Edited by
James G. Brennan and
Alistair S. Grandison

Food Processing Handbook

Volume 1

Related Titles

Rychlik, Michael (ed.)

Fortified Foods with Vitamins

Analytical Concepts to Assure Better and Safer Products

2011

ISBN: 978-3-527-33078-2

Moscicki, Leszek (ed.)

Extrusion-Cooking Techniques

Applications, Theory and Sustainability

2011

ISBN: 978-3-527-32888-8

Peinemann, K.-V., Pereira Nunes, S., Giorno, L. (eds.)

Membrane Technology

Volume 3: Membranes for Food Applications

2010

ISBN: 978-3-527-31482-9

Rijk, R., Veraart, R. (eds.)

Global Legislation for Food Packaging Materials

2010

ISBN: 978-3-527-31912-1

Stanga, Mario

Sanitation

Cleaning and Disinfection in the Food Industry

2010

ISBN: 978-3-527-32685-3

Popping, Bert, Diaz-Amigo, Carmen, Hoenicke, Katrin (eds.)

Molecular Biological and Immunological Techniques and Applications for Food Chemists

2010

ISBN: 978-0-470-06809-0

Wintgens, J. N. (ed.)

Coffee: Growing, Processing, Sustainable Production

2nd updated edition

2009

ISBN: 978-3-527-32286-2

Eßlinger, Hans Michael (ed.)

Handbook of Brewing

Processes, Technology, Markets

2009

ISBN: 978-3-527-31674-8

Chen, X. D., Mujumdar, A. S. (eds.)

Drying Technologies in Food Processing

ISBN: 978-1-4051-5763-6

Evans, J. (ed.)

Frozen Food Science and Technology

ISBN: 978-1-4051-5478-9

Heredia, N. L., Wesley, I. V. (eds.)

Microbiologically Safe Foods

2009

ISBN: 978-0-470-05333-1

Stadler, Richard H., Lineback, David R.

Process-Induced Food Toxicants

Occurrence, Formation, Mitigation, and Health Risks

2009

ISBN: 978-0-470-07475-6

Ziegler, H. (ed.)

Flavourings

Production, Composition, Applications, Regulations

2007

ISBN: 978-3-527-31406-5

Heller, K. J. (ed.)

Genetically Engineered Food

Methods and Detection

2006

ISBN: 978-3-527-31393-8

Edited by James G. Brennan and Alistair S. Grandison

Food Processing Handbook

2nd edition

Volume 1

WILEY-VCH Verlag GmbH & Co. KGaA

The Editors

James G. Brennan, MSc FIFST
16 Benning Way
Wokingham, Berks RG40 1XX
United Kingdom

Dr. Alistair S. Grandison
Department of Food and Nutritional Sciences
University of Reading
Whiteknights
Reading RG6 6AP
United Kingdom

Library of Congress Card No.: applied for

British Library Cataloguing-in-Publication Data
A catalogue record for this book is available from the British Library.

Bibliographic information published by the Deutsche Nationalbibliothek
The Deutsche Nationalbibliothek lists this publication in the Deutsche Nationalbibliografie; detailed bibliographic data are available on the Internet at <http://dnb.d-nb.de>.

Cover Design Adam-Design, Weinheim
Typesetting Laserwords Private Limited, Chennai, India
Printing and Binding Fabulous Printers Pte Ltd, Singapore

Printed in Singapore
Printed on acid-free paper

ISBN: 978-3-527-32468-2
ePDF ISBN: 978-3-527-63438-5
ePub ISBN: 978-3-527-63437-8
Mobi ISBN: 978-3-527-63439-2
oBook ISBN: 978-3-527-63436-1

Contents

Content of Volume 2

Preface to the Second Edition

In this second edition of Food Processing Handbook the chapters in the first edition have been retained and revised by including information on recent developments in each field and updating the reference lists. Some of the most notable changes are: the inclusion of a new section on ohmic heating in the Chapter on thermal processing (Chapter 2); extending the packaging chapter to cover intelligent packaging (Chapter 8); explaining the calculation of greenhouse gas emissions (carbon footprints) and providing a case study in the chapter on environmental aspects of food processing (Chapter 19). The original chapter entitled Baking, Extrusion and Frying has been split into three individual chapters providing extended coverage of these three important processes (Chapters 12, 13, and 14). Several new topics have been added to reflect recent trends and concerns in the food industry. These include chapters on: traceability in food processing and distribution (Chapter 16); hygienic design of food processing plant (Chapter 17); process realisation (Chapter 21); microscopy techniques and image analysis for the quantitative evaluation of food microstructure (Chapter 22); nanotechnology in the food sector (Chapter 23) and fermentation and the use of enzymes (Chapter 24). These changes have necessitated dividing the book into two volumes, the first consisting of the more basic food preservation processes and packaging, while volume 2 includes other manufacturing processes and other considerations relating to safety and sustainable manufacturing.

It is hoped that this much extended edition will be of interest to scientists and engineers involved in food manufacture and research and development in industry, and to staff and students participating in food related courses at undergraduate and postgraduate levels.

James G. Brennan,
Alistair S. Grandison

Preface to the First Edition

There are many excellent texts available which cover the fundamentals of food engineering, equipment design, modelling of food processing operations etc. There are also several very good works in food science and technology dealing with the chemical composition, physical properties, nutritional and microbiological status of fresh and processed foods. This work is an attempt to cover the middle ground between these two extremes. The objective is to discuss the technology behind the main methods of food preservation used in today's food industry in terms of the principles involved, the equipment used and the changes in physical, chemical, microbiological and organoleptic properties that occur during processing. In addition to the conventional preservation techniques, new and emerging technologies, such as high pressure processing and the use of pulsed electric field and power ultrasound are discussed. The materials and methods used in the packaging of food, including the relatively new field of active packaging, are covered. Concerns about the safety of processed foods and the impact of processing on the environment are addressed. Process control methods employed in food processing are outlined. Treatments applied to water to be used in food processing and the disposal of wastes from processing operations are described.

Chapter 1 covers the postharvest handling and transport of fresh foods and preparatory operations, such as cleaning, sorting, grading and blanching, applied prior to processing. Chapters 2, 3 and 4 contain up-to-date accounts of heat processing, evaporation, dehydration and freezing techniques used for food preservation. In Chapter 5, the potentially useful, but so far little used process of irradiation is discussed. The relatively new technology of high pressure processing is covered in Chapter 6, while Chapter 7 explains the current status of pulsed electric field, power ultrasound, and other new technologies. Recent developments in baking, extrusion cooking and frying are outlined in Chapter 8. Chapter 9 deals with the materials and methods used for food packaging and active packaging technology, including the use of oxygen, carbon dioxide and ethylene scavengers, preservative releasers and moisture absorbers. In Chapter 10, safety in food processing is discussed and the development, implementation and maintenance of HACCP systems outlined. Chapter 11 covers the various types of control systems applied in food processing. Chapter 12 deals with environmental issues including the impact of packaging wastes and the disposal of refrigerants. In Chapter 13, the various treatments applied to water to be used in food processing are described and the physical, chemical and biological treatments applied to food processing wastes are outlined. To complete the picture, the various separation techniques used in food processing are discussed in Chapter 14 and Chapter 15 covers the conversion operations of mixing, emulsification and size reduction of solids.

The editor wishes to acknowledge the considerable advice and help he received from former colleagues in the School of Food Biosciences, The University of Reading, when working on this project. He also wishes to thank his wife, Anne, for her support and patience.

Reading, August 2005 *James G. Brennan*

List of Contributors

Araya Ahromrit
Assistant Professor
Department of Food Technology
Khon Kaen University
Khon Kaen 40002
Thailand

Paul Ainsworth
Manchester Metropolitan
University
Retired Professor of Food
Technology
and Director of the Manchester
Food Research Centre
Old Hall Lane
Manchester M14 6HR
UK

Liliana Alamilla-Beltrán
National School of Biological
Sciences-National Polytechnic
Institute
Department of Food Science and
Technology
Carpio y Plan de Ayala s/n Sto.
Tomás 11340
Mexico City

Pedro Bouchon
Pontificia Universidad Católica
de Chile
Department of Chemical and
Bioprocess Engineering
P.O. Box 306
Santiago 6904411
Chile

James G. Brennan
16 Benning Way
Wokingham
Berkshire
RG40 1XX
UK

Stanley P. Cauvain
BakeTran
1 Oakland close
Freeland
Witney
OX 29 8AX
UK

José Jorge Chanona-Pérez
National School of Biological
Sciences-National Polytechnic
Institute
Department of Food Science and
Technology
Carpio y Plan de Ayala s/n Sto.
Tomás 11340
Mexico City

Dimitris Charalampopoulos
University of Reading
Department of Food and
Nutritional Sciences
PO Box 226
Reading RG6 6AP
UK

Brian P.F. Day
8 Cavanagh Close
Hoppers Crossing
VIC 3029
Australia

Ali Abd El-Aal Bakr
Food Science and Technology
Department
Faculty of Agriculture
Minufiya University
Shibin El-Kom
Egypt

Alistair S. Grandison
University of Reading
Department of Food and
Nutritional Sciences
P.O. Box 226
Whiteknights
Reading RG6 6AP
UK

Gustavo Fidel Gutierrez-López
National School of Biological
Sciences-National Polytechnic
Institute
Department of Food Science and
Technology
Carpio y Plan de Ayala s/n Sto.
Tomás 11340
Mexico City

Tony Hasting
37 Church Lane
Sharnbrook
Bedford
MK44 1HT
UK

Soojin Jun
University of Hawaii at Manoa
College of Tropical Agriculture
and Human Resources
Department of Human Nutrition
Food and Animal Sciences
1955 East West Rd. 302F
Honolulu, HI 96822
USA

Ashok S. Khare
University of Reading
Department of Food and
Nutritional Sciences
P.O. Box 226
Whiteknights
Reading RG6 6AP
UK

Christopher J. Kirby
Pharmaterials Ltd.
Unit B
5 Boulton Road
Reading RG2 0NH
UK

Christopher Knight
Head of Agriculture
Campden BRI
Chipping Campden
Glos. GL55 6LD
UK

Kevan G. Leach
Leach Associates Ltd.
Edgecumbe Lodge
Greenway Park
Chippenham
Wilts SN15 1QG
UK

Craig Leadley
Campden BRI
Food Manufacturing
Technologies
Chipping Campden
Gloucestershire GL55 6LD
UK

Dave A. Ledward
University of Reading
Department of Food and
Nutritional Sciences
Whiteknights
Reading RG6 6AP
UK

Michael J. Lewis
University of Reading
Department of Food and
Nutritional Sciences
P.O. Box 226
Whiteknights
Reading RG6 6AP
UK

José Mauricio Pardo
Universidad de la Sabana
Ingenieria de Produccion
Agroindustrial
A. A. 140013 Chia
Colombia

Angélica Gabriela Mendoza-Madrigal
National School of Biological
Sciences-National Polytechnic
Institute
Department of Food Science and
Technology
Carpio y Plan de Ayala s/n Sto.
Tomás 11340
Mexico City

Niharika Mishra
Agricultural and Biological
Engineering
717 W. Cherry lane
Apt # 2
State College, PA 16803
USA

Keshavan Niranjan
University of Reading
Department of Food and
Nutritional Sciences
P.O. Box 226
Whiteknights
Reading RG6 6AP
UK

Margaret F. Patterson
Agri-Food and biosciences Institute
Newforge Lane
Belfast BT9 5PX
Northern Ireland
UK

Maria de Jesús Perea-Flores
National School of Biological Sciences-National Polytechnic Institute
Department of Food Science and Technology
Carpio y Plan de Ayala s/n Sto. Tomás 11340
Mexico City

Nigel Rogers
Avure Technologies AB
Quintusvägen 2
Vasteras
SE 72166
Sweden

Gary Tucker
Head of Baking & Cereal Processing Department
Campden BRI
Chipping Campden
Glos, GL55 6LD
UK

Carol A. Wallace
University of Central Lancashire
International Institute of Nutritional Sciences and Applied Food Safety Studies
School of Sport
Tourism and the Outdoors
Preston
Lancashire PR1 2HE
UK

Jorge Welti-Chanes
Technological Institute of Advanced Studies of Monterrey
Food and Biotechnology Unit
Av. Eugenio Garza Sada 2501 Sur
Col. Tecnológico
64849 Monterrey
N.L.
Mexico

R. Andrew Wilbey
The University of Reading
Department of Food and Nutritional Sciences
Whiteknights
Reading RG6 6AP
UK

1
Postharvest Handling and Preparation of Foods for Processing

Alistair S. Grandison

1.1
Introduction

Food processing is seasonal in nature, both in terms of demand for products and availability of raw materials. Most crops have well-established harvest times – for example, the sugar beet season lasts for only a few months of the year in the United Kingdom, so beet sugar production is confined to the autumn and winter, yet demand for sugar is continuous throughout the year. Even in the case of raw materials that are available throughout the year, such as milk, there are established peaks and troughs in volume of production, as well as variations in chemical composition. Availability may also be determined by less predictable factors, such as weather conditions, which may affect yields or limit harvesting. In other cases demand is seasonal, for example, ice cream or salads are in greater demand in the summer, whereas other foods are traditionally eaten in the winter months, or even at more specific times, such as Christmas or Easter.

In an ideal world, food processors would like a continuous supply of raw materials, whose composition and quality are constant and whose prices are predictable. Of course this is usually impossible to achieve. In practice, processors contract ahead with growers to synchronize their needs with raw material production.

The aim of this chapter is to consider the properties of raw materials in relation to food processing, and to summarize important aspects of handling, transport, storage, and preparation of raw materials prior to the range of processing operations described in the remainder of this book. The bulk of the chapter will deal with solid agricultural products including fruits, vegetables, cereals, and legumes, although many considerations can also be applied to animal-based materials such as meat, eggs, and milk.

Food Processing Handbook, Second Edition. Edited by James G. Brennan and Alistair S. Grandison.

1.2 Properties of Raw Food Materials and Their Susceptibility to Deterioration and Damage

The selection of raw materials is a vital consideration to the quality of processed products. The quality of raw materials can rarely be improved during processing, and while sorting and grading operations can aid by removing oversize, undersize, or poor-quality units, it is vital to procure materials whose properties most closely match the requirements of the process. Quality is a wide-ranging concept and is determined by many factors. It is a composite of those physical and chemical properties of the material which govern its acceptability to the "user." The latter may be the final consumer, or more likely in this case, the food processor. Geometric properties, color, flavor, texture, nutritive value, and freedom from defects are the major properties likely to determine quality.

An initial consideration is selection of the most suitable cultivars in the case of plant foods (or breeds in the case of animal products). Other preharvest factors (such as soil conditions, climate, and agricultural practices), harvesting methods and postharvest conditions, maturity, storage, and postharvest handling also determine quality. These considerations, including seed supply and many aspects of crop production, are frequently controlled by the processor or even the retailer.

The timing and method of harvesting are determinants of product quality. Manual labor is expensive, therefore mechanized harvesting is introduced where possible. Cultivars most suitable for mechanized harvesting should mature evenly, producing units of nearly equal size that are resistant to mechanical damage. In some instances, the growth habits of plants (e.g., pea vines, fruit trees) have been developed to meet the needs of mechanical harvesting equipment. Uniform maturity is desirable as the presence of over-mature units is associated with high waste, product damage, and high microbial loads, while under-maturity is associated with poor yield, lack of flavor and color, and hard texture. For economic reasons, harvesting is almost always a "once over" exercise, hence it is important that all units reach maturity at the same time. The prediction of maturity is necessary to coordinate harvesting with processors' needs, as well as to extend the harvest season. It can be achieved primarily from knowledge of the growth properties of the crop combined with records and experience of local climatic conditions.

The "heat unit system," first described by Seaton [1] for peas and beans, can be applied to give a more accurate estimate of harvest date from sowing date in any year. This system is based on the premise that growth temperature is the overriding determinant of crop growth. A base temperature, below which no growth occurs, is assumed, and the mean temperature of each day through the growing period is recorded. By summing the daily mean temperatures minus base temperatures on days where mean temperature exceeds base temperature, the number of "accumulated heat units" can be calculated. By comparing this with the known growth data for the particular cultivar, an accurate prediction of harvest date can be computed. In addition, by allowing fixed numbers of accumulated heat units between sowings, the harvest season can be spread, so that individual fields may be harvested at peak maturity. Sowing plans and harvest date are determined by negotiation between the growers and the processors, and the latter may even provide the equipment and labor for harvesting and transport to the factory.

An important consideration for processed foods is that it is the quality of the processed product, rather than the raw material, that is important. For minimally processed foods, such as those subjected to modified atmospheres, low dose irradiation, mild heat treatment, or some chemical preservatives, the characteristics of the raw material are a good guide to the quality of the product. For more severe processing, including heat preservation, drying, or freezing, the quality characteristics may change markedly during processing. Hence, those raw materials which are preferred for fresh consumption may not be most appropriate for processing. For example, succulent peaches with delicate flavor may be less suitable for canning than harder, less flavorsome cultivars, which can withstand rigorous processing conditions. Similarly, ripe, healthy, well-colored fruit may be perfect for fresh sale, but may not be suitable for freezing due to excessive drip loss while thawing. For example, Maestrelli [2] reported that different strawberry cultivars with similar excellent characteristics for fresh consumption, exhibited a wide range of drip loss (between 8 and 38%), and hence would be of widely different value for the frozen food industry.

1.2.1 Raw Material Properties

The main raw material properties of importance to the processor are geometry, color, texture, functional properties, and flavor.

1.2.1.1 Geometric Properties

Food units of regular geometry are much easier to handle and are better suited to high-speed mechanized operations. In addition, the more uniform the geometry of raw materials, the less rejection and waste will be produced during preparation operations such as peeling, trimming, and slicing. For example, potatoes of smooth shape with few and shallow eyes are much easier to peel and wash mechanically than irregular units. Smooth-skinned fruits and vegetables are much easier to clean and are less likely to harbor insects or fungi than ribbed or irregular units.

Agricultural products do not come in regular shapes and exact sizes. Size and shape are inseparable, but are very difficult to define mathematically in solid food materials. Geometry is, however, vital to packaging and controlling fill-in weights. It may, for example, be important to determine how much mass or how many units may be filled into a square box or cylindrical can. This would require a vast number of measurements to perform exactly, and thus approximations must be made. Size and shape are also important to heat processing and freezing, as they will determine the rate and extent of heat transfer within food units. Mohsenin [3] describes numerous approaches by which the size and shape of irregular food units may be defined. These include the development of statistical techniques based on a limited number of measurements and more subjective approaches involving visual comparison of units to charted standards. Uniformity of size and shape is also important to most operations and processes. Process control to give accurately and uniformly treated products is always simpler with more uniform materials. For example, it is essential that wheat kernel size is uniform for flour milling.

Specific surface (area/mass) may be an important expression of geometry, especially when considering surface phenomena, such as the economics of fruit peeling, or surface processes such as smoking and brining.

The presence of geometric defects, such as projections and depressions, complicate any attempt to quantify the geometry of raw materials, as well as presenting processors with cleaning and handling problems, and yield loss. Selection of cultivars with the minimum defect level is advisable.

There are two approaches to securing optimum geometric characteristics: first, the selection of appropriate varieties, and second, sorting and grading operations.

1.2.1.2 **Color**

Color and color uniformity are vital components of the visual quality of fresh foods, and play a major role in consumer choice. However, it may be less important in raw materials for processing. For low-temperature processes, such as chilling, freezing, or freeze drying, the color changes little during processing, and thus the color of the raw material is a good guide to suitability for processing. For more severe processing, the color may change markedly during the process. Green vegetables such as peas, spinach, or green beans change color on heating from bright green to a dull olive green. This is due to the conversion of chlorophyll to pheophytin. It is possible to protect against this by addition of sodium bicarbonate to the cooking water, which raises the pH. However, this may cause softening of texture, and the use of added colorants may be a more practical solution. Some fruits may lose their color during canning, while pears develop a pink tinge. Potatoes are subject to browning during heat processing due to the Maillard reaction. Therefore, some varieties are more suitable for fried products, where browning is desirable, than for canned products, in which browning would be a major problem.

Again there are two approaches: procuring raw materials of the appropriate variety and stage of maturity, and sorting by color to remove unwanted units.

1.2.1.3 **Texture**

The texture of raw materials is frequently changed during processing. Textural changes are caused by a wide variety of effects, including water loss, protein denaturation which may result in loss of water-holding capacity or coagulation, hydrolysis, and solubilization of proteins. In plant tissues, cell disruption leads to loss of turgor pressure and softening of the tissue, while gelatinization of starch, hydrolysis of pectin, and solubilization of hemicelluloses also cause softening of the tissues.

The raw material must be robust enough to withstand the mechanical stresses during preparation, for example, abrasion during cleaning of fruit and vegetables.

Peas and beans must be able to withstand mechanical podding. Raw materials must be chosen so that the texture of the processed product is correct, such as canned fruits and vegetables in which raw materials must be able to withstand heat processing without being too hard or coarse for consumption.

Texture is dependent on the variety as well as the maturity of the raw material, and may be assessed by sensory panels or commercial instruments. One widely recognized instrument is the tenderometer used to assess the firmness of peas. The crop would be tested daily and harvested at the optimum tenderometer reading. In common with other raw materials, peas at different maturities can be used for different purposes, so that peas for freezing would be harvested at a lower tenderometer reading than peas for canning.

1.2.1.4 Flavor

Flavor is a rather subjective property which is difficult to quantify. Flavor quality of horticultural products is influenced by genotype and a range of pre- and postharvest factors [4]. Optimizing maturity/ripeness stage in relation to flavor at the time of processing is a key issue. Again, flavors are altered during processing, and following severe processing, the main flavors may be derived from additives. Hence, the lack of strong flavors may be the most important requirement. In fact, raw material flavor is often not a major determinant as long as the material imparts only those flavors which are characteristic of the food. Other properties may predominate. Flavor is normally assessed by human tasters, although sometimes flavor can be linked to some analytical test, such as sugar/acid levels in fruits.

1.2.1.5 Functional Properties

The functionality of a raw material is the combination of properties which determine product quality and process effectiveness. These properties differ greatly for different raw materials and processes, and may be measured by chemical analysis or process testing.

For example, a number of possible parameters may be monitored in wheat. Wheat for different purposes may be selected according to protein content. Hard wheat with 11.5–14% protein is desirable for white bread, and some whole wheat breads require even higher protein levels (14–16%) [5]. On the other hand, soft or weak flours with lower protein contents are suited to chemically leavened products with a lighter or more tender structure. Hence protein levels of 8–11% are adequate for biscuits, cakes, pastry, noodles, and similar products. Varieties of wheat for processing are selected on this basis, and measurement of protein content would be a good guide to process suitability. Furthermore, physical testing of dough using a variety of rheological testing instruments may be useful in predicting the breadmaking performance of individual batches of wheat flours [6]. A further test is the Hagberg Falling Number which measures the amount of α-amylase in flour or wheat [7]. This enzyme assists in the breakdown of starch to sugars, and high levels give rise to a weak bread structure. Hence, the test is a key indicator of wheat baking quality and is routinely used for bread wheat, and often determines the price paid to the farmer.

Similar considerations apply to other raw materials. Chemical analysis of fat and protein in milk may be carried out to determine its suitability for manufacturing cheese, yoghurt, or cream.

1.2.2
Raw Material Specifications

In practice, processors define their requirements in terms of raw material specifications for any process on arrival at the factory gate. Acceptance of, or price paid for, the raw material depends on the results of specific tests. Milk deliveries would be routinely tested for hygienic quality, somatic cells, antibiotic residues, extraneous water, as well as possibly fat and protein content. A random core sample is taken from all sugar beet deliveries and payment is dependent on the sugar content. For fruits, vegetables, and cereals, processors may issue specifications and tolerances to cover the size of units, the presence of extraneous vegetable matter, foreign bodies, levels of specific defects (e.g., surface blemishes, insect damage), and so on, as well as specific functional tests. Guidelines for sampling and testing many raw materials for processing in the United Kingdom are available from Campden BRI (*www.campden.co.uk*).

Increasingly, food processors and retailers may impose demands on raw material production which go beyond the properties described above. These may include "environmentally friendly" crop management schemes in which only specified fertilizers and insecticides are permitted, or humanitarian concerns, especially for food produced in developing countries. Similarly animal welfare issues may be specified in the production of meat or eggs. Another important issue is the growth of demand for organic foods in the United Kingdom and western Europe, which obviously introduces further demands on production methods that are beyond the scope of this chapter.

1.2.3
Deterioration of Raw Materials

All raw materials deteriorate following harvest, by some of the following mechanisms:

- **Endogenous enzymes:** Postharvest senescence and spoilage of fruit and vegetables occurs through a number of enzymic mechanisms, including oxidation of phenolic substances in plant tissues by phenolase (leading to browning); sugar–starch conversion by amylases; postharvest demethylation of pectic substances in fruits and vegetables leading to softening tissues during ripening and firming of plant tissues during processing.
- **Chemical changes:** These include deterioration in sensory quality by lipid oxidation; non-enzymic browning; and breakdown of pigments such as chlorophyll, anthocyanins, and carotenoids.
- **Nutritional changes:** Breakdown of ascorbic acid is an important example.

- **Physical changes:** These include dehydration and moisture absorption.
- **Biological changes:** Examples are the germination of seeds and sprouting.
- **Microbiological contamination:** Both the organisms themselves and their toxic products lead to deterioration of quality, as well as posing safety problems.

1.2.4 Damage to Raw Materials

Damage may occur at any point from growing through to the final point of sale. It may arise through external or internal forces.

External forces result in mechanical injury to fruits and vegetables, cereal grains, eggs, and even bones in poultry. They occur due to rough handling as a result of careless manipulation, poor equipment design, incorrect containerization, and unsuitable mechanical handling equipment. The damage typically results from impact and abrasion between food units, or between food units and machinery surfaces and projections, excessive vibration or pressure from overlying material. Increased mechanization in food handling must be carefully designed to minimize this.

Internal forces arise from physical changes such as variation in temperature and moisture content, and may result in skin cracks in fruits and vegetables, or stress cracks in cereals.

Either form of damage leaves the material open to further biological or chemical damage including enzymic browning of bruised tissue, or infestation of punctured surfaces by molds and rots.

1.2.5 Improving Processing Characteristics through Selective Breeding and Genetic Engineering

Selective breeding for yield and quality has been carried out for centuries in both plant and animal products. Until the twentieth century, improvements were made on the basis of selecting the most desirable looking individuals, while more systematic techniques have been developed more recently, based on greater understanding of genetics. The targets have been to increase yield as well as aiding factors of crop or animal husbandry such as resistance to pests and diseases, suitability for harvesting, or development of climate-tolerant varieties (e.g., cold-tolerant maize or drought-resistant plants) [8]. Raw material quality, especially in relation to processing, has become increasingly important. There are many examples of successful improvements in processing quality of raw materials through selective plant breeding including:

- improved oil percentage and fatty acid composition in oilseed rape;
- improved milling and malting quality of cereals;
- high sugar content and juice quality in sugar beets;

- development of specific varieties of potatoes for the processing industry, based on levels of enzymes and sugars, producing appropriate flavor, texture and color in products, or storage characteristics;
- Brussels sprouts which can be successfully frozen.

Similarly, traditional breeding methods have been used to improve yields of animal products such as milk and eggs, as well as improving quality – for example, fat/lean content of meat. Again the quality of raw materials in relation to processing may be improved by selective breeding. This is particularly applicable to milk, where breeding programs have been used at different times to maximize butterfat and protein content, and would thus be related to the yield and quality of fat- or protein-based dairy products. Furthermore, particular protein genetic variants in milk have been shown to be linked with processing characteristics, such as curd strength during manufacture of cheese [9]. Hence, selective breeding could be used to tailor milk supplies to the manufacture of specific dairy products.

Traditional breeding programs will undoubtedly continue to produce improvements in raw materials for processing, but the potential is limited by the gene pool available to any species. Genetic engineering extends this potential by allowing the introduction of foreign genes into an organism, with huge potential benefits. Again many of the developments have been aimed at agricultural improvements such as increased yield, or introducing herbicide, pest, or drought resistance. Other developments have aimed to improve the nutritional quality of foods. For example, transgenic "Golden" rice as a rich source of vitamin A; cereal grains with increased protein quantity and quality; oilseeds engineered to contain higher levels of omega-3 fatty acids. However, there is enormous potential in genetically engineered raw materials for processing [10]. The following are some examples which have been demonstrated:

- Tomatoes which do not produce pectinase and hence remain firm while color and flavor develop, producing improved soup, paste, or ketchup.
- Potatoes with higher starch content, which take up less oil and require less energy during frying.
- Canola (rape seed) oil tailored to contain high levels of lauric acid to improve emulsification properties for use in confectionery, coatings, or low-fat dairy products; high levels of stearate as an alternative to hydrogenation in manufacture of margarine; and high levels of polyunsaturated fatty acids for health benefits.
- Wheat with increased levels of high molecular weight glutenins for improved breadmaking performance.
- Fruits and vegetables containing peptide sweeteners such as thaumatin or monellin.
- "Naturally decaffeinated" coffee.

There is, however, considerable opposition to the development of genetically modified foods in the United Kingdom and elsewhere, due to fears of human health risks and ecological damage, discussion of which is beyond the scope of this book. It therefore remains to be seen if, and to what extent, genetically modified raw materials will be used in food processing.

1.3 Storage and Transportation of Raw Materials

1.3.1 Storage

Storage of food is necessary at all points of the food chain from raw materials, through manufacture, distribution, retailers, and final purchasers. Today's consumers expect a much greater variety of products, including non-local materials, to be available throughout the year. Effective transportation and storage systems for raw materials are essential to meet this need.

Storage of materials whose supply or demand fluctuate in a predictable manner, especially seasonal produce, is necessary to increase availability. It is essential that processors maintain stocks of raw materials, therefore storage is necessary to buffer demand. However, storage of raw materials is expensive for two reasons: stored goods have been paid for and may therefore tie up quantities of company money, and secondly, warehousing and storage space are expensive. All raw materials will deteriorate during storage. The quantities of raw materials held in store and the times of storage vary widely for different cases, depending on the above considerations. The "just in time" approaches used in other industries are less common in food processing.

The primary objective is to maintain the best possible quality during storage, and hence avoid spoilage during the storage period. Spoilage arises through three mechanisms:

1) Living organisms such as vermin, insects, fungi, and bacteria – these may feed on the food and contaminate it.
2) Biochemical activity within the food leading to quality reduction, such as respiration in fruits and vegetables; staling of baked products; enzymic browning reactions; rancidity development in fatty food.
3) Physical processes, including damage due to pressure or poor handling; physical changes such as dehydration or crystallization.

The main factors that govern the quality of stored foods are temperature, moisture/humidity, and atmospheric composition. Different raw materials provide very different challenges.

Fruits and vegetables remain as living tissues until they are processed and the main aim is to reduce respiration rate without damage to the tissue. Storage times vary widely between types. Young tissues such as shoots, green peas, and immature fruits have high respiration rates and shorter storage periods, while mature fruits and roots, and storage organs such as bulbs and tubers (e.g., onions, potatoes, sugar beets) respire much more slowly, and hence have longer storage periods.

Table 1.1 Storage periods of some fruits and vegetables under typical storage conditions.

Commodity	Temperature (°C)	Humidity (%)	Storage period
Garlic	0	70	6–8 mo
Mushrooms	0	90–95	5–7 d
Green bananas	13–15	85–90	10–30 d
Immature potatoes	4–5	90–95	3–8 wk
Mature potatoes	4–5	90–95	4–9 mo
Onions	−1 to 0	70–80	6–8 mo
Oranges	2–7	90	1–4 mo
Mangoes	5.5–14	90	2–7 wk
Apples	−1 to 4	90–95	1–8 mo
French beans	7–8	95–100	1–2 wk

Data from [13].

Some examples of conditions and storage periods of fruits and vegetables are given in Table 1.1. Many fruits (including bananas, apples, tomatoes, and mangoes) display a sharp increase in respiration rate during ripening, just before the point of optimum ripening, known as the "*climacteric*." The onset of the climacteric is associated with the production of high levels of ethylene, which is believed to stimulate the ripening process. Climacteric fruit can be harvested unripe and ripened artificially at a later time. It is vital to maintain careful temperature control during storage or the fruit will rapidly over-ripen. Non-climacteric fruits (e.g., citrus fruit, pineapples, strawberries) and vegetables do not display this behavior, and generally do not ripen after harvest. Quality is therefore optimal at harvest, and the task is to preserve quality during storage.

Harvesting, handling, and storage of fruit and vegetables are discussed in more detail by Thompson [11], while Nascimento Nunes [12] visually depicts the effects of time and temperature on the appearance of fruit and vegetables throughout postharvest life.

With meat storage the overriding problem is growth of spoilage bacteria, while avoiding oxidative rancidity. Cereals must be dried before storage to avoid germination and mold growth, and subsequently must be stored under conditions which prevent infestation with rodents, birds, insects, or molds.

Hence, very different storage conditions may be employed for different raw materials. The main methods employed in raw material storage are the control of temperature, humidity, and composition of atmosphere.

1.3.1.1 Temperature

The rate of biochemical reactions is related to temperature, such that lower storage temperatures lead to slower degradation of foods by biochemical spoilage, as well as reduced growth of bacteria and fungi. There may also be limited bacteriocidal effects at very low temperatures. Typical Q_{10} values for spoilage reactions are approximately 2, implying that spoilage rates would double for each 10 °C rise, or conversely that shelf life would double for each 10 °C reduction. This is an oversimplification as Q_{10} may change with temperature. Most insect activity is inhibited below 4 °C, although insects and their eggs can survive long exposure to these temperatures. In fact grain and flour mites can remain active and even breed at 0 °C.

The use of refrigerated storage is limited by the sensitivity of materials to low temperatures. The freezing point is a limiting factor for many raw materials, as the tissues will become disrupted on thawing. Other foods may be subject to problems at temperatures above freezing. Fruits and vegetables may display physiological problems that limit their storage temperatures, probably as a result of metabolic imbalance leading to a build-up of undesirable chemical species in the tissues. Some types of apples are subject to internal browning below 3 °C, while bananas become brown when stored below 13 °C, and many other tropical fruits display chill sensitivity. Less obvious biochemical problems may occur even where no visible damage occurs. For example, storage temperature affects starch/sugar balance in potatoes; in particular, below 10 °C a build up of sugar occurs, which is most undesirable for fried products. Examples of storage periods and conditions are given in Table 1.1, illustrating the wide ranges seen with different fruits and vegetables. It should be noted that predicted storage lives can be confounded if the produce is physically damaged, or by the presence of pathogens.

Temperature of storage is also limited by cost. Refrigerated storage is expensive, especially in hot countries. In practice, a balance must be struck incorporating cost, shelf life, and risk of cold injury. Slower growing produce such as onions, garlic, and potatoes can be successfully stored at ambient temperature and ventilated conditions in temperate climates.

It is desirable to monitor temperature throughout raw material storage and distribution.

Precooling to remove the "field heat" is an effective strategy to reduce the period of high initial respiration rate in rapidly respiring produce prior to transportation and storage. For example, peas for freezing are harvested in the cool early morning and rushed to cold storage rooms within 2–3 h. Other produce, such as leafy vegetables (lettuce, celery, cabbage) or sweetcorn, may be cooled using water sprays or drench streams. Hydrocooling obviously reduces water loss.

1.3.1.2 **Humidity**

If the humidity of the storage environment exceeds the equilibrium relative humidity (ERH) of the food, the food will gain moisture during storage, and vice versa. Uptake of water during storage is associated with susceptibility to growth of microorganisms, while water loss results in economic loss, as well as more specific problems such as cracking of seed coats of cereals, or skins of fruits and vegetables. Ideally the humidity of the store would equal the ERH of the food so that moisture is neither gained nor lost, but in practice a compromise may be necessary. The water activity (a_w) of most fresh foods (e.g., fruit, vegetables, meat, fish, milk) is in the range 0.98–1.00, but they are frequently stored at a lower humidity.

Some wilting of fruits or vegetable may be acceptable in preference to mold growth, while some surface drying of meat is preferable to bacterial slime. Packaging may be used to protect against water loss of raw materials during storage and transport (see Chapter 8).

1.3.1.3 Composition of Atmosphere

Controlling the atmospheric composition during storage of many raw materials is beneficial. The use of packaging to allow the development or maintenance of particular atmospheric compositions during storage is discussed in greater detail in Chapter 8.

With some materials the major aim is to maintain an oxygen-free atmosphere to prevent oxidation (e.g., coffee, baked goods), while in other cases adequate ventilation may be necessary to prevent anaerobic fermentation leading to off-flavors.

In living produce, atmosphere control allows the possibility of slowing down metabolic processes, hence retarding respiration, ripening, and senescence as well as the development of disorders. The aim is to introduce N_2 and remove O_2, allowing a build up of CO_2. Controlled-atmosphere storage of many commodities is discussed by Thompson [14]. The technique allows year-round distribution of apples and pears, where controlled atmospheres in combination with refrigeration can give shelf lives up to 10 months, much greater than by chilling alone. The particular atmospheres are cultivar specific, but are in the range 1–10% CO_2, 2–13% O_2 at 3 °C for apples and 0 °C for pears. Controlled atmospheres are also used during storage and transport of chill-sensitive crops, such as for transport of bananas, where an atmosphere of 3% O_2 and 5% CO_2 is effective in preventing premature ripening and the development of crown rot disease. Ethene (ethylene) removal is also vital during storage of climacteric fruit.

With fresh meat, controlling the gaseous environment is useful in combination with chilling. The aim is to maintain the red color by storage in high O_2 concentrations, which shifts the equilibrium in favor of high concentrations of the bright red oxymyoglobin pigment. At the same time, high levels of CO_2 are required to suppress the growth of aerobic bacteria.

1.3.1.4 Other Considerations

Odors and taints can cause problems, especially in fatty foods such as meat and dairy products, as well as less obvious commodities such as citrus fruits, which have oil in the skins. Odors and taints may be derived from fuels or adhesives and printing materials, as well as other foods (e.g., spiced or smoked products). Packaging and other systems during storage and transport must protect against contamination.

Light can lead to oxidation of fats in some raw materials (e.g., dairy products). In addition, light gives rise to solanine production and the development of green pigmentation in potatoes. Hence, storage and transport under dark conditions is essential.

1.3.2
Transportation

Food transportation is an essential link in the food chain, and is discussed in detail by Heap [15]. Raw materials, food ingredients, fresh produce, and processed products are all transported on a local and global level, by land, sea, and air. In the modern world, where consumers expect year-round supplies and non-local products, long-distance transport of many foods has become commonplace, and air transport may be necessary for perishable materials. Transportation of food is really an extension of storage; a refrigerated lorry is basically a cold store on wheels. However, transport also subjects the material to physical and mechanical stresses, and possibly rapid changes in temperature and humidity, which are not encountered during static storage. It is necessary to consider both the stresses imposed during the transport and those encountered during loading and unloading. In many situations transport is multimodal. Air or sea transport would commonly involve at least one road trip before and one road trip after the main journey. There would also be time spent on the ground at the port or airport where the material could be exposed to wide-ranging temperatures and humidities, or bright sunlight, and unscheduled delays are always a possibility. During loading and unloading, the cargo may be broken into smaller units where more rapid heat penetration may occur.

The major challenges during transportation are to maintain the quality of the food during transport, and to apply good logistics – in other words, to move the goods to the right place at the right time and in good condition.

1.4
Raw Material Cleaning

All food raw materials are cleaned before processing. The purpose is obviously to remove contaminants, which range from innocuous to dangerous. It is important to note that removal of contaminants is essential for protection of process equipment as well as the final consumer. For example, it is essential to remove sand, stones, or metallic particles from wheat prior to milling to avoid damaging the machinery. The main contaminants are:

- unwanted parts of the plant such as leaves, twigs, husks;
- soil, sand, stones, and metallic particles from the growing area;
- insects and their eggs;
- animal excreta, hairs, and so on;
- pesticides and fertilizers;
- mineral oil;
- microorganisms and their toxins.

Increased mechanization in harvesting and subsequent handling has generally led to increased contamination with mineral, plant, and animal contaminants, while there has been a general increase in the use of sprays, leading to increased chemical contamination. Microorganisms may be introduced preharvest from irrigation water, manure fertilizer, or contamination from feral or domestic animals, or postharvest from improperly cleaned equipment, wash waters, or cross-contamination from other raw materials.

Cleaning is essentially a separation process, in which some difference in physical properties of the contaminants and the food units is exploited. There are a number of cleaning methods available, classified into dry and wet methods, but a combination would usually be used for any specific material. Selection of the appropriate cleaning regime depends on the material being cleaned, the level and type of contamination and the degree of decontamination required. In practice a balance must be struck between cleaning cost and product quality, and an "acceptable standard" should be specified for the particular end-use. Avoidance of product damage is an important contributing factor, especially for delicate materials such as soft fruit.

1.4.1 Dry Cleaning Methods

The main dry cleaning methods are based on screens, aspiration, or magnetic separations. Dry methods are generally less expensive than wet methods and the effluent is cheaper to dispose of, but they tend to be less effective in terms of cleaning efficiency. A major problem is recontamination of the material with dust. Precautions may be necessary to avoid the risk of dust explosions and fires.

Screens Screens are essentially size separators based on perforated beds or wire mesh by which larger contaminants are removed from smaller food items (e.g., straw from cereal grains, or pods and twigs from peas). This is termed "*scalping*" (Figure 1.1a). Alternatively "de-dusting" is the removal of smaller particles (e.g., sand or dust) from larger food units (Figure 1.1b). The main geometries are rotary drums (also known as *reels* or *trommels*) and flatbed designs. Some examples are shown in Figure 1.2. Abrasion, either by impact during the operation of the machinery, or aided by abrasive disks or brushes, can improve the efficiency of dry screens. Screening gives incomplete separations and is usually a preliminary cleaning stage.

Aspiration This exploits the differences in aerodynamic properties of the food and the contaminants. It is widely used in the cleaning of cereals, but is also incorporated into equipment for cleaning peas and beans. The principle is to feed the raw material into a carefully controlled upward air stream. Denser material will fall, while lighter material will be blown away depending on the terminal velocity. *Terminal velocity* in this case can be defined as the velocity of upward air stream in which a particle remains stationary, and depends on the density and projected area of the particles (as described by Stokes' equation). By using different air velocities, it is possible to separate, say, wheat from lighter chaff (Figure 1.3) or denser small stones. Very accurate separations are possible, but large amounts of energy are

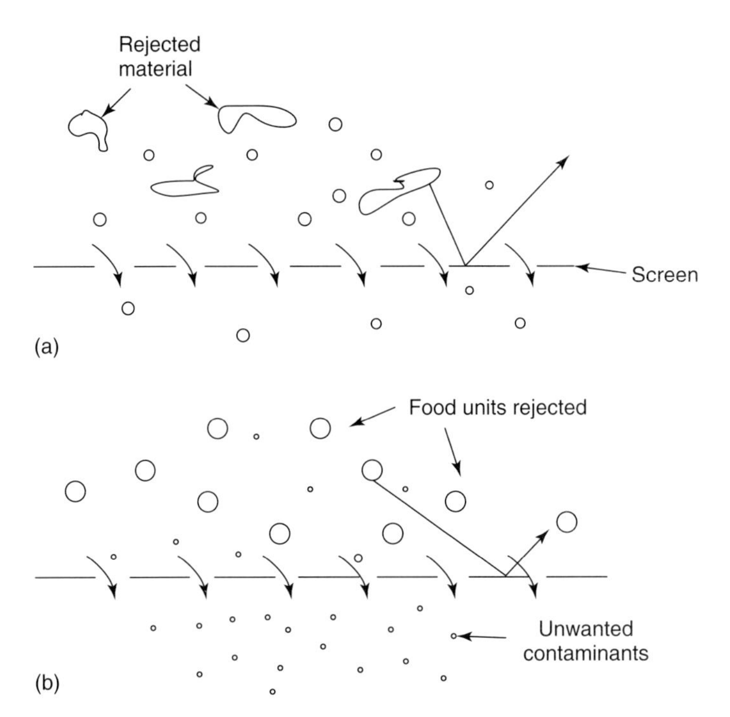

Figure 1.1 Screening of dry particulate materials: (a) scalping and (b) de-dusting.

required to generate the air streams. Obviously the system is limited by the size of raw material units, but is particularly suitable for cleaning legumes and cereals. Air streams may also be used simply to blow loose contaminants from larger items such as eggs or fruit.

Magnetic cleaning This is the removal of ferrous metal using permanent or electromagnets. Metal particles derived from the growing field or picked up during transport or preliminary operations constitute a hazard both to the consumer and to processing machinery (e.g. cereal mills). The geometry of magnetic cleaning systems can be quite variable: particulate foods may be passed over magnetized drums or magnetized conveyor belts, or powerful magnets may be located above conveyors. Electromagnets are easy to clean by turning off the power. Metal detectors are frequently employed prior to sensitive processing equipment as well as to protect consumers at the end of processing lines.

Electrostatic cleaning This can be used in a limited number of cases where the surface charge on raw materials differs from contaminating particles. The principle can be used to distinguish grains from other seeds of similar geometry but different surface charge, and has also been described for cleaning tea. The feed is conveyed on a charged belt and charged particles are attracted to an oppositely charged electrode according to their surface charge (Figure 1.4).

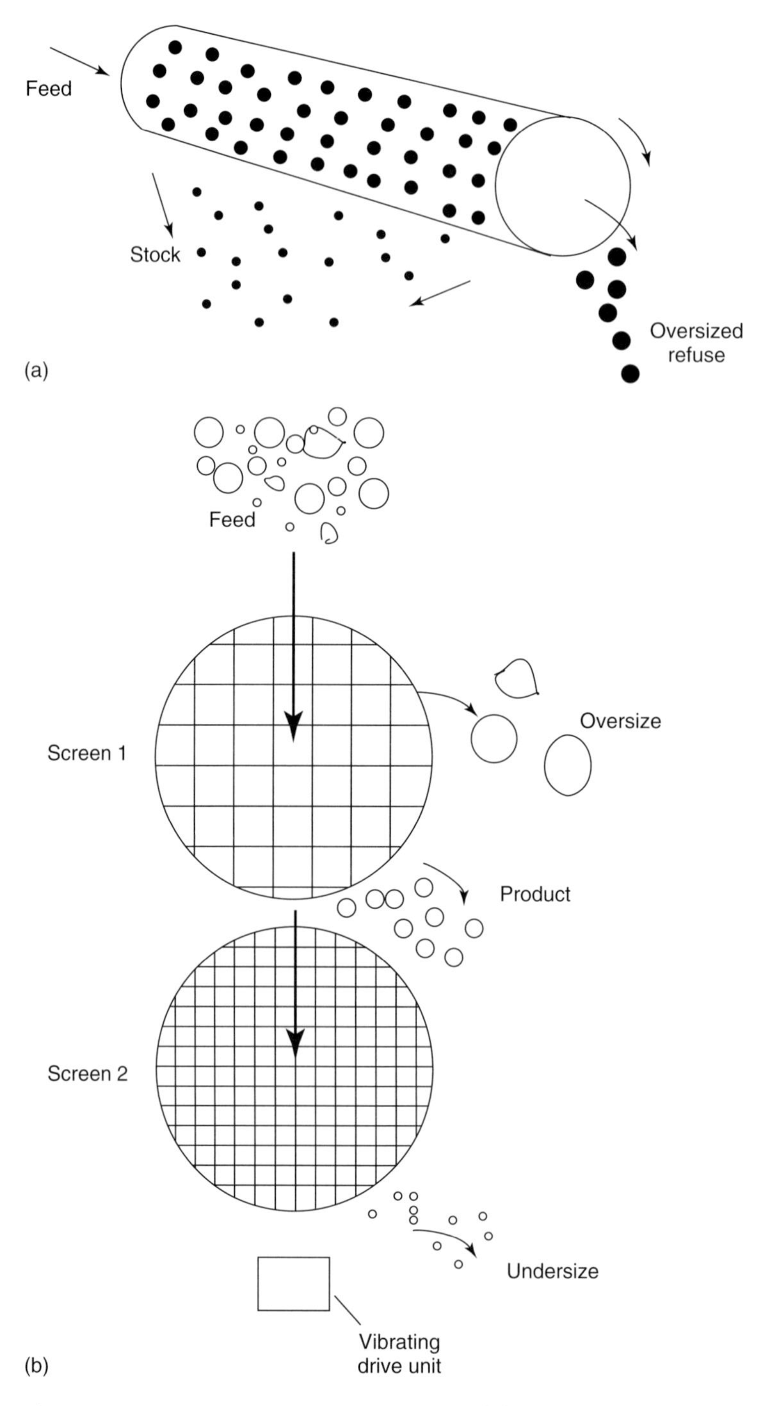

Figure 1.2 Screen geometries: (a) rotary screen and (b) principle of flatbed screen.

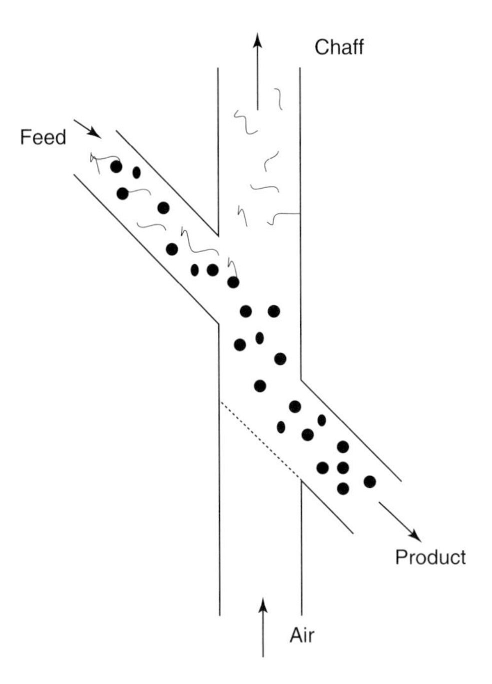

Figure 1.3 Principle of aspiration cleaning.

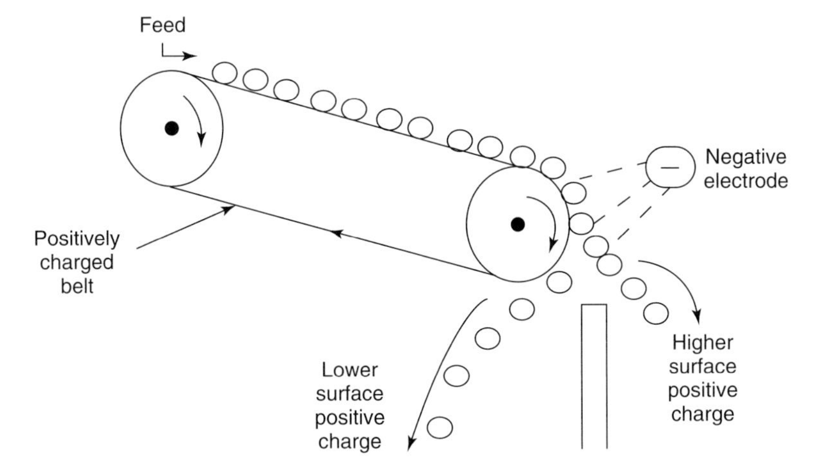

Figure 1.4 Principle of electrostatic cleaning.

1.4.2 Wet Cleaning Methods

Wet methods are necessary if large quantities of soil are to be removed, and are essential if detergents are used. They are, however, expensive as large quantities of high purity water are required, and the same quantity of dirty effluent is produced. Treatment and reuse of water can reduce costs. Employing the countercurrent principle can reduce water requirement and effluent volumes if accurately controlled. Sanitizing chemicals such as chlorine, citric acid, and ozone are commonly used in wash waters, especially in association with peeling and size reduction, where reducing enzymic browning may also be an aim [16]. Levels of 100–200 mg l^{-1} chlorine or citric acid may be used, although their effectiveness for decontamination has been questioned and they are not permitted in some countries.

Soaking is a preliminary stage in cleaning heavily contaminated materials such as root crops, permitting softening of the soil, and partial removal of stones and other contaminants. Metallic or concrete tanks or drums are employed, and these may be fitted with devices for agitating the water, including stirrers, paddles, or mechanisms for rotating the entire drum. For delicate produce such as strawberries or asparagus, or products which trap dirt internally (e.g., celery), sparging air through the system may be helpful. The use of warm water or including detergents improves cleaning efficiency, especially where mineral oil is a possible contaminant, but adds to the expense and may damage the texture.

Spray washing is very widely used for many types of food raw material. Efficiency depends on the volume and temperature of the water and time of exposure. As a general rule, small volumes of high-pressure water give the most efficient dirt removal, but this is limited by product damage, especially to more delicate produce. With larger food pieces it may be necessary to rotate the unit so that the whole surface is presented to the spray (Figure 1.5a). The two most common designs

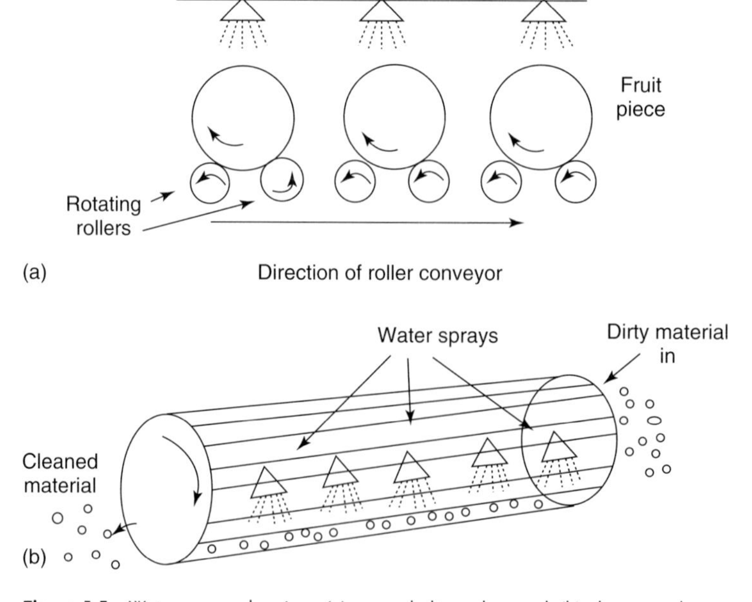

Figure 1.5 Water spray cleaning: (a) spray belt washer and (b) drum washer.

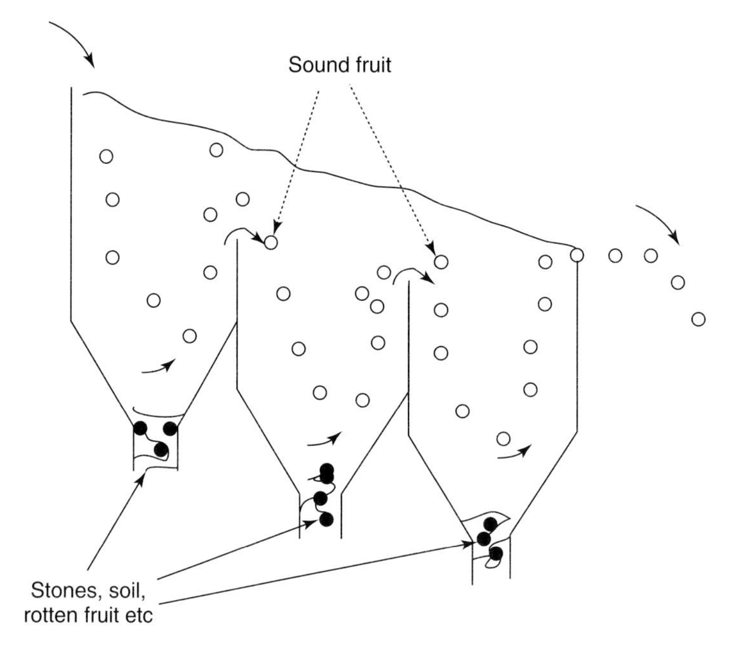

Figure 1.6 Principle of flotation washing.

are drum washers and belt washers (Figures 1.5). Abrasion may contribute to the cleaning effect, but again must be limited in delicate units. Other designs have included flexible rubber disks which gently brush the surface clean.

Flotation washing employs buoyancy differences between food units and contaminants. For instance, sound fruit generally floats, while contaminating soil, stones, or rotten fruits sink in water. Hence fluming fruit in water over a series of weirs gives very effective cleaning of fruit, peas, and beans (Figure 1.6). A disadvantage is high water use, thus recirculation of water should be incorporated.

Froth flotation is carried out to separate peas from contaminating weed seeds, and exploits surfactant effects. The peas are dipped in oil/detergent emulsion and air is blown through the bed. This forms a foam which washes away the contaminating material, and the cleaned peas can be spray washed.

Following wet cleaning it is necessary to remove the washing water. Centrifugation is very effective, but may lead to tissue damage, hence dewatering screens or reels are more common.

Prestorage hot water dipping has been used as an alternative to chemical treatments for preserving the quality of horticultural products. One recent development is the simultaneous cleaning and disinfection of fresh produce by a short hot water rinse and brushing (HWRB) treatment [17]. This involves placing the crops on rotating brushes and rinsing with hot water for 10–30 s. The effect is through a combination of direct cleaning action plus the lethal action of heat on surface pathogens. Fungicides may also be added to the hot water.

1.4.3
Peeling

Peeling of fruits and vegetables is frequently carried out in association with cleaning. Mechanical peeling methods require loosening of the skin using one of the following principles depending on the structure of the food and the level of peeling required [18]:

- **Steam** is particularly suited to root crops. The units are exposed to high-pressure steam for a fixed time and then the pressure is released causing steam to form under the surface of the skin, hence loosening it such that it can be removed with a water spray.
- **Lye** (1–2% alkali) solution can be used to soften the skin which can again be removed by water sprays. There is, however, a danger of damage to the product.
- **Brine** solutions can give a peeling effect but are probably less effective than the above methods.
- **Abrasion peeling** employs carborundum rollers or rotating the product in a carborundum-lined bowl followed by washing away the loosened skin. It is effective but here is a danger of high product loss by this method.
- **Mechanical knives** are suitable for peeling citrus fruits.
- **Flame peeling** is useful for onions in which the outer layers are burnt off and charred skin is removed by high-pressure hot water.

1.5
Sorting and Grading

Sorting and grading are terms which are frequently used interchangeably in the food processing industry, but strictly speaking they are distinct operations. Sorting is a separation based on a single measurable property of raw material units, while grading is "the assessment of the overall quality of a food using a number of attributes" [18]. *Grading of fresh produce* may also be defined as "sorting according to quality," as sorting usually upgrades the product.

Virtually all food products undergo some type of sorting operation. There are a number of benefits, including the need for sorted units in weight filling operations, and the aesthetic and marketing advantages in providing uniform-sized or uniform-colored units. In addition, it is much easier to control processes such as sterilization, dehydration, or freezing in sorted food units, and they are also better suited to mechanized operations such as size reduction, pitting, or peeling.

1.5.1
Criteria and Methods of Sorting

Sorting is carried out on the basis of individual physical properties. Details of principles and equipment are given in Saravacos and Kostaropoulos [19], Brennan *et al.* [20], and Peleg [21]. No sorting system is absolutely precise, and a balance is often struck between precision and flow rate.

Weight is usually the most precise method of sorting, as it is not dependent on the geometry of the products. Eggs, fruit, or vegetables may be separated into weight categories using spring-loaded, strain gauge, or electronic weighing devices incorporated into conveying systems. Using a series of tipping or compressed air blowing mechanisms set to trigger at progressively lesser weights, the heavier items are removed first followed by the next weight category, and so on. These systems are computer controlled and can additionally provide data on quantities and size distributions from different growers. An alternative system is to use the "catapult" principle where units are thrown into different collecting chutes, depending on their weight, by spring-loaded catapult arms. A disadvantage of weight sorting is the relatively long time required per unit and other methods are more appropriate with smaller items such as legumes or cereals, or if faster throughput is required.

Size sorting is less precise than weight sorting, but is considerably cheaper. As discussed in Section 1.2, the sizes and shapes of food units are difficult to define precisely. Size categories could include a number of physical parameters including diameter, length, or projected area. Diameter of spheroidal units such as tomatoes or citrus fruits is conventionally considered to be orthogonal to the fruit stem, while length is coaxial. Therefore rotating the units on a conveyor can make size sorting more precise.

Sorting into size categories requires some sort of screen, many designs of which are discussed in detail in Slade [22], Brennan *et al.* [20], and Fellows [18]. The main categories of screens are fixed-aperture and variable-aperture designs. Flatbed and rotary screens are the main geometries of fixed bed screen and a number of screens may be used in series or in parallel to sort units into several size categories simultaneously. The problem with fixed screens is usually contacting the feed material with the screen which may become blocked or overloaded. Fixed screens are often used with smaller particulate foods such as nuts or peas. Variable-aperture screens have either a continuous diverging or stepwise diverging apertures. These are much gentler and are commonly used with larger, more delicate items such as fruit. The principles of some sorting screens are illustrated in Figure 1.7.

Shape sorting is useful in cases where the food units are contaminated with particles of similar size and weight. This is particularly applicable to grain which may contain other seeds. The principle is that disks or cylinders with accurately shaped indentations will pick up seeds of the correct shape when rotated through the stock, while other shapes will remain in the feed (Figure 1.8).

Density can be a marker of suitability for certain processes. The density of peas correlates well with tenderness and sweetness, while the solids content of potatoes, which determines suitability for manufacture of crisps and dried products, relates to density. Sorting on the basis of density can be achieved using flotation in brine at different concentrations.

Photometric properties may be used as a basis for sorting. In practice this usually means color. Color is often a measure of maturity, presence of defects, or the degree of processing. Manual color sorting is carried out widely on conveyor belts

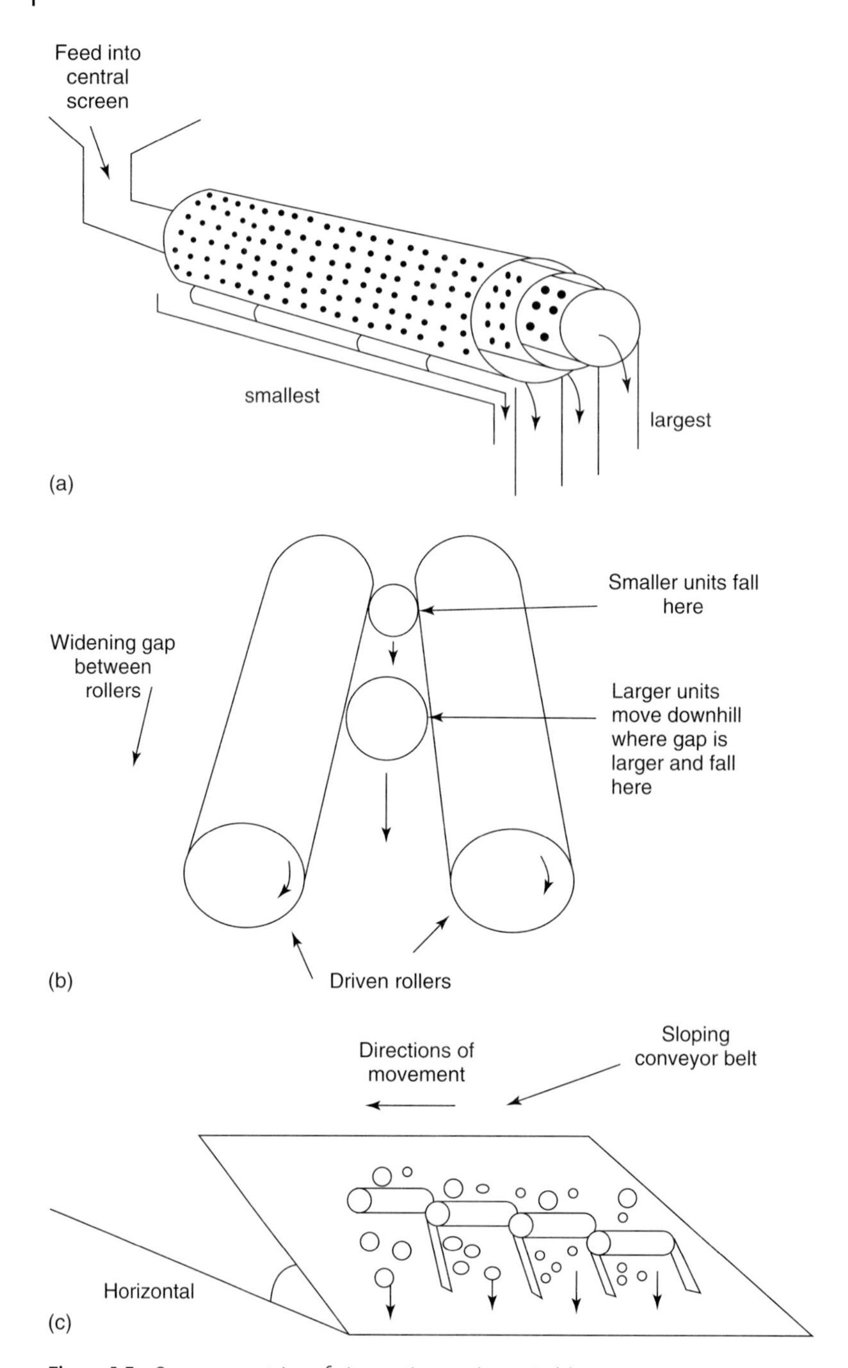

Figure 1.7 Some geometries of size-sorting equipment: (a) concentric drum screen; (b) roller size-sorter; and (c) belt and roller sorter.

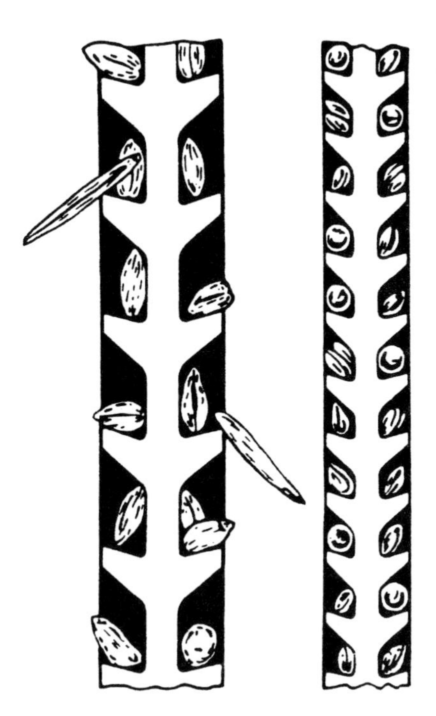

Figure 1.8 Cross section of disk separators for cleaning cereals.

or sorting tables, but is expensive. The process can be automated using highly accurate photocells which compare reflectance of food units to preset standards, and can eject defective or wrongly colored (e.g., blackened) units, usually by a blast of compressed air. This system is used for small particulate foods such as navy beans or maize kernels for canning, or nuts, rice, and small fruit (Figure 1.9). Extremely high throughputs have been reported (e.g., 16 t h^{-1}) [18]. By using more than one photocell positioned at different angles, blemishes on large units such as potatoes can be detected. Color sorting can also be used to separate materials which are to be processed separately such as red and green tomatoes. It is feasible to use transmittance as a basis for sorting, although as most foods are completely opaque very few opportunities are available. The principle has been used for sorting cherries with and without stones, and internal examination, or "candling," of eggs.

1.5.2 Grading

Grading is classification on the basis of quality (incorporating commercial value, end-use, and official standards [19]), and hence requires that some judgment on the acceptability of the food is made, based on simultaneous assessment of several properties, followed by separation into quality categories. Appropriate inspection belts or conveyors designed to present the whole surface to the operator, are frequently used. Trained manual operators are frequently used to judge the quality, and may use comparison to charted standards, or even plastic models. For

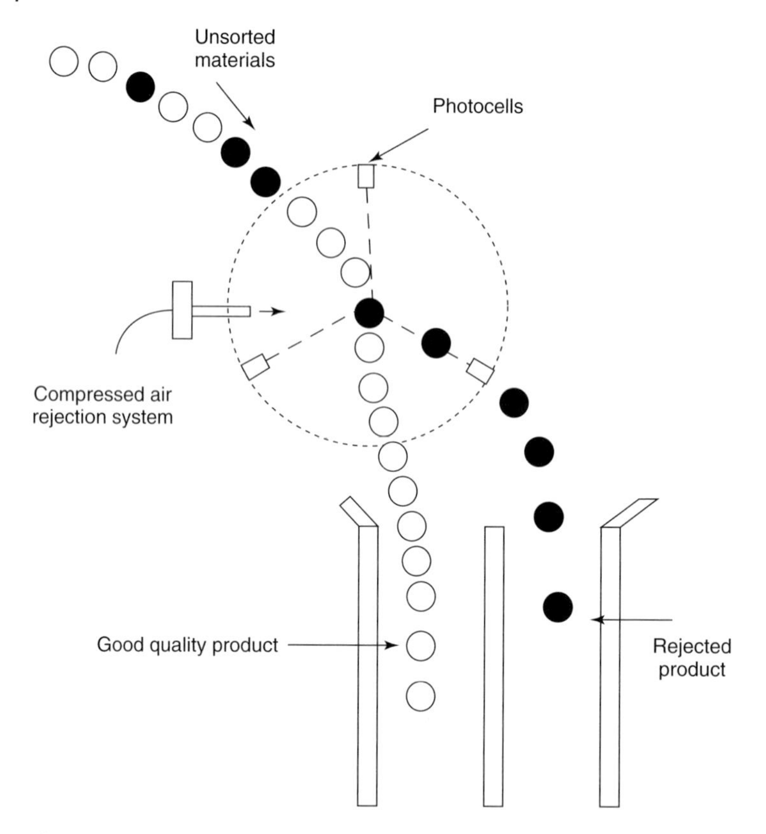

Figure 1.9 Principle of color sorter.

example, a fruit grader could simultaneously judge shape, color, evenness of color, and degree of russeting in apples. Egg "candling" involves inspection of eggs spun in front of a light so that many factors, including shell cracks, diseases, blood spots, or fertilization, can be detected. Apparently, experienced candlers can grade thousands of eggs per hour. Machine grading is only feasible where quality of a food is linked to a single physical property, and hence a sorting operation leads to different grades of material. Size of peas, for example, is related to tenderness and sweetness, therefore size sorting results in different quality grades.

Grading of foods is also the determination of the quality of a batch. This can be done by human graders who assess the quality of random samples of foods such as cheese or butter, or meat inspectors who examine the quality of individual carcasses for a number of criteria. Alternatively, batches of some foods may be graded on the basis of laboratory analysis.

There is increasing interest in the development of rapid, nondestructive methods of assessing the quality of foods, which could be applied to the grading and sorting of foods. Cubeddu *et al.* [23] and Nicolai *et al.* [24] have described potential application of advanced optical techniques to give information on both surface and internal properties of fruits, including textural and chemical properties. This could permit classification of fruit in terms of maturity, firmness, or the presence of defects, or even more specifically, noninvasive detection of chlorophyll, sugar, and acid levels. For example, Qin and Lu [25] were able to assess the ripeness of tomatoes and other fruit and vegetables from their absorption spectra in the visible and near infrared range, in particular, by the ratio of absorption coefficients of chlorophyll and anthocyanin. Another promising approach is the use of sonic techniques to measure the texture of fruits and vegetables [26, 27]. Similar applications of X-rays, lasers, infrared rays, and microwaves have also been studied [19].

An alternative approach to nondestructive testing is the Sinclair iQ firmness tester, which is based on the electrical response of the fruit to air bellows, which predicts the elastic response, which in turn reflects the ripeness. This equipment has been shown to relate very closely to standard penetrometer readings for a range of fruit [28] and is now available for commercial rapid online fruit grading. A recent patent [29] describes the use of nondestructive testing to assess size, Brix, maturity, and the presence of blemishes on fruit and hence to grade the fruit for the production of juice of preselected quality.

Numerous other miscellaneous mechanical techniques are available which effectively upgrade the material such as equipment for skinning and dehairing fish and meat, removing mussel shells, destemming and pitting fruit, and so on [19].

1.6 Blanching

Most vegetables and some fruits are blanched prior to further processing operations such as canning, freezing, or dehydration. Blanching is a mild heat treatment, but is not a method of preservation *per se*. It is a pretreatment usually performed between preparation and subsequent processing. Blanching consists of heating the food rapidly to a predetermined temperature, holding for a specified time, then either cooling rapidly or passing immediately to the next processing stage.

1.6.1 Mechanisms and Purposes of Blanching

Plant cells are discrete membrane-bound structures contained within semi-rigid cell walls. The outer or cytoplasmic membrane acts as a skin, maintaining turgor pressure within the cell. Loss of turgor pressure leads to softening of the tissue. Within the cell are a number of organelles including the nucleus, vacuole, chloroplasts, chromoplasts, and mitochondria. This compartmentalization is essential to the various biochemical and physical functions. Blanching causes cell death, and physical and metabolic chaos within the cells. The heating effect leads to enzyme destruction as well as damage to the cytoplasmic and other membranes, which become permeable to water and solutes. An immediate effect is the loss of turgor pressure. Water and solutes may pass into and out of the cells, a major consequence being nutrient loss from the tissue. Also cell constituents, which had previously been compartmentalized in subcellular organelles, become free to move and interact within the cell.

The major purpose of blanching is frequently to inactivate enzymes that would otherwise lead to quality reduction in the processed product. For example, with frozen foods, deterioration could take place during any delay prior to processing, during freezing, during frozen storage, or subsequent thawing. Similar considerations apply to processing, storage, and rehydration of dehydrated foods. Enzyme inactivation prior to heat sterilization is less important as the severe processing will destroy any enzyme activity, but there may be an appreciable time before the food is heated to sufficient temperature, so quality may be better maintained if enzymes are destroyed prior to heat sterilization processes such as canning.

It is important to inactivate quality-changing enzymes, that is, enzymes which will give rise to loss of color or texture, the production of off-odors and flavors or breakdown of nutrients. Many such enzymes have been studied including a range of peroxidases, catalases, and lipoxygenases. Peroxidase and to a lesser extent catalase are frequently used as indicator enzymes to determine the effectiveness of blanching. Although other enzymes may be more important in terms of their quality-changing effect, peroxidase is chosen because it is extremely easy to measure and it is the most heat resistant of the enzymes in question. More recent work indicates that complete inactivation of peroxidase may not be necessary and retention of a small percentage of the enzyme following blanching of some vegetables may be acceptable [30].

Blanching causes the removal of gases from plant tissues, especially intercellular gas. This is especially useful prior to canning where blanching helps achieve vacua in the containers, preventing expansion of air during processing, and hence reducing strain on the containers and the risk of misshapen cans. In addition, removing oxygen is useful in avoiding oxidation of the product and corrosion of the can. Removal of gases, along with the removal of surface dust, has a further effect in brightening the color of some products, especially green vegetables.

Shrinking and softening of the tissue is a further consequence of blanching. This is of benefit in terms of achieving filled weight into containers, so, for example, it may be possible to reduce the tin plate requirement in canning. It may also facilite the filling of containers. It is important to control the time/temperature conditions to avoid overprocessing leading to excessive loss of texture in some processed products. Calcium chloride addition to blanching water helps to maintain the texture of plant tissue through the formation of calcium pectate complexes. Some weight loss from the tissue is inevitable as both water and solutes are lost from the cells.

A further benefit is that blanching acts as a final cleaning and decontamination process. Selman [30] described the effectiveness of blanching in removing pesticide residues or radionuclides from the surface of vegetables, while toxic constituents naturally present (such as nitrites, nitrates, and oxalate) are reduced by leaching. Very significant reductions in microorganism content can be achieved, which is useful in frozen or dried foods, where surviving organisms can multiply on thawing or rehydration. It is also useful before heat sterilization if large numbers of microorganisms are present before processing.

1.6.2 Processing Conditions

It is essential to control the processing conditions accurately to avoid loss of texture (see Section 1.6.2), weight, color, and nutrients. All water-soluble materials, including minerals, sugars, proteins, and vitamins, can leach out of the tissue, leading to nutrient loss. In addition, some nutrient loss (especially ascorbic acid) occurs through thermal lability and, to a lesser extent, oxidation. Ascorbic acid is the most commonly measured nutrient with respect to blanching [30], as it covers all eventualities, being water soluble and hence prone to leaching from cells, thermally labile, as well as being subject to enzymic breakdown by ascorbic acid oxidase during storage. Wide ranges of vitamin C breakdown are observed depending on the raw material and the method and precise conditions of processing. The aim is to minimize leaching and thermal breakdown while completely eliminating ascorbic acid oxidase activity such that vitamin C losses in the product are restricted to a few percent. Generally steam blanching systems (see Section 1.6.3) give rise to lower losses of nutrients than immersion systems, presumably because leaching effects are less important.

Blanching is an example of unsteady state heat transfer involving convective heat transfer from the blanching medium and conduction within the food piece. Mass transfer of material into and out of the tissue is also important. The precise blanching conditions (time and temperature) must be evaluated for the raw material and will usually represent a balance between retaining the quality characteristics of the raw material and avoiding overprocessing. The following factors must be considered:

- fruit or vegetable properties, especially thermal conductivity, which will be determined by type, cultivar, degree of maturity, and so on;
- overall blanching effect required for the processed product, which could be expressed in many ways including: achieving a specified central temperature; achieving a specified level of peroxidase inactivation; retaining a specified proportion of vitamin C;
- size and shape of food pieces;
- method of heating and temperature of blanching medium.

Time/temperature combinations vary very widely for different foods and different processes and must be determined specifically for any situation. Holding times of 1–15 min at 70–100 °C are normal.

1.6.3 Blanching Equipment

Blanching equipment is described by Fellows [18]. The two main approaches in commercial practice are to convey the food through saturated steam or hot water.

Cooling may be with water or air. Water cooling may increase leaching losses but the product may also absorb water leading to a net weight gain. Air cooling leads to weight loss by evaporation but may be better in terms of nutrient retention.

Conventional steam blanching consists of conveying the material through an atmosphere of steam in a tunnel on a mesh belt. Uniformity of heating is often poor where food is unevenly distributed, and the cleaning effect on the food is limited. However, the volumes of wastewater are much lower than for water blanching. Fluidized bed designs and *individual quick blanching* (a three-stage process in which vegetable pieces are heated rapidly in thin layers by steam, held in a deep bed to allow temperature equilibration, followed by cooling in chilled air) may overcome the problems of nonuniform heating and lead to more efficient systems.

The two main conventional designs of hot water blancher are reel and pipe designs. In reel blanchers the food enters a slowly rotating mesh drum which is partly submerged in hot water. The heating time is determined by the speed of rotation. In pipe blanchers, the food is contacted with hot water recirculating through a pipe. The residence time is determined by the length of the pipe and the velocity of water.

There is much scope for improving energy efficiency and recycling water in either steam or hot water systems. Blanching may be combined with peeling and cleaning operations to reduce costs.

Microwave blanching has been demonstrated on an experimental scale but is too costly at present for commercial use.

1.7
Sulfiting of Fruits and Vegetables

Sulfur dioxide (SO_2) or inorganic sulfites (SO_3^{2-}) may be added to foods to control enzymic and nonenzymic browning, to control microbial growth, or as bleaching or reducing agents or antioxidants. The main applications are preserving or preventing discoloration of fruit and vegetables. The following sulfiting agents are permitted by European law: sulfur dioxide, sodium sulfite, sodium hydrogen sulfite, sodium metabisulfite, potassium metabisulfite, calcium sulfite, calcium hydrogen sulfite, and potassium hydrogen sulfite. However, sulfites have some disadvantages, notably dangerous side effects for asthmatics, and their use has been partly restricted by the US Food and Drug Administration.

Sulfur dioxide dissolves readily in water to form sulfurous acid (H_2SO_3), and the chemistry of sulfiting agents can be summarized as follows:

$$H_2SO_3 \Leftrightarrow H^+ + HSO_3^- \qquad pK_1 \approx 2$$

$$HSO_3^- \Leftrightarrow H^+ + SO_3^{2-} \qquad pK_1 \approx 7$$

Most foods are in the pH range 4–7, and therefore the predominant form is HSO_3^-. Sulfites react with many food components, including aldehydes, ketones, reducing sugars, proteins, and amino acids to form a range of organic sulfites [31]. It is not clear exactly which reactions contribute to the beneficial applications of sulfites in the food industry. It should be noted that some of the reactions lead to undesirable consequences, in particular, leading to vitamin breakdown. For example, Bender [32] reported losses of thiamin in meat products and fried potatoes when sulfiting agents were used during manufacture. On the other hand, the inhibitory effect of sulfiting agents on oxidative enzymes (e.g., ascorbic acid oxidase) may aid the retention of other vitamins, including ascorbic acid and carotene.

Sulfites may be used to inhibit and control microorganisms in fresh fruit and fruits used in the manufacture of jam, juice, or wine. In general, the antimicrobial action follows the order [31]: Gram-negative bacteria > Gram-positive bacteria > molds > yeasts. The mechanism(s) of action are not well understood although it is believed that undissociated H_2SO_3 is the active form, thus the treatment is more effective at low pH ($\leq$4).

A more widespread application is the inhibition of both enzymic and nonenzymic browning. Sulfites form stable hydroxysulfonates with carbonyl compounds, and hence prevent browning reactions by binding carbonyl intermediates such as quinones. In addition, sulfites bind reducing sugars, which are necessary for nonenzymic browning, and inhibit oxidative enzymes including polyphenoloxidase, which are responsible for enzymic browning. Therefore sulfite treatments can be used to preserve the color of dehydrated fruits and vegetables. For example, sun-dried apricots may be treated with gaseous SO_2 to retain their natural color [33], the product containing 2500–3000 ppm of SO_2. Sulfiting has commonly been used to prevent enzymic browning of many fruits and vegetables including peeled or sliced apple and potato, mushrooms for processing, grapes, and salad vegetables.

References

1. Seaton, H.L. (1955) Scheduling plantings and predicting harvest maturities for processing vegetables. *Food Technol.*, **9**, 202–209.
2. Maestrelli, A. (2000) Fruit and vegetables: the quality of raw material in relation to freezing, in *Managing Frozen Foods* (ed. C.J. Kennedy), Woodhead Publishing, Cambridge, pp. 27–55.
3. Mohsenin, N.N. (1989) *Physical Properties of Food and Agricultural Materials*, Gordon and Breach Science Publishers, New York.
4. Kader, A.A. (2008) Perspective: flavour quality of fruits and vegetables. *J. Sci. Food Agric.*, **88**, 1863–1868.
5. Chung, O.K. and Pomeranz, Y. (2000) Cereal processing, in *Food Proteins: Processing Applications* (eds S. Nakai and H.W. Modler), Wiley-VCH Verlag GmbH, Weinheim, pp. 243–307.
6. Nakai, S. and Wing, P.L. (2000) Breadmaking, in *Food Proteins: Processing Applications* (eds S. Nakai and H.W. Modler), Wiley-VCH Verlag GmbH, Weinheim, pp. 209–242.
7. Dobraszczyk, B.J. (2001) Wheat and flour, in *Cereals and Cereal Products: Chemistry and Technology* (eds D.A.V. Dendy and B.J. Dobraszczyk), Aspen, Gaithersburg, pp. 100–139.
8. Finch, H.J.S., Samuel, A.M., and Lane, G.P.F. (2002) *Lockhart and Wiseman's Crop Husbandry*, 8th edn, Woodhead Publishing, Cambridge.
9. Ng-Kwai-Hang, K.F. and Grosclaude, F. (2003) Genetic polymorphism of milk proteins, in *Advanced Dairy Chemistry, Proteins–Part B*, Vol. 1 (eds P.F. Fox and P.L.H. McSweeney), Kluwer Academic/Plenum, New York, pp. 739–816.

10. Nottingham, S. (1999) *Eat Your Genes*, Zed Books, London.
11. Thompson, A.K. (2003) *Fruit and Vegetables: Harvesting, Handling and Storage*, Blackwell Publishing, Oxford.
12. Nascimento Nunes, M.C. (2008) *Color Atlas of Postharvest Quality of Fruits and Vegetables*, Blackwell Publishing, Oxford.
13. Aked, J. (2002) Maintaining the post-harvest quality of fruits and vegetables, in *Fruit and Vegetable Processing: Improving Quality* (ed. W. Jongen), Woodhead Publishing, Cambridge, pp. 119–149.
14. Thompson, A.K. (1998) *Controlled Atmosphere Storage of Fruits and Vegetables*, CAB International, Wallingford.
15. Heap, R., Kierstan, M., and Ford, G. (1998) *Food Transportation*, Blackie, London.
16. Ahvenian, R. (2000) Ready-to-use Fruit and Vegetable. Fair-Flow Europe technical manual F-FE 376A/00, Fair-Flow, London.
17. Orea, J.M. and Gonzalez Urena, A. (2002) Measuring and improving the natural resistance of fruit, in *Fruit and Vegetable Processing: Improving Quality* (ed. W. Jongen), Woodhead Publishing, Cambridge, pp. 233–266.
18. Fellows, P.J. (2009) *Food Processing Technology: Principles and Practice*, 3rd edn, Woodhead Publishing, Cambridge.
19. Saravacos, G.D. and Kostaropoulos, A.E. (2002) *Handbook of Food Processing Equipment*, Kluwer Academic, London.
20. Brennan, J.G., Butters, J.R., Cowell, N.D., and Lilly, A.E.V. (1990) *Food Engineering Operations*, 3rd edn, Elsevier Applied Science, London.
21. Peleg, K. (1985) *Produce Handling, Packaging and Distribution*, AVI, Westport.
22. Slade, F.H. (1967) *Food Processing Plant*, Vol. 1, Leonard Hill, London.
23. Cubeddu, R., Pifferi, A., Taroni, P., and Torricelli, A. (2002) Measuring fruit and vegetable quality: advanced optical methods, in *Fruit and Vegetable Processing: Improving Quality* (ed. W. Jongen), Woodhead Publishing, Cambridge, pp. 150–169.
24. Nicolai, B.M., Beullens, K., Bobelyn, E., Peirs, A., Saeys, W., Theron, K.I., and Lammertyn, J. (2007) Nondestructive measurement of fruit and vegetable quality by means of NIR spectroscopy: a review. *Postharvest Biol. Technol.*, **46**, 99–118.
25. Qin, J. and Lu, R. (2008) Measurement of the optical properties of fruits and vegetables using spatially resolved hyperspectral diffuse reflectance imaging technique. *Postharvest Biol. Technol.*, **49**, 355–365.
26. Abbott, J.A., Affeldt, H.A., and Liljedahl, L.A. (2002) Firmness measurement in stored "Delicious" apples by sensory methods, Magness-Taylor, and sonic transmission. *J. Am. Soc. Hortic. Sci.*, **117**, 590–595.
27. Mizrach, A. (2008) Ultrasonic technology for quality evaluation of fresh fruit and vegetables in pre- and postharvest processes. *Postharvest Biol. Technol.*, **48**, 315–330.
28. Valero, C., Crisosto, C.H., and Slaughter, H. (2007) Relationship between non-destructive firmness measurements and commercially important ripening fruit stages for peaches, nectarines and plums. *Postharvest Biol. Technol.*, **44**, 248–253.
29. Evans, K., Garcia, S., Douglas-Mickey, J., Schroen, J.P., and Hitchcock, B. (2009) Method for producing juice having pre-selected properties and characteristics. US Patent 2,009,081,339 (A1).
30. Selman, J.D. (1987) The blanching process, in *Developments in Food Processing*, Vol. 4 (ed. S. Thorne), Elsevier Applied Science, London, pp. 205–249.
31. Chang, P.Y. (2000) Sulfites and food, in *Encyclopaedia of Food Science and Technology*, Vol. 4 (ed. F.J. Francis), John Wiley & Sons, Ltd, Chichester, pp. 2218–2220.
32. Bender, A.F. (1987) Nutritional changes in food processing, in *Developments in Food Processing*, Vol. 4 (ed. S. Thorne), Elsevier Applied Science, London, pp. 1–34.
33. Ghorpade, V.M., Hanna, M.A., and Kadam, S.S. (1995) *Handbook of Fruit Science and Technology: Production, Composition, Storage and Processing*, Marcel Dekker, New York, pp. 335–361.

2
Thermal Processing

Michael J. Lewis and Soojin Jun

2.1
Introduction

> "The examination of the "preserved" meats for the Navy has been resumed this week by the board of examiners appointed by the Admiralty, with the same results as last week. Eighteen cases, each containing from 10 lb were opened before one was found containing food fit for human sustenance. Subsequently, 306 canisters were opened, and the following results were arrived at: 264 cans unfit and 42 cans edible. The larger quantity was taken out to Spitalhead and thrown into the sea, and the remainder given to the poor.
> In many of the cases it appeared that the putridity had arisen from the atmosphere not being thoroughly expelled previously to the meat being put in, whilst in others there were indications of the animal either having died from disease, or of its having been slaughtered in a very inefficient manner. The stench arising from the examination of such a mass of putridity was so great, that it was impossible for the officials to carry out their duty without frequent and copious supplies of chloride of lime to the floor. Now and then a canister would emit such an odious stench as to cause all operations to be suspended for some minutes, and one was so overpowering that the examiners and their assistants had to beat a hasty retreat from the room."

Today there are a wide variety of thermally processed foods available to the consumer. Thankfully, we have moved a long way from the situation described above, during an inspection of canned meat at the naval dockyards in Portsmouth in 1861. However, even today incidents of this nature, although not quite so extreme, do still occur. So it is the aim of this chapter to review factors affecting the safety and quality of heat-preserved foods.

Thermal processing involves heating food, either in a sealed container or by passing it through a heat exchanger followed by packaging. It is important to ensure that the food is adequately heat treated and to minimize post-processing

Food Processing Handbook, Second Edition. Edited by James G. Brennan and Alistair S. Grandison.

contamination. The food should then be cooled quickly and it may require refrigerated storage or be stable at ambient temperature. The heating process can be either batch or continuous. In all thermal processes, the aim should be to heat and cool the product as quickly as possible. This has economic implications and may also lead to an improvement in quality. Heat or energy (in joules) is transferred from a high to a low temperature, the rate of heat transfer being proportional to the temperature difference. Therefore, high-temperature driving forces will promote heat transfer. SI units for rate of heat transfer (J s^{-1} or W) are mainly used but Imperial Units (Btu/h) may also be encountered [1, 2]. The heating medium is usually saturated steam or hot water. For temperatures above 100 °C, the steam and the hot water will be above atmospheric pressure. Cooling is achieved using either mains water or chilled water, brines, or glycol solutions. Regeneration is used in continuous processes to further reduce energy utilization, see Section 2.3.2.

2.1.1 Reasons for Heating Foods

Foods are heated for a number of reasons, the main one being to inactivate pathogenic or spoilage microorganisms. It may also be important to inactivate enzymes, to avoid the browning of fruit by polyphenol oxidases and minimize flavor changes resulting from lipase and proteolytic activity. The process of heating a food will also induce physical changes and chemical reactions, such as starch gelatinization, protein denaturation, or browning, which in turn will affect the sensory characteristics, such as color, flavor, and texture, either advantageously or adversely. For example, heating pretreatments are used in the production of evaporated milk to prevent gelation and age-thickening and for yoghurt manufacture to achieve the required final texture in the product. Heating processes may also alter the nutritional value of the food.

Thermal processes vary considerably in their intensity, ranging from mild processes such as thermization and pasteurization through to more severe processes such as in-container sterilization. The severity of the process will affect both the shelf life as well as other quality characteristics.

Foods that are heat treated can either be solid or liquid, so the mechanisms of conduction and convection may be involved. Since solid foods are poor thermal conductors of heat, it can take a long time to heat products such as meat or fish. However, for fluids there is more molecular mobility (compared to solids) and this improves heat transfer. Fluids range from those having a low viscosity (1–10 mPas), through to highly viscous fluids: the presence of particles (up to 25 mm in diameter) further complicates the process as it becomes necessary to ensure that both the liquid and solid phases are at least adequately and if possible equally heated. The presence of dissolved air in either of these phases is a problem as it becomes less soluble as temperature increases and will come out of solution. Air is a poor heat transfer fluid and for this reason hot air is rarely used as a heating medium.

Attention should be paid to removing air from steam, for example, venting of steam retorts and removing air from sealed containers (exhausting).

Heating is also involved in many other operations, which will not be covered in such detail in this chapter, such as evaporation and drying (see Chapter 3). It is also used for solids; in processing powders and other particulate foods: for example, extrusion, baking (see Chapter 9) and spice sterilization.

2.1.2 Safety and Quality Issues

The two most important issues connected with thermal processing are food safety and food quality. The major safety issue involves inactivating pathogenic microorganisms which are of public health concern. The World Health Organization estimates that there are over 100 million cases of food poisoning each year worldwide and that 1 million of these result in death. These pathogens will show considerable variation in their heat resistance. Some, such as *Campylobacter, Salmonella, Lysteria,* and, of more recent concern, *Escherichia coli* 0157, are heat labile and are inactivated by pasteurization. Others, such as *Bacillus cereus,* which may survive pasteurization and also grow at low temperatures, have greater heat resistance. The most heat-resistant pathogenic bacterial spore is *Clostridium botulinum.* As well as these major foodborne pathogens, it is important to inactivate microorganisms that cause food spoilage, such as yeasts, molds, and gas-producing and souring bacteria. Again there is considerable variation in their heat resistance, the most heat resistant being the spores of *Bacillus stearothermophilus.* The heat resistance of any microorganism will change as environmental conditions, for example, pH, water activity, or chemical composition change, and foods themselves provide a complex and variable environment. New microorganisms may also be encountered, such as *Bacillus sporothermodurans* (see Section 2.4.3.1). Therefore it is important to be aware of the type of microbial flora associated with all raw materials which are to be heat treated.

After processing, it is important to avoid contaminating the product; this is known generally as *post-processing contamination* and can cause problems in both pasteurization and sterilization. To avoid post-processing contamination, raw materials and finished product should not be allowed in close proximity of each other. Other safety issues are concerned with natural toxins, pesticides, herbicides, antibiotics and growth hormones, and environmental contaminants. Again, it is important that steps are taken to ensure that these do not finish up in the final product. Recently there have been some serious cases of strong allergic reactions, with some deaths, shown by some individuals to foods such as peanuts and shellfish. These are all issues which also need to be considered for heat-treated foods.

Quality issues revolve round minimizing chemical reactions and loss of nutrients and ensuring that sensory characteristics (appearance, color, flavor, and texture) are acceptable to the consumer. Quality changes, which may result from enzyme activity, must also be considered. There may also be conflicts between safety and quality issues. For example, microbial inactivation and food safety is increased by more severe heating conditions, but product quality will in general deteriorate. To summarize, it is important to understand reaction kinetics and how they relate to:

- microbial inactivation;
- chemical damage;
- enzyme inactivation; and
- physical changes.

2.1.3 Product Range

The products covered in this chapter include those that can be filled into containers and subsequently sealed and heat treated and those that can be processed by passing them through a continuous heat exchanger. This latter category includes milks and milk-based drinks, fruit and vegetable juices, purees, soups and sauces (both sweet and savory), and a range of products containing particulate matter, up to about 25 mm diameter. There are two distinct market sectors: (i) products given a mild heat treatment and then kept refrigerated throughout storage; and (ii) those that are sterilized and stored at ambient temperature. The relative importance of these two sectors varies from country to country for different products and from country to country for the same product. For example, in England and Wales, pasteurized milk, which is stored chilled, accounts for about 87% of liquid milk consumption, while ultra high temperature (UHT) and sterilized milk accounts for about 10%. In other countries, however, UHT milk accounts for a much greater proportion of milk consumed (e.g., France 78% and Germany 58%), so it is important to note that there are regional differences and preferences [3].

In general, heat processing eliminates the need for further additives to extend the shelf life, although additives may help improve the sensory characteristics or make processes less susceptible to fouling. In addition to reactions taking place during the heat treatment, chemical, enzymatic, and physical changes will continue to take place during storage. Microorganisms which survive the heat treatment may also grow if conditions are favorable. Pasteurized products are normally kept refrigerated during storage to retard microbial growth and low temperatures must be maintained throughout the cold chain. On the other hand, sterilized products are not normally refrigerated and are stored at ambient temperature. This may vary considerably throughout the world, ranging from below 0 °C to above 50 °C. All the reactions mentioned above are temperature dependent and considerable changes may take place during the storage period. One example is browning of milk and milk products, which is very significant after four months of storage at 40 °C. Changes during storage are discussed in more detail for pasteurized products in Section 2.4.2 and for sterilized products in Section 2.4.3.

2.2 Reaction Kinetics

2.2.1 Microbial Inactivation

All thermal processes involve three distinct periods: a heating period, a holding period, and a cooling period. In theory, all three may contribute to the reactions taking place, although in situations where heating and cooling are rapid, the holding period is the most significant. To determine the overall effect procedures are needed to evaluate each of these periods individually. By far the easiest to deal with is the holding period, as this takes place at constant temperature. It needs to be established how reaction rates are affected by changes in temperature during heating and cooling. To simplify the analysis, microbial inactivation is first measured at constant temperature. This is usually followed by observing how microbial inactivation changes with temperature.

2.2.2 Heat Resistance at Constant Temperature

When heat inactivation studies are carried out at constant temperature, it is often observed that microbial inactivation follows first order reaction kinetics, that is, the rate of inactivation is directly proportional to the population. This can be illustrated by plotting the log of the population against time (Figure 2.1) and finding that there is a straight line relationship.

The heat resistance of an organism is characterized by its *decimal reduction time* (D_T), which is defined as the time required to reduce the population by 90% or by 1 order of magnitude or one log cycle, that is, from 10^4 to 10^3, at a constant temperature T.

Every microorganism will have its own characteristic heat resistance and the higher its D value, the greater will be its heat resistance. Heat resistance is also affected by a wide range of other environmental factors, such as pH, water activity, and the presence of other solutes, such as sugars and salts.

The extent of microbial inactivation is measured by the number of decimal reductions which can be achieved. This is given by $\log(N_0/N)$, where N_0 is the initial population and N is the final population. It is determined from the following equation:

$$\log\left(\frac{N_0}{N}\right) = \frac{\text{Heating time}}{D_T} \tag{2.1}$$

There are two important aspects associated with first order reaction kinetics: (i) it is not theoretically possible to achieve 100% reduction and (ii) for a specific heat treatment, the final population will increase as the initial population increases. For example, for an organism with a D_{70} value of 10 s, heating for 10 s at 70 °C will achieve a 90% reduction in the population, 20 s heating will achieve $2D$ (99%) reduction, 30 s will achieve $3D$ (99.9%) and 60 s $6D$ or (99.9999%) reduction.

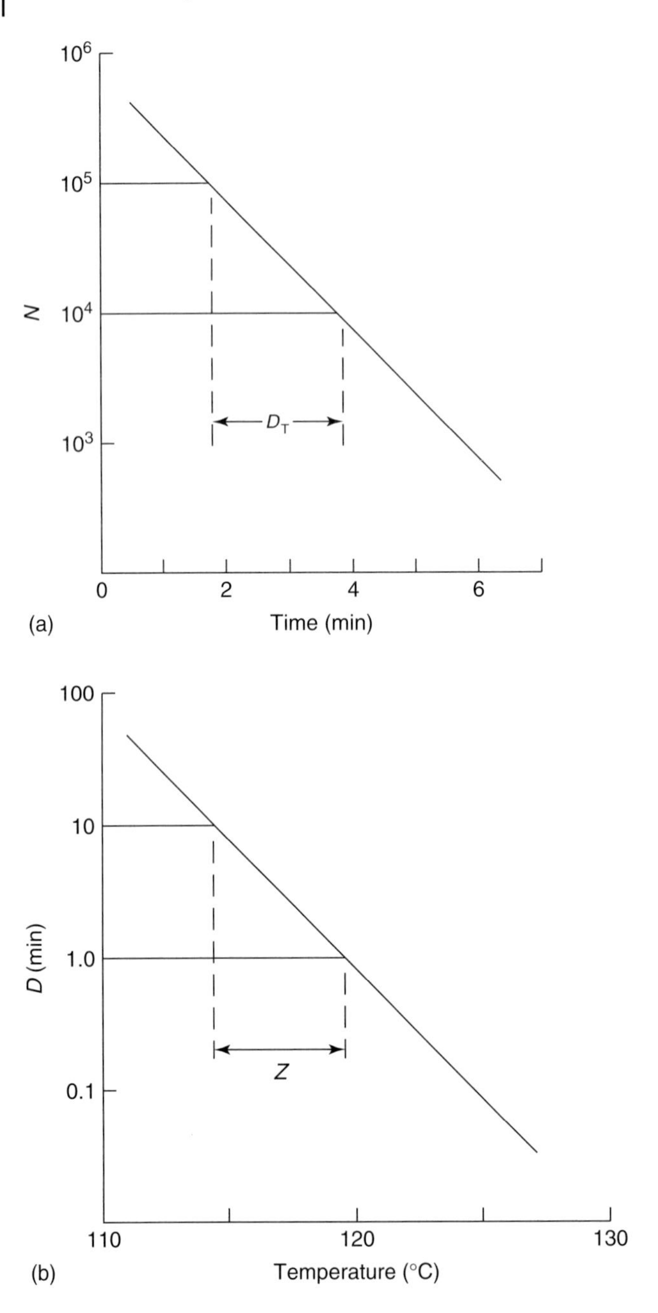

Figure 2.1 (a) Relationship between the population of microorganisms and time at a constant heating temperature and (b) relationship between the decimal reduction time and temperature, to determine the z value. From Lewis [1] with permission.

Although it is not theoretically possible to achieve 100% reduction, in practical terms this may appear to be the case. For sterilization processes, the term *commercial sterility* is used, rather than absolute sterility, to indicate that there will always be a small chance that one or more microorganisms will survive the heat treatment. Increasing the severity of the heating process, for example, by prolonging the time period or by using higher temperatures, will reduce the chance of finding survivors of any particular bacteria. Thus, if the initial population had been 10^6 ml^{-1}, after 80 s heating, the final population would be 10^{-2} ml^{-1}. It may be difficult to imagine a fraction of an organism, but another way of expressing this is 1 organism/100 ml. Thus, if the sample had been packaged in 1-ml portions, it should be possible to find one surviving microorganism in every 100 samples analyzed; that is, 99 samples would be free of microorganisms. The same heat treatment given to a raw material with a lower count might give one surviving microorganism in every 10 000 ml. Thus if 10 000 (1 ml) samples were analyzed, 9999 would be free of viable microorganisms. Note that just one surviving bacteria in a product or package can give rise to a spoiled product or package. In practice, deviations from first order reaction kinetics are often encountered and the reasons for this are discussed in more detail by Gould [4].

For pasteurization processes the temperature range of interest is 60–90 °C. Sterilization is a more severe process and temperatures in excess of 100 °C are needed to inactivate heat-resistant spores. Sterilization processes can be either in a sealed container, 110–125 °C for 10–120 min, or by continuous flow techniques, using temperatures in the range 135–142 °C for several seconds. In both cases, the amount of chemical reaction is increased and the flavor is different to pasteurized milk. The product will have a shelf life of up to six months at ambient storage conditions.

2.3 Temperature Dependence

As mentioned, most processes do not take place at constant temperature, but involve heating and cooling periods. Therefore, although it is easy to evaluate the effect of the holding period on the heat resistance and the lethality (i.e., the number of decimal reductions) of a process (see Equation 2.1), it is important to appreciate that the heating and cooling may also contribute to the overall lethality. It can easily be demonstrated that reaction rates increase as temperature increases and microbial inactivation is no exception. Food scientists use a parameter known as the *z value*, to describe temperature dependence. This is based on the observation that over a limited temperature range there is a linear relationship between the log of the decimal reduction time and the temperature (see Figure 2.1).

The z value for inactivation of any particular microorganism is the temperature change which results in tenfold change in the decimal reduction time. For most heat-resistant spores it is about 10 °C, whereas the z value for vegetative bacteria is considerably lower, usually between 4 and 8 °C. A low z value implies that the

Table 2.1 Values of D and z for microbial inactivation, enzyme inactivation and some chemical reactions.

Microbe	D_{121} (°C)	z (°C)
Bacillus stearothermophilus NCDO 1096, milk	181.0	9.43
B. stearothermophilus FS 1518, conc. milk	117.0	9.35
B. stearothermophilus FS 1518, milk	324.0	6.7
B. stearothermophilus NCDO 1096, milk	372.0	9.3
B. subtilis 786, milk	20.0	6.66
B. coagulans 604, milk	60.0	5.98
B. cereus, milk	3.8	35.9
Clostridium sporogenes PA 3679, conc. milk	43.0	11.3
C. botulinum NCTC 7272	3.2	36.1
C. botulinum (canning data)	13.0	10.0
Proteases inactivation	0.5–27.0 min at 150 °C	32.5–28.5
Lipases inactivation	0.5–1.7 min at 150 °C	42.0–25.0
Browning	–	28.2; 21.3
Total whey protein denaturation, 130–150 °C	–	30.0
Available lysine	–	30.1
Thiamin (B_1) loss	–	31.4–29.4
Lactulose formation	–	27.7–21.0

From Lewis and Heppell [7] with permission.

reaction in question is very temperature sensitive. In general, microbial inactivation is very temperature sensitive, with inactivation of vegetative bacteria being more temperature sensitive than that of heat-resistant spores.

In contrast to microbial inactivation, chemical reaction rates are much less temperature sensitive, having higher z values (20–40 °C) (Table 2.1). This is also the case for many heat-resistant enzymes, although heat-labile enzymes such as alkaline phosphatase or lactoperoxidase are exceptions to this rule. This difference for chemical reactions and microbial inactivation has some important implications for quality improvement when using higher temperatures for shorter times, and this is discussed in more detail in Section 2.4.2.1.

The relationship between D values at two different temperatures and the z value is given by:

$$\log\left(\frac{D_1}{D_2}\right) = \left[\frac{(\theta_2 - \theta_1)}{z}\right] \tag{2.2}$$

An alternative way of using z values is for comparing processes. For example, if it is known that a temperature of 68 °C is effective for 10 min and the z value is 6 °C; then equally effective processes would also be; 62 °C for 100 min or 74 °C for 1 min, that is, 1 min at θ is equivalent to 0.1 min at $(\theta + z)$ or 10 min at $(\theta - z)$.

Thus Equation 2.2 can be rewritten, replacing decimal reduction time (D) by the processing time (t), as:

$$\log\left(\frac{t_1}{t_2}\right) = \left[\frac{(\theta_2 - \theta_1)}{z}\right] \tag{2.3}$$

In the context of milk pasteurization, if the holder process (63 °C/30 min) is regarded as being equivalent to the high-temperature short-time (HTST) process of 72 °C for 15 s, the z value would come out to be about 4.3 °C.

Another approach is to use this concept of equivalence, together with a reference temperature. For example, 72 °C is used for pasteurization and 121.1 and 135 °C for sterilization processes (see later). Perhaps best known are the standard lethality tables, used in the sterilization of low-acid foods.

Thus, the lethality at any experimental temperature (T) can be compared to that at the reference temperature (θ), using the following equation:

$$\log L = \frac{(T - \theta)}{z} \tag{2.4}$$

For a standard reference temperature of 121.1 °C and an experimental temperature of 118 °C, using $z = 10$ °C, then $L = 0.490$. Thus 1 min at 118 °C would be equivalent to 0.490 min at 121.1 °C. Note that a temperature drop of 3 °C will halve the lethality. Some D and z values for heat-resistant spores and chemical reactions are given in Table 2.1.

Q_{10} is another parameter used to measure temperature dependence. It is defined as the ratio of reaction rate at $T + 10$ to that at T, that is, the increase in reaction rate caused by an increase in temperature of 10 °C.

It is interesting to note that, in spite of the many deviations reported, this log-linear relationship still forms the basis for thermal process calculations in the food industry. Gould [4] has surmised that there is a strong view that this relationship remains at least a very close approximation of the true thermal inactivation kinetics of spores. Certainly, the lack of major problems when sterilization procedures are properly carried out according to the above principles has provided evidence over many years that the basic rationale, however derived, is sound, even though it may be cautious.

2.3.1 Batch and Continuous Processing

A process such as pasteurization can be done batch-wise or continuously. Batch processing involves filling the vessel, heating, holding, cooling, emptying the vessel and filling into containers, and cleaning the vessel. Holding times may be up to 30 min. An excellent account of batch pasteurization is provided by Cronshaw [5]. Predicting the heating and cooling times involves unsteady state heat transfer and illustrates the exponential nature of the heat transfer process. The heating time is determined by equating the rate of heat transfer from the heating medium to the rate at which the fluid absorbs energy. Thus

$$UA\,(\theta_{\mathrm{h}} - \theta) = mc\frac{\mathrm{d}\theta}{\mathrm{d}t} \qquad (2.5)$$

which on integration becomes:

$$t = \frac{mc}{UA}\ln\left[\frac{(\theta_{\mathrm{h}} - \theta_{\mathrm{I}})}{(\theta_{\mathrm{h}} - \theta_{\mathrm{f}})}\right] \qquad (2.6)$$

where m is mass (kg); c is specific heat (J kg^{-1} K^{-1}); A is surface area (m^2); U is the overall heat transfer coefficient (OHTC) (W m^{-2} K^{-1}), and t is the heating time (s) required to raise the temperature from the initial temperature (θ_{I}) to the final temperature (θ_{f}), using a heating medium temperature θ_{h}.

The dimensionless temperature ratio represents the ratio of the initial temperature driving force to the approach temperature. The concept of approach temperature (i.e., how close the product approaches the heating or cooling medium temperature) is widely used in continuous heat exchangers. Heating and cooling times can be long.

Batch processes were easy to operate, flexible, able to deal with different size batches and different products, and this is still the case; also, if well mixed, there is no distribution of residence times, which is a problem with continuous processes. Heating and cooling rates are slower, however, and the operation is more labor intensive; it will involve filling, heating, holding, cooling, emptying, cleaning, and disinfecting, which may take up to 2 h. Post-processing contamination should be avoided in the subsequent packaging operations.

For both pasteurization and sterilization the alternative is a continuous process. Some advantages of continuous processes are as follows:

- Foods can be heated and cooled more rapidly compared with in-container processes: this improves the economics of the process and the quality of the product.
- There are none of the pressure constraints which apply to heating products in sealed containers. This allows the use of higher temperatures and shorter times, which results in less damage to the nutrients and improved sensory characteristics, such as appearance, color, flavor, and texture.
- Continuous processes provide scope for energy savings through using the hot fluid to heat the incoming fluid; this is known as *regeneration* and can save both heating and cooling costs (see Section 2.3.2).

Heating processes can be classified as direct or indirect. The most widely used is indirect heating, where the heat transfer fluid and the liquid food are separated by a barrier; for in-container sterilization this will be the wall of the bottle and for continuous processes, the heat exchanger plate or tube wall. In direct processes, steam is the heating medium and the steam comes into direct contact with the product (see Section 2.3.2).

The mechanisms of heat transfer are *conduction* in solids and *convection* in liquids. Thermal conductivity (W m^{-1} K^{-1}) is the property which measures the rate of heat transfer due to conduction. Metals are good conductors of heat. Although stainless steel has a much lower value (~20 W m^{-1} K^{-1}) than both copper (~400 W m^{-1} K^{-1}) and aluminum (~220 W m^{-1} K^{-1}), it is much higher than that of glass (~0.5 W m^{-1} K^{-1}). In general, foods are poor conductors of heat (≅0.5 W m^{-1} K^{-1}) and this can be a problem when heating particulate systems (see Section 2.5).

Heat transfer by convection is the predominant mechanism in fluids and is inherently faster than conduction heating. The efficiency of heat transfer by convection is measured by the *heat film coefficient*. Condensing steam has a much higher heat film coefficient (as well as a high latent heat of vaporization) than hot water, which in turn is higher than that of hot air. Heat transfer by convection is inherently faster than heat transfer by conduction.

In the indirect process, there are three resistances to the transfer of heat from the bulk of a hot fluid to the bulk of a cold fluid, two due to convection and one due to conduction. The overall heat transfer coefficient (U) provides a measure of the efficiency of the heat transfer process and takes into account all three resistances. It can be calculated from:

$$\frac{1}{U} = \frac{1}{h_1} + \frac{1}{h_2} + \frac{L}{k} \tag{2.7}$$

where h_1 and h_2 are the heat film coefficients (W m^{-2} K^{-1}) for the hot fluid and cold fluids respectively, L is heat exchanger wall thickness (m), and k is the thermal conductivity of the plate or tube wall (W m^{-1} K^{-1}).

The higher the value of U, the more efficient is the heat exchange system. Each of the terms in Equation 2.7 represents a resistance. The highest of the individual terms is known as the *limiting resistance*. This is the one that controls the overall rate of heat transfer. Thus, to improve the performance of a heat exchanger, it is best to focus on the limiting resistance.

The basic design equation for a heat exchanger is as follows:

$$Q = UA\Delta\theta_{\mathrm{m}} \tag{2.8}$$

where Q is the duty or rate of heat transfer (J s^{-1}), $\Delta\theta_m$ (°C or K) is the log mean temperature difference, and A is the surface area (m^2).

The duty (Q) is obtained from the following expression:

$$Q = m'c\Delta\theta \tag{2.9}$$

where m' is the mass flow rate (kg s^{-1}), c is the specific heat capacity (J kg^{-1} K^{-1}), and $\Delta\theta$ is the change in product temperature (K or °C).

In a continuous heat exchanger, the two fluids can either flow in the same direction (co-current) or in opposite directions (counter-current). Counter-current is the preferred direction as it results in a higher mean temperature driving force and a closer approach temperature.

One of the main practical problems associated with indirect heating is *fouling*. This is the formation of deposits on the wall of the heat exchanger, which will introduce one or two additional resistances to heat transfer and lead to a reduction in U. Fouling may be the result of deposits from the food or deposits from the service fluids in the form of sediment from steam, hardness from water, or microbial films from cooling water. Fouling may result in a decrease in product temperature and eventually in the product being underprocessed. A further problem is a reduction in the cross-sectional area of the flow passage, which will lead to a higher pressure drop. Fouled deposits will also need to be removed at the end of the process as they may serve as a breeding ground for bacteria, particularly thermoduric bacteria. For example, for milk, fouling becomes more of a problem as the processing temperature increases and the acidity of the milk increases.

Fouling is discussed in more detail by Lewis and Heppell [7] and strategies for minimizing fouling for milk products are discussed by Lewis and Deeth [8]. Heat exchangers also need to be cleaned and disinfected after their use (see Section 2.5.3).

2.3.2 Continuous Heat Exchangers

The viscosity of the product is a major factor that affects the choice of the most appropriate heat exchanger and the selection of pumps. The main types of indirect heat exchanger for fluids such as milk, creams, and liquid egg are the *plate heat exchanger* and the *tubular heat exchanger*. As product viscosity increases, this will give rise to a higher pressure drop, which might cause a problem in the cooling section, especially when phase transitions such as gelation or crystallization take place. For more viscous products, such as starch-based desserts, a scraped surface heat exchanger may be used, see Section 2.5.

One of the main advantages of continuous systems over batch systems is that energy can be recovered in terms of regeneration. The layout for a typical regeneration section is shown in Figure 2.2. The hot fluid (pasteurized or sterilized) can be used to heat the incoming fluid, thereby saving on heating and cooling costs. The regeneration efficiency (RE) is defined as follows:

$$\mathrm{RE} = \frac{\text{amount of heat supplied by regeneration}}{\text{amount of heat required assuming no regeneration}} \times 100 \qquad (2.10)$$

Regeneration efficiencies up to 95% can be obtained, which means that a pasteurized product requiring heating to 72 °C would be heated up to almost 68 °C by regeneration. Although high regeneration efficiencies result in considerable savings in energy, they necessitate the use of larger surface areas because of the lower temperature driving force and there is a slightly higher capital cost for the heat exchanger. This also means that the heating and cooling rates are also slower, and the transit times longer, which may affect the quality, especially in UHT processing.

Plate heat exchangers (Figure 2.3) are widely used for both pasteurization and sterilization processes. They have a high OHTC and are generally more compact than tubular models. Their main limitation is pressure, with an upper limit of about 20 bar. The normal gap width between the plates is between 2.5 and 5 mm but wider gaps are available for viscous liquids. The narrower the gap, the more pressure it will take and wide gap plates are not in regular use for UHT treatment

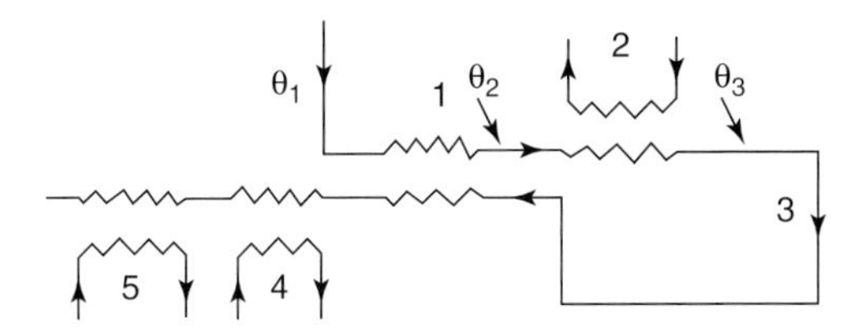

Figure 2.2 Heat exchanger sections for a high-temperature short-time (HTST) pasteurizer: 1, regeneration; 2, hot water section; 3, holding tube; 4, mains water cooling; and 5, chilled water cooling. From Lewis [1] with permission.

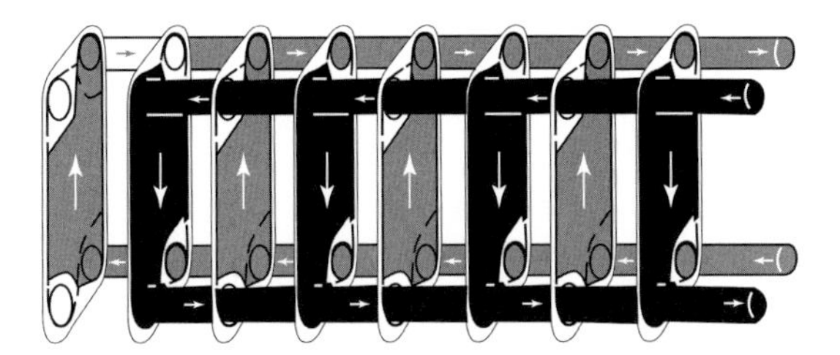

Figure 2.3 Flow through a plate heat exchanger, by courtesy of APV.

of low-acid foods. In general, a plate heat exchanger is the cheapest option and the one most widely used for low-viscosity fluids. However, maintenance costs may be high, as gaskets may need replacing at regular intervals and the integrity of the plates needs to be checked at regular intervals, especially those in the regeneration section, where a cracked or leaking plate may allow raw product, for example, raw milk, to contaminate already pasteurized milk or sterilized milk.

Tubular heat exchangers (Figure 2.4) have a lower OHTC than plates and generally occupy a larger space. They have slower heating and cooling rates with a longer transit time through the heat exchanger. In general they have fewer seals and provide a smoother flow passage for the fluid. A variety of tube designs are available to suit different product characteristics. These include single tubes with an outer jacket, double or multiple concentric tubes or shell, and tube types. Most UHT plants use a multi-tube design. They can withstand higher pressures than plate heat exchangers. Although, they are still susceptible to fouling, high pumping pressures can be used to overcome the flow restrictions. Thus, tubular heat exchangers give longer processing times with viscous materials and with products that are more susceptible to fouling.

For products containing fat, such as milk and cream, homogenization (see Chapter 11) must be incorporated to prevent fat separation. This may be upstream or downstream. For UHT processes, downstream homogenization requires the process to be achieved under aseptic conditions and provides an additional risk of recontaminating the product. Upstream homogenization removes the need to operate aseptically, but is thought to produce a less stable emulsion.

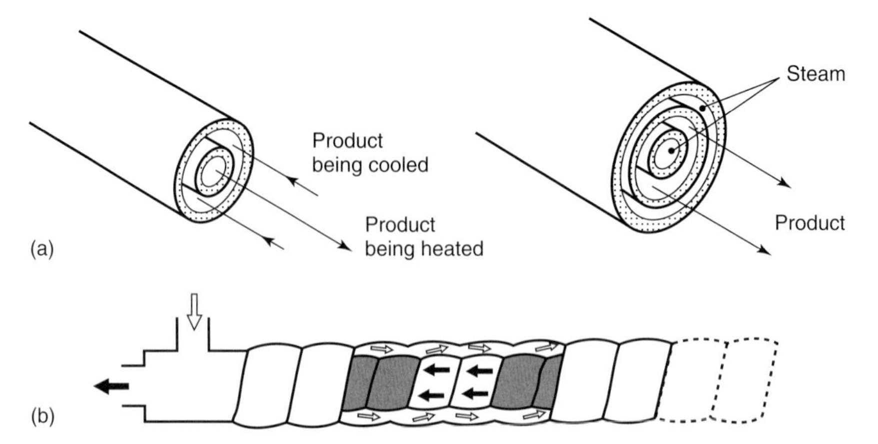

Figure 2.4 Types of concentric tube heat exchangers: (a) plain wall and (b) corrugated spiral wound. From Lewis and Heppell [7] with permission.

2.3.2.1 Direct Heating

In direct processes, the product is preheated up to a temperature of 75 °C, often by regeneration, before being exposed to culinary steam to achieve a temperature of 140–145 °C. The steam should be free of chemical contaminants and saturated to avoid excessive dilution of the product (which is between about 10 and 15%). This heating process is very rapid. The product is held for a few seconds in a holding tube. Added water is removed by flash cooling, which involves a sudden reduction in pressure to bring the temperature of the product down to between 75 and 80 °C. This sudden fall in temperature is accompanied by the removal of some water vapor. There is a direct relationship between the fall in temperature and the amount of water removed. The final temperature and hence the amount of water removed is controlled by the pressure (vacuum) in the flash cooling chamber. This cooling process, like the heating, is very rapid.

As well as removing the added water (as vapor), flash cooling removes other volatile components, which in the case of UHT milk gives rise to an improvement in the flavor. Direct processes employ a short, sharp, heating profile and result in less chemical damage compared to an equivalent indirect process of similar holding time and temperature. They are also less susceptible to fouling and will give long processing times, but their regeneration efficiencies are usually below 50%. Direct processes are usually employed for UHT rather than pasteurization processes.

There are two principal methods of bringing the steam and the food liquid together. Steam can either be injected into the liquid (injection processes) or liquid can be injected into the steam (infusion). There is a school of thought that claims that infusion is less severe than injection since the product has less contact with hot surfaces. However, direct experimental evidence is scant. There is no doubt that direct processes (both injection and infusion) produce a less intense cooked flavor than any indirect process, although claims that direct UHT milk is indistinguishable from pasteurized milk are not always borne out. Successful operation depends upon maintaining a steady vacuum, as the flash cooling vessel operates at the boiling point of the liquid. If the pressure fluctuates, the boiling point will also fluctuate and this will lead to boiling over if the pressure suddenly drops for whatever reason. Thus maintaining a steady vacuum is a major control point in the operation of these units. Note that some indirect UHT plants may incorporate a de-aeration unit, which operates under the same principle. The effects of heating and cooling profiles are compared for UHT products in Section 2.4.3.2.

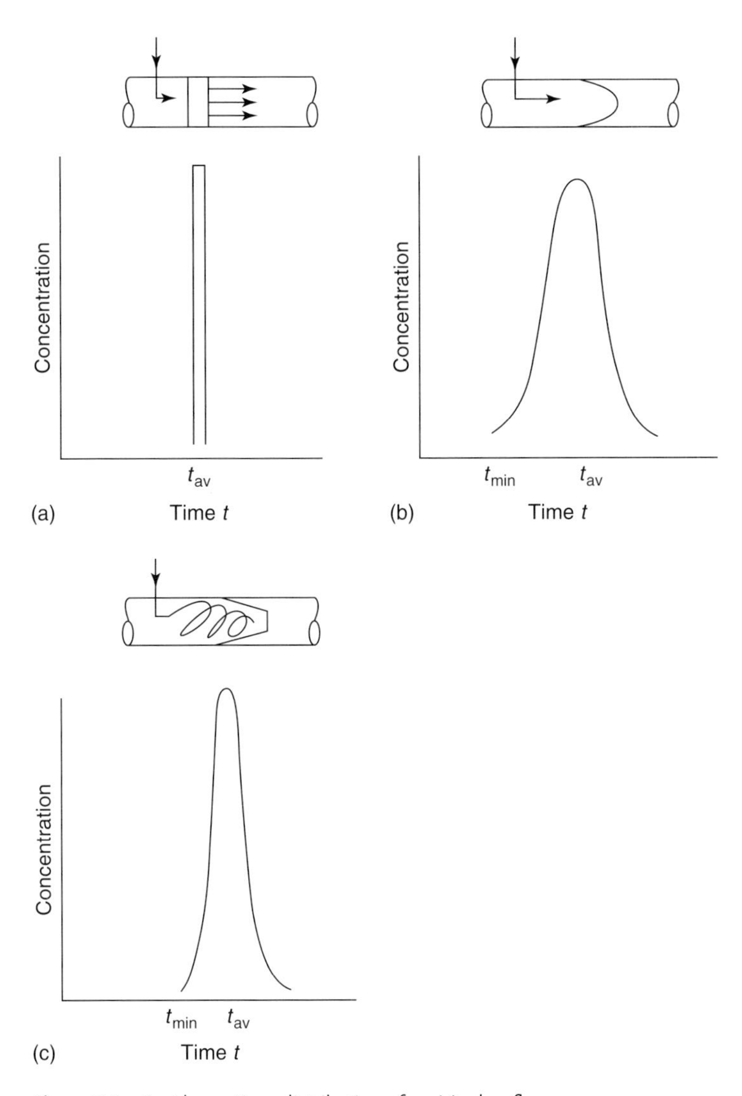

Figure 2.5 Residence time distributions for: (a) plug flow, (b) streamline flow, and (c) turbulent flow. From Lewis [1] with permission.

In continuous processes there is a distribution of residence times. It is important to know whether the flow is streamline or turbulent as this will influence heat transfer rates and the distribution of residence times within the holding tube and also the rest of the plant. This can be established by evaluating the Reynolds number (Re):

$$\mathrm{Re} = \frac{vD\rho}{\mu} = \frac{4Q\rho}{\Pi\mu D} \tag{2.11}$$

where v is average fluid velocity (m s^{-1}), ρ is fluid density (kg m^{-3}), D is pipe diameter (m), μ is fluid viscosity (Pas), and Q is volumetric flow rate (m^3 s^{-1}).

Note that the average residence time (based upon the average velocity) can be determined from t_{av} = volume tube/volumetric flow rate.

For viscous fluids, the flow in the holding tube is likely to be streamline, that is, Re would be less than 2000 and there will be a wide distribution of residence times. For Newtonian fluids the minimum residence time will be half the average residence time. Turbulent flow (Re > 4100) will result in a narrower distribution of residence times, with a minimum residence time of 0.83 times the average residence time. Figure 2.5 illustrates residence time distributions for three situations, namely, plug flow, streamline flow, and turbulent flow. Plug flow is the ideal situation with no spread of residence times, but for both streamline and turbulent flow, the minimum residence time should be greater than the stipulated residence time, to avoid underprocessing. Residence time distributions and their implications for UHT processing are discussed in more detail by Lewis and Heppell [5] and Burton [23].

2.4 Heat Processing Methods

The main types of heat treatment – pasteurization and sterilization – are covered below. But first we discuss a lesser used process, used specifically for raw milk, known as *thermization*.

2.4.1 Thermization

Thermization is a mild process which is designed to increase the keeping quality of raw milk. It is used mainly when it is known that it may not be possible to use raw milk immediately for conversion to other products, such as cheese or milk powder. The aim is to reduce the presence of psychrotropic bacteria, which can release heat-resistant protease and lipase enzymes into the milk. These enzymes will not be inactivated during pasteurization and may give rise to off-flavors if the milk is used for cheese or milk powders. Temperatures used are 58–68 °C for 15 s. Raw milk thus treated can be stored at a maximum of 8 °C for up to three days [8]. It is usually followed later by a more severe heat treatment.

Thermized milk should show the presence of alkaline phosphatase to distinguish it from pasteurized milk. To the author's knowledge, there is no equivalent process for other types of foods.

2.4.2 Pasteurization

Pasteurization is a mild heat treatment that is used on a wide range of different types of food products. The two primary aims of pasteurization are to remove pathogenic bacteria from foods, thereby preventing disease, and to remove spoilage (souring) bacteria to improve its keeping quality. It largely stems from the discovery by Louis Pasteur in 1857 that souring in milk could be delayed by heating milk to between 122 and 142 °F (50.0 and 61 °C), although it was not firmly established that causative agents of spoilage and disease were microorganisms until later in that century. However, even earlier than this, foods were being preserved in sealed containers by a process of sterilization, so for some considerable time foods were being supplied for public consumption without an understanding of the mechanism of preservation involved. In fact, the first stage in the history of pasteurization between 1857 and the end of the nineteenth century might well be called the *medical stage*, as the main history in heat treating milk came chiefly from the medical profession interested in infant feeding. By 1895 it was recognized that a thoroughly satisfactory product can only be secured where a definite quantity of milk is heated for a definite length of time at a definite temperature.

By 1927, North and Park [9] had established a wide range of time–temperature conditions for inactivating tubercle bacilli, ranging from 130 °F for 60 min up to 212 °F for 10 s. The effectiveness of heat treatments was determined by inoculating into guinea pigs samples of milk which had been heavily infected with tubercle organisms and then subjected to different time/temperature combinations and noting those conditions which did not kill the animals. The use of alkaline phosphatase as an indicator was first investigated in 1933 and is still standard practice. Pasteurization is now accepted as the simplest method to counter milkborne pathogens and has become commonplace, although there are still some devotees of raw milk.

The International Dairy Federation [10] definition of pasteurization is as follows: "pasteurization is a process applied to a product with the objective of minimizing possible health hazards arising from pathogenic microorganisms associated with the product (milk) which is consistent with minimal chemical, physical, and organoleptic changes in the product." This definition is also applicable to products other than milk, including, creams, ice cream mix, eggs, fruit juices, fermented products, soups, and other beverages.

In the early days of pasteurization, milk only had a short shelf life as domestic refrigeration was not widespread, but the introduction of refrigeration in the 1940s had an almost immediate impact on keeping quality. Initially there was considerable resistance expressed by the general public to the introduction of milk pasteurization, not dissimilar to that now being encountered by irradiation [11].

Pasteurization does not inactivate all microorganisms; those that survive pasteurization are termed *thermodurics* and those that survive a harsher treatment at 80–100 °C for 30 min are termed *spore formers*. Traditionally, pasteurization was carried out in a batch process – the Holder process – at 63 °C for 30 min, but this was replaced by the introduction and acceptance of continuous HTST processes.

2.4.2.1 **HTST Pasteurization**

The HTST process was first investigated in the late 1920s. It was approved for milk in the United States in 1933 and approval for the process was granted in the United Kingdom in 1941. Continuous operations offer a number of advantages over batch processes, such as faster heating and cooling rates, shorter holding times and regeneration, which saves both heating and cooling costs and contributes to the low processing costs incurred in thermal processing operations, compared to many novel techniques. Scales of operations on continuous heat exchangers range between 500 and 50 000 $l\,h^{-1}$, with experimental models down to 50 $l\,h^{-1}$. Continuous processing introduces some additional complications which have since been resolved; these include flow control, flow diversion, and distribution of residence times.

The schematic for the flow of fluid through the heat exchanger and the heat-exchange sections are shown in Figures 2.2 and 2.6. The fluid first enters the regeneration section, where it is heated from θ_1 to θ_2 by the fluid leaving the holding tube. It then enters the main heating section, where it is heated to the pasteurization temperature θ_3. It then passes through the holding tube. The tube is constructed such that the minimum residence time exceeds the stipulated residence time and this can be determined experimentally at regular intervals. It then passes back into the regeneration section, where it is cooled to θ_4. This is followed by further cooling sections, employing mains water and chilled water. The mains water section is usually dispensed with where it may heat the product rather than cool it, for example, at high regeneration efficiencies or high mains water temperatures.

As regeneration efficiency increases, the size and capital cost of the heat exchanger increases and heating rate decreases, which may affect quality, but this is more noticeable in UHT sterilization [5]. Heating profiles tend to become more linear at high regeneration efficiencies. Other features are a float-controlled balance tank, to ensure a constant head to the feed pump and a range of screens and filters to remove any suspended debris from the material. In most pasteurizers, one pump is used. It is crucial that the flow rate remains constant, despite any disturbances in feed composition temperature or changes in the system characteristics. The two most common options are a centrifugal pump with a flow controller or a positive displacement pump. If the product is to be homogenized, the homogenizer itself is a positive pump and is sized to control the flow rate.

In the majority of pasteurizers, the final heating process is provided by a hot water set. Steam is used to maintain the temperature of the hot water at a constant value, somewhere between 2 and 10 °C higher than the required pasteurization temperature. Electrical heating can be used, typically in locations where it would be

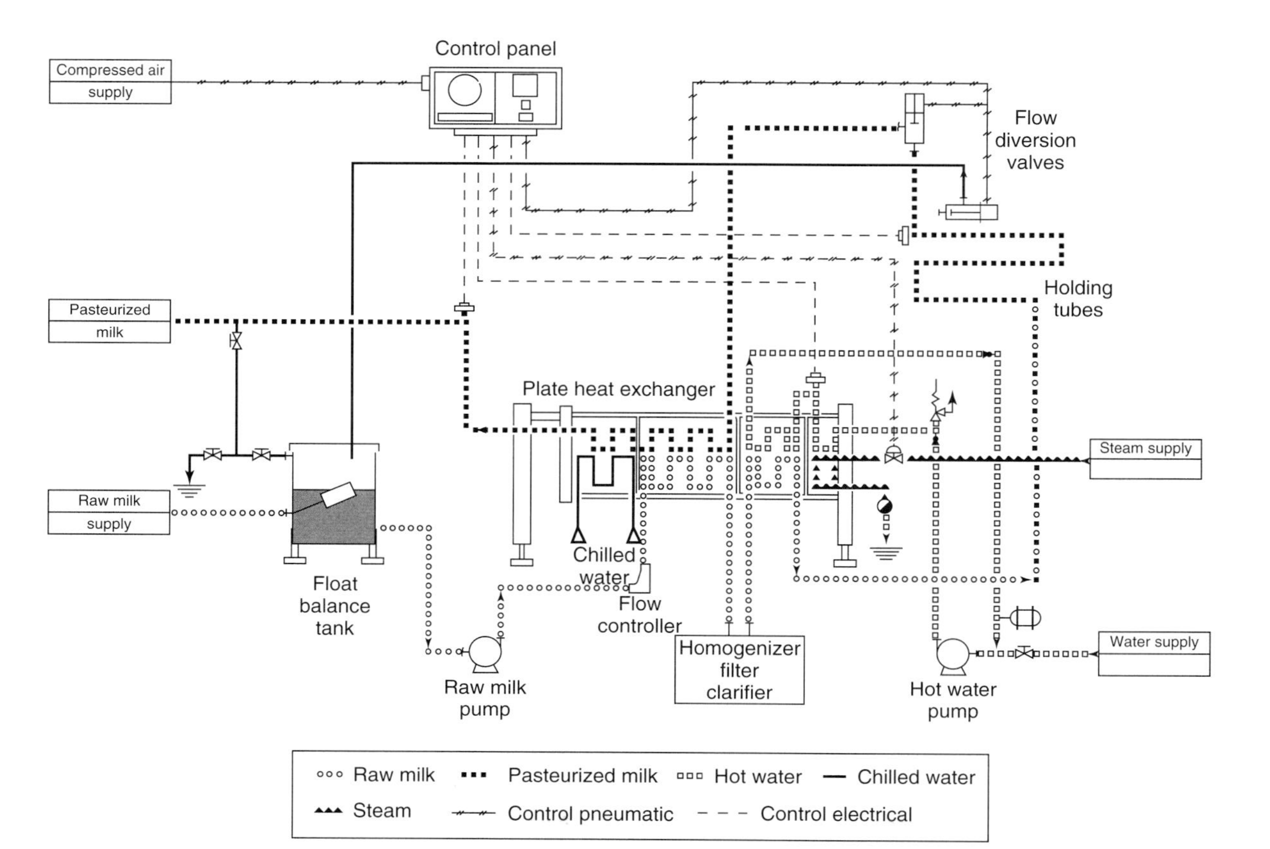

Figure 2.6 Typical milk pasteurization system. From Pearse [18] with permission.

costly or difficult to install a steam generator (boiler). The holding system is usually a straightforward holding tube, with a temperature recording probe in the beginning and a flow diversion valve at the end. The position of the temperature probe in the holding tube is one aspect for consideration. When positioned at the beginning, there is more time for the control system to respond to underprocessed fluid as it passes through the tube, but it will not measure the minimum temperature obtained, as there will be a reduction in temperature due to heat loss as the fluid flows through the tube. This could be reduced by insulating the holding tube, but it is not generally considered to be a major problem on commercial pasteurizers.

Ideally, the temperature control should be within $\pm 0.5\ ^{\circ}$C. Note that a temperature error of $1\ ^{\circ}$C, will lead to a reduction of about 25% in the process lethality (calculated for $z = 8\ ^{\circ}$C). From the holding tube, in normal operation, it goes back into regeneration, followed by final cooling. The pasteurized product may be packaged directly or stored in bulk tanks.

In principle, a safe product should be produced at the end of the holding tube. However, there may be an additional contribution to the total lethality of the process from the initial part of the cooling cycle. Thereafter, it is important to prevent recontamination, both from dirty pipes and from any recontamination with raw feed. Should this occur it is known as post-pasteurization contamination, which can be a major determinant of keeping quality. Failures of pasteurization processes have resulted from both causes. The most serious incidents have caused food poisoning outbreaks and have arisen where pasteurized milk has been recontaminated with raw milk containing pathogens. Such contamination may arise for a number of reasons, all of which involve a small fraction of raw milk not going through the holding tube (Figure 2.7), One explanation lies with pinhole leaks or cracks in the plates, which may appear with time due to corrosion. In plate heat exchangers, the integrity of the plates needs to be tested for these. This is most critical in the regeneration section; where there is a possibility of contamination from raw to treated, that is, from high to low pressure. An additional safeguard is the incorporation of an extra pump, to ensure that the pressure on the pasteurized side is higher than that on the raw side, but this further complicates the plant. In some countries this requirement may be incorporated into the heat treatment regulations. Another approach is the use of double-walled plates, which

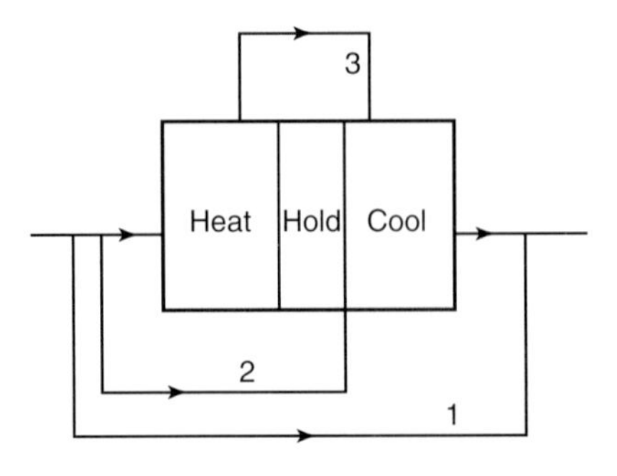

Figure 2.7 Bypass routes in a commercial pasteuriser: 1, via cleaning routes; 2, via flow diversion route; and 3, via regeneration section (e.g., pinhole leak in plates). From Lewis and Heppell [7] with permission.

also increases the heat transfer area by about 15–20% due to the air gap. These and other safety aspects have been discussed by Sorensen [13].

Pinholes in the heating and cooling sections could also lead to product dilution or product contamination in the hot water or chilled water sections, which may result in unexpected microbiological hazards. It is important that plates are regularly pressure-tested and the product tested by depression of freezing point to ensure that it has not been diluted. Similar problems may arise from leaking valves, either in recycle or detergent lines.

The introduction of more sensitive instrumentation for detecting phosphatase activity has made it easier to detect whether pasteurized milk has been recontaminated with raw milk. It is claimed that raw milk contamination as low as 0.01% can be detected (Section 2.4.2). Miller Jones [14] has documented a major pasteurization failure in America in which over 16 000 people were infected with salmonella and 10 died. The cause was believed to be due to post-processing contamination caused by a section of the plant which was not easy to drain and clean, hence leading to recontamination of the already pasteurized milk.

Some further considerations of the engineering aspects are provided by Kessler [34], the Society of Dairy Technology [16], and Hasting [12]. Fouling is not considered to be as much of a problem as that found in UHT processing. However, with longer processing times and poorer quality raw materials it may have to be accounted for and some products such as eggs may be more prone to fouling. One important aspect is a reduction in residence time due to fouling in the holding tube [12].

The trend is for HTST pasteurization plants to run for much longer periods (16–20 h), before cleaning and shutdown. Again, monitoring phosphatase activity at regular intervals is useful to ensure uniform pasteurization throughout and for detecting more subtle changes in plant performance, which may lead to a better estimation of when cleaning is required. However, it has also been suggested that there is an increase in thermophilic bacteria due to an accumulation of such bacteria in the regenerative cooling section arising toward the end of such long processing runs.

Other important issues are fouling, cleaning, and disinfecting, which are all paramount to the economics of the process. A recent review of pasteurization has been provided by Lewis [17].

2.4.2.2 Tunnel (Spray) Pasteurizers

Tunnel or spray pasteurizers are widely used in the beverage industry for continuous heating and cooling of products in sealed containers. They are ideal for high-volume throughput. Examples of such products are soft and carbonated drinks, juices, beers, and sauces. Using this procedure, post-processing contamination should be very much reduced, the major cause being defective seams on the containers. There are three main stages in the tunnel: heating, holding, and cooling, and in each stage water at the appropriate temperature is sprayed onto the container. Since heating rates are not as high as for plate or tubular heat exchangers, these processes are more suited to longer time/lower temperature processes. The total transit time may be about 1 h, with holding temperatures between 60 and 70 °C for about 20 min. Pearse [18] cites 60 °C for 20 min being a proven time/temperature profile. There is some scope for regeneration in these units.

2.4.2.3 Extended Shelf Life Products

There may be a requirement to further increase the shelf life of heat-treated products, either for the convenience of the processor and the consumer or to provide additional protection against temperature abuse, and more severe heating conditions can be used. However, for milk products it is important to avoid the development of cooked flavor, which would result from higher temperatures. There are some early references to time/temperature conditions required to induce a cooked flavor [5]. It is my experience that this occurs at a temperature of about 85–90 °C for 15 s. Therefore one approach is to use temperatures above 100 °C for shorter times. Wirjantoro and Lewis [19] showed that milk heated to 115 °C for 2 s had a much better keeping quality than milk heated at either 72 °C for 15 s or 90 °C for 15 s. There is no doubt that temperatures in the range 115–120 °C for 1–5 s are more effective than temperatures below 100 °C for extending the shelf life of refrigerated products.

A second approach is to use small amounts of a bacteriocin, such as nisin. The addition of small amounts of nisin (40 IU ml^{-1}) has been found to be effective in reducing microbial growth following heat treatment at 72 °C/15 s and more so at 90 °C/15 s. It was particularly effective at inhibiting *Lactobacillus* at both temperatures. Results for milk heat treated at 117 °C for 2 s with 150 IU ml^{-1} nisin were even more spectacular. Such milks have been successfully stored for over 150 days at 30 °C with only very low levels of spoilage [20]. Local regulations would need to be checked to establish whether nisin is a permitted additive in milk and milk-based beverages.

As a word of caution, it is important to eliminate post-pasteurization contamination, as nisin is not effective against Gram-negative contaminants, such as pseudomonads [20, 21]. A further strategy is to store pasteurized products at 2 °C, rather than 5–7 °C. This would further increase keeping quality but it may not be practicable to do this.

2.4.3 Sterilization

2.4.3.1 In-Container Processing

Sterilization of foods by the application of heat can either be in sealed containers or by continuous flow techniques. Traditionally it is an in-container process, although there have been many developments in container technology since the process was first commercialized at the beginning of the nineteenth century. Whatever the process, the main concerns are with food safety and quality. The most heat-resistant pathogenic bacterium, *Clostridium botulinum*, will not grow below a pH of 4.5. On this basis, the simplest classification is to categorize foods as either as acid foods (pH $<$ 4.5) or low-acid foods (pH $>$ 4.5). Note that a broader classification has been used for canning: low acid, pH > 5.0; medium acid, pH 4.5–5.0; acid, pH 4.5–3.7 and high acid, <3.7. However, as mentioned earlier, the main concern is with foods where the pH is greater than 4.5. For such foods the minimum recommended process is to achieve $12D$ reductions for *C. botulinum*. This is known as the *minimum botulinum cook* and requires heating at 121 °C for 3 min, measured at the slowest heating point. The evidence for this producing a safe process for sterilized foods is provided by the millions of units of heat-preserved foods consumed worldwide each year, without any botulinum-related problems.

The temperature of 121.1 °C (250 °F) is taken as a reference temperature for sterilization processes. This is used in conjunction with the z value for *C. botulinum*, which is taken as 10 °C, to construct the standard lethality tables (Table 2.2). Since lethalities are additive, it is possible to sum the lethalities for a process and determine the total integrated lethal effect, which is known as the *F_0 value*. In mathematical terms, F_0 is defined as the total integrated lethal effect, that is,

$$F_0 = \int L \, dt \tag{2.12}$$

It is expressed in minutes at a reference temperature of 121.1 °C, using the standard lethality tables derived for a z value of 10 °C.

For canned products, F_0 is determined by placing a thermocouple at the slowest heating point in the can and measuring the temperature throughout the sterilization process. This is known as the *general method* and is widely used to evaluate the microbiological severity of an in-container sterilization process. Other methods are available based on knowing the heating and cooling rates of the products.

The F_0 values recommended for a wide range of foods are given in Table 2.3. It can be seen from these values that some foods need well in excess of the minimum botulinum cook, that is, an F_0 of 3, with values ranging between 4 and 18, to achieve commercial sterility. This is because there are some other bacterial spores which are more heat resistant than *C. botulinum*, the most heat resistant of these being the thermophile *Bacillus stearothermophilus*, which has a decimal reduction time of about 4 min at 121 °C. Also of recent interest is the mesophilic spore-forming bacteria, *Bacillus sporothermodurans* [22]. Some heat-resistance values for other important spores are summarized in Lewis and Heppell [7] and more detailed compilations are given by Burton [23], Holdsworth [24], and Walstra *et al.* [25]. Such heat-resistant spores may cause food spoilage, either through the production of acid (souring) or through the production of gas. Again, such spores will not grow below a pH of 4.5 and many of them are inhibited at higher pH values than this. *Bacillus stearothermophilus*, for example, which causes flat-sour spoilage, will not continue to grow below a pH of about 5.2.

The severity of the process (F_0 value) selected for any food depends upon the nature of the spoilage flora associated with the food, the numbers likely to be present in that food, and to a limited extent on the size of the container, since more organisms will go into a larger container. Such products are termed *commercially sterile*, the target spoilage rate being less than 1 in 10 000 containers. It should be

Table 2.2 Lethality values, using a reference temperature of 121.1 °C, $z = 10$ °C. Lethality values are derived from $L = (T - \theta)/z$), where L is number of minutes at reference temperature equivalent to 1 min at experimental temperature, T is experimental temperature, θ is reference temperature (121.1 °C).

Processing temperature		Lethality
(°C)	(°F)	(L)
110	230.0	0.078
112	237.2	0.195
114	237.2	0.195
116	240.8	0.309
118	244.4	0.490
120	248.0	0.776
121	249.8	0.977
121.1	250.0	1.000
122	251.6	1.230
123	253.4	1.549
124	255.2	1.950
125	257.0	2.455
126	258.8	3.090
127	260.6	3.890
128	262.4	4.898
129	264.2	6.166
130	266.0	7.762
131	267.8	9.772
132	269.6	12.30
133	271.4	15.49
134	273.2	19.50
135	275.0	24.55
136	276.8	30.90
137	278.6	38.90
138	280.4	48.98
139	282.2	61.66
140	284.0	77.62
141	285.8	97.72
142	287.6	123.0
143	289.4	154.9
144	291.2	195.0
145	293.0	245.5
146	294.8	309.2
147	296.6	389.0
148	298.4	489.8
149	300.2	616.6
150	302.0	776.24

From Lewis and Heppell [7] with permission.

Table 2.3 F_0 values which have been successfully used for products on the UK market.

Product	Can size(s)	F_0 values
Baby foods	Baby food	3–5
Beans in tomato sauce	All	4–6
Peas in brine	Up to A2	6
	A2–A10	6–8
Carrots	All	3–4
Green beans in brine	Up to A2	4–6
	A2–A10	6–8
Celery	A2	3–4
Mushrooms in brine	A1	8–10
Mushrooms in butter	Up to A1	6–8
Meats in gravy	All	12–15
Sliced meat in gravy	Ovals	10
Meat pies	Tapered, flat	10
Sausages in fat	Up to 1 lb	4–6
Frankfurters in brine	Up to 16Z	3–4
Curries, meats, and vegetables	Up to 16Z	8–12
Poultry and game, whole in brine	A2.5–A10	15–18
Chicken fillets in jelly	Up to 16 oz	6–10
"Sterile" ham	1, 2 lb	3–4
Herrings in tomato	Ovals	6–8
Meat soups	Up to 16Z	10
Tomato soup, not cream of	All	3
Cream soups	A1–16Z	4–5
	Up to A10	6–10
Milk puddings	Up to 16Z	4–10
Cream	4–6 oz	3–4

From Brennan *et al.* [66] with permission.

remembered that canning and bottling operations are high-speed operations, with the production of up to 50 000 containers every hour from one single product line. It is essential that each of those containers is treated exactly the same and that products treated on every subsequent day are also subjected to the same conditions.

The philosophy for ensuring safety and quality in thermal processing is to identify the operations where hazards may occur (critical control points) and devise procedures for controlling these operations to minimize the hazards (see Chapter 15). Of crucial importance is the control of all those factors which affect heat penetration into the product and minimize the number of heat-resistant spores entering the can prior to sealing. It is also important to ensure that the closure (seal) is airtight, thereby eliminating post-processing contamination.

Since it is not practicable to measure the temperature in every can, the philosophy for quality assurance involves verifying that the conditions used throughout the canning process lead to the production of a product which is commercially sterile and ensuring that these conditions are reproduced on a daily basis. Processing conditions such as temperature and time are critical control points. Others are raw material quality (especially counts of heat-resistant spores), and controlling all factors affecting heat penetration, including filling temperature, size of headspace, ratio of solids to liquid, liquid viscosity, and venting procedures. It is also essential to reduce post-processing contamination by seal integrity, cooling water chlorination, and avoiding handling wet cans (after processing), and drying them quickly. Large pressure differentials between inside and outside of container should also be avoided. Drying cans quickly after cooling and reducing manual handling of cans are very important for minimizing post-processing contamination.

In-container sterilization involves the integration of a number of operations, all of which will contribute to the overall effectiveness of the process. These are summarized in Figure 2.8 [26].

Types of container A wide range of containers are available for in-container processes, the most common being the can. There have been many developments since the can's inception as mild steel coated with tin plate; these include soldered or welded cans, two-piece and three-piece cans; tin-free steel and aluminum. There are also many different lacquers to prevent chemical interactions between the metal and the foods (see Chapter 8). These modifications have been to effect cost reductions and provide greater convenience. Cans are able to tolerate reasonable pressure differentials.

Glass bottles and jars are also common; sterilized milk was traditionally produced in glass bottles but products in glass need to be heated and cooled more slowly to avoid breakage. There are also flexible pouches, plastic trays and bottles, metal, and glass containers. All these materials will have different wall thicknesses, different thermal conductivities, and different surface area-to-volume ratios, all of which can influence heat transfer rates and thus the quality of the final heat-treated product.

Supply of raw materials It is essential that there is a supply of raw materials of the right quality and quantity. Contracts between the supplier and the food processor should protect both their interests. For fruit and vegetables, appropriate varieties should be selected for canning, as they must be able to withstand the heat treatment without undue softening or disintegration. Also important are their sensory characteristics, spore counts, and chemical contaminants. Food to be processed should be transported as quickly as possible to the processing factory.

Preliminary operations Preliminary operations will depend upon the type of food and could include: inspection, preparation, cleaning, peeling, destoning, and size reduction. Where it is used, water quality is important and there will be considerable waste to be disposed of. Note that some pre-prepared vegetables may have been sulfited (see Chapters 1 and 20) and for sterilization in metal cans, these should be avoided as sulfite may stain and strip tin-plate.

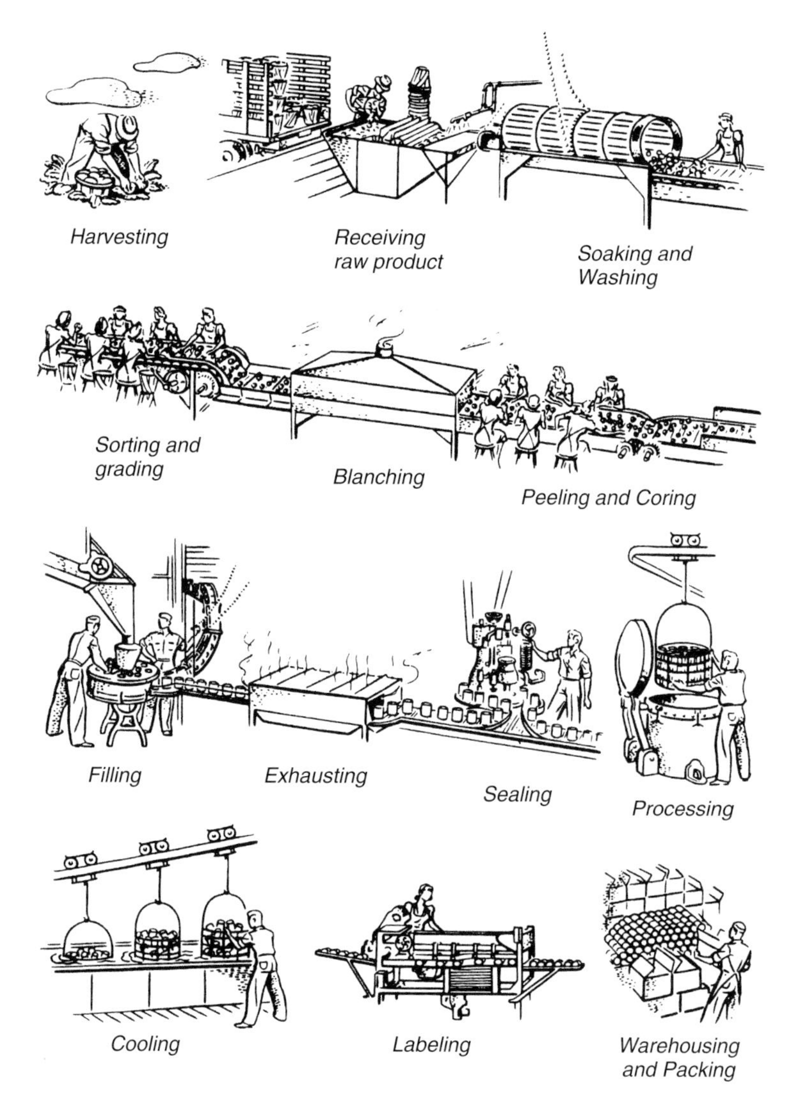

Figure 2.8 The canning process. From Jackson and Shinn [26] with permission.

Some of the processes are as follows:

- **Blanching:** This is an important operation using hot water or steam (see Chapter 1). Different products will have different time/temperatures combinations. Blanching inactivates enzymes and removes intracellular air, thereby helping to minimize the internal pressure generated on heating. It also increases the density of food and softens cell tissue, which facilitate filling and it further cleans the product as well as removing vegetative organisms. It may lead to some thermal degradation of nutrients and some leaching losses for hot-water blanching. An excellent detailed review on blanching is given by Selman [27].

- **Filling:** This is important both for the product and for any brine, syrup, or sauce that may accompany it. It is essential to achieve the correct filled and drained weights and headspace. When using hot-filling, the filling temperature must be controlled as variations will lead to variations in the severity of the overall sterilization process. Sauces, brines, and syrups may be used; their composition may be covered by Codes of Practice. One of the main reasons for their use is to improve heat transfer.
- **Exhausting:** This involves the removal of air prior to sealing, helping to prevent excessive pressure development in the container during heat treatment, which may damage the seal. Four methods are available for exhausting: mechanical vacuum, thermal exhausting, hot-filling, and steam-flow closing.
- **Sealing:** This produces an air-tight (hermetic) seal. For cans this is achieved by a double rolling process and the integrity of the seal is checked by visual inspection and by tearing down the seal and looking at the overlap and the tightness. Can seamers can handle from 50 to 2000 cans per minute (see Chapter 8).

Containers are sterilized in retorts, which are large pressure vessels (Figures 2.9 and 2.10). Batch and continuous retorts are available and the heating medium is

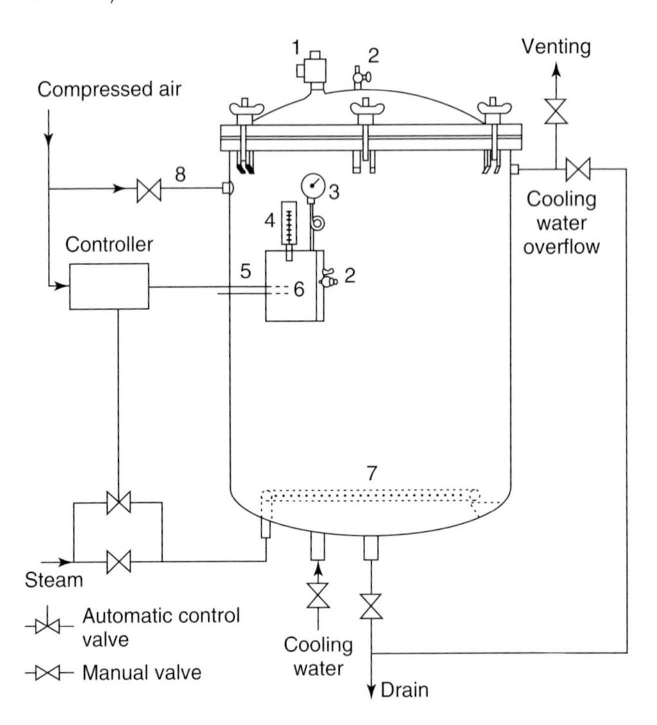

Figure 2.9 A vertical batch retort equipped for cooling under air pressure: 1, safety valve; 2, valve to maintain a steam bleed from retort during processing; 3, pressure gauge; 4, thermometer; 5, sensing element for controller; 6, thermometer box; 7, steam spreader; and 8, air inlet for pressure cooling. From Brennan *et al.* [66] with permission of the authors.

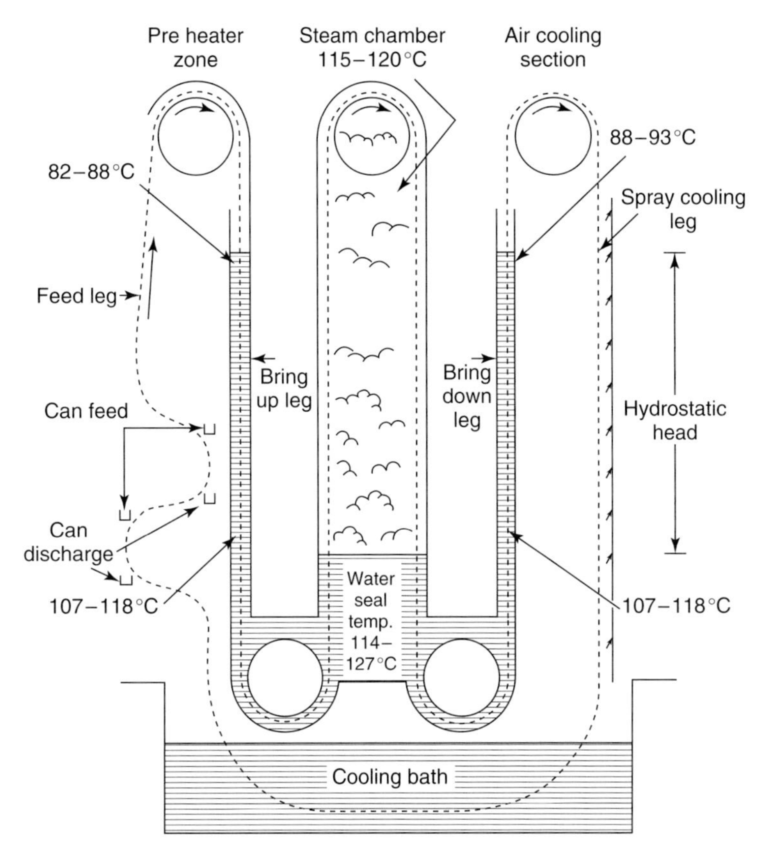

Figure 2.10 Diagram showing the principle of the hydrostatic sterilizer. From Brennan *et al.* [66] with permission of the authors.

either steam, pressurized hot water, or steam/air mixtures. For steam there is a fixed relationship between its pressure and temperature, given by the steam tables [1, 24].

The system should include an accurate temperature recording system and an indicating thermometer. A steam pressure gauge should also be incorporated as this will act indirectly as a second temperature monitoring device. Discrepancies between temperature and pressure readings will suggest that there may be some air in the steam or that the instruments are incorrect [7]. Venting involves the removal of air from the retort and venting conditions need to be established for each individual retort. Every product will have its own unique processing time and temperature and these would have been established to ensure that the appropriate F_0 value is achieved for that product. Ensuring $12D$ reductions for *Clostridium botulinum* (safety) will ensure that a food is safe but more stringent conditions may be required for commercial sterility. The processing time starts when the temperature in the retort reaches the required processing temperature.

Cooling is a very important operation and containers should be cooled as quickly as possible down to a final temperature of 35–40 °C to dry seams. As the product cools, the pressure inside the can will fall and it is important to ensure that the pressure in the retort falls at about the same rate. This is achieved by using a combination of cooling water and compressed air to avoid a sudden fall in pressure caused by steam condensation.

Water quality is important and it should be free of pathogenic bacteria. This can be assured by chlorination but an excessive amount should be avoided as this may cause container corrosion. It is also important avoid too much manual handling of wet cans to reduce the levels of post-processing contamination.

The containers are then labeled and stored. A small proportion may be incubated at elevated temperatures to observe for blown containers.

Quality assurance Strict monitoring of all these processes is necessary to ensure that in-container sterilization provides food which is commercially sterile. The target spoilage rate is <1 in 10^4 containers. There should be strict control of raw material quality, control of all factors affecting heat penetration, final product assessment (filled and drained weights, sensory characteristics, and regular seal evaluation). Use should be made of Hazard Analysis Critical Control Points (HACCPs) and advice given by the Institute of Food Science and Technology in *Food and Drink – Good Manufacturing Practice* [28].

One of the main problems with canning is that there is considerable heat damage to the nutrients and changes in the sensory characteristics; this can be assessed by the cooking value [17], where a reference temperature of 100 °C is used and a z value of 20–40 °C (typically about 33 °C). Also summarized are the z values for the sensory characteristics which range between 25 and 47 °C. Further information on canned food technology is provided by the following excellent reference works: Stumbo [29], Hersom and Hulland [30], Jackson and Shinn [26], Rees and Bettison [31], Footitt and Lewis [32], and Holdsworth [24].

It is noteworthy that most fruit and other acidic products, for example, pickles and fermented products, will need a less harsh heat treatment, and processing conditions of 100 °C for 10–20 min would suffice.

2.4.3.2 UHT Processing

More recently, continuous sterilization processes have been introduced. UHT or aseptic processing involves the production of a commercially sterile product by pumping the product through a heat exchanger. To ensure a long shelf life the sterile product is packed into pre-sterilized containers in a sterile environment (see Chapter 8). An airtight seal is formed, which prevents reinfection, in order to provide a shelf life of at least three months at ambient temperature. It has also been known for a long time that the use of higher temperatures for shorter times will result in less chemical damage to important nutrients and functional ingredients within foods, thereby leading to an improvement in product quality [15].

Sterilization of the product is achieved by rapid heating to temperatures about 140 °C, holding it for several seconds followed by rapid cooling. Ideally, heating and cooling should be as quick as possible. Indirect and direct heating methods are available (see Section 2.3). UHT products are in a good position to be able to improve the quality image of heat-processed, ambient stable foods.

Safety and spoilage considerations From a safety standpoint, the primary objective is the production of commercially sterile products, with an extended shelf life. The main concern is inactivation of the most heat-resistant pathogenic spore, namely *Clostridium botulinum*. The safety criteria used for UHT processing should be based upon those well established for canned and bottled products. The minimum F_0 value for any low-acid food should be 3. Fruit juices and other acidic products will require a less stringent process or may be heat treated in their containers, in either tunnels or oven-type equipment.

The time/temperature conditions required to achieve the minimum botulinum cook can be estimated at UHT temperatures. At a temperature of 141 °C, a time of 1.8 s would be required. There is experimental evidence to show that the data for botulinum can be extended up to 140 °C [33]. For UHT products, an approximate value of F_0 can be obtained from the holding temperature (T, °C) and minimum residence time (t, s) (Equation 2.13). In practice the real value will be higher than this estimated value because of the lethality contributions from the end of the heating period and the beginning of the cooling period as well as some additional lethality from the distribution of residence times.

$$F_0 = 10^{\frac{(T-121.1)}{10}} \cdot \frac{t}{60} \tag{2.13}$$

Therefore, the botulinum cook should be a minimum requirement for all low-acid foods, even those where botulinum has not been a problem (e.g., for most dairy products). In the United Kingdom there are statutory heat treatment regulations for some UHT products:

- milk 135 °C for 1 s;
- cream 140 °C for 2 s;
- milk-based products 140 °C for 2 s; and
- ice-cream mix 148.9 °C for 2 s.

In some cases, lower temperatures and longer times can be used, provided it can be demonstrated that the process renders the product free from viable microorganisms and their spores. If no guidelines are given, recommended F_0 values for similar canned products would be an appropriate starting point (typically 6–10 for dairy products). Sufficient pressure must also be applied in order to achieve the required temperature. A working pressure in the holding tube in excess of 1 bar over the saturated vapor pressure, corresponding to the UHT temperature, has been suggested.

UHT processes, like canned products, will also be susceptible to post-processing contamination. This will not usually give rise to a public health problem, although contamination with pathogens cannot be ruled out. However, high levels of spoiled product will not improve its quality image, particularly if not detected before it is released for sale. Contamination may arise from the product being reinfected in the cooling section of the plant, or in the pipelines leading to the aseptic holding or buffer tank or the aseptic fillers. This is avoided by heating all points downstream of the holding tube at 130 °C for 30 min. The packaging material may have defects, or the seals may not be airtight, or packaging may be damaged during subsequent handling. Any of these could result in an increase in spoilage rate.

Process characterization: safety and quality aspects As in other thermal processes, the requirements for safety and quality conflict, as a certain amount of chemical change will occur during adequate sterilization of the food. Therefore, it is important to ask what is the meant by quality and what is the scope for improving it. One aspect of quality that has already been discussed is reducing microbial spoilage. A second important aspect is minimizing chemical damage and reducing nutrient loss. In this aspect, UHT processing offers some distinct advantages over in-container sterilization. Chemical reactions are less temperature sensitive so the use of higher temperatures, combined with more rapid heating and cooling rates help to reduce the amount of chemical reaction. This has been well documented by Kessler [34, 35] and more recently by Browning *et al.* [36]. For example, reactions such as color changes, hydroxymethyl furfural formation, thiamin loss, whey protein denaturation, and lactulose formation will all be higher for in-container sterilization than for UHT processes.

There is a choice of indirect heat exchangers available, such as plates, tubular, and scraped surface, as well as direct steam injection or infusion plants, all of which will heat products at different rates and shear conditions. To better understand the quality of products produced from a UHT process knowledge of the temperature/time profile for the product is required. Some examples of such profiles for a number of different UHT process plants are shown in Figure 2.11. There are considerable differences in the heating and cooling rates for indirect processes and between the direct and indirect processes due to steam injection and flash cooling. Because of these differences, similar products processed on different plants may well be different in quality.

Two other parameters introduced for UHT processing of dairy products, but which could be more widely used for other UHT products, are B^* and C^* values [34]. The reference temperature used for this (135 °C) is much closer to UHT processing temperatures than that used for F_0 (121 °C) or cooking value (100 °C) estimations. B^* is a microbial parameter used to measure the total integrated lethal effect of a process. A process given as $B^* = 1$ would be sufficient to produce nine decimal reductions of mesophilic spores and would be equivalent to 10.1 s at 135 °C. C^* is a parameter to measure the amount of chemical damage taking place during the process. A process giving a $C^* = 1$ would cause 3% destruction of thiamin and would be equivalent to 30.5 s at 135 °C. Again, the aim in most cases is to obtain a high B^* and a low C^* value.

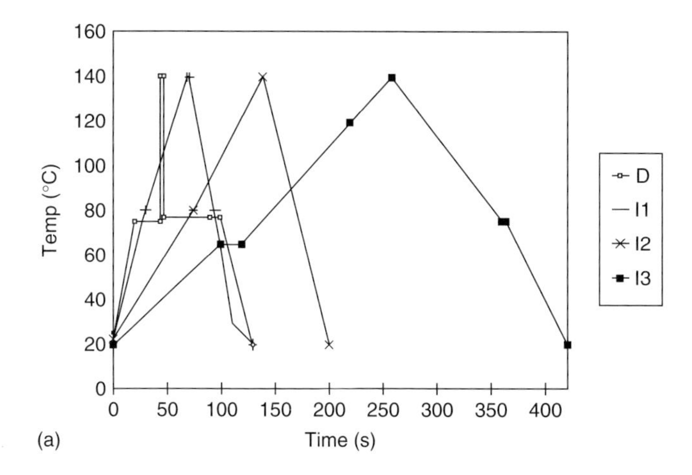

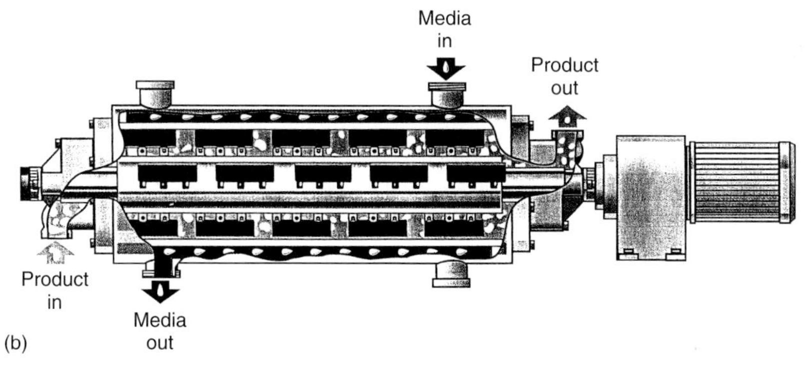

Figure 2.11 (a) Temperature/time profiles for different UHT plants. From Lewis and Heppell [7] with permission. (b) Cutaway view of a horizontal scraped surface heat exchanger. By courtesy of APV.

Calculations of B^* and C^* based on the minimum holding time (t, s) and temperature (°C) are straightforward:

$$B^* = 10^{\frac{(T-135.0)}{10.5}} \cdot \frac{t}{10.1} \tag{2.14}$$

and

$$C^* = 10^{\frac{(T-135.0)}{31.4}} \cdot \frac{t}{30.5} \tag{2.15}$$

Browning *et al.* [36] evaluated a standard temperature/holding time combination of 140 °C for 2 s for heating at cooling times from 1 to 120 s. A more detailed recent overview is given by Tran *et al.* [6], who have determined temperature/time profiles for over 20 UHT plants in Australia, and then used this data to evaluate a wide range of chemical and microbiological parameters (including B^* and C^*).

It is noticeable that the amount of chemical change increased significantly as the heating and cooling times increased and that the longer heating and cooling times gave rise to quite severe microbiological processing conditions, that is, high B^* and F_0 values. At a heating period of about 8 s, the amount of chemical damage done during heating and cooling exceeds that in the holding tube. It is this considerable increase in chemical damage that will be more noticeable in terms of decreasing the quality of the product. However, this may be beneficial in circumstances where a greater extent of chemical damage is required; for example, for inactivating enzymes or for heat inactivation of natural toxic components (e.g., trypsin inhibitor in beans) or softening of vegetable tissue. Differences in temperature/time processes arise because of the use of different heat exchangers and the extent of energy saving by regeneration.

Chemical damage could be further reduced by using temperatures in excess of 145 °C. The best solution would be the direct process, with its accompanying rapid heating and cooling. Steam is mixed with the product, pre-heated to about 75 °C, by injection or infusion. The steam condenses and becomes an ingredient in the product. Steam utilization is between about 10 and 15% (mass/mass). There are special requirements for the removal of impurities from the steam, such as water droplets, oil, and rust. Heating is almost instantaneous. The condensed steam is removed by flash cooling, if required. It will also remove heat-induced volatile components, for example, hydrogen sulfide and other low molecular weight sulfur-containing compounds thought to be responsible for the initial cooked flavor in milk. There is also a reduction in the level of dissolved oxygen, which may improve the storage stability of the product. Advantages of this process are reduced chemical damage and a less intense cooked flavor for many products.

One problem would be the very short holding times required, and the control of such short holding times. In theory it should be possible to obtain products with very high B^* and low C^* values at holding times of about 1 s. For indirect processes, the use of higher temperatures may be limited by fouling considerations and it is important to ensure that the heat stability of the formulation is optimized. It may be worth while developing simple tests to assess heat stability. The alcohol stability test is one such test which is useful for milk products. Generally direct systems give longer processing runs than indirect processes.

Raw material quality and other processing conditions In terms of controlling the process, the following aspects will also merit some attention. Aspects of raw material quality relate to an understanding of the physical properties of the food, through to spore loadings and chemical composition. Of particular concern would be high levels of heat-resistant spores and enzymes in the raw materials, as these could lead to increased spoilage and stability problems during storage. Dried products such as milk other dairy powders, cocoa, other functional powders, and spices are examples to be particularly careful with. Quality assurance programs must ensure that such poor quality raw materials are avoided. The product formulation is also important, including the nature of the principal ingredients, the levels of sugar, starch, salt as well as the pH of the mixture, particularly if there is appreciable amounts of protein. Some thought should also be given to water quality, particularly the mineral content.

Reproducibility in metering and weighing ingredients is also important, as is ensuring that powdered materials are properly dissolved or dispersed and that there are no clumps, which may protect heat-resistant spores. Homogenization conditions may be important: is it necessary to homogenize and if so at what pressures? Should the homogenizer be positioned upstream or downstream of the holding tube? Will two-stage homogenization offer any advantages? Homogenization upstream offers the advantage of breaking down any particulate matter to facilitate heat transfer, as well as avoiding the need to keep the homogenizer sterile during processing. All of these aspects will influence both the safety and the quality of the products.

It is important to ensure that sterilization and cleaning procedures have been properly accomplished. The plant should be sterilized downstream of the holding tube at 130 °C for 30 min. Cleaning should be adequate (detergent concentrations and temperatures) to remove accumulated deposits and the extent of fouling should be monitored, if possible. Steam barriers should be incorporated if some parts of the equipment are to be maintained sterile, while other parts are being cleaned.

All the important experimental parameters should be recorded. This will help ensure that any peculiarities can be properly investigated. Regular inspection and maintenance of equipment are important, particularly eliminating leaks. All staff involved with the process should be educated in order to understand the principles and be encouraged to be diligent and observant. With experience, further hazards will become apparent and methods for controlling them introduced. The overall aim should be to further reduce spoilage rates and to improve the quality of the product.

It is recognized that UHT processing is more complex than conventional thermal processing [28]. The philosophy of UHT processing should be based upon preventing and reducing microbial spoilage by understanding and controlling the process. One way of achieving this by using the principles of HACCPs [37] (see also Chapter 15). The hazards of the process are identified and procedures have been adapted to control them. An acceptable initial target spoilage rate of less than 1 in 10^4 should be aimed for. Such low spoilage rates require very large numbers of samples to be taken to verify that the process is being performed and controlled at the desired level. Initially a new process should be verified by 100% sampling. Once it is established that the process is under control, sampling frequency can be reduced and sampling plans can be designed to detect any spasmodic failures. More success will result from targeting high-risk occurrences, such as start-up, shut-down, and product change-overs. Thus, holding time and temperature are perhaps the two most critical parameters. Recording thermometers should be checked and calibrated regularly, and accurate flow control is crucial (as for pasteurization).

2.5
Special Problems with Viscous and Particulate Products

Continuous heat processing of viscous and particulate products provides some special problems. For viscous products it may be possible to use a tubular heat exchanger, but it is more probable that a scraped surface heat exchanger will be required (see Figure 2.11b). This incorporates a scraper blade which continually sweeps product away from the heat transfer surface. These heat exchangers are mechanically more complex, with seals at the inlet and outlet ends of the scraper blade shaft. Overall heat transfer coefficients will be low and the flow will be streamline with a consequent increase in the spread of residence times and the fluid being heated may also be non-Newtonian. Heating and cooling rates will be fairly slow and the process is generally more expensive to run, because of the higher capital and maintenance costs and less scope for regeneration. They may be used for pasteurization or sterilization processes.

Two specialist uses of this equipment are for freezing ice cream and for margarine and low-fat spread manufacture. In most cases increasing the agitation speed will improve heat transfer efficiency by increasing the overall heat transfer coefficient. However, when cooling products, some may become very viscous, for example, due to product crystallization, and higher agitation speeds may create additional frictional heat, making the product warmer rather than cooling. This is known as *viscous dissipation*. Scraped surface heat exchangers can also handle particulate systems, up to 25 mm diameter.

Problems arise with particulate systems because the solid phase conducts heat slowly so it will take longer to sterilize the particles compared to the liquid. Also, determination of the heat film coefficient is difficult due to the uncertainty in the relative velocity of the solid with respect to the liquid. There will also be problems determining the residence time distribution of the solid particles.

A second approach is to have a selective holding tube system, whereby larger particles are held up in the holding tube for a longer period of time. A third approach involves heating the solid and liquid phases separately and recombining them. This is the feature of the Jupiter heating system, as discussed in more detail by Lewis and Heppell [7].

A novel approach is ohmic heating, in which particulate material is pumped through a nonconducting tube containing electrodes (see Section 2.6). Dielectric heating, particularly microwaves, may also be used for pasteurization and sterilization of foods in continuous processes. For foods containing solid particles, there are some advantages in terms of being able to generate heat within the particles. Factors affecting the rate of heating include the field strength and frequency of the microwave energy and the dielectric loss factor of the food. However, dielectric heating processes are much more complex compared to well-established HTST and UHT processes and the benefits that might result have to be weighed against the increased costs. One major drawback is that of nonuniform heating. This is a critical aspect in pasteurization and sterilization processes, where it is a requirement that all elements of the food reach a minimum temperature for a minimum time (e.g., 70 °C for 2 min to inactivate *E. coli* 0157). If part of the food only reaches 65 °C, even though other parts may well be above 70 °C, serious under-processing will occur. Identifying with greater certainty where the slower heating points are will help to improve matters. Brody [38] provides further reading on microwave food pasteurization.

2.6 Ohmic Heating

2.6.1 Introduction

Ohmic heating, joule heating, or electrical resistance heating [39] refer to a process whereby materials are heated by passage of an electric current. The technology itself dates back to the 1800s and was revived in 1980s. Since the mid-1990s, commercial installations have been introduced into the market [40]. Ohmic heating is an attractive food-processing technology because of its technical simplicity, uniform, rapid heating patterns, high energy efficiency, and versatile applicability for liquids or particulate mixtures.

The electrical energy is almost entirely dissipated within the heated materials during ohmic heating, hence yielding very high energy transfer efficiency close to 100% [41, 42], whereas that of microwave heating is only 65% at best. Ohmic heating can be used for shear-sensitive foods or food mixtures containing high percentage of particulates [43] and heat-sensitive nutrients. Volumetric heat generation resulting in uniform and rapid heating can produce foods which retain better quality attributes and nutrients than conventional thermal processes. Ohmic heating has been used for several food-processing applications such as blanching, sterilization, pasteurization, thawing, dehydration and extraction, and commercial products such as diced, sliced fruits, and pasteurized liquid egg have appeared in the market.

Some basic principles and important factors of ohmic heating in food processing applications will now be outlined.

2.6.2 Fundamental Principles of Ohmic Heating

Heat generation by the passage of an electric current during ohmic heating is governed by Ohm's law:

$$Q = VI = IR^2 \tag{2.16}$$

If the voltage gradient is known, the heat can be calculated from voltage gradient and electrical conductivity:

$$Q = |\Delta V|^2 \sigma \tag{2.17}$$

Knowledge of electrical conductivity of food materials is critical for uniform and efficient ohmic heating. The electrical conductivity of foods $\sigma(S/m)$ can be determined by:

$$\sigma = \frac{I}{V} \cdot \frac{L}{A} \tag{2.18}$$

where V is applied voltage, I is electric current (A), L is distance between two electrodes (m), and A is the area of electrode (m^2). The electrical conductivities of

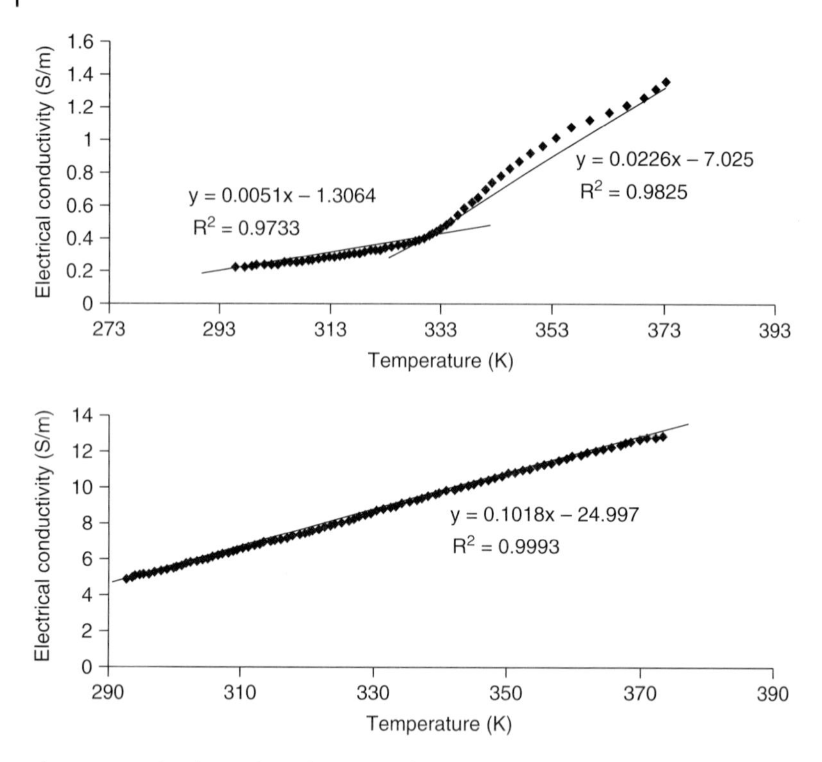

Figure 2.12 The electrical conductivities of potato (a) and 3% NaCl solution (b).

most food materials are dependent on their temperature values (Figure 2.12) [44]:

$$\sigma(T) = \sigma_{\text{ref}}(1 + mT) \tag{2.19}$$

where, σ_{ref} is the electrical conductivity at a reference temperature and m is a temperature coefficient. Foods containing high amount of ionic materials such as salts have high electrical conductivity, whereas foods composed of nonionized constituents and having high viscosity, such as honey and syrups, will thermally lag other components in the system.

2.6.2.1 **Electrochemical Reaction on Electrodes**

The application of alternating current (AC) at low frequencies during ohmic heating induces electrochemical reactions at electrode/solution interfaces, including electrode corrosion, migration of metal ions, and generation of O_2, H_2, and free radicals [45]. The overall electrolysis reaction which produces O_2 and H_2 at the electrodes is given by:

$$2H_2O \leftrightarrow 2H_2 + O_2 \tag{2.20}$$

A generalized anodic corrosion reaction is described by Samaranayake and Sastry [46]:

$$\text{M} \leftrightarrow \text{M}^{n+} + ne^-(n = 1, 2, 3\ldots) \tag{2.21}$$

Figure 2.13 The ohmically treatable pouch with two electrode tabs.

Metal ions migrating into the foods are considered as contaminants and toxic substances. These ions can also form various complexes with food constituents and catalyze undesired chemical reactions such as lipid oxidation [47]. Oxygen and free radicals adversely affect the food constituents such as lipids or vitamins through oxidative reactions. Electrochemical reactions occurring at the interfaces between electrodes and solution are greatly influenced by physical and chemical properties of electrodes. Other critical factors include pH of solution, frequency, pulse width, and pulse delay time. Studies on various electrode materials, such as titanium, stainless steel, platinized titanium, and graphite, at different pH values demonstrated that platinized titanium was relatively inert to electrochemical corrosion [46]. It is also known that the migration of ions such as Fe, Cr, Ni, Mn, and Mo from stainless steel electrodes into foods in a reusable pouch (Figure 2.13) is minimized when high-frequency pulsed waveform is used [48].

2.6.2.2 Heating Pattern of Multiphase Food in Ohmic System

Ohmic heating has been widely applied for the processing of liquid and particulate–liquid foods [49]. In general, ohmic heating is suitable for processing multiphase foods when the electrical conductivities of the solids and liquids are similar [50]. When multiphase food consisting of components with different electrical conductivity is ohmically treated, some portions of the food may not be not uniformly heated and there exists the danger of underprocessing. In a mixture where the aqueous medium has a higher electrical conductivity than surrounding particulates, the medium will be heated faster than the particulates. Nonuniform heating can be reduced by increasing the electrolytic contents in food particles by additional pretreatment such as soaking or blanching of particles in salt solution. The ohmic heating pattern of multiphase foods with such pretreatment methods was more uniform than that of ohmic heating without any pretreatment. Heating patterns are also affected by rheological properties of solutions [43, 51], often taking into consideration the arrangement of solids around particles and fluid flow fields around the particles [40].

2.6.2.3 Modeling of Ohmic Heating

In ohmic heating, knowledge of temperature distribution is important in process optimization, design or establishing process efficiency, and safety. It is difficult to experimentally measure temperature at different locations in the ohmic heating process, especially in the case of flowing particulates. Prediction of heating patterns has to rely on mathematical modeling which combines electrical, heat, and momentum transfer equations. Several modeling studies have investigated the factors influencing ohmic heating patterns, such as particle size and shape [52], orientation relative to electric field [53, 54], liquid viscosity [55], electrical conductivity of particulates and liquid [54, 56], and their volume fractions [57]. A comprehensive review of models for ohmic heating was given by Marcotte [43]. Various models have been developed, ranging from a single particle in a static heater [56, 58] to systems containing multiple particles in static or continuous flow ohmic heaters [55, 57, 59–64].

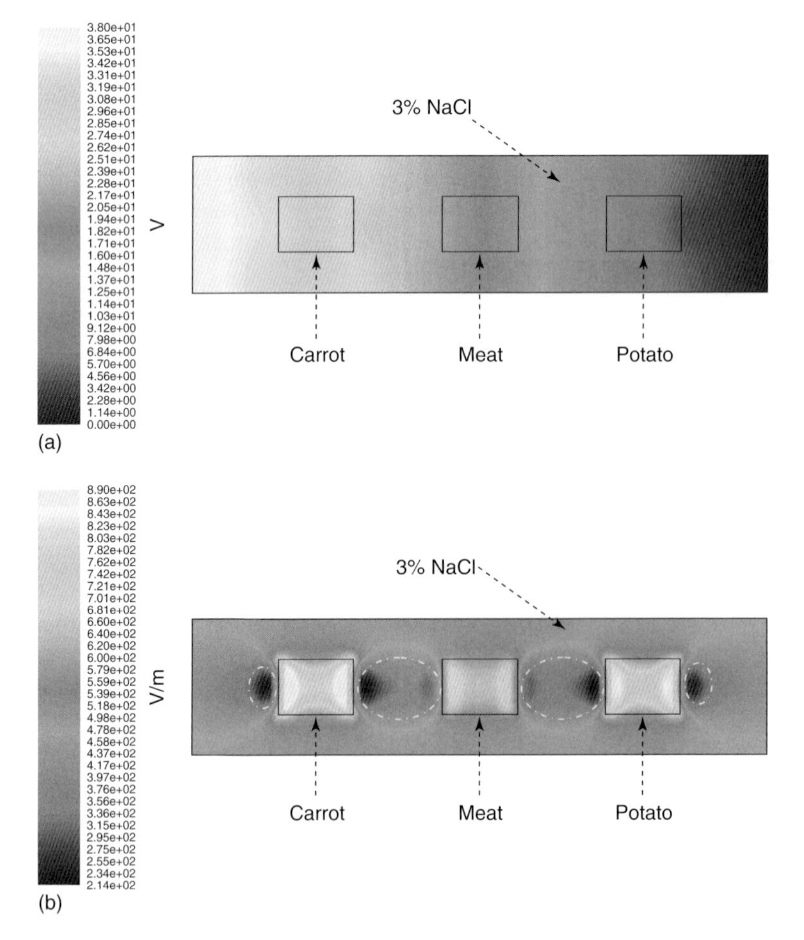

Figure 2.14 Simulated potential (a) and electric field strength (b).

Electric field distribution, which is important to estimate the heating rate, can be computed by using either Laplace's equation [56] or circuit analogy approach [57]. In the first approach, the continuity equation of electric currents for steady state is given by:

$$\nabla[\sigma(T)\nabla V] = 0 \tag{2.22}$$

where, σ is the electrical conductivity which relies on the function of temperature (T), V is the applied voltage. In contrast, the circuit analogy approach calculated the equivalent resistances based on electrical conductivity of solid particulates and liquids, cross section and length of incremental section, then the voltage was determined from the resulting resistance and electric current [57]. When the electric field distribution is interactively coupled with heat and momentum transfer equations, the temperature distribution can be obtained by solving those simultaneous equations. Examples of simulated electric potential, electrical field strength distributions and heating patterns obtained by Shim *et al.* [65] for multicomponent food mixtures by computational fluid dynamics simulation are shown in Figures 2.14 and 2.15. In particular, estimated overshoots and interruptions of electrical field

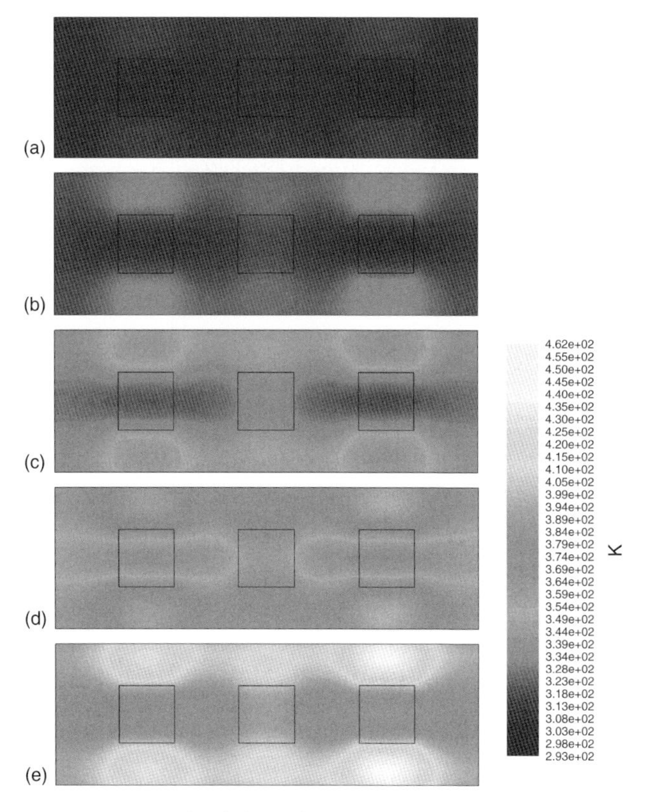

Figure 2.15 Simulated thermal patterns of solid and liquid foods inside the ohmic cell after heating for 50 s (a), 100 s (b), 150 s (c), 200 s (d), and 250 s (e).

strength caused by the existence of solid particles (Figure 2.14b) closely associated with localized hot and cold spots are key concerns.

In conclusion, ohmic heating is an attractive technology for thermal processing applications of foods. Despite some existing challenges, such as nonuniform heating, ion migration into the foods, and the identification of cold spots, the technology has been better understood and remarkably improved due to continued research efforts in food properties, process modeling, and electrochemical reactions vs. field characteristics. Future researches are required to investigate the combination of ohmic heating with other technologies and to develop in-depth understanding in modeling and food property changes under various process conditions.

2.7
Filling Procedures

For pasteurization and extended shelf-life products, clean filling systems are used and for UHT products aseptic systems are required. There are a number of aseptic packaging systems available. They all involve putting a sterile product into a sterile container in an aseptic environment. Pack sizes range from individual portions (14 ml), retail packs (125 ml to 1 l), through to bag-in-the-box systems up to 1000 l. The sterilizing agent is usually hydrogen peroxide (35% at about 75–80 °C); the contact time is short and the residual hydrogen peroxide is decomposed using hot air. The aim is to achieve 4*D* process for spores. Superheated steam has been used for sterilization of cans in the Dole process. Irradiation may be used for plastic bags (see Chapter 5).

Since aseptic packaging systems are complex, there is considerable scope for packaging faults to occur, which will lead to spoiled products. Where faults occur, the spoilage microorganisms would be more random and would include those microorganisms which would be expected to be inactivated by UHT processing; these often result in blown packages.

Packages should be inspected regularly to ensure that they are airtight, again focusing upon those more critical parts of the process, for example, start-up, shut-down, product changeovers, and for carton systems reel splices and paper splices. Sterilization procedures should be verified. The seal integrity of the package should be monitored as well as the overall microbial quality of packaging material itself. Care should be taken to minimize damage during subsequent handling. All these could result in an increase in spoilage rate.

2.8
Storage

UHT products are commonly stored at room (ambient temperature) and good-quality products should be microbiologically stable. Nevertheless, chemical reactions and physical changes will take place which will change the quality of the product. Particularly relevant are oxidation reactions and Maillard browning, both of which can lead to a deterioration in the sensory characteristics of the product. These have been discussed in more detail by Burton [23] and Lewis and Heppell [7].

References

1. Lewis, M.J. (1990) *Physical Properties of Foods and Food Processing Systems*, Woodhead Publishers, Cambridge.
2. Lewis, M.J. (1993) Physical properties of dairy products, in *Modern Dairy Technology*, Vol. 2 (ed. R.K. Robinson), Elsevier Applied Science, London, pp. 331–380.
3. Milk Marketing Board (1994) EC Facts and Figures, Milk Marketing Board, Thames Ditton, UK.
4. Gould, G.W. (1989) Heat induced injury and inactivation, in *Mechanisms of Action of Food Preservation Procedures* (ed. G.W. Gould), Elsevier Applied Science, London, pp. 11–42.
5. Cronshaw, H.B. (1947) *Dairy Information*, Dairy Industries Ltd, London.
6. Tran, H., Datta, N., Lewis, M., and Deeth, H. (2008) Predictions of some product parameters based on the processing conditions of ultra-high-temperature milk plants. *Int. Dairy J.*, **18**, 939–944.
7. Lewis, M.J. and Heppell, N. (2000) *Continuous Thermal Processing of Foods: Pasteurization and UHT Sterilization*, Aspen Publishers, Gaithersburg, MD.
8. Lewis, M.J. and Deeth, H.C. (2009) Heat treatment of milk, in *Milk Processing and Quality Management* (ed. A.Y. Tamine), SDT, John Wiley & Sons, Ltd, Chichester, pp. 168–204.
9. North, C.E. and Park, W.H. (1927) Standards for milk pasteurization. *Am. J. Hyg.*, **7**, 147–173.
10. International Dairy Federation (IDF) (1986) Monograph on Pasteurised Milk, Bulletin Number 200.
11. Satin, M. (1996) *Food Irradiation – A Guidebook*, Technomic Publishing Company Ltd, Lancaster.
12. Hasting, A.P.M. (1992) Practical considerations in the design, operation and control of food pasteurisation processes. *Food Control*, **3**, 27–32.
13. Sorensen, K.R. (1996) APV Heat Exchanger; AS introduces a new security system for pasteuriser installations, in IDF Heat Treatments and Alternative Methods, IDF/FILS No. 9602, pp. 179–183.
14. Miller Jones, J. (1992) *Food Safety*, Egan Press, St Paul, Minnesota.
15. International Dairy Federation (IDF) (1984) The Thermization of Milk, Bulletin Number 182.
16. Society of Dairy Technology (SDT) (1983) *Pasteurizing Plant Manual*, SDT, Huntingdon.
17. Lewis, M.J. (2010) Improving pasteurised and extended shelf-life milk, in *Improving the Safety and Quality of Milk: Milk Production and Processing*, Vol. 1 (ed. M.W. Griffiths), Woodhead Publishing, Cambridge, pp. 278–301.
18. Pearse, M.A. (1993) Pasteurization of liquid products, in *Encyclopedia of Food Science, Food Technology and Nutrition*, Academic Press, London, pp. 3441–3450.
19. Wirjantoro, T.I. and Lewis, M.J. (1996) Effect of nisin and high temperature pasteurisation on the shelf-life of whole milk. *J. Soc. Dairy Technol.*, **4**, 99–102.
20. Wirjantoro, T.I., Lewis, M.J., Grandison, A.S., Williams, G.C., and Delves-Broughton, J. (2001) The effect of nisin on the keeping quality of reduced heat treated (RHT) milks. *J. Food Prot.*, **64**, 213–219.
21. Phillips, J.D., Griffiths, M.W., and Muir, D.D. (1983) Effect of nisin on the shelf-life of pasteurized double cream. *J. Soc. Dairy Technol.*, **36**, 17–21.
22. Hammer, P., Lembke, F., Suhren, G., and Heeschen, W. (1996) Characterisation of heat resistant mesophilic Bacillus species affecting the quality of UHT milk, in *Proceedings of the IDF Symposium on Heat Treatments and Alternative Methods, Vienna, Austria, 6–8 September*

1995, IDF/FIL No. 9602, International Diary Federation, Brussels, pp. 9–25.
23. Burton, H. (1988) *UHT Processing of Milk and Milk Products*, Elsevier Applied Science, London.
24. Holdsworth, S.D. (1997) *Thermal Processing of Packaged Foods*, Blackie Academic and Professional, London.
25. Walstra, P., Guerts, T.J., Noomen, A., Jellema, A., and van Boekel, M.A.J.S. (2001) *Dairy Technology: Principles of Milk Properties and Processes*, Marcel Dekker, New York.
26. Jackson, J.M. and Shinn, B.M. (eds) (1979) *Fundamentals of Food Canning Technology*, AVI, Westport, CT.
27. Selman, J.D. (1987) The blanching process, in *Developments in Food Preservation*, Vol. 4 (ed. S. Thorne), Elsevier Applied Science, London, pp. 205–249.
28. Institute of Food Science and Technology (IFST) (1991) *Food and Drink – Good Manufacturing Practice: A Guide to its Responsible Management*, 3rd edn, IFST, London.
29. Stumbo, C.R. (1965) *Thermobacteriology in Food Processing*, Academic Press, New York.
30. Hersom, A.C. and Hulland, E.D. (1980) *Canned Foods*, Churchill Livingstone, Edinburgh.
31. Rees, J.A.G. and Bettison, J. (1991) *Processing and Packaging of Heat Preserved Foods*, Blackie, Glasgow.
32. Footitt, R.J. and Lewis, A.S. (1995) *The Canning of Fish and Meat*, Blackie Academic and Professional, Glasgow.
33. Gaze, J.E. and Brown, K.L. (1988) The heat resistance of spores of *Clostridium botulinum* 213B over the temperature range 120 to 140°C. *Int. J. Food Sci. Technol.*, **23**, 373–378.
34. Kessler, H.G. (1981) *Food Engineering and Dairy Technology*, Verlag A Kessler, Freising.
35. Kessler, H.G. (1989) Effect of thermal processing of milk, in *Developments in Food Preservation*, Vol. 5 (ed. S. Thorne), Chapman and Hall, London, pp. 91–130.
36. Browning, E., Lewis, M.J., and MacDougall, D. (2001) Predicting safety and quality parameters for UHT-processed milks. *Int. J. Dairy Technol.*, **54**, 111–120.
37. International Commission on Microbiological Specifications for Foods (ICMSF) (1988) *Microorganisms in Foods 4: Application of the Hazard Analysis Critical Control Point (HACCP) System to Ensure Microbiological Safety*, Blackwell Scientific Publications, Oxford.
38. Brody, A.L. (1992) Microwave food pasteurisation. *Food Technol. Int. Eur.*, 67–72.
39. Herrick, J.P., Sastry, S.K., Clyde, G.F., and Wedral, E.R. (2000) On-demand direct electrical resistance heating system and method thereof. US Patent 6,130,990.
40. Sastry, S. (2008) Ohmic heating and moderate electric field processing. *Food Technol. Int.*, **14** (5), 419–422.
41. Jun, S. and Sastry, S.K. (2005) Modeling and optimizing of pulsed ohmic heating of foods inside the flexible package. *J. Food Proc. Eng.*, **28** (4), 417–436.
42. Salengke, S. (2000) Electrothermal effects of ohmic heating on biomaterials: Temperature monitoring, heating of solid–liquid mixtures, and pre-treatment effects on drying rate and oil uptake. PhD Thesis, The Ohio State University, Ohio.
43. Marcotte, M. (1999) Ohmic heating of viscous liquid foods. PhD Dissertation, McGill University, Canada.
44. Palaniappan, S. and Sastry, S.K. (1991) Electrical conductivity of selected solid foods during ohmic heating. *J. Food Proc. Eng.*, **14**, 221–236.
45. Ghnimi, S., Flach-Malaspina, N., Dresch, M., Delaplace, G., and Maingonnat, J.F. (2008) Design and performance evaluation of an ohmic heating unit for thermal processing of highly viscous liquids. *Chem. Eng. Res. Des.*, **86**, 626–632.
46. Samaranayake, C.P. and Sastry, S.K. (2005) Electrode and pH effects on electrochemical reactions during ohmic heating, **577**, 125–135.
47. Samaranayake, C.D. (2003) Electrochemical reactions during Ohmic heating. PhD Thesis, The Ohio State University, Ohio.

48. Jun, S., Sastry, S.K., and Samaranayake, C. (2007) Migration of electrode components during ohmic heating of foods in retort pouches. *Innovative Food Sci. Emerg. Technol.*, **8** (2), 237–243.
49. Piette, G., Buteau, M.L., de Halleux, D., Chiu, L., Raymond, Y., Ramaswamy, H.S., and Dostie, M. (2004) Ohmic cooking of processed meats and its effects on product quality. *J. Food Sci.*, **69** (2), 71–78.
50. Wang, W.C. and Sastry, S.K. (1993) Salt diffusion into vegetable tissue as a pretreatment for ohmic heating: determination of parameters and mathematical model verification. *J. Food Eng.*, **20**, 311–323.
51. Khalaf, W.G. and Sastry, S.K. (1996) Effect of fluid viscosity on the ohmic heating rate of solid-liquid mixtures. *J. Food Eng.*, **27** (2), 145–158.
52. de Alwis, A.A.P. and Fryer, P.J. (1988) Preliminary experiments on heat transfer during ohmic heating of foods. Proceedings of the 2nd UK National Heat Transfer Conference, Glasgow, IMechE, Vol. 1, pp. 229–239.
53. de Alwis, A.A.P., Halden, K., and Fryer, P.J. (1989) Shape and conductivity effects in the ohmic heating of fluids. *Chem. Eng. Res. Des.*, **67**, 159–168.
54. Sastry, S.K. and Palaniappan, S. (1992) Influence of particle orientation on the effective electrical resistance and ohmic heating rate of a liquid-particle mixture. *J. Food Proc. Eng.*, **15**, 213–227.
55. Fryer, P.J., de Alwis, A.A.P., Koury, E., Stapley, A.G.F., and Zhang, L. (1993) Ohmic processing of solid-liquid mixtures: heat generation and convection effects. *J. Food Eng.*, **18**, 102–125.
56. de Alwis, A.A.P. and Fryer, P.J. (1990) A finite-element analysis of heat generation and transfer during ohmic heating of food. *Chem. Eng. Sci.*, **45** (6), 1547–1559.
57. Sastry, S.K. and Palaniappan, S. (1992) Mathematical modeling and experimental studies on ohmic heating of liquid-particle mixtures in a static heater. *J. Food Proc. Eng.*, **15**, 241–261.
58. de Alwis, A.A.P. and Fryer, P.J. (1990) The use of direct resistance heating in the food industry. *J. Food Eng.*, **11**, 3–27.
59. Sastry, S.K. (1992) A model for heating of liquid-particle mixtures in a continuous flow ohmic heater. *J. Food Proc. Eng.*, **15**, 263–278.
60. Sastry, S.K. and Li, Q. (1993) Models for ohmic heating of solid-liquid food mixtures in: "Heat Transfer in Food Processing". Presented at the 29th National Heat Transfer Conference of the American Society of Mechanical Engineers (ASME), Atlanta, Georgia, August 8–11, Volume 254, pp. 25–33.
61. Zhang, L., Liu, S., Pain, J.P., and Fryer, P.J. (1992) Heat transfer and flow of particle-liquid food mixtures. I. *ChemE. Symp. Ser.*, **126**, 79–88.
62. Zhang, L. and Fryer, P.J. (1995) A comparison of alternative formulation for the prediction of electrical heating rates of solid-liquid food materials. *J. Food Proc. Eng.*, **18**, 85–97.
63. Sastry, S.K. and Salengke, S. (1998) Ohmic heating of solid-liquid mixtures: a comparison of mathematical models under worst-case heating conditions. *J. Food Proc. Eng.*, **21**, 441–458.
64. Orangi, S. and Sastry, S.K. (1998) A numerical investigation of electroconductive heating in solid-liquid mixtures. *Int. J. Heat Mass Trans.*, **41** (14), 2211–2220.
65. Shim, J.Y., Lee, S.H., and Jun, S. (2010) Modeling of ohmic heating patterns of multiphase food products using computational fluid dynamics codes. *J. Food Eng.* **99** (2), 136–141.
66. Brennan, J.G., Butters, J.R., Cowell, N.D., and Lilly, A.E.V. (1990) *Food Engineering Operations*, 3rd edn, Elsevier Applied Science, London.

3
Evaporation and Dehydration

James G. Brennan

3.1
Evaporation (Concentration, Condensing)

3.1.1
General Principles

Most food liquids have relatively low solids contents. For example, whole milk contains approximately 12.5% total solids, fruit juice 12%, sugar solution after extraction from sugar beet 15%, solution of coffee solutes after extraction from ground roasted beans 25%. For various reasons, some of which are discussed in Section 3.1.5, it may be desirable to increase the solids content of such liquids. The most common method used to achieve this is to "boil off" or evaporate some of the water by the application of heat. Other methods used to concentrate food liquids are freeze concentration and membrane separation. These are discussed in Chapters 9 and 10, respectively. If evaporation is carried out in open pans at atmospheric pressure, the initial temperature at which the solution boils will be some degrees above 100 °C, depending on the solids content of the liquid. As the solution becomes more concentrated, the evaporation temperature will rise. It could take from several minutes to a few hours to attain the solids content required.

Exposure of the food liquid to these high temperatures for these lengths of time is likely to cause changes in the color and flavor of the liquid. In some cases such changes may be acceptable, or even desirable, for example, when concentrating sugar solutions for toffee manufacture or when reducing gravies. However, in the case of heat-sensitive liquids such as milk or fruit juice, such changes are undesirable. To reduce heat damage, the pressure above the liquid in the evaporator may be reduced below atmospheric by means of condensers, vacuum pumps, or steam ejectors. Since a liquid boils when the vapor pressure it exerts equals the external pressure above it, reducing the pressure in the evaporator lowers the temperature at which the liquid will evaporate. Typically, the pressure in the evaporator will be in the range 7.5–85.0 kPa absolute, corresponding to evaporation temperatures in the range 40–95 °C. The use of lower pressures is usually uneconomic. This is known as *vacuum evaporation*. The relatively low

Food Processing Handbook, Second Edition. Edited by James G. Brennan and Alistair S. Grandison.

evaporation temperatures that prevail in vacuum evaporation mean that reasonable temperature differences can be maintained between the heating medium, saturated steam, and the boiling liquid, while using relatively low steam pressures. This limits undesirable changes in the color and flavor of the product. For aqueous liquids, the relationship between pressure and evaporation temperature may be obtained from thermodynamic tables and psychrometric charts. Relationships are available for estimating the evaporating temperatures of nonaqueous liquids at different pressures [1, 2].

Another factor that affects the evaporation temperature is known as the boiling point rise (BPR) or boiling point elevation (BPE). The boiling point of a solution is higher than that of the pure solvent at the same pressure. The higher the soluble solids content of the solution, the higher its boiling point. Thus, the initial evaporation temperature will be some degrees above that corresponding to pressure in the evaporator, depending on the soluble solids content of the feed. However, as evaporation proceeds and the concentration of the soluble solids increases, the evaporation temperature rises. This is likely to result in an increase in changes in the color and flavor of the product. If the temperature of the steam used to heat the liquid is kept constant, the temperature difference between it and the evaporating liquid decreases. This reduces the rate of heat transfer and hence the rate of evaporation. To maintain a constant rate of evaporation, the steam pressure may be increased. However, this is likely to result in a further decrease in the quality of the product. Data on the BPR in simple solutions and some more complex foods is available in the literature in the form of plots and tables. Relationships for estimating the BPR with increase in solids concentrations have also been proposed [1, 3]. BPR may range from <1 to 10 °C in food liquids. For example, the BPR of a sugar solution containing 50% solids is about 7 °C.

In some long-tube evaporators (see Section 3.1.2.3), the evaporation temperature increases with increase in the depth of the liquid in the tubes in the evaporator, due to hydrostatic pressure. This can lead to overheating of the liquid and heat damage. This factor has to be taken into account in the design of evaporators and in selecting the operating conditions, in particular, the pressure of the steam in the heating jacket.

The viscosity of most liquids increases as the solids content increases during evaporation. This can lead to a reduction in the circulation rates and hence the rates of heat transfer in the heating section of the evaporator. This can influence the selection of the type of evaporator for a particular liquid food. Falling film evaporators (see Section 3.1.2.3) are often used for moderately viscous liquids. For very viscous liquids, agitated thin-film evaporators are used (see Section 3.1.2.5). Thixotropic (or time-dependent) liquids, such as concentrated tomato juice, can pose special problems during evaporation. The increase in viscosity can also limit the maximum concentration attainable in a given liquid.

Fouling of the heat transfer surfaces may occur in evaporators. This can result in a decrease in the rate of heat transfer and hence the rate of evaporation. It can also necessitate expensive cleaning procedures. Fouling must be taken into account in the design of evaporators and in the selection of the type of evaporator for a given duty. Evaporators that feature forced circulation of the liquid or agitated thin films are used for liquids that are susceptible to fouling.

Some liquids are prone to foaming when vigorously boiling in an evaporator. Liquids that contain surface active foaming agents, such as the proteins in skimmed milk, are liable to foam. This can reduce rates of heat transfer and hence rates of evaporation. It may also result in excessive loss of product by entrainment in the vapor leaving the heating section. This, in turn, can cause contamination of the cooling water to spray condensers and lead to problems in the disposal of that effluent. In some cases, antifoaming agents may be added to the feed to reduce foaming. Care must be taken not to infringe any regulations by the addition of such aids.

Volatile aroma and flavor compounds may be lost during vacuum evaporation, resulting in a reduction in the organoleptic quality of products such as fruit juices or coffee extract. In the case of fruit juices, this loss may be partly offset by adding some of the original juice, known as "*cutback juice*," to the concentrate. Alternatively, the volatiles may be stripped from the vapor, concentrated, and added to the concentrated liquid [1, 3–6].

3.1.2 Equipment Used in Vacuum Evaporation

A single-effect vacuum evaporator has the following components:

- A heat exchanger, known as a *calandria*, by means of which the necessary sensible and latent heat is supplied to the feed to bring about the evaporation of some of the liquid. Saturated steam is the usual heating medium but hot water and other thermal fluids are sometimes used. Tubular and plate exchangers of various designs are widely used. Other, more sophisticated designs are available, including agitated thin-film models, expanding flow chambers and centrifugal exchangers.
- A device to separate the vapor from the concentrated liquid phase. In vacuum evaporators, mechanical devices such as chambers fitted with baffles or meshes and cyclone separators are used to reduce entrainment losses.
- A condenser to convert the vapor back to a liquid and a pump, steam ejector, or barometric leg to remove the condensate, thus creating and maintaining the partial vacuum in the system.

Most evaporators are constructed in stainless steel except where there are extreme corrosion problems.

The following paragraphs describe various types of evaporators used in the food industry.

3.1.2.1 Vacuum Pans

A hemispherical pan equipped with a steam jacket and sealed lid, connected to a vacuum system, is the simplest type of vacuum evaporator in use in industry. The heat transfer area per unit volume is small and so the time required to reach the desired solids content can run into hours. Heating occurs by natural convection. However, an impeller stirrer may be introduced to increase circulation and reduce fouling. Small pans have a more favorable heat transfer area to volume ratio. They are useful for frequent changes of product and for low or variable throughputs. They are used in jam manufacture, the preparation of sauces, soups, and gravies and in tomato pulp concentration.

3.1.2.2 **Short Tube Vacuum Evaporators**

This type of evaporator consists of a calandria made up of a bundle of short vertical tubes surrounded by a steam jacket, located near the bottom of a large vessel (Figure 3.1). The tubes are typically 25–75 mm in diameter and 0.5–2.0 m long. The liquid being concentrated normally covers the calandria. Steam condensing on the outside of the tubes heats the liquid, causing it to rise by natural convection. Some of the water evaporates and flows to the condenser. The liquid circulates down through the larger, cooler tube in the center of the bundle, known as the *downcomer*.

This type of evaporator is suitable for low- to moderate-viscosity liquids, which are not very heat sensitive. With viscous liquids heat transfer rates are low, hence residence times are relatively long and there is a high risk of fouling. Sugar solutions, glucose, and malt extract are examples of products concentrated in this type of evaporator. It can also be used for crystallization operations. For this application an impeller may be located in the downcomer to keep the crystals in suspension.

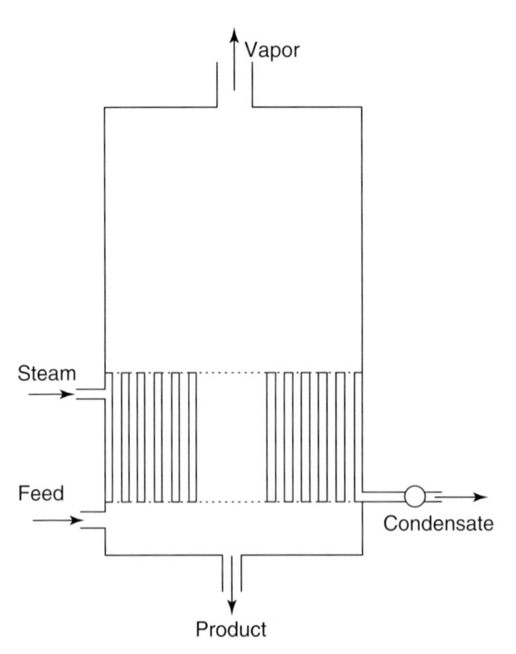

Figure 3.1 Vertical short tube evaporator. From Brennan *et al.* [1] with permission.

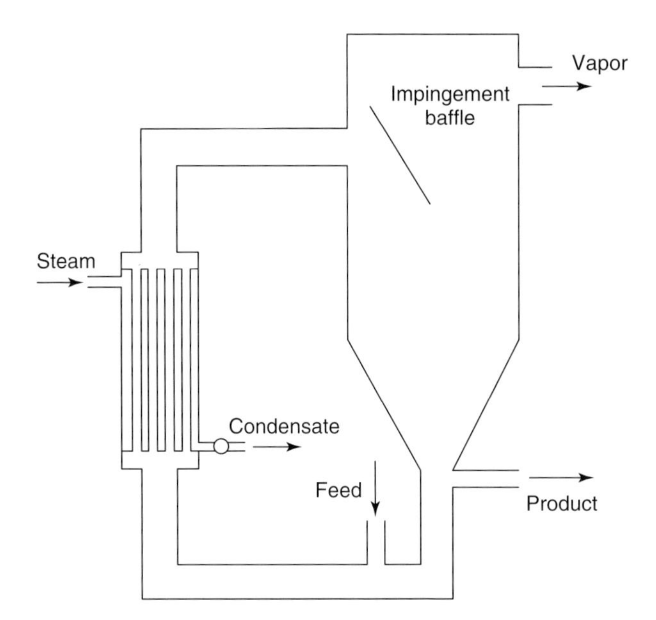

Figure 3.2 Natural circulation evaporator. From Brennan *et al.* [1] with permission.

In another design of a short tube evaporator the calandria is external to the separator chamber and may be at an angle to the vertical (Figure 3.2). The liquid circulates by natural convection within the heat exchanger and also through the separation chamber. The liquid enters the separation chamber tangentially. A swirling flow pattern develops in the chamber, generating centrifugal force, which assists in separating the vapor from the liquid. The vigorous circulation of the liquid results in relatively high rates of heat transfer. It also helps to break up any foam that forms. The tubes are easily accessible for cleaning. Such evaporators are also used for concentrating sugar solutions, glucose, and malt and more heat-sensitive liquids such as milk, fruit juices, and meat extracts.

A pump may be introduced to assist in circulating more viscous liquids. This is known as *forced circulation*. The choice of pump will depend on the viscosity of the liquid. Centrifugal pumps are used for moderately viscous materials, while positive displacement pumps are used for very viscous liquids.

3.1.2.3 **Long-Tube Evaporators**

These consist of bundles of long tubes, 3–15 m long and 25–50 mm in diameter, contained within a vertical shell into which steam is introduced. The steam condensing on the outside of the tubes provides the heat of evaporation. There are three patterns of flow of the liquid through such evaporators:

- In the *climbing film evaporator* the preheated feed is introduced into the bottom of the tubes. Evaporation commences near the base of the tubes. As the vapor expands, ideally, it causes a thin film of liquid to rise rapidly up the inner walls of the tubes around a central core of vapor. In practice, slugs of liquid and vapor bubbles also rise up the tubes. The liquid becomes more concentrated as it rises. At the top, the liquid–vapor mixture enters a cyclone separator. The vapor is drawn off to a condenser and pump, or into the heating jacket of another calandria in a multiple-effect system (see Section 3.1.3). The concentrated liquid may be removed as product, recycled through the calandria, or fed to another calandria. The residence time of the liquid in the tubes is relatively short. High rates of heat transfer are attainable in this type of evaporator, provided there are relatively large temperature differences between the heating medium and the liquid being concentrated. However, when these temperature differences are low, the heat transfer rates are also low. This type of evaporator is suitable for low-viscosity, heat-sensitive liquids such as milk and fruit juices.

- In the *falling film evaporator* the preheated feed is introduced at the top of the tube bundle and distributed to the tubes so that a thin film of the liquid flows down the inner surface of each tube, evaporating as it descends. Uniform distribution of the liquid so that the inner surfaces of the tubes are uniformly wetted is vital to the successful operation of this type of evaporator [7]. From the bottom of the tubes, the liquid–vapor mixture passes into a centrifugal separator and from there the liquid and vapor streams are directed in the same manner as in the climbing film evaporator. High rates of liquid flow down the tubes are attained by a combination of gravity and the expansion of the vapor, resulting in short residence times. These evaporators are capable of operating with small temperature differences between the heating medium and the liquid and can cope with viscous materials. Consequently, they are suitable for concentrating heat-sensitive foods and are very widely used in the dairy and fruit juice processing sections of the food industry today.
- A *climbing–falling film evaporator* is also available. The feed is first partially concentrated in a climbing film section and then finished off in a falling film section. High rates of evaporation are attainable in this type of plant.

3.1.2.4 Plate Evaporators

In these evaporators the calandria is a plate heat exchanger, similar to that used in pasteurizing and sterilizing liquids (see Chapter 2). The liquid is pumped through the heat exchanger, passing on one side of an assembly of plates, while steam passes on the other side. The spacing between plates is greater than that in pasteurizers to accommodate the vapor produced during evaporation. The liquid usually follows a climbing–falling film flow pattern. However, designs featuring only a falling film flow pattern are also available. The mixture of liquid and vapor leaving the calandria passes into a cyclone separator. The vapor from the separator goes to a condenser or into the heating jacket of the next stage, in a multiple-effect system. The concentrate is collected as product or goes to another stage. The advantages of plate evaporators include: high liquid velocities leading to high rates of heat transfer, short residence times, and resistance to fouling. They are compact and easily dismantled for inspection and maintenance. However, they have relatively high capital costs and low throughputs. They can be used for moderately viscous, heat-sensitive liquids such as milk, fruit juices, yeast, and meat extracts.

3.1.2.5 Agitated Thin-Film Evaporators

For very viscous materials and/or materials which tend to foul, heat transfer may be increased by continually wiping the boundary layer at the heat transfer surface. An agitated thin-film evaporator consists of a steam jacketed shell equipped with a centrally located, rotating shaft carrying blades which wipe the inner surface of the shell. The shell may be cylindrical and mounted either vertically or horizontally. Horizontal shells may be cone-shaped, narrowing in the direction of flow of the liquid. There may be a fixed clearance of 0.5–2.0 mm between the edge of the blades and the inner surface of the shell. Alternatively, the blades may float and swing out toward the heat transfer surface as the shaft rotates, creating a film of liquid with a thickness as little as 0.25 mm. Most of the evaporation takes place in the film that forms behind the rotating blades. Relatively high rates of heat transfer are attained and fouling and foaming are inhibited. However, these evaporators have relatively high capital costs and low throughputs. They are used as single-effect units, with relatively large temperature differences between the steam and the liquid being evaporated. They are often used as "finishers" when high solids concentrations are required. Applications include tomato paste, gelatin solutions, milk products, coffee extract, and sugar products.

3.1.2.6 Centrifugal Evaporators

In this type of evaporator a rotating stack of cones is housed in a stationary shell. The cones have steam on alternate sides to supply the heat. The liquid is fed to the undersides of the cones. It forms a thin film, which moves quickly across the surface of the cones, under the influence of centrifugal force, and rapid evaporation occurs. Very high rates of heat transfer, and so very short residence times, are attained. The vapor and concentrate are separated in the shell surrounding the cones. This type of evaporator is suitable for heat-sensitive and viscous materials. High capital costs and low throughputs are the main limitations of conical evaporators. Applications include fruit and vegetable juices and purees and extracts of coffee and tea.

3.1.2.7 Refractance Window Evaporator

In this relatively new type of evaporator, which operates at atmospheric pressure, the liquid to be concentrated flows countercurrent to heated water on either side of a thin plastic film. The water at a temperature of 95–98 °C provides the heat of evaporation by a combination of conduction and radiation. Due to evaporative cooling the product temperature seldom exceeds 70 °C. This heating method is also used to dehydrate liquid foods and the principle is described in more detail in Section 3.2.14.3. Refractance window evaporation has been used mainly to concentrate fruit juices and the quality of the concentrates compares favorably with those produced by falling film evaporators [8, 9].

3.1.2.8 Ancillary Equipment

Vapor–Concentrate Separators The mixture of vapor and liquid concentrate leaving the calandria needs to be separated and entrainment of droplets of the liquid in the vapor minimized. Entrained droplets represent a loss of product. They can also reduce the energy value of the vapor which would make it less effective in multiple-effect systems or when vapor recompression is being used (see Sections 3.1.3 and 3.1.4). Separation may be brought about by gravity. If sufficient headspace is provided above the calandria, as in Figure 3.1, the droplets may fall back into the liquid. Alternatively, this may occur in a second vessel, known as a *flash chamber*. Baffles or wire meshes may be located near the vapor outlet. Droplets of liquid impinge on these, coalesce into larger droplets and drain back into the liquid under gravity. The mixture of vapor and liquid may be directed tangentially, at high velocity, either by natural or forced circulation, into a cyclone separator. Centrifugal force is developed and the more dense liquid droplets impinge on the inner wall of the chamber, lose their kinetic energy and drain down into the liquid. This type of separator is used in most long-tube, plate, and agitated thin-film evaporators.

Condensers and Pumps The water vapor leaving the calandria contains some noncondensable gases which were in the feed or leaked into the system. The water vapor is converted back to a liquid in a condenser. Condensers may be of the indirect type, in which the cooling water does not mix with the water vapor. These are usually tubular heat exchangers. They are relatively expensive. Indirect condensers are used when volatiles are being recovered from the vapor or to facilitate effluent disposal. Direct condensers, known as *jet* or *spray condensers*, are more widely used. In these a spray of water is mixed with the water vapor to condense it. A condensate pump or barometric leg is used to remove the condensed vapor. The noncondensable gases are removed by positive displacement vacuum pumps or steam jet ejectors [1, 3, 5].

3.1.3 Multiple-Effect Evaporation

In a single-effect evaporator it takes 1.1–1.3 kg of steam to evaporate 1.0 kg of water. This is known as the *specific steam consumption* of an evaporator. The specific steam consumption of dryers used for liquid foods is much higher than this. A spray dryer has a specific steam consumption of 3.0–3.5 kg of steam to evaporate 1.0 kg of water. That of a drum dryer is slightly lower. Thus, it is common practice to concentrate liquid foods by vacuum evaporation before drying them in such equipment. It is also common practice to preheat the liquid to its evaporation temperature before feeding it to the evaporator. This improves the specific steam consumption. The vapor leaving a single-effect evaporator contains useful heat. This vapor may be put to other uses, for example, to heat water for cleaning. However, the most widely used method of recovering heat from the vapor leaving

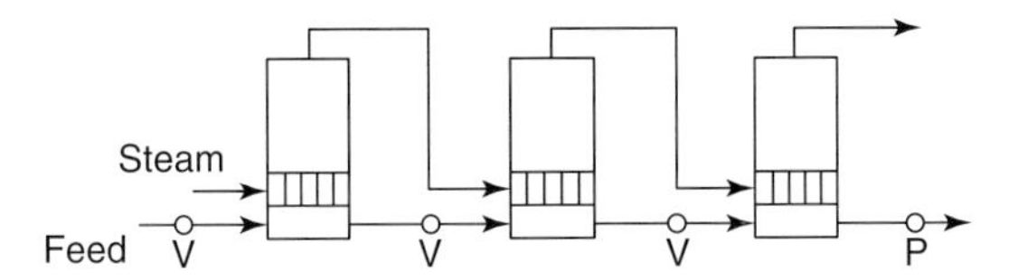

Figure 3.3 The principle of multiple-effect evaporation, with forward feeding (V - valve; P - product). Adapted from Brennan *et al.* [1] with permission.

an evaporator is *multiple-effect evaporation*, the principle of which is shown in Figure 3.3.

The vapor leaving the separator of the first evaporator, effect 1, enters the steam jacket of effect 2, where it heats the liquid in that effect, causing further evaporation. The vapor from effect 2 is used to heat the liquid in effect 3, and so on. The vapor from the last effect goes to a condenser. The liquid also travels from one effect to the next, becoming more concentrated as it does so. This arrangement is only possible if the evaporation temperature of the liquid in effect 2 is lower than the temperature of the vapor leaving effect 1. This is achieved by operating effect 2 at a lower pressure than effect 1. This principle can be extended to a number of effects. The pressure in effect 1 may be atmospheric or slightly below. The pressure in the following effects decreases with the number of effects. The specific steam consumption in a double effect evaporator is in the range 0.55–0.70 kg of steam per 1 kg of water evaporated, while in a triple effect it is 0.37–0.45 kg. However, the capital cost of a multiple-effect system increases with the number of effects.

The size of each effect is normally equivalent to that of a single-effect evaporator with the same capacity, working under similar operating conditions. Thus a three-effect unit will cost approximately three times that of a single effect. It must be noted that multiple-effect evaporation does not increase the throughput above that of a single effect. Its purpose is to reduce the steam consumption. Three to five effects are most commonly used. Up to seven effects are in use in some large milk and fruit juice processing plants.

The flow pattern in Figure 3.3 is known as *forward feeding* and is the most commonly used arrangement. Other flow patterns, including backward and mixed feeding, are also used. Each has its own advantages and limitations [1, 3, 5].

3.1.4 Vapor Recompression

This procedure involves compressing some or all of the vapor from the separator of an evaporator to a pressure that enables it to be used as a heating medium. The compressed vapor is returned to the jacket of the calandria. This reduces the amount of fresh steam required and improves the specific steam consumption of the evaporator. The vapor may be compressed thermally or mechanically.

In thermal vapor recompression (TVR), the vapor from the separator is divided into two streams. One stream goes to the condenser or to the next stage of a multiple-effect system. The other enters a steam jet compressor fed with fresh high pressure steam. As it passes through the jet of the compressor the pressure of the fresh steam falls and it mixes with the vapor from the evaporator. The vapor mix then passes through a second converging–diverging nozzle where the pressure increases. This high-pressure mixture then enters the jacket of the calandria.

In mechanical vapor recompression (MVR), all the vapor from the separator is compressed in a mechanical compressor which can be driven by electricity, a gas turbine, or a steam turbine. The compressed vapor is then returned to the jacket of the calandria. Both methods are used in industry. MVR is best suited to large capacity duties. Both TVR and MVR are used in conjunction with multiple-effect evaporators. Vapor recompression may be applied to one or more of the effects. A specific steam consumption of less than 0.10 kg steam per 1 kg of water evaporated is possible. A seven-effect falling film system, with TVR applied to the second effect, the compressed vapor being returned to the first effect, was reported by Pisecky [5]. See also [1, 3, 5, 6, 10–13].

3.1.5
Applications for Evaporation

The purposes for which evaporation is used in the food industry include: to produce concentrated liquid products (for sale to the consumer or as ingredients to be used in the manufacture of other consumer products), to preconcentrate liquids for further processing and to reduce the cost of transport, storage, and in some cases packaging, by reducing the mass and volume of the liquid.

3.1.5.1 Concentrated Liquid Products

Evaporated (Unsweetened Condensed) Milk The raw material for this product is whole milk which should of good microbiological quality. The first step is standardization of the composition of the raw milk so as to produce a finished product with the correct composition. The composition of evaporated milk is normally 8% fat and 18% solids not fat (snf), but this may vary from country to country. Skim milk and/or cream is added to the whole milk to achieve the correct fat:snf ratio. To prevent coagulation during heat processing and to minimize age thickening during storage, the milk is stabilized by the addition of salts, including phosphates, citrates, and bicarbonates, to maintain pH 6.6–6.7. The milk is then heat treated. This is done to reduce the microbiological load and to improve its resistance to coagulation during subsequent sterilization. The protein is denatured and some calcium salts are precipitated during this heat treatment. This results in stabilization of the milk.

The usual heat treatment is at 120–122 °C for several minutes, in tubular or plate heat exchangers. This heat treatment also influences the viscosity of the final product, which is an important quality attribute. The milk is then concentrated by vacuum evaporation. The evaporation temperature is in the range 50–60 °C and is usually carried out in multiple-effect, falling film evaporator systems. Plate and centrifugal evaporators may also be used. Two or three effects are common, but up to seven effects have been used. The density of the milk is monitored until it reaches a value that corresponds to the desired final composition of the evaporated milk. The concentrated milk is then homogenized in two stages in a pressure homogenizer, operating at 12.5–25.0 MPa. It is then cooled to 14 °C. The concentrated milk is then tested for stability by heating samples to sterilization temperature. If necessary, more stabilizing salts, such as disodium or trisodium phosphates, are added to improve stability. The concentrated milk is then filled into containers, usually tinplate cans, and sealed by double seaming. The filled cans are then sterilized in a retort at 110–120 °C for 15–20 min. If a batch retort is used, it should have a facility for continually agitating the cans during heating, to ensure that any protein precipitate formed is uniformly distributed throughout the cans. The cans are then cooled and stored at not more than 15 °C.

As an alternative to in-package heat processing, the concentrated milk may be UHT treated at 140 °C for about 3 s (see Chapter 2) and aseptically filled into cans, cartons, or "bag in box" containers.

Evaporated milk has been used as a substitute for breast milk, with the addition of vitamin D, in cooking and as a coffee whitener. A low-fat product may be manufactured using skimmed or semi-skimmed milk as the raw material. Skimmed milk concentrates are used in the manufacture of ice cream and yoghurt. Concentrated whey and buttermilk are used in the manufacture of margarine and spreads.

Sweetened Condensed Milk This product consists of evaporated milk to which sugar has been added. It normally contains about 8% fat, 20% snf, and 45% sugar. Because of the addition of the sugar, the water activity of this sweetened concentrate is low enough to inhibit the growth of spoilage and pathogenic microorganisms. Consequently, it is shelf stable, without the need for sterilization. In the manufacture of sweetened condensed milk, whole milk is standardized and heat treated in a similar manner to evaporated milk. At this point granulated sugar may be added, or sugar syrup may be added at some stage during evaporation. Usually, evaporation is carried out in a two- or three-stage, multiple-effect, falling film evaporator at 50–60 °C. Plate and centrifugal evaporators may also be used. If sugar syrup is used, it is usually drawn into the second effect evaporator. The density of the concentrate is monitored during the evaporation. Alternatively, the soluble solids content is monitored, using a refractometer, until the desired composition of the product is attained. The concentrate is cooled to about 30 °C. Finely ground lactose crystals are added, while it is vigorously mixed. After about 60 min of mixing, the concentrate is quickly cooled to about 15 °C. The purpose of this seeding procedure is to ensure that, when the supersaturated solution of lactose crystallizes out, the crystals formed will be small and not cause grittiness in the product. The cooled product is then poured into cans, cartons, or tubes and sealed in an appropriate manner.

Sweetened condensed milk is used in the manufacture of other products such as ice cream and chocolate. If it is to be used for these purposes, it is packaged in larger containers such as "bag in box" systems, drums, or barrels [14, 15].

Concentrated fruit and vegetable juices are also produced for sale to the consumer (see Section 3.1.5.3).

3.1.5.2 Evaporation as a Preparatory Step to Further Processing

One important application for evaporation is the preconcentration of liquids which are to be dehydrated by spray drying, drum drying, or freeze drying. As discussed in Section 3.1.3, the specific steam consumption of such dryers is greater than that of single-effect evaporators and much greater than that of multiple-effect systems, particularly if MVR or TVR is incorporated into one or more of the effects. Whole milk, skimmed milk, and whey are examples of liquids that are preconcentrated prior to drying. Equipment similar to that used to produce evaporated milk (see Section 3.1.5.1) is used and the solids content of the concentrate is in the range 40–55%.

Instant coffee Beverages such as coffee and tea are also available in powder form, so-called instant drinks. In the production of instant coffee, green coffee beans are cleaned, blended, and roasted. During roasting, the color and flavor develop. Roasting is usually carried out continuously. Different types of roasted beans, light, medium, and dark, are produced by varying the roasting time. The roasted beans are then ground in a mill to a particle size to suit the extraction equipment, usually in the range 1000–2000 μm. The coffee solubles are extracted from the particles, using hot water as the solvent. Countercurrent, static bed, or continuous extractors are used (see Chapter 9). The solution leaving the extractor usually contains 15–28% solids. After extraction, the solution is cooled and filtered. This extract may be directly dried by spray drying or freeze drying. However, it is more usual to concentrate the solution to about 60% solids by vacuum evaporation. Multiple-effect falling film systems are commonly used.

The volatile flavor compounds are stripped from the solution, before or during the evaporation, in a similar manner to that used when concentrating fruit juice (see Section 3.1.5.3) and added back to the concentrate before drying. Coffee powder is produced using a combination of spray drying and fluidized bed drying (see Section 3.2.6). Alternatively, the concentrated extract may be frozen in slabs, the slabs broken up and freeze dried in a batch or continuous freeze dryer (see Section 3.2.7 [15–17]).

Granulated sugar Vacuum evaporation is used in the production of granulated sugar from sugar cane and sugar beet. Sugar juice is expressed from sugar cane in roller mills. In the case of sugar beet, the sugar is extracted from sliced beet, using heated water at 55–85 °C, in a multistage, countercurrent, static bed, or moving bed extractor (see Chapter 9). The crude sugar juice, from either source, goes through a series of purification operations. These include screening and carbonation. Lime is added and carbon dioxide gas is bubbled through the juice. Calcium carbonate crystals are formed. As they settle, they carry with them a lot of the insoluble impurities in the juice. The supernatant is taken off and filtered. Carbonation may be applied in two or more stages during the purification of the juice.

The juice may be treated with sulfur dioxide to limit nonenzymic browning. This process is known as *sulfitation*. The treated juice is again filtered. Various type of filters are used in processing sugar juice, including plate and frame, shell and tube, and rotary drum filters (see Chapter 9). The purified juice is concentrated up to 50–65% solids by vacuum evaporation. Multiple-effect systems are employed, usually with five effects. Vertical short-tube, long-tube, and plate evaporators are used. The product from the evaporators is concentrated further in vacuum pans or single-effect short tube evaporators, sometimes fitted with an impeller in the downcomer. This is known as *sugar boiling*. Boiling continues until the solution becomes supersaturated. It is then "shocked" by the addition of a small amount of seeding material, finely ground sugar crystals, to initiate crystallization. Alternatively, a slurry of finely ground sugar crystals in isopropyl alcohol may be added at a lower degree of supersaturation. The crystals are allowed to grow, under carefully controlled conditions, until they reach the desired size and number.

The slurry of sugar syrup and crystals is discharged from the evaporator into a temperature-controlled tank, fitted with a slow-moving mixing element. From there it is fed to filtering centrifugals, or basket centrifuges (see Chapter 9) where the crystals are separated from the syrup and washed. The crystals are dried in heated air in rotary dryers (see Section 3.2.3.7) and cooled. The dry crystals are conveyed to silos or packing rooms. The sugar syrup from the centrifuges is subjected to further concentration, seeding, and separation processes, known as *second* and *third boilings*, to recover more sugar in crystal form [1, 18–21].

3.1.5.3 The Use of Evaporation to Reduce Transport, Storage, and Packaging Costs

Concentrated fruit and vegetable juices Many fruit juices are extracted, concentrated by vacuum evaporation, and the concentrate frozen on one site, near the growing area. The frozen concentrate is then shipped to several other sites where it is diluted, packaged, and sold as chilled fruit juice. Orange juice is the main fruit juice processed in this way [19]. The fruit is graded, washed, and the juice extracted using specialist equipment described in [20, 21]. The juice contains about 12.0% solids at this stage. The extracted juice is then "finished." This involves removing bits of peel, pips, pulp, and rag from the juice by screening and/or centrifugation. The juice is then concentrated by vacuum evaporation.

Many types of evaporator have been used, including high-vacuum, low-temperature systems. These had relatively long residence times, up to 1 h in some cases. To inactivate enzymes, the juice was pasteurized in plate heat exchangers before being fed to the evaporator. One type of evaporator, in which evaporation took place at temperatures as low as 20 °C, found use for juice concentration. This operated on a heat pump principle. A refrigerant gas condensed in the heating jacket of the calandria, releasing heat, which caused the liquid to evaporate. The liquid refrigerant evaporated in the jacket of the condenser, taking heat from the water vapor, causing it to condense [1].

Modern evaporators for fruit juice concentration work on a high-temperature, short-time (HTST) principle. They are multiple-effect systems, comprising up to seven falling film or plate evaporators. The temperatures reached are high enough to inactivate enzymes, but the short residence times limit undesirable changes in the product. They are operated with forward flow or mixed flow feeding patterns. Some designs of this type of evaporator are known as thermally accelerated short time evaporators or TASTEs.

Volatile compounds that contribute to the odor and flavor of fruit juices are lost during vacuum evaporation, resulting in a concentrate with poor organoleptic qualities. It is common practice to recover these volatiles and add them back to the concentrate. Volatiles may be stripped from the juice, prior to evaporation, by distillation under vacuum. However, the most widely used method is to recover these volatiles after partial evaporation of the water. The vapors from the first effect of a multiple-effect evaporation system consist of water vapor and volatiles. When these vapors enter the heating jacket of the second effect, the water vapor condenses first and the volatiles are taken from the jacket and passed through a distillation column, where the remaining water is separated off and the volatiles are concentrated [22–25].

Orange juice is usually concentrated up to 65% solids, filled into drums or "bag in box" containers and frozen in blast freezers. Unfrozen concentrated juice may also be transported in bulk in refrigerated tankers or ships' holds. Sulfur dioxide may be used as a preservative. Concentrated orange juice may be UHT treated and aseptically filled into "bag in box" containers or drums. These concentrates may be diluted back to 12% solids, packed into cartons or bottles and sold as chilled orange juice. They may also be used in the production of squashes, other soft and alcoholic drinks, jellies, and many other such products.

A frozen concentrated orange juice may also be marketed as a consumer product. This usually has 42–45% solids and is made by adding fresh juice, "cutback juice," to the 65% solids concentrate, together with some recovered volatiles. This is filled into cans and frozen. Concentrate containing 65% solids, with or without added sweetener, may be pasteurized and hot-filled into cans, which are rapidly cooled. This is sold as a chilled product. UHT-treated concentrate may also be aseptically filled into cans or cartons for sale to the consumer from the ambient shelf.

Juices from other fruits, including other citrus fruits, pineapples, apples, grapes, blackcurrants, and cranberries, may be concentrated. The procedures are similar to those used in the production of concentrated orange juice. When clear concentrates are being produced, enzymes are used to precipitate pectins, which are then separated from the juice.

Vacuum evaporation is also applied to the production of concentrated tomato products. Tomatoes are chopped and/or crushed and subjected to heat treatment, which may be the hot break or cold break process [26, 27]. Skin and seeds are removed and the juice extracted in a cyclone separator. The juice is then concentrated by vacuum evaporation. For small-scale production, vacuum pans may be employed. For larger throughputs, two- or three-effect tubular or plate evaporators may be used. If highly concentrated pastes are being produced, agitated thin-film evaporators may be used as finishers.

Glucose syrup, skimmed milk, and whey are among other food liquids that may be concentrated, to reduce weight and bulk, and so reduce transport and storage costs.

3.2
Dehydration (Drying)

3.2.1
General Principles

Dehydration is the oldest method of food preservation practiced by humans. For thousands of years we have dried and/or smoked meat, fish, fruits, and vegetables, to sustain us during out-of-season periods in the year. Today the dehydration section of the food industry is large and extends to all countries of the globe. Drying facilities range from simple sun or hot air dryers to high-capacity, sophisticated spray-drying or freeze-drying installations. A very large range of dehydrated foods is available and makes a significant contribution to the convenience food market.

In this chapter the terms *dehydration* and *drying* are used interchangeably to describe the removal of most of the water normally present in a foodstuff, by evaporation or sublimation, as a result of the application of heat. The main reason for drying a food is to extend its shelf life beyond that of the fresh material, without the need for refrigerated transport and storage. This goal is achieved by reducing the available moisture, or water activity (a_W; see Section 3.2.17) to a level which inhibits the growth and development of spoilage and pathogenic microorganisms, reducing the activity of enzymes and the rate at which undesirable chemical changes occur. Appropriate packaging is necessary to maintain the low a_W during storage and distribution.

Drying also reduces the weight of the food product. Shrinkage, which often occurs during drying, reduces the volume of the product. These changes in weight and volume can lead to substantial savings in transport and storage costs and, in some cases, the costs of packaging. However, dehydration is an energy-intensive process and the cost of supplying this energy can be relatively high compared to other methods of preservation.

Changes detrimental to the quality of the food may also occur during drying. In the case of solid food pieces, shrinkage can alter the size and shape of the pieces. Changes in color may also occur. When the food pieces are rehydrated, their color and texture may be significantly inferior to those of the fresh material. Dry powders may be slow to rehydrate. Changes in flavor may occur during drying solid or liquid foods, as a result of losing volatile flavor compounds and/or the development of cooked flavors. A reduction in the nutritional value of foods can result from dehydration. In particular, loss of vitamins C and A may be greater during drying than in canning or freezing.

Dehydration is usually described as a simultaneous heat and mass transfer operation. Sensible and latent heat must be transferred to the food to cause the water to evaporate. Placing the food in a current of heated air is the most widely used method of supplying heat. The heat is transferred by convection from the air to the surface of the food and by conduction within the food. Alternatively, the food may be placed in contact with a heated surface. The heat is transferred by conduction to the surface of the food in contact with the heated surface and within the food. There is limited use of radiant, microwave (MW), and radiofrequency (RF) energy in dehydration. Freeze drying involves freezing the food and removal of the ice by sublimation. This is usually achieved by applying heat, by conduction or radiation, in a very low-pressure environment. In osmotic drying food pieces are immersed in a hypertonic solution. Water moves from the food into the solution under the influence of osmotic pressure.

3.2.2 Drying Solid Foods in Heated Air

When a wet material is placed in a current of heated air, heat is transferred to its surface, mainly by convection. The water vapor formed is carried away from the drying surface in the air stream. Consider a model system in which a wet material in the form of a thin slab, consisting of an inert solid wetted with pure water, is placed in a current of heated air, flowing parallel to one of its large faces. The temperature, humidity, and velocity of the air are maintained constant. It is assumed that all the heat is transferred to the solid from the air by convection and that drying takes place from one large face only. If the rate of change of moisture content is plotted against time, as in Figure 3.4, the drying curve may be seen to consist of a number of stages or periods.

Period A–B is a settling down or *equilibration period*. The surface of the wet solid comes into equilibrium with the air. This period is usually short compared with the total drying time. Period B–C is known as the *constant rate period*. Throughout this period the surface of the solid is saturated with water. As water evaporates from the surface, water from within the solid moves to the surface, keeping it in a saturated state. The rate of drying during this period remains constant. So also does the surface temperature, at a value corresponding to the wet bulb temperature of the air. By considering heat and mass transfer across the drying surface, a model for the prediction of the duration of the constant rate period of drying can be developed as in Equation 3.1:

$$t_c = \frac{(W_0 - W_c)\rho_s L_s l}{h_c(\theta_a - \theta_s)} \tag{3.1}$$

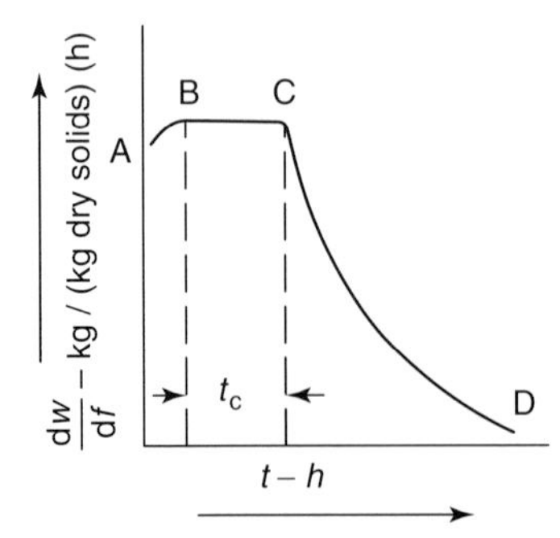

Figure 3.4 Drying curve for a wet solid in heated air at constant temperature, humidity, and velocity. Adapted from Brennan *et al.* [1] with permission.

where t_c is the duration of the constant rate period, W_0 is the initial moisture content of the wet solid (dry weight basis, dwb), W_c is the moisture content at the end of the constant rate period (dwb), ρ_s is the density of the material, L_s is the latent heat of evaporation at θ_s, l is the thickness of the slab, h_c is heat transfer coefficient for convection heating, θ_a is the dry bulb temperature of the air, and θ_s is the wet bulb temperature of the air.

From Equation 3.1 it can be seen that the main factors that influence the rate of drying during the constant rate period are the air temperature and humidity and the area of the drying surface. The air velocity also has an influence, as the higher this is the greater the value of h_c. As long as this state of equilibrium exists, high rates of evaporation may be maintained, without the danger of overheating the solid. This is an important consideration when drying heat-sensitive foods. Some foods exhibit a constant rate period of drying. However, it is usually short compared to the total drying time. Many foods show no measurable constant rate period.

As drying proceeds, at some point (represented by C in Figure 3.4) the movement of water to the surface is not enough to maintain the surface in a saturated condition. The state of equilibrium at the surface no longer holds and the rate of drying begins to decline. Point C is known as the *critical point* and the period C–D is the *falling rate period.* From point C on, the temperature at the surface of the solid rises and approaches the dry bulb temperature of the air as drying nears completion. Hence, it is toward the end of the drying cycle that any heat damage to the product is likely to occur. Many research workers claim to have identified two or more falling rate periods where points of inflexion in the curve have occurred. There is no generally accepted explanation for this phenomenon. During the falling rate period, the rate of drying is governed by factors which affect the movement of moisture within the solid. The influence of external factors, such as the velocity of the air, is reduced compared to the constant rate period. In the dehydration of solid food materials, most of the drying takes place under falling rate conditions.

Numerous mathematical models have been proposed to represent drying in the falling rate period. Some are empirical and were developed by fitting relationships to data obtained experimentally. Others are based on the assumption that a particular mechanism of moisture movement within the solid prevails. The best known of these is based on the assumption that moisture migrates within the solid by diffusion as a result of the concentration difference between the surface and the center of the solid. It is assumed that Fick's second law of diffusion applies to this movement. A well-known solution to this law is represented by Equation 3.2:

$$\frac{W - W_e}{W_c - W_e} = \frac{8}{\pi^2}\left[\exp\left\{-Dt\left(\frac{\pi}{2l}\right)^2\right\}\right] \tag{3.2}$$

where W is the average moisture content at time t (dwb), W_e is the equilibrium moisture content (dwb), W_c is the moisture content at the start of the falling rate period (dwb), D is the liquid diffusivity, and l is the thickness of the slab.

In Equation 3.2 it is assumed that the value of D is constant throughout the falling rate period. However, many authors have reported that D decreases as the moisture content decreases. Some authors who reported the existence of two or more falling rate periods successfully applied Equation 3.2 to the individual periods but used a different value of D for each period. Many other factors may change the drying pattern of foods. Shrinkage alters the dimensions of food pieces. The presence of cell walls can affect the movement of water within the solids. The density and porosity of the food material may change during drying. The thermal properties of the food material, such as specific heat and thermal conductivity, may change with a change in moisture content. As water moves to the surface, it carries with it any soluble material, such as sugars and salts. When the water evaporates at the surface, the soluble substances accumulate at the drying surface. This can contribute to the formation of an impervious dry layer at the surface, which impedes drying. This phenomenon is known as *case hardening*. The diffusion theory does not take these factors into account and so has had only limited success in modeling falling rate drying.

Many more complex models have been proposed, which attempt to take some of these changes into account, see [28] as an example [1, 3–5, 7, 29–35].

3.2.3
Equipment Used in Hot Air Drying of Solid Food Pieces

3.2.3.1 Cabinet (Tray) Dryer

This is a multipurpose, batch-operated hot air dryer. It consists of an insulated cabinet, equipped with a fan, an air heater, and a space occupied by trays of food. It can vary in size from a bench-scale unit holding one or two small trays of food to a large unit taking stacks of large trays. The air may be directed by baffles to flow the across surface of the trays of food or through perforated trays and the layers of food, or both ways. The moist air is partly exhausted from the cabinet and partly recycled by means of dampers. Small cabinet dryers are used in laboratories, while larger units are used as industrial dryers, mainly for drying sliced or diced fruits and vegetables. A number of large cabinets may be used in parallel, with a staggered loading sequence, to process relatively large quantities of food, up to 20 000 t day^{-1} of raw material [1, 3, 29, 32, 33, 37].

3.2.3.2 Tunnel Dryer

This type of dryer consists of a long, insulated tunnel. Tray loads of the wet material are assembled on trolleys which enter the tunnel at one end. The trolleys travel the length of the tunnel and exit at the other end. Heated air also flows through the tunnel, passing between the trays of food and/or through perforated trays and the layers of food. The air may flow parallel to and in the same direction as the trolleys, as shown in Figure 3.5. This is known as a *concurrent tunnel*.

Other designs featuring countercurrent, concurrent–countercurrent, and cross-flow of air are available. Each pattern of airflow has its advantages and limitations. The trolleys may move continuously through the tunnel. Alternatively, the movement may be semi-continuous. As a trolley full of fresh material is introduced into one end of the tunnel, a trolley full of dried product exits at the other end. Tunnels

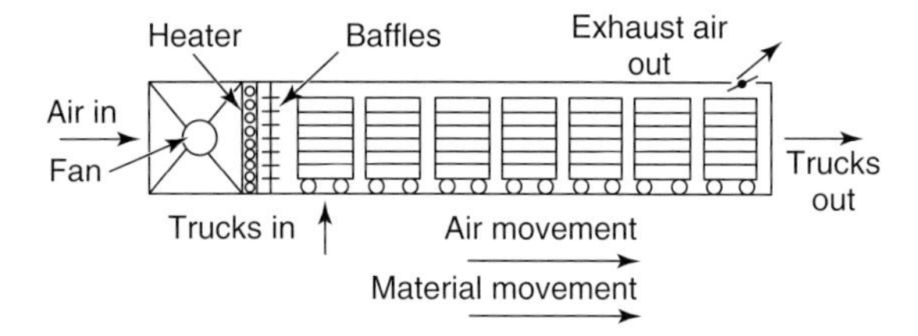

Figure 3.5 Principle of concurrent tunnel dryer. From Brennan *et al.* [1] with permission.

may be up to 25 m in length and about 2 m × 2 m in cross-section. Tunnel dryers are mainly used for drying sliced or diced fruits and vegetables [1, 3, 29, 33, 36, 37].

3.2.3.3 **Conveyor (Belt) Dryer**

In this type of dryer the food material is conveyed through the drying tunnel on a perforated conveyor made of hinged, perforated metal plates or wire or plastic mesh. The heated air usually flows through the belt and the layer of food, upward in the early stages of drying and downward in the later stages. The feed is applied to the belt in a layer 75–150 mm deep. The feed must consist of particles that form a porous bed allowing the air to flow through it. Conveyors are typically 2–3 m wide and up to 50 m long. The capacity of a conveyor dryer is much less than that of a tunnel dryer, occupying the same floor space. As shrinkage occurs during drying, the thickness of the layer of food on the conveyor becomes less. Thus the belt is being used less efficiently as drying proceeds.

The use of multistage conveyor drying is common. The product from the first conveyor is redistributed, in a thicker layer, on the second conveyor. This may be extended to three stages. In this way, the conveyor is used more efficiently, compared to a single-stage unit. On transfer of the particles from one stage to the next, new surfaces are exposed to the heated air, improving the uniformity of drying. The air temperature and velocity may be set to different levels in each stage. Thus, good control may be exercised over the drying, minimizing heat damage to the product. However, even when using two or more stages, drying in this type of dryer is relatively expensive. Consequently, they are often used to remove moisture rapidly, in the early stages of drying, and the partly dried product is dried to completion in another type of dryer. Diced vegetables, peas, sliced beans, and grains are examples of foods dried in conveyor dryers [1, 3, 29, 33, 36–38].

3.2.3.4 **Bin Dryer**

This is a throughflow dryer, mainly used to complete the drying of particulate material partly dried in a tunnel or conveyor dryer. It takes the form of a vessel fitted with a perforated base. The partly dried product is loaded into the vessel to up to 2 m deep. Dry, but relatively cool air, percolates up through the bed slowly, completing the drying of the product over an extended period, up to 36 h. Some migration of moisture between the particles occurs in the bin. This improves the uniformity of moisture content in the product [1, 3, 29, 37, 38].

3.2.3.5 **Fluidized Bed Dryer**

This is another throughflow, hot air dryer which operates at higher air velocities than the conveyor or bin dryer. In this type of dryer heated air is blown up through a perforated plate which supports a bed of solid particles. As the air passes through the bed of particles, a pressure drop develops across the bed. As the velocity of the air increases, the pressure drop increases. At a particular air velocity, known as the *incipient velocity*, the frictional drag on the particles exceeds the weight of the particles. The bed then expands, the particles are suspended in the air and the bed starts to behave like a liquid, with particles circulating within the bed. This is what is meant by the term *fluidized bed*. As the air velocity increases further, the movement of the particles becomes more vigorous. At some particular velocity, particles may detach themselves from the surface of the bed temporarily and fall back onto it. At some higher velocity, known as the *entrainment velocity*, particles are carried away from the bed in the exhaust air stream.

Fluidized bed dryers are operated at air velocities between the incipient and entrainment values. The larger and more dense the particles are, the higher the air velocity required to fluidize them. Particles in the size range 20 μm to 10 mm in diameter can usually be fluidized. The particles must not be sticky or prone to mechanical damage. Air velocities in the range 0.2–5.0 $m\,s^{-1}$ are used. There is very close contact between the heated air and the particles, which results in high rates of heat transfer and relatively short drying times.

Fluidized beds may be operated on a batch or continuous basis. Batch units are used for small-scale operations. Because of the mixing which occurs in such beds, uniform moisture contents are attainable. Continuous fluidized bed dryers used in the food industry are of the plug flow type shown in Figure 3.6. The feed enters the bed at one end and the product exits over a weir at the other. As the particles dry, they become lighter and rise to the surface of the bed and are discharged over the weir. However, if the size, density, and moisture content of the feed particles are not uniform, the moisture content of the product may also not be uniform.

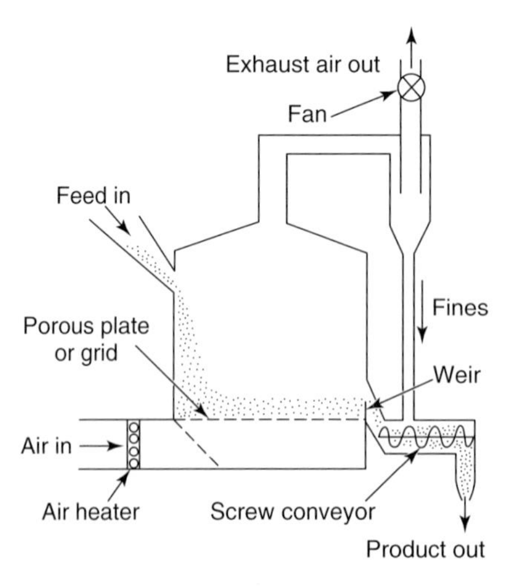

Figure 3.6 Principle of continuous fluidized bed dryer, with fines recovery system. From Brennan *et al.* [1] with permission.

The use of multistage fluidized bed dryers is common. The partly dried product from the first stage is discharged over the weir onto the second stage, and so on. Up to six stages have been used. The temperature of the air may be controlled at a different level in each stage. Such systems can result in savings in energy and better control over the quality of the product as compared with a single-stage unit. Fluidized beds may be mechanically vibrated. This enables them to handle particles with a wider size range than a standard bed. They can also accommodate sticky products and agglomerated particles better than a standard bed. The air velocities needed to maintain the particles in a fluidized state are less than in a stationary bed. Such fluidized beds are also known as *vibrofluidizers*.

Another design of the fluidized bed is known as the *spouted bed dryer* (Figure 3.7). Part of the heated air is introduced into the bottom of the bed in the form of a high velocity jet. A spout of fast-moving particles is formed in the center of the bed. On reaching the top of the bed, the particles return slowly to the bottom of the bed in an annular channel surrounding the spout. Some of the heated air flows upward through the slow-moving channel, countercurrent to the movement of the particles. High rates of evaporation are attained in the spout, while evaporative cooling keeps the particle temperature relatively low. Conditions in the spout are close to constant rate drying (see Section 3.2.3). Drying of the particles is completed

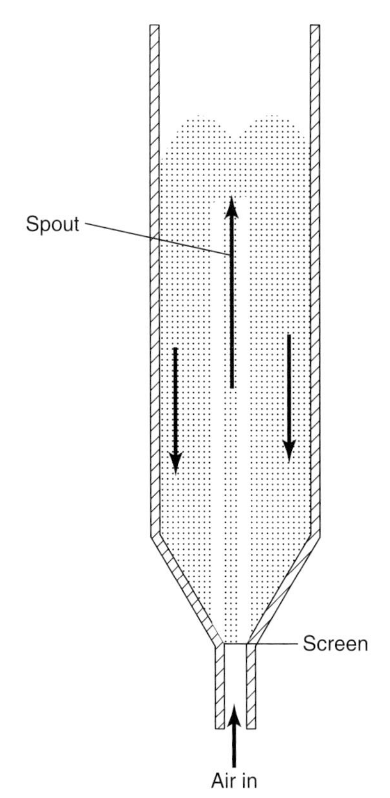

Figure 3.7 Principle of spouted bed dryer.

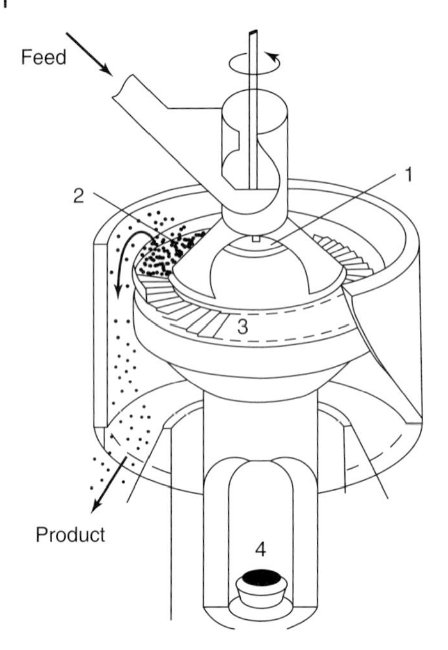

Figure 3.8 Principle of toroidal bed dryer: 1. Particle distributor, 2. Bed of particles, 3. Louvre plates, 4. Air heater.

in the annular channel. This type of dryer can handle larger particles than the conventional fluidized bed.

In some spouted bed dryers the air is introduced tangentially into the base of the bed and a screw conveyor is located at its center to control the upward movement of the particles. Such dryers are suitable for drying relatively small particles.

In the toroidal bed dryer the heated air enters the drying chamber through blades or louvers, creating a fast-moving, rotating bed of particles (Figure 3.8). High rates of heat and mass transfer in the bed enable rapid drying of relatively small particles. Peas, sliced beans, carrots and onions, grains, and flours are some foods that have been dried in fluidized bed dryers [1, 3, 5, 29, 33, 39–42].

3.2.3.6 **Pneumatic (Flash) Dryer**

In this type of dryer the food particles are conveyed in a heated air stream through ducting of sufficient length to give the required drying time. It is suitable for relatively small particles and high air velocities are used, in the range 10–40 m s^{-1}. The dried particles are separated from the air stream by cyclone separators or filters. The ducting may be arranged vertically or horizontally. Single vertical dryers are used mainly for removing surface moisture. Drying times are short, in the range 0.5–3.5 s and they are also known as *flash dryers*. When internal moisture is to be removed, longer drying times are required. Horizontal pneumatic dryers, or vertical dryers, consisting of a number of vertical columns in series, may be used for this purpose. The ducting may be in the form of a closed loop. The particles travel a number of times around the loop until they reach the desired moisture content. The dried particles are removed from the air stream by means of a cyclone. Fresh air is introduced continuously into the loop through a heater and humid air is continuously expelled from it. This type of dryer is known as a *pneumatic ring dryer*. Grains and flours are the main products dried in pneumatic dryers [1, 3, 5, 29, 43].

3.2.3.7 Rotary Dryer

The most common design of rotary dryer used for food application is known as the *direct rotary dryer*. This consists of a cylindrical shell, set at an angle to the horizontal. The shell rotates at 4–5 rpm. Wet material is fed continuously into the shell at its raised end and dry product exits over a weir at the lower end. Baffles or flights are fitted to the inner surface of the shell. These lift the material up as the shell rotates and allow it to fall down through a stream of heated air, that may flow concurrent or countercurrent to the direction of movement of the material. The feed material must consist of relatively small particles, which are free-flowing and reasonably resistant to mechanical damage. Air velocities in the range 1.5–2.5 $m\,s^{-1}$ are used and drying times are from 5 to 60 min.

One particular design of rotary dryer is known as the *louvered dryer*. The wall of the shell is made up of overlapping louver plates, through which the heated air is introduced. As the shell rotates, the bed of particles is gently rolled within it, causing mixing to take place and facilitating uniform drying. The movement is less vigorous than in the conventional rotary dryer and causes less mechanical damage to the particles. Grains, flours, cocoa beans, sugar, and salt crystals are among the food materials dried in rotary dryers [29, 44–47].

3.2.4 Drying of Solid Foods by Direct Contact with a Heated Surface

When a wet food material is placed in contact with a heated surface, usually metal, the sensible and latent heat is transferred to the food and within the food mainly by conduction. Most of the evaporation takes place from the surface, which is not in contact with the hot surface. The drying pattern is similar to that which prevails during drying in heated air (see Section 3.2.2). After an equilibration period, there a constant rate period, during which water evaporates at a temperature close to its boiling point at the prevailing pressure. The rate of drying will be higher than that in heated air at the same temperature. When the surface begins to dry out, the rate of drying will decrease and the temperature at the drying surface will rise and approach that of the heated metal surface, as drying nears completion. If drying is taking place at atmospheric pressure, the temperature at the drying surface in the early stages will be in excess of 100 °C. There is little or no evaporative cooling as is the case in hot air drying. The temperature of the heated metal surface will need to be well above 100 °C in order to achieve reasonable rates of drying and low product moisture content.

Exposure of a heat-sensitive food to such high temperatures for prolonged periods, up to several hours, is likely to cause serious heat damage. If the pressure above the evaporating surface is reduced below atmospheric, the evaporation temperature will be reduced and hence the temperature of the heated metal surface may be reduced. This is the principle behind *vacuum drying*.

If it is assumed that drying takes place from one large face only and that there is no shrinkage, the rate of drying, at any time t, may be represented by the following equation:

$$\frac{\mathrm{d}w}{\mathrm{d}t} = \frac{(W_0 - W_\mathrm{f})M}{t} = \frac{K_\mathrm{c}A(\theta_\mathrm{w} - \theta_\mathrm{e})}{L_\mathrm{e}} \tag{3.3}$$

where $\mathrm{d}w/\mathrm{d}t$ = rate of drying (rate of change of weight) at time t, W_0 is the initial moisture content of the material (dwb), W_f is the final moisture content of the material (dwb), M is the mass of dry solids on the surface, t is the drying time, K_c is the overall heat transfer coefficient for the complete drying cycle, A is the drying area, θ_w is the temperature of the heated metal surface (wall temperature), θ_e is the evaporating temperature, and L_e is the latent heat of evaporation at θ_e [1, 10, 25, 42, 43].

3.2.5 Equipment Used in Drying Solid Foods by Contact with a Heated Surface

3.2.5.1 Vacuum Cabinet (Tray or Shelf) Dryer

This dryer consists of a vacuum chamber connected to a condenser and vacuum pump. The chamber is usually cylindrical and has one or two access doors. It is usually mounted in a horizontal position. The chamber is equipped with a number of hollow plates or shelves, arranged horizontally. These shelves are heated internally by steam, hot water, or some other thermal fluid, which is circulated through them. A typical drying chamber may contain up to 24 shelves, each measuring 2.0 × 1.5 m. The food material is spread in relatively thin layers on metal trays. These trays are placed on the shelves, the chamber is sealed and the pressure reduced by means of the condenser and vacuum pump. Absolute pressures in the range 5–30 kPa are created, corresponding to evaporation temperatures of 35–80 °C. Drying times can range from 4 to 20 h, depending on the size and shape of the food pieces and the drying conditions. The quality of vacuum-dried fruits or vegetables is usually better than that of air-dried products. However, the capital cost of vacuum shelf dryers is relatively high and the throughput low, compared to most types of hot air dryer [1, 5, 29, 37, 47–49].

3.2.5.2 Double Cone Vacuum Dryer

This type of dryer consists of a hollow vessel in the shape of a double cone. It rotates about a horizontal axis and is connected to a condenser and vacuum pump. Heat is applied by circulating steam or a heated fluid through a jacket fitted to the vessel. The pressures and temperatures used are similar to those used in a vacuum shelf dryer (see Section 3.2.5.1). It is suitable for drying particulate materials, which are tumbled within the rotating chamber as heat is applied through the jacket. Some mechanical damage may occur to friable materials due to the tumbling action. If the particles are sticky at the temperatures being used, they may form lumps or balls or even stick to the wall of the vessel. Operating at low speeds or intermittently may reduce this problem. This type of dryer has found only limited application in the food industry [29, 47–49].

3.2.6
Freeze Drying (Sublimation Drying, Lyophilization) of Solid Foods

This method of drying foods was first used in industry in the 1950s. The process involves three stages: (i) freezing the food material, (ii) subliming the ice (primary drying), and (iii) removal of the small amount of water bound to the solids (secondary drying or desorption). Freezing may be carried out by any of the conventional methods including blast, immersion, plate, or liquid gas freezing (see Chapter 4). Blast freezing in refrigerated air is most often used. It is important to freeze as much of the water as possible. This may be difficult in the case of material with a high soluble solids content, such as concentrated fruit juice. As the water freezes in such a material, the soluble solids content of remaining liquid increases, and so its freezing point is lowered. At least 95% of the water present in the food should be converted to ice to attain successful freeze drying.

Ice will sublime when the water vapor pressure in the immediate surroundings is less than the vapor pressure of ice at the prevailing temperature. This condition could be attained by blowing dry, refrigerated air across the frozen material. However, this method has proved to be uneconomic on a large scale. On an industrial scale, the vapor pressure gradient is achieved by reducing the total pressure surrounding the frozen food to a value lower than the ice vapor pressure. The vapor pressure of ice at $-20\,^{\circ}\mathrm{C}$ is about 135 Pa, absolute. Industrial freeze drying is carried out in vacuum chambers operated at pressures in the range 13.5–270.0 Pa, absolute. The main components of a batch freeze dryer are a well-sealed vacuum chamber fitted with heated shelves, a refrigerated condenser, and a vacuum pump or pumps (Figure 3.9).

The refrigerated condenser removes the water vapor formed by sublimation. The water vapor freezes on to the condenser, thus maintaining the low water vapor pressure in the chamber. The vacuum pump(s) remove the noncondensable gases. The heated shelves supply the heat of sublimation. Heat may be applied from above the frozen food by radiation or from below by conduction or from both directions.

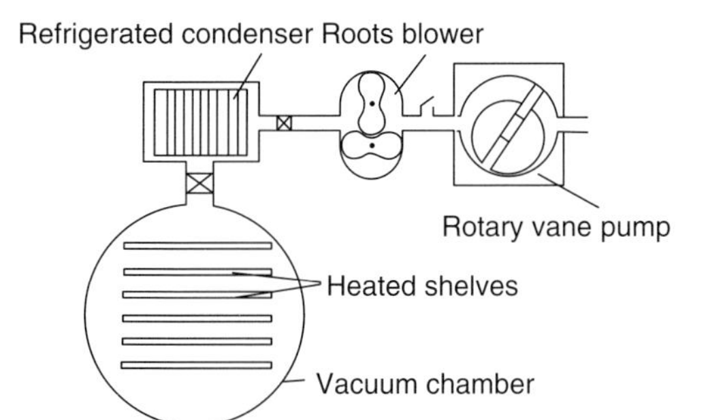

Figure 3.9 Main components of a batch freeze dryer. From Brennan [29] with permission.

Once sublimation starts, a dry layer will form on the top surface of the food pieces. Together, the rate at which water vapor moves through this dry layer and the rate at which heat travels through the dry layer and/or the frozen layer determine the rate of drying. If a slab-shaped solid is being freeze dried from its upper surface only and if the heat is supplied only from above, through the dry layer, a state of equilibrium will be attained between the heat and mass transfer. Under these conditions, the rate of drying (rate of change of weight, dw/dt), may be represented by the following model:

$$\frac{dw}{dt} = \frac{Ak_d(\theta_d - \theta_i)}{L_s l} = \frac{Ab(p_i - p_d)}{l} = A\rho_s(W_0 - W_f)\frac{dl}{dt} \quad (3.4)$$

where A is the drying area normal to the direction of flow of the vapor, k_d is the thermal conductivity of the dry layer, θ_d is the temperature at the top surface of the dry layer, θ_i is the temperature at the ice front, L_s is the heat of sublimation, l is the thickness of the dry layer, b is the permeability of the dry layer to water vapor, p_i is the water vapor pressure at the ice front; p_d is the water vapor pressure at the top surface of the dry layer, ρ_s is the density of the dry layer, W_0 is the initial moisture content of the material (dwb), W_f is the final moisture content of the material (dwb), and dl/dt is the rate of change of thickness of the dry layer. When heat is supplied from below, or from both above and below, the equilibrium between the heat and mass transfer no longer exists. More complex models representing these situations are available in the literature [33, 50–52].

The advantages of freeze drying as compared with other methods lie in the quality of the dried product. There is no movement of liquid within the solid during freeze drying. Thus, shrinkage does not occur and solutes do not migrate to the surface. The dried product has a light porous structure, which facilitates rehydration. The temperature to which the product is exposed is lower than in most other methods of drying, so heat damage is relatively low. There is good retention of volatile flavor compounds during freeze drying. However, some damage to the structure may occur during freezing which can result in some structural collapse on rehydration and a poor texture in the rehydrated product. Some denaturation of proteins may occur due to pH changes and concentration of solutes during freezing.

Freeze-dried food materials are usually hygroscopic, prone to oxidation and fragile. Relatively expensive packaging may be necessary, compared to other types of dried foods. The capital cost of freeze-drying equipment is relatively high and so are the energy costs. It is the most expensive method of drying food materials [1, 3, 5, 29, 33, 50–57].

3.2.7 Equipment Used in Freeze Drying Solid Foods

3.2.7.1 Cabinet (Batch) Freeze Dryer

This is a batch-operated dryer. The vacuum chamber is cylindrical, mounted horizontally, and fitted with doors front and back. The vacuum is created and maintained by a refrigerated condenser backed up by vacuum pumps (see Figure 3.9).

Once the frozen food is sealed into the chamber, the pressure must be reduced rapidly to avoid melting of the ice. Once the vacuum is established, the low pressure must be maintained throughout the drying cycle, which may range from 4 to 12 h. The vacuum system must cope with the water vapor produced by sublimation and noncondensable gases, which come from the food or through leaks in the system. The refrigerated condenser, which may be a plate or coil, is located inside the drying chamber, or in a smaller chamber connected to the main chamber by a duct. The water vapor freezes onto the surface of the plate or coil. The temperature of the refrigerant must be below the saturation temperature, corresponding to the pressure in the chamber. This temperature is usually in the range -10 to $-50\,^{\circ}$C. As drying proceeds, ice builds up on the surface of the condenser. This reduces its effectiveness. If it is not to be defrosted during the drying cycle, then a condenser with a large surface area is required. Defrosting may be carried out during the cycle by having two condensers. These are located in separate chambers, each connected to the main drying chamber via a valve. The condensers are used alternately. While one is isolated from the drying chamber and is defrosting, the other is connected to the main chamber and is condensing the water vapor. The roles are reversed periodically during the drying cycle.

Usually two vacuum pumps are used in series to cope with the noncondensable gases. The first may be a Roots blower or an oil-sealed rotary pump. The second is a gas-ballasted, oil-sealed rotary pump. In the early days of industrial freeze drying, multistage steam ejectors were used to evacuate freeze-drying chambers. These could handle both the water vapor and noncondensable gases. However, most of these have been replaced by the system described above, because of their low energy efficiency.

Heat is normally supplied by means of heated shelves. The trays containing the frozen are placed between fixed, hollow shelves, which are heated internally by steam, heated water, or other thermal fluids. Heat is supplied by conduction from below the trays and by radiation from the shelf above. Plates and trays are designed to ensure good thermal contact. Ribbed or finned trays are used to increase the area of the heated surface in contact with granular solids, without impeding the escape of the water vapor. The use of microwave heating in freeze drying is discussed in Section 3.2.9 [1, 3, 5, 29, 50–57].

3.2.7.2 **Tunnel (Semi-continuous) Freeze Dryer**

This is one type of freeze dryer capable of coping with a wide range of piece sizes, from meat and fish steaks down to small particles, while operating on a semi-continuous basis. It is comprised of a cylindrical tunnel, 1.5–2.5 m in diameter. It is made up of sections. The number of sections depends on the throughput required. Entry and exit locks are located at each end of the tunnel. Gate valves enable these locks to be isolated from the main tunnel. To introduce a trolleyload of frozen material into the tunnel, the gate valve is closed, air is let into the entry lock, the door is opened and the trolley pushed into the lock. The door is then closed and the lock is evacuated by a dedicated vacuum system. When the pressure is reduced to that in the main tunnel, the gate valve is opened and the trolley enters the main tunnel. A similar procedure is in operation at the dry end of the tunnel, to facilitate the removal of a trolleyload of dried product.

Fixed heater places are located in the tunnel. The food material, in trays or ribbed dishes, passes between these heater plates as the trolley proceeds through the tunnel. Since both the vacuum and heat requirements decrease as drying proceeds, the vacuum and heating systems are designed to match these requirements. The main body of the tunnel is divided into zones by means of vapor restriction plates. Each zone is serviced by its own vacuum and heating system. Each section of the tunnel can process 3–4 t of frozen material per 24 h [29, 55, 57].

3.2.7.3 **Continuous-Freeze Dryers**

There are a number of designs of continuous-freeze dryers suitable for processing granular materials. In one design, a stack of circular heater plates is located inside a vertical, cylindrical vacuum chamber. The frozen granules enter the top of the chamber alternately through each of two entrance locks, and fall onto the top plate. A rotating central vertical shaft carries arms which sweep the top surface of the plates. The arm rotating on the top plate pushes the granules outward and over the edge of the plate, onto the plate below, which has a larger diameter than the top plate. The arm on the second plate pushes the granules inward toward a hole in the center of the plate, through which they fall onto the third plate, which has the same diameter as the top plate. In this way, the granules travel down to the bottom plate in the stack. From that plate, they fall through each of two exit locks alternately and are discharged from the chamber.

In another design, the frozen granules enter at one end of a horizontal, cylindrical vacuum chamber via an entrance lock, onto a vibrating deck. This carries them to the other end of the chamber. They then fall onto a second deck, which transports them back to the front end of the chamber, where they fall onto another vibrating deck. In this way, the granules move back and forth in the chamber until they are dry. They are then discharged from the chamber through a vacuum lock. Heat is supplied by radiation from heated platens located above the vibrating decks. Another design is similar to the above but conveyor belts, rather than vibrating decks, move the granules back and forth within the chamber [33, 48].

3.2.7.4 **Vacuum Spray Freeze Dryer**

A prototype of this freeze dryer, developed for instant coffee and tea is described by Mellor [44]. The preconcentrated extract is sprayed into a tall cylindrical vacuum chamber, which is surrounded by a refrigerated coil. The droplets freeze by evaporative cooling, losing about 15% of their moisture. The partially dried particles fall onto a moving belt conveyor in the bottom of the chamber and are carried through a vacuum tunnel where drying is completed by the application of radiant heat. The dry particles are removed from the tunnel by means of a vacuum lock [29, 55, 57].

The production of instant coffee is the main industrial application of freeze drying. Freeze-dried fruits, vegetables, meat, and fish are also produced, for inclusion in ready meals and soups. However, they are relatively expensive compared to similar dried foods produced by hot air or vacuum drying.

3.2.8
Drying by the Application of Radiant (Infrared) Heat

When thermal radiation is directed at a body, it may be absorbed and its energy converted into heat, or reflected from the surface of the body, or transmitted through the material. It is the absorbed energy that can provide heat for the purposes of drying. Generally, in solid materials all the radiant energy is absorbed in a very shallow layer beneath the surface. Thus radiant drying is best suited to drying thin layers or sheets of material or coatings. Applications include textiles, paper, paints, and enamels. In the case of food materials, complex relationships exist between their physical, thermal, and optical properties. These in turn influence the extent to which radiant energy is absorbed by foods. The protein, fat, and carbohydrate components of foods have their own absorption patterns. Water in liquid, vapor, or solid form also has characteristic absorption patterns that influence the overall the absorption of radiant energy. It is very difficult to achieve uniform heating of foods by radiant heat.

Control of the heating rate is also a problem. Infrared heating is not normally used in the food industry for the removal of water in bulk from wet food materials. It has been used to remove surface moisture from sugar or salt crystals and small amounts of water from low-moisture particles such as breadcrumbs and spices. These are conveyed in thin layers beneath infrared heaters. Shortwave lamps are used for very heat-sensitive materials. Longwave bar heaters are used for more heat-resistant materials. Radiant heating is used in vacuum dryers and freeze dryers, usually in combination with heat transferred by conduction from heated shelves. The term *micronization* is used to describe high IR heat intensity processing which is applied to grains to make them more digestible to animals and in brewing to assist in the malting process [29, 58, 59].

3.2.9
Drying by the Application of Dielectric Energy

There is some confusion in the literature regarding the terminology when describing dielectric and microwave energy. In this section, the term *dielectric* is used to represent both the radiofrequency and microwave bands of the electromagnetic spectrum. Radiofrequency energy is in the frequency range 1–200 MHz and microwave from 300 MHz to 300 GHz. By international agreement, specific frequencies have been allocated for industrial use. These are: radiofrequency 27.12 and 13.56 MHz, microwave 2450 MHz and a band within the range 896–915 MHz. These are known as industrial, scientific, and medical (ISM) bands.

Dielectric heating is used for cooking, thawing, melting, and drying. The advantages of this form of heating over more conventional methods are that heat generation is rapid and occurs throughout the body of the food material. This is known as *volumetric heating*. Water is heated more rapidly than the other components in the food. This is an added advantage when it is used for drying foods. The depth to which electromagnetic waves penetrate into a material depends on their frequency and the characteristics of each material. The lower the frequency and so the longer the wavelength, the deeper the penetration. The energy absorbed by a wet material exposed to electromagnetic waves in the dielectric frequency range depends on a characteristic of the material known as its *dielectric loss factor*, which depends on the distribution of dipoles in the material. The higher the loss factor, the more energy is absorbed by the material.

The loss factor of a material is dependent on its moisture content, temperature and, to some extent, its structure. The loss factor of free water is greater than that of bound water. The loss factors of both free and bound water are greater than that of the dry matter. Heating both by radiofrequency and microwave methods is mainly due to energy absorbed by water molecules. However, the mechanism whereby this energy is absorbed is different for the two frequency ranges. In radiofrequency heating, heat is generated by the passage of an electric current through the water. This is due to the presence of ions in the water, which give it a degree of electrical conductivity. In microwave heating, dipolar molecules in the water are stressed by the alternating magnetic field and this results in the generation of heat. At frequencies corresponding to radiofrequency heating, the conductivity and hence the energy absorbed increases with increasing temperature. However, in the case of microwave frequencies, the loss factor and so the energy absorbed, decreases with increasing temperature.

At both radiofrequency and microwave frequencies, the rapid generation of heat within the material leads to the rapid evaporation of water. This gives rise to a total pressure gradient which causes a rapid movement of liquid water and water vapor to the surface of the solid. This mechanism results in shorter drying times and lower material temperatures, compared to hot air or contact drying. Drying is uniform, as thermal and concentration gradients are comparatively small. There is less movement of solutes within the material and overheating of the surface is less likely than is the case when convected or conducted heat is applied. There is efficient use of energy as the water absorbs most of the heat. However, too high a heating rate can cause scorching or burning of the material. If water becomes entrapped within the material, rupture of solid pieces may occur, due to the development of high pressure within them.

Equipment for the generation of radiofrequencies and microwaves is described in the literature [50, 53, 54]. The equipments for applying radiofrequencies and microwaves differ. A basic radiofrequency (platen) applicator consists of two metal plates, between which the food is placed or conveyed. The plates are at different electrical voltages. This is also known as a *throughfield applicator* and is used mainly for relatively thick objects. In a stray- or fringefield applicator, a thin layer of material passes over electrodes, in the form of bars, rods, or plates, of alternating polarity. In a staggered, throughfield applicator, bars are located above and below the product to form staggered throughfield arrays. This type is suitable for intermediate thickness products. In a basic batch microwave applicator, microwaves are directed from the generator into a metal chamber via a waveguide or coaxial cable. The food is placed in the chamber. To improve the uniformity of heating, the beam of microwaves may be disturbed by a mode stirrer, which resembles a slow turning fan. This causes reflective scattering of the waves.

Alternatively, the food may be placed on a rotating table in the chamber. In one continuous applicator, known as a *leaky waveguide applicator*, microwaves are allowed to leak in a controlled manner from slots or holes cut in the side of a waveguide. A thin layer of product passes over the top of the slots. In the slotted waveguide applicator, the product is drawn through a slot running down the center of the waveguide. This is also suitable for thin layers of material.

Dielectric heating is seldom used as the main source of heat for drying wet food materials. It is mainly used in conjunction with heated air.

Dielectric heating may be used to preheat the feed to a hot air dryer. This quickly raises the temperature of the food and causes moisture to rise to the surface. The overall drying time can be reduced in this way. It may also be applied in the early stages of the falling rate period of drying, or toward the end of the drying cycle, to reduce the drying time. It is more usual to use it near the end of drying. Radiofrequency heating is used in the later stage of the baking of biscuits (postbaking) to reduce the moisture content to the desired level. This significantly shortens the baking time compared to completing the baking in a conventional oven. A similar procedure has been used for breakfast cereals. Microwave heating is used, in combination with heated air of high humidity, to dry pasta. Cracking of the product is avoided and the drying time is shortened from 8 to 1 h, compared to drying to completion in heated air. Microwave heating has been used after frying of potato chips and crisps to attain the desired moisture content without darkening the product if the sugar content of the potatoes is high. It has also been used as a source of heat when vacuum drying pasta.

The use of microwave heating in freeze drying has been the subject of much research. Ice absorbs energy more rapidly than dry matter, which is an advantage. However, ionization of gases can occur in the very low-pressure conditions and this can lead to plasma discharge and heat damage to the food. Using a frequency of 2450 MHz can prevent this. If some of the ice melts, the liquid water will absorb energy so rapidly that it may cause solid food particles to explode. Very good control of the microwave heating is necessary to avoid this happening. Microwave heating in freeze drying has not yet been used on an industrial scale, mainly because of the high cost involved [1, 5, 29, 60–64].

3.2.10 Electrohydrodynamic Drying (EHD)

In electrohydrodynamic drying (EHD) electric fields of high intensity and frequency of 50 or 60 Hz are applied to the food being dried. This generates ionized forms of air constituents within the food. An ionic wind is created. The water molecules orient themselves in the direction of the field, resulting in a loss of water vapor. An EHD dryer consists of an earthed plate or mesh on which the food pieces are located. An electrode is positioned above the food. This electrode consists of a number of sharply pointed needles and is connected to a high voltage transformer. The transformer is in turn connected to a high voltage regulator which is supplied by a 110 V, 60 Hz source. In general, the greater the voltage applied the higher the drying date. However, the voltage must be controlled to avoid sparkover. The gap between the electrodes and plate is usually adjustable. A flow of air may be directed across the food pieces. To date, EHD drying has only been used experimentally. Among the foods tested are potato, apple, radish, and spinach. Results suggest that EHD enhances the drying rate, reduces shrinkage, and produces a product with better color, texture, and nutritional value than that produced by conventional convective drying [65, 66].

3.2.11 Osmotic Dehydration

When pieces of fresh fruits or vegetables are immersed in a sugar or salt solution that has a higher osmotic pressure than the food, water passes from the food into the solution under the influence of the osmotic pressure gradient; and the water activity of the food is lowered. This method of removing moisture from food is known as *osmotic dehydration* (drying). This term is misleading as the end product is seldom stable and further processing is necessary to extend its shelf life. *Osmotic concentration* would be a more accurate description of this process.

During osmosis the cell walls act as semi-permeable membranes, releasing the water and retaining solids. However, these membranes are not entirely selective and some soluble substances, such as sugars, salts, organic acids, and vitamins, may be lost from the cells, while solutes from the solution may penetrate into the food. Damage that occurs to the cells during the preparation of the pieces, by slicing or dicing, will increase the movement of soluble solids. The solutes, which enter the food from the solution, can assist in the reduction of the water activity of the food. However, they may have an adverse effect on the taste of the end product.

In the case of fruits, sugars with or without the addition of salt are used to make up the osmotic solution, also known as the *hypertonic solution*. Sucrose is commonly used, but fructose, glucose, glucose/fructose, and glucose/polysaccharide mixtures and lactose have been used experimentally, with varying degrees of success. The inclusion of 0.5–2.0% of salt in the sugar solution can increase the rate of osmosis. Some other low molecular weight compounds such as malic acid and lactic acid have been shown to have a similar effect. Sugar solutions with initial concentrations in the range 40–70% are used. In general, the higher the solute concentration the greater the rate and extent of drying. The higher the sugar concentration the more sugar will enter the food. This may result in the product being unacceptably sweet. The rate of water loss is high initially, but after 1–2 h it reduces significantly. It can take days before equilibrium is reached. A typical processing time to reduce the food to 50% of its fresh weight is 4–6 h.

In the case of vegetables, sodium chloride solutions in the range 5–20% are generally used. At high salt concentrations, the taste of the end product may be adversely affected. Glycerol and starch syrup have been used experimentally as solutes for osmotic drying of vegetables.

In general, the higher the temperature of the hypertonic solution the higher the rate of water removal. Temperatures in the range 20–70 °C have been used. At the higher temperatures, there is a danger that the cell walls may be damaged. This can result in excessive loss of soluble material, such as vitamins, from the food. Discoloration of the food may occur at high temperature. The food may be blanched, in water or in the osmotic solution, to prevent browning. This may affect the process in different ways. It can speed up water removal in some large fruit pieces, due to relaxation of the structural bonds in the fruit. In the case of some small fruit pieces, blanching may reduce water loss and increase the amount of solute entering into the fruit from the solution. Other pretreatments to enhance mass transfer have been investigated including the application of high-intensity electric field pulses, irradiation, microwaves, and high pressure. Carrying out osmotic dehydration under vacuum can also increase the rate of water transfer. In pulsed vacuum osmotic dehydration (PVOD), a vacuum is applied for a short period at the start of the process, after which the rest of the osmotic drying is carried out at atmospheric pressure. This technique improves the rate of mass transfer.

Applying edible coatings to food pieces before osmotic drying can increase dehydration and reduce solute uptake. Substances that may be used as coatings include pectin and its derivatives, low methoxy pectins, sodium alginate, some cellulose, and starch derivatives. In general, the lower the weight ratio of food to solution the greater the water loss and solids gain. Ratios of 1 : 4 to 1 : 5 are usually employed. The smaller the food pieces the faster the process, due to the larger surface area. However, the smaller the pieces the more cell damage is likely to occur when cutting them and, hence, the greater the amount of soluble solids lost from the food. Promoting movement of the solution relative to the food pieces should result in faster osmosis. However, vigorous mixing is likely to lead to cell damage. Delicate food pieces may remain motionless in a tank of solution. Some improvement in the rate of drying may be obtained by recirculating the solution through the tank by means of a pump. In large-scale installations, the food pieces may be contained within a basket immersed in the tank of solution. The basket is vibrated by means of an eccentric drive. Alternatively, the food pieces may be packed into a tall vessel and the solution pumped through the porous bed of solids.

Reuse of the hypertonic solution is desirable to make osmotic drying an economic process. Insoluble solids may be removed by filtration and the solution concentrated back to its original soluble solids content by vacuum evaporation. Discoloration may limit the number of times the solution can be reused. A mild heat treatment may be necessary to inactivate microorganisms, mainly yeasts, that may build up in the solution.

As stated above, the products from osmotic drying are usually not stable. In the case of fruits and vegetables, the osmosed products have water activities in the range 0.90–0.95. Consequently, further processing is necessary. Drying in heated air, vacuum drying, or freeze drying may be employed to stabilize such products. Alternatively, they may be frozen [29, 33, 67–71].

3.2.12
Sun and Solar Drying

In this section the term "*sun drying*" is used to describe the process whereby some or all of the energy for drying of foods is supplied by direct radiation from the sun. The term "*solar drying*" is used to describe the process whereby solar collectors are used to heat air, which then supplies heat to the food by convection.

For centuries, fruit, vegetables, meat, and fish have been dried by direct exposure to the sun. The fruit or vegetable pieces were spread on the ground on leaves or mats while strips of meat and fish were hung on racks. While drying in this way, the foods were exposed to the vagaries of the weather and to contamination by insects, birds, and animals. Drying times were long and spoilage of the food could occur before a stable moisture content was attained. Covering the food with glass or a transparent plastic material can reduce these problems. A higher temperature can be attained in such an enclosure compared to those reached by direct exposure to the sun. Most of the incident radiation from the sun will pass through such transparent materials. However, most radiation from hot surfaces within the enclosure will be of longer wavelength and so will not readily pass outwards through the transparent cover. This is known as the "*greenhouse effect*" and it can result in shorter drying times compared with those attained in uncovered food exposed to sunlight.

A transparent plastic tent placed over the food spread on a perforated shelf raised above the ground is the simplest form of covered sun dryer. Warm air moves by natural convection through the layer of food and contributes to the drying. A simple sun dryer of sturdy construction is shown in Figure 3.10.

The capacity of such a dryer may be increased by incorporating a solar collector. The warm air from the collector passes up through a number of perforated shelves supporting layers of food and is exhausted near the top of the chamber. A chimney may be fitted to the air outlet to increase the rate of flow of the air. The taller the chimney, the faster the air will flow. If a power supply is available, a fan may be incorporated to improve the airflow still further. Heating by gas or oil flames may be used in conjunction with solar drying. This enables heating to continue when sunlight is not available. A facility for storing heat may also be incorporated into solar dryers. Tanks of water and beds of pebbles or rocks may be heated via a solar

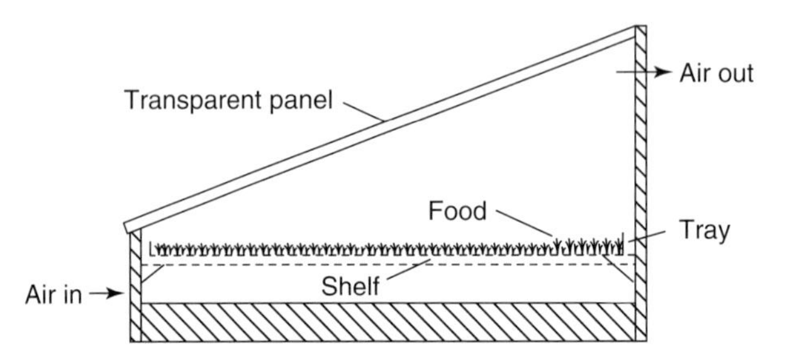

Figure 3.10 Simple sun dryer.

collector. The stored heat may then be used to heat the air entering the drying chamber. Drying can proceed when sunlight is not available. Heat-storing salt solutions or adsorbents may be used instead or water or stones. Quite sophisticated solar drying systems, incorporating heat pumps, are also available [3, 29, 33, 72–76].

3.2.13 Drying Food Liquids and Slurries in Heated Air

3.2.13.1 Spray Drying

This is the method most commonly used to dry liquid foods and slurries. The feed is converted into a fine mist or spray – a process known as *atomization* and the spray-forming device is an *atomizer*. The droplet size is usually in the range 10–200 μm, although for some applications larger droplets are produced. The spray is brought into contact with heated air in a large drying chamber. Because of the relatively small size of the droplets, a very large surface area is available for evaporation of the moisture. Also, the distance that moisture has to migrate to the drying surface is relatively short. Hence, the drying time is relatively short, usually in the range 1–20 s. Evaporative cooling at the drying surface maintains the temperature of the droplets close to the wet bulb temperature of the drying air; that is, most of the drying takes place under constant rate conditions (see Section 3.2.2). If the particles are removed quickly from the drying chamber once they are dried, heat damage is limited. Hence, spray drying can be used to dry relatively heat-sensitive materials.

The main components of a single-stage spray dryer are shown in Figure 3.11.

Inlet fan A draws air in through filter B and then through heater C into drying chamber D. Pump E delivers the feed from tank F to the atomizer G. This converts the feed into a spray which then contacts the heated air in drying chamber D, where drying takes place. Most of the dry powder is removed from the chamber through valve H and pneumatically conveyed through duct I to a storage bin. The air leaves the chamber through duct J and passes through one or more air/powder separators K to recover the fine powder carried in the air. This powder may be added to the main product stream through valve M or returned to the wet zone of the drying chamber through duct N.

The air may be heated indirectly through a heat exchanger using oil or gas fuel or steam. In recent years, natural gas has been used to heat the air directly. There is still some concern about possible contamination of the food with nitrate and nitrite compounds, in particular *N*-nitrosodimethylamine, which has been shown to be harmful. The use of low NO_x (nitrogen oxide) burners reduces this problem. However, the quality of the air should be monitored when direct heating is used.

It is important that the droplets produced by the atomizer are within a specified size range. If the droplets vary too much in size, drying may not be uniform. The drying conditions must be set so that the larger droplets reach the desired moisture content. This may result in smaller droplets being overexposed to the heated air. Droplet size can affect some important properties of the dry powder, such as its rehydration behavior and flow properties. There are three types of atomizer:

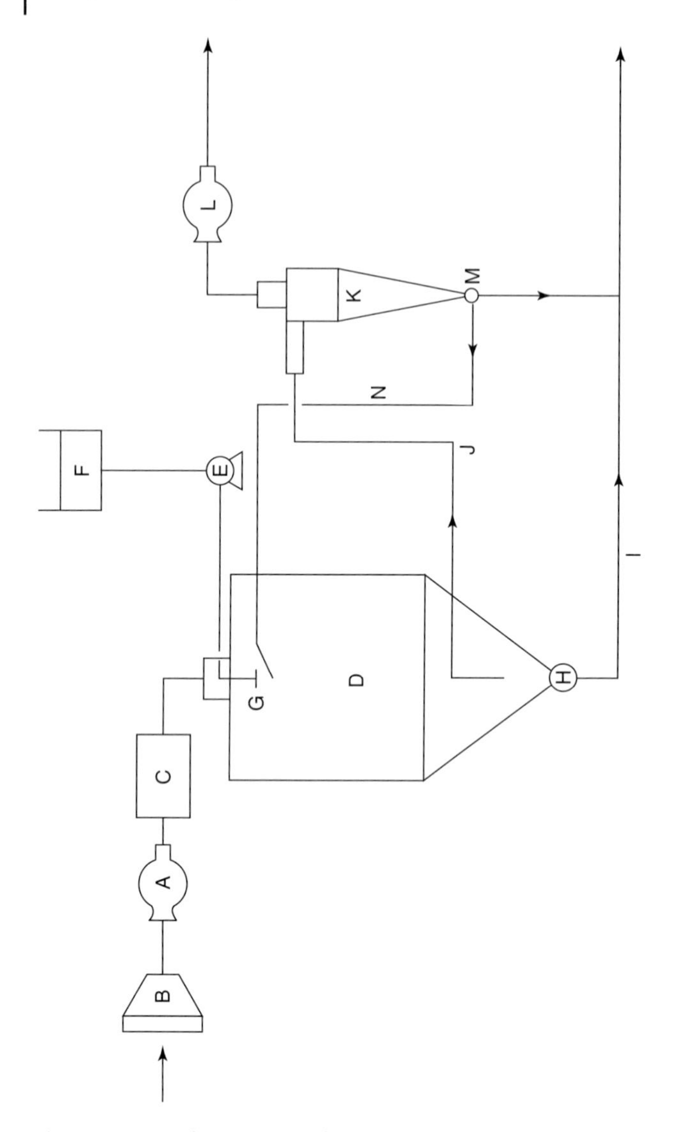

Figure 3.11 Single-stage spray drying system. From Brennan [29] with permission.

centrifugal atomizer, pressure nozzle, and two-fluid nozzle. A centrifugal atomizer consists of a disk, bowl, or wheel on the end of a rotating shaft. The liquid is fed onto the disk near its center of rotation. Under the influence of centrifugal force, it moves out to the edge of the disk and is spun off, initially in the form of threads, which then break up into droplets. There are many designs of disk in use. An example is shown in Figure 3.12. Disk diameters range over 50–300 mm and they rotate at speeds in the range 50 000–10 000 rpm. They are capable of producing uniform droplets and can handle viscous feeds and are not subject to blocking or abrasion by insoluble solid particles in the feed.

Figure 3.12 Centrifugal atomizer. By courtesy of Niro A/S.

A pressure nozzle features a small orifice, with a diameter in the range 0.4–4.0 mm, through which the feed is pumped at high pressure in the range 5.0–50.0 MPa. A grooved core insert, sited before the orifice, imparts a spinning motion to the liquid, producing a hollow cone of spray. Pressure nozzles are capable of producing droplets of uniform size, if the pumping pressure is maintained steady. However, they are subject to abrasion and/or blocking by insoluble solid particles in the feed. They are best suited to handling homogeneous liquids of relatively low viscosity.

A two-fluid nozzle, also known as a *pneumatic nozzle*, features an annular opening through which a gas, usually air, exits at high velocity. The feed exits through an orifice concentric with the air outlet. A Venturi effect is created and the liquid is converted into a spray. The feed-pumping pressure is lower than that required in a pressure nozzle. Such nozzles are also subject to abrasion and blocking if the feed contains insoluble solid particles. The droplets produced by two-fluid nozzles are generally not as uniform in size as those from the other two types of atomizer, especially when handling high-viscosity liquids. They are also best suited to handling homogeneous liquids.

There are numerous designs of drying chamber used in industry. Three types are shown in Figure 3.13. Figure 3.13a depicts a tall, cylindrical tower with a conical base. Both heated air and feed are introduced at the top of the chamber and flow concurrently down through the tower. This design is best suited to drying relatively large droplets of heat-sensitive liquids, especially if the powder particles are sticky and tend to adhere to the wall of the chamber. The wall of the conical section of the chamber may be cooled to facilitate removal of the powder. Figure 3.13b depicts another concurrent chamber, but featuring a shorter cylindrical body. The air enters tangentially at the top of the chamber and follows a downward, spiral flow path. The feed is introduced into the top of the chamber through a centrifugal atomizer. Because of the spiral flow pattern followed by the air, particles tend to be thrown against the wall of the chamber. Consequently, this type of chamber is best suited to drying foods that are not very heat sensitive or sticky. The chamber depicted in Figure 3.13c features a mixed flow pattern. The heated air is directed upward initially and contacts the spray of liquid from the centrifugal atomizer. The air containing the droplets then travels down to the bottom of the chamber.

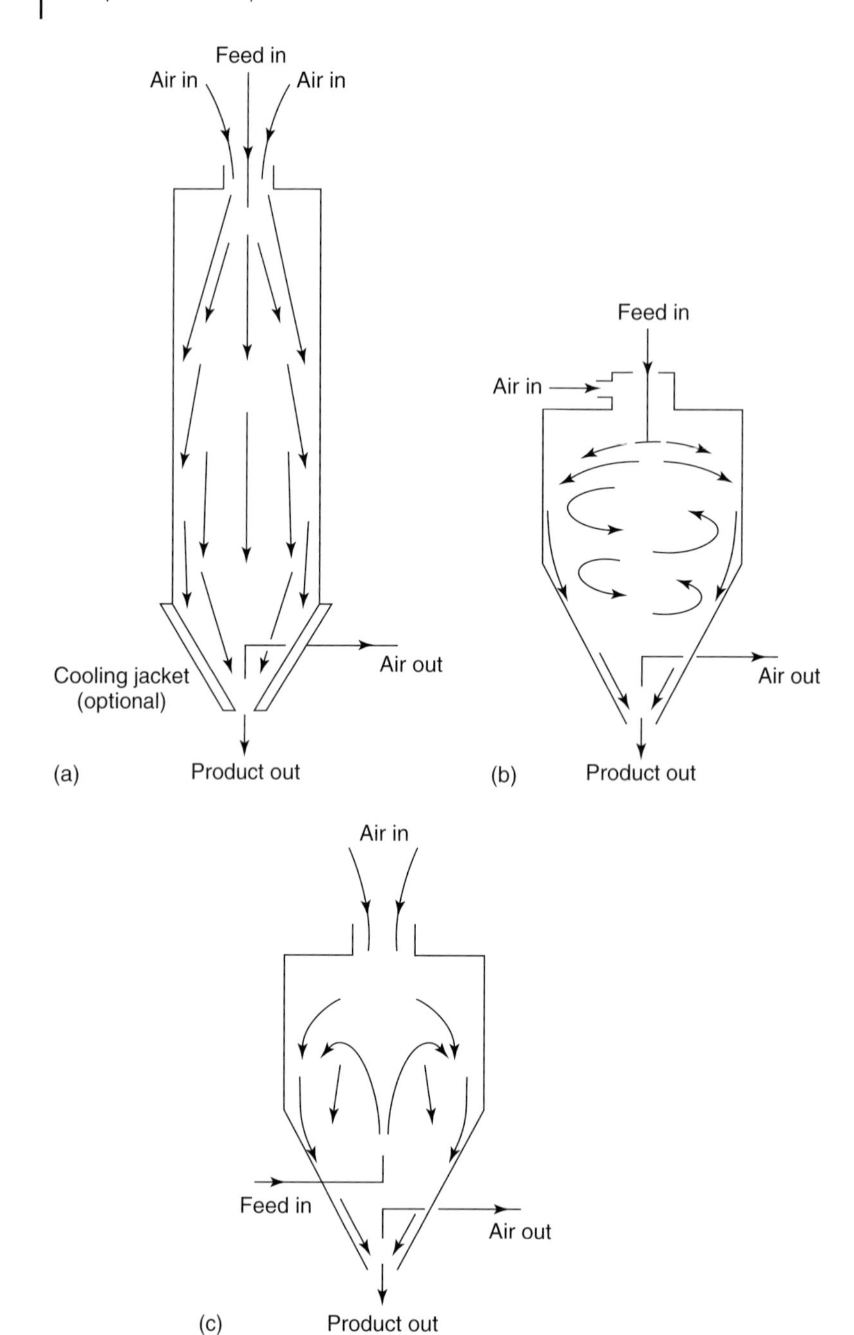

Figure 3.13 Different designs of spray drying chambers: (a) concurrent with straight-line flow path, (b) concurrent with spiral flow path, and (c) mixed flow. From Brennan [77] with permission.

The risk of heat damaging the product is greater than in concurrent chambers. This design is not widely used for food dehydration.

Many other designs of spray drying chambers are used in industry, including flat-bottomed cylindrical and horizontal box-like versions.

In industrial spray dryers, most of the dry powder is removed from the bottom of the chamber through a rotary valve or vibrating device. Some powders containing high amounts of sugar or fat may tend to adhere to the wall of the chamber. Various devices are used to loosen such deposits. Pneumatically operated hammers, which strike the outside wall of the chamber, may be used. Brushes, chains, or air brooms may sweep the inner surface of the chamber. The temperature of the wall of the chamber may be reduced by removing some of the insulation or by drawing cool air through a jacket covering some or all of the chamber wall. This can also reduce product build-up on the wall.

The exhaust air from the chamber carries with it some of the fine particles of product. These need to be separated from the air, as they represent valuable product and/or may contaminate the environment close to the plant. Large dry cyclone separators are often used, singly or in pairs in series, for this purpose. Fabric filters are also employed for this duty. Powder particles may be washed out of the exhaust air with water or some of the liquid feed and recycled to the drying chamber. Such devices are known as *wet scrubbers*. An advantage of this method is that heat may also be recovered from the air and used to preheat or preconcentrate the feed. Electrostatic precipitators could also be used for recovering fines from the air leaving the drying chamber. However, they are not widely employed in the food industry. Combinations of the above separation methods may be used. The air, after passing through the cyclones, may then pass through a filter or scrubber to remove very fine particles and avoid contaminating the outside environment. Fines recovered by cyclones or filters may be added to the main stream of product. However, powders containing large amounts of small particles, less than 50 μm in diameter, are difficult to handle and may have poor rehydration characteristics. When added to a hot or cold liquid, fine particles tend to form clumps which float on the surface of the liquid and are difficult to disperse. It is now common practice to recycle the fines back into the wet zone of the dryer where they collide with droplets of the feed to form small agglomerates. Such agglomerates disperse more readily than individual small particles when added to a liquid.

The powder is generally removed from the spray drying chamber before it reaches its final moisture content and drying is completed in another type of dryer. Vibrating fluidized bed dryers are most often used as secondary dryers (see Section 3.2.3.5). Some agglomeration of the powder particles occurs in this second stage, which can improve its rehydration characteristics. A second fluidized bed dryer may be used to cool the agglomerated particles. Some spray dryers permit multistage drying in one unit. In the integrated fluidized bed dryer, a fluidized bed in the shape of a ring is located at the base of the drying chamber. Most of the drying takes place in the main drying chamber. Drying is completed in the inbuilt fluidized bed [1, 3, 5, 11, 29, 33, 77–82].

3.2.14
Drying Liquids and Slurries by Direct Contact With a Heated Surface

3.2.14.1 Drum (Roller, Film) Dryer

The principle of this type of dryer is that the feed is applied in a thin film to the surface of a rotating hollow cylinder, heated internally, usually with steam under pressure. As the drum rotates, heat is transferred to the food film, mainly by conduction, causing moisture to evaporate. The dried product is removed by means of a knife which scrapes the surface of the drum. The pattern of drying is similar to that of solids drying in heated air drying (i.e., there are constant and falling rate periods – see Section 3.2.2). However, the temperature to which the food is exposed is usually higher than in heated air drying, as there is not the same degree of evaporative cooling. Drying times are relatively short, usually in the range 2–30 s.

In a single drum dryer, the feed may be applied to the drum by different means. The drum may dip into a trough containing the feed (Figure 3.14a). This feed system is most suited to low-viscosity liquids. For more viscous liquids and slurries, unheated rollers may apply the feed to the drum (Figure 3.14b). A multiple roller system is used for high-viscosity liquids and pastes (Figure 3.14c). The layer of feed becomes progressively thicker as it passes beneath successive feed rollers.

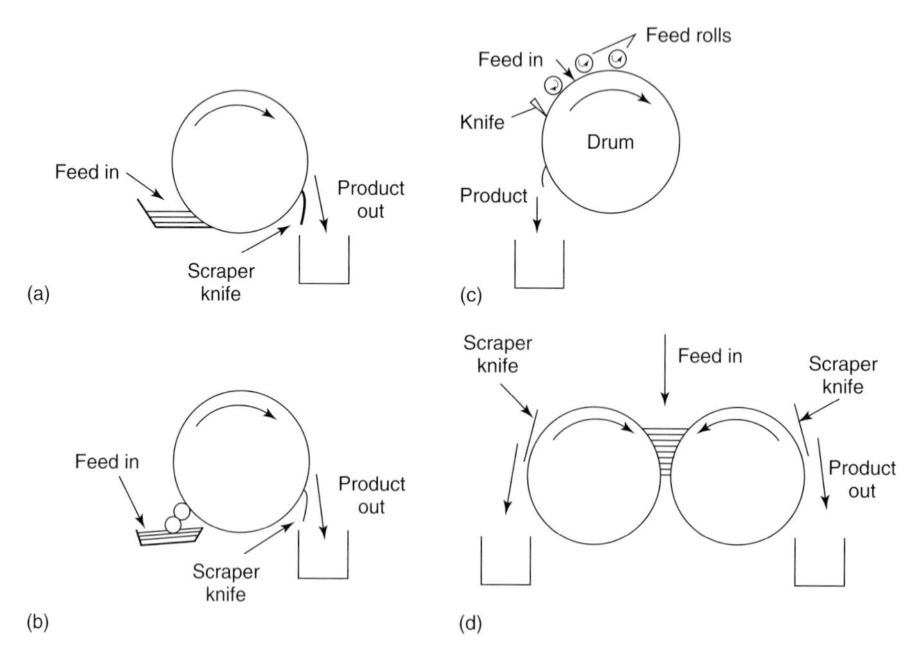

Figure 3.14 Different designs of drum dryers: (a) single drum with dip feed, (b) single drum with unheated roller feed, (c) single drum with multiroller feed, and (d) double drum. Adapted from Brennan [29] with permission.

The feed may also be sprayed or splashed onto the drum surface. A double drum dryer (Figure 3.14d) consists of two drums rotating toward each other at the top. The clearance between the drums is adjustable. The feed is introduced into the trough formed between the drums and a film is applied to the surface of both drums as they rotate. This type of drum dryer is suitable for relatively low-viscosity liquids that are not very heat sensitive.

Drums are 0.15–1.50 m in diameter and 0.2–3.0 m long. They rotate at speeds in the range 3–20 rpm. They may be made from a variety of materials. For food applications, chromium-plated cast iron or stainless steel drums are mostly used. The drum surface temperature is usually in the range 110–165 °C. As the food approaches these temperatures toward the end of drying, it is likely to suffer more heat damage than in spray drying. For very heat-sensitive materials, one or two drums may be located inside a vacuum chamber operated at an absolute pressure from just below atmospheric down to 5 kPa. The drum is heated by vacuum steam or heated water, with a surface temperature in the range 100 to 35 °C. Provision must be made for introducing the feed into and taking the product out of the vacuum chamber and for adjusting knives, feed rollers, and so on, from outside the chamber. Such vacuum drum dryers are expensive to purchase and maintain and are not widely used in the food industry [1, 3, 5, 29, 47, 83].

3.2.14.2 **Vacuum Band (Belt) Dryer**

In this type of vacuum dryer a metal belt or band, moving in a clockwise direction, passes over a heated and cooled drum inside a vacuum chamber (see Figure 3.15).

The band may be continuous and made of stainless steel or fine stainless steel wire mesh. Alternatively, it may be made of hinged, stainless steel plates. The feed, in the form of a viscous liquid or paste, is applied to the band by means of a roller, after entering the vacuum chamber via a valve. The cooled product is scraped off the band and removed from the chamber through a rotary valve, a sealed screw or into two vacuum receivers, working alternately. Radiant heaters or heated platens, in contact with the belt, supplement the heat supplied by the heated drum. The

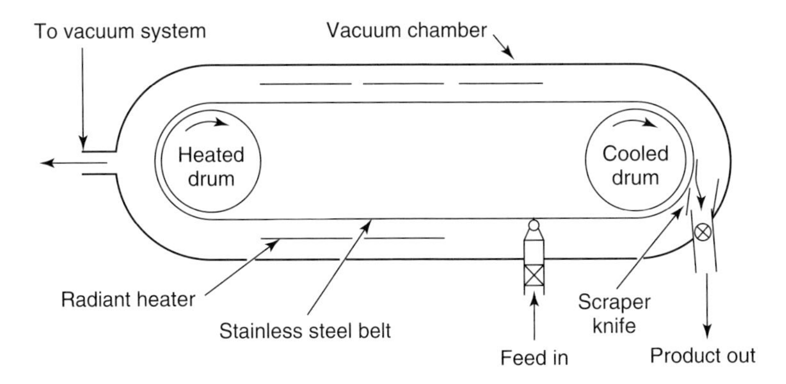

Figure 3.15 Principle of vacuum band (belt) dryer. From Brennan *et al.* [1] with permission.

chamber pressure and band temperature are in similar ranges to those used in vacuum drum dryers (see Section 3.2.13.1). A number of bands, positioned one above the other, may be located in one vacuum chamber. Vacuum band dryers are expensive to purchase and maintain and are only used for very heat-sensitive materials that can bear the high costs involved [1, 5, 29, 33, 47].

3.2.14.3 Refractance Window Drying System

In this drying system, which operates at atmospheric pressure, the wet material is applied as a thin film, usually 0.5–1.0 mm thick, to a thin plastic conveyor belt which passes over heated water circulating in a shallow trough. The belt is made of material that is relatively transparent to infrared radiation and is in contact with the heated water. The water is maintained at a temperature of 95–97 °C and is recycled by means of a pump via an insulated heating tank. Since the belt is very thin it rapidly reaches the temperature of the water. In the early stages of drying heat is transferred to the food material by a combination of conduction and radiation. In the later stages conduction predominates. Evaporative cooling occurs and the product temperature is typically 60–70 °C. Drying is rapid and drying times of 4–6 min are attainable. Beyond the drying section the belt passes over a trough of cooling water to reduce the temperature of the dry product below its glass transition temperature and facilitate its removal from the belt by a doctor blade. The technology is relatively inexpensive and the equipment easy to operate and maintain. The thermal efficiency of this system is relatively high at 52–77%.

Among the products dried by this system are herbs, spices, vegetable and fruit purees, scrambled egg, and nutritional supplements. The quality of the dried products is said to be high, in many cases comparable to that of freeze-dried materials [84–86].

3.2.15 Other Methods Used for Drying Liquids and Slurries

Concentrated liquids have been dried by heated air on conveyor dryers, using a technique known as *foam mat drying*. The liquid is made into a foam by the addition of a small amount, 1% or less, of a foaming agent, such as soya protein, albumin, fatty acid esters of sucrose and glycerol monostearate, and the incorporation of air or other gases by injection or mixing. The foam is spread in thin layers or strips on a wire mesh belt and conveyed through the dryer. Relatively rapid drying can be achieved, of the order of 1 h in air at 100 °C, yielding a porous dry product with good rehydration properties [25, 62].

Liquids have been dried in spouted beds of inert particles. Metal and glass spheres and various plastic particles have been used as inert material. These have been fluidized with heated air to form spouted beds (see Section 3.2.3.5). The liquid is sprayed onto the particles and is heated by convection from the heated air and conduction from the hot particles, causing moisture to evaporate. When dry, the film is released from the particles by abrasion and impact between particles and separated from the exhaust air by a cyclone separator [56].

Liquids, usually in concentrated form, have been dried in vacuum cabinet dryers (see Section 3.2.5.1). A technique known as *vacuum puff drying* has been used in this type of dryer and in vacuum band dryers. The concentrated liquid is spread in a thin layer on trays and placed in the vacuum chamber. When the vacuum is drawn, bubbles of water vapor and entrapped air form within the liquid and expand and the liquid froths up to form a foam. By careful control of pressure, temperature, and viscosity of the liquid, it may be made to expand to occupy a space up to 20 times that of the original material. When heat is applied, the foam dries rapidly to form a porous dry product, with good rehydration properties [29].

Concentrated liquids are freeze dried. The concentrate is frozen into slabs, which are broken up into pieces and dried in batch or continuous freeze dryers (see Section 3.2.6).

3.2.16
Applications of Dehydration

The following are examples of the many foods which are preserved in dehydrated form.

3.2.16.1 Dehydrated Vegetable Products

Many vegetables are available in dehydrated form. A typical process involves:

- wet and dry cleaning;
- peeling, if necessary, by mechanical, steam, or chemical methods;
- slicing, dicing, or shredding;
- blanching;
- sulfiting, if necessary;
- dewatering;
- drying;
- conditioning, if used;
- milling or kibbling, if used;
- screening; and
- packaging.

Most vegetables are *blanched* prior to drying. See Chapter 1 for a discussion of the purposes and methods of blanching. Many vegetables are *sulfured* or *sulfited* prior to drying. Again see Chapter 1 for a discussion of this step in the process. Many different types of dryers are used for drying vegetables, including: cabinets, single, or two-stage tunnels, conveyor, and fluidized bed dryers (see Section 3.2.3). Air inlet temperatures vary from product to product but are usually in the range 50–110 °C. Tunnel, conveyor, and fluidized bed dryers may be divided into a number of drying zones, each zone being controlled at a different temperature, to optimize the process. If the drying is not completed in the main dryer, the product may be finish dried or conditioned in a bin dryer, supplied with dry air at 40–60 °C.

Among the vegetables dried as outlined above are green beans, bell peppers, cabbage, carrot, celery, leeks, spinach, and swedes. Some vegetables, such as garlic, mushrooms, green peas, and onions, are not sulfited. Herbs such as parsley, sage, and thyme may be dried without blanching or sulfiting. Vegetables may be dried in vacuum cabinet dryers and in freeze dryers, to yield products of superior quality to those produced by air drying. However, such products will be more expensive.

Vegetable purees may be dried. Cooked and pureed carrot and green peas may be drum dried to produce a flaked product. Very finely divided, cooked carrot or green peas may be spray dried to a fine powder [25, 34, 64, 65].

A number of dried potato products are available. Dehydrated, diced potato is produced by a process similar to that outlined above. After blanching the potato pieces should be washed with a water spray to remove gelatinized starch from their surfaces. In addition to sulfite, calcium salts may be added to increase the firmness of the rehydrated dice. Cabinet, tunnel, or conveyor dryers may be used to dry the potato. Conveyor dryers are most widely used. In recent years, fluidized bed dryers have been applied to this duty. Finish drying in bins is often practiced. Potato flakes are produced by drum drying cooked, mashed potatoes. Two-stage cooking is followed by mashing or ricing. Sufficient sulfite is mixed with the mash to give 150–200 ppm in the dried product. An emulsion is made up containing, typically, monoglyceride emulsifier, sodium acid pyrophosphate, citric acid, and an antioxidant and mixed into the mash. In some cases milk powder may also be added. The mash is then dried on single drum dryers equipped with a feed roll and up to four applicator rolls. Steam at 520–560 kPa, absolute, is used to heat the drums. After drying, the dried sheet is broken up into flakes. Potato flour is made from poor-quality raw potatoes which are cooked, mashed, drum dried, and milled. Potato granules may also be produced from cooked, mashed potatoes. After cooking, the potato slices are carefully mashed. Some dry granules may be "added back" to the mash, which then has sulfite added to give 300–600 ppm in the dried product. A second granulation stage then follows. It is important that as little as possible rupture of cells occurs at this stage. The granules are cooled to 15.5–26.5 °C and held at that temperature for about 1 h. During this "conditioning" some retrogradaton of the starch occurs. A further gentle granulation then takes place and the granules are dried in pneumatic and/or fluidized bed dryers to a moisture content of 6–7% (wet weight basis, wwb) [29, 37, 85–89].

Sliced tomatoes may be sun dried. The slices are exposed to the fumes of burning sulfur in a chamber or dipped in or sprayed with sulfite solution before sun drying. In recent years, there has been an increase in demand for sun-dried tomatoes, which are regarded as being of superior quality to those dried by other means. Tomato slices are also air dried in cabinet or tunnel dryers to a moisture content of 4% (wwb). The dried slices tend to be hygroscopic and sticky. They are usually kibbled or milled into flakes for inclusion in dried soup mixes or dried meals.

Tomato juice may be spray dried. The juice is prepared by the "hot or cold break process" (see Section 3.1.5.3) and concentrated by vacuum evaporation up to 26–48% total solids content, depending on the preparation procedure, before it is spray dried. The powder is hygroscopic and sticky when hot and tends to adhere to the wall of the drying chamber. A tall drying chamber downward concurrent flow may be used. Alternatively, a chamber with a shorter body featuring a downward concurrent spiral flow path may be used. The wall of the chamber is fitted with a jacket through which cool air is circulated, to reduce wall deposition. An air inlet temperature in the range 140–150 °C is used and the product has a moisture content of 3.5% (wwb) [15, 29, 87–91].

3.2.16.2 **Dehydrated Fruit Products**

Many fruits are sun dried, including pears, peaches, and apricots. However, hot air drying is also widely applied to fruits such as apple slices, apricot halves, pineapple slices, and pears in halves or quarters. A typical process for such fruits involves:

- washing;
- grading;
- peeling/coring, if required;
- trimming, if required;
- sulfiting;
- cutting;
- resulfiting;
- drying;
- conditioning, if required;
- packing.

Fruits are not usually blanched. However, a procedure known as *dry–blanch–dry* is sometimes used for some fruits such as apricots, peaches, and pears. After an initial drying stage, in which the fruit is reduced to about half its initial weight, it is steam blanched for 4–5 min. It is then further dried and conditioned down to its final moisture content. Fruits are usually sulfured by exposure to the fumes of burning sulfur.

Cabinet or tunnel dryers are most commonly used at the drying stage(s). Air inlet temperatures in the range 50–75 °C are used. Tunnel dryers may be divided into two or three separate drying zones, each operating at a different air temperature.

Grapes are sun dried on a large scale to produce raisins or sultanas. They are also dried in heated air in cabinet or tunnel dryers. The grapes are dipped in a hot 0.25% NaOH bath, washed and heavily sulfured. prior to drying down to a moisture content of 10–15% (wwb). They are usually conditioned in bins or sweat boxes to attain a uniform moisture content. Fumigation may be necessary during conditioning to kill any infestation. The "golden bleached" appearance of the raisins or sultanas is due to the high SO_2 content which is in the range 1500–2000 ppm.

Whole plums are dried in cabinet or tunnel dryers to produce prunes. The fruits are not sulfured. They are dried down to a moisture content in the range 16–19% (wwb) but may be "stabilized" by rehydration in steam to a moisture content of 20–22% (wwb).

Fruit purees may be drum dried to produce a flake or powder product. Bananas may treated with SO_2, pulped, pasteurized, further treated with SO_2, homogenized and dried on a single drum dryer to a moisture content of 4% (wwb). Other fruit purees may be drum dried, including apricot, mango, and peach. Some of these products are hygroscopic and sticky. Additives, such as glucose syrup, may have to be mixed into the puree to facilitate removal of the product from the drum and its subsequent handling.

Fruit juices may be spray dried. Concentrated orange juice is an example. To avoid the powder sticking to the wall of the drying chamber, additives are used. The most common one is liquid glucose with a dextrose equivalent in the range 15–30. It may be added in amounts up to 75% of the concentrate, calculated on a solids basis. Other additives such as skim milk powder and carboxymethyl celluloses have been used. However, these limit the uses to which the dry powder can be put. It is important that the concentrated juice is well homogenized, if it contains insoluble solid particles. Both centrifugal and nozzle atomizers are used. Tall chambers featuring downward concurrent flow patterns are favored. The dried powder may be cooled on a fluidized bed to facilitate handling. Volatiles separated from the juice before or after concentration may be added to the powder to enhance its flavor. Other fruit juices which may be spray dried include lemon, mango, peach, and strawberry [29, 37, 85, 86, 92, 93].

3.2.16.3 **Dehydrated Dairy Products**

Skim or separated milk powder is mainly produced by spray drying. The raw whole milk is centrifuged to yield skim milk with 0.05% fat. The skim milk is then heat treated. The degree of heat treatment at this stage determines whether the powder produced is classed as low-heat, medium-heat, or high-heat powder. The more severe the heat treatment, the lower the amount of soluble whey proteins (albumin and globulin) that remain in the powder. Low-heat powder is used in recombined milk products such as cheese and baby foods. Medium-heat powder is used in the production of recombined concentrated milk products. High-heat powder is mainly used in the bakery and chocolate industries. The milk is then concentrated to a total solids content in the range 40–55%, by multiple-effect evaporation. Falling film evaporators are most widely used in recent years.

Various designs of spray drying chamber and atomizer can be used for drying skim milk, as it not a difficult material to dry. Air inlet and out temperatures used are in the ranges 180–230 and 80–100 °C, respectively. The moisture content of the powder is in the range 3.5–4.0% (wwb). It is common practice to recycle the fine powder from the separators into the wet zone of the dryer. The powder may be removed from the drying chamber at a moisture content of 5–7% (wwb) and the drying completed in a vibrated fluidized bed dryer. This promotes some agglomeration of the powder particles and improves its rehydration characteristics. Instant milk powder may be produced by rewetting the powder particles, usually with steam, mixing them to promote agglomeration and redrying them down to a stable moisture content. This can be carried out in a fluidized bed. The powder may be cooled in a second fluidized bed to facilitate handling.

Whole milk powder is also produced by spray drying in a similar manner to skim milk. Air inlet and outlet temperatures are usually in the ranges 175–200 and 75–95 °C, respectively. Whole milk powder is rather sticky when hot and can form a deposit on the chamber wall. Hammers, which are located outside the chamber and tap the wall of the chamber at intervals, may be used to assist in the removal of the powder. Whole milk powder may be agglomerated. However, this may not be as effective as in the case of skim milk powder. Fat may migrate to the surface of the particles. This reduces their wettability. The addition of small quantities of lecithin, a surface active agent, to whole milk powder, can improve its rehydration characteristics, particularly in cold liquids.

Other dairy products that are spray dried are buttermilk and whey. Whey powder is hygroscopic and difficult to handle, as the lactose is in a amorphous state. A crystallization process before and/or after drying can alleviate this problem.

Relatively small amounts of milk and whey are drum dried. Double drum dryers are usually used. The dry products are mainly used for animal feed. However, because of its good water-binding properties, drum-dried milk is used in some precooked foods [11, 13–15, 29, 77, 78, 94].

3.2.16.4 Instant Coffee and Tea

These products are produced by spray drying or freeze drying. The extract from ground roasted beans (see Section 3.1.5.2) is preconcentrated by vacuum evaporation before drying. Tall spray drying chambers, featuring a downward, concurrent flow pattern, and a nozzle atomizer, are favored as the powder is sticky. Air inlet and outlet temperatures in the ranges 250–300 and 105–115 °C, respectively, are used. Fine powder from the separators is recycled into the wet zone of the dryer. Agglomeration may be attained in a fluidized bed. As an alternative to spray drying, the preconcentrated extract may be frozen into slabs, the slabs broken into pieces and freeze dried in batch or continuous equipment.

Instant tea may also be produced by spray drying the extract from the leaves. Similar equipment to that used for instant coffee is employed but at lower air inlet temperatures (200–250 °C). The concentrated extract may be freeze dried in a similar manner to coffee extract [15–17].

3.2.16.5 Dehydrated Meat Products

Cooked minced meat may be hot air dried in cabinet, conveyor, fluidized bed, and rotary dryers, down to a moisture content of 4–6% (wwb). Chicken, beef, lamb, and pork may be dehydrated in this way. Chicken meat is the most stable in dried form, while pork is the least stable. The main cause of deterioration in such dried products is oxidation of fat leading to rancidity. Cooked minced meat may also be dried in vacuum cabinet dryers to give better quality products, than hot air-dried meat, but at a higher cost. Both raw and cooked meat in the form of steaks, slices, dice, or mince may be freeze dried down to a moisture content of 1.5–3.0% (wwb), at a still higher cost. Dehydrated meat products are mainly used as ingredients in dried soup mixes, sauces, and ready meals. They are also used in rations for troops and explorers [29, 37, 95].

3.2.16.6 Dehydrated Fish Products

The traditional methods of extending the shelf life of fish are salting and smoking. Salting could be regarded as a form of osmotic drying whereby salt is introduced into the flesh of fish to reduce its water activity (see Section 3.2.17). Some water may evaporate during or after salting. Smoking involves exposure of the fish to smoke from burning wood. This may be done at relatively low temperature, about 30 °C, and is known as *cold smoking*. A moisture loss of 10–11% may occur during smoking. Cold-smoked fish products can be chilled, which gives them a shelf life of about 7 days. Hot smoking is carried out at temperatures up to 120 °C. Hot-smoked products may have a sufficiently low water activity to be stable without refrigeration. Unsalted and unsmoked fish may be hot air dried in cabinet and tunnel dryers, using relatively low air temperatures, down to 30 °C. However, this is not widely practiced commercially.

Freeze-dried fish products are also available. Because of the high cost of freeze drying, only relatively expensive fish products, such as prawns and shrimps, are dried in this way. Dried fish products are used as ingredients in soup mixes and ready meals. Many fishery byproducts are produced in dried form, including fish hydrolysate, fish meal, and fish protein concentrate [29, 96–98].

3.2.17 Stability of Dehydrated Foods

When considering the stability of dehydrated foods it is not the total moisture content that is critical but rather the amount of moisture that is available to support microbial growth, enzymic, and chemical activity. It is generally accepted that a proportion of the total moisture present in a food is strongly bound to individual sites on the solid components and an additional amount is less firmly bound, but is still not readily available as a solvent for various soluble food components. In studying the availability of water in food, a fundamental property known as *water activity*, a_w, is measured. This property is defined by the expression:

$$a_w = \frac{p_v}{p_w} \tag{3.5}$$

where p_v is the water vapor pressure exerted by a solution or wet solid and p_w is the vapor pressure of pure water at the same temperature. This expression also describes the relative humidity of an air–water vapor mixture. A plot of moisture content as a function of water activity, at a fixed temperature, is known as a *sorption isotherm* (Figure 3.16). Isotherms may be prepared either by adsorption, that is, placing a dry material in contact with atmospheres of increasing relative humidity, or by desorption, that is, placing a wet material in contact with atmospheres of decreasing relative humidity. Thus two different curves may be obtained from the same material. This hysteresis effect is typical of many foods.

Food isotherms are often divided into three regions, denoted by A, B, and C in Figure 3.16. In region A, water molecules are strongly bound to specific sites on the solid. Such sites may be hydroxyl groups in polysaccharides, carbonyl, and amino groups in proteins and others on which water is held by hydrogen bonding, ion–dipole bonds, or other strong interactions. This bound water is regarded as being unavailable as a solvent and hence does not contribute to microbial, enzymic or chemical activity. It is in the a_w range 0–0.35, and is known as the *monomolecular*

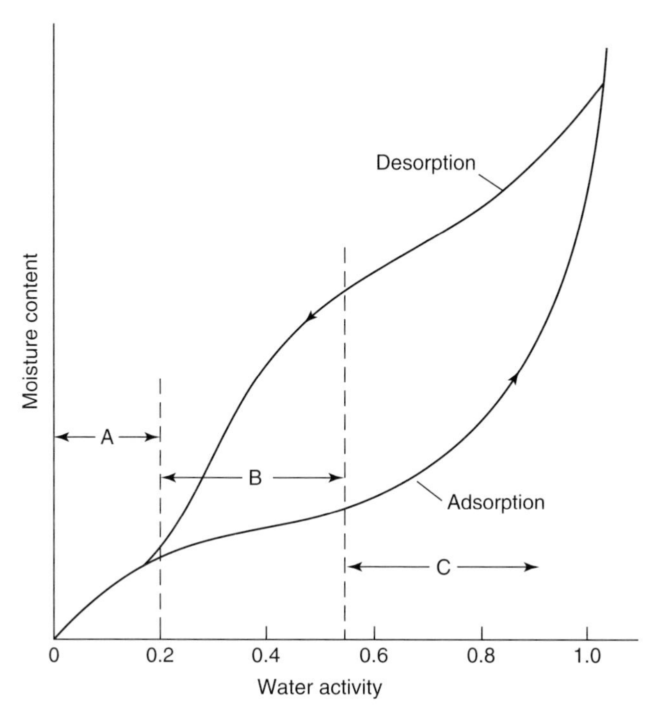

Figure 3.16 Adsorption and desorption isotherms showing hysteresis. From Brennan *et al.* [1].

or *monolayer value*. Monolayer moisture content in foods is typically in the range 0.05–0.11 (dry weight basis, dwb). Above region A, water may still be bound to the solid but less strongly than in region A. Region B is said to consist of a multilayer region and region C is one in which structural and solution effects account for the lowering of the water vapor pressure. However, this distinction is dubious, as these effects can occur over the whole isotherm. Above region A, weak bonding, the influence of capillary forces in the solid structure and the presence of soluble solids in solution all have the effect of reducing the water vapor pressure of the wet solid. All these effects occur at a moisture content below 1.0 (dwb). Most foods exhibit a water vapor pressure close to that of pure water when the moisture content is above 1.0 (dwb). Temperature affects the sorption behavior of foods. The amount of adsorbed water at any given values of a_w decreases with increase in temperature. Knowledge of the sorption characteristics of a food is useful in predicting its shelf life. In many cases the most stable moisture content corresponds to the monolayer value (Figure 3.17).

In many foods the rate of oxidation of fat is minimum in the a_w range 0.20–0.40. The rate of nonenzymic browning is highest in the a_w range 0.40–0.60. Below this range, reaction is slow due to the lack of mobility of the water. Hydrolytic reactions are also most rapid in the a_w range 0.40–0.70. The activity of enzymes starts to increase above the monolayer region and, at an a_w value above 0.80, it accelerates

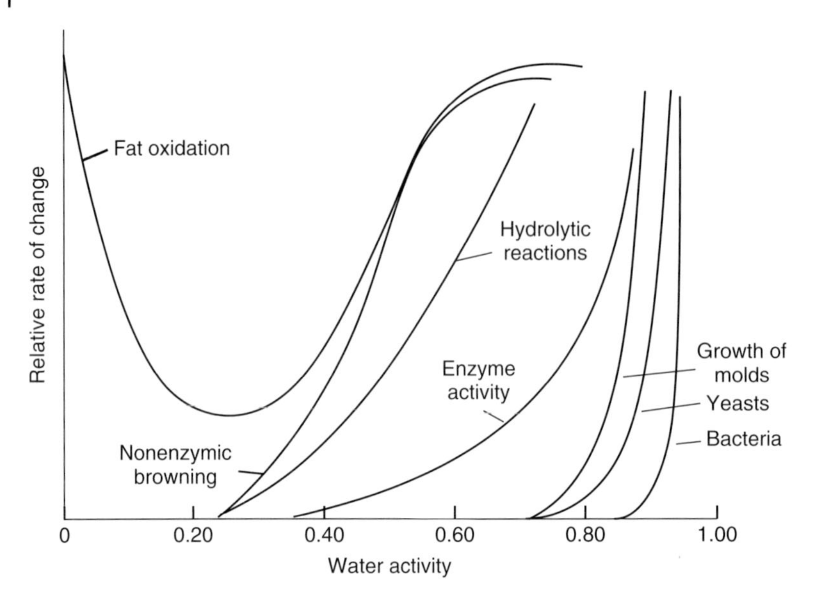

Figure 3.17 Influence of water activity on the stability of foods. Adapted from Labuza [99].

rapidly. Most molds will not grow below an a_w value of 0.70, yeasts below 0.80. Most bacteria will not grow below an a_w of 0.90. Halophilic (salt-loving) bacteria can grow at lower a_w.

It is important to note that many other factors influence the activity of microorganisms, including temperature, pH, and availability of oxygen and nutrients, and these may affect their behavior at different a_w levels [1, 10, 29, 99–101].

References

1. Brennan, J.G., Butters, J.R., Cowell, N.D., and Lilly, A.E.V. (1990) *Food Engineering Operations*, 3rd edn, Elsevier Applied Science, London.
2. Haywood, R.W. (1990) *Thermodynamic Tables in SI (Metric) Units*, 3rd edn, Cambridge University Press, Cambridge.
3. Berk, Z. (2009) *Food Process Engineering and Technology*, Academic Press, Amsterdam.
4. Ramaswamy, H. and Marcotte, M. (2006) *Food Processing Principles and Applications*, Taylor & Francis, Boca Raton.
5. Fellows, P. (2009) *Food Processing Technology*, 3rd edn, Woodhead Publishing, Cambridge.
6. Burkart, A. and Wiegand, B. (1987) Quality and economy in evaporator technology, in *Food Technology International Europe* (ed. A. Turner), Sterling Publications International, London, pp. 35–39.
7. Singh, R.P. and Heldman, D.R. (1993) *Introduction to Food Engineering*, 2nd edn, Academic Press, London.
8. Nindo, C.I., Tang, J., Powers, J.R., and Bolland, K. (2004) Energy consumption during Refractance Window evaporation of selected berry juices. *Int. J. Energy Res.*, **28**, 1089–1100.
9. Nindo, C.I., Powers, J.R., and Tang, J. (2007) Influence of refractance window evaporation on quality of juices from small fruits. *LWT Food Sci. Technol.*, **40**, 1000–1007.

10. Morison, K.R. and Hartel, R.W. (2007) Evaporation and freeze concentration, in *Handbook of Food Engineering*, 2nd edn (eds D.R. Heldman and D.B. Lund), Marcel Dekker, New York, pp. 495–552.
11. Pisecky, J. (1995) Evaporation and spray drying, in *Handbook of Industrial Drying*, 2nd edn, Vol. 1 (ed. A.S. Mujumdar), Marcel Dekker, New York, pp. 715–742.
12. Walstra, P., Guerts, T.J., Jellema, A., and van Boekel, M.A.J.S. (1999) *Dairy Technology, Principles of Milk Properties and Processes*, Marcel Dekker, New York.
13. Knipschild, M.E. and Anderson, G.G. (1994) Drying of milk and milk products, in *Modern Dairy Technology*, Vol. 1 (ed. R.K. Robinson), Chapman & Hall, London, pp. 159–254.
14. Early, R. (1998) Milk concentrates and milk powders, in *The Technology of Dairy Products*, 2nd edn (ed. R. Early), Blackie Academic & Professional, Glasgow, pp. 228–300.
15. Masters, K. (1991) *Spray Drying Handbook*, 5th edn, Longman Scientific and Technical, New York.
16. Clarke, R.J. (1987) Extraction, in *Coffee Technology*, Vol. 2 (eds R.J. Clarke and R. Macrae) Elsevier Applied Science, London, pp. 109–145.
17. Clarke, R.J. (1987) Drying, in *Coffee Technology*, Vol. 2 (eds R.J. Clarke and R. Macrae), Elsevier Applied Science, London, pp. 147–199.
18. McGinnis, R.A. (1971) *Beet-Sugar Technology*, 2nd edn, Beet Sugar Development Foundation, Fort Collins.
19. Meade, G.P. and Chen, J.C.P. (1977) *Cane Sugar Handbook*, 10th edn, John Wiley & Sons, Inc., New York.
20. Perk, C.G.M. (1973) *The Manufacture of Sugar from Sugarcane*, C.G.M. Perk, Durban.
21. Hugot, E. (1986) *Handbook of Cane Sugar Engineering*, 3rd edn, Elsevier, Amsterdam.
22. Rao, M.A. and Vitali, A.A. (1999) Fruit juice concentration and preservation, in *Handbook of Food Preservation* (ed. M.S. Rahman), Marcel Dekker, New York, pp. 217–258.
23. Nelson, P.E. and Tressler, D.K. (1980) *Fruit and Vegetable Juice Processing Technology*, 3rd edn, AVI Publishing Company, Westport.
24. Rebeck, H.M. (1990) Processing of citrus juices, in *Production and Packaging of Non-Carbonated Fruit Juices and Fruit Beverages* (ed. P.R. Ashurst), Blackie Academic and Professional, Glasgow, pp. 221–252.
25. Kale, P.N. and Adsule, P.G. (1995) Citrus, in *Handbook of Fruit Science and Technology* (eds D.K. Salunkhe and S.S. Kadam), Marcel Dekker, New York, pp. 39–65.
26. Goose, P.G. and Binsted, R. (1973) *Tomato Paste and Other Tomato Products*, Food Trade Press, London.
27. Madhavi, D.L. and Salunkhe, D.K. (1998) Tomato, in *Handbook of Vegetable Science and Technology*. Marcel Dekker, New York, pp. 171–201.
28. Wang, N. and Brennan, J.G. (1994) A mathematical model of simultaneous heat and moisture transfer during drying of potato. *J. Food Eng.*, **24**, 47–60.
29. Brennan, J.G. (1994) *Food Dehydration – A Dictionary and Guide*, Butterworth-Heinemann, Oxford.
30. Okos, M.R., Narsimhan, G., Singh, R.K., and Weitnauer, A.C. (1992) Food dehydration, in *Handbook of Food Engineering* (eds D.R. Heldman and D.B. Lund), Marcel Dekker, New York, pp. 437–562.
31. Mujumdar, A.S. (1997) Drying fundamentals, in *Industrial Drying of Foods* (ed. C.G.J. Baker), Blackie Academic & Professional, London, pp. 7–30.
32. Sokhansanj, S. and Jayes, D.S. (1995) Drying of foodstuffs, in *Handbook of Industrial Drying*, 2nd edn (ed. A.S. Mujumdar), Marcel Dekker, New York, pp. 589–625.
33. Barbosa-Canovas, G.V. and Vega-Mercado, H. (1996) *Dehydration of Foods*, Chapman & Hall, New York.
34. Toledo, R.T. (1991) *Fundamentals of Food Process Engineering*, 2nd edn, Van Nostrand Reinhold, New York.
35. Chen, X.D. (2008) Food drying fundamentals, in *Drying Technologies in Food*

Processing (eds X.D. Chen and A.S. Mujumdar), Blackwell Publishing Ltd., Oxford, pp. 1–55.
36. Brown, A.H., Van Arsdel, W.B., Lowe, E., and Morgan, A.I. Jr., (1973) Air drying and drum drying, in *Food Dehydration*, 2nd edn, Vol. 1 (eds W.B. Van Arsdel, M.J. Copley, and A.I. Morgan Jr., AVI Publishing Company, Westport, pp. 82–160.
37. Greensmith, M. (1998) *Practical Dehydration*, 2nd edn, Woodhead Publishing, Cambridge.
38. Sturgeon, L.F. (1995) Conveyors dryers, in *Handbook of Industrial Drying*, 2nd edn, Vol. 1 (ed. A.S. Mujumdar), Marcel Dekker, New York, pp. 525–537.
39. Hovmand, S. (1995) Fluidised bed drying, in *Handbook of Industrial Drying*, 2nd edn, Vol. 1 (ed. A.S. Mujumdar) Marcel Dekker, New York, pp. 195–248.
40. Pallai, E., Szentmarjay, T., and Mujumdar, A.S. (1995) Spouted bed drying, in *Handbook of Industrial Drying*, Vol. 1 (ed. A.S. Mujumdar), 2nd edn, Marcel Dekker, New York, pp. 453–488.
41. Brennan, J.G. (2003) Fluidized bed drying, in *Encyclopedia of Food Science and Nutrition*, 2nd edn (eds B. Caballero, L.C. Trugo, and P.M. Finglas), Academic Press, London, pp. 1922–1929.
42. Bahu, R.E. (1997) Fluidized bed dryers, in *Industrial Drying of Foods* (ed. C.G.J. Baker), Blackie Academic and Professional, London, pp. 65–88.
43. Kisakurek, B. (1995) Flash drying, in *Handbook of Industrial Drying*, 2nd edn, Vol. 1 (ed. A.S. Mujumdar), Marcel Dekker, New York, pp. 503–524.
44. Kelly, J.J. (1995) Rotary drying, in *Handbook of Industrial Drying*, 2nd edn, Vol. 1 (ed. A.S. Mujumdar), Marcel Dekker, New York, pp. 161–184.
45. Barr, D.J. and Baker, C.G.J. (1997) Specialized drying systems, in *Industrial Drying of Foods* (ed. C.G.J. Baker), Blackie Academic and Professional, London, pp. 179–209.
46. Mani, S. and Sokhansan, S. (2008) Rotary drum dryers, in *Food Drying Science and Technology* (eds Y.H. Hui, C. Clary, M.M. Farid, O.O. Fasina, A. NoomHorm, and J. Welti-Chanes), DEStech Publications Inc., Lancaster, PA, pp. 99–126.
47. Oakley, D. (1997) Contact dryers, in *Industrial Drying of Foods* (ed. C.G.J. Baker), Blackie Academic and Professional, London, pp. 115–133.
48. Anonymous (1992) Dryers: technology and engineering, in *Encyclopedia of Food Science and Technology*, Vol. 1 (ed. Y.H. Hui), John Wiley & Sons, Ltd, Chichester, pp. 619–656.
49. Noomhorm, A. and Ahmad, I. (2008) Vacuum drying, in *Food Drying Science and Technology* (eds Y.H. Hui, C. Clary, M.M. Farid, O.O. Fasina, A. Noomhorm, and J. Welti-Chanes), DEStech Publications, Inc., Lancaster, PA, pp. 203–214.
50. Mellor, J.D. (1978) *Fundamentals of Freeze-Drying*, Academic Press, New York.
51. Alzamora, S.M., Vergrara, F., and Welti-Chanes, J. (2008) *Food Drying Science and Technology*, DEStech Publications, Inc., Lancaster, PA, pp. 403–416.
52. Ratti, C. (2008) Freeze and vacuum drying of foods, in *Drying Technologies in Food Processing* (eds X.D. Chen and A.S. Mujumdar), Blackwell Publishing Ltd., Oxford, pp. 225–251.
53. Athanasios, I., Liapis, I., and Bruttini, R. (1995) Freeze drying, in *Handbook of Industrial Drying*, Vol. 1 (ed. A.S. Mujumdar), 2nd edn, Marcel Dekker, New York, pp. 309–344.
54. Snowman, J.W. (1997) Freeze dryers, in *Industrial Drying of Foods* (ed. C.G.J. Baker), Blackie Academic and Professional, London, pp. 134–155.
55. Lombrana, J.I. (2009) Fundamentals and tendencies in freeze-drying of foods, in *Advances in Food Dehydration* (ed. C. Ratti), CRC Press, Boca Raton, pp. 209–235.
56. Dalgleish, J.M.N. (1990) *Freeze-Drying for the Food Industries*, Elsevier Applied Science, London.
57. Lorentzen, J. (1975) Industrial freeze drying plants for food, in *Freeze Drying and Advanced Food Technology* (eds S.A. Goldblith, L. Rey, and H.H. Rothmayr), Academic Press, London, pp. 429–443.

58. Ratti, C. and Mujumdar, A.S. (1995) Infrared drying, in *Handbook of Industrial Drying*, 2nd edn, Vol. 1 (ed. A.S. Mujumdar), Marcel Dekker, New York, pp. 567–588.
59. Cenkowski, S., Arntfield, S.D., and Scanlon, M.G. (2008) Far infrared dehydration and processing, in *Food Drying Science and Technology* (eds Y.H. Hui, C. Clary, M.M. Farid, O.O. Fasina, A. Noomhorm, and J. Welti-Chanes), DEStech Publications, Inc., Lancaster, PA, pp. 157–202.
60. Edgar, R.H. (2001) Consumer, commercial, and industrial microwave ovens and heating systems, in *Handbook of Microwave Technology for Food Applications* (eds A.K. Datta and R.C. Anantheswaran), Marcel Dekker, New York, pp. 215–277.
61. Schiffmann, R.F. (1995) Microwave and dielectric drying, in *Handbook of Industrial Drying*, 2nd edn, Vol. 1 (ed. A.S. Mujumdar), Marcel Dekker, New York, pp. 345–372.
62. Jones, P.L. and Rowley, A.T. (1997) Dielectric dryers, in *Industrial Drying of Foods* (ed. C.G.J. Baker), Blackie Academic & Professional, London, pp. 156–178.
63. Schiffmann, R.F. (2001) Microwave processes for the food industry, in *Handbook of Microwave Technology for Food Applications* (eds A.K. Datta and R.C. Anantheswaran), Marcel Dekker, New York, pp. 215–299.
64. Li, H. and Ramaswamy, H.S. (2008) Microwave drying, in *Food Drying Science and Technology* (eds Y.H. Hui, C. Clary, M.M. Farid, O.O. Fasina, A. Noomhorm, and J. Welti-Chanes), DEStech Publications, Inc., Lancaster, PA, pp. 127–156.
65. Bajgai, T.R., Raghavan, V., Hashinaga, F., and Ngadi, M.O. (2006) Electrohydrodynamic drying – a concise overview. *Drying Technol.*, **24**, 905–910.
66. Ahmedou, S.A.O., Rouaud, O., and Havet, M. (2009) Assessment of the electrohydrodynamic drying process. *Food Bioprocess Technol.*, **2**, 240–247.
67. Lewicki, P.P. and Das Gupta, D.K. (1995) Osmotic dehydration of fruits and vegetables, in *Handbook of Industrial Drying*, 2nd edn, Vol. 1 (ed. A.S. Mujumdar), Marcel Dekker, New York, pp. 691–713.
68. Shi, J. and Xue, S.J. (2009) Application and development of osmotic dehydration technology in food processing, in *Advances in Food Dehydration* (ed. C. Ratti), CRC Press, Boca Raton, pp. 187–208.
69. Dhingra, D., Singh, J., Patil, R.T., and Uppal, D.S. (2008) Osmotic dehydration of fruits and vegetables: a review. *J. Food. Sci. Technol.*, **45**, 209–217.
70. Falade, K.O. and Igbeka, J.C. (2007) Osmotic dehydration of tropical fruits and vegetables. *Food Rev. Int.*, **23**, 373–405.
71. Shi, J. (2008) Osmotic dehydration of foods, in *Food Drying Science and Technology* (eds Y.H. Hui, C. Clary, M.M. Farid, O.O. Fasina, A. Noomhorm, and J. Welti-Chanes), DEStech Publications, Inc., Lancaster, PA, pp. 275–300.
72. Bolin, H.R. and Salunkhe, D.K. (1982) Fluid dehydration by solar energy. *CRC Crit. Rev. Food Sci. Technol.*, **16**, 327–354.
73. Brennan, J.G. (1989) Dehydration of foods, in *Water and Food Quality* (ed. T.M. Hardman), Elsevier Applied Science, London, pp. 33–70.
74. Imrie, L. (1997) Solar dryers, in *Industrial Drying of Foods* (ed. C.G.J. Baker), Blackie Academic & Professional, London, pp. 210–241.
75. Lopez-Malo, A. and Rios-Casas, L. (2008) Solar assisted drying of foods, in *Food Drying Science and Technology* (eds Y.H. Hui, C. Clary, M.M. Farid, O.O. Fasina, A. Noomhorm, and J. Welti-Chanes), DEStech Publications Inc., Lancaster, PA, pp. 83–98.
76. Bansal, P.K. and Chung, K.Y. (2008) Food drying equipment and design, in *Food Drying Science and Technology* (eds Y.H. Hui, C. Clary, M.M. Farid, O.O. Fasina, A. Noomhorm, and J. Welti-Chanes), DEStech Publications, Inc., Lancaster, PA, pp. 358–402.
77. Brennan, J.G. (2003) Spray drying, in *Encyclopedia of Food Science and Nutrition*, 2nd edn (eds B. Caballero, L.C.

Trugo, and P.M. Finglas), Academic Press, London, pp. 1929–1938.

78. Lawley, R. (2001) *LFRA Microbiology Handbook – Dairy Products*, 2nd edn, Leatherhead Food RA, Leatherhead.
79. Masters, K. (1997) Spray dryers, in *Industrial Drying of Foods* (ed. C.G.J. Baker), Blackie Academic & Professional, London, pp. 90–114.
80. Filkova, I. and Mujumdar, S.S. (1995) Industrial spray drying systems, in *Handbook of Industrial Drying*, 2nd edn, Vol. l (ed. A.S. Mujumdar), Marcel Dekker, New York, pp. 263–308.
81. Bhandari, B. (2008) *Food Drying Science and Technology*, DEStech Publications, Inc., Lancaster, PA, pp. 215–248.
82. Bhandari, B.R., Patel, K.C., and Chen, X.D. (2008) Spray drying of food materials – process and product characteristics, in *Drying Technologies in Food Processing*, Blackwell Publishing Ltd., Oxford, pp. 113–159.
83. Moore, J.G. (1995) Drum dryers, in *Handbook of Industrial Drying*, 2nd edn, Vol. 1 (ed. A.S. Mujumdar), Marcel Dekker, New York, pp. 249–262.
84. Nindo, C.I. and Tang, J. (2007) Refractance window dehydration technology: a novel contact drying method. *Drying Technol.*, **25**, 37–48.
85. Nindo, C.I., Feng, H., Shen, G.Q., Tang, J., and Kang, D.H. (2003) Energy utilization and microbial reduction in a new film drying system. *J. Food. Proc. Preserv.*, **27**, 117–136.
86. Nindo, C.I., Wang, S.W., Tang, J., and Powers, J.R. (2003) Evaluation of drying technologies for retention of physical and chemical quality of green asparagus. *LWT Food Sci. Technol.*, **36**, 507–516.
87. Salumkhe, D.K., Bolin, H.R., and Reddy, N.R. (1991) *Storage, Processing and Nutritional Quality of Fruits and Vegetables*, 2nd edn, Vol. 2, CRC Press, Boca Raton.
88. Jayaraman, K.S. and Das Gupta, D.K. (1995) Drying of fruits and vegetables, in *Handbook of Industrial Drying*, 2nd edn, Vol. 1 (ed. A.S. Mujumdar), Marcel Dekker, New York, pp. 643–690.
89. Luh, B.S. and Woodruff, J.G. (1975) *Commercial Vegetable Processing*, AVI Publishing Company, Westport.
90. Feinberg, B. (1973) Vegetables, in *Food Dehydration*, 2nd edn, Vol. 2 (eds W.B. Van Arsdel, M.J. Copley, and A.I. Morgan), AVI Publishing Company, Westport, pp. 1–82.
91. Talburt, W.F. and Smith, O. (1975) *Potato Processing*, 3rd edn, AVI Publishing Company, Westport.
92. Woodroof, J.G. and Luh, B.S. (1975) *Commercial Fruit Processing*, AVI Publishing Company, Westport.
93. Nury, F.S., Brekke, J.E., and Bolin, H.R. (1973) Fruits, in *Food Dehydration*, 2nd edn, Vol. 2 (eds W.B. Van Arsdel, M.J. Copley, and A.I. Morgan), AVI Publishing Company, Westport, pp. 158–198.
94. Kelly, P.M. (2008) *Food Drying Science and Technology*, DEStech Publications, Inc., Lancaster, PA, pp. 693–720.
95. Collignan, A., Santchurn, S., and Zakhia, N. (2008) *Food Drying Science and Technology*. DEStech Publications, Inc., Lancaster, PA, pp. 721–744.
96. Aitken, A., Mackies, I.M., Merritt, J.H., and Windsor, M.L. (1981) *Fish Processing and Handling*, 2nd edn, Her Majesty's Stationery Office, Edinburgh.
97. Windsor, M. and Barlow, S. (1981) *Introduction to Fishery By-products*, Fishing News Books, Farnham.
98. Peralta, J.P. (2008) in *Food Drying Science and Technology* (eds Y.H. Hui, C. Clary, M.M. Farid, O.O. Fasina, A. Noomhorm, and J. Welti-Chanes), DEStech Publications, Inc., Lancaster, PA, pp. 745–775.
99. Labuza, T.P. (1977) The properties of water in relationship to water binding in foods: a review. *J. Food Process. Preserv.*, **1**, 176–190.
100. Bhandari, B.R. and Adhikari, B.P. (2008) *Drying Technologies in Food Processing*, Blackwell Publishing Ltd., Oxford, pp. 55–89.
101. Barbosa-Canovas, G.V., Fontana, A.J. Jr., Schmidt, S.J., and Labuza, T.P. (2007) *Water Activity in Foods: Fundamentals and Applications*, Blackwell Publishing, Oxford.

4
Freezing

José Mauricio Pardo and Keshavan Niranjan

4.1
Introduction

To describe the refrigeration process as a mere temperature decrease in the product is simplistic. Many studies in the literature show the influence of this preservation process on important quality attributes of food, such as texture, color, flavor, and nutrient content [1–4].

It is clear that refrigeration affects biological materials in various ways, depending on their chemical composition, microstructure, and physical properties. Additionally, processing parameters such as cooling method used, cooling rate, final temperature, and pretreatment type play an important role in defining the quality of food [3]. Finally, the preservation of the cold chain plays an important role in the shelf life of the food products [5, 6].

In this chapter, the term *refrigeration* covers both chilling and freezing, which can be distinguished on the basis of the final temperature to which a material is cooled and the type of heat removed. In chilling, only sensible heat is extracted, whereas freezing involves the crystallization of water which requires the removal of latent heat and, therefore, the expenditure of more energy and time to complete the process.

This chapter is broadly divided into three sections: the first relates to the refrigeration equipment used to generate low temperatures and the mechanisms by which the heat removed from the food is transferred to a "heat sink," the second section deals with the kinetics of this process, and the third section addresses the effect of refrigeration on food quality.

4.2
Refrigeration Methods and Equipment

In earlier days, low temperatures were achieved by using ice obtained from high mountains, the polar regions of the earth, or that saved during winter. In the Roman Empire, the use of natural ice was widespread; food, water, and beverages

Food Processing Handbook, Second Edition. Edited by James G. Brennan and Alistair S. Grandison.

placed in isolated cabinets were cooled by the latent heat required to melt the ice (about 333 $kJ\ kg^{-1}$). Nowadays, industrial cooling systems can be divided into four main categories: (i) plate contact systems, (ii) gas contact systems, (iii) immersion and liquid contact systems, and (iv) cryogenic freezing systems.

The first three systems involve indirect cooling, that is, the food and the refrigerant are not brought into direct contact. An inherent advantage of indirect heat removal is that the final temperature of the product can be easily controlled and, therefore, these methods can be used for both chilling and freezing. However, cryogenic methods are only used to freeze biological materials.

4.2.1 Plate Contact Systems

In this kind of equipment the food is placed in contact with a cold surface, and the temperature difference transfers heat from the food. It is common to find plate freezers in which the sample is positioned between two cold metallic plates. Pressure is applied to the plates in order to ensure good contact between the cold surfaces and the object to be cooled. It is clear that the presence of regular shapes in the material produce better results in this type of equipment. Moreover, since high levels of overall heat transfer coefficients can be achieved, plate refrigerators are commonly used in production lines for solid or packaged food.

Other types of contact surfaces involve conveyor belts and rotating drums. In the former, the food (usually packaged) is placed on a temperature-controlled metallic belt conveyor and is expected to reach the desired low temperature at the end of the conveyor run. The refrigeration time is therefore controlled by the speed of the belt. The drum refrigerator, in contrast, is used to cool or freeze high-viscosity liquids by introducing them at the top of the rotating refrigerated drum and scraping the product off as the drum rotates through about 270°.

Both drum and belt refrigerators use secondary refrigerants such as brine in order to cool down their metallic surfaces, while the plate type uses direct evaporation of a refrigerant to lower its temperature.

4.2.2 Gas Contact Refrigerators

In this type of equipment, a cold gas (usually air) flows through the food, absorbing heat from it. Depending on the gas velocity and temperature, such a system can be used either for freezing or for cold storage. Even though air blast refrigeration systems are widely used for food processing [1], it does not mean that this kind of equipment is suitable for all types of materials.

A commonly used configuration is the refrigeration tunnel, where food is transported through a cold chamber on a conveyor. Cold air, forced inside the tunnel, establishes contact with the food and causes the heat transfer. Higher air velocities can achieve higher heat transfer rates, but there are limits [7]. Above a certain range of velocities, heat transfer through the food controls the overall rate;

and this should be considered in order to avoid unnecessary wastage of energy. In the case of particulate or granular materials, high speed air can be used to fluidize the material, which can result in a faster temperature drop [1]. On the other hand, impingement coolers in which cold gas jets (air or nitrogen usually) are forced to strike the food surface, have been found to increase by three times heat transfer coefficients found in traditional air tunnels [8]. This technology is recommended for products with high surface area and low weight, such as hamburger patties and pizzas.

The above procedures are sometimes combined with vacuum cooling, in which the food is introduced in a chamber where the pressure is reduced. Due to a drop in pressure, a fraction of the water present in the product vaporizes (usually surface water) carrying with it latent heat which reduces the temperature rapidly [9]. It is usually used to reduce field heats of leafy vegetables, however, applications in the fish and beef industries can be found as well [10]. Although vacuum cooling is rapid, it is more expensive than other refrigeration methods. Therefore, this technique is more appropriate for highly priced products where food safety considerations dominate.

Commercial pressures have aroused interest around spray freezing procedures, where atomized droplets are placed in contact with cold environments causing rapid nucleation, small crystals, and low phase separation. There are mainly two spray freezing methods which are classified according to the medium in which the droplets are immersed: spraying into gas and spraying into liquid. In the second one, microparticles are produced by spraying an aqueous solution or suspension through a nozzle directly into a cryogenic liquid [11]. Because of the influence of freezing rate, microbial survival is higher after spray freezing than that observed in traditional air freezing procedures. Thus spray freezing has valuable potential to preserve microorganisms [12].

4.2.3
Immersion and Liquid Contact Refrigeration

In this method, the product is brought into contact with a fast-flowing chilled liquid. It therefore tends to attain the temperature of the liquid rapidly. Chilled water, brines (salt and sugar solutions), and other type of liquids such as alcohol or ethylene glycol solutions are commonly used. The food (sometimes unpacked) can be brought into contact with the flowing liquid in two different ways: by immersion or by spraying. In spray contacting, the liquid is normally sprayed over the food, which makes heat transfer more efficient. Furthermore, less liquid is used, cutting costs and lowering environmental impact. Although immersion chilling is a cost and time effective procedure, its industrial implementation is not widespread because it is difficult to control mass transfer from the liquid to the sample and because of its effects on the product characteristics [13, 14].

Recently, some interest has arisen around high-pressure assisted freezing (HPAF) and high-pressure shift freezing (HPSF) methods. Both are based on the fact that freezing point changes with pressure. In HPAF water solidification

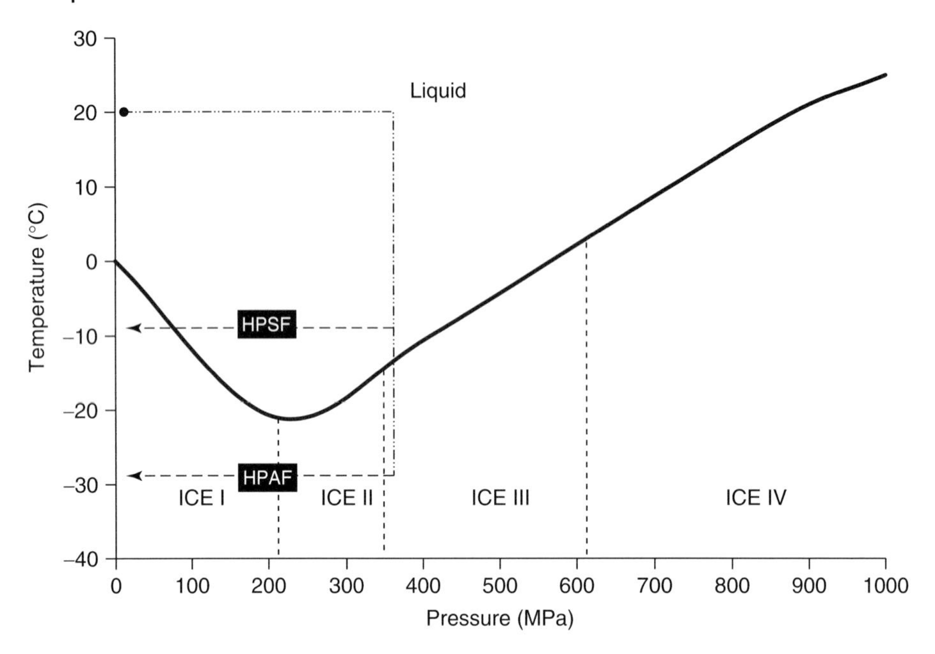

Figure 4.1 Phase diagram of water. Possible paths for high-pressure assisted freezing (HPAF) and high-pressure shift freezing (HPSF).

is produced while pressure around the product is high. Therefore the procedure involves introducing the sample into a vessel, increasing pressure, and afterwards reducing temperature until phase change is provoked; finally the pressure is reduced until atmospheric pressure is recovered. On the other hand, phase change during HPSF is produced during pressure reduction, as can be seen in Figure 4.1. Some characteristics of these freezing methods are that they lead to uniform and rapid nucleation [15], reduced microbial population on the sample surface [16], and low enzyme inactivation [17]. In addition, the texture of food samples, especially vegetables, is improved when compared with that of vegetables frozen at atmospheric pressure [16].

Although the pressure equipment is much more expensive than that used for atmospheric freezing, the improved final product characteristics could lead to creation of a niche in the vegetable industries where the advantages of high-pressure freezing have been observed.

4.2.4 Cryogenic Freezing

This type of refrigeration differs from other procedures because it does not depend on external low-temperature production systems. Low temperatures are produced because of the phase change of the cryogenic liquids themselves which are brought into contact with the food in freezing cabinets. The process is very rapid but very expensive. In some cases, cryogenic freezing is combined with air contact freezing as a two-stage process, in order to increase freezing rates while reducing process costs. In this process, also known as the *cryomechanical process*, a cold, hard crust is produced on the material by cryogenic freezing, before sending it to a cold chamber to finish off the solidification process.

The most common cryogenic liquid used is nitrogen, which boils at −195.8 °C. It is odorless, colorless, and chemically stable, but high costs restrict its use to high value products.

4.3 Low Temperature Production

Low temperatures are commonly achieved by mechanical refrigeration systems and by the use of cryogenic fluids. However, other ways have been studied and used. Some of these are described in Table 4.1 [18].

4.3.1 Mechanical Refrigeration Cycle

This method takes advantage of the latent heat needed for a refrigerant to change phase. The refrigerant – which is a fluid evaporating at very low temperatures – is circulated in an evaporation–condensation cycle and the energy it absorbs during evaporation is used to transfer heat from the refrigeration chamber (which contains the food) to a heat sink.

Basic components of typical refrigeration system are depicted in Figure 4.2 (a compressor, two heat exchangers, an expansion engine, and a refrigerant).

The most efficient refrigeration cycle is known as *Carnot cycle*, which consists of two steps: (i) frictionless and adiabatic compression and expansion (constant entropy) and (ii) heat rejection and absorption with a refrigerant at constant temperature.

It should be pointed out that frictionless compressions and expansions are highly ideal; therefore, the Carnot cycle is only useful to describe a perfect process. It is commonly used to evaluate the extent to which a real system deviates from ideal behavior.

From Figure 4.2, it can be seen that the cycle follows a specific path. The four stages of the cycle can be described as follows:

- **Compression (points 1–2)**: Initially, the refrigerant is in a gaseous state (point 1). Work is done on it by the compressor, when its pressure and temperature are elevated, resulting in a superheated gas with increased enthalpy (point 2).
- **Condensation (points 2–3)**: The superheated gas enters a heat exchanger, commonly referred to as a *high temperature exchanger* or *condenser*. Using either air-cooled or water-cooled atmospheres, the refrigerant gives up heat to the

Table 4.1 Some non-traditional methods of producing low temperatures.

Name	Description	Applications
Peltier cooling effect	In 1834, Peltier observed the inverse thermocouple effect: if electricity is forced into a thermocouple circuit, one of the metals cools down	Home freezers, water coolers and air conditioning systems in the former USSR and USA since 1949
Vortex cooling effect	In 1931, G. Ranqe discovered the vortex effect in which the injection of air into a cylinder at a tangent produces a spinning expansion of the air accompanied by the simultaneous production of cold and warm air streams	Cooling chocolate
Acoustic cooling	A hollow cylindrical tube filled with helium and xenon is used in combination with a 300 Hz speaker and a Helmholtz resounder. Sound waves compress the gas mixture and it is heated up. During decompression the gas cools down and absorbs heat from the structure. Inside the system, the noise is high (180 db). A well-insulated system is not more noisy than any mechanical system	Expensive home freezers have been produced with this environmentally friendly system
Cooling with gadoline	Gadoline is an element that increases its temperature while exposed to magnetic fields; and when exposure ceases, it reduces its temperature to a value lower than the initial	A prototype refrigerator has been built
Cooling with hydrogen	Dr. Feldman from Thermal Electric Devices has used hydrogen to refrigerate. The "HyFrig" device uses metal hydrides that are capable of absorbing large amounts of hydrogen (gas) and produce a significant decrease in temperature when the hydrogen is eliminated	Vehicle air conditioning

From Salvadori and Mascheroni [8].

surroundings and condenses to form a saturated liquid. By the time all the refrigerant has liquefied, its temperature may fall below the condensation point and a subcooled liquid may be obtained.

- **Expansion (points 3–4)**: The liquefied refrigerant enters the expansion engine, which separates the high- and low-pressure regions of the system. As the refrigerant passes through the engine, it experiences a pressure drop together with a drop in temperature. A mixture of liquid and gas leaves this process.

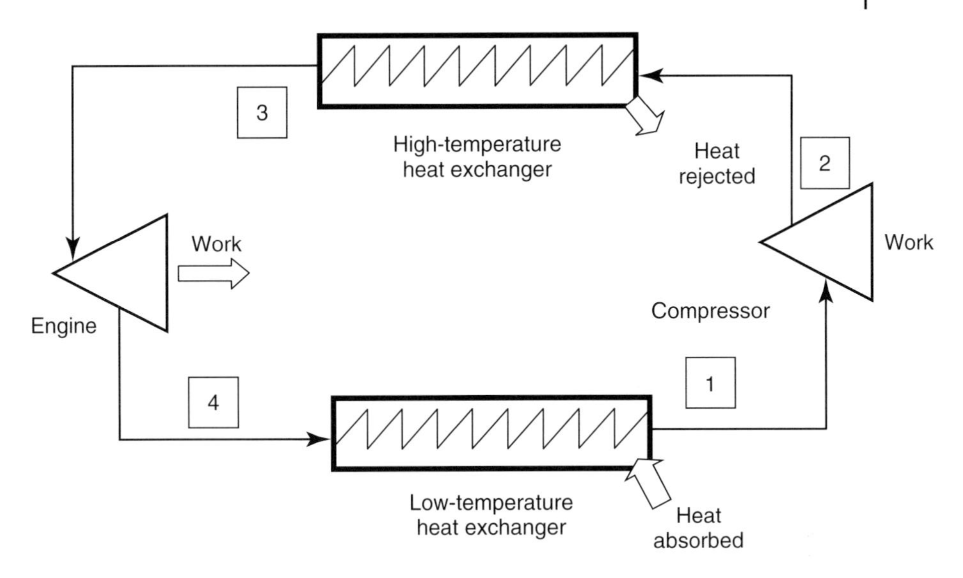

Figure 4.2 Basic components of a mechanical refrigeration cycle.

- **Evaporation (points 4–1)**: Due to the energy received in the heat exchanger, also known as the *low temperature exchanger* or *evaporator*, the refrigerant evaporates. The saturated vapor may gain more energy and become superheated before entering the compressor and continuing the cycle.

4.3.1.1 **The Pressure and Enthalpy Diagram**

The basic refrigeration cycle shown in Figure 4.2 allows a glimpse of the changes undergone by the refrigerant (energy absorption and rejection). However, other charts and diagrams are more useful while quantifying these changes. Using pressure–enthalpy charts, it is possible to estimate the changes in refrigerant properties as it goes through the process. Such charts are extremely useful in designing a refrigeration system.

The most common charts depict variations in enthalpy (x axis) with pressure (y axis), and are known as *p-h diagrams*. However, there are other useful charts such as T-s diagrams that show variation of entropy (x axis) with temperature (y axis). Figures 4.3 and 4.4 show the Carnot refrigeration cycle on a p-h diagram and T-s diagram, respectively. The domes depicted in these figures represent the saturation points. To the left of the dome is the subcooled liquid region; and to the right lies the superheated vapor region. The region beneath the dome represents two-phase (gas–liquid) mixtures. From Figures 4.3 and 4.4 it is clear that condensation (points 2–3) and evaporation (points 4–1) occur under constant temperature and pressure conditions, while expansion (points 3–4) and compression (points 1–2) occur under constant entropy conditions, as discussed earlier.

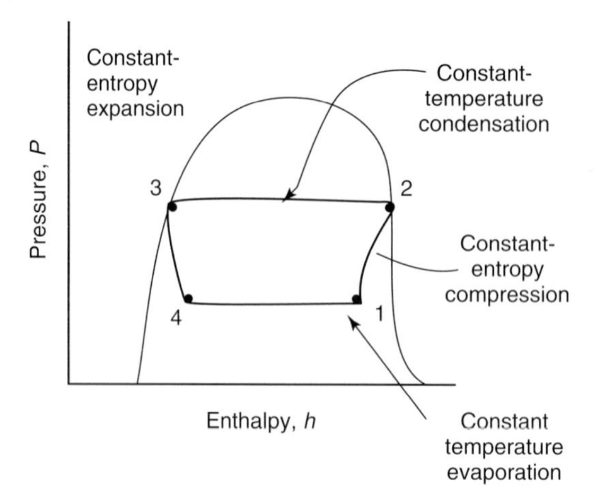

Figure 4.3 Pressure–enthalpy (p-h) diagram of the Carnot cycle.

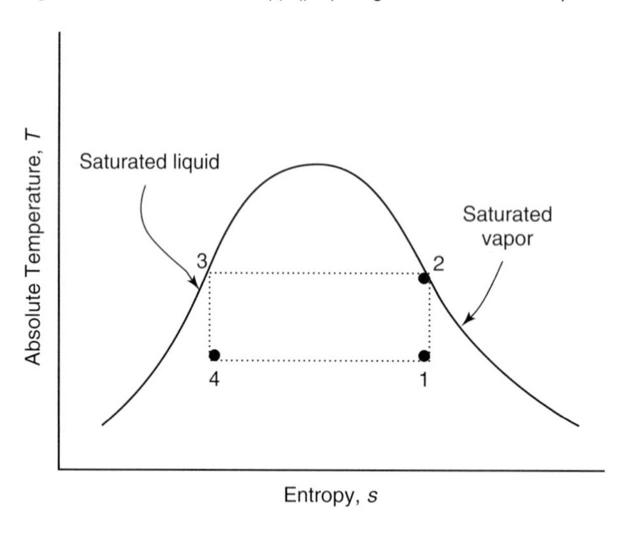

Figure 4.4 Temperature–entropy (T-s) diagram of the Carnot cycle.

4.3.1.2 **The Real Refrigeration Cycle (Standard Vapor Compression Cycle)**

If attempts are made to operate equipment under the conditions proposed by the Carnot cycle, severe mechanical problems are encountered, such as continuous valve breakdown, lack of good lubrication in the compression cycle, and so on. Therefore a real refrigeration cycle should be adapted, considering the mechanical possibilities. The main differences between the ideal Carnot cycle and that achieved in reality are described in the following paragraphs.

- **Compression: wet vs. dry**: The compression process observed in the ideal cycle (Figure 4.3, points 1–2) is commonly described as wet compression because it starts with a mixture of gas and vapor (point 1), and occurs completely in

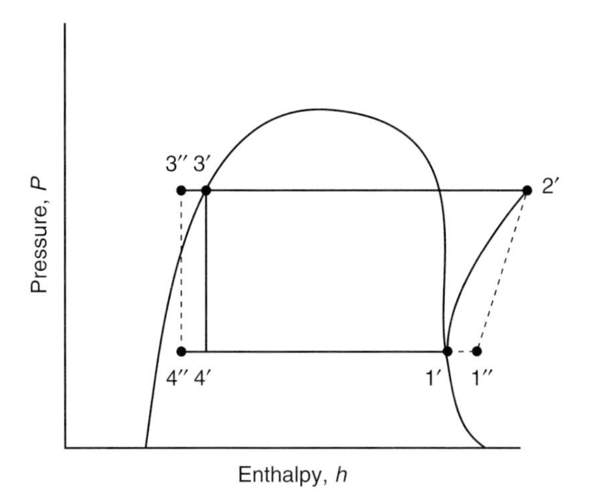

Figure 4.5 Standard refrigeration cycle and modified standard cycle.

the two-phase region (beneath the dome). The presence of a liquid phase can diminish lubrication effectiveness of some compressors. Moreover, droplets can also damage valves. Therefore, a real compression process should be carried out in the dry region shown in Figure 4.5 (process 1′–2′).

- **Expansion: engine vs. valve**: Expansion process 3–4 described in the ideal cycle (see Figure 4.3) assumes that the engine extracts energy from the refrigerant and uses it to reduce the compressor work. Practical problems include finding a suitable engine, controlling it and transferring its power to the compressor; therefore, it is uncommon to find expansion engines in refrigeration systems nowadays. Instead, the pressure drop is accomplished by a throttle valve (also called the *expansion valve*). Due to this change, the energy loss during expansion is negligible, thus $h_3 = h_4$. The changes made from cycle 1–2–3–4 to cycle 1′–2′–3′–4′ (see Figure 4.5) result in a cycle known as the *standard vapor compression cycle*. Even this standard cycle has been modified further, in order to adapt it to more efficient conditions, for example, cycle 1″–2′–3″–4″, where the vapor leaves the evaporator superheated and the liquid leaves the condenser subcooled.

4.3.2 Equipment for a Mechanical Refrigeration System

As mentioned earlier in this chapter, the basic components of a refrigeration system are the heat exchangers (evaporator and condenser), the compressor, the expansion valve, and the refrigerant.

4.3.2.1 Evaporators

Inside this heat exchanger, the liquid refrigerant absorbs heat from the air in the cold room and vaporizes. Depending on their design, evaporators can be classified into two types: direct expansion and indirect expansion evaporators. In the first type, the refrigerant vaporizes inside the coils which are in direct contact with either the atmosphere or the object being cooled. In contrast, indirect contact evaporators use a carrier fluid (secondary refrigerant) that transfers heat from the atmosphere or the object that is being refrigerated to the evaporator coils. Although this type of evaporator costs more, it is useful when several locations are to be refrigerated with a single refrigeration system.

The most common type of industrial evaporator is that in which the air is forced through a row of fins. This air absorbs heat from food and returns to the evaporator where it rejects the heat gained. The heat transfer within such an evaporator can be described as a typical multilayer heat transfer with resistances in series. Heat must move from the air to the outer wall of the coil (convective resistance), then to the inner wall of the coil (conductive resistance), and finally to the refrigerant (convective resistance). These resistances can be defined as follows:

$$R_{cv} = \frac{1}{HA_o} \tag{4.1}$$

$$R_{cd} = \frac{\Delta x}{KA_{mean}} \tag{4.2}$$

where R_{cv} and R_{cd} are the convection and conduction resistances, respectively, H and K are heat transfer coefficients for convection and conduction, and A is the related heat transfer area. It is clear from these equations that an increase in the heat transfer area will reduce the resistances, which is desirable. Additionally, it is known that the higher resistance is found in the outer layer of the coils and therefore it is common to see fins installed in the evaporator, in order to extend the contact area and decrease the resistance to heat transfer.

4.3.2.2 **Condensers**

Three main types of condensers can be found in refrigeration systems: air cooled (Figure 4.6a), water cooled (Figure 4.6b), and evaporative (Figure 4.6c). The purpose of each piece of equipment is to reject the heat absorbed by the refrigerant and transfer it to another medium such as air or water. Similar to evaporators, air-cooled condensers also use fins or plates to increase heat transfer efficiency. It is common to find fans attached to the condenser in order to increase air flow and therefore heat transfer rate. This type of equipment is easy and inexpensive to maintain and is therefore found in household appliances.

In water-cooled condensers, the refrigerant flows inside tubes. Double pipe condensers have been used for many years in the food industry, although shell and tube condensers are more common nowadays.

Evaporative condensers operate like cooling towers. Water is pumped from a reservoir at the base of the condenser and sprayed onto the coils. Water evaporation extracts heat from the coils and a fast flowing air stream takes away the vapor, favoring heat and mass transfer. The water that fails to evaporate returns to the pan and is recycled with the pump. These types of systems are very efficient, but they require a lot of space.

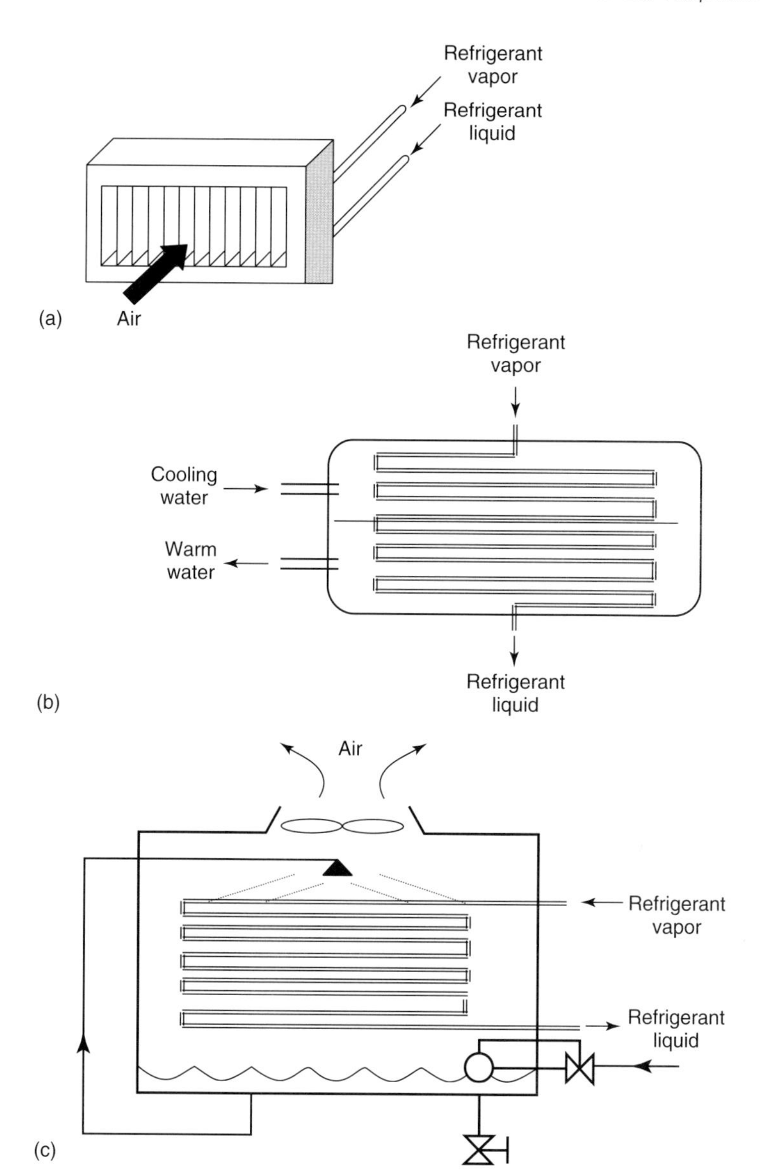

Figure 4.6 Three different types of condensers: (a) air cooling, (b) water cooling, and (c) evaporative.

4.3.2.3 **Compressors**

Refrigeration systems can be classified into three types depending on the method employed for compression:

- **Mechanical compression**. The most widely used system is the mechanical compression system, which is described in detail below.

- **Thermal compression (absorption-desorption).** This method is preferred when low-cost/low-pressure steam or waste heat is available. It is usually used to chill water (7–10 °C) and for capacities in the range from 300 kW to 5 MW.
- **Pressure difference (or ejectors).** These systems are used for similar applications as thermal compression systems and have lower initial and maintenance costs. However, they are not as common as thermal compression systems.

Mechanical compression compressors can be divided into two groups: positive displacement compressors (PDCs) and dynamic compressors. PDC increase the pressure by reducing the volume. Typical compressors in this group are the reciprocating compressors, which dominate in the range up to 300 kW applications and tend to be the first choice due to their lower costs. Rotary compressors also belong to the PDC group. Screw type rotary compressors are found in the range 300–500 kW. Other rotary compressors such as the Vane [19] are found in applications in low-capacity equipment.

In contrast, centrifugal compressors, which belong to the dynamic group, are found in equipment in the range 200–10 000 kW. These compressors have impellers turning at high speed and impart energy to the refrigerant. Some of this kinetic energy gained by the refrigerant is converted to pressure energy. These types of devices are more expensive than the reciprocating variety. However, they tend to be more economical to maintain. Further, they are generally small in size and are recommended when space is limited and long running periods are desired with minimum maintenance.

4.3.2.4 Expansion Valves

These are essentially flow-controlling devices that separate the high-pressure zone from the low-pressure zone in a refrigeration system. Common types are:

- **Manually operated.** These are used to set the volume of refrigerant flowing from the high to the low pressure side.
- **Automatic valves.** This group can be divided into float, thermostatic, and constant pressure valves. The float valve is the most cost effective of the three due to its simplicity and low maintenance costs, although it can only be used in flooded-type condensers. Thermostatic valves are the most widely used in the refrigeration industry.

4.3.2.5 Refrigerants

These can be classified into two groups: primary and secondary refrigerants. In the primary group belong those that vaporize and condense as they absorb and reject heat. Secondary refrigerants, in contrast, are heat transfer fluids commonly known as *carriers*. The choice of refrigerant is not easy, and there is a long list of commercial options. Desirable characteristics of these fluids are summarized in Table 4.2.

The refrigerants used in earlier days tended to be easily and naturally found: air, ammonia, CO_2, SO_2, ether, and so on. Ammonia is the only refrigerant from

Table 4.2 Desirable characteristics of modern refrigerants.

Safe	**Nontoxic; not explosive; nonflammable**
Environmentally friendly	Ozone friendly; no greenhouse potential; leaks should be easily detected; easily disposable
Low cost	Low cost per unit of mass; high latent heat/cost ratio
System compatible	Chemical stability; noncorrosive
Good thermodynamic and physical properties	High latent heat; low freezing temperature; high critical temperature (higher than ambient); low viscosity

this list that is still being used by the industry [20], but recently interest has been shown in CO_2. Ammonia shows several thermodynamically desirable properties, such as a high latent heat of vaporization. In addition, it has economic and environmental advantages, although its irritating and intoxicating effects are major drawbacks. Other types of refrigerants became important in the twentieth century due to their efficiency and safety considerations. Halocarbons such as R12 (Freon, dichlorofluormethane), R22 (chlorodifluormethane), R30 (methylene chloride), and others are still being used in air conditioning (building and car) and small refrigeration systems. During the 1970s, it was postulated that chlorofluorocarbons (CFCs) have damaging effects on the ozone layer. Because of their high stability in the lower atmosphere, they tend to migrate to the upper atmosphere where the chlorine portion splits and reacts with the ozone. The use of this kind of refrigerant started to decline in the 1990s under the influence of global environmental agreements such as the Montreal Treaty [5].

Hydrofluorocarbons (HFCs) are compounds containing carbon, hydrogen, and fluorine. Some of these chemicals are accepted by industry and scientists as alternatives to CFCs and hydrochlorofluorocarbons. The most commonly used HFCs are HFC 134a and HFC 152a. Because HFCs contain no chlorine, they do not directly affect stratospheric ozone and are therefore classified as substances with low ozone depletion potential. Although it is believed that HFCs do not deplete ozone, these compounds have other adverse environmental effects, such as infrared absorptive capacity. Concern over these effects may make it necessary to regulate the production and use of these compounds at some point in the future. Such restrictions have been proposed in the Kyoto Protocol. More information on refrigerant characteristics can be found elsewhere [21].

4.3.3 Common Terms Used in Refrigeration System Design

There are some useful expressions to define the capacity and efficiency of a refrigeration system and its components. Some of them are based on the variation of refrigerant energy and therefore how pressure–enthalpy diagrams are useful tools for the design of refrigeration systems.

Table 4.3 Basic sources of heat to be considered in a cold room design.

Source of heat	Considerations
Heat introduced through ceiling, floor, and walls	Conduction of heat through walls and insulation materials
Heat transferred by the food material	Sensible heat; respiration heat (applied to refrigerated vegetables); latent heat (when freezing occurs)
Heat transferred by people	Working time; number of people
Heat transferred by engines working inside the cold room	Working time; number of engines; total power
Heat transferred by illumination bulbs	Working time; number of bulbs; total power
Heat introduced by air renewal	Number of times the door is opened; other programmed changes of air

4.3.3.1 Cooling Load

The cooling load is defined as the rate of heat removal from a given space. As pointed out earlier in this chapter, ice was the main source of low temperatures when refrigeration was first commercialized; and therefore a typical unit for cooling load – known as a *tonne of refrigeration* – is equivalent to the latent heat of fusion of 1 t of ice in 24 h, which is equivalent to 3.52 kW. In designing a refrigeration system, several other sources of heat, in addition to the demand by the food itself, should be considered. Table 4.3 summarizes some of the most relevant factors.

4.3.3.2 Coefficient of Performance

The coefficient of performance (COP) is expressed as the ratio of the refrigeration effect obtained to the work done in order to achieve it. It is calculated by dividing the amount of heat absorbed by the refrigerant as it flows through the evaporator by the heat equivalent of the work done by the compressor. Thus in enthalpy values:

$$\mathrm{COP} = \frac{h_2 - h_1}{h_2 - h_3} \tag{4.3}$$

The COP value in a real refrigeration cycle is always less than that for the Carnot cycle. Therefore, industry is constantly striving to improve COP. It should be pointed out that methods to increase COP should be studied for every refrigerant separately.

4.3.3.3 Refrigerant Flow Rate

Refrigerant flow rate is useful for design purposes and depends on the cooling load in the following way:

$$\dot{m} = \frac{Q}{h_1 - h_4} \tag{4.4}$$

where is the refrigerant flow rate (kg s^{-1}) and Q is the total cooling load (kW).

4.3.3.4 Work Done by the Compressor

The energy used for raising the enthalpy of the refrigerant is:

$$W_c = \dot{m}(h_2 - h_1) \tag{4.5}$$

An efficiency term should be introduced into the above equation when calculating the work done by the compressor.

4.3.3.5 Heat Exchanged in the Condenser and Evaporator

As heat exchange in both condenser and evaporator occur at constant pressure and temperature, the amount of heat rejected or gained by the refrigerant in each process can be estimated as follows:

$$Q_c = \dot{m}(h_2 - h_3) \tag{4.6}$$

$$Q_c = \dot{m}(h_1 - h_4) \tag{4.7}$$

4.4 Freezing Kinetics

As explained earlier, from a thermodynamic point of view, freezing is a more complex process than cooling due to the phase change involved:

- cooling and undercooling of the liquid sample;
- nucleation;
- growth;
- further cooling of the frozen material.

The state diagram shown in Figure 4.7 is useful to describe a typical water freezing process which follows the cooling path A–B–C–D–E at atmospheric pressure. Starting at room temperature (point A), the sample is taken to point B without phase change (undercooling) due to an energy barrier that must be surmounted before nucleation starts. The formation of stable nuclei is governed by the net free energy of formation, which is the summation of the surface and volume energy terms. Higher undercooling favors the formation of a higher number of nuclei, which could have an effect on the final crystal size distribution. Due to the latent heat liberated, a temperature shift would be expected (points B–C in Figure 4.7). Once stable ice nuclei are formed, they continue to grow. Progress in crystal growth (points C–D in Figure 4.7) depends on the supply of water molecules from the liquid phase to the nuclei (mass transfer) and the removal of latent heat (heat transfer). After all of the water is frozen, the temperature decreases again (points D–E in Figure 4.7).

There are some differences between the phase diagrams of food and water. Because of the interaction of water with solutes present in the food, a freezing point depression occurs. In addition, ice crystallization causes a progressive concentration of the solution that remains unfrozen. Thus, as more ice is formed, the amorphous matrix increases its solid content and viscosity and reduces its freezing point, as

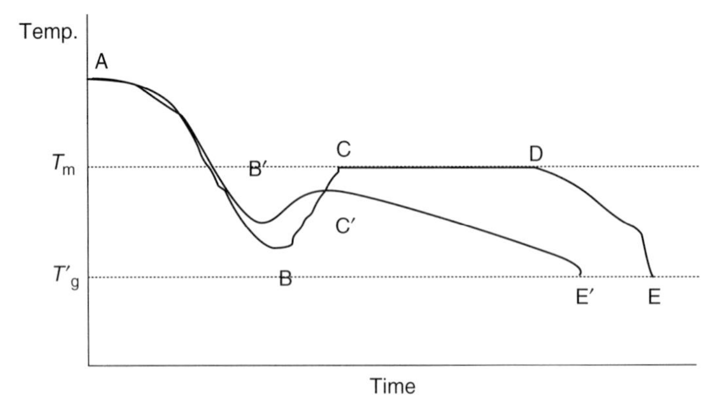

Figure 4.7 A time–temperature curve during typical freezing. Path A–B–C–D–E corresponds to the freezing of water; and path A–B′–C′–E′ corresponds to that for a solution.

seen from the path C′–D′. This concentration and ice formation continues until a temperature is reached at which no more water will freeze; this temperature is known as the *glass transition temperature* of the amorphous concentrated solution (T_g [22], point E in Figure 4.7). At this temperature, the concentrated solution vitrifies due to the high viscosity of this matrix (10^{12} Pa); and molecular movement is negligible (diffusion rates fall to a few micrometers per year) and therefore in food materials a fraction of unfrozen water is unavoidable.

Both water and temperature affect molecular mobility, because of their plasticizing capacity. Therefore, diffusion coefficients increase when water content or temperature is increased. Williams *et al.* [23] observed that the decrease in viscosity above T_g for various glass-forming substances could be modeled with what became the well-known WLF equation:

$$\log \frac{\mu}{\mu_o} = \frac{c_1(T - T_o)}{c_2 + T - T_o} \tag{4.8}$$

where μ and μ_o are the viscosities at temperatures T and T_o, and the so-called universal constants c_1 and c_2 are equal to 17.44 and 51.6, respectively. The validity of this equation over a temperature range of $T_g < T < (T_g + 100)$ has been tested and related to other physical changes such as collapse, crispiness, crystallization, ice formation, and deteriorative reactions. However, the validity of the so-called universal constants has been questioned.

4.4.1
Formation of the Microstructure during Solidification

The overall driving force for solidification is a complex balance of thermodynamic and mass transfer factors, which dominate in different parts of the process [24]. Crystal growth tends to be parallel and opposite to the direction of heat transfer;

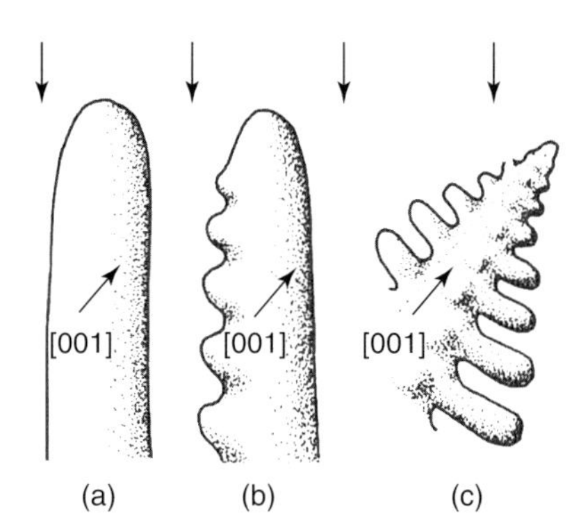

Figure 4.8 Different ice crystal shapes: (a) represents a cell which is only possible in unidirectional freezing, (b) this shape is known as *dendritic cell*, and (c) this shape is described as dendrite. After [17].

and therefore finger-like dendrites can be formed if unidirectional heat transfer is imposed [25].

Under unidirectional freezing, the most important crystal form is the dendrite [25]. Under directional cooling, cell-like structures can grow (Figure 4.8). However, anisotropy in heat and mass transfer properties tends to favor dendritic cells and dendrite growth (Figure 4.8b,c). The formation of a dendrite begins with the breakdown of the planar solid–liquid interface. Dendritic tips reject solute in all directions and therefore spaces between dendrites accumulate the rejected solute, favoring the formation of cell-type dendrites (Figure 4.8a). Secondary branches form afterwards if the conditions of the media allow.

A combination of heat and mass transfer governs the movement of the freezing front. During unidirectional cooling, heat transfer tends to be faster than mass transfer due to the high thermal conductivity of ice and low mass diffusion coefficients. Therefore, solute diffusion will be the limiting factor of growth and an undercooling in the tip region will be observed. This supercooling is known as *constitutional undercooling* and has been used to develop models on the primary spacing between dendrites, that is, the space between the tips of two dendrites [25–28]. All these models have, in common, the following general expression that correlates interdendritic spacing with freezing kinetics:

$$L \propto R^{\mathrm{a}} G^{\mathrm{b}} \tag{4.9}$$

where L is the mean interdendritic distance, R is the freezing front rate, G is the temperature gradient between the ice front and the freezing plate, and a, b are constants.

4.4.2 Mathematical Models for Freezing Kinetics

The *kinetics of freezing* has been widely studied and mathematically defined as a heat conduction process with phase change. The estimation of freezing times requires analysis of conductive heat flow through frozen and unfrozen layers, in addition to the heat transfer from the sample to the environment [29]. Numerical calculations of freezing times can be obtained using finite element or finite difference methods.

However, the work required for data preparation and computing time makes it impractical for the design engineer. Therefore, the greater part of the research effort to date has been in the development of semi-analytical/empirical food freezing time prediction methods that make use of simplifying assumptions. The most common analyses are based on either Plank's or Neumann's models, which are exact solutions valid for unidirectional freezing under the assumption of an isothermal phase change. While Plank's model assumes quasi-steady state heat transfer, the Neumann's model is more generally applicable and is based on unsteady state conduction through frozen and unfrozen layers. In addition to analytical methods, approximate analytical procedures are reviewed by Singh and Mannaperuma [30].

4.4.2.1 **Neumann's Model**

Under conditions of unidirectional freezing, it has been established that the position of the freezing front varies with time as follows [26, 27]:

$$x = c\sqrt{t} \tag{4.10}$$

where x is the position of the ice front, t is time, and c is a kinetic constant. This equation is valid under the assumption of a planar freezing front.

The equations following from Neumann's model are listed in Table 4.4. They can be solved analytically to result in a relationship similar to Equation 4.10:

$$x = 2\delta\sqrt{a_1 t} \tag{4.11}$$

where a_1 is the thermal diffusivity of the frozen zone and δ represents the Neumann dimensionless characteristic number.

4.4.2.2 **Plank's Model**

Plank's analytical solution is recognized as the first equation proposed for predicting freezing times. It is based on two assumptions: (i) the sample is initially at its freezing point and (ii) there is constant temperature in the unfrozen region. These assumptions imply that each layer in the unfrozen region remains at a constant temperature until the freezing front reaches it. In addition, Plank's model uses convective heat transfer as a boundary condition. Plank's model has been applied

Table 4.4 Basic heat equations involved in Neumman's model.

	Frozen layer	**Liquid layer**
Conservation equation	$\frac{\partial T_1}{\partial t} = a_1 \frac{\partial^2 T_1}{\partial x^2} 0 < x < s$	$\frac{\partial T_2}{\partial t} = a_2 \frac{\partial^2 T_2}{\partial x^2} s < x < x_1$
Initial conditions and boundary conditions	$T_1 = T_f, x = 0, t > 0$ $T_1 = T_m, x = s, t > 0$	$T_2 = T_0$ for all $x, t = 0$ $T_2 = T_0, x \to \infty, t > 0$ $T_2 = T_m, x = s, t > 0$
At the ice front	$-K_1 \frac{\partial T_1}{\partial x} + k_2 \frac{\partial T_2}{\partial x} + \rho_1 \Delta H_f \frac{ds}{dt}, x = s$	

to different basic geometries and the general solution is as follows:

$$t = \frac{\rho \Delta H}{T_f - T_a}\left(\frac{PD}{h} + \frac{RD^2}{K}\right) \tag{4.12}$$

where P and R are geometric constants with values which change according to the shape of the sample. For a slab $P = 1/2, R = 1/8$; for a cylinder $P = 1/4,\ R = 1/8$; and for a sphere $P = 1/6,\ R = 1/24$.

4.4.2.3 Cleland's Model

This model [29–31] estimates freezing times of food samples based on the following assumptions: (i) the conditions of the surrounding are constant, (ii) the sample is found initially at a uniform temperature, (iii) the final temperature has a fixed value, and (iv) Newton's cooling law describes the heat transfer at the surface.

Freezing time can be estimated using the following equation:

$$t_{\mathrm{slab}} = \frac{R}{h}\left[\frac{\Delta H_1}{\Delta T_1} + \frac{\Delta H_2}{\Delta T_2}\right]\left(1 + \frac{N_{\mathrm{Bi}}}{2}\right) \tag{4.13}$$

This equation is valid for Biot numbers between 0.02 and 11, Stefan numbers between 0.11 and 0.36 and Plank numbers between 0.03 and 0.61. Table 4.5 contains the equations used to estimate $\Delta H_1, \Delta H_2, \Delta T_1, \Delta T_2$.

4.4.2.4 Pham's Model

Pham [29, 32] proposed an estimation method similar to Plank's equation in which sensible heat effects are considered by calculating precooling, phase change, and subcooling times separately. In addition, Pham suggested the use of a mean freezing point, which is assumed to be 1.5 K below the initial freezing point of the food. Pham's freezing time estimation method is stated in terms of the volume and surface area of the food item and is therefore applicable to food items of any shape. This method is given as:

$$t_i = \frac{Q_i}{hA_s\Delta t_{\mathrm{mi}}}\left(1 + \frac{N_{\mathrm{Bi}}}{k_i}\right) \tag{4.14}$$

Table 4.6 contains other equations and variables needed to solve Equation 4.14. In this table N_{bil} and N_{bis} are the Biot numbers for the unfrozen and frozen phase, respectively; Q_1, Q_2, Q_3 are the heats of precooling, phase change, and subcooling respectively; t_c, t_{fm}, and t_c are temperatures related to the thermal center at the

Table 4.5 Equations and variables involved in Cleland's model.

$\Delta H_1 = C_u(T_i - T_3)$	C_u: unfrozen volumetric specific heat capacity J/(m^3 K)
$\Delta H_2 = H_L + C_f(T_3 - T_f)$	C_f: frozen volumetric heat capacity J/(m^3 K)
$\Delta T_1 = \frac{(T_i + T_3)}{2} - T_a$	H_L: latent heat of fusion kJ kg^{-1}
$T_3 = 1.8 + 0.263T_f + 0.105T_a$	T_i: initial temperature
$\Delta T_2 = T_3 - T_a$	T_a: chamber temperature

Table 4.6 Equations and variables involved in Pham's model.

Process	Equations and variables
Precooling $i = 1$	$k = 6$ $Q_i = C_f\,(t_i - t_{fm})\,V$ $N_{Bi} = N_{Bil} + N_{Bis}/2$ $\Delta t_{m1} = \dfrac{(t_i - t_m) - (t_{fm} - t_m)}{\ln\left[\dfrac{t_i - t_m}{t_{fm} - t_m}\right]}$
Phase change $i = 2$	$k = 4$ $Q_i = L_f V$ $N_{Bi} = N_{Bis}$ $\Delta t_{m2} = t_{fm} - t_m$
Subcooling $i = 3$	$k = 6$ $Q_i = C_s(t_{fm} - t_c)V$ $N_{Bi} = N_{Bis}$ $\Delta T_{m3} = \dfrac{(t_{fm} - t_m) - (t_o - t_m)}{\ln\left[\dfrac{t_{fm\cdot} - t_m}{t_o - t_m}\right]}$

end of the process, the mean freezing point, the mean final temperature; V is the volume of the food item.

4.5 Effects of Refrigeration on Food Quality

Market trends show that consumers are paying more attention to flavor and nutritional attributes of food rather than to texture and color characteristics; and therefore refrigerated products are gaining more importance in the marketplace due to the ability of this technology to maintain the two above-mentioned quality factors [33]. Foods gain from refrigerated storage (frozen and chilled) due to the positive effect of low temperatures on molecular movement, microbial growth, and chemical reaction rates. At temperatures marginally above zero, quality is well preserved for short periods (days or weeks). However, for longer periods, frozen storage is well suited because reactions continue at very low rates and microbial growth is virtually stopped to a point where the microbial population can be reduced [1].

During freezing, ice crystals can be formed in the space between cells and intracellular water can migrate, provided the cooling rate is slow enough [24]. This movement of water can produce irreversible changes in cell size. In addition, it can damage membranes, causing the loss of water and enzymes that are responsible for color and odor changes during thawing. Therefore rapid freezing, the use of cryoprotectants such as sugar and pretreatments such as blanching can improve the quality of the frozen product because cell wall damage and enzyme activity can both be reduced. Reid [34] suggests four processes that can help to explain the damages of vegetable tissue during freezing: cold (temperatures above 0 °C), solute concentration, dehydration, and ice crystal injuries. These injuries also apply to animal tissues. However, their relative importance is different: while solute concentration causes more damage to the latter, crystal injuries have a bigger effect on texture and therefore are more important in vegetable tissues.

References

1. Rahman, S. (2003) *Handbook of Food Preservation*, Marcel Dekker, New York.
2. Jeremiah, L. (ed.) (1995) *Freezing Effects on Food Quality*, Marcel Dekker, New York.
3. Gomez, F.G. and Sjoholm, I. (2004) Applying biochemical and physiological principles in the industrial freezing of vegetables: a case study on carrots. *Trends Food Sci. Technol.*, **15** (1), 39–43.
4. Jul, M. (1984) *The Quality of Frozen Foods*, Academic Press, London.
5. Coulomb, D. (2008) Refrigeration and cold chain serving the global food industry and creating a better future: two key IIR challenges for improved health and environment. *Trends Food Sci. Technol.*, **19** (8), 413–417.
6. Montanari, R. (2008) Cold chain tracking: a managerial perspective. *Trends Food Sci. Technol.*, **19** (8), 425–431.
7. Incropera, F. and Dewitt, D. (1996) *Fundamentals of Heat and Mass Transfer*, 4th edn, John Wiley & Sons, Inc., New York.
8. Salvadori, V.O. and Mascheroni, R.H. (2002) Analysis of impingement freezers performance. *J. Food Eng.*, **54** (2), 133–140.
9. Sun, D.W. and Zheng, L.Y. (2006) Vacuum cooling technology for the agri-food industry: past, present and future. *J. Food Eng.*, **77** (2), 203–214.
10. Cheng, Q.F. and Sun, D.W. (2007) Effects of combined water cooking-vacuum cooling with water on processing time, mass loss and quality of large pork ham. *J. Food Process Eng.*, **30** (1), 51–73.
11. Rogers, T.L., Nelsen, A.C., Hu, J.H., Brown, J.N., Sarkari, M., Young, T.J., Johnston, K.P., and Williams, R.O. (2002) A novel particle engineering technology to enhance dissolution of poorly water soluble drugs: spray-freezing into liquid. *Eur. J. Pharm. Biopharm.*, **54** (3), 271–280.
12. Volkert, M., Ananta, E., Luscher, C., and Knorr, D. (2008) Effect of air freezing, spray freezing, and pressure shift freezing on membrane integrity and viability of lactobacillus rhamnosus GG. *J. Food Process Eng.*, **87** (4), 532–540.
13. Blanda, G., Cerretani, L., Cardinali, A., Barbieri, S., Bendini, A., and Lercker, G. (2009) Osmotic dehydrofreezing of strawberries: polyphenolic content, volatile profile and consumer acceptance. *LWT Food Sci. Technol.*, **42** (1), 30–36.
14. Ribero, G.G., Rubiolo, A.C., and Zorrilla, S.E. (2007) Influence of immersion freezing in NaCl solutions and of frozen storage on the viscoelastic behavior of mozzarella cheese. *J. Food Sci.*, **72** (5), E301–E307.
15. Li, B. and Sun, D.W. (2002) Novel methods for rapid freezing and thawing of foods – a review. *J. Food Eng.*, **54** (3), 175–182.
16. Fernandez, P.P., Prestamo, G., Otero, L., and Sanz, P.D. (2006) Assessment of cell damage in high-pressure-shift frozen broccoli: comparison with market samples. *Eur. Food Res. Technol.*, **224** (1), 101–107.
17. Van Buggenhout, S., Messagie, I., Van der Plancken, I., and Hendrickx, M. (2006) Influence of high-pressure-low-temperature treatments on fruit and vegetable quality

related enzymes. *Eur. Food Res. Technol.*, **223** (4), 475–485.

18. James, S. (2001) Rapid chilling of food – a wish or a fact. *Bull. Int. Inst. Refrig.*, **3**, 13.
19. Pearson, A. (2001) New developments in industrial refrigeration. *ASHRAE J.*, **43**, 54–60.
20. Ayub, Z. (2006) Industrial refrigeration and ammonia enhanced heat transfer. *J. Enhanc. Heat Transf.*, **13** (2), 157–173.
21. Calm, J.M. (2006) Comparative efficiencies and implications for greenhouse gas emissions of chiller refrigerants. *Int. J. Refrig.*, **29** (5), 833–841.
22. Slade, L. and Levine, H. (1991) Beyond water activity: recent advances based on an alternative approach to the assessment of food quality and safety. *Crit. Rev. Food Sci. Nutr.*, **30**, 115–360.
23. Williams, M., Landel, R., and Ferry, J.D. (1955) The temperature dependence of relaxation mechanisms in amorphous polymers and other glass-forming liquids. *J. Am. Chem. Soc.*, **77**, 3701–3707.
24. Sahagian, M. and Douglas, G. (1995) Fundamental aspects of the freezing process, in *Freezing Effects on Food Quality* (ed. L. Jeremiah), Marcel Dekker, New York, pp. 1–50.
25. Kurz, W. and Fisher, D. (1987) *Fundamentals of Solidification*, Trans Tech Publications, Aedermansdorf.
26. Ueno, S., Do, G.S., Sagara, Y., Kudoh, K., and Higuchi, T. (2004) Three-dimensional measurement of ice crystals in frozen dilute solution. *Int. J. Refrig.*, **27** (3), 302–308.
27. Woinet, B., Andrieu, J., and Laurant, M. (1998) Experimental and theoretical study of model food freezing. Part I. Heat transfer modelling. *J. Food Eng.*, **35**, 381–393.
28. Pardo, J., Suess, F., and Niranjan, K. (2002) An investigation into the relationship between freezing rate and mean ice crystal size for coffee extracts, *Trans. IChemE*, **80** (Part C), 176–182.
29. Fricke, B.A. and Becker, B.R. (2001) Evaluation of thermophysical property models for foods. *HVAC R Res.*, **7** (4), 311–330.
30. Singh, R.P. and Mannaperuma, J.D. (1990) Developments in food freezing, in *Biotechnology and Food Process Engineering* (eds H. Schwartzberg and A. Roa), Marcel Dekker, New York, pp. 309–358.
31. Cleland, D.J., Cleland, A.C., and Earle, R.L. (1987) Prediction of freezing and thawing times for multi-dimensional shapes by simple formulae: I regular shapes. *Int. J. Refrig.*, **10**, 156–164.
32. Pham, Q.T. (1984) An extension to Plank's equation for predicting freezing times of foodstuffs of simple shapes. *Int. J. Refrig.*, **7**, 377–383.
33. Kadel, A. (2001) Recent advances and future research needs in postharvest technology of fruits. *Bull. Int. Inst. Refrig.*, 3–13.
34. Reid, D.S. (1993) Basic physical phenomena of the freezing and thawing of animal and vegetable tissue, in *Frozen Food Technology* (ed. C.P. Mallet), Blackie Academic & Professional, London, pp. 1–19.

5
Irradiation

Alistair S. Grandison

5.1
Introduction

Irradiation has probably been the subject of more controversy and adverse publicity prior to its implementation than any other method of food preservation. In many countries, including the United Kingdom, irradiation has been viewed with suspicion by the public, largely as a result of adverse and frequently misinformed reporting by the media. In the United Kingdom this has resulted in the paradoxical situation where irradiation of many foods is permitted, but not actually carried out. On the other hand, many non-food items such as pharmaceuticals, cosmetics, medical products, and plastics are routinely irradiated. Yet the process provides a means of improving the quality and safety of certain foods, while causing minimal chemical damage.

Irradiation of food is not a new idea. Since the discovery of X-rays in the late nineteenth century, the possibility of controlling bacterial populations by radiation has been understood. Intensive research on food irradiation has been carried out for over 50 years, and the safety and "wholesomeness" of irradiated food have been established to the satisfaction of most scientists.

5.2
Principles of Irradiation

Irradiation literally means exposure to radiation. In practice, three types of radiation may be used for food preservation: gamma (γ) rays, X-rays, or high-energy electron beams (β particles). These are termed *ionizing radiations*. Although the equipment and properties differ, the three radiation types are all capable of producing ionization and excitation of the atoms in the target material, but their energy is limited so that they do not interact with the nuclei to induce radioactivity. Gamma rays and X-rays are part of the electromagnetic spectrum, and are identical in their physical properties, although they differ in origin.

Food Processing Handbook, Second Edition. Edited by James G. Brennan and Alistair S. Grandison.

The energy of constituent particles or photons of ionizing radiations is expressed in electron volts (eV), or more conveniently in megaelectronvolts (1 MeV $= 1.602 \times 10^{-13}$ J). One electronvolt is equal to the kinetic energy gained by an electron on being accelerated through a potential difference of 1 V. An important, and sometimes confusing distinction exists between radiation energy and dose. When ionizing radiations penetrate a food, energy is absorbed. This is the "absorbed dose" and is expressed in grays (Gy), where 1 Gy is equal to an absorbed energy of 1 J kg^{-1}. Thus, while radiation energy is a fixed property for a particular radiation type, absorbed dose will vary in relation to intensity of radiations, exposure time, and composition of the food.

Gamma rays are produced from radioisotopes and hence they have fixed energies. In practice, cobalt-60 (Co-60) is the major isotope source. This isotope is specifically manufactured for irradiation and is not a nuclear waste product. An alternative radioisotope is cesium-137 (Cs-137), which is a byproduct of nuclear fuel reprocessing, and is used very much less widely than Co-60.

Electron beams and X-rays are machine sources that are powered by electricity. Hence they exhibit a continuous spectrum of energies depending on the type and conditions of the machinery. They hold a major advantage over isotopes in that they can be switched on and off, and can in no way be linked to the nuclear industry.

The mode of action of ionizing radiation can be considered in three phases:

- The primary physical action of radiation on atoms.
- The chemical consequences of these physical actions.
- The biological consequences to living cells of the food, or contaminating organisms.

5.2.1 Physical Effects

Although the energies of the three radiation types are comparable, and the results of ionization are the same, there are differences in their mode of action as shown in Figure 5.1.

High-energy electrons interact with the orbital electrons of the medium, giving up their energy. The orbital electrons are either ejected from the atom entirely, resulting in ionization, or moved to an orbital of higher energy, resulting in excitation. Ejected (secondary) electrons of sufficient energy can go on to produce further ionizations and excitations in surrounding atoms (Figure 5.1a).

X-rays and γ-rays can be considered to be photons, and the most important interaction is by the Compton effect (Figure 5.1b,c). The incident photons eject electrons from atoms in the target material, giving up some of their energy and changing direction. A single photon can give rise to many Compton effects, and can penetrate deeply into the target material. Ejected electrons which have sufficient energy (approximately 100 eV) go on in turn to cause many further excitations and ionizations. Such electrons are known as *delta rays*. It has been calculated that a

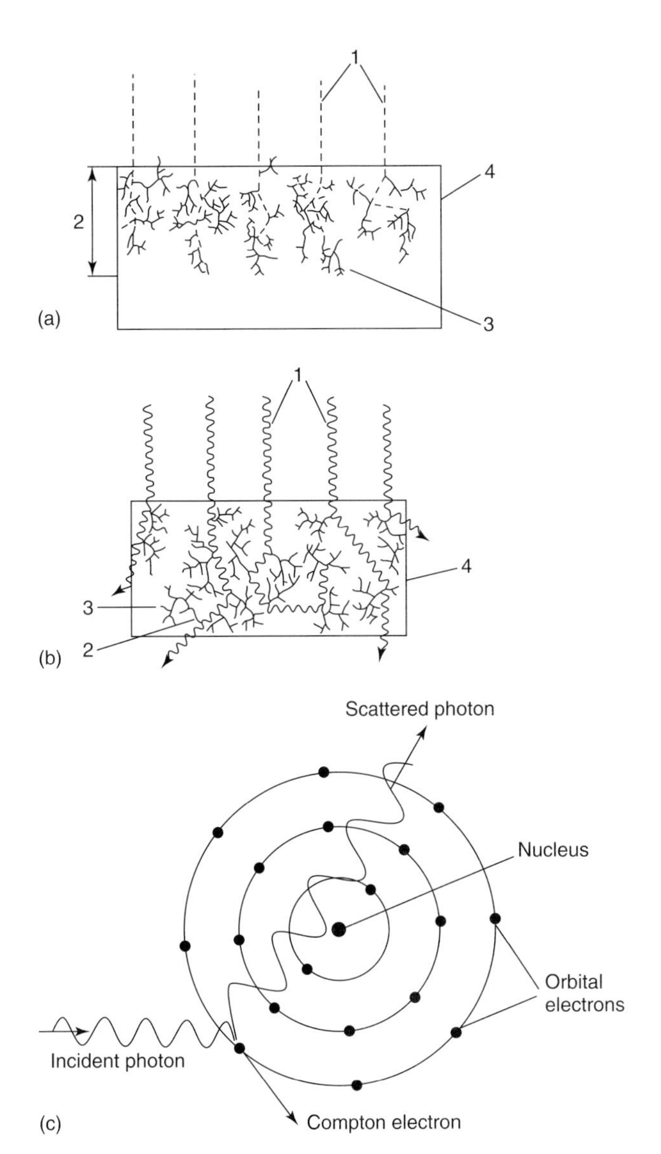

Figure 5.1 Interaction of radiation with matter. (Adapted from Diehl [2].) (a) Electron radiation; 1 – primary electron beam, 2 – depth of penetration, 3 – secondary electrons, and 4 – irradiated medium. (b) Gamma or X-radiation; 1 – gamma or X-ray photons, 2 – Compton electrons, 3 – secondary electrons, and 4 – irradiated medium. (c) Compton effect.

single Compton effect can result in 30 000–40 000 ionizations and 45 000–80 000 excitations [1].

The radiation-induced chemical changes produced by gamma or X-ray photons, or by electron beams are exactly the same because ionizations and excitations are ultimately produced by high-energy electrons in both cases. However, an important

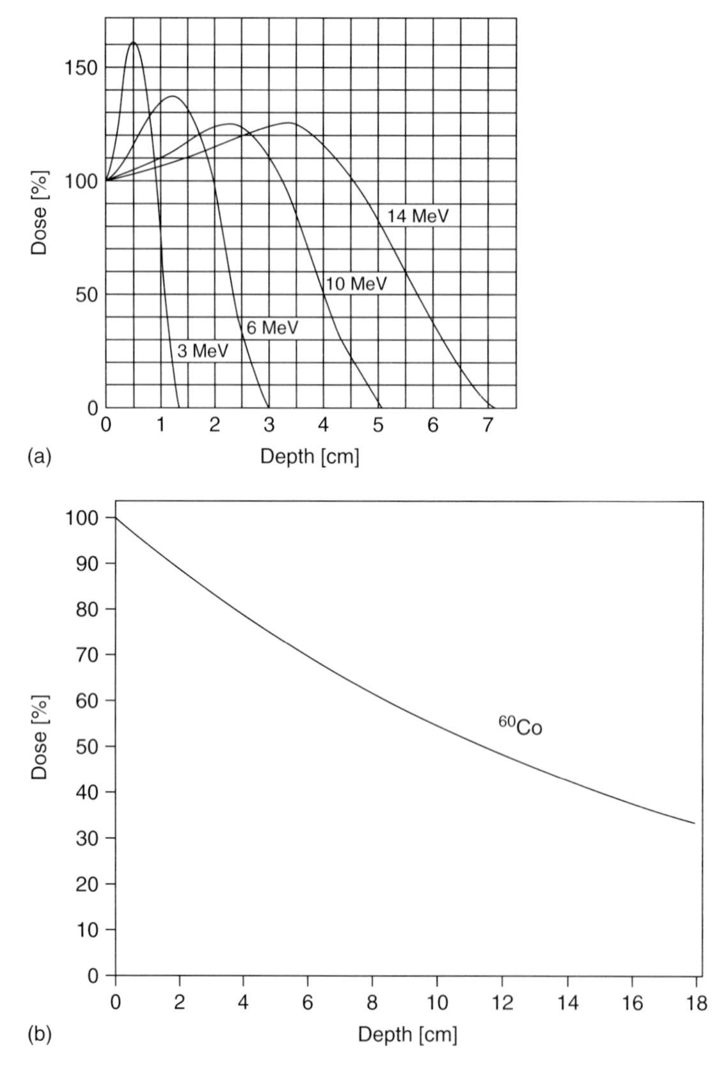

Figure 5.2 Depth–dose distribution in water from one side: (a) electrons of different energy and (b) gamma radiation. (Adapted from Diehl [2].)

difference between photons and high-energy electrons is the depth of penetration into the food. Electrons give up their energy within a few centimeters of the food surface depending on their energy, whereas gamma or X-rays penetrate much more deeply, as illustrated in Figure 5.1a,b. Depth–dose distribution curves for electrons and gamma rays are shown in Figure 5.2 for irradiation of water, which acts as a reasonable model for foods. The difference in shape of the curves between the two radiation types reflects the difference in mechanism of action. With electrons it can be seen that effective dose is greatest at some distance into the food as

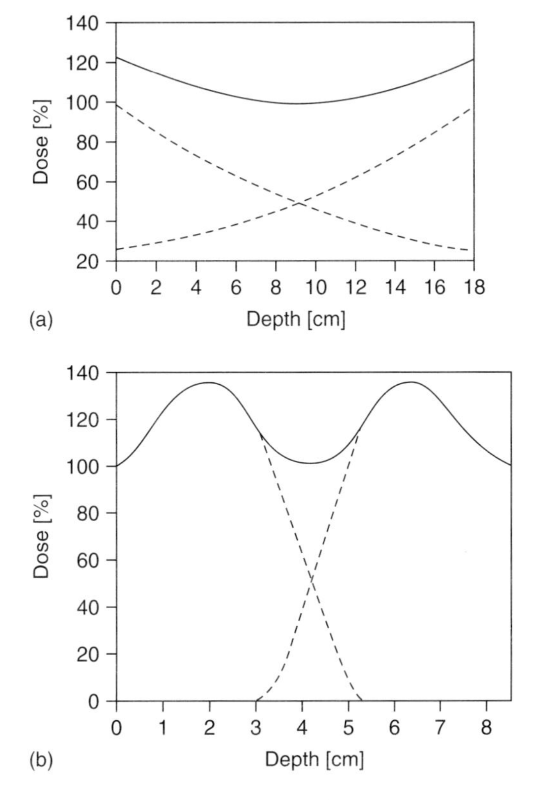

Figure 5.3 Depth–dose distribution in water from two sides: (a) gamma radiation and (b) 10 MeV electrons. Dashed lines indicate dose distribution for one-sided irradiation. (Adapted from Diehl [2].)

more secondary electrons are produced at that distance. The primary electrons lose their energy by interacting with water, so that the practical limit for electron irradiation is about 4 cm using the maximum permitted energy (10 MeV). Gamma rays, however, become depleted continuously as they penetrate the target and the effective depth is much greater. X-rays follow a similar pattern to Co-60 radiations. Two-sided irradiation permits treatment of thicker packages of food (Figure 5.3), but it is still quite limited for electrons (approximately 8 cm).

The net result of these primary physical effects is a deposition of energy into the material, giving rise to excited molecules and ions. These effects are nonspecific, with no preference for any particular atoms or molecules. Hall *et al.* [3] estimated that the timescale of these primary effects is 10^{-14} s. It is an essential feature that radiations do not have sufficient energy to interact with the nuclei, dislodging protons or neutrons, otherwise radioactivity could be induced in the target atoms. For this reason energies are limited to 5 and 10 MeV for X-rays and electrons respectively. Cobalt-60 produces γ-rays of fixed energy well below that required to interfere with the nucleus.

5.2.2
Chemical Effects

The secondary chemical effects of irradiation result from breakdown of the excited molecules and ions and their reaction with neighboring molecules, giving a cascade of reactions. The overall process by which these reactions produce stable end products is known as "*radiolysis.*" These reactions are considered in detail elsewhere [2], and only a brief summary is given here. The ions and excited molecules contain abnormal amounts of energy, which is lost through a combination of physical processes (fluorescence, conversion to heat, or transfer to neighboring molecules) and chemical reactions. This occurs irrespective of whether they are free or molecular components. The primary reactions include isomerization and dissociation within molecules, and reactions with neighboring species. The new products formed include *free radicals,* that is atoms or molecules with one or more unpaired electrons, which are available to form a chemical bond, and are thus highly reactive. Because most foods contain substantial quantities of water, and contaminating organisms contain water in their cell structure, the radiolytic products of water are particularly important. These include hydrogen and hydroxy radicals (H•, •OH), e_{aq}^-, H_2, H_2O_2, and H_3O^+. These species are highly chemically reactive and will react with many substances, although not water molecules.

Hall *et al.* [3] estimated the timescale of the secondary effects to be within 10^{-2} s. In most foods free radicals have a short lifetime ($<10^{-3}$ s). However, in dried or frozen foods, or foods containing hard components such as bones or shells, the free radicals may persist for longer.

The major components of foods and contaminating organisms, such as proteins, carbohydrates, fats, and nucleic acids, as well as minor components such as vitamins, are all chemically altered to some extent following irradiation. This can be through direct effects of the incident electrons or Compton electrons. In aqueous solutions, reactions occur through secondary effects by interaction with the radiolytic products of water. The relative importance of primary and secondary effects depends on the concentration of the component in question.

The chemical changes are important in terms of their effects on living food contaminants whose elimination is a major objective of food irradiation. However, it is also essential to consider effects on the components of foods inasmuch as this may affect their quality (e.g., nutritional status, texture, off-flavors). A further consideration is effects on living foods such as fruits and vegetables, where delaying ripening or senescence is the goal.

5.2.3
Biological Effects

The major purpose of irradiating food is to cause changes in living cells. These can either be contaminating organisms such as bacteria or insects, or cells of living foods such as raw fruits and vegetables. Ionizing radiation is lethal to all forms of life, the lethal dose being inversely related to the size and complexity

Table 5.1 Approximate lethal doses of radiation for different organisms.

Organism	Lethal dose (kGy)
Mammals	0.005–0.01
Insects	0.01–1
Vegetative bacteria	0.5–10
Sporulating bacteria	10–50
Viruses	10–200

Data from Pollard [4].

of the organism (Table 5.1). The exact mechanism of action on cells is not fully understood, but the chemical changes described above are known to alter cell membrane structure, reduce enzyme activity, reduce nucleic acid synthesis, affect energy metabolism through phosphorylation, and produce compositional changes in cellular DNA. The latter is believed to be far and away the most important component of activity but membrane effects may play an additional role.

The DNA damage may be caused by a direct effect whereby ionizations and excitations occur in the nucleic acid molecules themselves. Alternatively, the radiations may produce free radicals from other molecules, especially water, which diffuse to, and cause damage to the DNA through indirect effects. Direct effects predominate under dry conditions such as when dry spores are irradiated, while indirect effects are more important under wetter conditions, such as within the cell structures of fruits or vegetative bacterial cells.

It should be noted that remarkably little chemical damage to the system is required to cause cell lethality, because DNA molecules are enormous compared to other molecules in the cell, and thus present a large target. Yet only a very small amount of damage to the molecule is required to render the cell irreparably damaged. This is clearly illustrated in two separate studies reported by Diehl [2], which calculated that during irradiation of cells:

- a radiation dose of 0.1 kGy would damage 2.8% of DNA molecules, which would be lethal to most cells, while enzyme activity would only fall by 0.14%, which is barely detectable, and only 0.005% of amino acid molecules would be affected, which is completely undetectable [4]; and
- a dose of 10 kGy would affect 0.0072% of water molecules and 0.072% of glucose molecules, but a single DNA molecule (10^9 Da) would be damaged at about 4000 locations, including 70 double-strand breaks [5].

This is fortuitous for food irradiation, as the low doses required for efficacy of the process result in only minimal chemical damage to the food.

The phenomenon also explains the variation in lethal doses between organisms (Table 5.1). The target presented by the DNA in mammalian cells is considerably larger than that of insect cells, which in turn are much larger than the genome of bacterial cells; consequently, radiation sensitivity follows the same pattern. On the other hand, viruses have a much smaller nucleic acid content, and the high doses required for their elimination make irradiation an unlikely treatment procedure. Fortunately, it is unusual for foodborne viruses to cause health problems in humans.

5.3 Equipment

5.3.1 Isotope Sources

Cobalt-60 is the major isotope source for commercial irradiation. It is manufactured in specific reactors and over 80% of the world supply is produced in Canada. Cobalt-60 is produced from nonradioactive Co-59, which is compressed into small pellets and fitted into stainless steel tubes or rods a little larger than pencils. These are bombarded with neutrons in a nuclear reactor over a period of about one year to produce highly purified Co-60, which decays in a controlled manner to stable Ni-60 with the emission of γ-rays with energies of 1.17 and 1.33 MeV, and a half-life of about 5.2 years. The Co-60 is water-insoluble and thus presents minimal risk for environmental contamination.

Cesium-137 is an alternative possibility, but is much less widely used than Co-60. It decays to stable Ba-137, emitting a γ-photon (0.662 MeV) with a half-life of 30 years. It is unlikely to gain more importance, and future discussion of radioisotopes will be confined to Co-60.

In practice, a radioactive "source" is comprised of a number of Co-60 tubes arranged into the appropriate geometric pattern (e.g., plaques or cylinders). The short half-life of Co-60 means that the source will be depleted by approximately 1% per month. This depletion must be considered in calculations of dose, and requires that tubes be replaced periodically. In practice, individual rods are replaced in rotation at intervals, maintaining the total source energy at a fairly constant level. An individual rod may be utilized for 3–4 half-lives when activity has decayed to about 10% of its original level. When not in use the source is normally held under sufficient depth of water to completely absorb radiation to the surface so that personnel may safely enter the radiation cell area to load and unload radionuclides.

Examples of irradiation plants are shown in Figures 5.4 and 5.5. Designs can be batch or continuous, the latter being more appropriate for large-scale processing. As the source emits radiations in all directions and the rate of emission cannot be controlled, it is essential to control product movement past the source in the most efficient way possible, to make best use of the radiations. Another aim is to achieve the lowest possible dose uniformity ratio (see Section 5.3.3). For this reason the containerized products may follow a complex path around the source, often two or more product units in depth, and turning the units to effect two-sided irradiation. An example of product progress around a source is shown in Figure 5.6. The dose

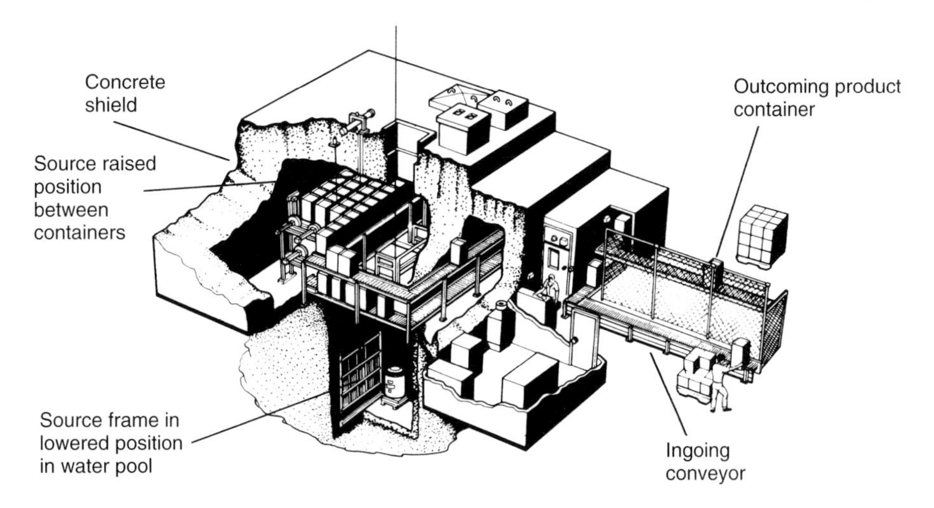

Figure 5.4 Commercial automatic tote-box gamma (Co-60) irradiator. (Courtesy of Atomic Energy of Canada Ltd.)

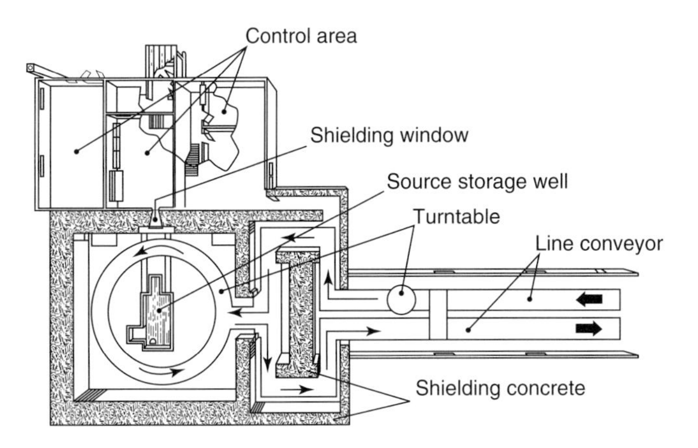

Figure 5.5 Plan of commercial potato irradiator. (Courtesy of Kawasaki Heavy Industries.)

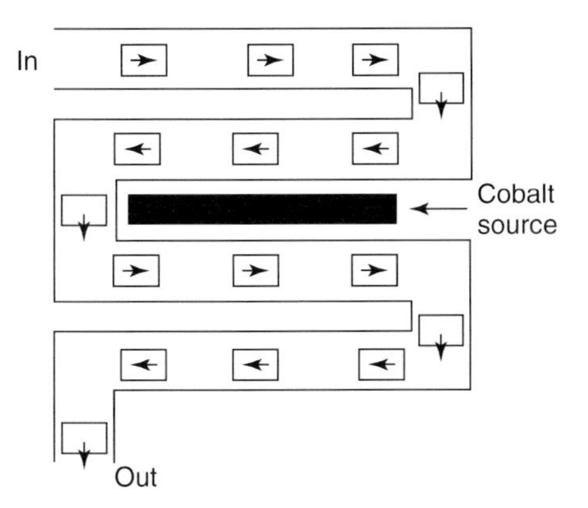

Figure 5.6 Example of progress of containerized product around a gamma irradiation source.

rates provided by isotope sources are generally low, so that irradiation may take around 1 h to complete. Therefore, product movement is often sequential with a finite time allowed in each position without movement, to allow absorption of sufficient radiation.

5.3.2 Machine Sources

Both electron and X-ray machines use electrons, which are accelerated to speeds approaching the speed of light by the application of energy from electric fields in an evacuated tube. The resulting electron beams possess a considerable amount of kinetic energy.

The main designs of electron irradiator available are the Dynamitron, which will produce electron energies up to 4.5 MeV, or linear accelerators for higher energies. In either case, the resulting beam diameter is only a few millimeters or centimeters. To allow an even dose distribution in the product it is necessary to scan the beam using a scanning magnet which creates an alternating magnetic field (analogous to the horizontal scan of a television tube), which moves the beam back and forwards at 100–200 Hz. An even, fan-shaped field of emitted electrons is created, through which the food is conveyed. A simplified diagram of an electron beam machine is shown in Figure 5.7. As described in Section 5.2.1, the low penetration of electrons requires that this technology is limited to surface treatment of food or to foods of limited thickness (8 cm maximum with two-sided irradiation).

When electrons strike a target they produce X-rays, which can be utilized to give greater penetration depth, but suffer from the disadvantage of low conversion efficiency. The efficiency of conversion depends on the energy of the electrons and the atomic number of the material of the target. In practice, therefore, X-rays are produced by firing high-energy electrons at a heavy metal target plate as shown schematically in Figure 5.8. Even with 10 MeV electrons and a tungsten (atomic number 74) plate, the efficiency of conversion is only 32%, hence cooling water must be applied to the converter plate.

Both electrons and X-rays deliver much higher dose rates than isotope sources, so that processing is complete in a matter of seconds. Also the beams may be directed, so that the complex transport of packages around the source is not required.

5.3.3 Control and Dosimetry

As with any processing technology, radiation treatment must be controlled and validated, and hence the level of treatment must be monitored in a quantifiable form. The subject is considered in detail by Ehlermann [6]. The basic parameter of importance is the absorbed dose, although information on dose rate may be important in some circumstances. A knowledge of absorbed dose is necessary, both to conform to legal limits, and to ensure that the advisory technological considerations have been met: in other words, whether the food has received

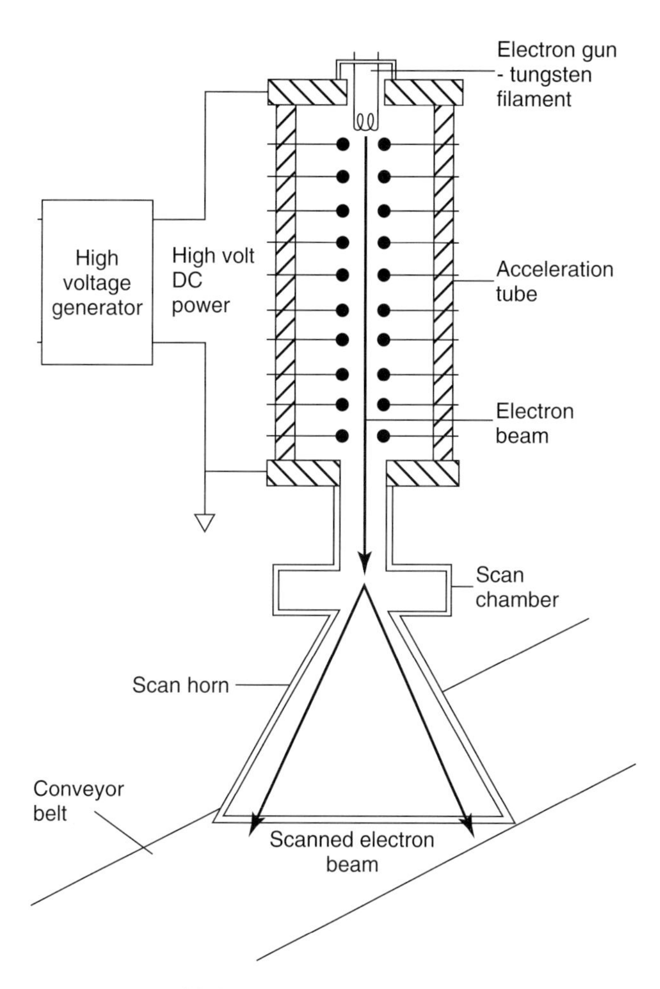

Figure 5.7 Simplified construction of electron beam machine. (Courtesy of Leatherhead Food RA, UK.)

sufficient treatment without exceeding either the legal limit or the threshold dose for production of sensory impairment. The dose absorbed by a food depends on the magnitude of the radiation field, the absorption characteristics of the food and the exposure time.

A further consideration is the dose uniformity ratio, that is, the ratio between the maximum and minimum dose absorbed by different parts of a food piece or within a food container. Radiation processing results in a range of absorbed doses in the product, in the same way as heating results in a range of temperatures. Dose is normally expressed in terms of average dose absorbed by the food during the treatment. Dose distribution will depend partly on the type of radiation used, but will be affected by the geometry of individual food units and the way in which the food is packaged and loaded into containers for processing. The ideal uniformity

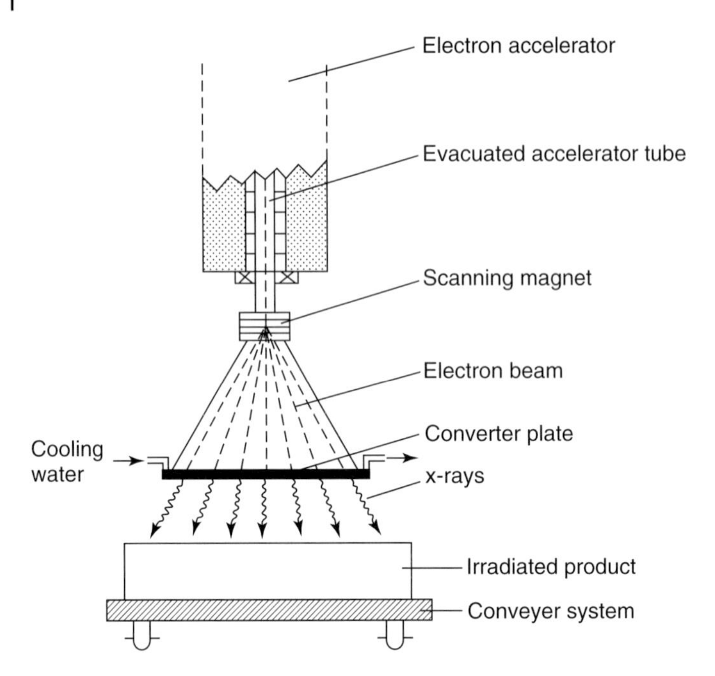

Figure 5.8 Electron accelerator with X-ray converter. (From Diehl [2], with permission.)

ratio of 1 is not achievable in any type of plant, a value of about 1.5–2.0 being a rough guideline for practical applications.

Validation of dose and quality assurance (as well as optimizing plant performance) can be carried out using dosimeters. Routine dosimetry is carried out by attaching dosimeters to packages and then reading on completion of treatment. The dosimeters are usually in the form of plastic strips whose absorption characteristics change in a linear fashion in relation to a given irradiation dose. Radiation-induced changes are measured with a spectrophotometer at the appropriate wavelength. It is essential that routine dosimeters are ultimately related to a primary standard at a specialized national standards laboratory, such as the National Physical Laboratory in the United Kingdom. The primary standard is the actual energy absorbed by water as determined by calorimetry, that is, measurement of the temperature rise, and hence heat absorbed during irradiation.

Chemical dosimetry systems such as the Fricke system can be used as reference systems to ensure the reliability of routine systems. This system is based on the conversion of ferrous to ferric ions in acidic solutions, measured with a spectrophotometer, which is highly accurate but too complex for routine use.

A recent development has been to apply computer processing techniques, such as Monte-Carlo simulation and computed tomography, to generate three-dimensional dose distributions during irradiation of food products and hence allow better control of radiation processing [7].

5.4
Safety Aspects

The safety of consuming irradiated foods is no longer in serious question, and is discussed in other parts of this chapter. However, the safe operation of irradiation facilities warrants further consideration. In particular, precautions must be taken to protect workers, the public, and the environment from accidental exposure. These considerations must encompass the irradiation facilities as well as the transport and disposal of the radioactive materials associated with gamma plants. There are more than 170 gamma plants and 600 electron beam facilities in operation worldwide [2], but only a small fraction of these are used for treatment of food. As with any industrial process, there is the potential for accidents, but radiation processing is one of the most strictly regulated and controlled industries with an excellent safety record [8].

Any design of plant must incorporate a "cell" to contain the radiation source. Commercial plants usually have walls constructed of concrete, while lead may be employed in smaller plants. Gamma sources are held under water when not in use. Irradiators are designed with overlapping protection to prevent leaking of radiation to workers or the public (see Figures 5.4 and 5.5). It is of paramount importance that personnel cannot enter the cell when irradiation is taking place as exposure for even a few seconds could deliver a lethal dose. Precautions against uncontrolled entry, such as interlocks, are therefore an essential feature. Transportation of radioactive source materials is carried out in special casks, which are designed to survive the most severe accidents and disasters. Waste radioactive materials are returned to the manufacturer and either reused or stored until harmless. The quantity of waste is, in fact, very small. The North American Food Irradiation Processing Alliance [9] report that in 40 years of transporting radioactive materials there has never been a problem with the materials, and during 35 years of operating irradiation facilities, there has never been a fatality. This is an excellent record compared to other industries such as those involving shipping toxic materials or crude oil and petroleum products.

It should be remembered that electron and X-ray equipment carries no problems of transporting, storing, or disposal of radioactive materials.

5.5
Effects on the Properties of Food

The radiation doses suitable for food systems cause very little physicochemical damage to the actual food, but potential effects on the major components will be considered.

The mineral content of foods is unaffected by irradiation, although the status of minerals could be changed (e.g., the extent of binding to other chemicals). This could feasibly affect their bioavailability.

The major effect of radiation on carbohydrates concerns the breakdown of the glycosidic bond in starch, pectin, or cellulose to give smaller carbohydrates. This can lead to a reduction in the viscosity of starches, or a loss of texture in some foods, for example leading to softening of some fruits.

Irradiation of proteins at doses up to 35 kGy causes no discernible reduction in amino acid content [10], and hence no reduction in their protein nutritional quality. Changes in secondary and tertiary structures are also minimal, although there may be isolated examples where functionality is impaired. For example, whipping quality of egg whites is impaired following irradiation. As stated previously, effects on enzyme action are very limited and are probably undetectable at the doses likely to be used for foods.

Insignificant changes in the physical (viscosity, melting point) and many chemical properties (iodine value, peroxide number, etc.) of lipids are produced by irradiation up to 50 kGy [10]. The major concern with fat-containing foods is the acceleration of autoxidation when irradiated in the presence of O_2. Unsaturated fatty acids are converted to hydroperoxyl radicals, which form unstable hydroperoxides, which break down to a range of mainly carbonyl compounds. Many of the latter have low odor thresholds and lead to rancid off-flavors. It should be noted that the same end products are found following long storage of unirradiated lipids. This is obviously a concern when irradiating foods containing significant quantities of unsaturated fats, and irradiation and subsequent storage of these foods in the absence of oxygen is advisable.

Loss of vitamins during processing is another concern and has been studied in great detail in a variety of foods [11, 12]. Inactivation results mainly from their reaction with radiolytic products of water, and is dependent on the chemical structure of the vitamin. The degree of inactivation is also determined by the composition of the food as other food components can act as "quenchers," in competition for the reactive products. This has led to an overestimation of vitamin losses in the literature where irradiation of pure solutions of vitamins has been studied. Another consideration is the role of post-irradiation storage where, in common with other processing methods, further breakdown of some vitamins may occur. Hence, losses of vitamins during irradiation vary from vitamin to vitamin and from food to food, and are obviously dose dependent. In addition, subsequent storage or processing of food must be considered.

According to Stewart [1], the sensitivity of the water-soluble vitamins to irradiation follows the order: vitamin B1 > vitamin C > vitamin B6 > vitamin B2 > folate = niacin > vitamin B12, while the sensitivity of fat-soluble vitamins follows the order: vitamin E > carotene > vitamin A > vitamin D > vitamin K.

As a general rule, vitamin losses during food irradiation are quite modest compared to those during other forms of processing. In fact, many studies have demonstrated 100% retention of individual vitamins following low- or medium-dose treatments.

5.6 Detection Methods for Irradiated Foods

There is a clear need to be able to distinguish between irradiated and non-irradiated foods. This would permit proof of authenticity of products labeled as irradiated, or conversely, of unlabeled products which had been irradiated. In addition, it would be useful to be able to estimate accurately the dose to which a food had been treated. Solving these issues would improve consumer confidence in the technology and would benefit international trade in irradiated foods, especially where legislation differs between countries. However, as discussed in Section 5.5, physicochemical changes occurring during food irradiation are minimal, and thus detection is difficult. The changes measured may not be specific to radiation processing, and may alter during subsequent storage. Also the compositional and structural differences between foods, and the different doses required for different purposes, mean that it is unlikely that a universal detection method is possible. Various approaches have been studied in many different foods with some success, and the subject has been reviewed in full by Stewart [13] and Diehl [2]. Methods currently available depend on physical, chemical, biological, and microbiological changes occurring during irradiation, and are summarized only briefly.

Electron spin resonance (ESR) can detect free radicals produced by radiation, but these are very short lived in high moisture foods (see Section 5.2.2). However, in foods containing components with high dry matter, such as bones, shells, seeds, or crystalline sugars, the free radicals remain stable and may be detected. ESR has been successfully demonstrated in some meats and poultry, fish and shellfish, berries, nuts, stone fruit spices, and dried products.

Luminescence techniques are also very promising, and are based on the fact that excited electrons become trapped in some materials during irradiation. The trapped energy can be released and measured as emitted light, either by heat in thermoluminescence (TL), or by light in photostimulated luminescence (PSL). In fact the luminescence is produced in trapped mineral grains rather than the actual food, but successful tests have been demonstrated in a wide range of foods where mineral grains can be physically separated.

Other physical principles, including viscosity changes to starch, changes in electrical impedance of living tissues and near infrared reflectance, hold some promise, but suffer from errors due to variation between unprocessed samples, dose thresholds, and elapsed time from processing.

The most promising chemical methods are the detection of long-chain hydrocarbons and 2-alkylcyclobutanones which are formed by radiolysis of lipids, and are hence limited to foods with quite high fat contents. Numerous other potentially useful chemical changes have been studied with limited success.

DNA is the main target for ionizing radiation, and hence it is logical that DNA damage should be an index for detection. The "comet assay" is used to detect the presence of tails of fragmented DNA produced by irradiation, on electrophoresis gels, as opposed to more distinct nuclei from non-irradiated samples. It is limited by the fact that cooking and other processing produces DNA damage, but is nonetheless promising, as it is applicable to most foods.

Microbiological methods have looked at changes in populations of microorganisms which may have resulted from irradiation. They are nonspecific, but may act as useful screening procedures for large numbers of samples.

A further development is the use of antibody assays in which antibodies are raised to products of radiolysis and incorporated into enzyme-linked immunosorbent assays. These tests are rapid and specific, but not yet in routine use.

5.7 Applications and Potential Applications

In practice, the application of irradiation is limited by legal requirements. Approximately 40 countries have cleared food irradiation within specified dose limits for specific foods. This does not mean, however, that the process is carried out in all these countries, or indeed that irradiated foods are freely available within them, the United Kingdom being a prime example of a country where irradiation of many foods is permitted, but none is actually carried out. The picture is further complicated by trade agreements, labeling requirements, and legal questions, which will not be pursued here. Legislation generally requires that irradiated foods, or foods containing irradiated ingredients, be labeled appropriately. The "radura" symbol (Figure 5.9) is accepted by many countries.

One general principle has been to limit the overall average dose to 10 kGy, which essentially means that radappertization of foods is not an option. This figure was adopted by the WHO/FAO Codex Alimentarius Commission in 1983 [14, 15] as a result of extensive research studying the nutritional, toxicological, and microbiological properties of foods irradiated up to this level. There was never any implication, however, that larger doses were unsafe, and there have been exceptions to this generalization. South Africa has permitted average doses up to 45 kGy to be used in the production of shelf-stable meat products and several countries

Figure 5.9 Radura symbol indicates that a food has been irradiated.

have permitted doses greater than 10 kGy for treatment of spices. A recent joint FAO/IAEA/WHO study group [16] concluded that doses greater than 10 kGy can be considered safe and nutritionally adequate, and it would not be surprising to see more widespread national clearances of higher doses in the future.

Two basic purposes can be achieved by food irradiation:

- extension of storage life; and
- prevention of foodborne illness.

It is difficult to classify the applications of food irradiation as the process may be acting through different mechanisms in different foods or at different doses. Some foods are much more suitable for irradiation than others, and the factors determining shelf life vary between foods. This section will therefore consider the general effects and mechanisms of irradiation, followed by a brief overview of applications in the major food classes.

5.7.1 General Effects and Mechanisms of Irradiation

5.7.1.1 Inactivation of Microorganisms

When a population of microorganisms is irradiated, a proportion of the cells will be damaged or killed, depending on the dose. In a similar way to heat treatment, the number of surviving organisms decreases exponentially as dose is increased. A common measure of the radiation sensitivity in bacteria is the D_{10} value, which is the dose required to kill 90% of the population. Figure 5.10 shows typical survival

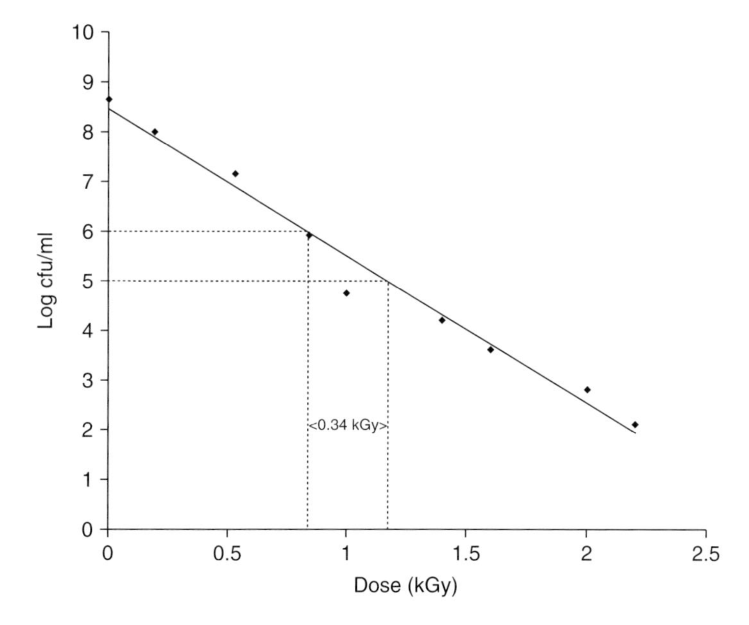

Figure 5.10 Survival curve for electron irradiated *Escherichia coli*.

curve is shown for *Escherichia coli* (data from Fielding *et al.*, [17]) showing a D_{10} value of 0.34 kGy.

Radiation resistance varies widely among different species of bacteria, yeasts, and molds. Bacterial spores are generally more resistant than vegetative cells (see Table 5.1), which is at least partly due to their lower moisture content. Vegetative cells may contain 70% water, whereas spores contain less than 10% water. Hence indirect damage to DNA through the radiolytic products of water is much less likely in spores. An extremely important concept determining the radiation sensitivity of different species and genera of bacteria is their widely varying ability to repair DNA. D_{10} values in the range 0.03–10 kGy have been reported. Some organisms have developed highly efficient repair mechanisms, but fortunately these species are not pathogenic and have no role in food spoilage.

Sensitivity of microorganisms to radiation is also related to environmental conditions. Temperature of irradiation, a_w, pH and the presence of salts, nutrients, or toxins such as organic acids, all exert a great effect on microbial populations after irradiation. The mechanisms of effect may be through modification of the lethality of the applied dose, for example, in dry or frozen environments the effectiveness may be reduced because of suppression of indirect effects caused by radiolytic products of water. Alternatively, the environmental conditions will undoubtedly affect the ability of the organisms to repair themselves, and their ability to reproduce subsequent to treatment. Generally, reducing a_w, by freezing, water removal or addition of osmotically active substances, will increase the resistance of microorganisms, as the secondary effects due to radiolytic products of water will be reduced. The presence of food components generally reduces sensitivity as they can be considered to compete with microorganisms for interaction with the radiolytic products of water. pH also affects lethality in two ways: (i) radiolysis of water is pH sensitive, and (ii) pH changes will affect the general functioning of cells, including the efficiency of DNA repair mechanisms. In addition, recovery of surviving cells following irradiation is pH dependent [17].

The following radiation treatments are aimed at inactivation of microorganisms.

- **Radappertization**: This aims to reduce the number and/or activity of microorganisms to such a level that they are undetectable. Properly packaged radappertized foods should keep indefinitely, without refrigeration. Doses in the range 25–50 kGy are normally required.
- **Radicidation**: This aims to reduce the number of viable spore-forming pathogenic bacteria to an undetectable level. Doses of 2–8 kGy are normally required.
- **Radurization**: This treatment is sufficient to enhance the keeping quality of foods through a substantial reduction in the numbers of viable specific spoilage organisms. Doses vary with the type of food and level of contamination, but are often in the range 1–5 kGy.

5.7.1.2 **Inhibition of Sprouting**

The shelf life of tuber and bulb crops, such as potatoes, yams, garlic, and onions, may be extended by irradiation at low dose levels. Sprouting is the major sign of deterioration during storage of these products, and occurs after a time lag (dormant period) after harvest. The duration of the dormant period differs between different crops, different agricultural practices and different storage conditions, but is usually a number of weeks.

It is believed that the inhibitory effect of irradiation on sprouting results from a combination of two metabolic effects [18]. First, irradiation impairs the synthesis of endogenous growth hormones such as gibberellin and indolyl-3-acetic acid, which are known to control dormancy and sprouting. Secondly, nucleic acid synthesis in the bud tissues, which form the sprouts, is thought to be suppressed. Treatments in the range 0.03–0.25 kGy are effective, depending on the commodity, while higher doses may cause deterioration of the tissue.

5.7.1.3 Delay of Ripening and Senescence

Living fruits and vegetables may be irradiated to extend shelf life by delaying the physiological and biochemical processes leading to ripening. The mechanisms involved are complex and not well understood, and it is probable that different mechanisms predominate in different cultivars.

In some fruits, ripening is associated with a rapid increase in rate of respiration and associated quality changes (flavor, color, texture, etc.) known as the *climacteric*, which more or less coincides with eating ripeness. This represents the completion of maturation, and is followed by senescence. Ripening is triggered by ethylene, which is then produced autocatalytically by the fruit. Climacteric fruit such as tomatoes, bananas, or mangoes, can either be harvested at full ripeness for immediate consumption, or harvested before the climacteric for storage and transport before ripening (either through endogenous or exogenous ethylene). Irradiation can be used either to delay senescence in fully ripe fruit, or to extend the preclimacteric life of unripe fruit. Applied doses of radiation are usually limited to 2 kGy, and often much less, due to radiation injury to the fruit, leading to discoloration or textural damage.

Vegetables and non-climacteric fruit (e.g., citrus, strawberries, cherries) are usually fully mature at harvest, and respiration rate declines steadily thereafter. Irradiation can be used to delay the rate of senescence in these products in some instances, although other purposes such as control of fungi or sprout inhibition may be more important.

5.7.1.4 Insect Disinfestation

Insects can cause damage to food as well as leading to consumer objections. Fortunately insects are sensitive to irradiation (Table 5.1) and can be controlled by doses of 0.1–1.0 kGy. These doses may not necessarily cause immediate lethality, but will effectively stop reproduction and egg development.

5.7.1.5 Elimination of Parasites

Relatively few foodborne parasites afflict humans. The two major groups are single-celled protozoa and intestinal worms (helminths), which can occur in meats, fish, fruit, and vegetables. Irradiation treatment is feasible although not widespread.

5.7.1.6 Miscellaneous Effects on Food Properties and Processing

Although chemical changes to food resulting from irradiation are very limited, it is possible that irradiation could produce beneficial changes to the eating or processing quality of certain foods. There have been reports of improvements in the flavor of some foods following processing, but these are not particularly well substantiated. Chemical changes could result in textural changes in the food. The most likely examples are depolymerization of macromolecules such as starch, which could lead to altered baking performance or changes in drying characteristics. Irradiation may cause cellular injury in some fruits giving rise to easier release of cell contents, and hence increased juice recovery from berry fruits.

Recent research [19] has indicated the potential of using irradiation technology to reduce or eliminate toxic or undesirable compounds in food. This includes structural denaturation of proteins responsible for food allergies, nitrosamine and nitrite reduction, degradation of biogenic amines produced during fermentations, and reduction in phytic acid levels in various foods. Similarly, the technology could be used to modify the colour of foods such as vegetable oils by breaking down chlorophyll.

5.7.1.7 Combination Treatments

Combining food processes is a strategy that permits effective processing while minimizing the severity of treatment. The benefits of combining low-dose irradiation with heat, low temperatures, high pressures, modified atmosphere packaging, or chemical preservatives have been described [20, 21]. In some cases an additive effect of combining processes may be obtained, but synergistic combinations are sometimes observed.

The major drawback of these strategies is the economics of such complex processing regimes.

5.7.2 Applications to Particular Food Classes

A brief outline of the potential application of radiation processing of the major food classes is given. It should be reiterated that the actual practices adopted by different countries usually result from local legislation and public attitudes rather than technical feasibility and scientific evaluation of product quality and safety.

5.7.2.1 Meat and Meat Products

The application of irradiation for control of bacteria and parasites in meat and poultry is discussed in detail by Molins [22]. A major problem with these products is the development of undesirable flavors and odors, which have variously been described as "wet dog" or "goaty." The precise chemical nature of the irradiated flavor is unclear despite much research, but lipid oxidation is a contributing factor. The phenomenon is dose dependent, species dependent, and can be minimized by irradiating at low (preferably subzero) temperatures or under vacuum or in an oxygen-free atmosphere. It is notable that the flavors and odors are transient, and have been shown in chicken and different meats to be reduced or disappear following storage for a period of days. The sensory problems may also be masked by subsequent cooking. The phenomenon clearly limits applications and applied doses even though there is no suggestion of toxicological problems. Appropriate dose–temperature combinations that yield acceptable flavor, while achieving the intended purpose, must be evaluated for any application.

Radiation sterilization of meat and meat products was the objective of much research in the 1950s and 1960s, largely aimed at military applications. The high doses required (e.g., 25–75 kGy) are usually prohibitive on legislative grounds, but the process is technically feasible if combined with vacuum and very low temperatures (say $-40\,^{\circ}C$) during processing [23]. Radiation-sterilized shelf-stable precooked meals incorporating meat and poultry have been produced for astronauts and military purposes, and are produced commercially in South Africa for those taking part in outdoor pursuits.

Effective shelf life extensions of many fresh, cured, or processed meats and poultry have been demonstrated with doses in the range 0.5–2.5 kGy. Increases in shelf life of two to three times or more, without detriment to the sensory quality, are reported [22]. Irradiation cannot, however, make up for poor manufacturing practice. For example, Roberts and Weese [24] demonstrated that quality and shelf life extension of excellent initial quality ground beef patties ($<10^2$ aerobic CFU g^{-1}) was much better than lower initial quality (10^4 CFU g^{-1}) materials when irradiated at the same dose. The former were microbiologically acceptable for up to 42 days at $4\,^{\circ}C$. Also, subsequent handling of irradiated meats or poultry, which may give rise to recontamination following treatment, should be avoided. It is preferable to irradiate products already packaged for retail. Packaging considerations are important for both preventing recontamination and inhibiting growth of contaminating microorganisms. Vacuum or modified atmosphere packaging may be useful options. It is important to note that such products continue to require refrigeration following treatment.

Elimination of pathogens in vegetative cell form is a further aim of irradiating meat and poultry products, and may go hand in hand with shelf life extension. Many studies have reported effective destruction of different pathogens in meat – reviewed by many authors including Farkas [25] and Lee *et al.* [26]. Much work has focused on *Campylobacter* spp. in pork and *Salmonella* spp. in chicken, but recent outbreaks of food poisoning attributed to *Escherichia coli* 0157:H7 in hamburgers and other meats have gained attention and accelerated FDA approval of treatment of red meat in the United States. Fan [27] describes the introduction of irradiated ground beef into the National School Lunch Program in the United States.

5.7.2.2 Fish and Shellfish

Both finfish and shellfish are highly perishable unless frozen on board fishing vessels or very shortly after harvesting. If unfrozen, their quality depends on rapid ice cooling. Irradiation can play a role in preservation and distribution of unfrozen fish and shellfish, with minimal sensory problems. The subject is reviewed extensively by Kilgen [28] and Nickerson *et al.* [29]. The mechanism of action is by reducing the microbial load (hence extending refrigerated shelf life), inactivating parasites and reducing pathogenic organisms. Low- and medium-dose irradiation (up to 5 kGy) has been studied as a means of preserving many varieties of finfish and commercially important shellfish, and shelf life extensions of up to one month under refrigerated storage, have been reported in many species, without product deterioration. Onboard irradiation of eviscerated, fresh, iced fish has been suggested.

Problems have been noted with fatty fish, such as flounder and sole, which must be irradiated and stored in the absence of O_2 to avoid rancidity. Salmon and trout exhibit bleaching of the carotenoid pigments following irradiation, which may limit the application [10]. Combination treatments involving low-dose irradiation (1 kGy) with either chemical dips (e.g., potassium sorbate or sodium tripolyphosphate) or elevated CO_2 atmospheres may be effective for treatment of fresh fish [30].

Strategies for irradiation treatment of frozen fish products, dried fish, fish paste, and other fish products have been investigated [30].

5.7.2.3 Fruits and Vegetables

There are a number of purposes for irradiating fresh fruits and vegetables, including extending shelf life by delay of ripening and senescence, control of fungal pathogens which lead to rotting, and insect disinfestation [31]. Irradiation can also be used to ensure the hygienic quality of minimally processed fruits and vegetables [32]. Radurization and radicidation of processed fruits and vegetables have also been studied. The unusual property of fruits and vegetables is that they consist of living tissue, hence irradiation may effect life changes on the product and such changes may not be immediately apparent, but result in delayed effects.

The main problems associated with these products are textural problems which result from radiation-induced depolymerization of cellulose, hemicellulose, starch, and pectin, leading to softening of the tissue. The effect is dose related and hence severely limits applied doses. Nutritional losses at such doses are minimal. Other disorders include discoloration of skin, internal browning, and increased susceptibility to chilling. Tolerance to radiation varies markedly; for example, some citrus fruits can withstand doses of 7.5 kGy, whereas avocados may be sensitive to 0.1 kGy.

The conditions used for irradiating any fruit or vegetable cultivars are very specific depending on the intended purpose, and the susceptibility of the tissue to irradiation damage. Some specific examples are discussed in detail by Urbain [10] and Thomas [18]. Combination treatments involving low-dose irradiation and hot water or chemical dips, or modified atmospheres, may be effective.

5.7.2.4 Bulbs and Tubers

Potatoes, sweet potatoes, yams, ginger, onions, shallots, and garlic may be treated by irradiation to produce effective storage life extensions due to inhibition of sprouting as described in Section 5.7.1.2. A general guideline for dose requirements is 0.02–0.09 kGy for bulb crops and 0.05–0.15 kGy for tubers

[33]. These low doses have no measurable detrimental effect on nutritional quality, and are too low to produce significant reductions in microbiological contaminants.

5.7.2.5 **Spices and Herbs**

Spices and herbs are dry materials which may contain large numbers of bacterial and fungal species, including organisms of public health significance. Small quantities of contaminated herbs and spices could inoculate large numbers of food portions, and hence decontamination is essential. Chemical fumigation with ethylene oxide is now banned in many countries on account of its toxic and potentially carcinogenic properties, and radiation treatment offers a viable alternative [30]. Fortunately, being dry products, herbs and spices are resistant to ionizing radiation, and can usually tolerate doses up to 10 kGy, in fact higher doses are permitted in some countries. FDA regulations currently permit irradiation of dry or dehydrated spices and seasonings up to 30 kGy in the United States. In general, doses in the range 3–10 kGy are employed, which gives a reduction in aerobic viable count to below 10^3 or 10^4, which is considered equivalent to chemical fumigation. Individual treatments are discussed in detail by Farkas [34].

5.7.2.6 **Cereals and Cereal Products**

Lorenz [35] reviewed the application of radiation to cereals and cereal products. Insects are the major problem during storage of grains and seeds. Disinfestation is therefore the main purpose of irradiation of cereal grain (e.g., wheat, maize, rice, barley). This can be achieved with doses of 0.2–0.5 kGy, with minimal change to the properties of flour or other cereal products [30].

Chemical fumigants, such as methyl bromide gas, are considered a health risk and could be phased out in favor of irradiation disinfestation of grain.

Radurization of flour for bread making, at a dose of 0.75 kGy, to control the "rope" defect caused by *Bacillus subtilis* gives rise to a 50% increase in the shelf life of the resulting bread. However, higher doses lead to reduced bread quality. Alternatively, finished loaves and other baked goods may be irradiated to increase storage life by suppression of mold growth, with a dose of 5 kGy.

5.7.2.7 **Other Miscellaneous Foods**

Radiation processing of most foods has been investigated at some point. Some foods which do not appear in the above categories are described below.

Milk and dairy products are very susceptible to radiation-induced flavor changes, even at very low doses. Significant application to dairy products is therefore unlikely, with the exception of mold-ripened soft cheeses (e.g., Brie and Camembert) produced from raw milk.

Control of *Salmonella* in eggs would be a very beneficial application. However, irradiation of whole eggs causes weakening of yolk membrane and loss of yolk appearance. There may be potential for irradiation of liquid egg white, yolk, or other egg products.

Nuts contain high levels of oil and are again susceptible to lipid oxidation. Low doses to control insects, or sprouting are feasible, but higher doses required to control microbial growth, are less likely.

Ready meals, such as those served on airlines or in hospitals and other institutions, offer a potential application for radiation treatments. The major problem is the variety of ingredients, each with different characteristics and radiation sensitivities. However, careful selection of ingredients and use of low doses and combination treatments may lead to some practical applications.

References

1. Stewart, E.M. (2001) Food irradiation chemistry, in *Food Irradiation: Principles and Applications* (ed. R. Molins), John Wiley & Sons, Ltd, Chichester, pp. 37–76.
2. Diehl, J.F. (1995) *Safety of Irradiated Foods*, Marcel Dekker, New York.
3. Hall, K.L., Bolt, R.O., and Carroll, J.G. (1963) Radiation chemistry of pure compounds, in *Radiation Effects on Organic Materials* (eds R.O. Bolt and J.G. Carroll), Academic Press, New York, pp. 63–125.
4. Pollard, E.C. (1966) Phenomenology of radiation effects on microorganisms, in *Encyclopedia of Medical Radiology*, vol. 2 (2) (ed. A. Zuppinger), Springer-Verlag, New York, pp. 1–34.
5. Brynjolffson, A. (1981) *Combination Processes in Food Irradiation*, Proceedings Series, International Atomic Energy Agency, Vienna, pp. 367–373.
6. Ehlermann, D. (2001) Process control and dosimetry in food irradiation, in *Food Irradiation: Principles and Applications* (ed. R. Molins), John Wiley & Sons, Ltd, Chichester, pp. 387–413.
7. Kim, J., Moriera, R.G., Huang, Y., and Castell-Perez, M.E. (2007) 3-D dose distributions for optimum radiation planning in complex foods. *J. Food Eng.*, **79**, 312–321.
8. IAEA (1992) *Radiation Safety of Gamma and Electron Irradiation Facilities*, Safety Series, vol. 107, International Atomic Energy Agency, Vienna.
9. Food Irradiation Processing Alliance (2001) Food Irradiation: Questions and Answers, Technical Document. *http://www.fipa.us/q%26a.pdf* (accessed January 2011).
10. Urbain, W.M. (1986) *Food Irradiation*, Academic Press, London.
11. World Health Organization (1994) *Safety and Nutritional Adequacy of Irradiated Food*, World Health Organization, Geneva.
12. Thayer, D.W., Fox, J.B., and Lakrotz, L. (1991) Effects of ionising radiation on vitamins, in *Food Irradiation* (ed. S. Thorne), Elsevier Applied Science, London, pp. 285–325.
13. Stewart, E.M. (2001) Detection methods for irradiated foods, in *Food Irradiation: Principles and Applications* (ed. R. Molins), John Wiley & Sons, Ltd, Chichester, pp. 347–386.
14. Codex Alimentarius Commission (1984) Codex General Standard for Irradiated Foods, CAC/vol. XV, E-1, CODEX STAN 106-1983, Joint FAO/WHO Food Standards Programme, FAO, Rome.
15. Codex Alimentarius Commission (1984) Recommended International Code of Practice for the Operation of Radiation Facilities Used for the Treatment of Foods, CAC/vol. XV, E-1, CAC/RCP 19-1979 (rev. 1), Joint FAO/WHO Food Standards Programme, FAO, Rome.
16. World Health Organization (1999) High Dose Irradiation: Wholesomeness of Food Irradiated with Doses above 10 kGy, Technical report series No. 890, World Health Organization, Geneva.
17. Fielding, L.M., Cook, P.E., and Grandison, A.S. (1994) The effect of electron beam irradiation and modified pH on the survival and recovery of

Escherichia coli. *J. Appl. Bacteriol.*, **76**, 412–416.
18. Thomas, P. (2001) Irradiation of fruits and vegetables, in *Food Irradiation: Principles and Applications* (ed. R. Molins), John Wiley & Sons, Ltd, Chichester, pp. 213–240.
19. Byun, M.Y., Jo, C., and Lee, J.-W. (2006) Potential applications of ionizing radiation, in *Food Irradiation Research and Technology* (eds C.H. Sommers and X. Fan), Blackwell, Oxford, pp. 249–262.
20. Campbell-Platt, G. and Grandison, A.S. (1990) Food Irradiation and combination processes. *Radiat. Phys. Chem.*, **35**, 253–257.
21. Patterson, M. (2001) Combination treatments involving food irradiation, in *Food Irradiation: Principles and Applications* (ed. R. Molins), John Wiley & Sons, Ltd, Chichester, pp. 313–327.
22. Molins, R.A. (2001) Irradiation of meats and poultry, in *Food Irradiation: Principles and Applications* (ed. R. Molins), John Wiley & Sons, Ltd, Chichester, pp. 131–191.
23. Thayer, D.W. (2001) Development of irradiated shelf-stable meat and poultry products, in *Food Irradiation: Principles and Applications* (ed. R. Molins), John Wiley & Sons, Ltd, Chichester, pp. 329–345.
24. Roberts, W.T. and Weese, J.O. (1998) Shelf-life of ground beef patties treated by gamma radiation. *J. Food Prot.*, **61**, 1387–1389.
25. Farkas, J. (1987) Decontamination, including parasite control, of dried, chilled and frozen foods by irradiation. *Acta Aliment.*, **16**, 351–384.
26. Lee, M., Sebranek, J.G., Olson, D.G., and Dickson, J.S. (1996) Irradiation and packaging of fresh meat and poultry. *J. Food Prot.*, **59**, 62–72.
27. Fan, X. (2006) Irradiated ground beef for the national school lunch program, in *Food Irradiation Research and Technology* (eds C.H. Sommers. and X. Fan), Blackwell, Oxford, pp. 237–248.
28. Kilgen, M.B. (2001) Irradiation processing of fish and shellfish products, in *Food Irradiation: Principles and Applications* (ed. R. Molins), John Wiley & Sons, Ltd, Chichester, pp. 193–211.
29. Nickerson, J.F.R., Licciardello, J.J., and Ronsivalli, L.J. (1983) Radurization and radicidation: fish and shellfish, in *Preservation of Food by Ionizing Radiation* (eds E.S. Josephson and M.S. Peterson), CRC Press, Boca Raton, FL, pp. 12–82.
30. Wilkinson, V.M. and Gould, G.W. (1996) *Food Irradiation: A Reference Guide*, Butterworth-Heinemann, Oxford.
31. Kader, A.A. (1986) Potential applications of ionising radiation in postharvest handling of fresh fruits and vegetables. *Food Technol.*, **40**, 117–121.
32. International Atomic Energy Agency (2007) *Use of Irradiation to Ensure the Hygienic Quality of Fresh, Pre-cut Fruits and Vegetables and Other Minimally Processed Foods of Plant Origin*, International Atomic Energy Agency, Vienna.
33. Thomas, P. (2001) Radiation treatment for control of sprouting, in *Food Irradiation: Principles and Applications* (ed. R. Molins), John Wiley & Sons, Ltd, Chichester, pp. 241–271.
34. Farkas, J. (2001) Radiation decontamination of spices, herbs, condiments and other dried food ingredients, in *Food Irradiation: Principles and Applications* (ed. R. Molins), John Wiley & Sons, Ltd, Chichester, pp. 291–312.
35. Lorenz, K. (1975) Irradiation of cereal grains and cereal grain products. *CRC Crit. Rev. Food Sci. Nutr.*, **6**, 317–382.

6
High Pressure Processing

Margaret F. Patterson, Dave A. Ledward, Craig Leadley, and Nigel Rogers

6.1
Introduction

In recent years there has been a consumer-driven trend toward better tasting and additive-free foods with a longer shelf life. Some of the scientific solutions, such as genetic modification and gamma irradiation, have met with consumer resistance. However, high pressure processing (HPP) is one technology that has the potential to fulfill both consumer and scientific requirements.

The use of HPP has, for over 50 years, found applications in diverse non-food industries. For example, it has made a significant contribution to aircraft safety and reliability, currently being used to treat turbine blades to eliminate minute flaws in their structure. These microscopic imperfections can lead to cracks and catastrophic failure in highly stressed aero engines. HPP gives a several-fold increase in component reliability, leading to a longer engine life and, ultimately, lower flight cost to the public.

Although the effect of high pressure on food has been known for just over 100 years [1], the technology remained within the R&D environment until relatively recently (Table 6.1). Early R&D pressure vessels were generally regarded as unreliable, costly, and had a very small usable vessel volume. Hence, the prospect of "scaling up" the R&D design to a full production system was commercially difficult. Equally, the food industry, particularly in Europe, has mainly focused upon cost reductions, restructuring, and other programs, often to the neglect of emerging technologies.

A situation had to develop where a need was created that could not be fully satisfied by current technology. This occurred on a small scale in Japan with the desire to produce delicate, fresh, quality, long shelf life fruit-based products for a niche market. The HPP products were, and are today, produced on small- to medium-sized machines at a premium price to satisfy that particular market need.

In the United States, issues relating to food poisoning outbreaks, notably with unprocessed foods such as fruit juices and oysters, led to action by the US Food and Drug Administration (FDA). They attempted to regulate the situation by requiring a significant reduction in the natural microbial levels of a fresh food.

Food Processing Handbook, Second Edition. Edited by James G. Brennan and Alistair S. Grandison.

Table 6.1 The history of high pressure processing food products.

Year	Events
1895	Royer (France) used high pressure to kill bacteria experimentally
1899	Hite (USA) used high pressure for food preservation
1980	Japan started producing high pressure jams and fruit products
1990	Avomex (USA) began to produce high pressure guacamole from avocados with a fresh taste and extended shelf life
2000	Mainland Europe began producing and marketing fresh fruit juices (mainly citrus) and delicatessen-style cooked meats. High pressure self-shucking oysters, poultry products, fruit juices, and other products were marketed in the USA
2001	HPP fruit pieces given approval for sale in the UK. Launch of the first HPP fruit juices in the UK
2009	Over 140 commercial HPP facilities around the world, processing over 100 different foods, including meat, seafood, fruit and vegetable, and ready-to-eat products
	The FDA issue a letter of no objection for shelf-stable mashed potato treated by pressure-assisted thermal processing (PATP)

Foods that did not achieve this reduction were required to be labeled with a warning notice. This meant that producers of freshly squeezed orange juice, for example, marketing their product as wholesome and healthy were faced with a dilemma. If they reverted to the established method of reducing microbial counts, that is, by heating the juice, then the product became of inferior quality in all respects. If they continued to sell the juice "untreated," they faced market challenges with a product essentially labeled to say that it could be harmful to the consumer. Another example involved the producers of avocado-based products who knew that there was a huge demand for fresh quality product. However, the very short shelf life of the product coupled with the opportunity for harmful microbes to flourish with time effectively restricted this market opportunity. Oysters are another case in point. They have traditionally been a niche market with romantic connotations, but food poisoning outbreaks associated with *Vibrio* spp. in the oyster population caused serious illness and even death, which effectively devastated the market.

In all these cases, the producers needed to look for a process that reduced the numbers of spoilage or harmful microorganisms but left the food in its natural, fresh state. HPP has been shown to be successful in achieving this aim.

Today, a wide range of pressure-treated products are commercially available around the world, including fruit juice, guacamole, cooked and ready-to-eat meats, oysters, and other shellfish.

High pressure processing of food is the application of high pressure to a food product in an isostatic manner. This implies that all atoms and molecules in the food are subjected to the same pressure at exactly the same time, unlike heat processing where temperature gradients are established. The second key feature of HPP, arising from Le Chatelier's principle, indicates that any phenomenon that results in a volume decrease is enhanced by an increase in pressure. Thus, hydrogen bond formation is favored by the application of pressure while some of the other weak linkages found in proteins are destabilized. However, covalent bonds are unaffected.

For food applications, "high" pressure can be generally considered to be up to 600 MPa for most food products (600 MPa = 6000 bar = 6000 atm ≅ 87 000 psi). With increasing pressure the food reduces in overall size in proportion to the pressure applied during treatment, but retains its original shape. Hence, a delicate food such as a grape can be subjected to 600 MPa of isostatic pressure and emerge apparently unchanged, although the different rates and extents of compressibility of the gaseous (air), liquid, and solid phases may lead to some physical damage. Pressure kills microorganisms, including pathogens and spoilage organisms, producing a high-quality food with a significantly longer and safer chilled shelf life. The conventional way to do this is to process the food by heat, but this may also damage the organoleptic and visual quality of the food, whereas HPP does not.

In summary, the advantages of HPP are as follows:

- The fresh taste and texture of products such as fruit juices, shellfish, cooked meats, dips, sauces, and guacamole is retained.
- Microbiological safety and shelf life are increased by inactivation of pathogens and spoilage organisms and also some enzymes.
- Innovative foods can be produced, such as gelled products and modification of the properties of existing foods, for example, milk with improved foaming properties.
- There are savings in labor compared with more traditional techniques, for example, self-shucking oysters.
- Energy consumption is low.
- There is minimal heat input, thus retaining fresh-like quality in many foods.
- There is minimal effluent.
- It leads to uniform isostatic pressure and adiabatic temperature distribution throughout the product, unlike thermal processing.

The current disadvantages of HPP are as follows:

- Initial outlay on equipment remains high (in the region of US$0.6–1.8 million for a typical production system). However, numerous companies have justified this cost by offsetting it against new product opportunities, cost-savings in end-of-shelf life costs, removal of preservatives, and supported by the relatively low running cost of the HPP equipment. In 2008, there were over 140 commercial HPP food facilities around the world, many of them in the United States.
- There was initial uncertainty about the ease of gaining approval for new products under the European "novel foods" directive (May 1997). However, it is now understood that for all food products commercially available within the European Union before May 1997, the use of HPP does not make them "substantially different" and HPP technology itself is not a novel process. Therefore, the earlier

concerns expressed about HPP coming under the general novel food regulations are unfounded. Companies thinking about investing in HPP for improvement of existing foods or the creation of new ones based upon already known constituents do not need to apply for any special approvals.

High pressure has many effects on the properties of the food ingredients themselves as well as on the spoilage organisms, food-poisoning organisms, and enzymes. In addition to preserving a fresher taste than most other processing technologies, HPP can affect the texture of foods such as cheese and the foaming properties of milk. This chapter looks at how these effects can make HPP foods more marketable, less labor intensive to produce and generally more attractive to the producer, retailer, and consumer alike.

6.2
Effect of High Pressure on Microorganisms

The lethal effect of high pressure on bacteria is a result of a number of different processes taking place simultaneously. In particular, damage to the cell membrane and inactivation of key enzymes, including those involved in DNA replication and transcription, are thought to play a key role in inactivation [2]. The cell membranes are generally regarded as a primary target for damage by pressure [3]. The membranes consist of a bilayer of phospholipids with a hydrophilic outer surface (composed of fatty acids) and an inner hydrophobic surface (composed of glycerol). Pressure causes a reduction in the volume of the membrane bilayers and the cross-sectional area per phospholipid molecule [4]. This change affects the permeability of the membrane, which can result in cell damage or death. The extent of the pressure inactivation achieved depends on a number of interacting factors as discussed below. These factors have to be considered when designing process conditions to ensure the microbiological safety and quality of HPP foods.

6.2.1
Bacterial Spores

Bacterial spores can be extremely resistant to high pressure, just as they are resistant to other physical treatments, such as heat and irradiation. However, low/moderate pressures are more effective than higher pressures. It was concluded that inactivation of spores is a two-step process involving pressure-induced germination [5]. This has led to the suggestion that spores could be killed by applying pressure in two stages. The first pressure treatment would germinate or activate the spores while the second treatment would kill the germinated spores [6].

Temperature has a profound effect on pressure-induced germination. In general, the initiation of germination increases with increasing temperature over a specified temperature range [7, 8]. Applying a heat treatment before or after pressurization can also enhance spore kill [9]. A small proportion of a spore population will remain resistant to pressure-induced germination [9]. These superdormant spores could be a major limitation for high pressure sterilization, especially in low-acid foods.

In recent years the application of high temperature along with high pressure has been used to develop shelf-stable foods.

Many published approaches for high pressure sterilization utilize the fact that when pressure is applied to a foodstuff, a rapid temperature rise due to compression is observed. By pre-heating a product and then applying pressure, the compression heating in conjunction with the applied pressure could be sufficient to inactivate bacterial spores and produce an ambient shelf-stable product. This process would essentially be a high-temperature short-time process with an additional contribution to lethality from the applied pressure. What makes this approach unique is that it could be applied equally well to a liquid or a solid food product. An additional benefit is that when the pressure is released, the product rapidly cools to its temperature at the start of the pressure cycle. High pressure in combination with elevated temperature offers advantages over other volumetric heating techniques such as ohmic heating or microwave heating because it delivers rapid heating and cooling.

In February 2009, the US National Center for Food Safety and Technology announced that the FDA had accepted a filing for a "pressure-assisted sterilization process" [10]. This is thought to be the first filing of its kind anywhere in the world. Despite this positive development, a number of technical challenges still need to be addressed before pressure-assisted thermal sterilization (PATS) finds widespread commercial use. These principally relate to the need to control temperature variation within the processing vessel so as to ensure that products receive a consistent minimum process regardless of spatial position within the vessel.

6.2.2 Vegetative Bacteria

Pressure treatments normally used for foods (up to 600 MPa) can give a significant reduction in numbers of both pathogenic and spoilage vegetative bacteria. Gram-positive bacteria, especially cocci such as *Staphylococcus aureus*, tend to be more pressure resistant than Gram-negative rods, such as *Salmonella* spp. However, there are exceptions to this general rule. For example, certain strains of *Escherichia coli* O157:H7 are relatively resistant to pressure [11].

One proposed explanation for the difference in pressure response between Gram-positive and Gram-negative bacteria is that the more complex cell membrane structure of Gram-negative bacteria makes them more susceptible to environmental changes brought about by pressure [12].

6.2.3 Yeasts and Molds

Yeasts are generally not associated with foodborne disease. They are, however, important in food spoilage due to their ability to grow in low water activity (a_w) products and to tolerate relatively high concentrations of organic acid preservatives. Yeasts are thought to be relatively sensitive to pressure, although ascospores are more resistant than vegetative bacteria. The age of ascospores can also affect the response to pressure, with older ascospores being more resistant [13].

Unlike yeasts, certain molds are toxigenic and may present a safety problem in foods. In one study, a range of molds including *Byssochlamys nivea*, *B. fulva*, *Eupenicillium* sp., and *Paecilomyces* sp. were exposed to a range of pressures (300–800 MPa) in combination with a range of temperatures (10–70 °C) [14]. The vegetative forms were inactivated within a few minutes using 300 MPa at 25 °C, but ascospores were more resistant. A treatment of 800 MPa at 70 °C for 10 min was required to reduce a starting inoculum of $<10^6$ ascospores ml^{-1} of *B. nivea* to undetectable levels. A treatment of 600 MPa at 10 °C for 10 min was sufficient to reduce a starting inoculum of 10^7 ascospores ml^{-1} of *Eupenicillium* to undetectable levels.

Information on the effect of pressure on preformed mycotoxins is limited. Brâna [15] reported that patulin, a mycotoxin produced by several species of *Aspergillus*, *Penicillium*, and *Byssochlamys* could be degraded by pressure. The patulin content in apple juice decreased by 42, 53, and 62% after 1 h treatment at 300, 500, and 800 MPa, respectively, at 20 °C.

6.2.4 Viruses

There is relatively little information on high pressure inactivation of viruses, compared to the information available on vegetative bacteria. Studies in culture media suggest that a treatment of 600 MPa for 600 s was sufficient to give reductions in viral infectivity titers of >5 $\log_{10}$ for feline calicivirus (a surrogate for norovirus), >3 $\log_{10}$ reduction for hepatitis A virus, and ≤ 2 $\log_{10}$ for poliovirus [16]. Hepatitis A virus, present in contaminated oysters, was also inactivated by pressure. Reductions of >1, >2, and >3 $\log_{10}$ plaque forming units (PFUs) were observed for 1 min treatments at 350, 375, and 400 MPa respectively, within a temperature range of 8.7–10.3 °C [17]. Inactivation of murine norovirus in oyster tissue has been demonstrated, with a 5 min treatment at 5 °C being sufficient to inactivate 4.05 $\log_{10}$ PFU [18]. These results would suggest that the pressure treatments necessary to kill vegetative bacterial pathogens would also be sufficient to cause significant inactivation of human virus particles.

6.2.5 Parasites

There is some evidence that *Toxoplasma gondii* cysts are relatively sensitive to pressure. A treatment of 30–400 MPa for <90 s was sufficient to render ground pork containing viable tissue cysts of the VEG strain of *T. gondii* nonviable in a mouse bioassay study [19].

6.2.6
Strain Variation within a Species

Pressure resistance varies not only between species but also within a species. For example, Linton *et al.* [20] reported a 4 log difference in resistance of pathogenic *E. coli* strains to treatment at 600 MPa for 15 min at 20 °C. Alpas *et al.* [21] also demonstrated variability in pressure resistance within strains of *Listeria monocytogenes, Salmonella* spp., *S. aureus*, and *E. coli* O157:H7. However, they found that the range of pressure differences within a species decreased when the temperature during the pressure treatment was increased from 25 to 50 °C. This finding would be helpful in a commercial situation, where the combination of pressure and mild heat could be used to enhance the lethal effect of the treatment.

6.2.7
Stage of Growth of Microorganisms

Vegetative bacteria tend to be most sensitive to pressure when treated in the exponential phase of growth and most resistant in the stationary phase of growth [22]. When bacteria enter the stationary phase they can synthesize new proteins that protect the cells against a variety of adverse conditions , such as high temperature, high salt concentrations, and oxidative stress. It is not known if these proteins can also protect bacteria against high pressure but this may explain the increase in resistance in the stationary phase.

6.2.8
Magnitude and Duration of the Pressure Treatment

Pressure is similar to heat, in that there is a threshold below which no inactivation occurs. This threshold varies depending on the microorganism. Above the threshold, the lethal effect of the process tends to increase as the pressure increases but not necessarily as the time increases. This can lead to inactivation curves with very definite "tails." These nonlinear inactivation curves have been reported by many workers [11, 12, 23].

Several theories to explain the tailing effect have been proposed. The phenomenon may be independent of the mechanisms of inactivation but be due to population heterogeneity such as clumping or genetic variation. Alternatively, tailing may be a normal feature of the mechanism of resistance (adaptation and recovery) [24]. In practice, the nonlogarithmic inactivation curves make it difficult to calculate accurate D and z values.

6.2.9
Effect of Temperature on Pressure Resistance

The temperature during pressure treatment can have a significant effect on microbial resistance. As a general rule, pressure treatments carried out below 20 °C or above the growth range for the microorganism result in greater inactivation. The inactivation of a murine norovirus in growth medium was enhanced when pressure was applied at an initial temperature of 5 °C; a 5 min treatment of 350 MPa at 30 °C inactivated 1.15 $\log_{10}$ PFU, while the same treatment at 5 °C resulted in a reduction of 5.56 $\log_{10}$ PFU [18]. The simultaneous application of pressure (up to 700 MPa) with mild heating (up to 60 °C) was more lethal than either treatment alone in inactivating pathogens such as *E. coli* O157:H7 and *Staph. aureus* in milk and poultry meat [25].

6.2.10
Substrate

The composition of the substrate can significantly affect the response of microorganisms to pressure and there can be significant differences in the levels of kill achieved with the same organism on different substrates. For example, *E. coli* O157:H7 treated under the same conditions of 700 MPa for 30 min at 20 °C resulted in a 6 log reduction in numbers in phosphate-buffered saline, a 4 log reduction in poultry meat and a <2 log reduction in UHT milk [11]. The reasons for these effects are not clear but it may be that certain food constituents like proteins and carbohydrates can have a protective effect on the bacteria and may even allow damaged cells to recover more readily.

There is evidence that the a_w and pH of foods can significantly affect the inactivation of microorganisms by pressure. A reduction in a_w appears to protect microorganisms from pressure inactivation. Oxen and Knorr [26] reported that in sucrose solution ($a_w \sim 0.98$), the pressure inactivation (at 200–400 MPa) of *Rhodotorula rubra* was independent of pH at pH 3.0–8.0. However, at a_w values below 0.94 there was a protective effect, irrespective of solute (glucose, sucrose, fructose, or sodium chloride). Most microorganisms are more susceptible to pressure at lower pH values and the survival of pressure-damaged cells is less in acidic environments. This can be of commercial value, such as in the pressure treatment of fruit juices where, in the high acid conditions, pathogens, such as *E. coli* O157:H7, which may survive the initial pressure treatment, will die within a relatively short time during cold storage [27].

6.2.11
Combination Treatments Involving Pressure

High pressure processing can be successfully used in combination with other techniques to enhance its preservative action and/or reduce the severity of one or all of the treatments. It is possible that this hurdle approach will be used in many of the commercial applications of HPP technology. Beneficial combination treatments include the use of pressure (usually <15 MPa) with carbon dioxide to improve the microbial quality of chicken, egg yolk, shrimp, orange juice [28], and fermented vegetables [29]. Pressure combined with irradiation has been proposed to improve the microbial quality of lamb meat [30] and poultry meat [31]. Pressure combined with heat and ultrasound has been successful in inactivating *B. subtilis* spores [32]. Several studies have also reported synergism between HPP and bacteriocins, mainly nisin, for the inactivation of pathogens in a range of foods, including cheese [33], sausage [34], and cooked ham [35].

The commercial value of these combinations still has to be assessed, given that one of the advantages of HPP is that it can be regarded as a "natural" minimal processing technology.

6.2.12 Effect of High Pressure on the Microbiological Quality of Foods

HPP is already used commercially to enhance the microbiological quality of certain food products. Fruit products have been most extensively studied. These products are acidic so, in terms of their microbiology, pathogens are generally not so important, but spoilage microorganisms, particularly yeasts and mold, are of concern. The limiting factor for shelf life of such products is often the action of enzymes, particularly those which can cause browning, although the problem may be at least partially overcome by blanching or adding an oxygen scavenger such as ascorbic acid. Most fruit products are given a treatment of 400–500 MPa for up to a few minutes. Pressure treatment of vegetables tends to be less successful due to their higher pH and the potential presence of spores, which can be very pressure resistant. In addition, the quality of some vegetables deteriorates as a result of pressure processing, which can also be a limiting factor [36]. However, it should be noted that one of the most successful commercial products available to date is HPP guacamole. Research has shown that clostridial spores cannot outgrow in this product and it can have a shelf life of around one month at 4 °C without modification of color, texture, or taste.

The use of HPP to improve the microbiological quality of meat, fish, and dairy products has been investigated by a number of workers. These products tend to have a more neutral pH and provide a rich growth medium for most microorganisms, with pathogens being of particular concern. The need to ensure microbiological safety is one of the main drivers for pressure treatment of cooked and ready-to-eat meats, where the treatment is used as an additional control measure for any *L. monocytogenes* which may be present [37]. Research has also shown that pressure processing can be successful, in terms of improving microbiological safety and quality, for the treatment of pork [38], minced beef [39], duck foie gras (liver pâté) [40], fish [41], ovine milk [42], and liquid whole egg [43]. In many cases, the authors also comment on the ability of the pressure treatment to maintain or enhance sensory, nutritional, or functional quality compared to conventional processing methods.

In all cases, the optimum treatment conditions need to be carefully defined and thoroughly tested to ensure food safety is not compromised. This may include extensive inoculation studies, under standardized conditions and using the most resistant strains of pathogens to ensure that the product will be microbiologically safe during its shelf life.

6.3
Ingredient Functionality

It is now well established [44, 45] that changes in protein structure and functionality occur during high pressure treatment. Studies carried out on volume changes in proteins, have shown that the main targets of pressure are hydrophobic and electrostatic interactions [46, 47]. Hydrogen bonding, which stabilizes the α helical and β pleated sheet forms of proteins, is almost pressure insensitive. At the pressures used in food processing, covalent bonds are unaffected [48] but at pressures of about 300 MPa sulfhydryl groups may oxidize to S–S bonds in the presence of oxygen.

It is readily apparent from the above that pressure and temperature do not normally work synergistically with respect to protein unfolding, since the weak linkages that are most labile to heat, that is, hydrogen bonds, are stabilized or only marginally affected by pressure, while the bonds most labile to pressure (electrostatic and hydrophobic interactions) are far less temperature sensitive. However, in the presence of oxygen, both increasing temperature and pressure encourage disulfide bond formation and/or interchange. Such considerations help to explain the effects of pressure and temperature on the phase diagram for most, if not all, native/denatured protein systems (Figure 6.1), where, for example, it is seen that, up to a certain temperature, pressure stabilizes the protein against heat denaturation. On removal from the denaturing environments, the proteins will, if free to do so, refold to a native-like structure. However, because the pressure/temperature dependences of many weak linkages differ so markedly, on pressure release the "reformed" structures often differ from that of the native structure, that is, conformational drift occurs. Thus the amount of α helix, β sheet, and random coil

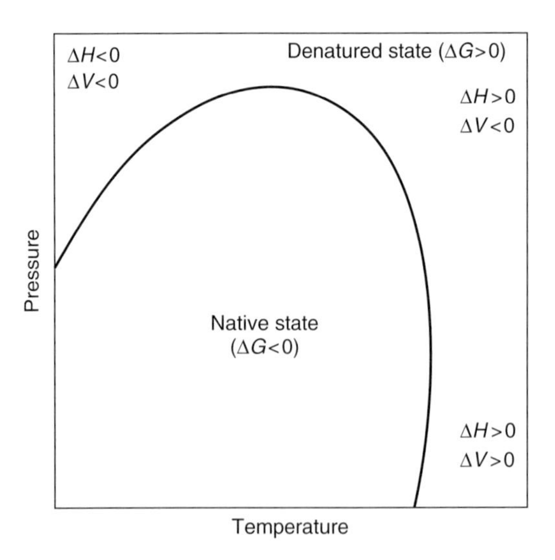

Figure 6.1 Phase diagram for native/denatured proteins. Most proteins denature in the range 40–80 °C.

present will vary, as also will such properties as surface hydrophobicity and charge. As these "renatured" pressure-treated proteins yield such different structures, it is not surprising that they exhibit marked differences in behavior to the native protein or its heat-denatured product. Thus their functional properties vary.

6.4 Enzyme Activity

Enzymes, being proteins, will at sufficiently high pressure undergo conformational changes and thus lose activity. Most enzymes of importance in food deterioration are relatively resistant to pressure and complete inactivation is difficult to achieve. Thus polyphenoloxidases (PPOs), the enzymes responsible for browning in fruits and vegetables, require pressures of 800 MPa, at room temperature or more to bring about complete inactivation. The degree of inactivation is usually dependent on pressure, temperature, and time and, not unnaturally, pH is a further very important factor [49].

A review by Ludikhuyze *et al.* [50] has summarized present knowledge regarding the combined effects of pressure and temperature on those enzymes that are related to the quality of fruits and vegetables. The enzymes of importance include the PPOs, pectin methylesterases, which induce cloud loss and consistency changes, and lipoxygenases, which are responsible for the development of off-flavors, loss of essential fatty acids and color changes. These authors discuss how complete characterization of the pressure/temperature phase diagram for inactivation of these enzymes has been achieved and also suggest how this information could be useful in generating integrated kinetic information with regard to process engineering associated with HPP treatment. Interestingly, these authors found that, for most of the enzymes so far studied, the reaction kinetics were first order, the only exception being pectin methylesterase which only gave fractional conversion and thus the kinetics were difficult to resolve. As discussed in Section 6.3, these studies showed that most enzymes (proteins) have maximum stability to pressure at temperatures around 25–40 °C.

In these model systems, the loss of activity is invariably due to a change in the conformation of the protein, that is, unfolding/denaturation, which tends to be irreversible. However, although the loss of enzyme activity in the majority of cases is associated with major structural changes, this is not always the case. For example, the secondary and tertiary structure of papain is little affected by pressures up to 800 MPa [50], but a significant loss of activity is observed on treatment at these pressures. This loss in activity can be largely inhibited by applying pressure in the absence of oxygen, since treatment at 800 MPa in air causes a loss in activity of about 41%, but only 23% of the initial activity is lost on pressure treatment after flushing with nitrogen; and, after flushing with oxygen, the loss in activity is 78% at 800 MPa. Gomes *et al.* [51] suggested that, in this case, the loss in activity is related to specific thiol oxidation at the active site. The active site in papain contains both a cysteine group and a histidine group and, at pH 7, they exist as a relatively stable S^{-}–N^{+} ion pair. In an aqueous environment, pressure causes this linkage to rupture due to the associated decrease in volume due to electrostriction of the separated charges. However, steric considerations mean the S^- ion cannot form a disulfide linkage and thus, in the presence of oxygen, the ion oxidizes to the stable SO_3 [52].

Since many enzymes are relatively difficult to denature, it is not surprising that, when whole foods are subjected to pressure, the effects on enzyme activity are difficult to predict. Thus many fruits and vegetables, when subjected to pressure, undergo considerable browning since at pressures below that necessary to inactivate the enzyme some change occurs which makes the substrate more available [53]. For this reason, combined with the cost of the process, high pressure is an effective means of preventing enzymic browning only if applied to the food with appropriate control. Although the application of high pressure may well accelerate the activity of PPO in some fruits and vegetables, it can be controlled and thus in the commercial manufacture of avocado paste (guacamole) treatment at 500 MPa for a few minutes is adequate to extend the color shelf life of the product so that it has a shelf life at chill temperatures of eight weeks compared to a few hours under the same conditions if not subjected to pressure. From the previous discussion, it is apparent that 500 MPa does not fully inactivate PPO but it brings about sufficient decrease in activity to permit the extended shelf life. However, where modification of enzymic activity is required, pressure treatment may be of benefit as in optimizing protease activity in meat and fish products [54] or in modifying the systems that affect meat color stability. For example, Cheah and Ledward [55, 56] have shown that subjecting fresh beef to pressures of only 70–100 MPa leads to a significant increase in color stability due to some, as yet unidentified, modification of an enzyme-based system that causes rapid oxidization of the bright red oxymyoglobin to the brown oxidized metmyoglobin.

Enzymes, as well as being responsible for many color changes in fruit, vegetables, and meat systems, are also intimately related to flavor development in fruits and vegetables. Thus, lipoxygenase plays an important role in the genesis of volatiles [57], as this enzyme degrades linoleic and linolenic acids to volatiles such as hexanal and *cis*-3-hexenal. The latter compound transforms to *trans*-2-hexenal which is more stable. These compounds are thought to be the major volatile compounds contributing to the fresh flavor of blended tomatoes [58]. Tangwongchai *et al.* [59] reported that pressures of 600 MPa led to a complete and irreversible loss of lipoxygenase activity in cherry tomatoes when treated at ambient temperature. This loss of enzyme activity resulted in flavor differences between the pressure-processed tomato and the fresh product. Compared to unpressurized tomatoes, treatment at 600 MPa gave significantly reduced levels of hexanal, *cis*-3-hexenal, and *trans*-2-hexenal, all of which are important contributors to fresh tomato flavors. It is well established that high pressure can very satisfactorily maintain the flavor quality of fruit juices, as well as their color quality, but obviously with regard to the fresh fruit, differences become apparent and it is likely therefore that the technology will not be of benefit for some whole fruits, although its use for fruit juices cannot be disputed.

As well as being involved in flavor development in fruits and vegetables, enzymes are intimately involved in the textural changes that take place during growth and ripening. The enzymes primarily responsible are believed to be polygalacturonase and pectin methylesterase. Tangwongchai *et al.* [60] showed that, in whole cherry tomatoes, these two enzymes are affected very differently by pressure. Although a sample of purified commercial pectin methylesterase was partially inactivated at all pressures above 200 MPa, irrespective of pH, in whole cherry tomatoes no significant inactivation was seen even after treatment at 600 MPa, presumably because other components in the tomato offered protection, or the isoenzymes were different. Polygalacturonase was more susceptible to pressure, being almost totally inactivated after treatment at 500 MPa. It is interesting to note that these authors observed, both visually and by microscopy, that whole cherry tomatoes showed increasing textural damage with increasing pressures up to about 400 MPa. However, at higher pressures (500–600 MPa) there was less apparent damage than that caused by treatment at the lower pressures, the tomatoes appearing more like the untreated samples. These authors concluded that the textural changes in tomato induced by pressure involve at least two related phenomena. Initially, damage is caused by the greater compressibility of the gaseous phase (air) compared to the liquid and solid components, giving rise to a compact structure which on pressure release is damaged as the air rapidly expands, leading to increases in membrane permeability. This permits egress of water and the damage also enables enzymic action to increase, causing further cell damage and softening. The major enzyme involved in the further softening is polygalacturonase (which is inactivated above 500 MPa) and not pectin methylesterase (which in the whole fruit is barotolerant). Thus, at pressures above 500 MPa, less damage to the texture is seen.

From the brief overview above it is apparent that high pressure has very significant effects on the quality of many foods, especially fruits and vegetables, if enzymes are in any way involved in the development of color, flavor, or texture.

Although to date pressure has largely been concerned with the preservation of quality, either by inhibiting bacteria or inactivating enzymes, it does offer potential as a processing aid in assisting reactions that are pressure sensitive. An example that may have commercial application is that moderate pressures (300–600 MPa) cause significant increases in the activity of the amylases in wheat and barley flour, because the pressure-induced gelatinization of the starch makes it, the starch, more susceptible to enzymic attack [61]. Higher pressures lead to significant decreases in activity due to unfolding and aggregation of the enzymes. This technology thus has potential in producing glucose syrups from starch by a more energy-efficient process than heat.

The effects of pressure on enzyme activity suggest that it may well be a very effective processing tool for some industrial applications in the future.

6.5 Foaming and Emulsification

The structural changes undergone by a protein will affect its functionality. For example, pressure-treated β-lactoglobulin at 0.01% concentration has significantly

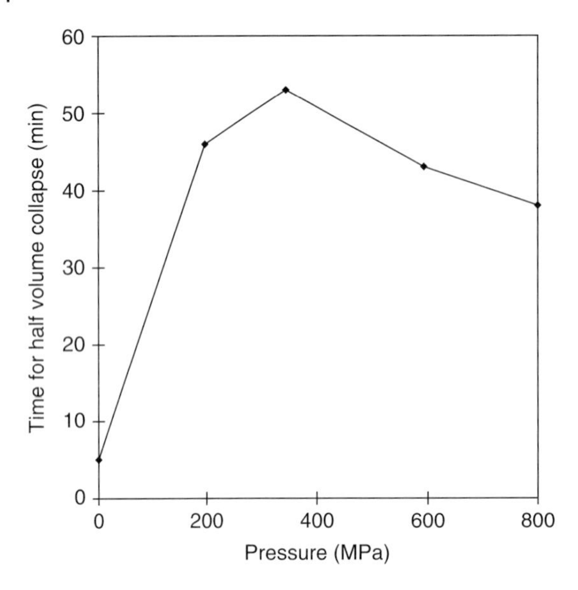

Figure 6.2 Effects of high pressure processing treatment on the foam stability of β-lactoglobulin at pH 7.0. The time for half volume collapse is plotted as a function of pressure applied for 20 min. From Galazka *et al.* [62].

improved foaming ability compared to its native counterpart (Figure 6.2). However, if pressure is applied to a relatively concentrated solution of β-lactoglobulin (0.4%), disulfide bond formation may lead to significant aggregation so that it is less useful as a functional ingredient. If disulfide bond-induced aggregation does occur on pressure treatment, then only dilute solutions should be treated, to avoid loss of functionality due to increased size. In addition, other factors that aid aggregation or disulfide bond formation, such as pressures above that necessary to cause unfolding, extended treatment times, alkaline pH, and the presence of oxygen, should be avoided so as to optimize functionality.

In addition to β-lactoglobulin, many studies have been reported on the effect of high pressure on the emulsifying and foaming properties of other water-soluble proteins, including ovalbumin, vegetable proteins such as soy and pea, and casein. One advantage of pressure treatment on proteins to improve or modify their functional properties is that the process is invariably easier to control than thermal processing, where the effects are rather drastic and less easy to control. For example, model emulsions prepared with high pressure-treated (<600 MPa) protein after homogenization show that pressurization induced extensive droplet flocculation which increased with protein concentration and severity of treatment [63]. However, it was also noted that moderate thermal processing (80 °C for 5 min) had a far greater effect than pressure treatment at 800 MPa for 40 min on the state of flocculation of ovalbumin-coated emulsion droplets. The increase in emulsion viscosity is due to the formation of a network from the aggregated dispersed oil droplets and denatured polymers in aqueous solution.

The level of pressure-induced modification can be controlled more efficiently by altering the intensity of high pressure treatment rather than by controlling the temperature in thermal processing. Thus, HPP can be viewed as a novel way of manipulating the microstructure of proteins such as ovalbumin, while maintaining the nutritional value and natural flavor of such compounds.

As with ovalbumin, emulsions made with pressure-treated 11S soy protein were found to have poorer emulsifying and stabilizing ability with respect to initial droplet size and creaming behavior than the native protein. This is probably due to the enhanced association of subunits and/or aggregation induced by the formation of intermolecular disulfide bridges via SH/–S–S interchange. As with ovalbumin, moderate heat treatment (80 °C for 2 min) had a far greater effect than high pressure treatment on the changes in emulsion stability and droplet size distribution.

Though only in its infancy, the ability of HPP to modify the structure and surface hydrophobicity of a protein does suggest that, as well as a preservation technique, HPP may well have commercial/industrial application for modifying the functional properties of potential emulsifiers and foaming agents.

6.6 Gelation

At sufficiently high concentrations and at the appropriate pH many proteins, especially if they have potential disulfide bond-forming abilities, will gel or precipitate, but the texture of the gels formed will be markedly different to their heat-set counterparts. Such pressure-set gels will normally contain a relatively high concentration of hydrogen-bonded structure(s) and thus will melt or partially melt on heating. In addition, they will be less able to hold water, that is, they will synerese and be much softer in texture and "glossier" in appearance. For example, myoglobin (a protein with no amino acids containing sulfur) will unfold at sufficiently high pressure on pressure release and then normally revert back to a soluble monomer or dimer; but at its isoelectric point (pH 6.9) it will precipitate. However, unlike the precipitate formed on heat denaturation, it will be relatively unstable and, because it is primarily stabilized by hydrogen bonds, will dissolve or melt on gently raising the temperature [62].

If, as suggested, pressure-set gels are stabilized, at least to some extent by hydrogen bonds, then it would be expected that heat treatment of such a system will destroy this network and enable a "heat" set gel to form. Such effects are apparent in both whey protein concentrate gels at pH 7 and in the toughness or hardness of fish and meat flesh, that is, myosin gels. For example, calorimetric studies on fish and meat myofibrillar proteins and whole muscle have clearly demonstrated the presence of a hydrogen-bonded network in pressure-treated muscle, myofibrillar protein and myosin [64, 65] that is destroyed on heat treatment. Just as the thermal stability of different myosins reflects the body temperature of the species, that from cod being less stable than that from beef or pork, so their relative pressure sensitivities also vary, with cod myosin denaturing at about 100–200 MPa at 20 °C, and that from turkey and pork only unfolding at pressures above 200 MPa. Thus, the simultaneous or sequential treatment of proteins with heat and pressure does raise the possibility of generating gels with interesting and novel textures. The likely mechanisms are discussed in more detail by Ledward [66].

As well as being able to modify the functional properties of water-soluble proteins such as β-lactoglobulin and myoglobin, and generate heat-sensitive gels in proteinaceous foods such as meat and fish, high pressures can also be used to texturize many insoluble plant proteins such as gluten and soya. An extensive study has been carried out by Apichartsrangkoon *et al.* [67, 68] on the use of various pressure/temperature treatments to texturize wheat gluten and soy protein and it can be seen from Figure 6.3 how pressure/temperature and time can be used to generate a range of soya gels of different rheological properties.

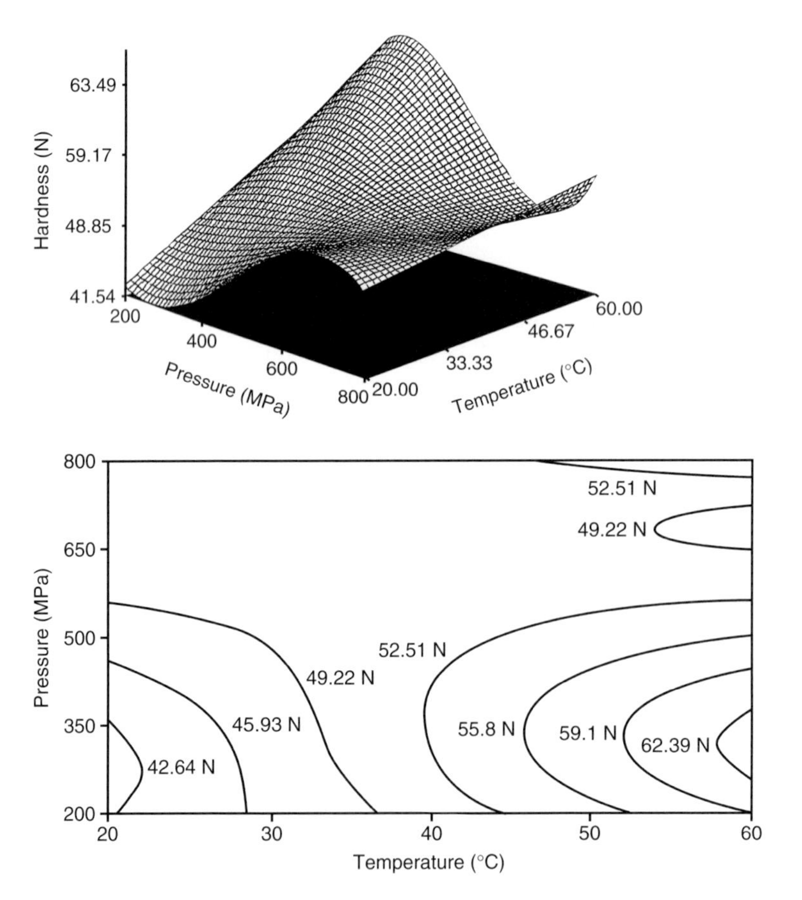

Figure 6.3 Response surface and contour plots of the hardness of soy protein gels prepared from a commercial soy concentrate mixed with 3.75 times its weight of water after treatment at temperatures of 20–60 °C and pressures up to 800 MPa for 50 min. From Apichartsrangkoon [69].

The above are but a few examples of the rapidly expanding literature on the use of pressure and/or temperature to generate gels with very different rheological properties. The potential of such technology in the development of new proteinaceous foods is very exciting.

From the above brief review it is readily apparent that, although our knowledge of the effects of pressure on proteins has advanced in recent years, a great deal more research is needed before our understanding approaches that of the effect of other parameters on proteins, such as temperature and pH.

6.7 Organoleptic Considerations

Since covalent bonds are unaffected by pressure, many of the small molecules that contribute to the color, flavor, or nutritional quality of a food are unchanged by pressure. This is a major advantage of the process and has led to its successful application to such products as guacamole, fruit juices, and many other fruit-based jams and desserts.

However, if the organoleptic quality of a food depends upon the structural or functional macromolecules, especially proteins, pressure may affect the quality. The most obvious case of this is with meat and fish products where, at pressures above ~100 MPa for coldwater fish and 300–400 MPa for meats and poultry, the myosin will unfold/denature and the meat or fish will take on a cooked appearance as these proteins gel. In addition, in red meats such as beef and lamb, myoglobin, the major heme protein responsible for the red color of the fresh product, will denature at pressures around 400 MPa and give rise to the brown hemichrome, further contributing to the cooked meat appearance. However, since the smaller molecules are not affected, the flavor of the fish or meat will be that of the uncooked product. On subsequent heat treatment, these foods will develop a typical cooked flavor.

Although in many circumstances the flavor of a food will remain unchanged after pressure treatment, Cheah and Ledward [55] and Angsupanich and Ledward [64] have shown that, under pressure, inorganic transition metals (especially iron) can be released from the transition metal compounds in both meat and fish. These may catalyze lipid oxidation and thus limit shelf life and may also contribute to the flavor when the product is subsequently cooked. This release of inorganic iron takes place at pressures above 400 MPa and does not appear to be a problem with cured meats or shellfish. This phenomenon may limit the usefulness of HPP for many uncured fish and meat products.

Since enzyme activity is affected by pressure, many foods whose flavor, texture, or color is dependent on enzymic reactions, may have their sensory properties modified on pressure treatment. Also, in multiphase foods, such as fruits and vegetables, pressure can give rise to significant textural changes both due to modification of the enzyme activity and because of the different rates and extents of compression and decompression of the aqueous, solid, and gaseous phases. This may lead to physical damage, as described above for cherry tomatoes.

6.8 Equipment for HPP

The key features of any HPP system used for food processing include: a pressure vessel, the pressure transmission medium, and a means of generating the pressure. The emergence of production-sized HPP equipment for food occurred more or less simultaneously in the United States and Europe in the mid-1990s.

For convenience, potable water is used as the pressurizing medium. The water is usually separated from the food by means of a flexible barrier or package forming the actual consumer pack. Products such as shellfish, however, are usually immersed directly into the water as beneficial effects upon yield can be achieved. Up to the minute details of HPP technology and applications can be found on *www.avure.com*

6.8.1 HPP Systems

These vessels, typically 35–680 l in capacity, use water as the pressurizing medium and accept prepackaged or unpackaged food products (Figure 6.4).

When a high water content food is subjected to isostatic pressure up to 600 MPa, it compresses in volume by 10–15%, depending on the food type and structure. As the pressurizing medium is water, the food packaging must be capable of preventing water ingress and accepting the volume reduction.

The packaging has to withstand the pressure applied but it should be noted that standard plastic bottles, vacuum packs, and plastic pouches are usually suitable. These flexible materials have essentially no resistance to volume under the effects of high pressure. Hence, a plastic bottle of fruit juice, having a small (standard) headspace, will compress by the volume of the headspace plus the compressibility of the juice. Provided that the pack is resilient enough to withstand the volume reduction imposed, then the pressurizing water will be kept out of the bottle. Once depressurized, the bottle and contents will revert to their original state, minus the spoilage organisms.

Figure 6.4 Illustration of a typical high pressure processing system.

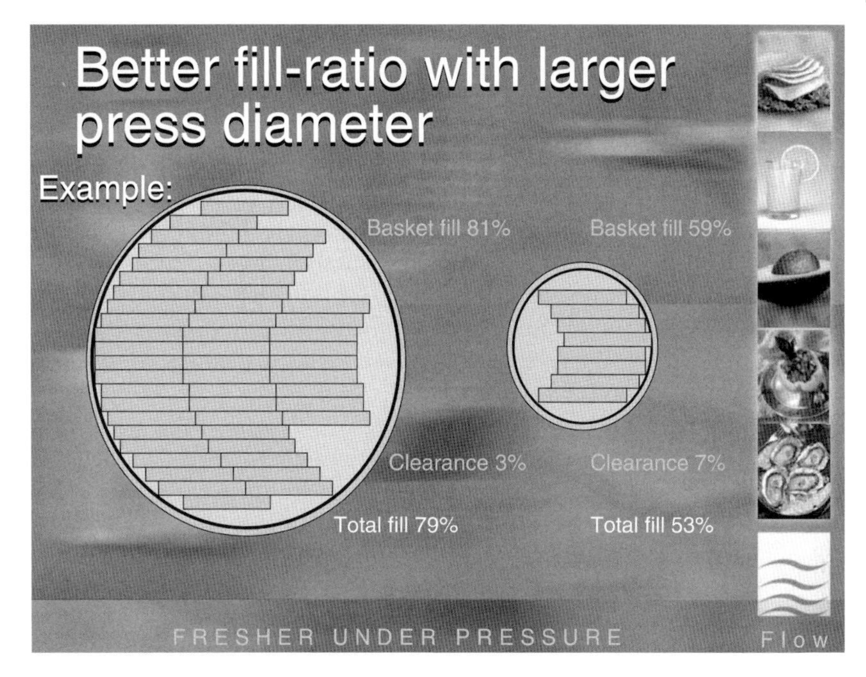

Figure 6.5 Schematic representation showing how a better fill ratio is attainable with a larger pressure vessel.

A large vessel size gives better utilization of the available space; and therefore the greatest production output per batch cycle. A 350 l vessel could be filled with a 350 l plastic bag of food, therefore obtaining the maximum use of available volume. Normally, however, the batch vessel will be used for a variety of prepacked consumer food products of different size and styles. Fruit juice bottles, for example, will not normally pack together without gaps and therefore the effective output of the vessel will reduce accordingly (Figure 6.5).

Batch systems are ideal where wide ranges of products are required and the products comprise solids and liquids combined. Cooked meats, stews, guacamole, fruits in juice, shellfish, and ready meals are typical batch HPP examples. Batch systems filled with consumer-type packs of food benefit from both the packaging and the food being HPP treated. This considerably aids quality assurance through distribution and shelf life and may reduce or simplify some production processes.

6.9 Pressure Vessel Considerations

In a commercial environment, a HPP system is likely to be run at least 8 h day^{-1} and so the design criteria are critical. "Standard" pressure vessels use a thick steel wall construction, where the strength of the steel parts alone is used to contain the pressure within. This is fine for a vessel expected to perform a low number of cycles. However, every time the pressure is applied, the steel vessel expands slightly as it takes up the strain. This imposes expansion stresses within the steel structure and, with time, can lead to the creation of microscopic cracks in the steel itself. If undetected, it is possible for the vessel to fail and rapidly depressurize, causing a safety hazard and destroying the vessel.

The technique used to create a long life and guaranteed "leak before break" vessel is known as "*wire winding*." In this, a relatively thin-walled pressure vessel (too thin to contain the working pressure of its own) is wrapped in steel wire. The wire wrapping, which can amount to several hundred kilometers of wire for a large vessel, is stretched and wrapped onto the vessel by a special machine and powered turntable.

The wire attempts to return to its unstretched state and in doing so exerts a compressive force upon the pressure vessel. The compressive force of the wire wrapping is engineered to slightly exceed the expansion force of the HPP water pressure. Hence, the thin-walled pressure vessel, even with 600 MPa water pressure within it, is actually still under compression from the wire windings. This means that the only forces within the steel of the pressure vessel are compressive and therefore cracks cannot propagate.

A wire-wound frame is made to hold in place "floating" end closures, so as to allow access to the vessel for loading and unloading. The frame construction is similar to that described for the vessel (Figure 6.6). A wire-wound vessel has theoretically unlimited life expectancy.

High-grade forged pressure vessel steel is not the preferred material for contact with foods and so the wire-wound vessel incorporates an inner liner of stainless steel.

6.9.1 High Pressure Pumps

Normal pumps cannot achieve the required pressures for HPP food applications. Therefore, a standard hydraulic pump is used to drive an intensifier pump, comprising a large piston driven back and forth by the hydraulic oil in a low pressure pump cylinder. The large piston has two smaller pistons connected to it, one each side, running in high pressure cylinders. The ratio of large and small piston areas and hydraulic pump pressure gives a multiplication of the pressure seen at the output of the high pressure cylinders. The small, high pressure pistons are pumping the potable water used as the pressurizing medium in the HPP food process.

The pressure and volume output from the intensifier is dependent upon the overall sizes of the pistons and hydraulic pressure. However, it is often more convenient to use several small intensifiers working "out of phase" to give a smooth pressure delivery and some degree of redundancy, rather than use a single, large intensifier.

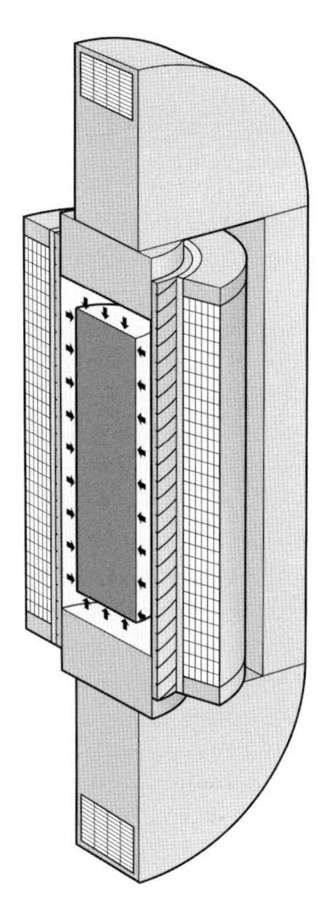

Figure 6.6 A wire-wound pressure vessel.

6.9.2
Control Systems

High pressure processing systems are mainly sequentially operated machines and so are controlled by standard product life cycles (PLCs). Verification of each sequence before proceeding to the next is essential in some cases, due to the very high pressures involved and the materials used for the sanitization of food processes. Many critical sequences will include "fail safe" instrumentation and logic.

The user interface forms an important part of any process and the demands today are for even the most complex machine or system to be operable by non-engineering operatives. The operator is therefore presented with a screen or monitor with the process parameters and operator requirements presented in a format similar to that of a home PC running the most popular software. System logic design and password protection means that any suitably trained person can carry out normal production and maintenance.

Fault conditions are automatically segregated into those that the operator can deal with and those that dictate either the automatic safety shutdown of the system (or part thereof) or a sequence halt. In every case, the fault and probable solution is available to the user.

6.10
Current and Potential Applications of HPP for Foods

France was the first country in the European Community to have HPP food products commercially available in 1994. Orange juice was the main product, although some lemon and grapefruit juice was also produced. Their motivation for moving into HPP was a desire to extend the shelf life of their fresh fruit juice, then only 6 days at chilled temperature. High pressure treatment allowed a shelf life of up to 16 days. This reduced logistical problems and transportation costs, without harming the sensory quality and vitamin content of the juice. HPP-processed fruit juice and fruit smoothies have also been available in the EU since 2002.

In Europe, HPP products are widely available and sold as branded and own-label products in many of the most popular supermarket chains. These include ready-to-eat meats, juices, lobster, ready meals, and fish.

HPP guacamole, with a chill shelf life of 35 days, produced by Fresherized Foods Inc., has been commercially available for many years in the United States. It is actually manufactured across the border in Mexico to take full advantage of low raw material costs and avoid the duty associated with the import of raw avocados to the United States. Its market share continues to grow and is based on the consumer preference for the "fresher" taste of guacamole processed in this manner. The same company now operates 12 commercial HPP systems and "toll-processes" for other companies.

"Gold Band" oysters from Motivatit Seafoods Inc. (*www.theperfectoyster.com*) are achieving top national awards in the United States for HPP shucked oysters. They report up to 75% yield increase with HPP compared to manual shucking processes and now "contract shuck" oysters for other famous oyster companies in the region. Shellfish shucking using HPP is also found in Australia, New Zealand, Canada, and Europe and is subject to international patents held by Avure Technologies Inc.

Throughout the world, many commercial added-value products now benefit from HPP, from guacamole to dips, wet salads, raw and cooked meats, sausages, tapas, fish, shellfish, lobster, crab, mussels, fruit and vegetable juices, purees, sauces, surimi, and many more. The estimated total market value of these products is approaching US$3 billion per annum, as of 2009.

The first high-temperature HPP product was approved by the US food authorities in mid 2009. This potato product combined the quality of a traditional pasteurized product with the ambient shelf life capability of a sterilized product. The combination of temperature and HPP has been proven to provide the necessary control of spores, notably *Clostridium botulinum*, and will form the basis of a series of products to fulfill the requirements of the US Army and their partners. The treatment is referred to as *pressure-assisted thermal sterilization* or pressure assisted thermal processing (PATP).

HPP is an eco-friendly technology with a small carbon footprint. It consumes relatively low power, produces negligible effluent, and uses normal potable water as the processing medium. In many cases, the potable water is at room temperature or already at a chilled temperature due to other process needs. HPP adds no chemicals or other additives to the food and takes away only the potentially harmful pathogens and spoilage microbes, leaving the original goodness and quality of the food untouched.

Modern, commercial size HPP systems can produce 2 or more tonnes of product per hour per system, depending upon the vessel utilization and packaging configuration. There is the opportunity to install multiple systems to match the required process production rate. The total operational cost, including capital cost, energy, labor, spares, and so on, ranges from €0.05 to 0.20 per kg produced.

Internal cost reductions, new market opportunities and new products can more than offset the capital cost of the equipment, as has been seen in many commercial application successes. Food producers and retailers should carefully consider the currently accepted high costs associated with short production runs of fresh quality, limited shelf life foods. HPP offers foods with quality, safety, and convenience, in keeping with today's consumer trend.

References

1. Hite, B.H. (1989) The effect of pressure in the preservation of milk. *Bull. West Virginia Univ. Agric. Exp. Stn*, **58**, 15–35.
2. Hoover, D.G., Metrick, K., Papineau, A.M., Farkas, D.F., and Knorr, D. (1989) Biological effects of high hydrostatic pressure on food microorganisms. *Food Technol.*, **43**, 99–107.
3. Morita, R.Y. (1975) Psychrophilic bacteria. *Bacteriol. Rev.*, **39**, 144–167.
4. Chong, G. and Cossins, A.R. (1983) A differential polarized fluorometric study of the effects of high hydrostatic pressure upon the fluidity of cellular membranes. *Biochemistry*, **22**, 409–415.
5. Sale, A.J.H., Gould, G.W., and Hamilton, W.A. (1970) Inactivation of bacterial spores by high hydrostatic pressure. *J. Gen. Microbiol.*, **60**, 323–334.
6. Ananth, E., Heinz, V., Schlter, O., and Knorr, D. (2001) Kinetic studies on high pressure inactivation of *Bacillus stearothermophilus* spores suspended in food matrices. *Innov. Food Sci. Emerging Technol.*, **2**, 261–272.
7. Roberts, C.M. and Hoover, D.G. (1996) Sensitivity of *Bacillus coagulans* spores to combinations of high hydrostatic pressure, heat, acidity and nisin. *J. Appl. Bacteriol.*, **81**, 363–368.
8. Mills, G., Earnshaw, R., and Patterson, M.F. (1998) Effects of high hydrostatic pressure on *Clostridium sporogenes* spores. *Lett. Appl. Microbiol.*, **26**, 227–230.
9. Gould, G.W. (1973) Inactivation of spores in food by combined heat and hydrostatic pressure. *Acta Aliment.*, **2**, 377–383.
10. National Center for Food Safety and Technology (2009) NCFST Receives Regulatory Acceptance of Novel Food Sterilization Process. Press release, *www.avure.com/archive/documents/Press-release/ncfst-receives-regulatory-acceptance-of-novel-food-sterilization-process.pdf* (accessed 18th May 2011).
11. Patterson, M.F., Quinn, M., Simpson, R., and Gilmour, A. (1995) Sensitivity of vegetative pathogens to high hydrostatic pressure treatment in phosphate-buffered saline and foods. *J. Food Prot.*, **58**, 524–529.
12. Shigehisa, T., Ohmori, T., Saito, A., Taji, S., and Hayashi, R. (1991) Effects of high pressure on characteristics of pork slurries and inactivation of microorganisms associated with meat and meat products. *Int. J. Food Microbiol.*, **12**, 207–216.
13. Chapman, B., Winley, E., Fong, A.S.W., Hocking, A.D., Stewart, C.M., and

Buckle, K.A. (2007) Ascospore inactivation and germination by high pressure processing is affected by ascospore age. *Innov. Food Sci. Emerging Technol.*, **8**, 531–534.
14. Butz, P., Funtenberger, S., Haberditzl, T., and Tauscher, B. (1996) High pressure inactivation of *Byssochlamys nivea* ascospores and other heat-resistant moulds. *Lebensm.-Wiss. Technol.*, **29**, 404–410.
15. Brâna, D., Voldrich, M., Marek, M., and Kamarád, J. (1997) Effect of high pressure treatment on patulin content in apple concentrate, in *High Pressure Research in the Biosciences* (ed. K. Heremans), Leuven University Press, Leuven, pp. 335–338.
16. Grove, S.F., Forsyth, S., Wan, J., Coventry, J., Cole, M., Stewart, C.M., Lewis, T., Ross, T., and Lee, A. (2008) Inactivation of hepatitis A virus, poliovirus and a norovirus surrogate by high pressure processing. *Innov. Food Sci. Emerging Technol.*, **9**, 206–210.
17. Kevin, R., Meade, G.K., Tezloff, R.C., and Kingsley, D.H. (2005) High-pressure inactivation of hepatitis A virus within oysters. *Appl. Environ. Microbiol.*, **71**, 339–343.
18. Kingsley, D.H., Holliman, D.R., Calci, K.R., Chen, H., and Flick, G. (2007) Inactivation of norovirus by high pressure processing. *Appl. Environ. Microbiol.*, **73**, 581–585.
19. Lindsay, D.S., Collins, M.V., Holliman, D., Flick, G.J., and Dubey, J.P. (2006) Effects of high-pressure processing on *Toxoplasma gondii* tissue cysts in ground pork. *J. Parasitol.*, **92**, 195–196.
20. Linton, M., McClements, J.M.J., and Patterson, M.F. (2001) Inactivation of pathogenic *Escherichia coli* in skimmed milk using high hydrostatic pressure. *Int.J. Food Sci. Technol.*, **2**, 99–104.
21. Alpas, H., Kalchayanand, N., Bozoglu, F., Sikes, T., Dunne, C.P., and Ray, B. (1999) Variation in resistance to hydrostatic pressure among strains of foodborne pathogens. *Appl. Environ. Microbiol.*, **65**, 4248–4251.
22. Mackey, B.M., Forestiere, K., and Isaacs, N. (1995) Factors affecting the resistance of *Listeria monocytogenes* to high hydrostatic pressure. *Food Biotechnol.*, **9**, 1–11.
23. Styles, M.F., Hoover, D.G., and Farkas, D.F. (1991) Response of *Listeria monocytogenes* and *Vibrio parahaemolyticus* to high hydrostatic pressure. *J. Food Sci.*, **56**, 1404–1497.
24. Earnshaw, R.G. (1995) High pressure microbial inactivation kinetics, in *High Pressure Processing of Foods* (eds D.A. Ledward, D.E. Johnston, R.G. Earnshaw and A.P.M. Hastings), Nottingham University Press, Nottingham, pp. 37–46.
25. Patterson, M.F. and Kilpatrick, D.J. (1998) The combined effect of high hydrostatic pressure and mild heat on inactivation of pathogens in milk and poultry. *J. Food Prot.*, **61**, 432–436.
26. Oxen, P. and Knorr, D. (1993) Baroprotective effects of high solute concentrations against inactivation of *Rhodotorula rubra*. *Lebensm.-Wiss. Technol.*, **26**, 220–223.
27. Linton, M., McClements, J.M.J., and Patterson, M.F. (1999) Survival of *Escherichia coli* O157:H7 during storage in pressure-treated orange juice. *J. Food Prot.*, **62**, 1038–1040.
28. Wei, C.I., Balaban, M.O., Fernando, S.Y., and Peplow, A.J. (1991) Bacterial effect of high pressure CO_2 treatment of foods spiked with *Listeria* or *Salmonella*. *J. Food Prot.*, **54**, 189–193.
29. Hong, S.I., Park, W.S., and Pyun, Y.R. (1997) Inactivation of *Lactobacillus* sp. from Kimchi by high pressure carbon dioxide. *Lebensm.-Wiss. Technol.*, **30**, 681–685.
30. Paul, P., Chawala, S.P., Thomas, P., and Kesavan, P.C. (1997) Effect of high hydrostatic pressure, gamma-irradiation and combined treatments on the microbiological quality of lamb meat during chilled storage. *J. Food Saf.*, **16**, 263–271.
31. Crawford, Y.J., Murano, E.A., Olson, D.G., and Shenoy, K. (1996) Use of high hydrostatic pressure and irradiation to eliminate *Clostridium sporogenes*

in chicken breast. *J. Food Prot.*, **59**, 711–715.

32. Raso, J., Palop, A., Pagan, R., and Condon, S. (1998) Inactivation of *Bacillus subtilis* spores by combining ultrasonic waves under pressure and mild heat treatment. *J. Appl. Microbiol.*, **85**, 849–854.

33. Rodriguez, E., Arques, J.L., Nuñez, M., Gaya, P., and Medina, M. (2005) Combined effect of high-pressure treatments and bacteriocin-producing lactic acid bacteria on inactivation of *Escherichia coli* O157:H7 in raw-milk cheese. *Appl. Environ. Microbiol.*, **71**, 3389–3404.

34. Chung, Y.K., Vurma, M., Turek, E.J., Chrism, G.W., and Tousef, A.E. (2005) Inactivation of barotolerant *Listeria monocytogenes* in sausage by combination of high-pressure processing and food grade additives. *J. Food Prot.*, **68**, 744–750.

35. Jofré, A., Garriga, M., and Aymerich, T. (2008) Inhibition of *Salmonella* sp., *Listeria monocytogenes* and *Staphylococcus aureus* in cooked ham by combining antimicrobials, high hydrostatic pressure and refrigeration. *Meat Sci.*, **78**, 53–59.

36. Arroyo, G., Sanz, P.D., and Prestamo, G. (1997) Effects of high pressure on the reduction of microbial populations in vegetables. *J. Appl. Microbiol.*, **82**, 735–742.

37. US Food and Drug Administration (2009) Listeria monocytogenes Action Plan. *http://www.fda.gov/Food/FoodSafety/FoodSafetyPrograms/ActionPlans/Listeriamoncytogenes ActionPlan/default.htm* (accessed January 2011).

38. Ananth, V., Dickson, J.S., Olson, D.G., and Murano, E.A. (1998) Shelf-life extension, safety and quality of fresh pork loin treated with high hydrostatic pressure. *J. Food Prot.*, **61**, 1649–1656.

39. Carlez, A., Rosec, J.P., Richard, N., and Cheftel, J.C. (1994) Bacterial growth during chilled storage of pressure-treated minced meat. *Lebensm.-Wiss. Technol.*, **27**, 48–54.

40. El Moueffak, A.C., Antoine, M., Cruz, C., Demazeau, G., Largeteau, A., Montury, M., Roy, B., and Zuber, F. (1995) High pressure and pasteurisation effect on duck foie gras. *Int.J. Food Sci. Technol.*, **30**, 737–743.

41. Carpi, G., Buzzoni, M.M., Gola, S., Maggi, A., and Rovere, P. (1995) Microbial and chemical shelf-life of high-pressure treated salmon cream at refrigeration temperatures. *Ind. Conserve,* **70**, 386–397.

42. Gervilla, R., Felipe, X., Ferragut, V., and Guamis, B. (1997) Effect of high hydrostatic pressure on *Escherichia coli* and *Pseudomonas fluorescens* strains in ovine milk. *J. Dairy Sci.*, **80**, 2297–2303.

43. Ponce, E., Pla, R., Mor-Mur, M., Gervilla, R., and Guamis, B. (1998) Inactivation of *Listeria innocua* inoculated in liquid whole egg by high hydrostatic pressure. *J. Food Prot.*, **72**, 119–122.

44. Balny, C., Masson, P., and Travers, F. (1989) Some recent aspects of the use of high pressure for protein investigations in solutions. *High Pressure Res.*, **2**, 1–28.

45. Heremans, K. (1989) From living systems to biomolecules, in *High Pressure and Biotechnology*, vol. 224 (eds C. Balny, R. Hayashi, K. Heremans, and P. Masson), Colloque INSERM, Paris, pp. 37–44.

46. Balny, C. and Masson, P. (1993) Effects of high pressure on proteins. *Food Rev. Int.*, **9**, 611–628.

47. Heremans, K. (1982) High pressure effects on proteins and other biomolecules. *Annu. Rev. Biophys. Bioeng.*, **11**, 1–21.

48. Mozhaev, V.V., Heremanks, K., and Frunk, J. (1994) Exploiting the effects of high hydrostatic pressure in biotechnological applications. *Trends Biotechnol.*, **12**, 493–501.

49. Galazka, V.B. and Ledward, D.A. (1998) High pressure effects on biopolymers, in *Functional Properties of Food Macromolecules*, 2nd edn (eds S.E. Hill, D.A. Ledward, and J.R. Mitchell), Aspen Press, Aspen, pp. 278–301.

50. Ludikhuyze, L., Van Loey, A., Indrawati, I., Denys, S., and Hendrickx, M. (2002) The effect of pressure processing on food quality related enzymes, from kinetic information to process engineering, in *Trends in High Pressure Bioscience* *and Biotechnology* (ed. R. Hayashi), Elsevier, London, pp. 517–524.

51. Gomes, M.R.A., Sumner, I.G., and Ledward, D.A. (1997) High Pressure Treatment of Papain. *J. Sci. Food Agric.*, **75**, 67–75.

52. Baker, E.N. and Drenth, J. (1987) The thiol proteases, structure and mechanism, in *Biological Macromolecules and Assemblies* (eds F.A. Jurnak and A. McPherson), John Wiley & Sons, Inc., New York, pp. 313–368.

53. Gomes, M.R.A. and Ledward, D.A. (1996) High pressure effects on some polyphenoloxidases. *Food Chem.*, **56**, 1–5.

54. Cheftel, J.C. and Culioli, J. (1997) Effects of high pressure on meat, a review. *Meat Sci.*, **46**, 211–236.

55. Cheah, P.B. and Ledward, D.A. (1997) Catalytic mechanisms of lipid oxidation in high pressure treated pork fat and meat. *J. Food Sci.*, **62**, 1135–1141.

56. Cheah, P.B. and Ledward, D.A. (1997) Inhibition of metmyoglobin formation in fresh beef by pressure treatment. *Meat Sci.*, **45**, 411–418.

57. Eskin, N.A.M., Grossman, S., and Pinsky, A. (1977) Biochemistry of lipoxygenase in relation to food quality. *Crit. Rev. Food Sci. Nutr.*, **22**, 1–33.

58. Kazeniac, S.J. and Hall, R.M. (1970) Flavour chemistry of tomato volatiles. *J. Food Sci.*, **35**, 519–530.

59. Tangwongchai, R., Ledward, D.A., and Ames, J.M. (2000) Effect of high pressure treatment on lipoxygenase activity. *J. Agric. Food Chem.*, **48**, 2896–2902.

60. Tangwongchai, R., Ledward, D.A., and Ames, J.M. (2000) Effect of high pressure on the texture of cherry tomato. *J. Agric. Food Chem.*, **48**, 1434–1441.

61. Gomes, M.R.A., Clark, R., and Ledward, D.A. (1998) Effects of high pressure on amylases and starch in wheat and barley flours. *Food Chem.*, **63**, 363–372.

62. Defaye, A.B., Ledward, D.A., MacDougall, D.B., and Tester, R.F. (1995) Renaturation of proteins subjected to high isostatic pressure. *Food Chem.*, **52**, 19–22.

63. Galazka, V.B., Ledward, D.A., and Varley, J. (1997) High pressure processing of β-lactoglobulin and bovine serum albumen, in *Food Colloids: Proteins, Lipids and Polysaccharides* (eds E. Dickenson and B. Bergenstahl), Royal Society of Chemistry, Cambridge, pp. 127–136.

64. Angsupanich, K. and Ledward, D.A. (1998) High pressure treatment effects on cod muscle. *Food Chem.*, **63**, 39–50.

65. Angsupanich, K., Edde, M., and Ledward, D.A. (1999) The effects of high pressure on the myofibrillar proteins of cod and turkey. *J. Agric. Food Chem.*, **47**, 92–99.

66. Ledward, D.A. (2000) Effects of pressure on protein structure. *High Press. Res.*, **19**, 1–10.

67. Apichartsrangkoon, A., Ledward, D.A., Bell, A.E., and Brennan, J.G. (1998) Physiochemical properties of high pressure treated wheat gluten. *Food Chem.*, **63**, 215–220.

68. Apichartsrangkoon, A., Ledward, D.A., Bell, A.E., and Schofield, J.D. (1999) Dynamic viscoelastic behaviour of high pressure treated wheat gluten. *Cereal Chem.*, **76**, 777–782.

69. Apichartsrangkoon, A. (1998) Effects of high pressure on rheological and chemical characteristics of plant proteins. PhD thesis, University of Reading.

7
Emerging Technologies for Food Processing

Liliana Alamilla-Beltrán, Jorge Welti-Chanes, José Jorge Chanona-Pérez, Ma de Jesús Perea-Flores, and Gustavo F. Gutierrez-López

7.1
Introduction

Food processing is an old practice and the early methods of food preservation were either physical or chemical. Physical methods include drying, canning, and refrigeration/freezing, while the chemical methods involve the use of additives and preparation of pickled, smoked, and cured meals. These traditional methods are still extensively applied, and their aim is to prevent undesirable biochemical changes in food products as well as to reduce microbiological growth, extend shelf life and to develop healthy and safe foods aiming to enhance the consumer's well-being. Innovative technologies of food preservation are also designed to extend the shelf life and stability of products, improve digestibility, enhance microbial food safety, and quality without affecting nutritional aspects; they may also be designed to develop foodstuffs with special functional and sensorial characteristics [1–3]. Emerging processing preservation techniques include: pulsed electric fields (PEFs), high hydrostatic pressures, low temperature plasmas, ultrasound power, high intensity pulsed light, oscillating magnetic fields, irradiation by X or gamma rays, microwave and radiofrequency heating, ohmic heating, and ultraviolet (UV) light pulses [1].

Emerging food technologies are processes that are in the research and development stage and which have significant potential to be commercialized or that currently represent only a small percentage of the market. A notable characteristic of many such technologies is the absence of heat inputs. Industry and university teams have given extensive attention to the development of these processes, trying to understand the effects of emerging processing technologies on different food materials [2].

In emerging thermal processes (ohmic, microwave, infrared radiation, and radiofrequency), conventional control techniques may be applied, providing processing time and temperature can be measured. Traditional evaluation methods of these variables cannot always be applied for emerging nonthermal processing (radiation, high hydrostatic pressure, electric pulses, ultrasound, and oscillating

Food Processing Handbook, Second Edition. Edited by James G. Brennan and Alistair S. Grandison.

magnetic waves). For example, when evaluating time–temperature curves to determine the heating time required for disinfection, it is necessary to evaluate and report radiation dose levels, pressure levels, degree and depth of penetration, electric power, and pulse durations. In addition, it is important to identify the most resistant pathogens during processing and to evaluate their inactivation kinetics [4].

It is also necessary to establish close relationships between users and food regulatory bodies to improve the acceptance of emerging technologies and the commercial success of the products by the end users [5]. "Food neophobia" is not necessarily an important factor for the acceptance of emerging technologies, and indeed the opposite is often observed, with consumers having a strong predilection for organic food [5, 6]. Consumers sometimes assert that if the food product is highly beneficial and satisfies their needs, the technology applied to produce it is less important. This reaction of consumers to accept food products produced by new food technologies has been associated with the rate of commercial implementation of processes [3].

In this chapter, some of the most promising emerging technologies for food processing are discussed.

7.2 Pulsed Electric Field Processing

Pulsed electric field processing is an emerging technology with a wide variety of applications in the food and biotechnology fields. It has been applied to the inactivation of bacteria, molds, and yeasts with promising results; other applications include the inactivation and modification of enzymes with no or minimum changes to the sensory, physicochemical, and nutritional characteristics of the product. The use of electric discharges to inactivate microorganisms and enzymes in food products was first attempted in the 1920s with the "ElectroPure Process" (ohmic heating process) applied to milk, but it was not until the 1960s that electric fields were widely used to inactivate a variety of microorganisms in model food systems [7]. The increase in the shelf life of products treated with PEF with "like-fresh" quality is the main reason for the development of this technology, aiming to improve or substitute the use of other preservation methods, mainly those using heat.

The PEF technique is based on the application of high voltage (10–80 $\mathrm{kV\,cm^{-1}}$) pulses for a short time (usually microseconds) to a food placed between two electrodes [8]. Although, during processing with PEF there is an increase in product temperature, PEF is considered to be a nonthermal method because the temperature and treatment time are not as elevated as in thermal treatments and, in addition, the increase in temperature can be controlled. In some cases it is desirable to combine PEF technology with mild heat and other preservation methods to enhance its lethal effects on microorganisms. This concept is known as *hurdle technology* and has the main advantage that it reduces the intensity of each one of the preservation methods used for a particular case.

7.2.1 PEF Treatment Chambers

Design of PEF treatment chambers is a key factor affecting microbial and enzymatic inactivation and is consequently one of the most important parameters in PEF. The treatment chamber should provide a uniform electric field to foods with minimum increase in temperature and the electrodes are designed to minimize the effect of electrolysis [9]. There are a great variety of treatment chambers designed at laboratory and pilot-plant scale, mainly by research laboratories and private industries, and it is also possible to find industrial-scale PEF treatment systems nowadays [10].

In general, PEF equipment consists of a pulse generator, fluid-handling and monitoring systems, and a treatment chamber housing two electrodes to generate a high voltage [10]. The treatment chambers are operated as static or continuous systems. Static chambers are classified as: U-shaped, parallel plate, disk-shaped, wire-cylinder, rod-rod, or sealed static. The U-shaped static treatment chamber has a polythene spacer (U-shaped) placed between the electrodes. This chamber design was proposed by Sale and Hamilton [11] and contains two carbon electrodes backed with brass blocks hollowed out for coolant flow (Figure 7.1). The maximum field strength that can be applied is 25 kV cm^{-1} due to the dielectric strength of air.

The parallel plate static treatment chamber uses flat electrodes separated by an insulating spacer, creating an area in which uniform electric field strength can be achieved. One disadvantage of this design is the field strength limitation due to surface tracking on the fluid or insulator that leads to arcing. Some parallel plate static treatment chambers have been designed at a laboratory scale by Dunn and Pearlman [12] (Figure 7.2) and by Grahl and Markl [13].

A disk-shaped static treatment chamber designed at Washington State University (WSU) consists of two round-edged, disk-shaped stainless steel electrodes polished to mirror surfaces that are held in position by insulating material that also forms

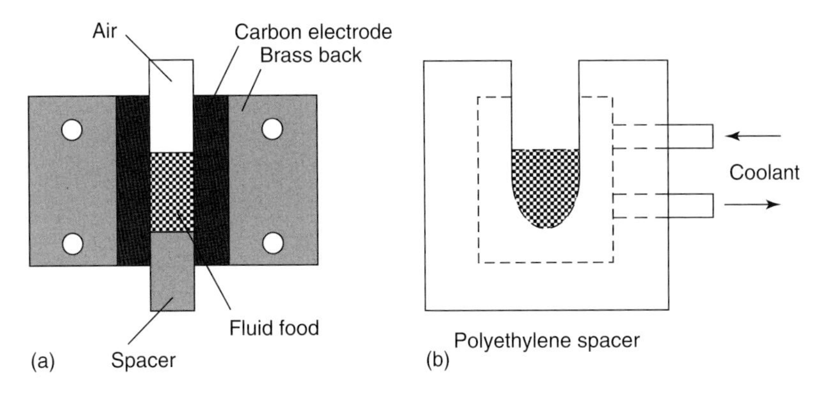

Figure 7.1 Static pulsed electric field treatment chamber. (a) Cut-away view showing the alignment of three parts. (b) U-shaped spacer and coolant connection [11].

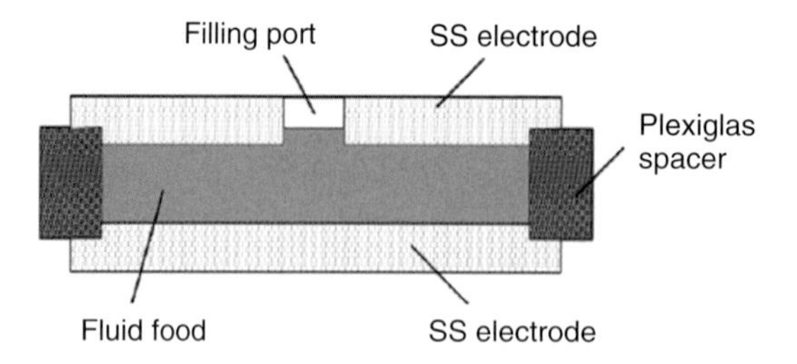

Figure 7.2 Cut-away cross-section of the static treatment chamber designed by Dunn and Pearlman [12].

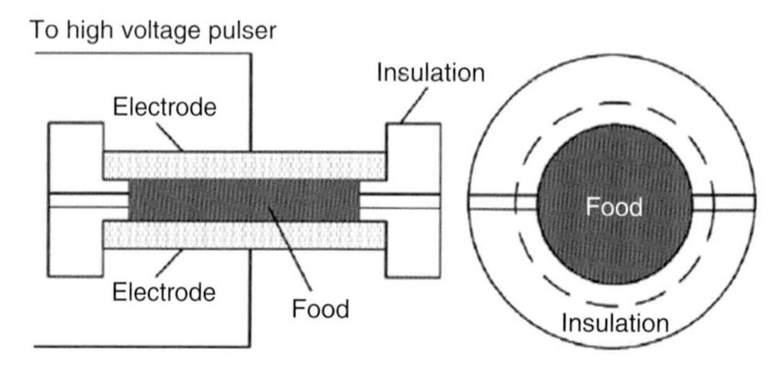

Figure 7.3 Schematic drawing of Washington State University static treatment chamber [14].

an enclosure containing the food [14]. The disk-shaped electrodes minimize the electric field and reduce the possibility of dielectric breakdown (Figure 7.3). The design incorporates a cooling system with water circulating through jackets.

7.2.2
Effects of PEF on Microorganisms

The effects of PEF treatment on microorganisms does not always results in lethal injury. The effectiveness of the PEF process depends on the intensity of the treatment conditions used [15]. A number of theories have been developed to explain the mechanisms of action of PEF on microorganisms. One of these is known as *electrical breakdown*. Electrical breakdown arises when an electric field increases the potential difference across the cell membrane, causing a reduction in the thickness of the membrane. When the applied potential difference reaches a critical value, repulsion between molecules produces pore formation in weak areas of the membrane [16] or enlarges existing pores. This causes an immediate flow of cellular materials and, consequently, membrane damage. Loss of cell materials or penetration into the cell of medium components causes microbial death. Pore formation can be permanent or temporary, depending on the intensity of the treatment applied. Breakdown of the membrane may be reversible if the pores are small in relation to the total surface of the membrane. If pores are formed across large areas of the membrane the injury is not reversible [17]. It is important to

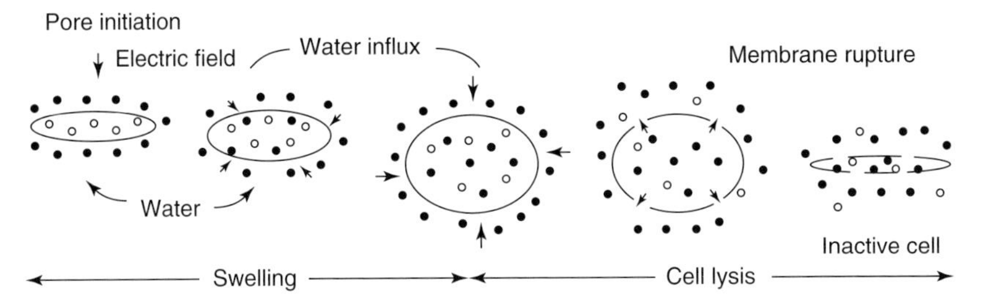

Figure 7.4 Microbial inactivation by electroporation caused by the pulsed electric field treatment. Adapted from Vega-Mercado *et al.* [7].

emphasize that the critical membrane potential depends on the size of the cell, the surface charge of the membrane, and the electrical conductivities of the membrane, cytoplasm, and suspending liquid medium [18, 19].

The principle of *electroporation* has also been used to explain the mechanism of action of PEF for inactivating microorganism. During electroporation, the lipid bilayer and proteins of the cell membrane are temporarily destabilized [20]. This destabilization causes changes in the conformation of lipids forming the bilayer; existing pores are enlarged and others are formed. These pores can conduct electric current, inducing heating that changes the lipid bilayer from a rigid gel to a liquid crystalline form. Once the semi-permeable nature of the membrane is disrupted, swelling, and eventual rupture of the cell is induced (Figure 7.4) [7, 21].

7.2.3 Factors Affecting the Ability of PEF to Inactivate Microorganisms

The effectiveness of PEF is related to factors associated with processing, microbiological issues, and characteristics of the product, a discussion of which is provided next.

7.2.3.1 Processing Factors

These include electric field strength ($kV\,cm^{-1}$), number of pulses, treatment time (generally in milliseconds or microseconds), pulse width (μs), pulse wave-form, process temperature (°C), frequency of repetitive pulses (Hz), and design and type of operation (batch or continuous) of the treatment chamber.

The electric field strength, defined as the voltage difference between two electrodes, divided by the distance between them [22], is one of the most important variables for microbial destruction by PEF. As electric field strength increases, the lethal effect of PEF rises due to transmembrane potential modification in the microbial cell [15, 17]. However, there are some studies indicating a threshold level ($3.0\,V\,\mu m^{-1}$) above which no differences in *Listeria innocua* inactivation are detected despite the increasing of the field strength [15].

As the number of pulses is increased, the level of microbial inactivation rises, but there is also a notable elevation of the temperature of the product [15].

Giner *et al.* [23] proved this effect of temperature elevation in a desalted aqueous solution of a commercial enzyme preparation treated with PEF. An increase in treatment time, defined as the number of pulses multiplied by the duration of pulse, raises the level of microbial inactivation achieved, but this inactivation could be due to a combination of PEF and temperature.

As the pulse width gets larger, the microbial inactivation achieved is greater due to a reduction in the field intensity required to produce a transmembrane potential large enough to initiate pore formation. Nevertheless, there is also a direct relationship between the pulse width and the degree of heating observed in the food, thus a balance must be found to achieve a high microbial inactivation while minimizing product heating [17].

The form of the pulse wave affects the level of inactivation achieved by PEF; square wave forms are more lethal and energy efficient than exponentially decaying pulses. Bipolar pulses are more effective in reducing the microbial load than monopolar pulses because of an additional stress to the cell membrane. Zhang *et al.* [22] evaluated the effect of square and exponential wave forms on the inactivation of *Saccharomyces cerevisiae* suspended in apple juice. Both wave forms were effective in microbial inactivation. Inactivation of *S. cerevisiae* treated with square wave pulses was larger than that with exponential decay pulses. Elez-Martínez *et al.* [24] observed an inverse relationship between pulse frequency and microbial inactivation.

The lethality of PEF increases with the temperature due to changes in membrane fluidity and permeability or because of an increase in conductivity of the product [15, 17]. A higher reduction is achieved in continuous PEF than in static systems because the distribution of the electric field is better in continuous systems [14].

7.2.3.2 Microorganism Factors

The type of microorganism, phase of growth, and microbial initial load are all important factors in the effectiveness of PEF. In terms of resistance to electric pulses, bacterial spores present the highest resistance, followed by Gram-positive bacteria, Gram-negative bacteria, and yeasts. Permeabilization studies on Gram-positive and Gram-negative bacteria indicate that differences in structure and composition of the cell wall could be responsible for the different behaviour observed when treated by PEF [25]. Microorganisms in the log phase of growth are generally more sensitive to PEF than those in the stationary phase. There is also evidence suggesting that initial concentration of microorganisms can affect the lethality of the process [20]. Opposite effects were reported by Álvarez *et al.* [26] who found that the inactivation of *S. senftenberg* was not a function of the initial microbial load.

7.2.3.3 Food Factors

Electrical conductivity, pH, presence of particles, particle size, and initial temperature all influence the lethality of PEF treatment in foods. Only foodstuffs capable of withstanding a high electric field can be treated by PEF. The electrical conductivity of the product to be treated is a very important parameter in PEF processing. This property depends on the physical structure and chemical composition of the food affecting its ionic strength which directly affects its conductivity. Conductivity

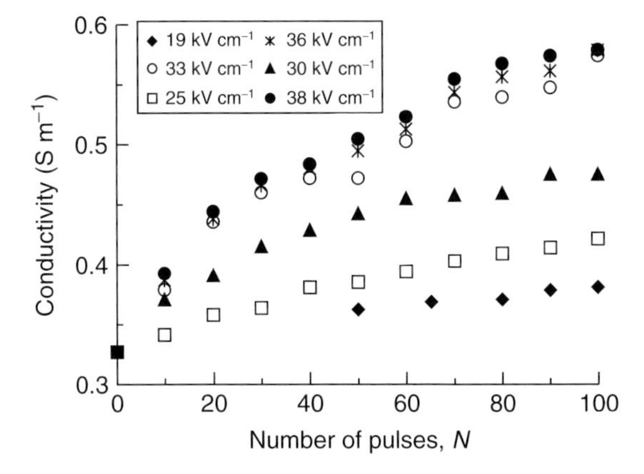

Figure 7.5 Effect of number of pulses in the conductivity of a desalted aqueous solution of a commercial enzyme preparation treated with pulsed electric field at different field strengths. Adapted from Giner *et al.* [23].

decreases as the ionic strength increases. Consequently, it is more difficult to build up electric field strength when the conductivity of the product is high [15]. Giner *et al.* [23] observed that the greater the field strength and the number of pulses, the higher the conductivity of a desalted aqueous solution of a commercial enzyme preparation treated with PEF (Figure 7.5).

There is an inverse relationship between the conductivity and the lethality of the PEF process [15, 27]. Foods with a high electrical conductivity are not suitable for processing with PEF because the peak electrical field across the chamber is reduced [17]. Liquid and homogeneous foods with a low electrical conductivity are the most appropriate for the continuous application of this technology, although solid foods not containing air bubbles can also be treated [8]. The effect of pH on microbial inactivation by PEF is not yet clear. Gómez *et al.* [28] found for *Listeria monocytogenes* in McIlvaine buffer, a direct relationship between microbial load and pH. The same tendency was obtained by Wouters *et al.* [15], for *Listeria innocua* and by Vega-Mercado *et al.* [7] for *E. coli*. The opposite effect, however, was found by Álvarez *et al.* [26] for *Salmonella senftenberg*. Heinz and Knorr [29] suggested that pH modifications down to pH 5.5 had minimal effect on the lethality of PEF processing of *B. subtilis*.

The presence of particles and their size in liquid foods causes processing problems since high energy inputs may be needed to inactivate microorganism attached to particles and there is a risk of dielectric breakdown of the food [17]. Particles should be smaller than the gap in the treatment chamber [8].

Broadly speaking, high initial temperatures in the product increase the inactivation effects of PEF in microorganisms [27, 30]. The process, microbiological, and product factors influencing PEF treatment are summarized in Table 7.1.

Table 7.1 Effects of different process, microbiological, and product factors on the inactivation of microorganisms by pulsed electric field.

		Low is better	High is better	Other considerations
Process factors	Electric field strength	–	X	–
	Number of pulses	–	X	–
	Treatment time	–	X	–
	Pulse width	–	X	–
	Pulse wave-form	–	–	Bipolar square pulses are better
	Pulse frequency	X	–	–
	Temperature	–	X	–
	Operation model of the treatment chamber	–	–	Continuous operation is better
Microbiological factors	Type of microorganism	–	–	Order of resistance to PEF: yeast < Gram-negative < Gram-positive < bacteria < bacterial spores
	Phase of growth	–	–	In the log phase microorganisms are more susceptible to PEF
	Initial load	–	–	There is no clear effect
Product factors	Electric conductivity	X	–	–
	pH	–	–	There is no clear effect
	Presence of particles (viscosity)	X	–	–
	Particle size	X	–	–
	Initial temperature of the foodstuff	–	X	–

7.2.4
Effects of PEF on Enzymes

Some reports indicate that PEF is effective for inactivating enzymes. Most studies on the effects of PEF on enzymes have been carried out on fruit juices and pectin methyl esterase (PME) is among the most widely investigated enzymes. The mechanism of enzyme inactivation by PEF has not been elucidated at a satisfactory level. However, the inactivation has been associated with the induction of conformational changes in the protein such as the reduction of the α-helix fractions [31], protein unfolding, breakdown of covalent bonds, denaturation, and redox reactions, such as those between sulfide groups and disulfide bonds [32].

Bendicho *et al.* [33] studied the effects of the type of operation (continuous or batch) on the activity of a protease from *Bacillus subtilis* suspended in simulated milk ultrafiltrate or milk. Products were processed in three different types of PEF equipment; two operated continuously and one in batch mode and operating conditions in each system were not the same. The protease of *B. subtilis* was minimally affected (approximately 10% of the activity was inhibited) after PEF treatment as compared with a low temperature for a long time or with a high temperature for a short time. However, it was also observed that the effect of the PEF treatment on enzyme activity varied according to the type of equipment used.

Schilling *et al.* [31] studied the nonthermal pasteurization of apple juice with PEF. The juice composition did not show changes, but browning was observed after some time, indicating residual enzymatic activity.

Aguiló-Aguayo *et al.* [34] compared a PEF treatment (35 kV cm^{-1} for 1500 μs using bipolar 4 μs pulses at 100 Hz) with a thermal treatment at 90 °C for 0.5–1 min on the residual activity of peroxidase, PME, and polygalacturonase from tomato juice. Higher efficiency in the inactivation of peroxidase was obtained by PEF compared to thermal treatments (97% vs. 90% and 79% after 1 min and 30 s respectively), while thermal treatments were found to be more efficient for the other enzymes.

The activity of lipase, glucose oxidase, and heat stable α-amylase was reduced by 70–80% after a 30-pulse PEF treatment at 13–87 kV cm^{-1}, 0.5 Hz, 2 μs pulse width at 20 °C. For the same treatment conditions, peroxidase and polyphenol oxidase showed a moderate reduction of 30–40%. Alkaline phosphate activity was reduced only by 5%. On the other hand, the activity of lysozyme and pepsin increased under a certain range of voltage (approximately 4–15 kV for lysozyme and 6–13 kV for pepsin) [35].

Giner *et al.* [36] utilized the PEF process for the inhibition of PME activity from tomato and a 93.8% reduction of the initial activity was achieved by using 400 pulses 0.02 ms long at 24 kV cm^{-1}.

Giner *et al.* [23] studied the inactivation of PME in a commercial enzyme preparation under PEF. Samples were exposed to monopolar exponentially decay wave form pulses for up to 463 μs at electric field intensities ranging from 19 to 38 kV cm^{-1}. As the treatment time and field strength increase, the residual activity of the PME enzyme decreases (Figure 7.6).

Yeom *et al.* [37] observed a reduction of about 90% in the activity of PME in orange juice after a treatment of 35 kV cm^{-1} for 59 μs. Recovery of the enzyme activity was not observed after 112 days of storage at 4, 22, and 37 °C.

When working with PME from commercial orange juice, Yeom *et al.* [38] observed, that PEF treatments at 25 kV cm^{-1}, were more effective when the juice temperature was increased (10–50 °C). When 50 °C was used, a 90% reduction in enzyme activity was achieved. The thermal treatment of the product at 50 °C did not cause a significant reduction in enzyme activity, so the inhibition can be attributed to the PEF process and not to the mild heat applied.

It is important to mention that factors affecting enzyme inactivation by PEF are, in general, the same as those for microbial inactivation.

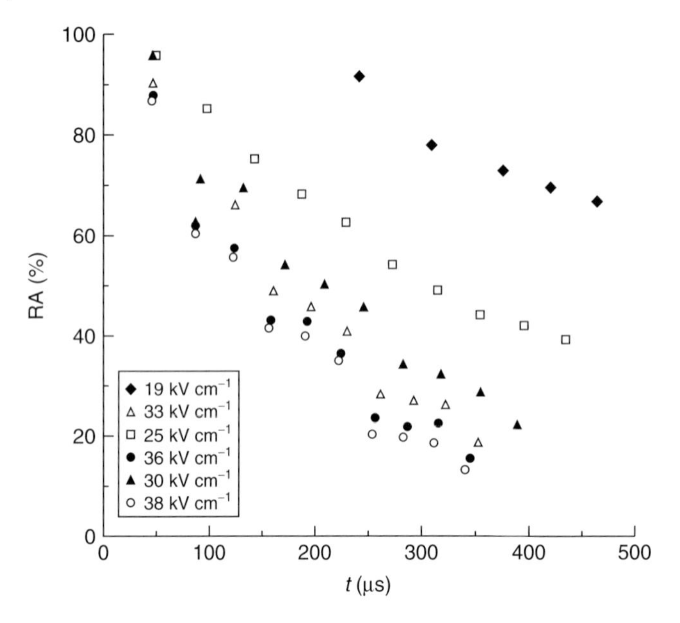

Figure 7.6 Residual activity (RA) of pectin methyl esterase in an aqueous solution of a commercial enzyme preparation versus total treatment time (t), due to PEF treatment at different field strengths [23].

7.2.5
Other Applications of PEF

In addition to the traditional applications of PEF in the food field, this nonthermal technology has found other practical uses. For example, in drying operations this process is very helpful since products treated by PEF can be dried more rapidly due to changes in texture that increase the effective diffusion coefficients. Lebovka *et al.* [39] compared the temperature dependencies of the moisture effective diffusion coefficient (D_{eff}) for intact, PEF-treated, and freeze-thawed potato tissues. It was shown that D_{eff} was sensitive to the pretreatment procedure, and although higher values of D_{eff} were observed for freeze-thawed pretreated samples than for PEF-treated ones, PEF processing led to higher D_{eff} values than those in untreated (intact) samples. When the product was pretreated at 50 °C the D_{eff} was comparable with that for PEF pretreated samples (Figure 7.7).

PEF has also been used for oil extraction. There are some results showing that PEF can be used as a pretreatment before oil separation to increase oil yield and content of functional food ingredients under gentle conditions [40]. Loginova *et al.* [41] demonstrated the benefits of PEF application (field strength of 100–600 V cm^{-1}, treatment time 10^{-3}–50 s, and temperature 20–80 °C) for enhancement of soluble matter extraction from chicory. The PEF pretreatment noticeably accelerated oil diffusion. PEF treatment could be promising for future industrial applications of "cold" soluble matter extraction from chicory roots.

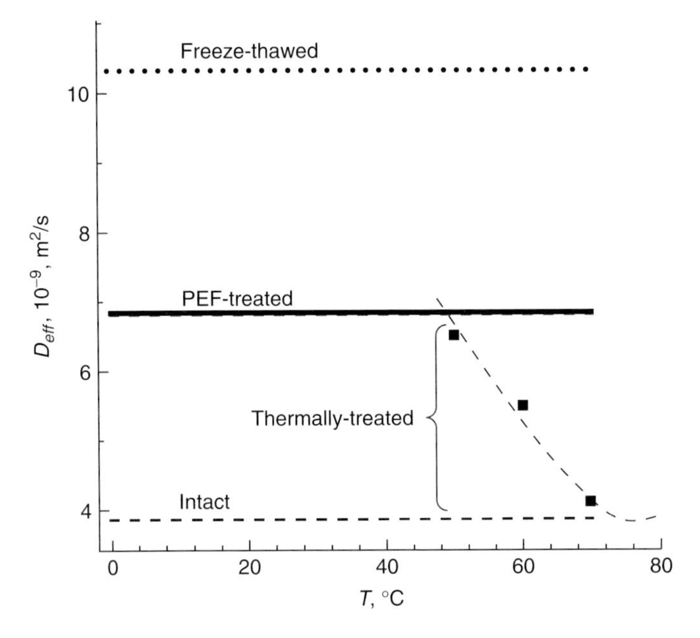

Figure 7.7 Effective diffusion coefficient D_{eff} versus temperature of thermal pretreatment (symbols), intact (dashed line), pulsed electric field treated (continuous line), and freeze-thawed (pinpoint line) pretreated potatoes. The drying temperature is 50 °C. (Adapted from Lebovka *et al.* [39].)

PEF also has some promising applications in the fruit juice extraction industry. Bazhal *et al.* [42] showed that the combination of extraction by mechanical expression and PEF treatment enhanced the juice yield in comparison with untreated samples.

PEF processing has also found biotechnological uses based on the capacity of PEF to form pores in membranes. Electroporation or electrofusion is used in biotechnology to insert foreign DNA into other cells to modify their cellular characteristics.

7.3
Ultrasound Power

Sound waves are mechanical vibrations that travel through solid materials in the form of transverse waves, and through liquids and gases in the form of longitudinal waves. Sound waves used in most engineering applications are longitudinal waves [43]. Three types of waves have been used in food areas: *longitudinal waves*, whose movement is in the direction of the particle displacement; *shear waves* which move perpendicular to the particle displacement; and *Rayleigh waves* which travel near to the surface of the material [44]. When ultrasonic waves strike the surface of an object, a perpendicular force is generated which results in a compression wave that moves through the material. When the ultrasound waves travel across the food material they produce cavitation, heating of the medium, microstirring at interfaces, and microchannels. Cavitation is a phenomenon in which the high intensity of the waves generates the growth and collapse of bubbles inside liquids.

If the force is parallel to the surface, it will produce a shearing wave [45]. Ultrasound produces changes in pressure (creating regions of alternating compression and expansion) and temperature that cause cavitation. In liquid foods, gas bubbles are formed and there are thinning of cell membranes, localized heating, and free radical production, which have a lethal effect on microorganisms [44, 45].

Ultrasound refers to sound waves having a frequency higher than the range audible to humans. Normal human hearing will detect sound frequencies ranging from 0.016 to 18 kHz [43, 46]. The lowest ultrasonic limit is 20 kHz (1 Hz = 1 cycle per second) and the upper limit of ultrasound frequencies for gases is 5 and 500 MHz for liquids and solids [43]. In water, the velocity of ultrasound waves changes 3 m s^{-1} per °C. The ultrasonic velocity of food materials is a nonlinear function of temperature [45, 46].

Ultrasonic waves are characterized by their frequency and wavelength. The product of these parameters is the speed of the wave through the medium:

$$c = \lambda f \tag{7.1}$$

where c is speed of sound, f is frequency, and λ is wavelength. The speed of sound increases with the density of the media. Some advantages of the use of ultrasound in the food industry include minimization of flavor loss, homogenization effects, and significant energy saving [47].

7.3.1 Applications of Ultrasound in the Food Industry

Applications of ultrasound have been divided into two categories: low-intensity ultrasound (LIU) and high-intensity ultrasound (HIU).

7.3.1.1 Low-Intensity Ultrasound

This is also known as "*diagnostic*" or "*high-frequency*" ultrasound, and involves low-amplitude sound waves [43, 44]. Low-power ultrasound uses very high frequencies of 100 kHz to 10 MHz or more, with low sound intensities of 100 mW cm^{-2} to 1 W cm^{-2}. It measures the velocity and attenuation of the wave.

In these range, the information is used for medical imaging (e.g., fetal scanning), chemical analysis, food quality assessment (ripeness, composition), and nondestructive tests (e.g., regular crack testing for aircraft structures) [43, 45]. A low-power ultrasound measurement system is composed of a transducer, a signal generator, a digitizer, and a measurement cell [43]. Some applications of low-power ultrasound in the food industry are in processing, quality assurance, and nondestructive food inspections.

Examples of the use of such technology are the identification of foreign bodies and analysis of droplet size in emulsions. The foreign bodies in foodstuffs of soft consistency are identified by immersing the sample in a water bath containing the ultrasound probe. The echo is preamplified and digitized. Parallel to this, the same identification is carried out in the food without foreign bodies. The echo signals of a sample without foreign body are subtracted from those of the studied sample including a foreign body. The results are processed by filtering, applying Fourier functions and objects are detected based on a signal-to-noise ratio [45]. Determinations include measurement of temperature and flowrate of liquid compounds; assessment of fish, meat, fruit, vegetables, and egg composition; physical characterization such as particle size and texture; creaming and sedimentation profiles; monitoring of phase transitions; study of gelation; fouling detection; texture, viscosity measurements; and the study of molecular properties [43–45].

7.3.1.2 High-Intensity Ultrasound

Also known as "*power ultrasound*," this uses lower frequencies (20–100 kHz) and produces sound intensities of high power which rank from 10 to 1000 W cm^{-2}, with amplitudes ranging from 5 to 50 μm [44].

This kind of ultrasound treatment has enough energy to break intermolecular bonds. Some applications in food processing operations include emulsification, homogenization, modification of viscosity, defoaming, extrusion, extraction (release of plant material), dispersion of aggregates, drying, filtration, separation, heat and mass transfer enhancement, separation of biological components, and modification and control of crystallization (as that for induction of the formation of smaller ice crystals in freezing), inactivation of microbes and enzymes by breaking the cell membranes due to the violence of cavitation and due to the formation of free radicals [43, 44, 46, 48].

7.3.2 Enzymes

Ultrasound power can inactivate enzymes. Inactivation of tomato PME based on cavitation effects, was observed at a rate of 0.004–0.020 mg of hydrogen peroxide $(l\,min)^{-1}$; at 50, 61, and 72 °C, the application of ultrasound increased PME inactivation compared with equivalent thermal treatments alone [49]. The shearing and compression effects of ultrasound cause denaturation of proteins, resulting in reduced enzyme activity. Short ultrasound treatments may increase the activity of enzymes, possibly by breaking down large molecular structures, making the active site more accessible for the substrates [50].

7.3.3 Microorganisms

Ultrasound has been used for bacterial inactivation by shear forces that change rapidly due to produced waves which are effective for destroying cells, particularly when combined with other treatments such as heating, pH modification, and chlorination. Microorganisms show a great sensitivity to ultrasound at elevated temperatures due to the weakening of the bacterial membrane, which enhances the effect of cavitation [51].

7.3.4 Fruits and Vegetables

Apple samples in hypertonic sucrose solutions subjected to ultrasound waves interchanged mass transfer at increased rates. The ultrasonic treatment at 10 W cm^{-2} (170 dB) at 20 kHz had a slightly greater effect on mass transfer than a well stirred solution. On the other hand, strong stirring produced a large amount of foam that would hinder the dewatering and the solute gain of the samples [52]. In the case of corn starch, the application of ultrasound disrupted the crystalline structure of starch granules before a reversible hydration of the amorphous phase, which resulted in destruction of the granular structure. This rupture increases the facility for water entrance to the corn starch granules, which leads to a higher water uptake and retention [46]. The application of ultrasonic pretreatment prior to air-drying of bananas showed that water diffusivity increased after the application of ultrasound and that the overall drying time was reduced by 11% [53]. The use of ultrasound as a pretreatment method in drying of mushrooms, Brussels sprouts, and cauliflower reduced the drying time to 3 and 10 min, respectively [54].

7.4 Other Technologies

7.4.1 High-Pressure Carbon Dioxide

Carbon dioxide at high pressure has been used as a nonthermal technology in food processing, for sterilization, and for enzyme inactivation. This technology is known as high-pressure carbon dioxide (HPCD) or dense phase carbon dioxide (DPCD) and can be an effective method to inactivate microorganisms by several orders of magnitude in liquid foods [55, 56]. The understanding of the microbial inactivation mechanisms could help in the explanation of the survival kinetics curves. The efficiency of contact between CO_2 and the microorganism may affect the kinetic behavior of the process.

In this technology the food comes in contact with either sub- or supercritical CO_2; pressurized CO_2 has the capacity to diffuse through the product and dissolve materials that result in bactericidal action. The CO_2 used has to be very pure and the efficacy of microbial inactivation is affected by pressure, temperature, and exposure time [55–57]. HPCD treatments may be carried out in batch, semi-batch or continuous apparatus which use CO_2 of 99.5% purity at pressures from 10 to 30 MPa, and temperatures from 32 to 42 °C.

A general process includes heating the pressure vessel to the required temperature, evacuating the air then introducing CO_2. The vessel is then pressurized to the required level, and maintained for the required treatment time [55, 57].

To date, HPCD preservation technology has not been implemented in the food industry on a commercial scale. Industrial application requires systematic data on the inactivation of important pathogenic and spoilage microorganisms, and knowledge of the influence of the process on the food ingredients and on the properties of the food matrix [56].

7.4.2 Ozonization

Ozone may be applied as an antimicrobial agent for food treatment, storage, and processing [58]. Currently, there are more than 3000 ozone-based water treatment installations all over the world and more than 300 potable water treatment plants in the United States [59, 60]. Ozone is a form of oxygen with three atoms. Ozone gas is blue whereas both liquid (−111.9 °C at 1 atm) and solid ozone (−192.7 °C) are opaque blue-black color. It is a relatively unstable gas at normal temperatures and pressures and is partially soluble in water, being the strongest disinfectant currently available for foods [60–62]. Ozone can be generated by exposure of air to electrical discharges as well as to electrochemical and UV radiation. Common electrical discharge methods have relatively low efficiencies (2–10%) and consume large amounts of electricity. The other two methods (electrochemical and UV treatments) are less cost effective [61].

The sensitivity of microorganisms to ozone depends on the treated product and the mode of application. The susceptibility of microorganisms to ozone varies with the physiological state of the culture, pH of the media, temperature, humidity, and presence of additives [61, 62]. Some products preserved with ozone include eggs during cold storage, fresh fruits and vegetables (grapes, broccoli florets) and fresh fish. Some examples of doses of ozone applied for sanitization purposes in the food industry are: 0.3 ppm in blackberries containing *Botrytis cinerea*; 0.2–0.4 ppm in poultry containing *Salmonella* sp. and Enterobacteriaceae; 5 ppm in dairy products containing *Alcaligens faecalus* and *P. fluorescens*; 0.111 $mg\,l^{-1}$ in fish containing *Enterococcus seriolicida and* 0.1–0.5 ppm in ground black peppers with *E. coli* [61–63].

7.4.3
Plasma Processing

Plasma is a neutral ionized gas, constituted by particles such as photons, electrons, positive and negative ions, atoms, free radicals, and excited or non-excited molecules interacting with each other. Nonthermal plasmas are obtained at lower pressures and power. Low temperature and medium pressurized plasmas are of industrial interest because they do not require extreme conditions for producing and handling. Examples of these plasmas are the corona discharge and the gliding arc discharge [64]. In the nonthermal plasma technique, a gas–liquid gliding arc discharge is generated between at least two metal electrodes with AC high voltage.

Air, oxygen, nitrogen, and argon are gases used in this technique. Treatment of gas phase and liquid phase pollution, removal organic compounds from aqueous solution and decolorization reaction are applications of plasma technique. The electrodes are either concentric metallic tubes or flat, plate-like electrodes. When a voltage is supplied to the electrodes, a discharge is produced between the two electrodes [61, 65]. Plasma may induce perforations (pores) in the membranes of microorganisms. Few UV photons are emitted by a gliding arc discharge but the plasma generated causes marked changes in the structure of DNA. Microorganisms have a very dense cytoplasm and cytoplasmic proteins may efficiently buffer the variations of pH. In case of *Erwinia carotovora atroseptica*, a decrease of the pH of the medium is not an essential factor for bacterial inactivation [64]. This technique is effective for the destruction of *Hafnia alvei*, *Erwinia carotovora carotovora*, *Erwinia chrysanthemi*, *Pectobacterium carotovorum carotovorum*, and *Dickeya chrysanthemi*, destroying the outer bacterial membrane which releases constitutive proteins and induces the dissociation of membrane-linked DNA [61, 64, 65].

7.5
Conclusions

Emerging technologies have attracted research interest in many areas, mainly within the food field. The fact that some of these technologies are considered to be nonthermal offers industry and consumers many advantages, including the destruction of microorganisms and enzymes while avoiding sensory and quality deterioration. Despite research done in this field, there are still many gaps to be filled for these technologies to be applied commercially. In the case of PEF, the gap to be filled is mainly in the design of commercial-scale equipment with continuous operation mode and, in the case of pulsed UV light, consumer acceptance must yet be achieved. Further research is necessary to develop safe processes effective in destroying microorganism and inactivating enzymes, avoiding or minimizing sensory and nutrient changes and with no ill-effects on consumers' health. The considerable research carried out on high hydrostatic pressure processing has already overcome these obstacles, and nowadays, the Food and Drug Administration (FDA) accepts it for commercial use. Much more research must be conducted on other emerging technologies before commercial-scale processes can be implemented.

References

1. Lund, D.B. (2002) Food engineering for the 21st century, in *Engineering and Food for the 21st Century* (eds J. Welti-Chanes, G. Barbosa-Cánovas, and J.M. Aguilera), CRC Press, Boca Raton, Florida, pp. 3–14.
2. Bimbenet, J.J. and Trystram, G. (2002) Trends in food engineering, in *Engineering and Food for the 21st Century* (eds J. Welti-Chanes, G. Barbosa-Cánovas and J.M. Aguilera), CRC Press, Boca Raton, Florida, pp. 15–34.
3. Wan, J., Coventry, J., Swiergon, P., Sanguansri, P., and Versteeg, C. (2009) Advances in innovative processing technologies for microbial inactivation and enhancement of food safety – pulsed electric field and low-temperature plasma. *Food Sci. Technol.*, **20** (9), 414–424.
4. Swartzel, K.R. (2002) Challenges for the process specialist in the 21st century, in *Engineering and Food for the 21st Century* (eds J. Welti-Chanes, G. Barbosa-Cánovas, and J.M. Aguilera), CRC Press, Boca Raton, Florida, pp. 35–40.
5. Sarkara, S. and Costa, A. (2008) Dynamics of open innovation in the food industry. *Trends Food Sci. Technol.*, **19**, 574–580.
6. Siegrist, M. (2008) Factors influencing public acceptance of innovative food technologies and products. *Trends Food Sci. Technol.*, **19**, 603–608.
7. Vega-Mercado, H., Pothakamury, U.R., Chang, F.J., Barbosa-Cánovas, G.V., and Swanson, B.G. (1996) Inactivation of *Escherichia coli* by combining pH, ionic strength and pulsed electric fields hurdles. *Food Res. Int.*, **29**, 117–121.
8. Charles, A.V., Nevárez, G.V., Zhang, Q.H., and Ortega, E. (2007) Comparison of thermal processing and pulsed electric fields treatment in pasteurization of apple juice. *Food Bioprod. Proc.*, **85** (2), 93–97.
9. Buta, P. and Tauscher, B. (2002) Emerging technologies: chemical aspects. *Food Res. Int.*, **35** (2–3), 279–284.
10. Huang, K. and Wang, J. (2009) Designs of pulsed electric fields treatment chambers for liquid foods pasteurization process: a review. *J. Food Eng.*, **95**, 227–239.
11. Sale, A.J.H. and Hamilton, W.A. (1967) Effects of high electric fields on microorganisms: I. Killing of bacteria and yeasts. *Biochim. Biophys. Acta*, **148** (3), 781–788.
12. Dunn, J.E. and Pearlman, J.S. (1987). Methods and apparatus for extending the shelf life of fluid food products. US Patent 4,695,472.
13. Grahl, T. and Markl, H. (1996) Killing of micro-organisms by pulsed electric field. *Appl. Microbiol. Biotechnol.*, **45**, 148–157.
14. Qin, B.L., Zhang, Q.H., Barbosa-Canovas, G.V., Swanson, B.G., and Pedrow, P.D. (1995) Pulsed electric field treatment chamber design for liquid food pasteurization using a finite element method. *Trans. ASAE (Am. Soc. Agric. Eng.)*, **38**, 557–565.
15. Wouters, P.C., Dutreux, N., Smelt, J.P.P.M., and Lelieveld, H.L.M. (1999) Effect of pulsed electric fields on inactivation kinetics of *Listeria innocua*. *Appl. Environ. Microbiol.*, **65** (12), 5364–5371.
16. Somolinos, M., Mañas, P., Condón, S., Pagán, R., and García, D. (2008) Recovery of *Saccharomyces cerevisiae* sublethally injured cells after pulsed electric fields. *Int. J. Food Microbiol.*, **125**, 352–356.
17. Leadley, C. and Williams, A. (2006) Pulsed electric field processing, power ultrasound and other emerging technologies, in *Food processing Handbook* (ed. J. Brennan), Wiley-VCH Verlag GmbH, Weinheim, pp. 201–236.
18. Gross, D. (1988) Electromobile surface charge alters membrane and potential

changes induced by applied electric fields. *Biophys. J.*, **54**, 879–884.
19. Lojewska, Z., Farkas, D.L., Ehrenberg, B., and Loew, L.M. (1989) Analysis of the effect of medium and membrane conductance on the amplitude and kinetics of membrane potentials induced by externally applied electric fields. *Biophys. J.*, **56**, 121–128.
20. Barbosa-Cánovas, G.V., Pierson, M.D., Zhang, Q.H., and Schaffner, D.W. (2000) Pulsed electric fields. *J. Food Sci.*, **2000**, 65–81.
21. Jeyamkondan, S., Jayas, D.S., and Holley, R.A. (1999) Pulsed electric field processing. A review. *J. Food Prot.*, **62**, 1088–1096.
22. Zhang, Q., Monsalve-González, A., Qin, B., Barbosa-Cánovas, G., and Swanson, B.G. (1994) Inactivation of *Saccharomyces cerevisiae* in apple juice by square-wave and exponential-decay pulsed electric fields. *J. Food Proc. Eng.*, **17** (4), 469–478.
23. Giner, J., Grouberman, P., Gimeno, V., and Martín, O. (2005) Reduction of pectinesterase activity in a commercial enzyme preparation by pulsed electric fields: comparison of inactivation kinetic models. *J. Sci. Food Agric.*, **85**, 1613–1621.
24. Elez-Martínez, P., Escolá -Hernández, J., Soliva-Fortuny, R.C., and Martín-Belloso, O. (2005) Inactivation of *Lactobacillus brevis* in orange juice by high- intensity pulsed electric fields. *Food Microbiol.*, **22**, 311–319.
25. García, D., Gómez, N., Mañas, P., Raso, J., and Pagán, R. (2007) Pulsed electric fields cause bacterial envelopes permeabilization depending on the treatment intensity, the treatment medium pH and the microorganism investigated. *Int. J. Food Microbiol.*, **113**, 219–227.
26. Álvarez, I., Raso, J., Palop, A., and Sala, F.J. (2000) Influence of different factors on the inactivation of Salmonella senftenberg by pulsed electric fields. *Int. J. Food Microbiol.*, **55**, 143–146.
27. Jayaram, S., Castle, G.S.P., and Margaritis, A. (1992) Kinetics of sterilization of *Lactobacillus brevis* cells by application of high voltage pulses. *Biotechnol. Bioeng.*, **40**, 1412–1420.
28. Gómez, N., García, D., Álvarez, I., Condon, S., and Raso, J. (2005) Modelling inactivation of *Listeria monocytogenes* by pulsed electric fields in media of different pH. *Int. J. Food Microbiol.*, **103**, 199–206.
29. Heinz, V. and Knorr, D. (2000) Effect of pH, ethanol addition and high hydrostatic pressure on the inactivation of *Bacillus subtilis* by pulsed electric fields. *Innovative Food Sci. Emerging Technol.*, **1** (2), 151–159.
30. Pothakamury, U.R., Vega, H., Zhang, Q., Barbosa-Cánovas, G.V., and Swanson, B.G. (1996) Effect of growth stage and processing temperature on the inactivation of *E. coli* by pulsed electric fields. *J. Food Prot.*, **59**, 1167–1171.
31. Schilling, S., Schmid, S., Jager, H., Ludwig, M., Dietrich, H., Toepfl, S., Knorr, D., Neidhart, S., Schieber, A., and Carle, R. (2008) Comparative study of pulsed electric field and thermal processing of apple juice with particular consideration of juice quality and enzyme deactivation. *J. Agric. Food Chem.*, **56**, 4545–4554.
32. Barsotti, L., Merle, P., and Cheftel, C. (1999) Food processing by pulsed electric fields I. Physical aspects. *Food Rev. Int.*, **15** (2), 163–180.
33. Bendicho, S., Marselles-Fontanet, A.R., Barbosa-Cánovas, G.V., and Martín-Belloso, O. (2005) High intensity pulsed electric fields and heat treatments applied to a protease from *Bacillus subtilis*. A comparison study of multiple systems. *J. Food Eng.*, **69**, 317–323.
34. Aguiló-Aguayo, I., Soliva-Fortuny, R., and Martín-Belloso, O. (2008) Comparative study on color, viscosity and related enzymes of tomato juice by high-intensity pulsed electric fields or heat. *Eur. Food Res. Technol.*, **227**, 599–606.
35. Ho, S.Y., Mittal, G.S., and Crosd, J.D. (1997) Effects of high field electric pulses on the activity of selected enzymes. *J. Food Eng.*, **31**, 69–84.
36. Giner, J., Gimeno, V., Espachs, A., Elez, P., Barbosa-Cánovas, G.V., and Martín, O. (2000) Inhibition of tomato

Licopersicon esculentum Mill./pectin methylesterase by pulsed electric fields. *Innovative Food Sci. Emerg. Technol.*, **1** (1), 57–67.

37. Yeom, H.W., Streaker, C.B., Zhang, Q.H., and Min, D.B. (2000) Effects of pulsed electric fields on the activities of microorganisms and pectin methyl esterase in orange juice. *J. Food Sci.*, **65** (8), 1359–1363.
38. Yeom, H.W., Zhang, Q.H., and Chism, G.W. (2002) Inactivation of pectin methyl esterase in orange juice by pulsed electric field. *J. Food Sci.*, **67** (6), 2154–2159.
39. Lebovka, N.I., Shynkaryk, N.V., and Vorobiev, E. (2007) Pulsed electric field enhanced drying of potato tissue. *J. Food Eng.*, **78**, 606–613.
40. Guderjan, M., Elez-Martínez, P., and Knorr, D. (2007) Application of pulsed electric fields at oil yield and content of functional food ingredients at the production of rapeseed oil. *Innovative Food Sci. Emerg. Technol.*, **8** (1), 55–62.
41. Loginova, K.V., Shynkaryk, M.V., Lebovka, I., and Vorobiev, E. (2010) Acceleration of soluble matter extraction from chicory with pulsed electric fields. *J. Food Eng.*, **96** (3), 374–379.
42. Bazhal, M.I., Lebovka, N.I., and Vorobiev, E. (2001) Pulsed electric field treatment of apple tissue during compression for juice extraction. *J. Food Eng.*, **50**, 129–139.
43. Feng, H. and Yang, W. (2006) Power ultrasound, in *Handbook of Food Science, Technology, and Engineering* (ed. H.Y. Hui), Taylor & Francis, New York, pp. 121-1–121-9.
44. Benedito, J., Sanjuan N., Cárcel, J.A., and Mulet, A. (2002) Applications of low-intensity ultrasonics in the dairy industry, in *Engineering and Food for the 21st Century* (eds J. Welti-Chanes, G. Barbosa-Cánovas, and J.M. Aguilera), CRC Press, Boca Raton, Florida, pp. 763–783.
45. Dolatowski, Z.J., Stadnik, J., and Stasiak, D. (2007) Application of ultrasound in food technology. *Acta Sci. Pol. Technol. Aliment.*, **6** (3), 89–99.
46. Rezěk-Jambrak, A., Herceg, Z., Sŭbarić, D., Babić, J., Brncić, M., Rimac, S., Bosiljkov, T., Ĉvek, D., Tripalo, B., and Gelo, J. (2009) Ultrasound effect on physical properties of corn starch. *Carbohydr. Polym.*, **79** (1), 91–100.
47. Earnshaw, R.G., Appleyard, J., and Hurst, R.M. (1995) Understanding physical inactivation progress: Combined preservation opportunities using heat, ultrasound and pressure. *Int. J. Food Microbiol.*, **28**, 197–219.
48. Angersbach, A., Heinz, V., and Knorr, D. (2000) Effects of pulsed electric fields on cell membranes in real food systems. *Innovative Food Sci. Emerg. Technol.*, **1** (2), 135–149.
49. Raviyan, P., Zhang, Z., and Feng, H. (2005) Ultrasonication for tomato pectinmethylesterase inactivation: effect of cavitation intensity and temperature on inactivation. *J. Food Eng.*, **70**, 189–196.
50. Fellows, P. (2009) *Food Processing Technology. Principles and Practice*, 3rd edn, Woodhead Publishing, Cambridge.
51. Patist, A. and Bates, D. (2008) Ultrasonic innovations in the food industry: from the laboratory to commercial production. *Innovative Food Sci. Emerg. Technol.*, **9**, 147–154.
52. Cárcel, J.A., Benedito, J., Rossello, C., and Mulet, A. (2007) Influence of ultrasound intensity on mass transfer in apple immersed in a sucrose solution. *J. Food Eng.*, **78**, 472–479.
53. Fernandes, F. and Rodrigues, S. (2007) Ultrasound as pre-treatment for drying of fruits: dehydration of banana. *J. Food Eng.*, **82**, 261–267.
54. Rezěk-Jambrak, A., Mason, T.J., Paniwnyk, B.L., and Lelas, V. (2007) Accelerated drying of button mushrooms, brussels sprouts and cauliflower by applying power ultrasound and its rehydration properties. *J. Food Eng.*, **81**, 88–97.
55. Ferrentino, G., Balabanb, M.O., Ferrari, G., and Polettoa, M. (2009) Food treatment with high pressure carbon dioxide: *Saccharomyces cerevisiae* inactivation kinetics expressed as a function of CO_2 solubility. *J. Supercrit. Fluids*, Article on line. doi: 10.1016/j.supflu.2009.07.005
56. Garcia-Gonzalez, L., Geeraerd, A.H., Elst, K., Van Ginneken, L., Van Impe,

J.F., and Devlieghere, F. (2009) Influence of type of microorganism, food ingredients and food properties on high-pressure carbon dioxide inactivation of microorganisms. *Int. J. Food Microbiol.*, **129**, 253–263.

57. Liao, H., Zhang, Y., Hu, X., Liao, X., and Wu, J. (2008) Behavior of inactivation kinetics of *Escherichia coli* by dense phase carbon dioxide. *Int. J. Food Microbiol.*, **126**, 93–97.
58. Kim, G.J., Yousef, A.E., and Dave, S. (1999) Application of ozone for enhancing safety and quality of foods: a review. *J. Food Prot.*, **62** (9), 1071–1087.
59. Rice, R.G., Overbeck, P., and Larson, K.A. (2000) *Proceedings on Small Drinking Water and Wastewater Systems*, NSF International, Ann Arbor, Michigan, p. 27.
60. Najafi, M.B.H. and Khodaparast, M.H.H. (2009) Efficacy of ozone to reduce microbial populations in date fruits. *Food Control*, **20**, 27–30.
61. Mahapatra, A.K., Muthukumarappan, K., and Julson, J.L. (2005) Applications of ozone, bacteriocins and irradiation in food processing: a review. *Crit. Rev. Food Sci. Nutr.*, **45**, 447–461.
62. Aguayo, E., Escalona, V.H., and Artés, F. (2006) Effect of cyclic exposure to ozone gas on physicochemical, sensorial and microbial quality of whole and sliced tomatoes. *Postharvest Biol. Technol.*, **39**, 169–177.
63. Emer, Z., Yesilcimen, M., and Ozdemir, M. (2008) Bactericidal activity of ozone against *Escherichia coli* in whole and ground black peppers. *J. Food Prot.*, **71** (5), 914–917.
64. Moreau, M., Orange, N., and Feuilloley, M.J.G. (2008) Non-thermal plasma technologies: new tools for bio-decontamination. *Biotechnol. Adv.*, **26**, 610–617.
65. Du, Ch., Shi, T., Sun, Y., and Zhuang, X. (2008) Decolorization of Acid Orange 7 solution by gas–liquid gliding arc discharge plasma. *J. Hazardous Mater.*, **154**, 1192–1197.

8
Packaging

James G. Brennan and Brian P.F. Day

8.1
Introduction

The main functions of a package are to contain the product and protect it against a range of hazards which might adversely affect its quality during handling, distribution, and storage. The package also plays an important role in marketing and selling the product. In this chapter only the protective role of the package will be considered. In this context the following definition may apply: "Packaging is the protection of materials by means of containers designed to isolate the contents, to some predetermined degree, from outside influences. In this way, the product is contained in a suitable environment within the package." The qualification "to some predetermined degree" is included in the definition as it is not always desirable to completely isolate the contents from the external environment.

Today most food materials are supplied to the consumer in a packaged form. Even foods which are sold unpackaged, such as some fruits and vegetables, will have been bagged, boxed, or otherwise crudely packaged at some stage in their distribution. A wide range of packaging materials is used for packaging foods including: papers, paperboards, fiberboards, regenerated cellulose films, polymer films, semi-rigid and rigid containers made from polymer materials, metal foil, rigid metals, glass, timber, textiles, and earthenware. Very often a combination of two or more materials is employed to package one product. It is important to look on packaging as an integral part of food processing and preservation. The success of most preservation methods depends on appropriate packaging, for example, to prevent microbiological contamination of heat-processed foods or moisture pick-up by dehydrated foods.

It is essential to consider packaging at an early stage in any product development exercise, not only for its technical importance, but also because of its cost implications. Packaging should not give rise to any health hazard to the consumer. No harmful substances should leach from the packaging material into the food. Packaging should not lead to the growth of pathogenic microorganisms when anaerobic conditions are created within the package. Packages should be convenient to use. They should be easy to open and resealable, if appropriate. The contents should be

Food Processing Handbook, Second Edition. Edited by James G. Brennan and Alistair S. Grandison.

readily dispensed from the container. Other examples of convenient packaging are "boil in the bag" products and microwavable packaging.

There are many environmental implications to packaging. In the manufacture of packaging materials, the energy requirements and the release of undesirable compounds into the atmosphere have environment implications. The use of multitrip containers and recycling of packaging materials can have positive influences on the environment. The disposal of waste packaging materials, particularly those that are not biodegradable, is a huge problem. These topics are outside the scope of this chapter but are covered elsewhere in the literature [1–5].

8.2 Factors Affecting the Choice of a Packaging Material and/or Container for a Particular Duty

8.2.1 Mechanical Damage

Fresh, processed, and manufactured foods are susceptible to mechanical damage. The bruising of soft fruits, the break up of heat-processed vegetables, and the cracking of biscuits are examples. Such damage may result from sudden impacts or shocks during handling and transport, vibration during transport by road, rail, and air, and compression loads imposed when packages are stacked in warehouses or large transport vehicles. Appropriate packaging can reduce the incidence and extent of such mechanical damage. Packaging alone is not the whole answer. Good handling and transport procedures and equipment are also necessary.

The selection of a packaging material of sufficient strength and rigidity can reduce damage due to compression loads. Metal, glass, and rigid plastic materials may be used for primary or consumer packages. Fiberboard and timber materials are used for secondary or outer packages. The incorporation of cushioning materials into the packaging can protect against impacts, shock, and vibration. Corrugated papers and boards, pulpboard, and foamed plastics are examples of such cushioning materials. Restricting movement of the product within the package may also reduce damage. This may be achieved by tight-wrapping or shrink-wrapping. Inserts in boxes or cases, or thermoformed trays may be used to provide compartments for individual items such as eggs and fruits.

8.2.2 Permeability Characteristics

The rate of permeation of water vapor, gases (O_2, CO_2, N_2, and ethylene), and volatile odor compounds into or out of the package is an important consideration in the case of packaging films, laminates, and coated papers. Foods with relatively high moisture contents tend to lose water to the atmosphere. This results in a loss of weight and deterioration in appearance and texture. Meat and cheese are typical examples of such foods. Products with relatively low moisture contents will tend to pick up moisture, particularly when exposed to a high humidity atmosphere. Dry powders such as cake mixes and custard powders may cake and lose their freeflowing characteristics. Biscuits and snack foods may lose their crispness. If the water activity of a dehydrated product is allowed to rise above a certain critical level, microbiological spoilage may occur. In such cases a packaging material with a low permeability to water vapor, effectively sealed, is required.

In contrast, fresh fruit and vegetables continue to respire after harvesting. They use up oxygen and produce water vapor, carbon dioxide, and ethylene. As a result, the humidity inside the package increases. If a high humidity develops, condensation may occur within the package when the temperature fluctuates. In such cases, it is necessary to allow for the passage of water vapor out of the package. A packaging material that is semi-permeable to water vapor is required in this case.

The shelf life of many foods may be extended by creating an atmosphere in the package which is low in oxygen. This can be achieved by vacuum packaging or by replacing the air in the package with carbon dioxide and/or nitrogen. Cheese, cooked and cured meat products, dried meats, egg, and coffee powders are examples of such foods. In such cases, the packaging material should have a low permeability to gases and be effectively sealed. This applies also when modified atmosphere packaging (MAP) is used (see Section 8.4).

If a respiring food is sealed in a gastight container, the oxygen will be used up and replaced with carbon dioxide. The rate at which this occurs depends on the rate of respiration of the food, the amount in the package and the temperature. Over a period of time, an anaerobic atmosphere will develop inside the container. If the oxygen content falls below 2%, anaerobic respiration will set in and the food will spoil rapidly. The influence of the level of carbon dioxide in the package varies from product to product. Some fruits and vegetables can tolerate, and may even benefit from, high levels of carbon dioxide while others do not. In such cases, it is necessary to select a packaging material that permits the movement of oxygen into and carbon dioxide out of the package at a rate that is optimum for the contents. Ethylene is produced by respiring fruits. Even when present in low concentrations, this can accelerate the ripening of the fruit. The packaging material must have an adequate permeability to ethylene to avoid this problem.

To retain the pleasant odor associated with many foods, such as coffee, it is necessary to select a packaging material that is a good barrier to the volatile compounds which contribute to that odor. Such materials may also prevent the contents from developing taints due to the absorption of foreign odors. It is worth noting here that films that are good barriers to water vapor may be permeable to volatiles.

In those cases where the movement of gases and vapors is to be minimized, metal and glass containers, suitably sealed, may be used. Many flexible film materials, particularly if used in laminates, are also good barriers to vapors and gases. Where some movement of vapors and/or gases is desirable, films that are semi-permeable to them may be used. For products with high respiration rates the packaging material may be perforated. A range of microperforated films is available for such applications.

In the case of an intact polymer film, the rate at which vapors and gases pass through it is specified by its "permeability" or "permeability constant," P, defined by the following relationship:

$$P = \frac{ql}{A(p_1 - p_2)} \tag{8.1}$$

where q is the quantity of vapor or gas passing through A, an area of the film in unit time, l is the thickness of the film, and p_1, p_2 are the partial pressures of the vapor or gas in equilibrium with the film at its two faces. The permeability of a film to water vapor is usually expressed as x g m^{-2} day^{-1} (i.e., per 24 h) and is also known as the water vapor transfer rate (WVTR). Highly permeable films have values of WVTR in the range from 200 to >800 g m^{-2} day^{-1}, while those with low permeability have values of 10 g m^{-2} day^{-1}. The permeability of a film to gases is usually expressed as x cm^3 m^{-2} day^{-1}. Highly permeable films have P values from 1000 to >25 000 cm^3 m^{-2} day^{-1}, while those with low permeability have values of 10 cm^3 m^{-2} day^{-1} or below. When stating the P value of a film, the thickness of the film and the conditions under which it was measured, mainly the temperature and (p_1, p_2), must be given.

8.2.3
Greaseproofness

In the case of fatty foods, it is necessary to prevent egress of grease or oil to the outside of the package, where it would spoil its appearance and possibly interfere with the printing and decoration. Greaseproof and parchment papers (see Section 8.3.1) may give adequate protection to dry fatty foods, such as chocolate and milk powder, while hydrophilic films or laminates are used with wet foods, such as meat or fish.

8.2.4
Temperature

A package must be able to withstand the changes in temperature which it is likely to encounter, without any reduction in performance or undesirable change in appearance. This is of particular importance when foods are heated or cooled in the package. For many decades metal and glass containers were used for foods which were retorted in the package. It is only in relatively recent times that heat-resistant laminates were developed for this purpose. Some packaging films become brittle when exposed to low temperatures and are not suitable for packaging frozen foods. The rate of change of temperature may be important. For example, glass containers have to be heated and cooled slowly to avoid breakage. The method of heating may influence the choice of packaging.

Many new packaging materials have been developed for foods which are to be processed or heated by microwaves.

8.2.5 Light

Many food components are sensitive to light, particularly at the blue and ultraviolet end of the spectrum. Vitamins may be destroyed, colors may fade, and fats may develop rancidity when exposed to such light waves. The use of packaging materials which are opaque to light will prevent these changes. If it is desirable that the contents be visible, for example, to check the clarity of a liquid, colored materials which filter out short wavelength light may be used. Amber glass bottles, commonly used for beer in the UK, perform this function. Pigmented plastic bottles are used for some health drinks.

8.2.6 Chemical Compatibility of the Packaging Material and the Contents of the Package

It is essential in food packaging that no health hazard to the consumer should arise as a result of toxic substances present in the packaging material leaching into the contents. In the case of flexible packaging films, such substances may be residual monomers from the polymerization process or additives such as stabilizers, plasticizers, coloring materials, and so on. To establish the safety of such packaging materials two questions need to be answered: (i) are there any toxic substances present in the packaging material and (ii) will they leach into the product? Toxicological testing of just one chemical compound is lengthy, complicated, and expensive, usually involving extensive animal feeding trials and requiring expert interpretation of the results. Such undertakings are outside the scope of all but very large food companies. In most countries there are specialist organizations to carry out such this type of investigation, for example, the British Industrial Biological Research Association (BIBRA) in the United Kingdom. Such work may be commissioned by governments, manufacturers of packaging materials, and food companies.

To establish the extent of migration of a chemical compound from a packaging material into a food product is also quite complex. The obvious procedure would be to store the food in contact with the packaging material for a specified time under controlled conditions and then to analyze the food to determine the amount of the specific compound present in it. However, detecting a very small amount of a specific compound in a food is a difficult analytical problem. It is now common practice to use simulants instead of real foods for this purpose. These are liquids or simple solutions which represent different types of foods in migration testing. For example, the European Commission specifies the following simulants:

- **Simulant A**: distilled water or equivalent (to represent low acid, aqueous foods);
- **Simulant B**: 3% (w/v) acetic acid in aqueous solution (to represent acid foods);
- **Simulant C**: 15% (w/v) ethanol in aqueous solution (to represent foods containing alcohol);
- **Simulant D**: rectified olive oil (to represent fatty foods).

The EC also specifies which simulants are to be used when testing specific foods. More than one simulant may be used with some foods. After being held in contact with the packaging material, under specified conditions, the simulant is analyzed to determine how much of the component under test it contains. Migration testing is seldom carried out by food companies. Specialist organizations mostly do this type of work, for example, in the United Kingdom Pira International.

Most countries have extensive legislation in place controlling the safety of flexible plastic packaging materials for food use. These include limits on the amount of monomer in the packaging material. There is particular concern over the amount of vinyl chloride monomer (VCM) in polyvinylchloride (PVC). The legislation may also include: lists of permitted additives which may be incorporated into different materials, limits on the total migration from the packaging material into the food and limits on the migration of specific substances, such as VCM. The types of simulants to be used in migration tests on different foods and the methods to be used for analyzing the simulants may also be specified. While the discussion above is concerned only with flexible films, other materials used for food packaging may result in undesirable chemicals migrating into foods. These include semi-rigid and rigid plastic packaging materials, lacquers, and sealing compounds used in metal cans, materials used in the closures for glass containers, additives and coatings applied to paper, board and regenerated cellulose films, wood, ceramics, and textiles.

Apart from causing a health hazard to the consumer, interaction between the packaging material and the food may affect the quality and shelf life of the food and/or the integrity of the package; and it should be avoided. An example of this is the reaction between acid fruits and tinplate cans. This results in the solution of tin in the syrup and the production of hydrogen gas. The appearance of the syrup may deteriorate and colored fruits may be bleached. In extreme cases, swelling of the can (hydrogen swelling) and even perforation may occur. The solution to this problem is to apply an acid-resistant lacquer to the inside of the can. Packaging materials that are likely to react adversely with the contents should be avoided, or another barrier substance should be interposed between the packaging material and the food [6–9].

8.2.7
Protection against Microbial Contamination

Another role of the package may be to prevent or limit the contamination of the contents by microorganisms from sources outside the package. This is most important in the case of foods that are heat-sterilized in the package, where it is essential that post-processing contamination does not occur. The metal can has dominated this field for decades and still does. The reliability of the double seam (see Section 8.3.8) in preventing contamination is one reason for this dominance. Some closures for glass containers are also effective barriers to contamination.

It is only in relatively recent times that plastic containers have been developed that not only withstand the rigors of heat processing but also whose heat seals are effective in preventing post-processing contamination. Effective seals are also necessary on cartons, cups, and other containers that are aseptically filled with UHT products. The sealing requirements for containers for pasteurized products and foods preserved by drying, freezing, or curing are not so rigorous. However, they should still provide a high level of protection against microbial contamination.

8.2.8
In-Package Microflora

The permeability of the packaging material to gases and the packaging procedure employed can influence the type of microorganisms that grow within the package. Packaging foods in materials that are highly permeable to gases is not likely to bring about any significant change in the microflora, compared to unpackaged foods. However, when a fresh or mildly processed food is packaged in a material that has a low permeability to gases and when an anaerobic atmosphere is created within the package as a result of respiration of the product or because of vacuum or gas packaging, the type of microorganisms that grow inside the package are likely to be different to those that would grow in the unpackaged food. There is a danger that pathogenic microorganisms could flourish under these conditions and result in food poisoning. Such packaging procedures should not be used without a detailed study of the microbiological implications, taking into account the type of food, the treatment it receives before packaging, the hygienic conditions under which it is packaged and the temperature at which the packaged product is to be stored, transported, displayed in the retail outlet, and kept in the home of the consumer.

8.2.9
Protection against Insect and Rodent Infestation

In temperate climates, moths, beetles, and mites are the insects that mainly infest foods. Control of insect infestation is largely a question of good housekeeping. Dry, cool, clean storage conditions, good ventilation, adequate turnaround of warehouse stocks, and the controlled use of fumigants or contact insecticides can all help to limit insect infestation. Packaging can also contribute, but an insect-proof package is not normally economically feasible, with the exception of metal and glass containers. Some insects are classified as penetrators, as they can gnaw their way through some packaging materials. Paper, paperboard, and regenerated cellulose materials offer little resistance to such insects. Packaging films vary in the resistance they offer. In general, the thicker the film the more resistant it is to penetrating insects. Oriented films are usually more resistant that unoriented forms of the same materials. Some laminates, particularly those containing foil, offer good resistance to penetrating insects. Other insects are classified as invaders as they enter through openings in the package. Good design of containers to eliminate as far as possible cracks, crevices, and pinholes in corners and seals can limit the ingress of invading insects. The use of adhesive tape to seal any such openings can help. The application of insecticides to some packaging materials is practiced to a limited extent, for example, to the outer layers of multiwall paper sacks. They may be incorporated into adhesives. However, this can only be done if regulations allow it.

Research has also been carried out into the use of natural substances to protect food packages from insects. Among the substances that have been investigated are extracts from neem, turmeric, cinnamon, and eucalyptus, which were coated onto or incorporated into paper and plastic packaging materials. Their effectiveness as insect repellents varied widely and between insect species [10–12].

Packaging does not make a significant contribution to the prevention of infestation by rodents. Only robust metal containers offer resistance to rats and mice. Good, clean storekeeping, provision of barriers to infestation, and controlled use of poisons, gassing, and trapping are the usual preventive measures taken to limit such infestation.

8.2.10 Taint

Many packaging materials contain volatile compounds which give rise to characteristic odors. The contents of a package may become tainted by absorption or solution of such compounds when in direct contact with the packaging materials. Food not in direct contact with the packaging material may absorb odorous compounds present in the free space within the package. Paper, paperboard, and fiberboard give off odors which may contaminate food. The cheaper forms of these papers and boards, which contain recycled material, are more likely to cause tainting of the contents. Clay, wax, and plastic coatings applied to such materials may also cause tainting. Storage of these packaging materials in clean, dry, and well-ventilated stores can reduce the problem. Some varieties of wood, such as cedar and cypress, have very strong odors which could contaminate foods. Most polymers are relatively odor-free, but care must be taken in the selection of additives used. Lacquers and sealing compounds used in metal and glass containers are possible sources of odor contamination. Some printing inks and adhesives give off volatile compounds when drying, which may give rise to tainting of foods. Careful selection of such materials is necessary to lessen the risk of contamination of foods in this way [9].

8.2.11 Tamper-Evident/Resistant Packages

There have been many reports in recent years of food packages being deliberately contaminated with toxic substances, metal, or glass fragments. The motive for this dangerous practice is often blackmail or revenge against companies. Another less serious, but none the less undesirable activity, is the opening of packages to inspect, or even taste, the contents and returning them to the shelf in the supermarket. This habit is known as *grazing*. There is no such thing as a tamper-proof package. However, tamper-resistant and/or tamper-evident features can be incorporated into packages. Reclosable glass or plastic bottles and jars are most vulnerable to tampering. Examples of tamper-evident features include: a membrane heat-sealed to the mouth of the container, beneath the cap, roll-on closures (see Section 8.3.9), polymer sleeves heat-shrunk over the necks and caps, breakable caps which are connected to a band by means of frangible bridges that break when the cap is opened and leave the band on the neck of the container [12–15].

8.2.12 Other Factors

There are many other factors to be considered when selecting a package for a particular duty. The package must have a size and shape which makes it easy to handle, store, and display on the supermarket shelf. Equipment must be available to form, fill, and seal (FFS) the containers at an acceptable speed and with an adequately low failure rate. The package must be aesthetically compatible with the contents. For example, the consumer tends to associate a particular type of package with a given food or drink. Good-quality wines are packaged in glass, whereas cheaper ones may be packaged in "bag in box" containers or plastic bottles.

The decoration on the package must be attractive. A look around a supermarket confirms the role of the well-designed package in attracting the consumer to purchase that product. The labeling must clearly convey all the information required to the consumer and comply with relevant regulations.

Detailed discussion of these factors is not included in this chapter but further information is available in the literature [14–20].

8.3 Materials and Containers Used for Packaging Foods

8.3.1 Papers, Paperboards, and Fiberboards

8.3.1.1 Papers

While paper may be manufactured from a wide range of raw materials, almost all paper used for food packaging is made from wood. Some papers and boards are made from repulped waste paper. Such materials are not used in direct contact with foods. The first stage in the manufacture of papers and boards is pulping. Groundwood pulp is produced by mechanical grinding of wood and contains all the ingredients present in the wood (cellulose, lignin, carbohydrates, resins, and gums). Paper made from this type of pulp is relatively weak and dull compared to the alternative chemical pulp. Chemical pulp is produced by digesting wood chips in an alkaline (sulfate pulp) or acid (sulfite pulp) solution, followed by washing. This pulp is a purer form of cellulose, as the other ingredients are dissolved during the digestion and removed by washing. Some mechanical pulp may be added to chemical pulp for paper manufacture, but such paper is not usually used in direct contact with foods. The first step in the paper-making process itself is known as *beating* or *refining*. A dilute suspension of pulp in water is subjected to controlled mechanical treatment in order to split the fibers longitudinally and produce a mass of thin fibrils. This enables them to hold together when the paper is manufactured, thus increasing the strength of the paper. The structure and density of the finished paper is mainly determined by the extent of this mechanical treatment.

Additives such as mineral fillers and sizing agents are included at this stage to impart particular properties to the paper. The paper pulp is subjected to a series of refining operations before being converted into paper. There are two types of equipment used to produce paper from pulp. In the Fourdrinier machine, a dilute suspension of the refined pulp is deposited on to a fine woven, moving and vibrating mesh belt. By a sequence of draining, vacuum filtration, pressing, and drying, the water content is reduced to 4–8% and the network of fibers on the belt is formed into paper. In the alternative cylinder machine, six or more wiremesh cylinders rotate, partly immersed in a suspension of cellulose fibers. They pick up fibers and deposit them in layers onto a moving felt blanket. Water is removed by a sequence of operations similar to those described above. This method is mainly used for the manufacture of boards where combinations of different pulps are used.

Types of papers used for packaging foods include the following:

- **Kraft paper**: Made from sulfate pulp, this is available unbleached (brown) or bleached. It is a strong multipurpose paper used for wrapping individual items or parceling a number of items together. It may also be fabricated into bags and multiwall sacks.
- **Sulfite paper**: Made from pulp produced by acid digestion, again it is a general purpose paper, not as strong as kraft. It is used in the form of sachets and bags.
- **Greaseproof paper**: Made from sulfite pulp which is given a severe mechanical treatment at the beating stage, this is a close-textured paper with greaseproof properties under dry conditions.
- **Glassine paper**: Produced by polishing the surface of greaseproof paper, this has some resistance to moisture penetration.
- **Vegetable parchment**: This is produced by passing paper made from chemical pulp through a bath of sulfuric acid, after which it is washed, neutralized, and dried. The acid dissolves the surface layers of the paper, decreasing its porosity. It has good greaseproof characteristics and retains its strength when wet better than greaseproof paper.
- **Tissue paper**: This light, open structured paper is used to protect the surface of fruits and provide some cushioning.
- **Wet-strength papers**: These have chemicals added which are crosslinked during the manufacturing process. They retain more of their strength when wet compared to untreated papers. They are not used in direct contact with food, but mainly for outside packaging.

- **Wax-coated papers**: Although these are heat-sealable and offer moderate resistance to water and water vapor transfer, the heat seals are relatively weak and the wax coating may be damaged by creasing and abrasion.
- **Other coatings**: Other coatings may be applied to papers to improve their functionality. These include many of the polymer materials discussed in Section 8.3.2. They may be used to increase the strength of paper, make it heat sealable, and/or improve its barrier properties.

These various types of papers may be used to wrap individual items or portions. Examples include waxed paper wraps for toffees and vegetable parchment paper wraps for butter and margarine. They may be made into small sachets or bags. Examples include sulfite paper sachets for custard powders or cake mixes and bags for sugar and flour. Kraft papers may be fabricated into multiwall paper sacks containing from two to six plies. They are used for fruits, vegetables, grains, sugar, and salt in quantities up to 25 kg. Where extra protection is required against water vapor, one or more plies maybe wax- or polyethylene (PE) coated. The outer layer may consist of wet-strength paper.

8.3.1.2 Paperboards

Paperboards are made from the same raw materials as papers. They are normally made on the cylinder machine and consist of two or more layers of different quality pulps with a total thickness in the range 300–1100 μm. The types of paperboard used in food packaging include the following:

- **Chipboard**: This is made from a mixture of repulped waste with chemical and mechanical pulp. It is dull gray in color and relatively weak. It is available lined on one side with unbleached, semi, or fully bleached chemical pulp. A range of such paperboards are available, with different quality liners. Chipboards are seldom used in direct contact with foods, but are used as outer cartons when the food is already contained in a film pouch or bag, for example, breakfast cereals.
- **Duplex board**: This is made from a mixture of chemical and mechanical pulp, usually lined on both sides with chemical pulp. It is used for some frozen foods, biscuits, and similar products.
- **Solid white board**: All plies are made from fully, bleached chemical pulp. It is used for some frozen foods, food liquids, and other products requiring special protection.
- Other paperboards are available that are coated with wax or polymer materials such as PE, polyvinylidene chloride (PVdC), and polyamides (PAs). These are mainly used for packaging wet or fatty foods.

Paperboards are mainly used in the form of cartons. Cartons are fed to the filling machine in a flat or collapsed form, where they are erected, filled, and sealed. The thicker grades of paperboards are used for set-up boxes, which come to the filling machine already erected. These are more rigid than cartons and provide additional mechanical protection.

8.3.1.3 **Molded Pulp**

Molded pulp containers are made from a waterborne suspension of mechanical, chemical or waste pulps, or mixtures of same. The suspension is molded into shape either under pressure (pressure injection molding) or vacuum (suction molding) and the resulting containers are dried. Such containers have good cushioning properties and limit in-pack movement, thus providing good mechanical protection to the contents. Trays for eggs and fruits are typical examples.

8.3.1.4 **Fiberboards**

Fiberboard is available in solid or corrugated forms. Solid fiberboard consists of a layer of paperboard, usually chipboard, lined on one or both faces with kraft paper. Solid fiberboard is rigid and resistant to puncturing. Corrugated fiberboard consists of one or more layers of corrugated material (medium) sandwiched between flat sheets of paperboard (linerboard), held in place by adhesive. The medium may be chipboard, strawboard, or board made from mixtures of chemical and mechanical pulp. The completed board may have one (single wall), two (double wall), or three (triple wall) layers of corrugations with linerboard in between. Four different flute sizes are available:

- A (104–125 flutes m^{-1}) is described as coarse and has good cushioning characteristics and rigidity.
- B (150–184 flutes m^{-1}) is designated as fine and has good crush resistance.
- C (120–145 flutes m^{-1}) is a compromise between these properties.
- E (275–310 flutes m^{-1}) is classed as very fine and is used for small boxes and cartons, when some cushioning is required.

Wax- and polymer-coated fiberboards are available. Fiberboards are usually fabricated into cases which are used as outer containers, to provide mechanical protection to the contents. Unpackaged products such as fruits, vegetables, and eggs are packaged in such containers. Inserts within the case reduce in-pack movement. Fiberboard cases are also used for goods already packaged in pouches, cartons, cans, and glass containers.

8.3.1.5 **Composite Containers**

So-called composite containers usually consist of cylindrical bodies made of paperboard or fiberboard with metal or plastic ends. Where good barrier properties are required, coated or laminated board may be used for the body or aluminum foil may be incorporated into it. Small containers, less than 200 mm in diameter, are referred to as *tubes* or *cans* and are used for foods such as salt, pepper, spices, custard powders, chocolate beverages, and frozen fruit juices. Larger containers, known as *fiberboard drums*, are used as alternatives to paper or plastic sacks or metal drums for products such as milk powder, emulsifying agents, and cooking fats [14, 17–23].

8.3.2
Wooden Containers

Outer wooden containers are used when a high degree of mechanical protection is required during storage and transport. They take the form of crates and cases. Wooden drums and barrels are used for liquid products. The role of crates has largely been replaced by shipping containers. Open cases find limited use for fish, fruits, and vegetables, although plastic cases are now widely used. Casks, kegs, and barrels are used for storage of wines and spirits. Oak casks are used for high-quality wines and spirits. Lower quality wines and spirits are stored in chestnut casks [14].

8.3.3
Textiles

Jute and cotton are woven materials which have been used for packaging foods. Sacks made of jute are used, to a limited extent, for fresh fruit and vegetables, grains, and dried legumes. However, multiwall paper sacks and plastic sacks have largely replaced them for such products. Cotton bags have been used in the past for flour, sugar, salt, and similar products. Again, paper and plastic bags are now mainly used for these foods. Cotton scrims are used to pack fresh meat. However, synthetic materials are increasingly used for this purpose [14].

8.3.4
Flexible Films

Nonfibrous materials in continuous sheet form up to 0.25 mm thick are termed *packaging films*. They are flexible, usually transparent, unless deliberately pigmented, and with the exception of regenerated cellulose, thermoplastic to some extent. This latter property enables many of them to be heat-sealed.

With the exception of regenerated cellulose, most films consist of a polymer, or a mixture of two or more polymers, to which are added other materials to give them particular functional properties, alter their appearance, or improve their handling characteristics. Such additives may include plasticizers, stabilizers, coloring materials, antioxidants, antiblocking, and slip agents.

- Extrusion is the method most commonly used to produce polymer films. The mixture of polymer and additives is fed into the extruder, which consists of a screw revolving inside a close-fitting, heated barrel. The combination of the heat applied to the barrel and that generated by friction melts the mixture, which is then forced through a die in the form of a tube or flat film. The extrudate is stretched to control the thickness of the film and rapidly cooled. By using special adaptors, it is possible to extrude two or more different polymers simultaneously. They fuse together to form a single web. This is known as *coextrusion*.
- Calendering is another techniques used to produce polymer films and sheets. The heated mixture of polymer and additives is squeezed between a series of heated rollers with a progressively decreasing clearance. The film formed

then passes over cooled rollers. Some PVC, ethylene-vinyl acetate (EVA), and ethylene-propylene copolymers are calendered.

- Solution casting is also used to a limited extent. The plastic material, with additives, is dissolved in a solvent, filtered and the solution cast through a slot onto a stainless steel belt. The solvent is driven off by heating. The resulting film is removed from the belt. Films produced in this way have a clear, sparkling appearance. Cellulose acetate and ethyl cellulose films are among those that are produced by solvent casting.
- Orientation is a process applied to some films in order to increase their strength and durability. It involves stretching the film in one (uniaxial orientation) or two directions at right angles to each other (biaxial orientation). This causes the polymer chains to line up in a particular direction. In addition to their improved strength, oriented films have better flexibility and clarity and, in some cases, lower permeability to water vapor and gases, compared to non-oriented forms of the same polymer film. Oriented films tear easily and are difficult to heat-seal. The process involves heating the film to a temperature at which it is soft before stretching it. Flat films are passed between heated rollers and then stretched on a machine known as a *tenter*, after which they are passed over a cooling roller. Films in the form of tubes are flattened by passing through nip rollers, heated to the appropriate temperature, and stretched by increasing the air pressure within the tube. When stretched to the correct extent they are cooled on rollers. Polyester, polypropylene (PP), low-density polyethylene (LDPE), and PA are the films that are mainly available in oriented form.
- Irradiation of some thermoplastic films can bring about crosslinking of the C–C bonds, which can increase their tensile strength, broaden their heating sealing range, and improve their shrink characteristics. PE is the film most widely irradiated, using an electron beam accelerator.

The following are brief details of the packaging films which are commonly used to package food.

8.3.4.1 Regenerated Cellulose

Regenerated cellulose (cellophane) differs from the polymer films in that it is made from wood pulp. Good-quality, bleached sulfite pulp is treated with sodium hydroxide and carbon disulfide to produce sodium cellulose xanthate. This is dispersed in sodium hydroxide to produce viscose. The viscose is passed through an acid-salt bath which salts out the viscose and neutralizes the alkali. The continuous sheet of cellulose hydrate produced in this way is desulfured, bleached, and passed through a bath of softener solution to give it flexibility. It is then dried in an oven. This is known as *plain (P) regenerated cellulose*. It is clear, transparent, not heat-sealable, and has been described as a transparent paper. It provides general protection against dust and dirt, some mechanical protection and is greaseproof. When dry it is a good barrier to gases, but becomes highly permeable when wet. Plain cellulose is little used in food packaging. Plain regenerated cellulose is mainly used coated with various materials which improve its functional properties. The most common coating material is referred to as "*nitrocellulose*" but is actually a mixture of nitrocellulose, waxes, resins, plasticizers, and some other agents. The following code letters are used to reflect the properties of coated regenerated cellulose films:

- A: anchored coating, that is, lacquer coating
- D: coated on one side only
- M: moistureproof
- P: uncoated
- Q: semi-moistureproof
- S: heat-sealable
- T: transparent
- X or XD: copolymer coated on one side only
- XX: copolymer coated on both sides.

The types of film most often used for food packaging include:

- MSAT: nitrocellulose-coated on both sides, a good barrier to water vapor, gases and volatiles, and heat-sealable;
- QSAT: nitrocellulose-coated on both sides, more permeable to water vapor than MSAT and heat-sealable;
- DMS: nitrocellulose-coated on one side only;
- MXXT: copolymer coated on both sides, very good barrier to water vapor, gases and volatiles, strong heat-seal;
- MXDT: copolymer coated on one side only.

The copolymer used in the X and XX films is a mixture of PVC and PVdC. The various coated films are used in the form of pouches and bags and as a component in laminates.

8.3.4.2 **Cellulose Acetate**

Cellulose acetate is made from waste cotton fibers which are acetylated and partially hydrolyzed. The film is made by casting from a solvent or extrusion. It is clear, transparent, and has a sparkling appearance. It is highly permeable to water vapor, gases, and volatiles. It is not much used in food packaging except as window material in cartons. It can be thermoformed into semi-rigid containers or as blister packaging.

8.3.4.3 **Polyethylene**

Polyethylene, commonly called *polythene*, is made in one of two ways. Ethylene is polymerized at high temperature and pressure in the presence of a little oxygen and the polymer converted into a film by extrusion. Alternatively, lower temperatures and pressures may be used to produce the polymer if certain alkyl metals are used as catalysts. The film is available in low (LDPE), medium (MDPE), and high (HDPE) density grades. The lower density grades are most widely used in food packaging. The main functional properties of LDPE are its strength, low permeability to water vapor and it forms a very strong heat seal. It is not a good barrier to gases, oils, or volatiles. It is used on its own in the form of pouches, bags, and sacks. It is also used for coating papers, boards, and plain regenerated cellulose and as a component in laminates. HDPE has a higher tensile strength and stiffness than LDPE. Its permeability to gases is lower and it can withstand higher temperatures. It is used for foods which are heated in the package, so-called "boil in the bag" items.

8.3.4.4 Polyvinyl Chloride

Polyvinyl chloride is made by chlorination of acetylene or ethylene followed by polymerization under pressure in the presence of a catalyst. The film can be formed by extrusion or calendering. It is a clear, transparent film which on its own is brittle. The addition of plasticizers and stabilizers to the polymer are necessary to give it flexibility. It is essential that PVC film used in food packaging contains only permitted additives to avoid any hazard to the consumer (see Section 8.3.6). It has good mechanical properties. Its permeability to water vapor, gases, and volatiles depends on the type and amount of plasticizers added to the polymer. The most common grade used for food packaging is slightly more permeable to water vapor than LDPE, but less permeable to gases and volatiles. It is a good grease barrier. It can be sealed by high-frequency welding. It can be orientated and as such is heat-shrinkable. Highly plasticized grades are available with stretch and cling properties. This is one form of "cling film" which is used for stretch-wrapping foods in industry and in the home.

8.3.4.5 Polyvinylidene Chloride

Polyvinylidene chloride is made by further chlorination of vinyl chloride in the presence of a catalyst, followed by polymerization. The polymer itself is stiff and brittle and unsuitable for use as a flexible film. Consequently, it is a copolymer of PVdC with PVC that is used for food packaging. Typically, 20% of vinyl chloride is used in the copolymer, although other ratios are available. The film is usually produced by extrusion of the copolymer. The properties of the copolymer film depend on the degree of polymerization, the properties of the monomers used, and the proportion of each one used. The copolymer film most widely used for food packaging has good mechanical properties, is a very good barrier to the passage of water vapor, gases, and volatiles and is greaseproof and heat-sealable. It can withstand relatively high temperatures such as those encountered during hot filling and retorting. It is available in oriented form which has improved strength and barrier properties and is highly heat-shrinkable. PVdC/PVC copolymer film is used for shrink-wrapping foods such as meat and poultry products and as a component in laminates.

8.3.4.6 Polypropylene

Polypropylene is produced by low-pressure polymerization of propylene in the presence of a catalyst. The film is normally extruded onto chilled rollers and is known as *cast polypropylene*. Its mechanical properties are good except at low temperature, when it becomes brittle. The permeability of cast PP to water vapor and gases is relatively low, comparable with HDPE. It is heat-sealable, but at a very high temperature, 170 °C. It is usually coated with PE or PVdC/PVC copolymer to facilitate heat-sealing. Cast PP is used in the form of bags or overwraps for applications similar to PE. Oriented polypropylene (OPP) has better mechanical properties than cast PP, particularly at low temperature, and is used in thinner gauges. It is a good barrier to water vapor but not gases. It is often coated with PP or PVdC/PVC copolymer to improve its barrier properties and to make it heat-sealable. It is normally heat-shrinkable. It is used in coated or laminated form to package a wide range of food products, including biscuits, cheese, meat, and coffee. It is stable at relatively high temperature and is used for in-package heat processing. A white opaque form of OPP, known as *pearlized film*, is also available. Copolymers of PP and PE are also available. Their functional properties tend to be in a range between PP and HDPE.

8.3.4.7 **Polyester**

Polyester film used in food packaging is polyethylene terephthalate (PET), which is usually produced by a condensation reaction between terephthalic acid and ethylene glycol and extruded. There is little use of the non-oriented form of PET but it widely used in the biaxially oriented form. Oriented PET has good tensile strength and can be used in relatively thin gauges. It is often used coated with PE or PVdC/PVC copolymer to increase its barrier properties and facilitate heat-sealing. It is stable over a wide temperature range and can be used for "boil in the bag" applications. Metallized PET is also available and has a very low permeability to gases and volatiles. Metallized, coextruded PE/PET is used for packaging snack foods.

8.3.4.8 **Polystyrene**

Polystyrene (PS) is produced by reacting ethylene with benzene to form ethylbenzene. This is dehydrogenated to give styrene which is polymerized at a relatively low temperature, in the presence of catalysts, to form PS. PS film is produced by extrusion. It is stiff and brittle with a clear sparkling appearance. In this form it is not useful as a food packaging film. Biaxially oriented polystyrene (BOPS) is less brittle and has an increased tensile strength, compared to the non-oriented film. BOPS has a relatively high permeability to vapors and gases and is greaseproof. It softens at about 80–85 °C, but is stable at low temperature, below 0 °C. It shrinks on heating and may be heat-sealed by impulse sealers. The film has few applications in food packaging, apart from wrapping of fresh produce. PS is widely used in the form of thermoformed semi-rigid containers and blow-molded bottles. For these applications it is coextruded with ethylene-vinyl alcohol (EVOH) or PVdC/PVC copolymer. PS is also used in the form of a foam for containers such as egg cartons, fruit trays, and containers for takeaway meals.

8.3.4.9 **Polyamides**

Polyamides, known generally as *nylons*, are produced by two different reactions. Nylon 6,6 and 6,10 are formed by condensation of diamines and dibasic acids.

The numbers indicate the number of carbon atoms in the diamine molecules followed by the number in the acid. Nylons 11 and 12 are formed by condensation of ω-amino acids. Here, the numbers indicate the total number of carbon atoms involved. The film may be extruded or solution cast. PA films are clear and attractive in appearance. As a group they are mechanically strong, but the different types do vary in strength. The permeability to water vapor varies from high, nylon 6, to low, nylon 12. They are good barriers to gases, particularly under dry conditions, volatiles, and greases. They are stable over a very wide temperature range. They can be heat-sealed but at a high temperature, 240 °C. They do absorb moisture and their dimensions can change by 1–2% as a result. Nylon films may be combined with other materials, by coating, coextrusion, or lamination, in order to facilitate heat-sealing and/or improve their mechanical and barrier properties. PE, ionomers, EVA, and ethylene-acrylic acid (EAA) (see below) are among such other materials. Different types of nylon may be combined as copolymers, for example, nylon 6/6,6 or 6/12. Biaxially oriented nylon films are also available. Their functional properties may be further modified by vacuum metallizing. Applications for nylon films include packaging of meat products, cheese, and condiments.

8.3.4.10 Polycarbonate

Polycarbonate (PC) is made by the reaction of phosgene or diphenyl carbonate with bisphenol A. The film is produced by extrusion or casting. It is mechanically strong and grease-resistant. It has a relatively high permeability to vapors and gases. It is stable over a wide temperature range, from −70 to 130 °C. It is not widely used for food packaging but could be used for "boil in the bag" packages, retortable pouches, and frozen foods.

8.3.4.11 Polytetrafluoroethylene

Polytetrafluoroethylene (PTFE) is made by the reaction of hydrofluoric acid with chloroform followed by pyrolysis and polymerization. The film is usually produced by extrusion. It is strong, has a relatively low permeability to vapors and gases and is grease-resistant. It is stable over a wide temperature range, −190 to 190 °C and has a very low coefficient of friction. It is not widely used in film form for packaging of foods but could be used for retortable packages and where a high resistance to the transfer of vapors and gases is required, for example, for freeze-dried foods. It is best known for its nonstick property and is used on heat sealers and for coating cooking utensils.

"Ionomers" are formed by introducing ionic bonds as well as the covalent bonds normally present in polymers such as PE. This is achieved by reacting with metal ions. Compared to LDPE, they are stiffer and more resistant to puncturing and have a higher permeability to water vapor and good grease resistance. They are most widely used as components in laminates with other films, such as PC or PET, for packaging cheese and meat products.

8.3.4.12 Ethylene-Vinyl Acetate Copolymers

Ethylene-vinyl acetate copolymers are made by the polymerization of PE with vinyl acetate. Compared to LDPE, they have higher impact strength, higher permeability to water vapor and gases, and are heat-sealable over a wider temperature range. EVA with other polymers such as ethylene ethyl acrylate (EEA) and EAA form a family of materials that may be used, usually in laminates with PE, PP, and other films, for food packaging. Care must be taken in selecting these materials as there are limitations on the quantity of the minor components which should be used for particular food applications. EVA itself has very good stretch and cling characteristics and can be used as an alternative to PVC for cling-wrap applications [14, 17, 18, 24, 25].

8.3.5 Metallized Films

Many flexible packaging films can have a thin coating, less than 1 μm thick, of metal applied to them. This was originally introduced for decorative purposes. However, it emerged that metallizing certain films increased their resistance to the passage of water vapor and gases, by up to 100%. Today metallized films are used extensively to package snack foods. The process involves heating the metal, usually aluminum, to temperatures of 1500–1800 °C in a vacuum chamber maintained at a very low pressure, about 10^{-4} torr (0.13 Pa). The metal vaporizes and deposits onto the film which passes through the vapor on a chilled roller. PET, PP, PA, PS, PVC, PVdC, and regenerated cellulose are available in metallized form [18, 25, 26].

8.3.6 Flexible Laminates

When a single paper or film does not provide adequate protection to the product, two or more flexible materials may be combined together in the form of a laminate. In this form the functional properties of the individual components complement each other to suit the requirements of a particular food product. The materials involved may include papers or paperboards, films, and aluminum foil. The paper or paperboard provides stiffness, protects the foil against mechanical damage and has a surface suitable for printing.

The film(s) contributes to the barrier properties of the laminate, provides a heat-sealable surface, and strengthens the laminate. The foil acts as a barrier material and has an attractive appearance. Laminates may be formed from paper–paper, paper–film, film–film, paper–foil, film–foil, and paper–film–foil combinations. The layers of a laminate may be bonded together by adhesive. When one or more of the layers is permeable to water vapor, an aqueous adhesive may be used. Otherwise, nonaqueous adhesives must be used. If one or more of the components is thermoplastic, it may be bonded to the other layer by passing them between heated rollers. A freshly extruded thermoplastic material, still in molten form, may be applied directly to another layer and thus bonded to it. Two or more thermoplastic materials may be combined to together by coextrusion (see Section 8.3.4). There are hundreds of combinations of different materials available. Examples include:

- vegetable parchment–foil for wrapping butter and margarine;
- MXXT regenerated cellulose–PE for vacuum-packed cheese, cooked, and cured meats;
- PET–PE for coffee, paperboard–foil–PE for milk and fruit juice cartons.

Retortable pouches may be made of a three-ply laminate typically consisting of PET–foil–PP or PET–foil–HDPE [14, 17, 18, 25].

8.3.7
Heat-Sealing Equipment

Many flexible polymer films are thermoplastic and heat-sealable. Nonthermoplastic materials may be coated with or laminated to thermoplastic material to facilitate heat-sealing. Heat-sealing equipment must be selected to suit the type of material being sealed. Nonthermoplastic materials such as papers, regenerated cellulose, and foil, which are coated with heat-sealable material, are best sealed with a hot bar or *resistance sealer*. The two layers of material are clamped between two electrically heated metal bars. The temperature of the bars, the pressure exerted by them, and the contact time all influence the sealing. Metal jaws with matching serrations are often used for coated regenerated cellulose films. The serrations stretch out wrinkles and improve the seal. In the case of laminates, smooth jaws meeting uniformly along their length are used. Alternatively, one of the jaws is made of resilient material, often silicone elastomer, which is not heated.

For continuous heat-sealing of coated material, heated rollers are used. Heated plates are used to seal wrapped items. For unsupported, thermoplastic materials, *impulse sealers* are used. In such sealers, the layers of film are clamped between jaws of resilient material, one or both of which has a narrow metal strip running the length of the jaw. An accurately timed pulse of low-voltage electricity is passed through the strip(s), heating it and fusing the two layers of material together. The jaws are held apart by unmelted material each side of the strip. This minimizes the thinning of the sealed area, which would weaken the seal. The jaws remain closed until the melted material solidifies. The jaws are coated with PTFE to prevent the film sticking to them. For continuous sealing of unsupported thermoplastic material a *band sealer* may be used. A pair of moving, endless metal belts or bands is heated by stationary, heated shoes. The shoes are so shaped that they touch the center of the bands only. This minimizes thinning of the seal. After passing between the heated shoes, the layers of material pass between pressure rollers and then between cooled shoes to solidify the melted film. A heated wire may be used to simultaneously cut and seal unsupported thermoplastic films.

Electronic sealing is used on relatively thick layers of polymer material with suitable electrical properties, mainly PVC and PVC/PVdC copolymers. The layers of film are placed between shaped electrodes and subjected to a high-frequency electric field. This welds the layers of material together. *Ultrasonic sealing* may be used to seal layers of film or foil together. This is particularly suited to uncoated, oriented materials that are difficult to seal by other methods [16–18, 25].

8.3.8
Packaging in Flexible Films and Laminates

Flexible films may be used to overwrap items of food such as portions of meat. The meat is usually positioned on a tray made of paperboard or foamed plastic, with an absorbent pad between it and the tray. The film is stretched over the meat and under the tray. It may be heat-sealed on a heated plate or held in position by clinging to itself. Films may also be made into preformed bags which are filled by hand or machine and sealed by heat or other means. Heavy-gauge material, such as PE, may be made into shipping sacks for handling large amounts, 25–50 kg, of foods such as grains or milk powder.

However, films and laminates are most widely used in the form of sachets or pillow packs. A *sachet* is a small square or rectangular pouch heat-sealed on all four edges (Figure 8.1), whereas a *pillow pack* is a pouch with a longitudinal heat seal and two end seals (Figure 8.2). These are formed, filled, and sealed by a sequential operation, known as a *form–fill–seal* system. FFS machines may operate vertically or horizontally. The principle of a vertical FFS machine for making sachets is shown in Figure 8.1. Vertical FFS machines can also make pillow packs (Figure 8.2). Vertically formed pillow and sachet packs may be used for either liquids or solids.

The principle of a horizontal FFS for sachets is shown in Figure 8.3. Such a system is used for both solid and liquid products. Horizontal FFS machines can also produce pillow packs (Figure 8.4). Systems like these are used for solid items

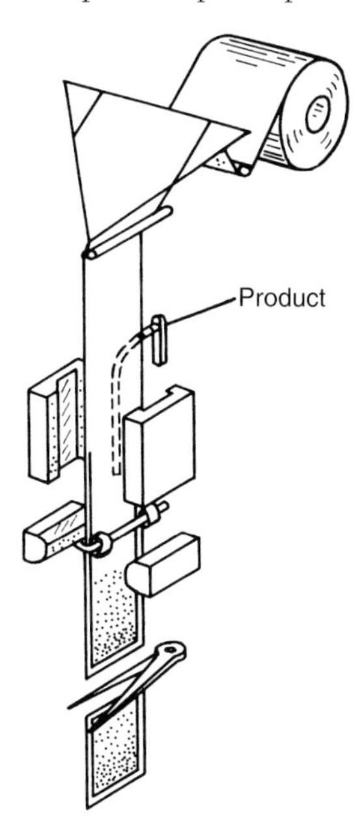

Figure 8.1 Vertical form-fill-seal machine for sachets. Adapted from Brennan *et al.* [16] with permission.

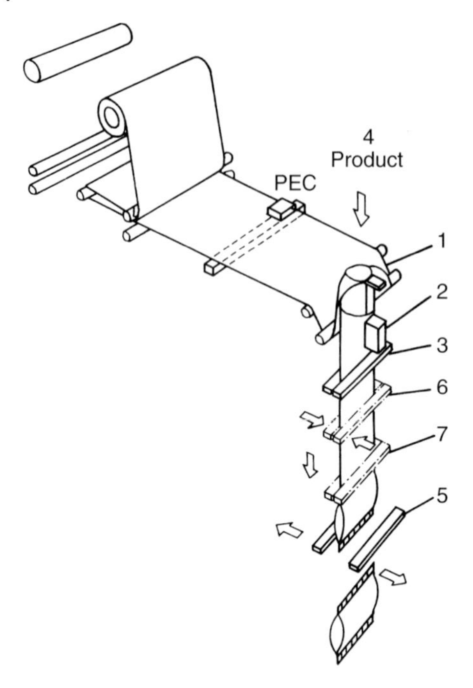

Figure 8.2 Vertical form-fill-seal machine for pillow packs. 1. Film from reel made into a tube over forming shoulder. 2. Longtitudinal seal made. 3. Bottom of tube closed by heat crimped jaws which move downwards drawing film from reel. 4. Predetermined quantity of product falls through collar into pouch. 5. Jaws open and return on top of stroke. 6. Jaws partially close and "scrape" product into pouch out of seal area. 7. Jaws close, crimp seal top of previous pouch and bottom of new one. Crimp-sealed container cut-off with knife. Adapted from Brennan *et al.* [16] with permission.

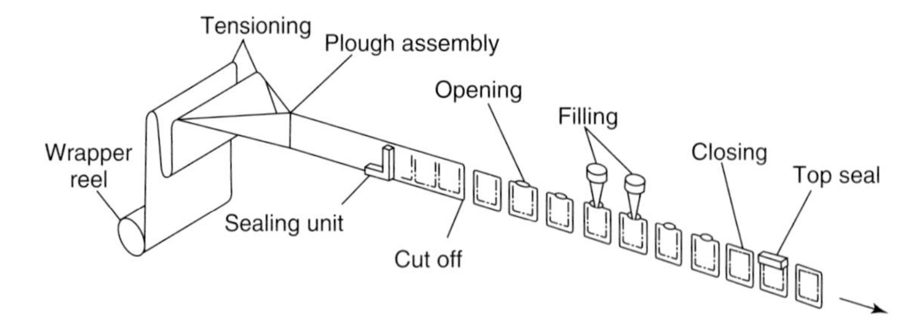

Figure 8.3 Horizontal form-fill-seal machine for sachets. Adapted from Brennan *et al.* [16] with permission.

such as candy bars or biscuits. Pillow packs are more economical than sachets in the use of packaging material. The packaging materials must be thin and flexible, have good slip characteristics and form a strong seal, even before cooling. Sachets are made from stiffer material and can be used for a wider range of product types. They are usually used in relatively small sizes, for example, for individual portions of sauce or salad dressing [14, 16–18, 25, 27].

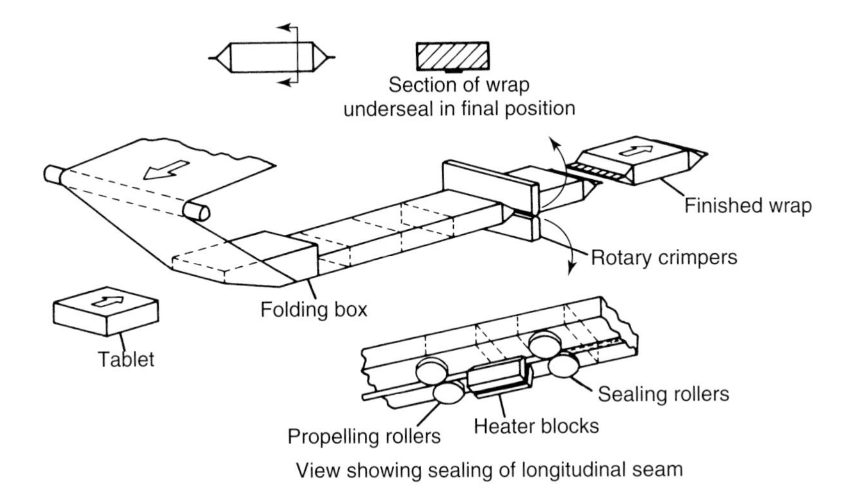

Figure 8.4 Horizontal form-fill-seal machine for pillow packs. Film is drawn from reel and formed into horizontal tube around product with continuous seal underneath formed by heater blocks and crimping rollers. Then, rotary heaters make the crimped end seals and cut-off produces individual packs. Adapted from Brennan *et al.* [16] with permission.

8.3.9 Rigid and Semi-rigid Plastic Containers

Many of the thermoplastic materials described above can be formed into rigid and semi-rigid containers, the most common being LDPE, HDPE, PVC, PP, PET, and PS, singly or in combinations. Acrylic plastics are also used for this purpose, including polyacrylonitrile and acrylonitrile-butadiene-styrene (ABS). Urea formaldehyde, a thermosetting material, is used to make screwcap closures for glass and plastic containers.

The following methods are used to convert these materials into containers:

8.3.9.1 Thermoforming

In thermoforming, a plastic sheet is clamped in position above a mold. The sheet is heated until it softens and then made to take up the shape of the mold by either (i) having an air pressure greater than atmospheric applied above the sheet, (ii) having a vacuum created below the sheet, or (iii) sandwiching the sheet between a male and female mold. The sheet cools through contact with the mold, hardens, and is ejected from the mold. Plastic materials that are thermoformed include PS, PP, PVC, HDPE, and ABS. Thermoforming is used to produce open-topped or wide-mouthed containers such as cups and tubs for yoghurt, cottage cheese or margarine, trays for eggs or fresh fruit and inserts in biscuit tins, or chocolate boxes.

8.3.9.2 Blow Molding

In blow molding, a mass of molten plastic is introduced into a mold and compressed air is used to make it take up the shape of the mold. The plastic cools, hardens, and is ejected from the mold. Blow molding is mainly used to produce narrow-necked containers. LDPE is the main material used for blow molding, but PVC, PS, and PP may also be processed in this way. Food applications include bottles for oils, fruit juices and milk, and squeezable bottles for sauces and syrups.

8.3.9.3 Injection Molding

In injection molding, the molten plastic from an extruder is injected directly into a mold, taking up the shape of the mold. On cooling, the material hardens and is ejected from the mold. Injection molding is used mainly used to produce wide-mouthed containers, but, narrow-necked containers can be injection molded in two parts which are joined together by a solvent or welding. PS is the main material used for injection molding, but PP and PET may also be processed in this way. Food applications include cups and tubs for cream, yoghurt, mousses as well as phials, and jars for a variety of uses.

8.3.9.4 Compression Molding

Compression molding is used from thermosetting plastics, such as urea formaldehyde. The plastic powder is held under pressure between heated male and female molds. It melts and takes up the shape of the mold after which it is cooled, the mold opened and the item ejected. The main application for this method is to produce screwcaps [14, 17, 18, 25, 28–30].

8.3.10 Metal Materials and Containers

The metal materials used in food packaging are aluminum, tinplate, and electrolytic chromium-coated steel (ECCS). Aluminum is used in the form of foil or rigid metal.

8.3.10.1 Aluminum Foil

Aluminum foil is produced from aluminum ingots by a series of rolling operations down to a thickness in the range 0.15–0.008 mm. Most foil used in packaging contains not less than 99.0% aluminum, with traces of silicon, iron, copper, and in some cases, chromium and zinc. Foil used in semi-rigid containers also contains up to 1.5% manganese. After rolling, foil is annealed in an oven to control its ductility. This enables foils of different tempers to be produced from fully annealed (dead folding) to hard, rigid material. Foil is a bright, attractive material, tasteless, odorless, and inert with respect to most food materials. For contact with acid or salty products, it is coated with nitrocellulose or some polymer material. It is mechanically weak, easily punctured, torn, or abraded. Coating or laminating it with polymer materials will increase its resistance to such damage. Relatively thin foil, less than 0.03 mm thick, will contain perforations and will be permeable to vapors and gases. Again, coating or laminating it with polymer material will improve its barrier properties. Foil is stable over a wide temperature range. Relatively thin, fully annealed foil is used for wrapping chocolate and processed cheese portions. Foil is used as a component in laminates, together with polymer materials and, in some cases, paper. These laminates are formed into sachets or pillow packs on FFS equipment (see Section 8.3.6). Examples of foods packaged in this way include dried soups, sauce mixes, salad dressings, and jams.

Foil is included in laminates used for retortable pouches and rigid plastic containers for ready meals. It is also a component in cartons for UHT milk and fruit juices. Foil in the range 0.040–0.065 mm thick is used for capping glass and rigid plastic containers. Plates, trays, dishes, and other relatively shallow containers are made from foil in the thickness range 0.03–0.15 mm and containing up to 1.5% manganese. These are used for frozen pies, ready meals, and desserts, which can be heated in the container.

8.3.10.2 **Tinplate**

Tinplate is the most common metal material used for food cans. It consists of a low-carbon, mild steel sheet, or strip, 0.50–0.15 mm thick, coated on both sides with a layer of tin. This coating seldom exceeds 1% of the total thickness of the tinplate. The structure of tinplate is more complex than would appear from this simple description and several detectable layers exist (Figure 8.5).

The mechanical strength and fabrication characteristics of tinplate depend on the type of steel and its thickness. The minor constituents of steel are carbon, manganese, phosphorus, silicon, sulfur, and copper. At least four types of steel, with different levels of these constituents, are used for food cans. The corrosion resistance and appearance of tinplate depend on the tin coating. The stages in the manufacture of tinplate are shown in Figure 8.6. These result in two types of tinplate, that is, single-reduced or cold-reduced (CR) and double-reduced (DR) electroplate. DR electroplate is stronger in one direction than CR plate and can be used in thinner gauges than the latter for certain applications.

The thickness of tinplate used for food can manufacture is at the lower end of the range given above. CR plate thickness may be as low as 0.17 mm and DR plate 0.15 mm. The amount of tin coating is now usually expressed as x g m^{-2}. This may be the same on both sides of the plate or a different coating weight may be applied to each side. The latter is known as *differentially coated plate.* In general, the more corrosive the product the higher the coating weight used. Coating weights range over 11.2–1.1 g m^{-2} (represented as E.11.2/11.2 to E.1.1/1.1) if the same

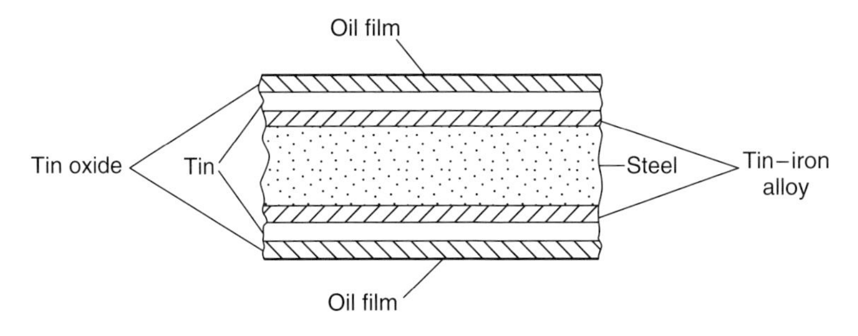

Figure 8.5 Structure of tinplate.

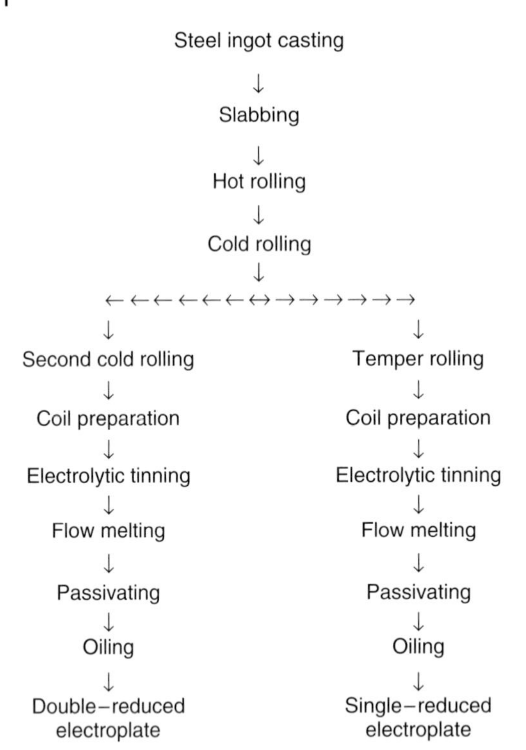

Figure 8.6 Stages in the manufacture of tinplate.

weight is applied to both sides. Differentially coated plate is identified by the letter D followed by the coating weights on each side. For example, D.5.6/2.8 plate has 5.6 g m^{-2} of tin on one side and 2.8 g m^{-2} on the other. Usually, the higher coating weight is applied to the side that will form the inside of the can.

Lacquer (enamel) may be applied to tinplate to prevent undesirable interaction between the product and the container. Such interactions arise with: (i) acid foods which may interact with tin dissolving it into the syrup and, in some cases, causing a loss of color in the product, (ii) some strongly colored products where anthocyanin color compounds react with the tin, causing a loss of color, (iii) sulfur-containing foods where the sulfur reacts with the tin, causing a blue-black stain on the inside of the can, and (iv) products sensitive to small traces of tin, such as beer. Lacquers can provide certain functional properties, such as a nonstick surface to facilitate the release of the contents of the can, for example, solid meats packs. A number of such lacquers are available, including natural, oleoresinous materials, and synthetic materials (Table 8.1). Cans may be made from prelacquered plate or the lacquer may be applied to the made-up can.

8.3.10.3 Electrolytic Chromium-Coated Steel

Electrolytic chromium-coated steel (ECCS), sometimes described as tin-free steel, is finding increasing use for food cans. It consists of low-carbon, mild CR, or DR steel coated on both sides with a layer of metallic chromium and chromium

Table 8.1 Main types of internal can lacquer (enamel).

General type of resin and compounds blended to produce it	Sulfide stain resistance	Typical uses	Comments
Oleo-resinous	Poor	Acid fruits	Good general purpose natural range at relatively low cost
Sulfur-resistant oleoresinous with added zinc oxide	Good	Vegetables, soups	Not for use with acid products
Phenolic (phenol or relatively low-substituted phenol with vegetables formaldehyde)	Very good	Meat, fish, soups	Good at cost but film thickness restricted by flexibility
Epoxy-phenolic (epoxy resins with phenolic resins)	Poor	Meat, fish, soups vegetables, beer, beverages (top coat)	Wide range of properties may be obtained by modifications
Epoxy-phenolic with zinc oxide	Good	Vegetables, soups (especially can ends)	Not for use with acid products; possible color change with green vegetables
Aluminized epoxyphenolic (metallic aluminum powder added)	Very good	Meat products	Clean but dull appearance
Vinyl solution (vinyl chloride-vinyl acetate copolymers)	Not applicable	Spray-on can bodies, roller coating on ends, as topcoat for beer and beverages	Free from flavor taints; not usually suitable for direct application to tinplate
Vinyl organosol or plastisol (high MW vinyl resins suspended in a solvent)	Not applicable	Beer and beverage topcoat on ends, drawn cans	As for vinyl solutions, but giving a thicker, tougher layer
Acrylic (acrylic resin usually pigmented white)	Very good when pigmented	Vegetables, soups, prepared foods containing sulfide stainers	Clean appearance
Polybutadiene (hydrocarbon resins)	Very good if zinc oxide added	Beer and beverage first coat, vegetables and soups with ZnO	Costs depend on country

From Morgan [31].

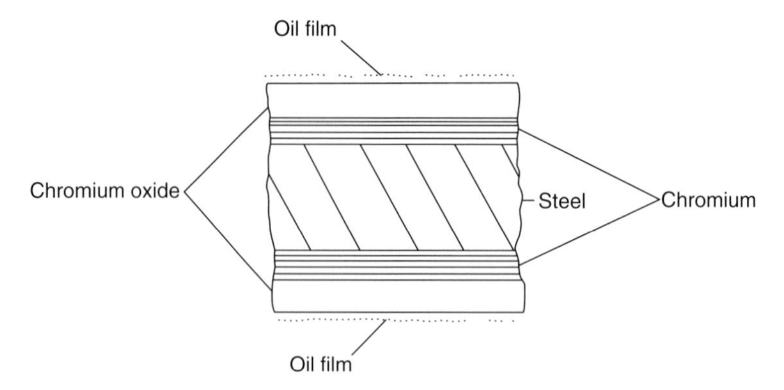

Figure 8.7 Structure of ECCS plate (tin-free steel).

sesqueoxide, applied electrolytically. It is manufactured by a similar process to that shown in Figure 8.6, but the flow melting and chemical passivation stages are omitted. A typical coating weight is 0.15 g m^{-2}, much lower than that on tinplate. ECCS is less resistant to corrosion than tinplate and is normally lacquered on both sides. It is more resistant to weak acids and sulfur staining than tinplate. It exhibits good lacquer adhesion and a range of lacquers, suitable for ECCS, is available. The structure of ECCS is shown in Figure 8.7.

8.3.10.4 **Aluminum Alloy**

Hard-temper aluminum alloy, containing 1.5–5.0% magnesium, is used in food can manufacture. Gauge for gauge, it is lighter but mechanically weaker than tinplate. It is manufactured in a similar manner to aluminum foil. It is less resistant to corrosion than tinplate and needs to be lacquered for most applications. A range of lacquers suitable for aluminum alloy is available, but the surface of the metal needs to be treated to improve lacquer adhesion.

8.3.10.5 **Metal Containers**

Metal cans are the most common metal containers used for food packaging. The traditional *three-piece can* (open or sanitary) is still very widely used for heat-processed foods. The cylindrical can body and two ends are made separately. One end is applied to the can body by the can maker, the other (the canners end) by the food processor after the can has been filled with product. The ends are stamped out of sheet metal, the edges curled in, and a sealing compound injected into the curl. The body blank is cut from the metal sheet, formed into a cylinder and the lapped side seam sealed by welding or by PA adhesive. Both ends of the cylindrical body are flanged in preparation for the application of the can end (Figure 8.8a).

The can end is applied to the body by means of double-seaming (Figure 8.9). The can body and end are clamped tightly between a chuck and a base plate. The chuck is made to rotate rapidly; and the can body and base rotate with the chuck. The first seaming roller moves in and engages with the chuck, forming mating hooks on the can body and end. The first seaming roller moves out and a second

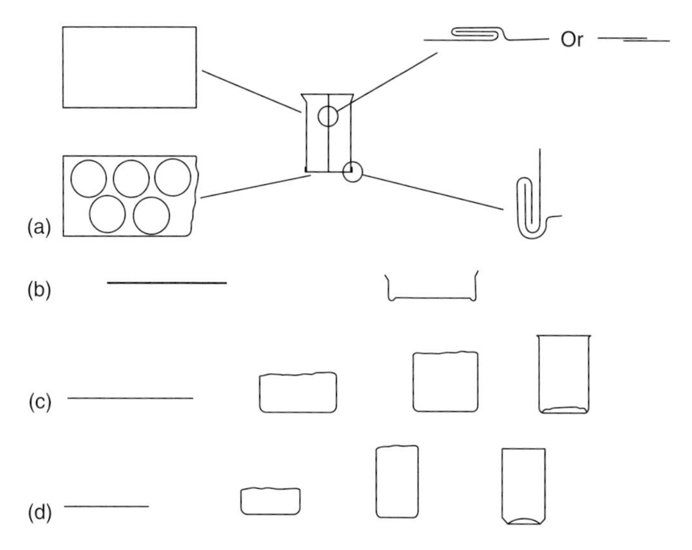

Figure 8.8 Schematic representation of the manufacture of: (a) three-piece can, (b) drawn can, (c) drawn and redrawn can, and (d) drawn and wall-ironed can. From Fellows [15] with permission of the authors.

roller moves in, tightens the hooks, and completes the seam. In some high-speed seaming machines and those used for noncylindrical cans, the can body, end and chuck remain stationary and the seaming rollers rotate on a carriage around them.

The *drawn can* is a type of two-piece container. The can body and base are made in one operation from a blank metal sheet by being pressed out with a suitable die. The open end of the body is flanged. The can end, manufactured as described above, is applied to the body by double-seaming after the can is filled with product. Because of the strain on the metal, DR cans are shallow with a maximum height : diameter ratio of 1 : 2 (Figure 8.8b).

The drawn and redrawn (DRD) can is another type of two-piece can. It is made by drawing a cup to a smaller diameter in a series of stages to produce a deeper container than the drawn can. The can end is applied to the filled can by double-seaming. DRD cans are usually relatively small, cylindrical, and have a height : diameter ratio of up to 1.2 : 1.0 (Figure 8.8c).

The drawn and wall-ironed (DWI) can is made from a disk of metal 0.30–0.42 mm thick. This is drawn into a shallow cup which is forced through a series of ironing rings of reducing internal diameter so that the wall of the cup gets thinner and higher. The top of the body is trimmed, flanged, and the end applied by double-seaming after filling the can (Figure 8.8d). Because of the very thin body wall, typically 0.10 mm thick, DWI cans are mainly used for packaging carbonated beverages. The internal pressure supports the thin wall.

The dimensions of cylindrical cans are usually specified in diameter and height, in that order. In many countries the units of diameter and height are millimeters. In the United Kingdom and United States inches and 16ths of an inch are used.

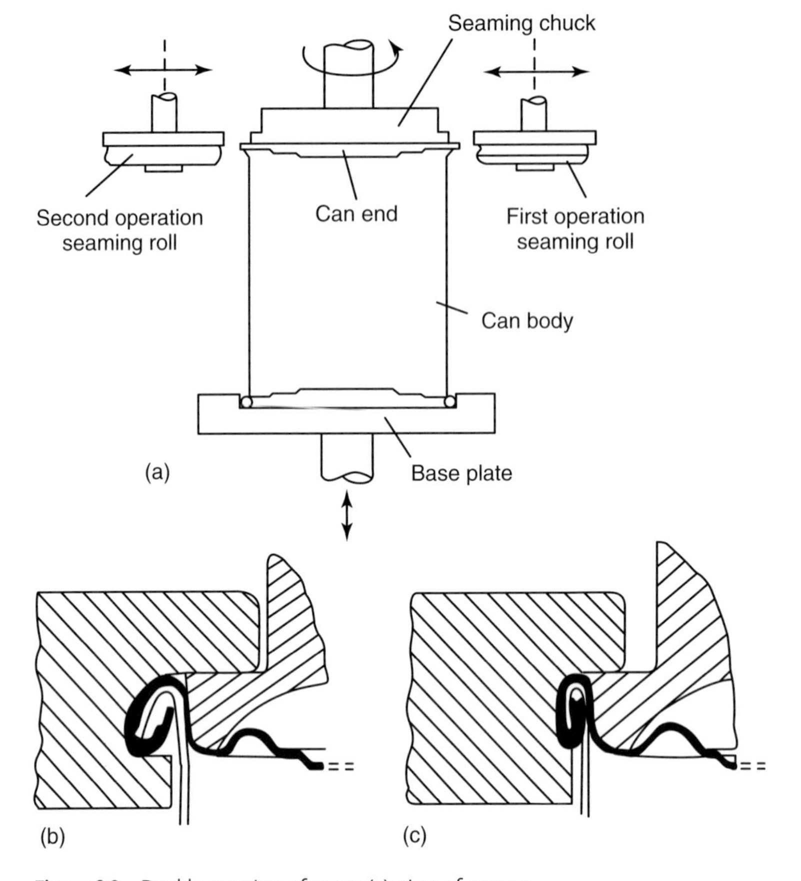

Figure 8.9 Double-seaming of cans: (a) view of seamer, (b) seam after completion of first operation, and (c) seam after completion of second operation. Adapted from Fellows [15] with permission of the authors.

Thus a can specified as 401/411 has a diameter of 41/16 inches and a height of 411/16 inches. In the case of rectangular or oval cans, two horizontal dimensions must be given.

Other metal containers used for packaging foods include [14, 15, 17, 18, 31–37]:

- cylindrical cans with a friction plug closure at the canners end, used for dry powders such as coffee and custard powders or for liquids such as syrups and jams;
- rectangular or cylindrical containers with push-on lids, often sealed with adhesive tape, used for biscuits and sweets;
- rectangular or cylindrical containers, incorporating apertures sealed with screw-caps, used for liquids such as cooking oils and syrups;
- metal drums used for beer and other carbonated drinks.

Table 8.2 Composition of a typical British glass.

Silica (from sand) (%)	72.0
Lime (from limestone) (%)	11.0
Soda (from synthetic sodium carbonate) (%)	14.0
Alumina (from aluminum minerals) (%)	1.7
Potash (as impurity) (%)	0.3

8.3.11 Glass and Glass Containers

In spite of the many developments in plastic containers, glass is still widely used for food packaging. Glass is inert with respect to foods, transparent, and impermeable to vapors, gases, and oils. Because of the smooth internal surface of glass containers, they can be washed and sterilized and used as multitrip containers, for example, milk and beer bottles. However, glass containers are relatively heavy compared to their metal or plastic counterparts, susceptible to mechanical damage and cannot tolerate rapid changes in temperature (low thermal shock resistance). Broken glass in a food area is an obvious hazard. The composition of a typical UK glass is shown in Table 8.2.

These ingredients, together with up to 30% recycled glass or cullet, are melted in a furnace at temperatures in the range 1350–1600 °C. The viscous mass passes into another chamber which acts as a reservoir for the forming machines. Two forming methods are used, that is, the blow and blow (B and B) process, which is used for narrow-necked containers, and the press and blow (P and B) process, which is used for wide-mouthed containers (Figure 8.10).

After forming, the containers are carried on a conveyor belt through a cooling tunnel, known as an *annealing lehr*. In this lehr, they are heated to just below the softening temperature of the glass, held at that temperature for about 5 min and then cooled in a controlled manner. This is done to remove any stresses in the glass that may have developed during forming and handling. These stresses would weaken the containers and make them less resistant to mechanical damage. As the containers leave the lehr, they are inspected for faults. To produce colored containers, coloring compounds such as metal oxides, sulfides, or selenides are included in the formulation. It is important that the dimensions and capacities of glass containers only vary within specified tolerance limits. Otherwise, breakages and hold ups may occur in the bottling plant and customer complaints may arise. When a delivery of new glass containers is received at the bottling plant, samples should be removed on a statistical basis and their dimensions and capacities measured and checked against specifications. In the case of cylindrical containers, the dimensions that are usually measured are the height, diameter, verticality (how truly vertical the container is), and ovality (how truly cylindrical it is). In the case of noncylindrical containers, other dimensions may be measured. The mechanical strength of glass containers, that is, their resistance to internal pressure, vertical

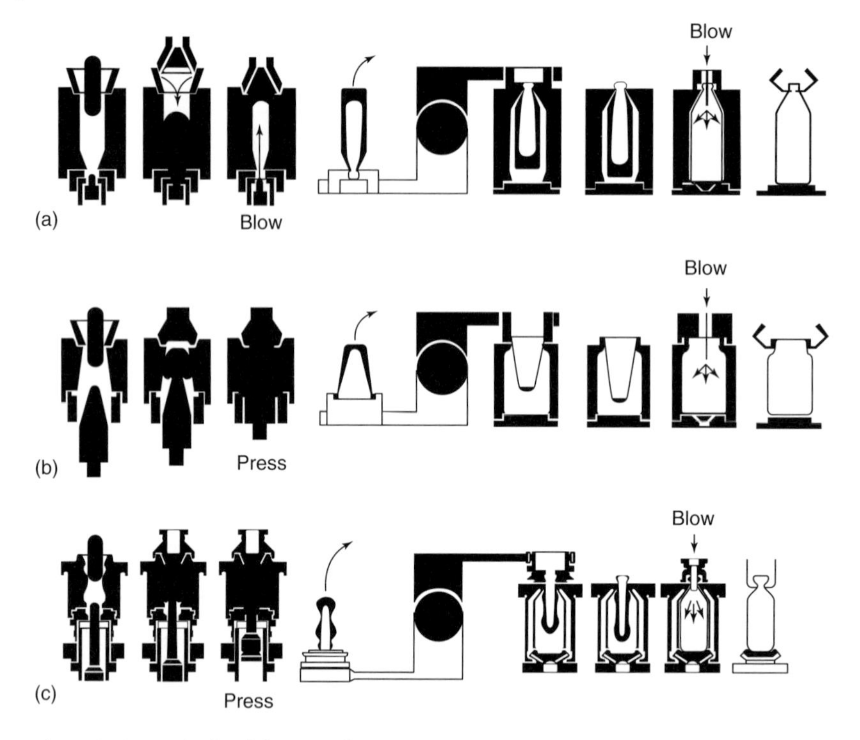

Figure 8.10 Methods of forming glass containers: (a) blow and blow forming, (b) press and blow forming for wide-mouthed containers, and (c) press and blow forming for narrow-necked containers. By courtesy of Rockware Glass Ltd.

loads, and impacts, increases with increasing thickness of the glass in the bodies and bases. The design of the container also influences its strength. Cylindrical containers are more durable than more complex shapes featuring sharp corners. The greater the radius of curvature of the shoulder, the more resistant the container is to vertical loads (Figure 8.11). The thickness of the glass in the base is usually greater than that in the body. The circle where the body joins the base is weak due to the change in thickness. The insweep (Figure 8.11) minimizes container-to-container contact in this weak area.

Glass containers become weaker with use, due to abrasion of the outer surface as a result of container-to-container contact or contact with other surfaces. Treating the surface with compounds of titanium or tin and replacement of the sodium ions at the surface with potassium ions can reduce this problem. The resistance of glass containers to sudden changes in temperature is reduced as the thickness of the glass increases. Thus, when designing glass containers which are to be subjected to heating or cooling, for example, when the product is to be sterilized or pasteurized in its bottle or jar, or if the container is to be hot-filled with product, a compromise has to be achieved between their mechanical strength and thermal

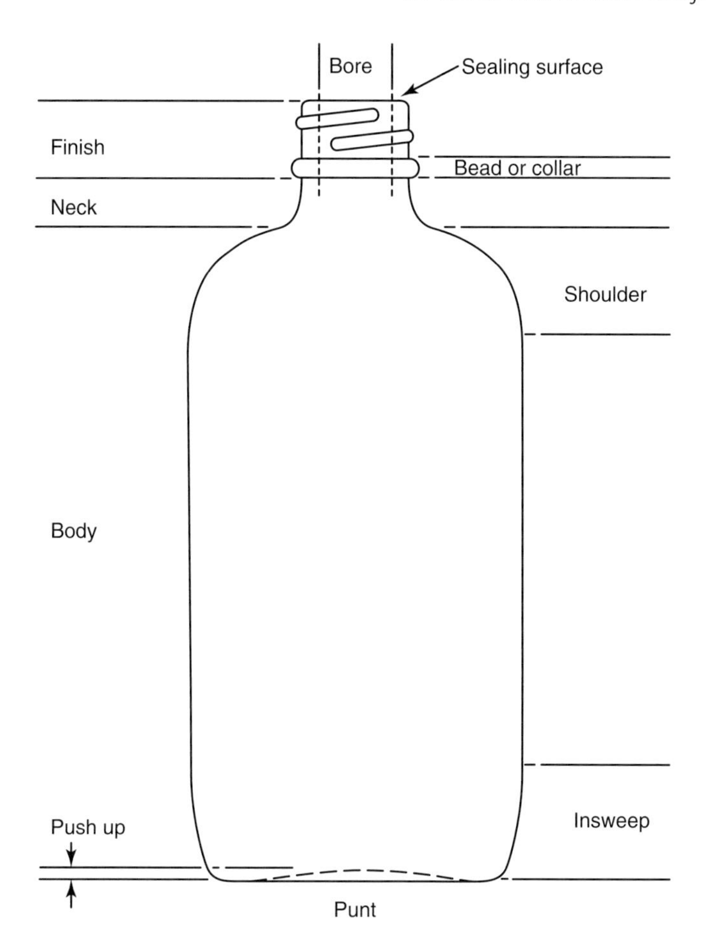

Figure 8.11 Important features of glass containers.

shock resistance. Heating and cooling should be carried out relatively slowly to avoid thermal damage to glass containers.

Glass containers are sealed by compressing a resilient disk, ring, or plug against the sealing surface of the container, and maintaining it in the compressed condition by means of a retaining cap. The resilient material may be cork, rubber, or plastic. The cap is made of metal or plastic. The cap may be screwed on, crimped on or pushed in, or onto the finish of the container. Rollon caps are used as tamper-evident closures. Different closures are effective when: (i) the pressure inside the container is close to atmospheric pressure (normal seal), (ii) the pressure inside the container is less than that outside (vacuum seal), and (iii) the pressure inside the container is higher than that outside (pressure seal). Pressure seals are necessary when packaging carbonated drinks.

Singletrip glass containers are used for liquids such as some beers, soft drinks, wines, sauces, salad dressings, and vinegars and for dry foods such as coffee and milk powders. Multitrip containers are used for pasteurized milk, some beers, and soft drinks. Products heated in glass containers include sterilized milk, beer, fruit juices, and pickled vegetables [14, 17, 18, 38–41].

8.4 Modified Atmosphere Packaging

Modified atmosphere packaging (MAP) is a procedure which involves replacing air inside a package with a predetermined mixture of gases prior to sealing it. Once the package is sealed, no further control is exercised over the composition of the in-package atmosphere. However, this composition may change during storage as a result of respiration of the contents and/or solution of some of the gas in the product.

Vacuum packaging is a procedure in which air is drawn out of the package prior to sealing but no other gases are introduced. This technique has been used for many years for products such as cured meats and cheese. It is not usually regarded as a form of MAP.

In MAP proper, the modified atmosphere is created by one of two methods. In the case of trays, the air is removed by a vacuum pump and the appropriate mixture of gases introduced prior to sealing. In the case of flexible packages, such as pouches, the air is displaced from the package by flushing it through with the gas mixture before sealing. In the case of horticultural products, a modified in-package atmosphere may develop as a result of respiration of the food. The concentration of oxygen inside the package will fall and that of carbon dioxide will rise. The equilibrium composition attained inside the package will largely depend on the rate of respiration of the food and the permeability of the packaging material to gases (see Section 8.2.2).

The gases involved in MAP, as applied commercially today, are carbon dioxide, nitrogen, and oxygen.

Carbon dioxide reacts with water in the product to form carbonic acid which lowers the pH of the food. It also inhibits the growth of certain microorganisms, mainly molds and some aerobic bacteria. Lactic acid bacteria are resistant to the gas and may replace aerobic spoilage bacteria in MAP meat. Most yeasts are also resistant to carbon dioxide. Anaerobic bacteria, including food poisoning organisms, are little affected by carbon dioxide. Consequently, there is a potential health hazard in MAP products from these microorganisms. Strict temperature control is essential to ensure the safety of MAP foods. Molds and some Gram-negative, aerobic bacteria, such as *Pseudomonas* spp, are inhibited by carbon dioxide concentrations in the range 5–50%. In general, the higher the concentration of the gas, the greater is its inhibitory power. The inhibition of bacteria by carbon dioxide increases as the temperature decreases. Bacteria in the lag phase of growth are most affected by the gas.

Nitrogen has no direct effect on microorganisms or foods, other than to replace oxygen, which can inhibit the oxidation of fats. As its solubility in water is low, it is used as a bulking material to prevent the collapse of MAP packages when the carbon dioxide dissolves in the food. This is also useful in packages of sliced or ground food materials, such as cheese, which may consolidate under vacuum.

Oxygen is included in MAP packages of red meat to maintain the red color, which is due to the oxygenation of the myoglobin pigments. It is also included in MAP packages of white fish, to reduce the risk of botulism.

Other gases have antimicrobial effects. Carbon monoxide will inhibit the growth of many bacteria, yeasts, and molds in concentrations as low as 1%. However, due to its toxicity and explosive nature, it is not used commercially. Sulfur dioxide has been used to inhibit the growth of molds and bacteria in some soft fruits and fruit juices. In recent years, there has been concern that some people may be hypersensitive to sulfur dioxide.

So called noble gases, such as argon, helium, xenon, and neon, have also been used in MAP of some foods. However, apart from being relatively inert, it is not clear what particular benefits they bring to this technology.

MAP packages are either thermoformed trays with heat-sealed lids or pouches. With the exception of packages for fresh produce, these trays and pouches need to be made of materials with low permeability to gases (CO_2, N_2, O_2). Laminates are used, made of various combinations of PET, PVdC, PE, and PA (nylons; see Section 8.3.4). The oxygen permeability of these laminates should be less than 15 cm^3 m^{-2} day^{-1} at a pressure of 1 atm (101 kPa). The following are some examples of modified atmosphere packages (Figure 8.12).

- **Meat products**: Fresh red meat packaged in an atmosphere consisting of 80% oxygen and 20% carbon dioxide or 70% oxygen, 20% carbon dioxide, and 10% nitrogen should have a shelf life of 7–12 days at $2 \pm 1\,^\circ C$. The meat is usually placed on an absorbent pad, contained in a deep tray with a heat-sealed lid. Poultry can be modified atmosphere packaged in a mixture of nitrogen and carbon dioxide. However, this is not widely practiced because of cost considerations. Cooked and cured meats may be packaged in a mixture of nitrogen and carbon dioxide.
- **Fish**: Fresh white fish, packaged in a mixture of 30% oxygen, 30% nitrogen, and 40% carbon dioxide, should have a shelf life of 10–14 days at a temperature of $0\,^\circ C$. Such packages should not be exposed to a temperature above $5\,^\circ C$, because of the risk of botulism. Fatty fish are packaged in mixtures of carbon dioxide and nitrogen.
- **Fruits and vegetables**: Respiration in such products leads to a build-up of carbon dioxide and a reduction in the oxygen content (see Section 8.2.2). Some build-up of carbon dioxide may reduce the rate of respiration and help to prolong the shelf life of the product. However, if the oxygen level is reduced to 2% or less, anaerobic respiration will set in and the product will spoil. The effect of the build-up of carbon dioxide varies from product to product. Some fruits and vegetables can tolerate high levels of this gas while others cannot. Each fruit or vegetable will have an optimum in-package gas composition which will result in a maximum shelf life. Selection of a packaging film with an appropriate permeability to water

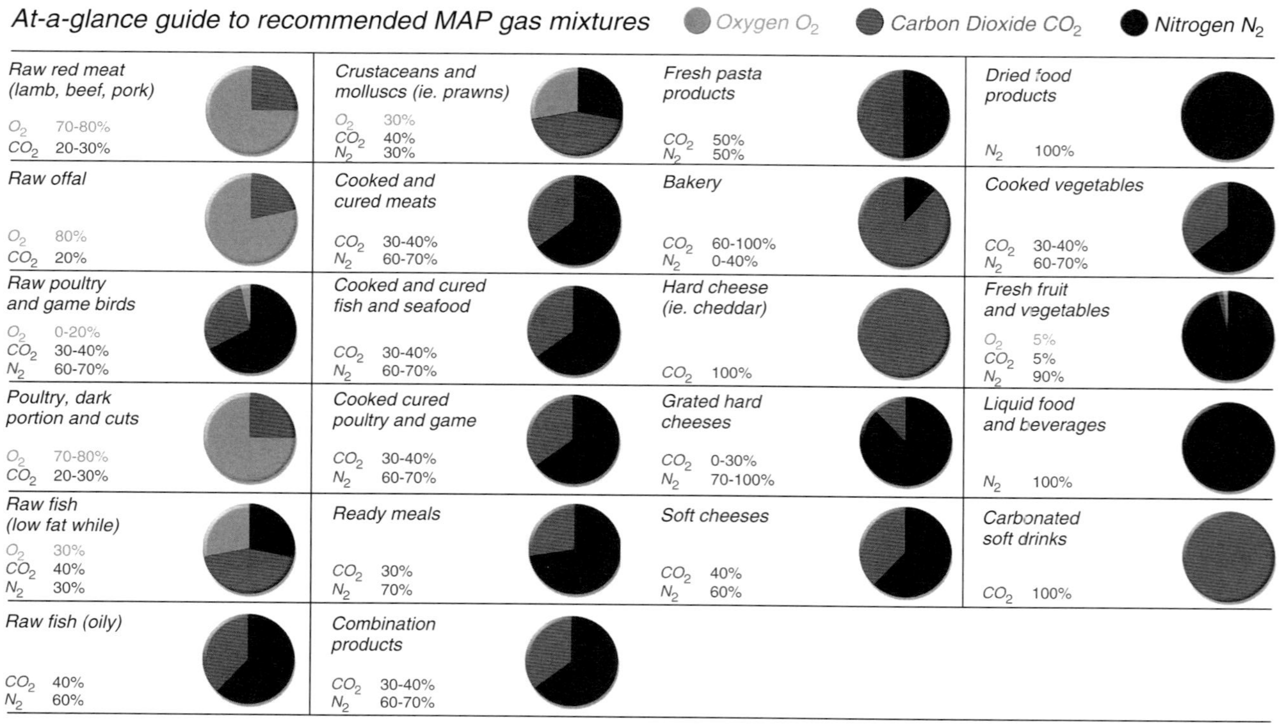

At-a-glance guide to recommended MAP gas mixtures

Oxygen O_2 · Carbon Dioxide CO_2 · Nitrogen N_2

Raw red meat (lamb, beef, pork) O_2 70-80% CO_2 20-30%	Crustaceans and molluscs (ie. prawns) O_2 30% CO_2 40% N_2 30%	Fresh pasta products CO_2 50% N_2 50%	Dried food products N_2 100%
Raw offal O_2 80% CO_2 20%	Cooked and cured meats CO_2 30-40% N_2 60-70%	Bakery CO_2 60-100% N_2 0-40%	Cooked vegetables CO_2 30-40% N_2 60-70%
Raw poultry and game birds O_2 0-20% CO_2 30-40% N_2 60-70%	Cooked and cured fish and seafood CO_2 30-40% N_2 60-70%	Hard cheese (ie. cheddar) CO_2 100%	Fresh fruit and vegetables O_2 5% CO_2 5% N_2 90%
Poultry, dark portion and cuts O_2 70-80% CO_2 20-30%	Cooked cured poultry and game CO_2 30-40% N_2 60-70%	Grated hard cheeses CO_2 0-30% N_2 70-100%	Liquid food and beverages N_2 100%
Raw fish (low fat while) O_2 30% CO_2 40% N_2 30%	Ready meals CO_2 30% N_2 70%	Soft cheeses CO_2 40% N_2 60%	Carbonated soft drinks CO_2 100%
Raw fish (oily) CO_2 40% N_2 60%	Combination products CO_2 30-40% N_2 60-70%		

Figure 8.12 Gas compositions commonly used in MAP food containers. By courtesy of Air Products.

vapor and gases can lead to the development of this optimum composition. For fruits with very high respiration rates, the package may need to be perforated. A range of microperforated films are available for such applications.

- **Cheese**: Portions of hard cheese may be packaged by flushing with carbon dioxide before sealing. The gas will be absorbed by the cheese, creating a vacuum. Cheese packaged in this way may have a shelf life of up to 60 days. To avoid collapse of the package, some nitrogen may be included with the carbon dioxide. Mold-ripened cheese may be packaged in nitrogen.
- **Bakery products and snack foods**: The shelf life of bread rolls, crumpets, and pita bread may be significantly increased by packaging in carbon dioxide or nitrogen/carbon dioxide mixtures. Nuts and potato crisps benefit by being modified atmosphere packaged in nitrogen.
- **Pasta**: Fresh pasta may be modified atmosphere packaged in nitrogen or carbon dioxide.
- **Other foods**: Pizza, quiche, lasagne, and many other prepared foods may benefit from MAP. It is very important to take into account the microbiological implications of modified atmosphere packaging such products. Maintenance of low temperatures during storage, distribution, in the retail outlet, and in the home is essential [42–53].

The term *Bio-MAP* has been used to represent MAP combined with some form of biological control. An example is the use of lactic acid bacteria in combination with MAP. Such bacteria grow well in an atmosphere rich in CO_2. If they are present in a modified atmosphere packaged food they will remain dormant as long as the temperature is maintained at an appropriately low level and their presence will not be evident to the consumer. However, if the temperature is allowed to rise, that is, if temperature abuse occurs, the lactic acid bacteria will initiate their growth themselves and outgrow any pathogenic microorganisms present, thus inhibiting or excluding them. As they grow the lactic acid bacteria produce lactic acid and other odorous compounds which will alert the consumer to a potential health hazard when the package is opened [54].

8.5 Aseptic Packaging

Chapter 2 discussed the advantages, in terms of product quality, to be gained by heat processing foods in bulk prior to packaging (UHT treatment) as compared with heat treating the packaged product. UHT-treated products have to be packaged under conditions which prevent microbiological contamination, that is, aseptically packaged. With some high-acid foods ($pH < 4.5$), it may be sufficient to cool the product after UHT treatment to just below 100 °C, fill it into a clean container, seal the container, and hold it at that temperature for some minutes before cooling it. This procedure will inactivate microorganisms that may have been in the container or entered during the filling operation and which might grow in the product. The filled container may need to be inverted for some or all of the holding period. However, in the case of low-acid foods (pH $>$ 4.5) this procedure would not be adequate to ensure the sterility of the product. Consequently for such products, aseptic filling must involve sterilizing the empty container or the material from which the container is made, filling it with the UHT-treated product and sealing it without it being contaminated with microorganisms.

In the case of rigid metal containers, superheated steam may be used to sterilize the empty containers and maintain a sterile atmosphere during the filling and sealing operations. Empty cans are carried on a stainless steel conveyor through a stainless steel tunnel. Superheated steam, at a temperature of approximately 260 °C, is introduced into the tunnel to sterilize the cans. They then move into an enclosed filling section, maintained sterile by superheated steam. They are sprayed on the outside with cool sterile water before being filled with the cooled UHT product. The filled cans move into an enclosed seaming section, which is also maintained in a sterile condition with superheated steam. The can ends are also sterilized with superheated steam and double-seamed onto the filled cans in the sterile seaming section. The filled and seamed cans then exit from the tunnel. The whole system has to be presterilized and the temperatures adjusted to the appropriate levels before filling commences. This aseptic filling procedure is known as the *Dole process* [51]. Glass containers and some plastic and composite containers may be aseptically filled by this method.

Cartons made from a laminate of paper/aluminum foil/PE are widely used for UHT products such as liquid milk and fruit juices. This type of packaging material cannot be sterilized by heat alone. A combination of heat and chemical sterilant is used. Treatments with hydrogen peroxide, peracetic acid, ethylene oxide, ionizing radiation, ultraviolet radiation, and sterile air have all been investigated. Hydrogen peroxide at a concentration of 35% in water and 90 °C is very effective against heat-resistant, spore-forming microorganisms, and is widely used commercially as a sterilant in aseptic packaging in laminates. FFS systems are available, an example being the Tetra Brik system, offered by Tetra Pak Ltd. (Figure 8.13).

The packaging material, a PE/paper/PE/foil/PE laminate, is unwound from a reel and a plastic strip is attached to one edge, which will eventually overlap the internal longitudinal seal in the carton. It then passes through a deep bath of hot hydrogen peroxide, which wets the laminate. As it emerges from the bath, the laminate passes between squeeze rollers, which express liquid hydrogen peroxide for return to the bath. Next, a high-velocity jet of hot sterile air is directed onto both sides of the laminate to remove residual hydrogen peroxide, as a vapor. The laminate, which is now sterile and dry, is formed into a tube with a longitudinal seal in an enclosed section which is maintained sterile by means of hot, sterile air under pressure. The product filling tube is located down the center of the laminate tube. The presterilized product is fed into the sterile zone near the bottom of the tube, which is heat-sealed. The air containing the vaporized hydrogen peroxide is collected in a cover and directed to a compressor where it is mixed with water, which washes out the residual hydrogen peroxide. The air is sterilized by heat and returned to the filling zone.

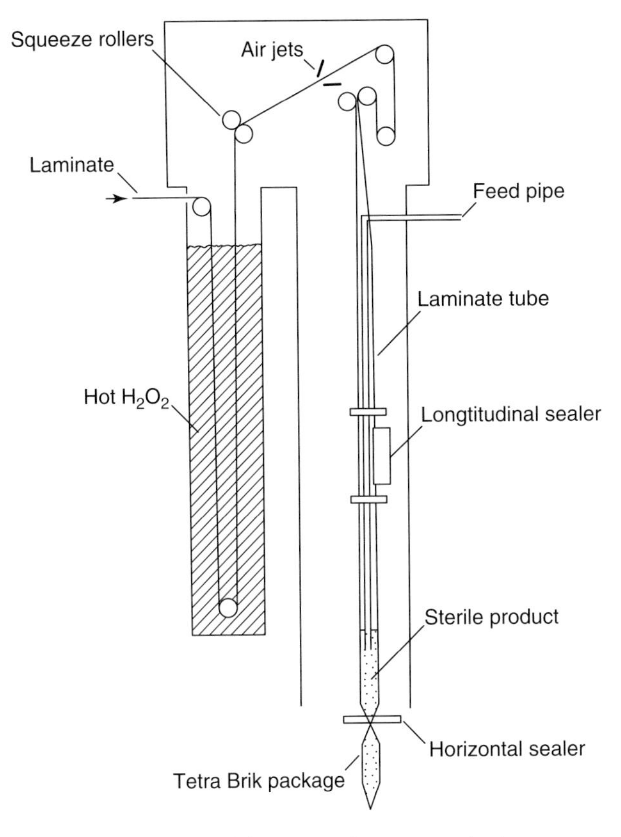

Figure 8.13 Principle of the Tetra Brik aseptic packaging system. From Fellows [15] with permission.

In another system, the laminate is in the form of carton blanks which are erected and then sterilized by a downward spray of hydrogen peroxide followed by hot sterile air. This completes the sterilization and removes residual hydrogen peroxide. The presterilized product is filled into the cartons and the top sealed within a sterile zone.

Similar systems are available to aseptically fill into preformed plastic cups. The lidding material is sterilized with hydrogen peroxide or infrared radiation before being heat-sealed onto the cups within a sterile zone. Thermoform filling systems are available to aseptically fill into polymer laminates. The web of laminate passes through a bath of hydrogen peroxide and then is contacted by hot sterile air which completes the sterilization, removes residual hydrogen peroxide and softens the laminate. The laminate is then thermoformed into cups and filled with presterilized product within a sterile zone. The sterilized lidding material is applied before the cups leave the sterile zone. Thermoforming systems are usually used to fill small containers, for example, for individual portions of milk, cream, and whiteners [55–62].

8.6 Active Packaging

8.6.1 Introduction

Active packaging is the term used to describe the inclusion of certain additives in a package in order to maintain and extend the shelf life of the product. The additives may be incorporated into the packaging material itself, or into a label attached to the inside of the package or contained within a sachet inside the package. Active packaging not only provides a barrier from outside influences but also plays an additional role in preservation of the product. Table 8.3 lists some of the food applications that have benefited from active packaging technology.

In the following sections selected examples of active packaging are discussed. More detailed information on active packaging systems available commercially can be found in the references provided.

8.6.2 Oxygen Scavengers

The presence of oxygen in a package of food can have an undesirable effect on the quality of the product and in some cases can shorten its shelf life. Oxidative rancidity can occur in cooked meats, cheeses, oily fish, nuts, and many other fatty foods, causing off-flavors and odors to develop. The staling of baked goods may be accelerated in the presence of oxygen. Some vitamins and pigments in foods may be degraded by oxidation. Aerobic spoilage microorganisms can grow in the presence of oxygen. Thus the removal of oxygen from packaged food can have a beneficial effect of the quality and shelf life of the product.

Vacuum packaging and MAP (see Section 8.4) are widely used to remove oxygen from packaged foods. However, these techniques typically result in 0.3–3.0% residual oxygen concentration in the package. The use of oxygen scavenging agents can reduce this figure to less than 0.01%. Such agents may be used alone or in combination with MAP. The most commonly used oxygen scavengers consist of various iron-based powders containing a variety of catalysts. These mixtures react with water from the food to produce a reactive metallic reducing agent that scavenges oxygen within the package, forming a stable oxide. The powders are sealed within a small sachet made of a material highly permeable to oxygen, which in turn is contained within the food package. Possible disadvantages of these metal-based scavengers are that they may impart a metallic taint to some foods, they may set off inline metal detectors, and they do not work at frozen food temperatures. Nonmetallic oxygen scavengers are also available. These include ascorbic acid which, in the presence of a metal such as copper, can remove oxygen. Immobilized enzymes can also be used for this purpose. Mixtures of glucose oxidase and catalase have been used, as has ethanol oxidase.

Table 8.3 Selected examples of active packaging systems.

Systems	Mechanisms	Food applications
Oxygen scavengers	1. Iron-based	Bread, cakes, cooked rice, biscuits, pizza, pasta, cheese, cured meats, cured fish, coffee, snack foods, dried foods, and beverages
	2. Metal/acid	
	3. Metal (e.g., platinum) catalyst	
	4. Ascorbate/metallic salts	
	5. Enzyme-based	
Carbon dioxide scavengers/emitters	1. Iron oxide/calcium hydroxide	Coffee, fresh meats, fresh fish, nuts, other snack food products, and sponge cakes
	2. Ferrous carbonate/metal halide	
	3. Calcium oxide/activated charcoal	
	4. Ascorbate/sodium bicarbonate	
Ethylene scavengers	1. Potassium permanganate	Fruit, vegetables, and other horticultural products
	2. Activated carbon	
	3. Activated clays/zeolites	
Preservative releasers	1. Organic acids	Cereals, meats, fish, bread, cheese, snack foods, fruit, and vegetables
	2. Silver zeolite	
	3. Spice and herb extracts	
	4. BHA/BHT antioxidants	
	5. Vitamin E antioxidant	
	6. Volatile chlorine dioxide/sulfur dioxide	
Ethanol emitters	1. Alcohol spray	Pizza crusts, cakes, bread, biscuits, fish, and bakery products
	2. Encapsulated ethanol	

(*continued overleaf*)

Table 8.3 (*Continued*)

Systems	Mechanisms	Food applications
Moisture absorbers	1. PVA blanket 2. Activated clays and minerals 3. Silica gel	Fish, meats, poultry, snack foods, cereals, dried foods, sandwiches, fruit, and vegetables
Flavor/odor adsorbers	1. Cellulose triacetate 2. Acetylated paper 3. Citric acid 4. Ferrous salt/ascorbate 5. Activated carbon/clays/zeolites	Fruit juices, fried snack foods, fish, cereals, poultry, dairy products, and fruit
Temperature control packaging	1. Non-woven plastics 2. Double-walled containers 3. Hydrofluorocarbon gas 4. Lime/water 5. Ammonium nitrate/water	Ready meals, meats, fish, poultry, and beverages

From Buchner [59].

Oxygen scavenging materials can now be incorporated into adhesive labels, laminated polymer films, and trays. This avoids the danger of sachets being accidentally consumed. They can also be incorporated into plastic cap liners of containers for noncarbonated and carbonated beverages to remove oxygen from the headspace, as such products are sensitive to oxygen when exposed to UV light. Oxygen scavenging technology has been applied to a variety of food products including cooked and cured meats, pizzas, cakes, breads, biscuits, coffee, tea, fresh pasta, dried milk and eggs, spices, herbs, snack foods, noncarbonated, and carbonated beverages [63–75].

8.6.3 Carbon Dioxide Scavengers

Carbon dioxide scavengers in sachet or label form have been used in a number of packaged food products in recent years. Fresh roasted or ground coffees release carbon dioxide during storage. This can cause the package to expand and in extreme cases burst. One solution to this is to vacuum pack the coffee and incorporate a one-way valve on the package to release the carbon dioxide. Another is to use carbon dioxide scavengers, usually in sachet form. Fresh fruit and vegetables with high respiration rates generate carbon dioxide. Within a pack this can cause undesirable changes in the produce which may include tissue damage, discoloration, and the development of off-flavors and odors. The pack may collapse as the carbon dioxide dissolves in moisture within it. The anaerobic atmosphere may encourage the growth of undesirable microorganisms. Carbon dioxide scavengers can reduce these problems.

The most commonly used scavenger is calcium hydroxide, which converts to calcium carbonate by reacting with the carbon dioxide in the presence of moisture. Potassium and sodium hydroxides have also been used, as has calcium oxide with activated charcoal. Dual-action oxygen and carbon dioxide scavengers are also available in sachet and label form [63–72].

8.6.4 Carbon Dioxide Emitters

High levels of carbon dioxide inside a food package can inhibit the growth of aerobic microorganisms and oxidative changes and thus prolong the shelf life of many food products. MAP (see Section 8.4) can be used to achieve these levels. However, the carbon dioxide can be absorbed by moisture or fat within the pack, thus reducing the concentration and possibly leading to the collapse of the package. The inclusion of a carbon dioxide generator in the pack can offset this effect. Dual-action oxygen scavengers/carbon dioxide emitters are available in sachet and label forms. They consist of ferrous carbonate and a metal halide catalyst or a mixture of ascorbic acid and sodium bicarbonate. Both these systems will absorb oxygen and generate an equal volume of carbon dioxide. Food applications for these systems include snack foods, such as potato crisps and nuts, bakery products, and dried meat and fish products [63–68, 70].

8.6.5
Ethylene Scavengers

The plant hormone ethylene is produced during respiration by most fruits and vegetables. Its presence inside a pack will accelerate the ripening of the produce, which in turn can lead to senescence, deterioration in the quality, and a shortening of the shelf life of the produce. The inclusion of ethylene scavengers in packages of fresh produce can reduce the concentration of the hormone in the pack, slow down ripening, and respiration and extend the shelf life of the produce.

Potassium permanganate ($KMnO_4$) is used as an ethylene scavenger. It is immobilized on an inert mineral such as alumina, perlite, or silica gel, to provide a large surface area, and sealed in sachets to avoid contact with the food, as it is highly toxic. It oxidizes ethylene to acetate and ethanol and in doing so changes color from purple to brown. This progressive color change is an indication of the residual scavenging capacity of the permanganate. Activated carbon together with catalysts such as bromine or palladium can be used to scavenge ethylene. The scavenger may be sealed in sachets or embedded into paper bags or fiberboard cases for produce packaging or storage. Finely ground minerals and clays such as pumice stone, zeolite, and oya stone have the ability to adsorb ethylene. These are embedded or blended into packaging films, mainly PE, which are used to package fresh produce. While these films have been shown to extend the shelf life of fruits and vegetables the exact mechanism whereby they do so is not clear [59]. Dual-action ethylene scavengers/moisture absorbers are also available [63–68, 74].

8.6.6
Ethanol Emitters

Ethanol is a well-known antimicrobial agent. It is very effective against molds but can also inhibit the growth of some bacteria. Its effectiveness against yeasts is not so clear [60]. It can be sprayed directly onto foods, such as bakery products, just before packaging and can increase their shelf life by inhibiting mold growth. As an alternative to spraying, ethanol emitters can be included in food packages. Ethanol is absorbed or encapsulated in a carrier material, sealed inside a sachet, which is inserted into the food package. On contact with moisture from the product, ethanol vapor is released into the headspace in a controlled manner. Silicon powder, for example, silicon dioxide, is also used as a carrier material. Since moisture plays a part in releasing the ethanol vapor, such emitters can only be used with high moisture products. Some sachets contain small quantities of flavoring material, such as vanilla, to mask the odor of alcohol. Ethanol emitters have been used with high ratio cakes and other high-moisture bakery products, mainly in Japan, and have resulted in substantial increases in the shelf lives of such products. Ethanol also exhibits an anti-staling capacity which can inhibit the hardening of the surface of some bakery products, such as sponge cakes. Dual-action oxygen scavenging/ethanol emitting sachets are also available [63–65, 67, 68, 70, 71].

8.6.7 Moisture Absorbers

The shelf lives of low-moisture foods can be extended by the use of moisture absorbers. These consist of silica gel, calcium oxide, activated clays, and minerals, which are sealed in moisture permeable sachets inserted into the food packages. Dried foods, snack foods, and cereals are the main applications for these sachets. They have also been used with some bakery products to maintain the crispness of the crust. They can be included in packages of fresh produce to absorb moisture vapor produced by respiration and thus reduce the in-pack humidity and condensation. Oxygen scavengers and odor absorbers may be included in sachets with the moisture absorbers.

Moisture absorbing pads are used under fresh meat, fish, and poultry to absorb unsightly drip that exudes from such products. They usually consist of two layers of a microporous nonwoven plastic film, such as PE or PP, with a layer of absorbent polymer between them. Polyacrylate salts and copolymers of carboxymethyl cellulose (CMC) and starch are examples of so-called superabsorbers as they can absorb up 500 times their own weight of water. One disadvantage of some of these pads is that red meat in contact with them will become brown due to lack of oxygen. Pads are available with a breathable surface that enables oxygen to contact the meat and thus retain its red color.

Absorbers for high-moisture foods which take up water vapor rather than liquid water are also available. These consist of two layers of film, highly permeable to water vapor, with one or more humectants sealed in between. Various humectants are used, an example being propylene glycol combined with a carbohydrate. Such materials are available in the form of sheets, rolls, pouches, or bags and can be used to wrap or package meat, fish, and poultry. The moisture content at the surface of the food is reduced by osmosis and this can result in an extension of its shelf life [63–65, 67, 68, 70, 71, 76].

8.6.8 Flavor/Odor Absorbers

Undesirable odors and flavors are produced when food materials break down. For example, volatile amines, such as trimethylamine, are produced when fish muscle deteriorates and these give rise to unpleasant odors. Since these compounds are alkaline they can be neutralized by acid compounds. Pouches and bags made from film containing a ferrous salt and an organic acid, such as citrate or ascorbate, are available and are claimed to reduce odors due to amines. Oxidation of fats and oils leads to the formation of aldehydes, which can produce off-flavors in high-fat foods. Inorganic sulfates and tocopherols, such as vitamin E, have the capacity to remove aldehydes and can be incorporated into packaging films. Sorbitol and cyclodextrin, incorporated into PET film, have been trialed as aldehyde scavengers. Another approach is to incorporate synthetic aluminosilicate zeolites into packaging materials, especially papers. These have a very porous structure and can absorb odorous gases, such as aldehydes. Snack foods, biscuits, cereals, dairy, meat, and fish products can benefit from the use of flavor/odor absorbers. One potential hazard associated with this technology is that if all odors were removed, the consumer might inadvertently eat unsafe food.

It is also possible to incorporate flavor and aroma compounds into packaging materials. These can be released inside the package to enhance the flavor/aroma of the product, but can also be released outside the package, for example, when it is on the supermarket shelf, to attract the consumer to that product [63–65, 67, 70, 71, 77].

8.6.9 Antioxidant Release

Antioxidants, such as butylated hydroxytoluene (BHT) and butylated hydroxyanisole (BHA), can be incorporated into packaging materials and released into the package to reduce the degree of oxidation of fatty products. While most of the antioxidant is lost some will be absorbed by the food. Alpha-tocopherol (vitamin E) can also be used for this purpose. Crisps, nuts, cereals, snack foods, and some processed meats are the main applications for this technology [63, 64, 70].

8.6.10 Antimicrobial Packaging

Since microbial spoilage of food usually begins at the surface, there has been considerable interest in recent years in the possibility of the incorporation of antimicrobial agents into packaging materials and their release onto the surface of the food to extend its shelf life. A great deal of research has been carried out on a wide range of antimicrobial agents, but very few have been applied commercially to date. The main reasons for this are safety considerations and legislative restrictions.

The agents can act in one of two ways, by direct contact between the packaging film and the food or by diffusion in a gaseous phase from the packaging material to the surface of the food. The antimicrobial agents may be incorporated into the film or coated onto its surface. Agents investigated include metals, and silver in particular. Silver ions have a strong antimicrobial capacity. They are used in the form of silver zeolites, which are applied to the surface or incorporated into the packaging film or container. They are released from the packaging material which needs to be in direct contact with the product. They are effective against bacteria, fungi, and yeasts.

Weak organic acids, such as acetic, citric, benzoic, proprionic, and sorbic acids, have an inhibitory effect on the growth of bacteria and fungi. Benzoic and proprionic acids have been used in waxes and edible coatings on some citrus fruits to inhibit fungal growth. Another antifungal agent, imazalil, has been used in PE film to limit mold growth on some vegetables and fruits.

Bacteriocins are proteinaceous compounds (usually peptides) which are produced by bacteria and have an inhibitory effect on the growth of certain microorganisms, notably *Listeria*, *Clostridium*, and lactic acid bacteria. Nisin, pediosin, piscicolin 126, and subtilosin A are examples of bacteriocins that have been investigated for potential use in packaging materials, with limited success. Nisin, attached to cellulose used as a casing for meat and poultry and has been shown to reduce the risk of *Listeria*. It had a similar effect on a number of products when incorporated into PE and corn zein [60, 77].

Triclosan (2,4,4′-trichloro-2′hydroxydiphenyl ether) has antibacterial capability and has been incorporated into a number of polymers used for food packaging, with limited success. Since the antimicrobial compounds are nonvolatile they are only effective when the film is in direct contact with the food.

Chitosan is a derivative of chitin, the exoskeletal glucosamine polymer of insects and shellfish, which is available in the form of a clear flexible film. It has an antimicrobial effect on some types of bacteria, yeasts, and fungi and has been investigated as a potential antimicrobial packaging material for foods, again with limited success.

Naturally occurring substances, with antimicrobial properties, have also been considered for use in antimicrobial packaging. Allylisothiocyanate (AIT), an oil from wasabi, has been incorporated into packaging films and labels and was found to be effective in inhibiting the growth of some pathogenic and spoilage bacteria. Essential oils extracted from herbs and spices have antimicrobial and antioxidant properties and have been investigated for use in packaging materials. Extracts from oregano, cinnamon, clove, basil, and rosemary have been studied, with some limited success. Possible contamination of the product with odors and flavors from these extracts has to be considered [63–65, 67, 68, 71, 72, 78–82].

8.6.11 Lactose and Cholesterol Removers

To reduce the lactose content of some foods, attempts have been made to incorporate the enzyme lactase into packaging materials. However, there are difficulties with this process. Attempts have also been made to incorporate cholesterol reductase into packaging materials to remove cholesterol from certain foods [60].

8.6.12 UV Light Absorbers

Many food materials are sensitive to UV light, which can cause a loss of vitamins, color, and flavor. Ultraviolet absorbers (UVAs) and hindered amine light (HAL) stabilizers can be added to clear packaging films and containers to reduce these problems. Such additives can also stabilize some packaging materials which are degraded by exposure to UV light [64].

8.6.13
Other Active Packaging Systems

Many other devices are discussed in the literature under the heading of active packaging. These include *microwave susceptors* used to promote crisping and browning of foods in microwave ovens; widgets to assist the formation of a head on canned beers and self-heating and cooling packs. These are discussed in more detail elsewhere [63, 64, 67, 68, 70, 71].

8.7
Intelligent Packaging

8.7.1
Introduction

The main roles of intelligent or smart packaging are to sense and inform. *Intelligent packaging* devices are capable of sensing and providing information about the function and properties of packaged food relevant to the quality and safety of the product, product authenticity, and traceability. They can also provide evidence of pack integrity and tampering. Such information can be provided throughout the food chain, from the point of manufacture to the point of consumption. In the following sections selected examples of intelligent packaging are discussed. More detailed information on these and other intelligent packaging devices can be found in the literature [64, 65, 70, 72, 83].

8.7.2
Time–Temperature Indicators (TTIs)

Time–temperature indicators (TTIs) can provide a record of the temperatures a product has been exposed to and the length of time it has been exposed to these temperatures. This record can last from the time of production for up to several months, depending on the anticipated shelf life of the product. It can provide information on the temperature history of the packaged product during transport and storage in warehouses, at the point of sale and in the home or catering outlet. Indicators are available for frozen and chilled foods and products transported and stored at ambient temperature. There are two main categories of TTIs. Partial history indicators only respond when a predetermined temperature threshold has been exceeded. They are used to indicate when the product has been exposed to unacceptably high temperatures which could lead to a health hazard and/or a deterioration in quality. Full history indicators provide a continuous record of the temperatures the product is exposed to throughout its life. Many TTIs indicate by means of a color change when a specified temperature has been exceeded. Thermochromic inks or liquid crystals are used to create the colors. Indicators may operate on the bases of mechanical, chemical, enzymatic, or microbiological changes. TTIs are available in the form of labels and tags. There are many patented TTI systems available. Further details can be found in the literature [64, 68, 70, 83, 84].

8.7.3 Quality Indicators and Sensors

Much research has been directed at developing quality indicators to warn the consumer if the quality of the product has fallen below an acceptable standard. Such indicators usually undergo a permanent change in color when activated. They detect chemical or microbial changes in the product or changes in gas composition within the pack. However, very few such indicators are in current use in the food industry.

8.7.3.1 Chemical Indicators

Some chemical indicators react to changes in the concentration of organic acids in products such as meat. An example is polyaniline, which is sensitive to changes in pH and changes color as the pH falls. This has been investigated as a quality indicator for meat and dairy products. Amine compounds are produced when fish products spoil. A pH-sensitive dye that changes color in the presence of such compounds has been used as a quality indicator for fish products.

8.7.3.2 Microbial Indicators

Microbiological indicators usually consist of labels containing microorganisms, often in a gel. As the microorganisms grow the indicators change color, thus indicating the microbiological status of the product. As these microorganisms are exposed to the same conditions as the product throughout its shelf life these indicators simulate the spoilage of the product. Another way of using a microbiological indicator involves applying a transparent label containing microorganisms in a gel over the barcode on the package. As the microorganisms increase in number the label becomes increasingly opaque. When the barcode can no longer be read, the product is no longer fit for consumption. Microbial indicators are mostly used on packaged chilled foods such as ready meals, salads, and sandwiches. More sophisticated systems involve the use of biosensors. A bioreceptor, which may consist of enzymes, microbes, antigens, or hormones, is incorporated into a label which is read by a transducer. The bioreceptor measures the microbial activity within the package and the transducer converts the biological signals into numbers.

8.7.3.3 Gas Concentration Indicators

The composition of the atmosphere inside a food package may change during its shelf life as a result of respiration of the product or microorganisms, or the permeation of gases into or out of the pack through the packaging material or though faults in the package. Gas concentration indicators can detect changes in the composition of the in-package atmosphere and such changes can give an indication as to the quality of the product. The gases usually monitored by indicators are oxygen, carbon dioxide, and hydrogen sulfide.

Oxygen indicators can be used in packaged products susceptible to oxidative rancidity, such as cooked meats and some bakery products. They contain a redox dye which changes color on contact with oxygen and may be in the form of a tablet, label, or printed on to the packaging material. They may also be used in MAP (see Section 8.4).

Carbon dioxide indicators can be used in MAP packages and in packages of fresh foods which produce carbon dioxide by respiration. In the latter case they can indicate if an anaerobic atmosphere is developing in the package, as this would be detrimental to the quality of such products. Such indicators consist of carbon dioxide scavengers (see Section 8.6.3) containing a redox dye sensitive to carbon dioxide.

Hydrogen sulfide indicators can be used to monitor the spoilage of meat and poultry products. When hydrogen sulfide reacts with myoglobin the latter changes color. This change forms the basis of most hydrogen sulfide indicators [64, 65, 70, 72, 83].

8.7.4 Radiofrequency Identification Devices (RFID)

A radiofrequency identification device (RFID) tag consists of a very small microchip connected to a tiny antenna. They are usually incorporated into labels, but may also be incorporated into packaging films and some containers. A reader, which emits radio waves, captures data from the tags, and the data are fed into a computer for analysis. Tags may be classified into two types. Passive chips have no battery and are powered by the energy from the reader. Active tags have their own power supply. Active tags can be read over a greater distance than passive ones and can be updated as required. Passive tags are smaller and cheaper and last longer than active tags. They are used mainly in retail packages, whereas active tags are attached to crates, cases, and pallets and used to track them through the distribution chain. A RFID tag can contain a unique identification number for the product and information on where and when it was made, the sources of the ingredients, nutritional information, and other useful data. The tags can be deactivated as the customer leaves the supermarket.

Although the application of RFID technology in the food industry is still somewhat limited, it has great potential to contribute to logistic control, and enhance food safety and security in the future [63, 64, 69, 70, 85, 86].

8.7.5 Other Intelligent Packaging Devices

Many other devices are discussed in the literature under the heading of intelligent packaging. These include temperature sensors, new types of barcodes, microwave doneness indicators, special inks to detect counterfeiting, and numerous devices to provide evidence of tampering [64, 65, 68, 72].

8.7.6 Consumer Attitudes, Safety, and Legal Aspects of Active and Intelligent Packaging

Consumer attitudes to active and intelligent packaging vary from country to country. Sachets containing active packaging materials, such as oxygen absorbers and ethanol emitters, have been generally accepted in the United States, Australia, Japan, and other eastern countries, but much less so in European countries. On the other hand, some intelligent packaging systems, such as TTIs, are more widely used in Europe. The reasons for these different attitudes are not clear. They may be partly due to cultural differences, but also to a lack of understanding of the functions and benefits of these systems. For example, there are still concerns about the possibility of the contents of active packaging sachets being accidentally ingested. If active and intelligent packaging is to become more widely accepted the consumer needs to be better informed about their purpose and use.

In conventional packaging, where the packaging material is in contact with the food, an important objective is that interaction between them should be kept to a minimum (Section 8.2.6). Packaging legislation worldwide embodies that objective. However, many of the active and intelligent packaging systems outlined above involve promoting deliberate interaction between the packaging and the food or the food environment. This poses additional difficulties in evaluating the safety of such systems and in developing appropriate legislation to apply to them. In the case of active packaging substances may migrate into the food. Such migration may be intended or unintended. Examples of intended migrants are ethanol, antioxidants, and antimicrobial agents. These would require regulatory approval before use. Unintended migrants could include metal compounds used as scavengers. These would have to be regulated also.

It is essential to consider the effects of any active packaging on the microbiological population in the package. For example, the use of oxygen scavengers may encourage the growth of anaerobic bacteria. Antimicrobial packaging has the potential to inhibit the growth of spoilage microorganisms without affecting the growth of pathogens. It is very important that all such potential hazards are fully investigated before applying active packaging to any particular food product.

Informative and accurate labeling is very important in the case of active and intelligent packaging. Sachets containing active materials and any other active or intelligent devices contained within the package should be clearly marked as not for consumption. Additional information on the label explaining the purpose of any inclusions in or attachments to the package would also be helpful to the consumer.

Environmental implications of active and intelligent packaging also need to be considered. In Europe, the Packaging Waste Directive (1994), covering the reuse, recycling, and recovery of energy from packaging materials, applies. European companies planning to apply active and intelligent packaging will have to comply with this directive.

Figure 8.14 Symbol to illustrate that the active or intelligent device should not be consumed.

From the brief discussion above it is clear that regulations governing conventional food packaging throughout the world have to be modified and extended to cover active and intelligent packaging. Many countries are in the process of doing so with varying success. For example, in Europe, following a multinational research project, the Regulation 1935/2004 on materials and articles intended to come in contact with food was introduced. This contains some general provisions on the safety of active and intelligent packaging. For example, whenever active or intelligent materials or articles may be mistaken as part of the food and a risk of ingestion exist, they shall be accompanied by the words "DO NOT EAT" and, where technically possible, the symbol shown in Figure 8.14 [60]. Much more detailed regulations are in the process of development and some are expected to be published soon [63–68, 70, 72, 83, 87, 88].

8.8 The Role of Nanotechnology in Food Packaging

In recent years there have been considerable developments in the use of nanoparticulates in food packaging materials to alter their properties. The incorporation of such particulates into polymer packaging materials can change their mechanical and permeability properties, enhance their active capabilities, improve their biodegradability, and impart intelligent functions to them. Such developments are discussed in more detail in Chapter 23 (Section 23.5).

References

1. Levy G.M. (ed.) (1993) *Packaging in the Environment*, Blackie Academic & Professional, London.
2. Lauzon, C. and Wood, G. (1995) *Environmentally Responsible Packaging – A Guide to Development, Selection and Design*, Pira International, Leatherhead.
3. Levy, G.M. (ed.) (2000) *Packaging Policy and the Environment*, Aspen Publishers, Gaithersburgh.
4. McCormack, T. (2000) Plastics packaging and the environment, in *Materials and Development of Plastic Packaging for the Consumer Market* (eds G.A. Giles and D.R. Bain), Sheffield Academic Press, Sheffield, pp. 152–176.
5. Dent, I.S. (2000) Recycling and reuse of plastics packaging for the consumer market, *Materials and Development of Plastic Packaging for the Consumer Market* (eds G.A. Giles and D.R. Bain), Sheffield Academic Press, Sheffield, pp. 177–202.

6. Ashby, R., Cooper, I., Harvey, S., and Tice, P. (1997) *Food Packaging Migration and Legislation*, Pira International, Leatherhead.
7. Watson, D.H. and Meah, M.N. (eds) (1994) *Chemical Migration from Food Packaging, Food Science Reviews*, vol. 2, Ellis Horwood, London.
8. Crosby, N.T. (1981) *Food Packaging Materials – Aspects of Analysis and Migration of Contaminants*, Applied Science Publishers, London.
9. Ackermann, P., Jagerstad, M., and Ohlsson, T. (eds) (1995) *Food Packaging Materials – Chemical Interactions*, Royal Society of Chemistry, Cambridge.
10. Highland, H.A. (1978) Insect resistance of food packages – a review. *J. Food Process. Preserv.*, 2, 123–130.
11. Wohlgemoth, R. (1979) Protection of stored foodstuffs against insect infestation by packaging. *Chem. Ind.*, May, 330–333.
12. Navarro, S., Zehavi, D., Angel, S., and Finkelman, S. (2007) Natural non-toxic insect repellent packaging materials, in *Intelligent and Active Packaging for Fruits and Vegetables* (ed. C.L. Wilson), Taylor & Francis Group, Boca Raton, FL, pp. 201–236.
13. Paine, F.A. (1989) *Tamper Evident Packaging – A Literature Review*, Pira, Leatherhead.
14. Paine, F.A. (ed.) (1991) *The Packaging User's Handbook*, Blackie and Sons, Glasgow.
15. Fellows, P.J. (2000) *Food Processing Technology*, 2nd edn, Woodhead Publishing, Cambridge.
16. Brennan, J.G., Butters, J.R., Cowell, N.D., and Lilly, A.E.V. (1990) *Food Engineering Operations*, 3rd edn, Elsevier Applied Science, London.
17. Paine, F. A. and Paine, H.Y. (eds) (1992) *A Handbook of Food Packaging*, 2nd edn, Blackie Academic & Professional, London.
18. Robertson, G.L. (1993) *Food Packaging, Principles and Practice*, Marcel Dekker, New York.
19. Paine, F. (1990) *Packaging Design and Performance*, Pira, Leatherhead.
20. DeMaria, K. (2000) *The Packaging Design Process*, Technomic Publishing Co., Lancaster.
21. Kirwan, M.J. (2003) Paper and paperboard packaging, in *Food Packaging Technology* (eds R. Coles, D. McDowell and M.J. Kirwan), Blackwell Publishing, Oxford, pp. 241–281.
22. Anonymous (1997) Paper and paperboard, in *The Wiley Encyclopedia of Packaging Technology*, 2nd edn (eds A.L. Brody and K.S. Marsh), John Wiley & Sons, Inc., New York. pp. 714–723.
23. Foster, G.E. (1997) Boxes, corrugated, in *The Wiley Encyclopedia of Packaging Technology*, 2nd edn (eds A.L. Brody and K.S. Marsh), John Wiley & Sons, Inc., New York, pp. 100–108.
24. Hernandez, R.J. (1997) Polymer properties, in *The Wiley Encyclopedia of Packaging Technology*, 2nd edn (eds A.L. Brody and K.S. Marsh), John Wiley & Sons, Inc., New York, pp. 738–764.
25. Kirwan, M.J. and Strawbridge, J.W. (2003) Plastics in food packaging, in *Food Packaging Technology*, (eds R. Coles, D. McDowell, and M.J. Kirwan), Blackwell Publishing, Oxford, pp. 174–240.
26. Bakish, R. (1997) Metallizing, vacuum, in *The Wiley Encyclopedia of Packaging Technology*, 2nd edn (eds A.L. Brody and K.S. Marsh), John Wiley & Sons, Inc., New York, pp. 629–638.
27. Anonymous (1997) Sealing, heat, in *The Wiley Encyclopedia of Packaging Technology*, 2nd edn (eds A.L. Brody and K.S. Marsh), John Wiley & Sons, Inc., New York. pp. 823–827.
28. Staines, G. (2000) Injection moulding, in *Development of Plastic Packaging for the Consumer Market* (eds G.A. Giles and D.H. Bain), Sheffield Academic Press, Sheffield, pp. 8–24.
29. Hind, V. (2001) Extrusion blow-moulding, in *Technology of Plastics Packaging for the Consumer Market* (eds D.A. Giles and D.R. Bain), Sheffield Academic Press, Sheffield, pp. 25–52.
30. Bain, D.R. (2001) Thermoforming technologies for the manufacture of rigid plastics packaging, in *Technology of Plastics Packaging for the Consumer Market* (eds D.A. Giles and D.R. Bain), Sheffield Academic Press, Sheffield, pp. 146–159.

31. Morgan, E. (1985) *Tinplate and Modern Canmaking Technology*, Pergamon Press, Oxford.
32. Page, B., Edwards, M., and May, N. (2003) Metal cans, in *Food Packaging Technology* (eds R. Coles, D. McDowell, and M.J. Kirwan), Blackwell Publishing, Oxford, pp. 120–151.
33. Britten, S.C. (1975) *Tin Versus Corrosion*, International Tin Research Institute, Middlesex, (ITRI Publication No. 510).
34. Good, R.H. (1988) Recent advances in metal can interior coatings, in *Food and Packaging Interactions* (ed. J.H. Horchkiss), American Chemical Society, Washington, DC, pp. 203–219.
35. Turner, T.A. (1998) *Canmaking – The Technology of Metal Protection and Decoration*, Blackie Academic & Professional, London.
36. Selbereis, J. (1997) Metal cans, fabrication, in *The Wiley Encyclopedia of Packaging Technology*, 2nd edn (eds A.L. Brody and K.S. Marsh), John Wiley & Sons, Inc., New York, pp. 616–629.
37. Kraus, F.J. (1997) Cans, steel, in *The Wiley Encyclopedia of Packaging Technology* (eds A.L. Brody and K.S. Marsh), 2nd edn, John Wiley & Sons, Inc., New York, pp. 144–154.
38. Girling, P.J. (2003) Packing of food in glass containers, in *Food Packaging Technology* (eds R. Coles, D. McDowell, and M.J. Kirwan), Blackwell Publishing, Oxford, pp. 152–173.
39. Moody, B.E. (1977) *Packaging in Glass*, Hutchinson and Benham, London.
40. Cavanagh, J. (1997) Glass container manufacturing, in *The Wiley Encyclopedia of Packaging Technology*, 2nd edn (eds A.L. Brody and K.S. Marsh), John Wiley & Sons, Inc., New York, pp. 475–484.
41. Tooley, F.V. (1974) *The Handbook of Glass Manufacture*, Ashlee Publishing, New York.
42. Zagory, D. (1997) Modified atmosphere packaging, in *The Wiley Encyclopedia of Packaging Technology*, 2nd edn (eds A.L. Brody and K.S., Marsh), John Wiley & Sons, Inc., New York, pp. 650–656.
43. Mullan, M. and McDowell, D. (2003) Modified atmosphere packaging in *Food Packaging Technology*, (eds R. Coles, D. McDowell, and M.J. Kirwan), Blackwell Publishing, Oxford, pp. 303–338.
44. Blakistone, B.A. (1998) Meats and poultry, in *Principles and Applications of Modified Atmosphere Packaging of Foods*, 2nd edn (ed. B.A. Blakistone), Blackie Academic and Professional, London, pp. 240–290.
45. Gill, C.O. (1995) MAP and CAP of fresh red meats, poultry and offal, in *Principles of Modified-Atmosphere and Sous Vide Product Packaging* (eds J.M. Fraber and K.L. Dodds), Technomic Publishing Co., Lancaster, pp. 105–136.
46. Davis, H.K. (1998) Fish and shellfish, in *Principles and Applications of Modified Atmosphere Packaging of Foods*, 2nd edn (ed. B.A. Blakistone), Blackie Academic and Professional, London, pp. 194–239.
47. Gibson, D.M. and Davis, H.K. (1995) Fish and shellfish products in sous vide and modified atmosphere packs, in *Principles of Modified-Atmosphere and Sous Vide Product Packaging* (eds J.M. Faber and K.L. Dodds), Technomic Publishing Co., Lancaster, pp. 150–174.
48. Garrett, E.H. (1998) Fresh-cut produce, in *Principles and Applications of Modified Atmosphere Packaging of Foods*, 2nd edn (ed. B.A. Blakistone), Blackie Academic and Professional, London, pp. 125–134.
49. Zagory, D. (1995) Principles and practice of modified atmosphere packaging of horticultural commodities, in *Principles of Modified-Atmosphere and Sous Vide Product Packaging* (eds J.M. Faber and K.L. Dodds), Technomic Publishing Co., Lancaster, pp. 175–206.
50. Seiler, D.A.L. (1998) Bakery products, in *Principles and Applications of Modified Atmosphere Packaging of Foods*, 2nd edn (ed. B.A. Blackistone), Blackie Academic and Professional, London, pp. 135–157.
51. Subramanian, P.J. (1998) Dairy foods, multi-component products, dried foods and beverages, in *Principles and Applications of Modified Atmosphere Packaging of Foods*, 2nd edn (ed. B.A. Blackistone), Blackie Academic and Professional, London, pp. 158–193.
52. Day, B.P.F. (2008) Modified atmosphere packaging (MAP), in *Food Biodeterioration and Preservation* (ed. G.S.

Tucker), Blackwell Publishing, Oxford, pp. 165–191.
53. Mensitieri, G. and Buonocore, G.G. (2008) A perspective on actual technologies and future developments. *New Food*, (3), 40–45.
54. Yuan, J.T.C. (2007) Bio-MAP: Modified atmosphere packaging with biological control for shelf-life extension, in *Packaging for Nonthermal Processing of Food*, (ed. J.H. Han), Blackwell Publishing, Oxford, pp. 53–65.
55. White, F.S. (1993) The Dole process, in *Aseptic Processing and Packaging of Particulate Foods* (ed. E.M.A. Willhoft), Blackie Academic and Professional, London, pp. 148–154.
56. Hersom, A.C. and Hulland, E.D. (1980) *Canned Foods: Thermal Processing and Microbiology*, 7th edn, Churchill Livingstone, Edinburgh.
57. Burton, H. (1988) *Ultra-High Temperature Processing of Milk and Milk Products*, Elsevier Applied Science Publishers, London.
58. Holdsworth, S.D. (1992) *Aseptic Processing and Packaging of Food Products*, Elsevier Applied Science Publishers, London.
59. Buchner, N. (1993) Aseptic processing and packaging of food particulates, in *Aseptic Processing and Packaging of Particulate Foods* (ed. E.M.A. Willhoft), Blackie Academic and Professional, London, pp. 1–22.
60. Joyce, D.A. (1993) Microbiological aspects of aseptic processing and packaging, in *Aseptic Processing and Packaging of Particulate Foods* (ed. E.M.A. Willhoft), Blackie Academic and Professional, London, pp. 155–180.
61. Wakabayashi, S. (1993) Aseptic packaging of liquid foods, in *Aseptic Processing and Packaging of Particulate Foods* (ed. E.M.A. Willhoft), Blackie Academic and Professional, London, pp. 181–187.
62. Jairus, R.D., David, R.H., Graves, R.H., and Carlson, V.R. (1996) *Aseptic Processing and Packaging of Food – A Food Industry Perspective*, CRC Press, London.
63. Brennan, J.G. and Day, B.P.F. (2006) Packaging, in *Food Processing Handbook*, (ed. J.G. Brennan), Wiley-VCH Verlag GmbH, Weinheim, pp. 291–350.
64. Potter, L., Campbell, A., and Cava, D. (2008) *Active and Intelligent Packaging–A Review*, Campden & Chorleywood Food Research Association Group, Chipping Campden.
65. Brody, A.L., Bugusu, B., Han, J.H., Sand, C.K., and McHugh, T.H. (2008) Innovative food packaging solutions. *J. Food Sci.*, **73**, 107–116.
66. Rooney, M.L. (2005) Introduction to active food packaging technologies, in *Innovations in Food Packaging* (ed. J.H. Han), Elsevier Academic Press, Amsterdam, pp. 64–79.
67. Day, B.P.F. (2003) Active packaging, in *Food Packaging Technology* (eds R. Coles, D. McDowell, and M.J. Kirwan), Blackwell Publishing, Oxford, pp. 282–302.
68. Lee, D.S., Yam, K.L., and Piergiovanni, L. (2008) *Food Packaging Science and Technology*, CRC Press, Boca Raton,pp 445–478.
69. Fellows, P.J. (2009) *Food Processing Technology – Principles and Practice*, 3rd edn, Woodhead Publishing Limited, Oxford, pp. 761–767.
70. Robertson, G.L. (2006) *Food Packaging Principles and Practice*, 2nd edn, Taylor & Francis, Boca Raton, pp. 285–311.
71. Day, B.P.F. (2001) Active packaging – a fresh approach. *J. Brand Technol.*, **1**, 32–41.
72. Ahvenainen, R. (2003) Active and intelligent packaging: an introduction, in *Novel Food Packaging Techniques* (ed. R. Ahvenainen), Woodhead Publishing Limited, Cambridge, pp. 5–21.
73. Rooney, M.L. (2005) Oxygen-scavenging packaging, in *Innovations in Food Packaging* (ed. J.H. Han), Elsevier Academic Press, Amsterdam, pp. 123–137.
74. Vermeiren, L., Heirlings, L., Devlieghere, F., and Debevere, J. (2003) Oxygen, ethylene and other scavengers, in *Novel Food Packaging Techniques* (ed. R. Avhenainen), Woodhead Publishing Limited, Cambridge, pp. 22–49.
75. Rooney, M.L. (2000) Applications of ZERO2 oxygen scavenging films for food and beverage products, *Proceedings of the International Conference on Active and Intelligent Packaging*, Campden & Chorleywood Food Research Association, Chipping Campden.

76. Powers, T. and Calvo, W.J. (2003) Moisture regulation, in *Novel Food Packaging Techniques* (ed. R. Avhenainen), Woodhead Publishing Limited, Cambridge, pp. 172–188.
77. Linssen, J.P.H., van Willage, R.W.G., and Dekker, M. (2003) Packaging–flavour interactions, in *Novel Food Packaging Techniques* (ed. R. Avhenainen), Woodhead Publishing Limited, Cambridge, pp. 144–171.
78. Han, J.H. (2005) Antimicrobial packaging systems, in *Innovations in Food Packaging* (ed. J.H. Han), Elsevier Academic Press, Amsterdam, pp. 80–107.
79. Lee, D.S. (2005) Packaging containing natural antimicrobial or antioxidative agents, in *Innovations in Food Packaging* (ed. J.H. Han), Elsevier Academic Press, Amsterdam, pp. 108–122.
80. Appendini, P. and Hotchkiss, J.H. (2002) Review of antimicrobial food packaging. *Innov. Food Sci. Emerg.*, **3**, 113–126.
81. Cha, D.S. and Chinnan, M.S. (2004) Bipolymer-based antimicrobial packaging: a review. *Crit. Rev. Food Sci. Nutr.*, **44**, 223–237.
82. Lopez, P., Sanchez, C., Battle, R., and Nerin, C. (2007) Development of flexible antimicrobial films using essential oils as active agents. *J. Agric. Food Chem.*, **55**, 8814–8824.
83. Han, J.H., Ho, C.H.L., and Rodrigues, E.T. (2005) Intelligent packaging, in *Innovations in Food Packaging* (ed. J.H. Han), Elsevier Academic Press, Amsterdam, pp. 138–155.
84. Taoukis, P.S. and Labuza, T.P. (2003) Time–temperature indicators (TTIs), in *Novel Food Packaging Techniques* (ed. R. Ahvenainen), Woodhead Publishing Limited, Cambridge, pp. 103–126.
85. Han, J.H., Hydamaka, A.W., and Zong, Y. (2007) Radio frequency identification systems for packaged foods, in *Packaging for Nonthermal Processing of Food* (ed. J.H. Han), Blackwell Publishing, Oxford, pp. 117–137.
86. Roberts, B. (2007) RFID temperature monitoring: trends, opportunities and challenges, in *Intelligent and Active Packaging for Fruits and Vegetables* (ed. C.L. Wilson), Taylor and Francis Group, Boca Raton, pp. 237–247.
87. de Kruijf, N. and Rijk, R. (2003) Legislative issues relating to active and intelligent packaging, in *Novel Food Packaging Techniques* (ed. R. Ahvenainen), Woodhead Publishing Limited, Cambridge, pp. 459–496.
88. Dainelli, D., Gontard, N., Spyropoulos, D., Zondervan-van den Beuken, E., and Tobback, P. (2008) Active and intelligent packaging: legal aspects and safety concerns. *Trends Food Sci. Technol.*, **19**, S103–S112.

Edited by
James G. Brennan and
Alistair S. Grandison

Food Processing Handbook

Volume 2

Related Titles

Rychlik, Michael (ed.)

Fortified Foods with Vitamins

Analytical Concepts to Assure Better and Safer Products

2011

ISBN: 978-3-527-33078-2

Moscicki, Leszek (ed.)

Extrusion-Cooking Techniques

Applications, Theory and Sustainability

2011

ISBN: 978-3-527-32888-8

Peinemann, K.-V., Pereira Nunes, S., Giorno, L. (eds.)

Membrane Technology

Volume 3: Membranes for Food Applications

2010

ISBN: 978-3-527-31482-9

Rijk, R., Veraart, R. (eds.)

Global Legislation for Food Packaging Materials

2010

ISBN: 978-3-527-31912-1

Stanga, Mario

Sanitation

Cleaning and Disinfection in the Food Industry

2010

ISBN: 978-3-527-32685-3

Popping, Bert, Diaz-Amigo, Carmen, Hoenicke, Katrin (eds.)

Molecular Biological and Immunological Techniques and Applications for Food Chemists

2010

ISBN: 978-0-470-06809-0

Wintgens, J. N. (ed.)

Coffee: Growing, Processing, Sustainable Production

2nd updated edition

2009

ISBN: 978-3-527-32286-2

Eßlinger, Hans Michael (ed.)

Handbook of Brewing

Processes, Technology, Markets

2009

ISBN: 978-3-527-31674-8

Chen, X. D., Mujumdar, A. S. (eds.)

Drying Technologies in Food Processing

ISBN: 978-1-4051-5763-6

Evans, J. (ed.)

Frozen Food Science and Technology

ISBN: 978-1-4051-5478-9

Heredia, N. L., Wesley, I. V. (eds.)

Microbiologically Safe Foods

2009

ISBN: 978-0-470-05333-1

Stadler, Richard H., Lineback, David R.

Process-Induced Food Toxicants

Occurrence, Formation, Mitigation, and Health Risks

2009

ISBN: 978-0-470-07475-6

Ziegler, H. (ed.)

Flavourings

Production, Composition, Applications, Regulations

2007

ISBN: 978-3-527-31406-5

Heller, K. J. (ed.)

Genetically Engineered Food

Methods and Detection

2006

ISBN: 978-3-527-31393-8

Edited by James G. Brennan and Alistair S. Grandison

Food Processing Handbook

2nd edition

Volume 2

WILEY-VCH Verlag GmbH & Co. KGaA

The Editors

James G. Brennan, MSc FIFST
16 Benning Way
Wokingham, Berks RG40 1XX
United Kingdom

Dr. Alistair S. Grandison
Department of Food and Nutritional Sciences
University of Reading
Whiteknights
Reading RG6 6AP
United Kingdom

Library of Congress Card No.: applied for

British Library Cataloguing-in-Publication Data
A catalogue record for this book is available from the British Library.

Bibliographic information published by the Deutsche Nationalbibliothek
The Deutsche Nationalbibliothek lists this publication in the Deutsche Nationalbibliografie; detailed bibliographic data are available on the Internet at <http://dnb.d-nb.de>.

Cover Design Adam-Design, Weinheim
Typesetting Laserwords Private Limited, Chennai, India
Printing and Binding Fabulous Printers Pte Ltd, Singapore

Printed in Singapore
Printed on acid-free paper

ISBN: 978-3-527-32468-2
ePDF ISBN: 978-3-527-63438-5
ePub ISBN: 978-3-527-63437-8
Mobi ISBN: 978-3-527-63439-2
oBook ISBN: 978-3-527-63436-1

Contents

Preface to the Second Edition

In this second edition of Food Processing Handbook the chapters in the first edition have been retained and revised by including information on recent developments in each field and updating the reference lists. Some of the most notable changes are: the inclusion of a new section on ohmic heating in the Chapter on thermal processing (Chapter 2); extending the packaging chapter to cover intelligent packaging (Chapter 8); explaining the calculation of greenhouse gas emissions (carbon footprints) and providing a case study in the chapter on environmental aspects of food processing (Chapter 19). The original chapter entitled Baking, Extrusion and Frying has been split into three individual chapters providing extended coverage of these three important processes (Chapters 12, 13, and 14). Several new topics have been added to reflect recent trends and concerns in the food industry. These include chapters on: traceability in food processing and distribution (Chapter 16); hygienic design of food processing plant (Chapter 17); process realisation (Chapter 21); microscopy techniques and image analysis for the quantitative evaluation of food microstructure (Chapter 22); nanotechnology in the food sector (Chapter 23) and fermentation and the use of enzymes (Chapter 24). These changes have necessitated dividing the book into two volumes, the first consisting of the more basic food preservation processes and packaging, while volume 2 includes other manufacturing processes and other considerations relating to safety and sustainable manufacturing.

It is hoped that this much extended edition will be of interest to scientists and engineers involved in food manufacture and research and development in industry, and to staff and students participating in food related courses at undergraduate and postgraduate levels.

James G. Brennan,
Alistair S. Grandison

Preface to the First Edition

There are many excellent texts available which cover the fundamentals of food engineering, equipment design, modelling of food processing operations etc. There are also several very good works in food science and technology dealing with the chemical composition, physical properties, nutritional and microbiological status of fresh and processed foods. This work is an attempt to cover the middle ground between these two extremes. The objective is to discuss the technology behind the main methods of food preservation used in today's food industry in terms of the principles involved, the equipment used and the changes in physical, chemical, microbiological and organoleptic properties that occur during processing. In addition to the conventional preservation techniques, new and emerging technologies, such as high pressure processing and the use of pulsed electric field and power ultrasound are discussed. The materials and methods used in the packaging of food, including the relatively new field of active packaging, are covered. Concerns about the safety of processed foods and the impact of processing on the environment are addressed. Process control methods employed in food processing are outlined. Treatments applied to water to be used in food processing and the disposal of wastes from processing operations are described.

Chapter 1 covers the postharvest handling and transport of fresh foods and preparatory operations, such as cleaning, sorting, grading and blanching, applied prior to processing. Chapters 2, 3 and 4 contain up-to-date accounts of heat processing, evaporation, dehydration and freezing techniques used for food preservation. In Chapter 5, the potentially useful, but so far little used process of irradiation is discussed. The relatively new technology of high pressure processing is covered in Chapter 6, while Chapter 7 explains the current status of pulsed electric field, power ultrasound, and other new technologies. Recent developments in baking, extrusion cooking and frying are outlined in Chapter 8. Chapter 9 deals with the materials and methods used for food packaging and active packaging technology, including the use of oxygen, carbon dioxide and ethylene scavengers, preservative releasers and moisture absorbers. In Chapter 10, safety in food processing is discussed and the development, implementation and maintenance of HACCP systems outlined. Chapter 11 covers the various types of control systems applied in food processing. Chapter 12 deals with environmental issues including the impact of packaging wastes and the disposal of refrigerants. In Chapter 13, the various treatments applied to water to be used in food processing are described and the physical, chemical and biological treatments applied to food processing wastes are outlined. To complete the picture, the various separation techniques used in food processing are discussed in Chapter 14 and Chapter 15 covers the conversion operations of mixing, emulsification and size reduction of solids.

The editor wishes to acknowledge the considerable advice and help he received from former colleagues in the School of Food Biosciences, The University of Reading, when working on this project. He also wishes to thank his wife, Anne, for her support and patience.

Reading, August 2005 *James G. Brennan*

List of Contributors

Araya Ahromrit
Assistant Professor
Department of Food Technology
Khon Kaen University
Khon Kaen 40002
Thailand

Paul Ainsworth
Manchester Metropolitan
University
Retired Professor of Food
Technology
and Director of the Manchester
Food Research Centre
Old Hall Lane
Manchester M14 6HR
UK

Liliana Alamilla-Beltrán
National School of Biological
Sciences-National Polytechnic
Institute
Department of Food Science and
Technology
Carpio y Plan de Ayala s/n Sto.
Tomás 11340
Mexico City

Pedro Bouchon
Pontificia Universidad Católica
de Chile
Department of Chemical and
Bioprocess Engineering
P.O. Box 306
Santiago 6904411
Chile

James G. Brennan
16 Benning Way
Wokingham
Berkshire
RG40 1XX
UK

Stanley P. Cauvain
BakeTran
1 Oakland close
Freeland
Witney
OX 29 8AX
UK

José Jorge Chanona-Pérez
National School of Biological
Sciences-National Polytechnic
Institute
Department of Food Science and
Technology
Carpio y Plan de Ayala s/n Sto.
Tomás 11340
Mexico City

Dimitris Charalampopoulos
University of Reading
Department of Food and
Nutritional Sciences
PO Box 226
Reading RG6 6AP
UK

Brian P.F. Day
8 Cavanagh Close
Hoppers Crossing
VIC 3029
Australia

Ali Abd El-Aal Bakr
Food Science and Technology
Department
Faculty of Agriculture
Minufiya University
Shibin El-Kom
Egypt

Alistair S. Grandison
University of Reading
Department of Food and
Nutritional Sciences
P.O. Box 226
Whiteknights
Reading RG6 6AP
UK

Gustavo Fidel Gutierrez-López
National School of Biological
Sciences-National Polytechnic
Institute
Department of Food Science and
Technology
Carpio y Plan de Ayala s/n Sto.
Tomás 11340
Mexico City

Tony Hasting
37 Church Lane
Sharnbrook
Bedford
MK44 1HT
UK

Soojin Jun
University of Hawaii at Manoa
College of Tropical Agriculture
and Human Resources
Department of Human Nutrition
Food and Animal Sciences
1955 East West Rd. 302F
Honolulu, HI 96822
USA

Ashok S. Khare
University of Reading
Department of Food and
Nutritional Sciences
P.O. Box 226
Whiteknights
Reading RG6 6AP
UK

Christopher J. Kirby
Pharmaterials Ltd.
Unit B
5 Boulton Road
Reading RG2 0NH
UK

Christopher Knight
Head of Agriculture
Campden BRI
Chipping Campden
Glos. GL55 6LD
UK

Kevan G. Leach
Leach Associates Ltd.
Edgecumbe Lodge
Greenway Park
Chippenham
Wilts SN15 1QG
UK

Craig Leadley
Campden BRI
Food Manufacturing
Technologies
Chipping Campden
Gloucestershire GL55 6LD
UK

Dave A. Ledward
University of Reading
Department of Food and
Nutritional Sciences
Whiteknights
Reading RG6 6AP
UK

Michael J. Lewis
University of Reading
Department of Food and
Nutritional Sciences
P.O. Box 226
Whiteknights
Reading RG6 6AP
UK

José Mauricio Pardo
Universidad de la Sabana
Ingenieria de Produccion
Agroindustrial
A. A. 140013 Chia
Colombia

Angélica Gabriela Mendoza-Madrigal
National School of Biological
Sciences-National Polytechnic
Institute
Department of Food Science and
Technology
Carpio y Plan de Ayala s/n Sto.
Tomás 11340
Mexico City

Niharika Mishra
Agricultural and Biological
Engineering
717 W. Cherry lane
Apt # 2
State College, PA 16803
USA

Keshavan Niranjan
University of Reading
Department of Food and
Nutritional Sciences
P.O. Box 226
Whiteknights
Reading RG6 6AP
UK

Margaret F. Patterson
Agri-Food and biosciences
Institute
Newforge Lane
Belfast BT9 5PX
Northern Ireland
UK

Maria de Jesús Perea-Flores
National School of Biological
Sciences-National Polytechnic
Institute
Department of Food Science and
Technology
Carpio y Plan de Ayala s/n Sto.
Tomás 11340
Mexico City

Nigel Rogers
Avure Technologies AB
Quintusvägen 2
Vasteras
SE 72166
Sweden

Gary Tucker
Head of Baking & Cereal
Processing Department
Campden BRI
Chipping Campden
Glos, GL55 6LD
UK

Carol A. Wallace
University of Central Lancashire
International Institute of
Nutritional Sciences and Applied
Food Safety Studies
School of Sport
Tourism and the Outdoors
Preston
Lancashire PR1 2HE
UK

Jorge Welti-Chanes
Technological Institute of
Advanced Studies of Monterrey
Food and Biotechnology Unit
Av. Eugenio Garza Sada 2501 Sur
Col. Tecnológico
64849 Monterrey
N.L.
Mexico

R. Andrew Wilbey
The University of Reading
Department of Food and
Nutritional Sciences
Whiteknights
Reading RG6 6AP
UK

9
Separations in Food Processing Part 1

James G. Brennan and Alistair S. Grandison

9.1
Introduction

Separations are vital to all areas of the food processing industry. Separations usually aim to remove specific components in order to increase the added value of the products, which may be the extracted component, the residue or both. Purposes include cleaning, sorting, and grading operations (see Chapter 1), extraction, and purification of fractions such as sugar solutions or vegetable oils, recovery of valuable components such as enzymes or flavor compounds, or removal of undesirable components such as microorganisms, agricultural residues, or radionuclides. Operations range from separation of large food units, such as fruits and vegetables measuring many centimeters, down to separation of molecules or ions measured in nanometers.

Separation processes always make use of some physical or chemical difference between the separated fractions; examples are size, shape, color, density, solubility, electrical charge, and volatility.

The separation rate is dependent on the magnitude of the driving force and may be governed by a number of physical principles involving concepts of mass transfer and heat transfer. Rates of chemical reaction and physical processes are virtually always temperature-dependent, such that separation rate will increase with temperature. However, high temperatures give rise to degradation reactions in foods, producing changes in color, flavor, and texture, loss of nutritional quality, protein degradation, and so on. Thus a balance must be struck between rate of separation and quality of the product. Separations may be classified according to the nature of the materials being separated, and a brief overview is given below.

9.1.1
Separations from Solids

Solid foods include fruits, vegetables, cereals, legumes, animal products (carcasses, joints, minced meat, fish fillets, shellfish), and various powders and granules. Their separation has been reviewed by Lewis [1] and can be subdivided as follows.

Food Processing Handbook, Second Edition. Edited by James G. Brennan and Alistair S. Grandison.

9.1.1.1 Solid–Solid Separations

Particle size may be exploited to separate powders or larger units using sieves or other screen designs, examples of which are given in Chapter 1. Air classification can be achieved using differences in aerodynamic properties to clean or fractionate particulate materials in the dry state. Controlled air streams will cause some particles to be fluidized in an air stream depending on the terminal velocity, which in turn is related primarily to size, but also to shape and density. Also in the dry state, particles can be separated on the basis of photometric (color), magnetic, or electrostatic properties. By suspending particles in a liquid, particles may be separated by settlement on the basis of a combination of size and density differences. Buoyancy differences can be exploited to separate products from heavy materials such as stones or rotten fruit in flotation washing, while surface properties can be used to separate peas from weed seeds in froth flotation.

9.1.1.2 Separation from a Solid Matrix

Plant materials often contain valuable components within their structure. In the case of oils or juices, these may be separated from the bulk structure by expression, which involves the application of pressure. Alternatively, components may be removed from solids by extraction (see Section 9.4), which utilizes the differential solubilities of extracted components in a second medium. Water may be used to extract sugar, coffee, fruit, and vegetable juices, and so on. Organic solvents are necessary in some cases, for example, hexane for oil extraction. Supercritical CO_2 may be used to extract volatile materials such as in the decaffeination of coffee. A combination of expression and extraction is used to remove 99% of the oil from oilseeds.

Water removal from solids plays an important role in food processing (see Chapter 3).

9.1.2 Separations from Liquids

Liquid foods include aqueous or oil-based materials, and frequently contain solids either in true solution or dispersed as colloids or emulsions.

9.1.2.1 Liquid–Solid Separations

Discrete solids may be removed from liquids using a number of principles. Conventional filtration (see Section 9.2) is the removal of suspended particles on the basis of particle size using a porous membrane or septum, composed of wire mesh, ceramics, or textiles. A variety of pore sizes and geometric shapes are available and the driving force can be gravity, upstream pressure (pumping), downstream pressure (vacuum), or centrifugal force. Using smaller pore sizes, microfiltration, ultrafiltration, and related membrane processes can be used to fractionate solids in true solution (see Chapter 10).

Density and particle size determine the rate of settlement of dispersed solids in a liquid according to Stokes' law. Settlement due to gravity is very slow, but is widely used in water and effluent treatment. Centrifugation subjects the dispersed particles to forces greatly exceeding gravity, which dramatically increases the rate of separation and is widely used for clarifying liquid food products. A range of geometries for batch and continuous processing are available (see Section 9.3).

9.1.2.2 Immiscible Liquids

Centrifugation is again used to separate immiscible liquids of different densities. The major applications are cream separation and the dewatering of oils during refining.

9.1.2.3 General Liquid Separations

Differences in solubility can be exploited by contacting a liquid with a solvent which preferentially extracts the component(s) of interest from a mixture. For example, organic solvents could be used to extract oil soluble components, such as flavor compounds, from an aqueous medium.

An alternative approach is to induce a phase change within the liquid, such that components are separated on the basis of their freezing or boiling points. Crystallization is the conversion of a liquid into a solid plus liquid state by cooling or evaporation (see Section 9.6). The desired fraction, solid or liquid, can then be collected by filtration or centrifugation. Alternatively, evaporation (see Chapter 3) is used to remove solvent or other volatile materials by vaporization. In heat-sensitive foods, this is usually carried out at reduced operating pressures and hence reduced temperature, frequently in the range 40–90 °C. Reverse osmosis (see Section 10.1) is an alternative to evaporation in which pressure rather than heat is the driving force. Ion exchange and electrodialysis (see Sections 10.2 and 10.3) are used to separate dissolved components in liquids, depending on their electrostatic charge.

9.1.3 Separations from Gases and Vapors

These separations are not common in food processing. Removal of solids suspended in gases is required in spray drying and pneumatic conveying is achieved by filter cloths, bag filters, or cyclones. Another possibility is wet scrubbing to remove suspended solids on the basis of solubility in a solvent (see Chapter 3).

9.2 Solid–Liquid Filtration

9.2.1 General Principles

In this method of separation the insoluble solid component of a solid–liquid suspension is separated from the liquid component by causing the latter to flow through a porous membrane, known as the *filter medium*, which retains the solid particles within its structure and/or as a layer on its upstream face. If a layer of solid particles does form on the upstream face of the medium it is known as the *filter cake*. The clear liquid passing through the medium is known as the *filtrate*. The flow of the liquid through the medium and cake may be brought about by means of gravity alone (gravity filtration), by pumping it through under pressure (pressure filtration), by creating a partial vacuum downstream of the medium (vacuum filtration) or by centrifugal force (centrifugal filtration). Once the filtration stage is complete, it is common practice to wash the cake free of filtrate. This is done to recover valuable filtrate and/or to obtain a cake of adequate purity. When filtering oil, the cake may be blown free of filtrate by means of steam. After washing, the cake may be dried with heated air.

In the early stages of a filtration cycle, solid particles in the feed become enmeshed in the filter medium. As filtration proceeds, a layer of solids begins to build up on the upstream face of the medium. The thickness of this layer, and so the resistance to the flow of filtrate, increases with time. The pressure drop, $-\Delta p_c$, across the cake at any point in time may be expressed as:

$$-\Delta p_c = \frac{\alpha \eta w V}{A^2} \left(\frac{dV}{dt} \right) \tag{9.1}$$

where η is the viscosity of the filtrate, w is the mass of solids deposited on the medium per unit volume of filtrate, V is the volume of filtrate delivered in time t, A is the filter area normal to the direction of flow of the filtrate, and α is the *specific cake resistance*. The specific cake resistance characterizes the resistance to flow offered by the cake and physically represents the pressure drop necessary to give unit superficial velocity of filtrate of unit viscosity through a cake containing unit mass of solid per unit filter area.

If a cake is composed of rigid, nondeformable solid particles, then α is independent of the pressure drop across the cake and is constant throughout the depth of the cake. Such a cake is known as *incompressible*. In the case of incompressible cakes, it is possible to calculate the value of α. In contrast, a *compressible* cake is made up of nonrigid deformable solid particles or agglomerates of particles. In such cakes, the value of α increases with increase in pressure and also varies throughout the depth of the cake, being highest near the filter medium. The relationship between α and $-\Delta p_c$ is often expressed as:

$$\alpha = \alpha_0 \left(-\Delta p_c \right)^s \tag{9.2}$$

where α_0 and s are empirical constants. The *compressibility coefficient* of the cake, s, is zero for an incompressible cake and rises toward 1.0 as the compressibility increases. In the case of compressible cakes, values of α must be obtained by experiment.

The filter medium also offers resistance to the flow of the filtrate and a pressure drop, $-\Delta p_m$, develops across it. This pressure drop may be expressed as:

$$-\Delta p_m = \frac{R_m \eta}{A} \left(\frac{dV}{dt} \right) \cdots \tag{9.3}$$

where R_m is known as the *filter medium resistance*. Values of R_m are determined experimentally. It is usual to assume that R_m is constant throughout any filtration cycle.

The total pressure drop across the cake and medium, $-\Delta p$, is obtained by adding the two pressure drops together, thus:

$$-\Delta p = -\Delta p_c - \Delta p_m = \frac{\eta}{A}\left(\frac{dV}{dt}\right)\left(\frac{\alpha w V}{A} + R_m\right) \quad (9.4)$$

or:

$$\frac{dV}{dt} = \frac{A(-\Delta p)}{\eta\left(\frac{\alpha w V}{A} + R_m\right)} \quad (9.5)$$

A filter cycle may be carried out by maintaining a constant total pressure drop across the cake and medium. This is known as *constant pressure filtration*. As the cake builds up during the cycle, the rate of flow of filtrate decreases. Alternatively, the rate of flow of filtrate may be maintained constant throughout the cycle, in which case the pressure increases as the cake builds up. This is known as *constant rate filtration*. A combination of constant rate and constant pressure filtration may also be employed by building up the pressure in the early stages of the cycle and maintaining it constant throughout the remainder of the cycle.

Equation 9.5 may be applied to both constant pressure and constant rate filtration. In the case of a compressible cake, a relationship such as Equation 9.2 needs to be used to account for the change in α with increase in pressure during constant rate filtration [2–11].

9.2.2 Filter Media

The main functions of the filter medium are to promote the formation of the filter cake and to support the cake once it is formed. Once the filter cake is formed, it becomes the primary filter medium. The medium must be strong enough to support the cake under the pressure and temperature conditions that prevail during the filtering cycle. It must be nontoxic and chemically inert with respect to the material being filtered. Filter media may be flexible or rigid. The most common type of flexible medium is a woven cloth, which may be made of cotton, wool, silk, or synthetic material. The synthetic materials used include nylon, polyester, polyacrylonitrile, polyvinylchloride, polyvinylidenechloride, polyethylene, and polytetrafluoroethylene. Such woven materials are available with different mesh counts, mesh openings, thread sizes, and weaves. Woven glass fiber and flexible metal meshes are also used as filter media. Nonwoven, flexible media are fabricated in the form of belts, sheets, or pads of various shapes. These tend to be used for filtering liquids with relatively low solids content. Most of the solids remain enmeshed within the depth of the media rather than forming a cake on the surface. Rigid media may be fixed or loose. Fixed rigid media are made in the forms of disks, pads, and cartridges.

They consist of rigid particles set in permanent contact with one another. They include ceramic and diatomaceous materials and foamed plastics made from polyvinylchloride, polyethylene, polypropylene, and other polymer materials. Perforated metal plates and rigid wire meshes are used for filtering relatively large particles. Loose rigid media consist of rigid particles that are merely in contact with each other but remain in bulk, loose form. They include sand, gravel, charcoal, and diatomaceous material arranged in the form of beds. All types of media are available with different pore sizes to suit particular filtration duties. Practical trials are the most reliable methods for selecting media for particular tasks [7–12].

9.2.3 Filter Aids

Filter aids are employed to improve the filtration characteristics of highly compressible filter cakes or when small amounts of finely divided solids are being filtered. They consist of hard, strong, inert incompressible particles of irregular shape. They form a porous, permeable rigid lattice structure which allows liquid to pass through but retains solid particles. They are usually applied in small amounts in the range 0.01–4.00% of the weight of the suspension. They may be applied in one of two ways. The filter medium may be precoated with a layer of filter aid prior to introducing the suspension to be filtered. This precoat, which is usually 1.5–3.0 mm thick, prevents the suspension particles from becoming enmeshed in the filter medium and reducing the flow of liquid. It may also facilitate the removal of the cake when filtration and washing are complete. Alternatively, the filter aid may be added to the suspension before it is introduced into the filter unit. It increases the porosity of the cake and reduces its compressibility. Sometimes a combination of precoating and premixing is used.

The materials most commonly used as filter aids include: diatomaceous material (which is made from the siliceous remains of tiny marine plants known as *diatoms*, known as *diatomite* and kieselguhr), expanded perlite (made from volcanic rock), charcoal, cellulose fibers, and paper pulp. These materials are available in a range of grades. Experimental methods are used to select the correct grade, amount, and method of application for a particular duty [7–11].

9.2.4 Filtration Equipment

Gravity filtration is not widely applied to food slurries, but is used in the treatment of water and waste disposal. These applications are covered in Chapter 20.

9.2.4.1 Pressure Filters

In pressure filters the feed is pumped through the cake and medium and the filtrate exits at atmospheric pressure. The following are examples of pressure filters used in processing of foods.

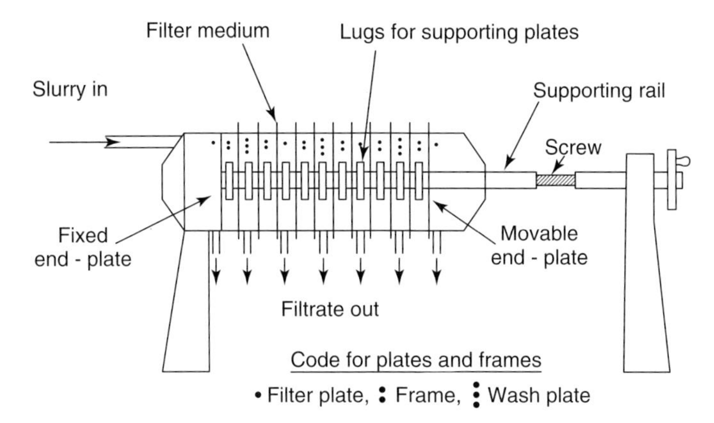

Figure 9.1 Schematic drawing of assembled plate-and-frame filter press. From Ref. [2] with permission of the authors.

Plate-and-frame filter press In this type of press grooved plates, covered on both sides with filter medium, alternate with hollow frames in a rack (Figure 9.1). The assembly of plates and frames is squeezed tightly together to form a liquid-tight unit. The feed is pumped into the hollow frames through openings in one corner of the frames (Figure 9.2).

The cake builds up in the frames and the filtrate passes through the filter medium onto the grooved surface of the plates, from where it exits via an outlet channel in each plate. When filtering is complete, wash liquid may be pumped through the press following the same path as the filtrate. Some presses are equipped with special wash plates (see Figure 9.2). Every second plate in the frame is a wash plate. During filtration, these act as filter plates. During washing, the outlets from

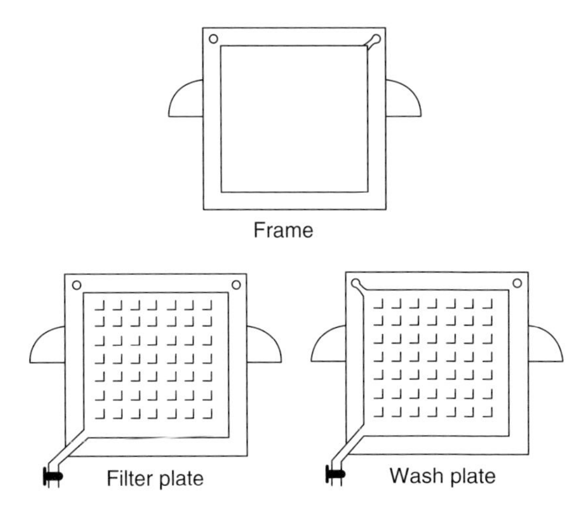

Figure 9.2 Schematic drawings of plates and frames. From Ref. [2] with permission of the authors.

the wash plates are closed and the wash liquid is pumped onto their surfaces via an inlet channel (see Figure 9.2). The wash liquid then passes through the full thickness of the cake and two layers of filter medium before exiting from the filter plates. This is said to achieve more effective washing than that attainable without the wash plates. After washing, the press is opened, the cake is removed from the frames, the filter medium is cleaned and the press is reassembled ready for the next run. This and other types of vertical plate filters are compact, flexible, and have a relatively low capital cost. However, labor costs and filter cloth consumption can be high.

Horizontal plate filter In this type of filter, the medium is supported on top of horizontal drainage plates which are stacked inside a pressure vessel. The feed is pumped in through a central duct, entering above the filter medium. The filtrate passes down through the medium onto the drainage plates and exits from them through an annular outlet. The cake builds up on top of the filter medium. After filtration, the feed is replaced by wash liquid which is pumped through the filter. After washing, the assembly of plates is lifted out of the pressure vessel and the cake removed manually. This type of filter is compact. The units are readily cleaned and can be sterilized if required. Labor costs can be high. They are used mainly for removing small quantities of solids and are known as *polishing filters*.

Shell-and-leaf filters A filter leaf consists of a wire mesh screen or grooved plate over which the filter medium is stretched. Leaves may be rectangular or circular in shape. They are located inside a pressure vessel or shell. They are either supported from the bottom or center or suspended from the top, inside the shell. The supporting member is usually hollow and acts as a takeaway for the filtrate. In horizontal shell-and-leaf filters, the leaves are mounted vertically inside horizontal pressure vessels (Figure 9.3).

As the feed slurry is pumped through the vessel, the cake builds up on the filter medium covering the leaves while the filtrate passes through the medium into the hollow leaf and then out through the leaf supports. Leaves may be stationary

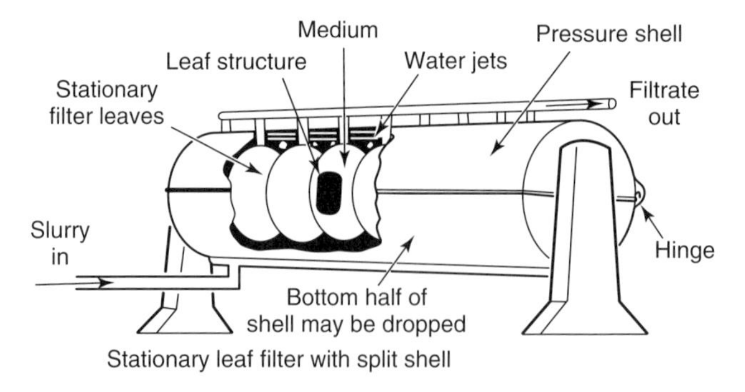

Figure 9.3 Schematic drawing of a shell-and-leaf filter. Adapted from Ref. [2] with permission of the authors.

or they may rotate about a horizontal axis. When filtering is stopped, washing is carried out by pumping wash liquid through the cake and leaves.

The cake may be removed by withdrawing the leaf assembly from the shell and cleaning the leaves manually. In some designs, the bottom half of the shell may be opened and the cake sluiced down with water jets. In vertical shell-and-leaf filters, rectangular leaves are mounted vertically inside a vertical pressure vessel. Shell-and-leaf filters are generally not as labor intensive as plate-and-frame presses but have higher capital costs. They are mainly used for relatively long filtration runs with slurries of low or moderate solids content.

Edge filters In this type of filter a number of stacks of rings or disks, known as *filter piles* or *packs*, are fixed to a header plate inside a vertical pressure vessel. Each pile consists of a number of disks mounted one above the other on a fluted vertical rod and held together between a boss and nut (Figure 9.4). The clearance between the disks is in the range 25–250 μm. Before filtration commences, a precoat of filter aid is applied to the edges of the disks. When the feed slurry is pumped into the pressure vessel, the cake builds up on top of the precoat of filter aid, while the filtrate passes between the disks and exits via the grooves on the supporting rod. Additional filter aid may be mixed with the feed. When filtering and washing are complete, the cake is removed by back flushing with liquid through the filtrate

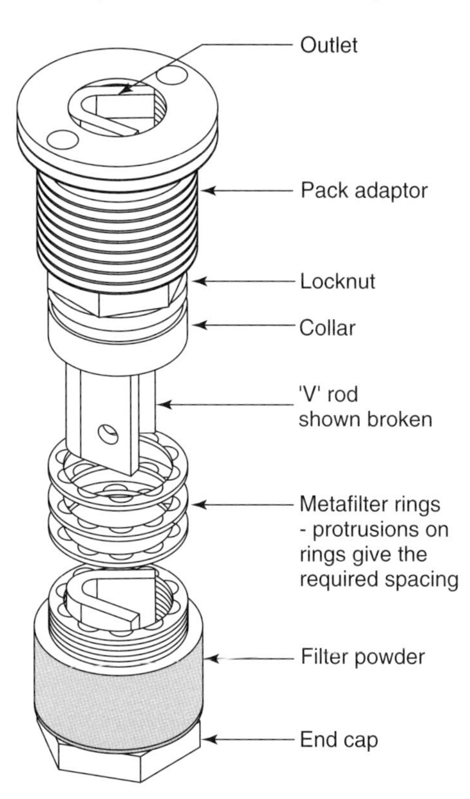

Figure 9.4 Metafilter Pack filter pile. By courtesy of Stella-Meta Ltd.

outlet and removing the cake in the form of a sludge through an outlet in the bottom of the pressure vessel. The disks may be made of metal or plastic. Edge filters have a relatively low labor requirement and use no filter cloth. They are used mainly for removing small quantities of fine solids from liquids.

9.2.4.2 Vacuum Filters

In vacuum filters a partial vacuum is created downstream of the medium and atmospheric pressure is maintained upstream. Most vacuum filters are operated continuously, as it is relatively easy to arrange continuous cake discharge under atmospheric pressure.

Rotary drum vacuum filters There are a number of different designs of this type of filter, one of which is depicted in Figure 9.5.

A cylindrical drum rotates about a horizontal axis partially immersed in a tank of the feed slurry. The surface of the drum is divided into a number of shallow compartments by means of wooden or metal strips running the length of the drum. Filter medium is stretched over the drum surface, supported on perforated plates or wire mesh. A pipeline runs from each compartment to a rotary valve located centrally at one end of the drum. Consider one of the compartments on the surface of the drum (shown shaded in Figure 9.5). As the drum rotates, this compartment becomes submerged in the slurry. A vacuum is applied to the compartment through the rotary valve. Filtrate is drawn through the medium and flows through the pipe to the rotary valve, from where it is directed to a filtrate receiver. The solids form a layer of cake on the outer surface of the medium. The cake increases in thickness as long as the compartment remains submerged in the slurry. As it emerges from the slurry, residual filtrate is sucked from the cake. Next the compartment passes beneath sprays of wash liquid. The washings are directed to a different receiver by means of the rotary valve. As the compartment passes from beneath the sprays,

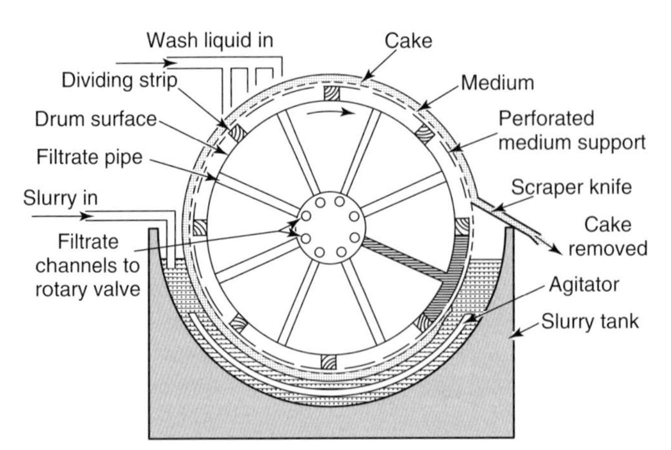

Figure 9.5 Principle of operation of a rotary drum vacuum filter. From Ref. [2] with permission of the authors.

residual wash liquid is sucked from the cake. Next, by means of the rotary valve, the compartment is disconnected from the vacuum source and compressed air introduced beneath the medium for a short period of time. This loosens the cake from the surface of the medium and facilitates its removal by means of a scraper knife. Many other designs of rotary drum filters are available featuring different methods of feeding the slurry onto the drum surface removing the cake from the medium.

Rotary drum vacuum filters incur relatively low labor costs and have large capacities for the space occupied. However, capital costs are high and they can only handle relatively free-draining solids. In common with all vacuum filters, they are not used to process hot and/or volatile liquids. For removing small quantities of fine solids from a liquid, a relatively thick layer (up to 7.5 cm) of filter aid may be precoated onto the medium. A thin layer of this precoat is removed together with the cake by the scraper knife.

Rotary vacuum disk filters In a disk filter, instead of a drum, a number of circular filter leaves, mounted on a horizontal shaft, rotate partially submerged in a tank of slurry. Each disk is divided into sections. Each section is covered with filter medium and is connected to a rotary valve, which controls the application of vacuum and compressed air to the section. Scraper knives remove the cake from each disk. Such disk filters have a larger filtering surface per unit floor area, compared to drum filters. However, cake removal can be difficult and damage to filter cloth excessive.

Other designs of continuous vacuum filters are available featuring moving belts, rotating tables, and other supports for the filter medium. These are used mainly for waste treatment rather than in direct food applications [2, 5–7, 9, 10].

9.2.4.3 **Centrifugal Filters (Filtering Centrifugals, Basket Centrifuges)**

In this type of filter, the flow of filtrate through the cake and medium is induced by centrifugal force. The slurry is fed into a rotating cylindrical bowl with a perforated wall. The bowl wall is lined on the inside with a suitable filter medium. Under the action of centrifugal force, the solids are thrown to the bowl wall where they form a filter cake on the medium. The filtrate passes through the cake and medium and leaves the bowl through the perforations in the wall.

Batch centrifugal filters The principle of this type of filter is shown in Figure 9.6. The cylindrical metal bowl in suspended from the end of a vertical shaft within a stationary casing. With the bowl rotating at moderate speed, slurry is fed into the bowl. A cake forms on the medium lining the inside of the perforated bowl wall and the filtrate passes through the perforations into the casing and out through a liquid outlet. The speed of the bowl is increased to recover most of the filtrate. Wash liquid may be sprayed onto the cake and spun off at high speed. The bowl is then slowed down, the cake cut out with an unloader knife or plough and removed through an opening in the bottom of the bowl. Cycle times vary over 3–30 min. Fully automated versions of these batch filters operate at a constant speed, about horizontal axes,

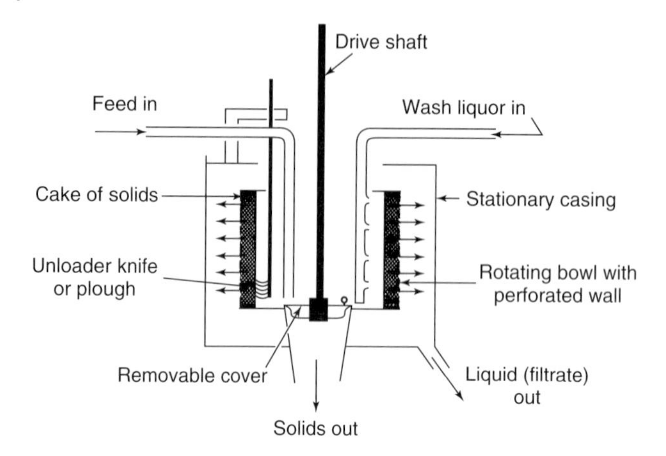

Figure 9.6 Batch centrifugal filter: from [2] with permission.

throughout a shorter cycle of 0.5–1.5 min. The feed and wash liquid are introduced automatically and the cake is cut out by a hydraulically operated knife.

Continuous centrifugal filters The principle of one type of continuous centrifugal filter is shown in Figure 9.7. A conical perforated bowl (basket) rotates about a vertical axis inside a stationary casing. The incline of the bowl causes the separation force to be split between vertical and horizontal elements resulting in the product moving upwards. The vertical force pushes the product up over the basket lip into the casing from where it is discharged. The horizontal element ensures that purging of the liquid phase takes place. This type of centrifuge is used for separating

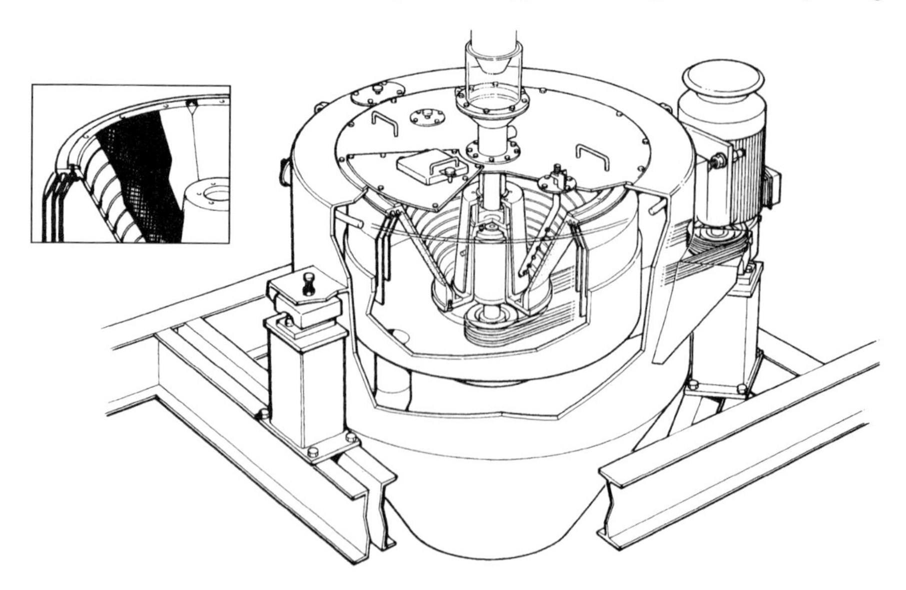

Figure 9.7 A continuous centrifugal filter. By courtesy of Broadbent Customer Services Ltd.

sugar crystals from syrup. Sliding of the product upwards and its discharge form the lip of the bowl, usually at high speed, is a relatively violent process and may damage the product, that is, fracture of crystals. Washing the solid phase while it is moving may limit its effectiveness.

Other types of continuous centrifugal filters feature reciprocating pushing devices, screw conveyors, or vibrating mechanisms to facilitate removal of the cake [2, 5, 6, 11].

A multistage version of this push-type centrifuge features a number of perforated bowls arranged in a telescopic fashion, rotating inside a stationary casing. Alternate bowls reciprocate while the rest do not. The cake progresses from one bowl to the next by being pushed off the end of the smaller bowl onto the inner surface of the next largest bowl until it is discharged from the open end of the largest bowl. In this type of centrifuge, buckling of the cake is less likely and washing is said to be more effective than in the single-stage machine.

Other types of continuous centrifugal filters feature cone-shaped perforated bowls and screw conveyors or vibrating mechanisms to facilitate removal of the cake [2, 5–7, 11].

9.2.5 Applications of Filtration in Food Processing

9.2.5.1 Edible Oil Refining

Filtration is applied at a number of stages in refining of edible oils. After extraction or expression, crude oil may be filtered to remove insoluble impurities such as fragments of seeds, nuts, cell tissue, and so on. For large-scale applications rotary filters are used. Plate-and-frame filters are used for smaller operations. Bleaching earths used in decolorizing oils are filtered off using rotary or plate filters. The catalysts used in hydrogenating fats and oils are recovered by filtration. Since hydrogenated fats have relatively high melting points, heated plate filters may be used. During winterization and fractionation of fats, after cooling, the higher melting point fractions are filtered off using plate-and-frame or belt filters [13–17].

9.2.5.2 Sugar Refining

The juice produced by extraction from sugar cane or sugar beet contains insoluble impurities. The juice is treated with lime to form a flocculent precipitate which settles to the bottom of the vessel. The supernatant liquid is filtered to produce a clear juice for further processing. Plate-and-frame presses, shell-and-leaf, and rotary drum vacuum filters are used. The settled "mud" is also filtered to recover more juice. Plate-and-frame presses or rotary drum vacuum filters are used for this duty. Filtration is also used at a later stage in the refining process to further clarify sugar juice. In the production of granulated sugar, purified sugar juice is concentrated up to 50–60% solids content by vacuum evaporation and seeded with finely ground sugar crystals to initiate crystallization. When the crystals have grown to the appropriate size, they are separated from the juice in batch or continuous centrifugal filters [18, 19].

9.2.5.3 Beer Production

During maturation of beer, a deposit of yeast and trub forms on the bottom of the maturation tank. Beer may be recovered from this by filtration using plate-and-frame presses, shell-and-leaf, or rotary drum vacuum filters. The beer is clarified by treatment with isinglass finings, centrifugation, or filtration. If filtration is used, the beer is first chilled and then filtered through plate-and-frame, horizontal plate, or edge filters. In the case of plate filters, the filter medium consists of sheets of cellulose, aluminum oxide, or zirconium oxide fibers, with added kieselguhr. Insoluble polyvinyl pyrrolidone may also be incorporated into the medium to absorb phenolic materials associated with beer haze. Edge filters are precoated with filter aid and more filter aid is usually added to the beer prior to filtration. Yeasts and bacteria may also be removed from beer by filtration. Although the pore sizes in the media are much larger than the microorganisms, the fibers hold the negatively charged microorganisms electrostatically. The pressure drop across these filters needs to be limited to avoid the microorganisms being forced off the media fibers. When a sterile product is desired, the sealed filter must be presterilized before use [20–22].

9.2.5.4 Wine Making

Wine is filtered at different stages of production: after racking, after decolorizing, and finally just before bottling. Plate-and-frame presses, shell-and-leaf filters, edge filters, and precoated rotary drum vacuum filters have been used. Filter media are mainly sheets made of cellulose incorporating filter aid material (mainly diatomaceous earth) which is bound into the cellulose sheets with bitumen. With edge and precoated drum filters, loose filter aid material is used. Sterile wine may be produced by filtration in presterilized equipment [22–25].

There are many other applications for filtration in the food industry, including the filtration of starch and gluten suspensions and the clarification of brines, sugar syrups, fruit juices, yeast, and meat extracts.

9.3 Centrifugation

9.3.1 General Principles

Centrifugation involves the application of centrifugal force to bring about the separation of materials. It may be applied to the separation of immiscible liquids and the separation of insoluble solids from liquids.

9.3.1.1 Separation of Immiscible Liquids

If two immiscible liquids, A and B, with different densities, are introduced into a cylindrical bowl rotating about a vertical axis, under the influence of centrifugal force, the more dense liquid A moves toward the wall of the bowl where it forms

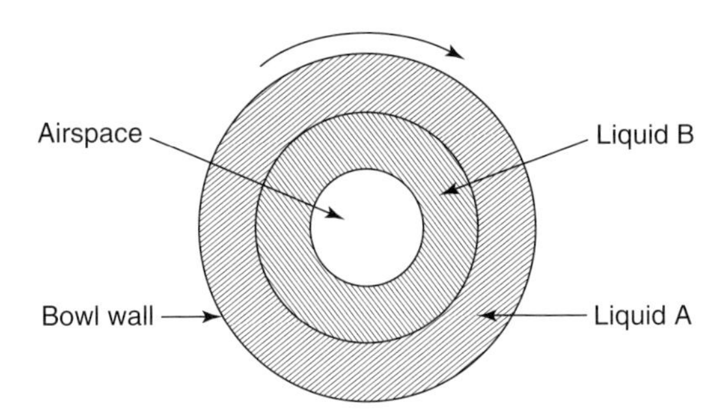

Figure 9.8 Separation of immiscible liquids in a cylindrical bowl (plan view). From Ref. [2] with permission of the authors.

an annular ring (Figure 9.8). The less dense liquid B is displaced toward the center of the bowl where it forms an inner annular ring. If the feed is introduced continuously into the bottom of the bowl through a vertical feed pipe, the liquids may be removed separately from each layer by a weir system, as shown in Figure 9.9. The more dense liquid A flows out over a circular weir of radius R_A and the less dense liquid B over a weir of radius R_B.

The interface between the two layers is known as the *neutral zone*. The position of this interface can influence the performance of the centrifuge. In the outer zone (A), light liquid is effectively stripped from a mass of dense liquid while, in the inner zone (B), dense liquid is more effectively stripped from a mass of light liquid. Thus, if the centrifuge is being used to strip a mass of dense liquid free of light liquid so that the dense phase leaves in as pure a state as possible, then the dwell time in zone A should be greater than that in zone B. For such a duty, the interface is best moved toward the center of rotation so that the volume of zone A exceeds that of zone B. In this situation, the light component is exposed to a relatively small centrifugal force for a short time, while the heavy component is exposed to a large force for a longer time. An example of such a duty is the separation of cream from milk, where the objective is to produce skim milk with as little fat

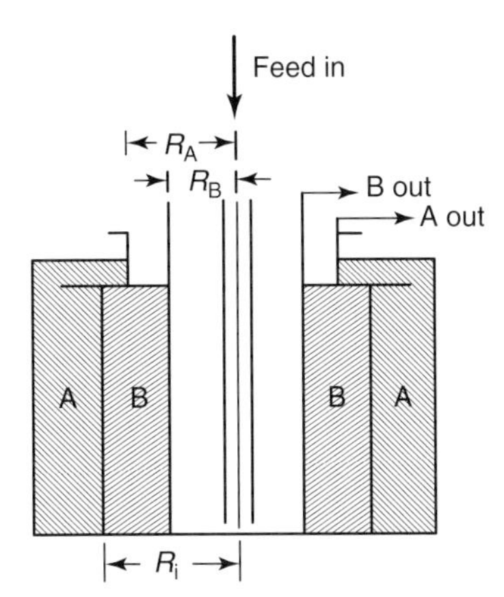

Figure 9.9 Separation of immiscible liquids in a cylindrical bowl with submerged weir (sectional view). From Ref. [2] with permission of the authors.

in it as possible. In contrast, if the duty is to strip a mass of light liquid free of dense liquid, that is, to produce a pure light phase, the interface is best moved out toward the bowl wall so that the volume occupied by zone B exceeds that of zone A. An example of such a duty would be the removal of small amounts of water from an oil. The actual change in position of the interface is quite small, in the order of 25–50 μm. However, it does affect the performance of the separator. The position of the interface can be changed by altering the radii of the liquid outlets. For example, if the radius of the light liquid outlet R_B is fixed, decreasing the radius of the dense liquid outlet R_A will move the interface toward the center of rotation. In practice, the radius of either liquid outlet and hence the position of the interface, is determined by fitting a ring with an appropriate internal diameter to the outlet. Such rings are known as *ring dams* or *gravity disks*. It has also been established that the best separation is achieved by introducing the feed to the bowl at a point near the interface. The density difference between the liquids needs to be 3% or more for successful separation.

Other factors that influence the performance of liquid–liquid centrifugal separators are bowl speed and the rate of flow of the liquids through them. In general, the higher the speed of the bowl, the better the separation. However, at very high speeds the viscosity of the oil phase may impede its flow through the centrifuge. The higher the rate of flow of the liquids through the bowl, the shorter the dwell time in the action zone; and so the less effective the separation is likely to be. For each duty a compromise needs to be struck between throughput and efficiency of separation.

9.3.1.2 **Separation of Insoluble Solids from Liquids**

If a liquid containing insoluble solid particles is fed into the bottom of a cylindrical bowl rotating about a vertical axis, under the influence of centrifugal force, the solid particles move toward to the bowl wall. If a particular solid particle reaches the bowl wall before being swept out by the liquid leaving through a central outlet in the top of the bowl (Figure 9.10), it remains in the bowl and thus is separated from the liquid. If it does not reach the bowl wall, it is carried out by the liquid. The fraction of the solid particles remaining in the bowl and the fraction passing out in the liquid depend on the rate of feed, that is, the dwell time in the bowl.

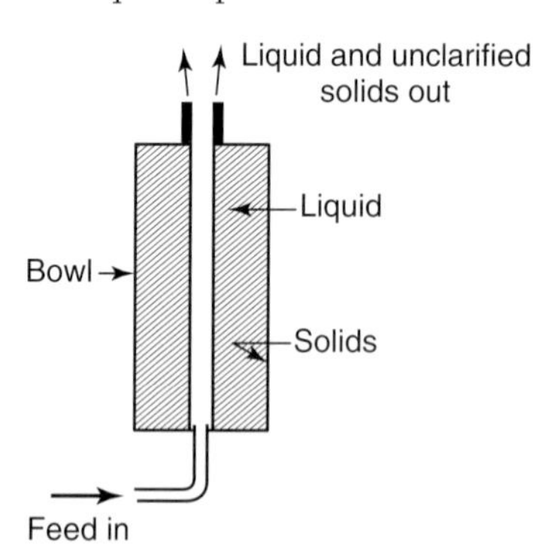

Figure 9.10 Principle of simple cylindrical centrifugal clarifier. From Ref. [2] with permission of the authors.

The following expression relates the throughput of liquid through a cylindrical bowl centrifuge to the characteristics of the feed and the dimensions and speed of the bowl:

$$q = 2\left[\frac{g\,(\rho_s - \rho_l)\,D_p^2}{18\eta}\right]\left[\frac{\omega^2 V}{2g\ln\left(\frac{R_2}{\left[\left(R_1^2+R_2^2\right)/2\right]^{1/2}}\right)}\right] \tag{9.6}$$

where q is the volumetric flow rate of liquid through the bowl, g is acceleration due to gravity, ρ_s is the density of the solid, ρ_l is the density of the liquid, D_p is the minimum diameter of a particle that will be removed from the liquid, η is the viscosity of the liquid, ω is the angular velocity of the bowl, V is the volume of liquid held in the bowl at any time, R_1 is the radius of the liquid outlet, and R_2 is the inner radius of the bowl. Note the quantities contained within the first set of square brackets relate to the feed material, while those within the second set refer to the centrifuge. This expression can be used to calculate the throughput of a specified feed material through a cylindrical bowl centrifuge of known dimensions and speed. It can also be used for scaling-up calculations. Alternative expressions for different types of bowl (see Section 9.3.2) can be found in the literature [2, 5–7, 26].

9.3.2
Centrifugal Equipment

9.3.2.1 Liquid–Liquid Centrifugal Separators

Tubular bowl centrifuge This type of centrifuge consists of a tall, narrow bowl rotating about a vertical axis inside a stationary casing. Bowl diameters range from 10 to 15 cm with length : diameter ratios of 4–8. The feed enters into the bottom of the bowl through a stationary pipe and is accelerated to bowl speed by vanes or baffles. The light and dense phases leave via a weir system at the top of the bowl and flow into stationary discharge covers. Depending on the duty it has to perform, a gravity disk of appropriate size is fitted to the dense phase outlet, as explained in Section 9.3.1.1. Bowl speeds range from 15 000 (large) to 50 000 rpm (small).

Disk bowl centrifuge In this type of centrifuge a relatively shallow, wide bowl rotates within a stationary casing. The bowl usually has a cylindrical body, with a diameter in the range 20–100 cm and a conical top (Figure 9.11). The bowl contains a stack of truncated metal cones, known as *disks*, which rotate with the bowl. The clearance between the disks is of the order of 50–130 μm. The disks have one or more sets of matching holes which form vertical channels in the stack. The feed is introduced to the bottom of the bowl, flows through these channels and enters the spaces between the disks. Under the influence of centrifugal force, the dense phase travels in a thin layer down the underside of the disks toward the bowl wall while the light phase, displaced toward the center of rotation, flows over the top

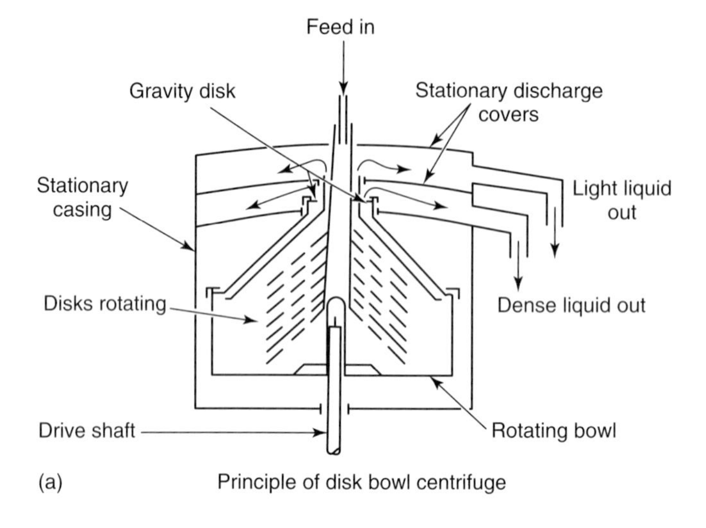

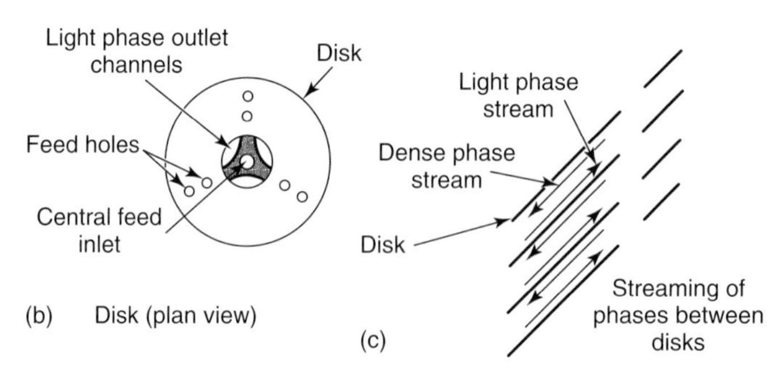

Figure 9.11 Principle of disk bowl centrifuge. From Ref. [2] with permission of the authors.

of the disks. Thus the space between each pair of disks is a minicentrifuge. The distance any drop of one liquid must travel to get into the appropriate stream is small compared to that in a tubular bowl machine or indeed an empty bowl of any design. In addition, in a disk machine there is considerable shearing at the interface between the two counter-current streams of liquid which contributes to the breakdown of emulsions. The phases leave the bowl through a weir system fitted with an appropriate gravity disk and flow into stationary discharge covers.

9.3.2.2 Solid–Liquid Centrifugal Separators

Both tubular bowl and disk bowl centrifuges can be used for solid–liquid separation, within certain limitations. For this type of duty, the dense phase outlet is closed off and the clear liquid exits the bowl through the central, light phase outlet. The solid particles which are separated from the liquid remain in the bowl, building up as a deposit on the wall of the bowl. Consequently, the centrifuges are operated on a batch principle and have to be stopped and cleaned at intervals. The tubular bowl machines have a relatively small solids capacity and are only suitable for handling feeds with low solids content, less than 0.5%. However, because of the high speeds they operate at, they are particularly suited to removing very fine solids. Disk bowl machines have up to five times the capacity for solids, compared to tubular bowl centrifuges. However, to avoid frequent cleaning, they are also used mainly with feeds containing relatively low solids content, less than 1.0%.

Solid bowl centrifuge (clarifier) For separating solid particles that settle relatively easily, a bowl similar in shape to that shown in Figure 9.6 may be used. However, the bowl wall is not perforated and no filter medium is used. The solids build up on the inside of the bowl wall and the clear liquid spills out over the top rim of the bowl into the outer casing. At intervals, the feed is stopped and the solids removed by means of a knife or plough and discharged through an opening in the bottom of the bowl. This type of clarifier can handle feeds with up to 2.0% solids content.

Nozzle-discharge centrifuge In this type of centrifuge there is provision for the continuous discharge of solids, in the form of a sludge, as well as the clear liquid. There are many different designs available. One design consists of a disk bowl machine with 2–24 nozzles spaced around the bowl. The size of the nozzles is in the range 0.75–2.00 mm, depending on the size of the solid particles in the feed. From 5 to 50% of the feed is continuously discharged in the form of a slurry through these nozzles. The slurry may contain up to 25% v/v solids. By recycling some of the slurry the solids content may be increased. Up to 75% of the slurry may be recycled, depending on its flowability, and the solids content increased up to 40%.

Self-opening centrifuge In this type of centrifuge the ports discharging the slurry open at intervals and the solids are discharged under a pressure of up to 3500 kN m^{-2}. The opening of the ports may be controlled by timers. Self-triggering ports are also available. The build-up of solids in the bowl is monitored and a signal is generated which triggers the opening of the ports. An example of one such centrifuge, used in the brewing industry, is shown in Figure 9.12.

The slurry discharged from these self-opening centrifuges usually has a higher solids content compared to that continuously discharged through open nozzles.

Decanting centrifuge Nozzle and valve discharge centrifuges can only handle feeds containing a few percent or less of solids. For feeds containing a higher percent of solids, decanting, or conveyor bowl centrifuges may be used. The principle of operation of one such centrifuge is shown in Figure 9.13. A solid bowl containing a screw conveyor rotates about a horizontal axis. The bowl and conveyor rotate in the same direction but at different speeds. The feed enters the bowl through the conveyor axis. The solids are thrown to the bowl wall and are conveyed to one end of the bowl, up a conical section, from where they are discharged. The clear liquid is discharged through an adjustable weir at the other end of the bowl. Such machines can handle feeds containing up to 90% (v/v) of relatively large solid particles. Particles 2 μm or less in diameter are normally not be removed from the

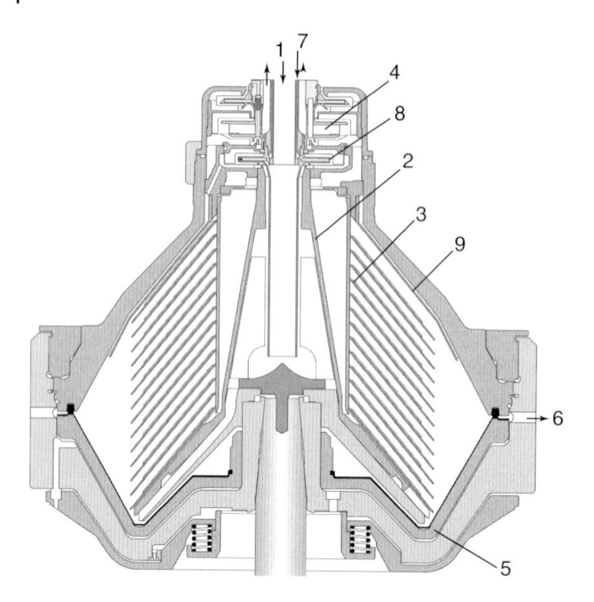

Figure 9.12 A self-triggering solids-ejecting clarifier centrifuge. By courtesy of Alfa Laval Ltd.

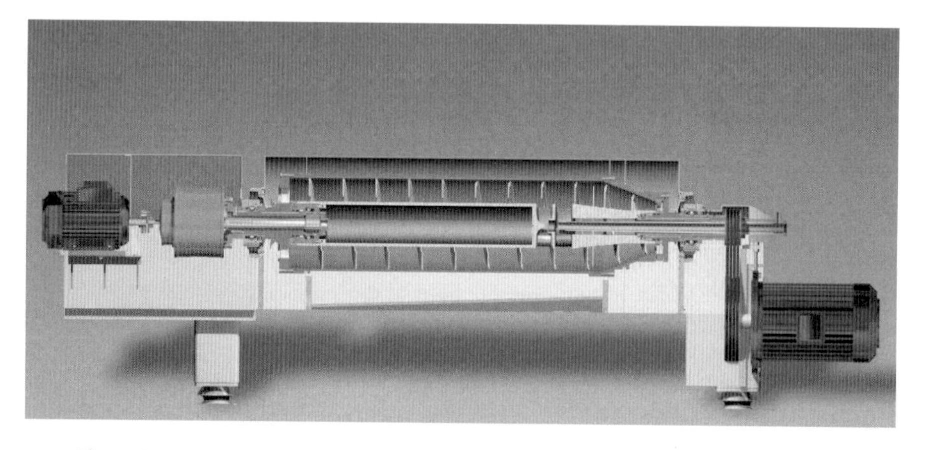

Figure 9.13 Cut away view – Alfa Laval Decanter. By courtesy of Alfa Laval Ltd.

liquid. Where necessary, the liquid discharged from decanting centrifuges may be further clarified in tubular or disk bowl centrifuges [2, 5–7, 11].

9.3.3
Applications for Centrifugation in Food Processing

9.3.3.1 Milk Products

Centrifugation is used in the separation of milk to produce cream and/or skim milk. Disk bowl centrifuges are generally used for this duty. They may be hermetically sealed and fitted with centripetal pumps. Milk is usually heated to between 40 and 50 °C prior to separation, to reduce its viscosity and optimize the density difference between the fat and aqueous phases. The fat content of the skim milk may be reduced to less than 0.05%. Although the process is continuous, insoluble solids present in the milk (dirt particles, casein micelles, microorganisms) build up as sludge in the centrifuge bowl. The bowl has to be cleaned out at intervals. Alternatively, nozzle or valve discharge centrifuges may be used, but with outlets for the cream and skim milk as well as the sludge. Fat may be recovered from whey and buttermilk by centrifugation [27].

9.3.3.2 **Edible Oil Refining**

In the early stages of oil refining, the crude oil is treated with water, dilute acid, or alkali to remove phosphatides and mucilaginous material. This process is known as *degumming*. Nozzle or valve discharge centrifuges are used to remove the gums after these treatments. In the case of acid-degumming, the degummed oil may be washed with hot water and the washings removed by centrifugation. The next step in oil refining is neutralization. The free fatty acids, phosphatides and some of the pigments are treated with caustic soda to form soapstock which is then separated from the oil by centrifugation, using nozzle or valve discharge centrifuges. The oil is then washed with hot water and the washings removed by centrifugation [14].

9.3.3.3 **Beer Production**

Centrifugation may be used as an alternative to filtration at various stages in the production of beer. Nozzle discharge centrifuges may be used for clarifying rough beer from fermenting vessels and racking tanks. Valve discharge centrifuges may be used for wort and beer clarification. Centrifuges used for the treatment of beer may be hermetically sealed to prevent the loss of carbon dioxide and the take-up of oxygen by the beer. Valve discharge centrifuges may also be used for the recovery of beer from fermenters and tank bottoms. Decanting centrifuges may be used for clarifying worts and beers containing relatively high contents of yeast or trub. They may also be used as an alternative to valve discharge machines to recover beer from fermenters and tank bottoms [20].

9.3.3.4 **Wine Making**

Centrifugation may be used instead of or in combination with filtration at various stages in the production of wine. Nozzle or valve discharge centrifuges are generally used. Applications include: the clarification of must after pressing, provided that the solids content is relatively low, the clarification of wine during fermentation to stabilize it by gradual elimination of yeast, the clarification of new wines after fermentation and before filtration, the clarification of new red wines before filling into barrels and the facilitation of tartrate precipitation for the removal of tartrate crystals [23–25].

9.3.3.5 Fruit Juice Processing

Centrifugation may be used for a variety of tasks in fruit juice processing. Valve discharge centrifuges are used to remove pulp and control the level of pulp remaining in pineapple and citrus juices. Centrifuged apple juice is cloudy but free from visible pulp particles. Tubular bowl centrifuges were originally used to clarify apple juice but more recently nozzle and valve discharge machine are used. The use of hermetically sealed centrifuges prevents excessive aeration of the juice. In the production of oils from citrus fruits centrifugation is applied in two stages. The product from the extractor contains an emulsion of 0.5–3.0% oil. This is concentrated up to 50–70% oil in a nozzle or valve discharge centrifuge. The concentrated emulsion is then separated in a second centrifuge to produce the citrus oil [28, 29].

There are many other applications for centrifugation in food processing, for example, tubular bowl machines for clarifying cider and sugar syrups and separating animal blood into plasma and hemoglobin, nozzle and valve discharge machines for dewatering starches, and decanting centrifuges for recovering animal and vegetable protein, separating fat from comminuted meat and separating coffee and tea slurries.

9.4 Solid–Liquid Extraction (Leaching)

9.4.1 General Principles

This is a separation operation in which the desired component, the solute, in a solid phase is separated by contacting the solid with a liquid, the solvent, in which the desired component is soluble. The desired component leaches from the solid into the solvent. Thus the compositions of both the solid and liquid phases change. The solid and liquid phases are subsequently separated and the desired component recovered from the liquid phase.

Solid–liquid extraction is carried out in single or multiple stages. A stage is an item of equipment in which the solid and liquid phases are brought into contact, maintained in contact for a period of time and then physically separated from each other. During the period of contact, mass transfer of components between the phases takes place and they approach a state of equilibrium. In an equilibrium or theoretical stage, complete thermodynamic equilibrium is attained between the phases before they are separated. In such a stage, the compositional changes in both phases are the maximum which are theoretically possible under the operating conditions. In practice, complete equilibrium is not reached and the compositional changes in a real stage is less than that attainable in an equilibrium stage. The *efficiency of a real stage* may be defined as the ratio of the compositional change attained in the real stage to that which would have been reached in an equilibrium stage under the same operating conditions. When estimating the number of stages required to carry out a particular task in a multistage system, the number of equilibrium stages is first estimated and the number of real stages calculated by dividing the number of equilibrium stages by the stage efficiency. Graphical and numerical methods are used to estimate the number of equilibrium stages required for a particular duty [2, 4–6]. After the period of contact, the solid–liquid mixture is separated into two streams: a "clear" liquid stream or overflow, consisting of a solution of the solute in the solvent and a "residue" stream or underflow, consisting of the insoluble solid component with some solution adhering to it. In an equilibrium stage, the composition of the overflow is the same as that of the solution leaving with the insoluble solid in the underflow. In a real stage, the concentration of solute in the overflow is less than that in the solution leaving with the insoluble solid in the underflow.

The extraction of the solute from a particle of solid takes place in three stages. The solute dissolves in the solvent. The solute in solution then diffuses to the surface of the particle. Finally, the solute transfers from the surface of the particle into the bulk of the solution. One or more of these steps can limit the rate of extraction. If the correct choice of solvent has been made, the solution of the solute in solvent is rapid and is unlikely to influence the overall rate of extraction. The rate of movement of the solute to the surface of the solid particle depends on the size, shape, and internal structure of the particle and is difficult to quantify. The rate of transfer of the solute from the surface of the solid particle to the bulk of the solution may be represented by the expression:

$$\frac{dw}{dt} = KA(C_S - C) \tag{9.7}$$

Here, dw/dt is the rate of mass transfer of the solute, A is the area of the solid–liquid interface, C_S and C are the concentration of the solute at the surface of the solid particle and in the bulk of the solution, respectively, and K is the mass transfer coefficient.

In a single-stage extraction unit where V is the total volume of the solution and is constant, then:

$$\frac{dw}{dt} = VdC$$

and so:

$$\frac{dC}{dt} = \frac{KA(C_S - C)}{V} \tag{9.8}$$

The main factors which influence the rate of extraction are as follows:

- **The solid–liquid interface area**: The rate of mass transfer from the surface of the particle to the bulk of the solution increases with increase in this area. Reducing the size of the solid particles increases this area and so increases the rate of mass transfer. In addition, the smaller the particle the shorter the distance the solute has to travel to reach the surface. This is likely to further speed up the extraction. However, very small particles may impede the flow of solvent through the bed of solid in an extractor and some particles may not come in contact with the solvent. In the case of cellular material, such as sugar beet the cell wall acts as a semipermeable membrane releasing sugar but retaining larger nonsugar molecules. Therefore, the beet is sliced rather than comminuted to increase the surface area for extraction but to limit cell wall damage.

- **Concentration gradient**: To ensure as complete extraction as possible, a gradient must be maintained between the concentration of solute at the surface of the solid particles and that in the bulk of the solution. In a single-stage extractor, as the phases approach equilibrium, this gradient decreases and so does the rate of extraction until it ceases. When this occurs, the solid may still contain a significant amount of solute and the solution may be relatively dilute, depending on the equilibrium conditions. This solution could be drained off and replaced with fresh solvent, resulting in further extraction of the solute. This could be repeated until the solute content of the solid reached a suitably low level. However, this would result in the production of a large volume of relatively dilute solution. The cost of recovering the solute from this solution increases as its solute content decreases. For example, the lower the concentration of sugar in the solution obtained after extraction of sugar beet, the more water has to be evaporated off before crystallization occurs. Multistage counter-current extraction systems enable a concentration gradient to be maintained even when the concentration of solute in the solid is low (Figure 9.14b). This results in more complete extraction compared with that attainable in single-stage or multistage concurrent systems (Figure 9.14a).

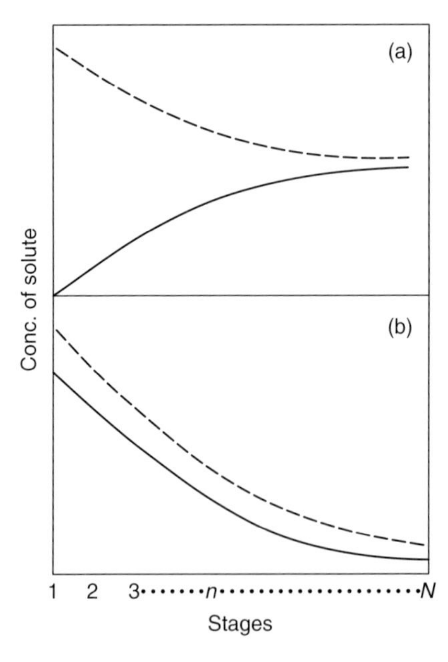

Figure 9.14 Graphical representation of concurrent and counter-current extraction systems. (a) Concurrent system and (b) counter-current system. Solid lines indicate solution, dashed lines indicate solid matter. From Ref. [2] with permission of the authors.

- **Mass transfer coefficient**: An increase in temperature increases the rate of solution of the solute in the solvent and also the rate of diffusion of solute through the solution. This is reflected in a higher value of K in Equations 9.7 and 9.8. Thus, the solvent is usually heated prior to and/or during extraction. The upper limit in temperature depends on the nature of the solids. For example, in the extraction of sugar from beet, too high a temperature can result in peptization of the beet cells and the release of nonsugar compounds into the solution (see Section 9.4.3.2). In the case of the extraction of solubles from ground roasted coffee beans, too high a temperature can result in the dried coffee powder having an undesirable flavor (see Section 9.4.3.3). Increasing the velocity and turbulence of the liquid as it flows over the solid particles can result in an increase in the value of K in Equations 9.7 and 9.8 and hence an increase in the rate of extraction. In some industries, when fine particles are being extracted, they are mechanically stirred. However, in most food applications, this is not the case as agitation of the solid can result in undesirable breakdown of the particles. In most food applications, the solvent is made to flow through a static bed of solids under the influence of gravity or with the aid of a pump. Alternatively, the solids are conveyed slowly, usually counter-current to the flow of solvent [2, 4–7, 30, 31].

9.4.2 Extraction Equipment

9.4.2.1 Single-Stage Extractors

A simple extraction cell consists of a tank fitted with a false bottom which supports a bed of the solids to be extracted. The tank may be open or closed. If extraction is to be carried under pressure, as in the case with extraction of ground roasted coffee (see Section 9.4.3.3), or if volatile solvents are used, as in the case of edible oil extraction (see Section 9.4.3.1), the tank is enclosed (see Figure 9.15). The solvent is sprayed over the top surface of the bed of solids, percolates down through the bed and exits via an outlet beneath the false bottom. The tank may be jacketed and/or a heater incorporated into the solvent feed line to maintain the temperature of the solution at the optimum level. Usually a pump is provided for recirculating the solution. The spent solid is removed manually or dumped through an opening in the bottom of the tank. In large cells, additional supports may be provided for the bed of solids to prevent consolidation at the bottom of the cell.

Single extraction cells are used for laboratory trials and for small-scale industrial applications. As discussed in Section 9.4.1, the bulked solution from such units is relatively dilute. If a volatile solvent is used, the overflow from the cell may be heated to vaporize the solvent, which is then condensed and recycled through the cell. In this way, a more concentrated solution of the solute may be obtained.

9.4.2.2 Multistage Static Bed Extractors

One method of applying multistage counter-current extraction is to use a number of single cells arranged in a circuit. Each cell contains a charge of solids. The

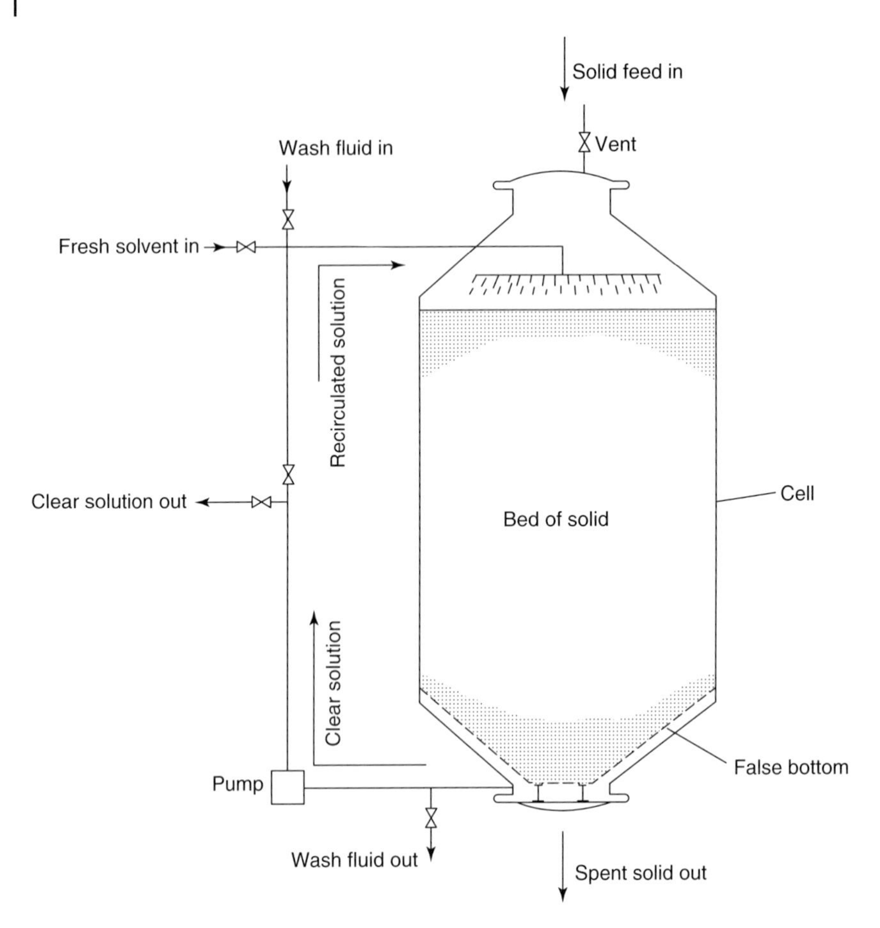

Figure 9.15 Single-stage, enclosed extraction cell. From Ref. [2] with permission of the authors.

solution from the preceding cell is sprayed over the surface of the bed of solids and percolates down through the bed, becoming more concentrated as it does so. The solution leaving from the bottom of the cell is introduced into the top of the next cell. A typical battery, as used for the extraction of sugar beet, contains 14 cells, as shown in Figure 9.16. At the time depicted in this figure, three of the cells are excluded from the circuit. Cells 10, 11, and 12 are being filled, washed, and emptied, respectively. The fresh water enters cell 13 and the concentrated sugar solution, or overflow, leaves from cell 9. When the beet in cell 9 is fully extracted, this cell is taken out of the circuit and cell 10 brought in to take its place. Fresh water then enters cell 14 and the concentrated sugar solution leaves from cell 10. By isolating cells in turn around the circuit, the principle of counter-current extraction may be achieved without physically moving the beet from one cell to the next. The number of cells in such a circuit may vary from 3 to 14.

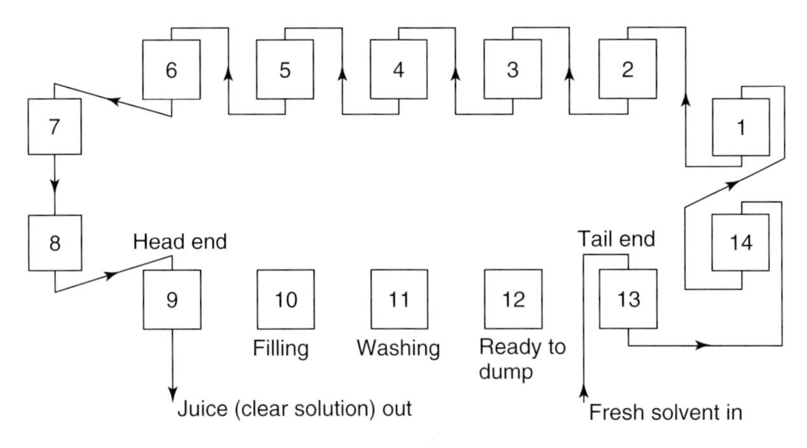

Figure 9.16 Multistage, counter-current extraction battery showing flow of solution at one particular point in time. From Ref. [2] with permission of the authors.

9.4.2.3 **Multistage Moving Bed Extractors**

There are many different designs of moving bed extractors available. They usually involve moving the solid gently from one stage to the next, counter-current to the flow of the solution. One type of continuous extractor consists of a trough set at a small angle to the horizontal containing two screw conveyors with intermeshing flights. The solvent is introduced at the elevated end of the trough. The solid is fed in at the other end and is carried up the slope by conveyors counter-current to the flow of the solution. The trough is enclosed and capable of withstanding high pressure. Extractors of this type are used for sugar beet and ground roasted coffee.

Another type, known as the *Bonotto extractor* is shown in Figure 9.17. It consists of a vertical tower divided into sections by horizontal plates. Each plate has an opening through which the solid can pass downwards from plate to plate; and each plate is fitted with a wiper blade which moves the solid to the opening. The holes are positioned 180° from each other in successive plates. The solid is fed onto the top plate. The wiper blade moves it to the opening and it falls onto the plate below and so on down the tower from plate to plate. Fresh solvent is introduced at the bottom of the tower and is pumped upwards counter-current to the solid. The rich solution leaves at the top of the tower and the spent solid is discharged from the bottom. This type of extractor is used for oil extraction from nuts and seeds.

Many other designs of continuous moving bed extractors are in use in industry. One design features moving perforated baskets which carry the solid through a stream of the solvent. In another design, the solid is conveyed by screw conveyors, with perforated blades, through vertical towers, counter-current to the flow of solvent [2, 4–6, 30, 32].

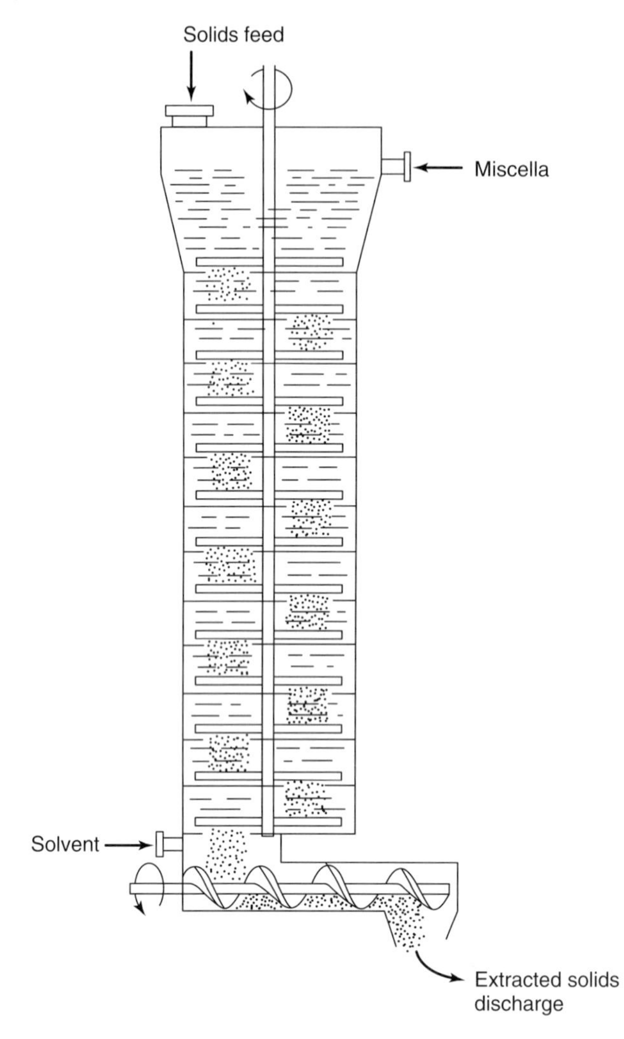

Figure 9.17 Bonotto extractor.

9.4.3 Applications of Solid–Liquid Extraction in Food Processing

9.4.3.1 Edible Oil Extraction

Solvent extraction may be used as an alternative to or in combination with expression to obtain oil from nuts, seeds, and beans. The most commonly used solvent is hexane. This is a clear hydrocarbon, derived from petroleum that boils at 68.9 °C. It is miscible with oil, immiscible with water and does not impart any objectionable odor or taste to the oil or spent solid. Hexane is highly flammable and so the plant must be vapor-tight and care must be taken to avoid the generation of sparks that might ignite the solvent. Other solvents have been investigated, including heptane and cyclohexane. Nonflammable solvents, such as trichloroethylene and carbon disulfide, have been studied but are toxic and difficult to handle. Various alcohols and supercritical carbon dioxide (see Section 9.4.4) have also been investigated. Various designs of moving bed extractors are used in large scale-oil extraction, including those described in Section 9.4.2. After extraction, the solution of oil in solvent is filtered, the solvent removed by vacuum evaporation followed by distillation and the solvent reused. Residual solvent may be removed from the spent solid by direct or indirect heating with steam and the resulting meal used for animal feed. Batch extractors featuring solvent recovery and reuse can be used for small-scale operations. Cottonseed, linseed, rapeseed, sesame, sunflower, peanuts, soybean, and corn germ may be solvent-extracted [2, 28].

9.4.3.2 **Extraction of Sugar from Sugar Beet**

Sugar is extracted from sugar beet using heated water as solvent. The beets are washed and cut into slices, known as *cossettes*. This increases the surface area for extraction and limits cell wall damage. Water temperature ranges from 55 °C in the early stages of extraction to 85 °C toward the end. Higher temperatures can cause peptization of the beet cells and release nonsugar compounds into the extract. Multistage static bed batteries, as depicted in Figure 9.17, are widely used. So also are various designs of moving bed extractors, including those described in Section 9.4.2.3. The solution leaving the extractor contains about 15% of dissolved solids. This is clarified by settling and filtration, concentrated by vacuum evaporation, seeded and cooled to crystallize the sugar. The crystals are separated from the syrup by centrifugation, washed, and air dried (see Section 3.1.5.2 [2, 33]).

9.4.3.3 **Manufacture of Instant Coffee**

A blend of coffee beans is roasted to the required degree, ground to the appropriate particle size range and extracted with heated water.

Extraction may be carried out in a multistage, counter-current static bed system consisting of five to eight cells. Each cell consists of a tall cylindrical pressure vessel as temperatures above 100 °C are used. Heat exchangers are located between the cells. Water at about 100 °C is introduced into the cell containing the beans that are almost fully extracted and then passes through the other cells, until the rich solution exits from the cell containing the freshly ground beans. The temperature of the solution increases up to a maximum of 180 °C as it passes through the battery of cells. In the later stages of extraction, some hydrolysis of insoluble carbohydrate material occurs, resulting in an increase in the yield of soluble solids. Higher temperatures may impart an undesirable flavor to the product due to excessive hydrolysis.

Continuous, counter-current extractors featuring screw conveyors within pressurized chambers (see Section 9.4.2.3) may be used instead of the static bed system. The rich solution leaving the extractor usually contains 15–28% solids. This may be fed directly to a spray drier. Alternatively, the solution may be concentrated up to 60% solids by vacuum evaporation (see Section 3.1.5.2). The volatiles may be stripped from the extract before or during evaporation and added back to the concentrated extract prior to dehydration either by spray drying or freeze drying [2, 34, 35].

9.4.3.4 Manufacture of Instant Tea

Dried, blended tea leaves may be extracted with heated water in a static bed system consisting of three to five cells. Water temperature ranges from 70 °C in the early stages of extraction to 90 °C in the later stages. The cells may be evacuated after filling with the dry leaves and the pressure brought back to atmospheric level by introducing gaseous carbon dioxide. This facilitates the flow of the water through the cells. Continuous tower or other moving bed extractors are also used to extract tea leaves. The rich solution coming from the extractor usually contains 2.5–5.0% solids. This is concentrated by vacuum evaporation to 25–50% solids. The volatile aroma compounds are stripped from the extract prior to or during evaporation and added back before dehydration by spray drying, vacuum drying, or freeze drying. An alternative process for producing instant tea from juice expressed from green leaves has been reported in the literature. The juice is fermented, steamed, centrifuged, and freeze dried. The residue left after pressing the leaves is used to produce tea granules. The instant tea was reported to be of good quality and cheaper to produce compared to that made from extracted leaves [36].

9.4.3.5 Fruit and Vegetable Juice Extraction

In recent years there has been considerable interest in using extraction instead of expression for recovering juices from fruits and vegetables. Counter-current screw extractors, some operated intermittently, have been used to extract juice with water. In some cases this results in higher yields of good quality compared to that obtained by expression [37].

9.4.4 The Use of Supercritical Carbon Dioxide as a Solvent

The critical pressure and temperature for carbon dioxide are 73.8 kPa and 31.06 °C, respectively. At pressures and temperatures above these values carbon dioxide exists in the form of a supercritical fluid (supercritical carbon dioxide; SC-CO_2). In this state it has the characteristics of both a gas and a liquid. It has the density of a liquid and can be used as a liquid solvent, but it diffuses easily like a gas. It is highly volatile, has a low viscosity, a high diffusivity, is nontoxic and nonflammable. These properties make it a very useful solvent for extraction.

However, the fact that SC-CO_2 has to be used at high pressure means that relatively expensive pressure-resistant equipment is required and running costs are also high. The solvent power of SC-CO_2 increases with increase in temperature and pressure. For the extraction of highly soluble compounds or for deodorization, pressures and temperatures close to the critical values may be used. When a single component is to be extracted from an insoluble matrix, so-called *simple extraction*, the highest pressure and temperature possible for each application should be used. The upper limit on temperature will depend on the heat sensitivity of the material.

The limit on pressure will be determined by the cost of the operation. When all soluble matter is to be extracted, so-called *total extraction*, high pressures and temperatures are again necessary.

The following are examples of the industrial application of SC-CO_2 extraction.

9.4.4.1 Hop Extract

A good quality hop extract, for use in brewing, may be obtained by extraction with SC-CO_2. A multistage counter-current static bed system, consisting of four extraction cells, is normally used. The SC-CO_2 percolates down through the hop pellets in each cell in turn. The solution of the extract in the SC-CO_2 leaving the battery is heated and the carbon dioxide evaporates, precipitating out the extract. The carbon dioxide is recompressed and cooled and condenses back to SC-CO_2 which is chilled to 7 °C and recycled through the extraction battery.

9.4.4.2 Decaffeination of Coffee Beans

SC-CO_2 may be used as an alternative to water or methylene chloride for the extraction of caffeine from coffee beans. The beans are moistened before being loaded into the extractor. SC-CO_2 is circulated through the bed of beans extracting the caffeine. The caffeine-laden SC-CO_2 passes to a scrubbing vessel where the caffeine is washed out with water. Alternatively, the caffeine may be removed by passing the caffeine-laden SC-CO_2 through a bed of activated charcoal. SC-CO_2 has also been used to decaffeinate tea.

9.4.4.3 Removal of Cholesterol from Dairy Fats

SC-CO_2 at 40 °C and 175 kPa has been used to remove cholesterol from butter oil in a packed column extractor. The addition of methanol as an entrainer increases the solubility of cholesterol in the fluid phase. The methanol is introduced with the oil into the column. Cholesterol can also be extracted from egg yolk and meat using SC-CO_2.

Many other potential applications for SC-CO_2 extraction have been investigated including: extraction of oils from nuts and seeds, extraction of essential oils from roots, flowers, herbs, and leaves, extraction of flavor compounds from spices and concentration of flavor compounds in citrus oils [2, 32, 38, 39, 58–61].

9.5 Distillation

9.5.1 General Principles

Distillation is a method of separation which depends on there being a difference in composition between a liquid mixture and the vapor formed from it. This difference in composition develops if the different components of the mixture have different vapor pressures or volatilities. In *batch distillation*, a given volume of liquid is heated and the vapors formed are separated and condensed to form a product. In batch distillation, the compositions of the liquid remaining in the still and the vapor collected change with time. Batch distillation is still used in some whisky distilleries. However, continuous distillation columns are used in most industrial applications of distillation.

Consider a liquid mixture consisting of two components with different volatilities. If such a mixture is heated under constant pressure conditions, it does not boil at a single temperature. The more volatile component starts to vaporize first. The temperature at which this commences is known as the *bubble point*. If a vapor consisting of two components with different volatilities is cooled, the less volatile component starts to condense first. The temperature at which this commences is known as the *dew point*. A diagram of the temperature composition for liquid–vapor equilibrium of a two-component mixture is presented in Figure 9.18.

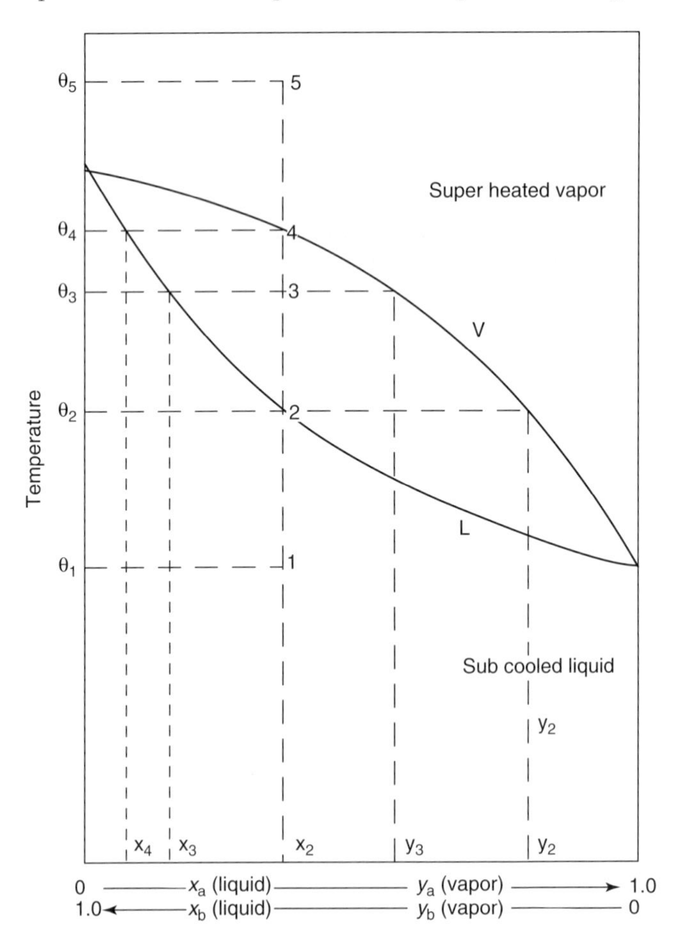

Figure 9.18 Temperature-composition diagram for liquid–vapor equilibrium for a two-component mixture.

The bottom "L" line in the phase envelope represents liquid at its bubble point and the top "V" line represents vapor at its dew point. The 0–1.0 composition axis refers to the more volatile component "a," x_a, and y_a are the mole fractions of "a" in the liquid and vapor phases, respectively, and x_b and y_b are the mole fractions of component "b," the less volatile component, in the liquid and vapor phases, respectively. The region below the bubble point curve represents subcooled liquid. The region above the dew point curve represents superheated vapor. Within the envelope, between the L and V lines, two phases exist. Saturated liquid and saturated vapor exist in equilibrium with each other.

If a liquid mixture at temperature θ_1 is heated until it reaches temperature θ_2, its bubble point, it will start to vaporize. The vapor produced at this temperature will contain mole fraction y_2 of component "a." Note that the vapor is richer in "a" than the liquid. As a result of the evaporation the liquid becomes less rich in component "a" and richer in "b" so its temperature rises further. At temperature θ_3 the liquid phase contains mole fraction x_3 of "a" and vapor phase y_3 of "a." Note that, at this temperature, the vapor is less rich in "a" than it was at its bubble point. When temperature θ_4 is reached, all of the liquid is evaporated and the composition of the vapor is the same as the original liquid, $y_4 = x_1$. A similar sequence of events occurs if we start with a superheated vapor at temperature θ_5 and cool it. When it reaches its dew point θ_4, it begins to condense and the liquid contains mole fraction x_4 of "a." If cooled further to θ_3, the liquid and vapor contain mole fractions x_3 and y_3 of "a," respectively. Thus both partial vaporization and partial condensation bring about an increase of the more volatile component in the vapor phase. A distillation column consists of a series of stages, or plates, on which partial vaporization and condensation takes place simultaneously.

The principle of a continuous distillation column, also known as a *fractionation* or *fractionating column*, is shown in Figure 9.19. The column contains a number of plates that are perforated to allow vapor rising from below to pass through them. Each plate is equipped with a weir over which the liquid flows and then through a downtake onto the plate below. The liquid contained in the reboiler at the bottom of the column is heated. When the liquid reaches its bubble point temperature, vapor is formed and this vapor bubbles through the liquid on the bottom plate. The vapor from the reboiler has a composition richer in the more volatile components than the liquid remaining in the reboiler. This vapor is at a higher temperature than the liquid on the bottom plate. Some of that vapor condenses and causes some of the liquid on the bottom plate to evaporate. This new vapor is richer in the more volatile components than the liquid on the bottom plate. This vapor in turn bubbles through the liquid on the plate above the bottom plate, causing some of it to evaporate and so on up the column. Thus, partial condensation and partial vaporization takes place on each plate. The vapor rising up the column becomes increasingly rich in the more volatile components while the liquid flowing down from plate to plate becomes richer in the less volatile components. If all the vapor leaving the top of the column is condensed and removed, then the liquid in the column becomes progressively less rich in the volatile components as does the vapor being removed at the top of the column.

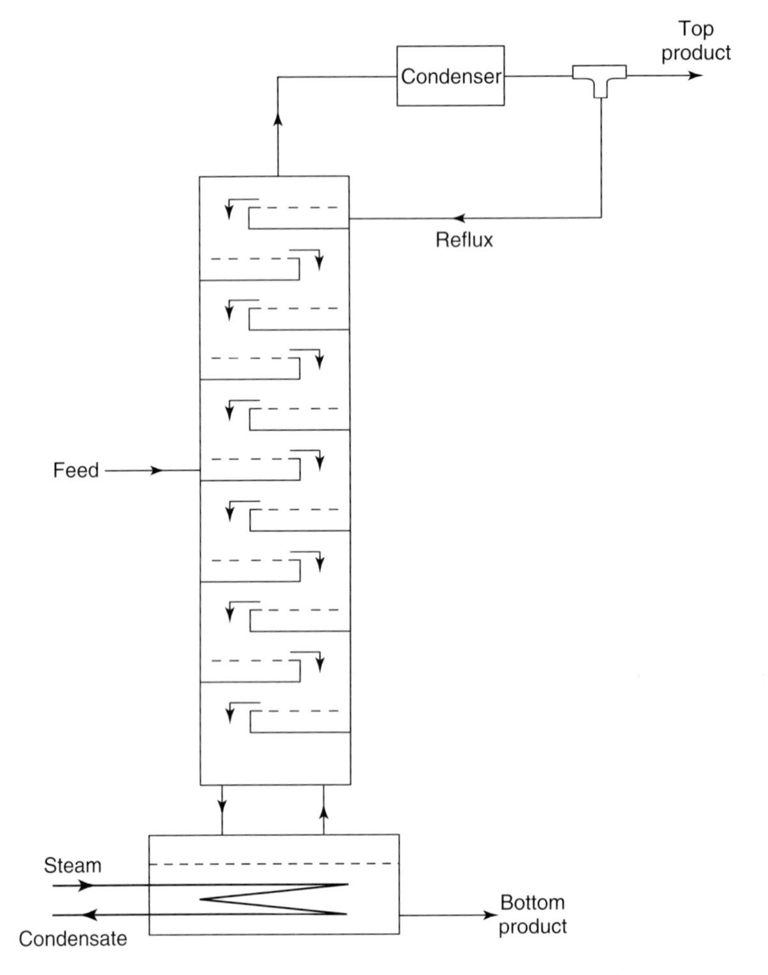

Figure 9.19 Principle of a continuous distillation (fractionating) column.

This is the equivalent of batch distillation. However, if some of the condensed vapor is returned to the column and allowed to flow down from plate to plate, the concentration of the volatile components in the column is maintained at a higher level. The condensed vapor returned to the column is known as *reflux*. If feed material is introduced continuously into the column, then a product rich in volatile components can be withdrawn continuously from the top of the column and one rich in the less volatile components from the reboiler at the bottom of the column. The feed is usually introduced onto a plate partway up the column.

Each plate represents a stage in the separation process. As is the case in solid–liquid extraction (see Section 9.4.1), the terms "equilibrium" or "theoretical plate" and "plate efficiency" may be applied to distillation. Graphical and numerical methods are used to estimate the number of equilibrium plates required to perform a particular duty and the number of real plates is calculated by dividing the number of equilibrium plates by the plate efficiency.

Steam distillation is applicable to mixtures which have relatively high boiling points and which are immiscible with water. If steam is bubbled through the liquid in the still, some of it will condense and heat the liquid to boiling point. Two liquid layers will form in the still. The vapor will consist of steam and the volatile vapor, each exerting its own vapor pressure. The mixture will boil when the sum of these pressures equals atmospheric pressure. Thus, the distillation temperature will always be lower than 100 °C at atmospheric pressure. By reducing the operating pressure, the distillation temperature will be reduced further and less steam will be used. Steam distillation may be used to separate temperature-sensitive, high boiling point materials from volatile impurities, or for removing volatile impurities with high boiling points from even less volatile compounds. The separation of essential oils from nonvolatile compounds dissolved or dispersed in water is an example of its application [5–7, 40–42].

9.5.2 Distillation Equipment

9.5.2.1 Pot Stills

Pot stills, used in the manufacture of good quality whisky, are usually made from arsenic-free copper. Apparently the copper has an influence on the flavor of the product. The still consists of a pot which holds the liquor to be distilled and which is fitted with a swan neck. Because of their shape, pot stills are sometimes known as *onion stills*. A lyne arm, a continuation of the swan neck, tapers toward the condenser. The condenser is usually a shell and tube heat exchanger also made of copper. Heat is applied to the pot by means of steam passing through coils or a jacket. There is very limited use of direct heating by means of a solid fuel furnace beneath the pot. Such stills operate on a batch principle and two or three may operate in series (see Section 9.5.3.1).

9.5.2.2 Continuous Distillation (Fractionating) Columns

A distillation column consists of a tall cylindrical shell fitted with a number of plates or trays. The shell may be made of stainless steel, monel metal, or titanium. As described in Section 9.5.1, the vapor passes upwards through the plates while the refluxed liquid flows across each plate, over a weir and onto the plate below via a downtake. There are many different designs of plate including the following examples.

Sieve plates These consist of perforated plates with apertures of the order of 5 mm diameter, spaced at about 10 mm centers. The vapor moving up through the perforations prevents the liquid from draining through the holes. Each plate is fitted with a weir and downtake for the liquid.

Bubble cap plates These are also perforated but each hole is fitted with a riser or "chimney" through which the vapor from the plate below passes. Each riser is covered by a bell-shaped cap, which is fastened to the riser by means of a spider or other suitable mounting. There is sufficient space between the top of the riser and the cap to permit the passage of the vapor. The skirt of the cap may be slotted or the edge of the cap may be serrated. The vapor rises through the chimney, is diverted downwards by the cap and discharged as small bubbles through the slots or from the serrated edge of the cap beneath the liquid. The liquid level is maintained at some 5–6 cm above the top of the slots in the cap by means of the weir. The bubbles of vapor pass through the layer of liquid, heat, and mass transfer occur and vapor, now richer in the more volatile components, leaves the surface of the liquid and passes to the plate above.

Valve plates These are also perforated but the perforations are covered by liftable caps or valves. The caps are lifted as the vapor flows upwards through the perforations, but they fall and seal the holes when the vapor flow rate decreases. Liquid is prevented from falling down through the perforations when the vapor flow rate drops. The caps direct the vapor horizontally into the liquid thus promoting good mixing.

Packed columns These may be used instead of plate columns. The cylindrical column is packed with an inert material. The liquid flows down the column in the form of a thin film over the surface of the packing material, providing a large area of contact with the vapor rising up the column. The packing may consist of hollow cylindrical rings or half rings, which may be fitted with internal cross pieces or baffles. These rings may be made of metal, various plastics, or ceramic materials. The packing is supported on perforated plates or grids. Alternatively, the column may be packed with metal mesh. The liquid flows through the packing in a zigzag pattern providing a large area of contact with the vapor [5, 6, 40–42].

Spinning cone column This consists of a vertical cylinder with a rotating shaft at its center. A set of inverted cones are fixed to the shaft and rotate with it. Alternately between the rotating cones is a set of stationary cones fixed to the internal wall of the cylinder. The feed material, which may be in the form of a liquid, puree, or slurry, is introduced into the top of the column. It flows by gravity down the upper surface of a fixed cone and drops onto the next rotating cone.

Under the influence of centrifugal force, it is spun into a thin film which moves outwards to the rim of the spinning cone and onto the next stationary cone below. The liquid, puree, or slurry moves from cone to cone to the bottom of the column. The stripping gas, usually nitrogen or steam, is introduced into the bottom of the column and flows upwards, counter-current to the feed material.

The thin film provides a large area of contact and the volatiles are stripped from the feed material by the gas or steam. Fins on the underside of the rotating cones create a high degree of turbulence in the rising gas or steam which improves mass transfer. They also provide a pumping action which reduces the pressure drop across the column. The gas or steam flows out of the top of the column and passes through a condensing system, where the volatile aroma compounds are condensed and collected in a concentrated form. This equipment is used to recover aroma compounds from fruits, vegetables, and their byproducts, tea, coffee, meat extracts, and some dairy products. It has also been used for the production of low alcohol beers and wines [43–46].

9.5.3 Applications of Distillation in Food Processing

9.5.3.1 Manufacture of Whisky

Whisky is a spirit produced by the distillation of a mash of cereals, which may include barley, corn, rye, and wheat, and is matured in wooden casks. There are three types of Scotch and Irish whisky: malt whisky produced from 100% malted (germinated) barley, grain whisky produced from unmalted cereal grains, and blended whisky which contains 60–70% grain whisky and 30–40% malt whisky.

Malting of barley is carried out by steeping the grain in water for 2–3 days and allowing it to germinate. The purpose of malting is the production of amylases which later convert grain starch to sugar. Malting is stopped by drying the grain down to a moisture content of 5% in a kiln. In traditional Scotch whisky production, the grain is dried over a peat fire which contributes to the characteristic flavor of the end product. Alternatively, the kiln may be heated indirectly by gas or oil or directly by natural gas. The dried grain is milled by means of corrugated roller mills, hammer mills, or attrition mills (see Section 11.3.2), in order to break open the bran layer without creating too much fines. The milled grain is mashed. In the case of malted barley, the grain is mixed with water at 63–68 °C for 0.5–1.5 h before filtering off the liquid, known as *wort*. In the production of grain whisky, the milled grain is cooked for 1.5 h at 120 °C to gelatinize the starch in the nonbarley cereals. It is cooled to 60–65 °C and 10–15% freshly malted barley is added to provide a source of enzymes for the conversion of the starch to the sugar maltose. The wort is filtered, transferred to a fermentation vessel, cooled to 20–25 °C, inoculated with one or more strains of the yeast *Saccharomyces cerevisiae* and allowed to ferment for 48–72 h. This produces a wash containing about 7% ethanol and many flavor compounds. The fermented wash is then distilled. In the batch process for the production of Scotch whisky, two pot stills are used. In the first, known as the *wash still*, the fermented mash is boiled for 5–6 h to produce a distillate, known as *low wines*, containing 20–25% (v/v) ethanol. This is condensed and transferred to a smaller pot still, known as the *low wines still*, and distilled to produce a spirit containing about 70% (v/v) ethanol. A crude fractionation is carried out in this second still.

The distillate first produced, known as the *foreshots*, contains aldehydes, furfurols, and many other compounds which are not used directly in the product. The foreshots are returned to the second still. The distillation continues and the distillate collected as product until a specified distillate strength is reached. Distillation continues beyond that point but the distillate, known as "*feints*", is returned to the second still. A similar process, but incorporating three pot stills, is used in the production of Irish whisky.

The continuous distillation of fermented mash to produce whisky is usually carried out using two distillation columns. The fermented wash is fed toward the top of the first column, known as the *beer column*. Alcohol-free stillage is withdrawn from the bottom of this column. The vapor from the top of the first column is introduced at the base of the second column, known as the *rectifier column*. The vapor from the top of this column is condensed and collected as product. The bottom product is returned as reflux to top of the first column.

The rectifier column contains mainly sieve plates but with some bubble cap plates near the top. If the columns are fabricated from stainless steel, a flat disk of copper mesh may be fitted near the top of the rectifier column to improve the flavor of the product.

Whiskies produced by batch or continuous distillation are matured in wooden barrels, usually oak, for periods ranging from 1 year to more than 18 years, depending on the type and quality of the whisky. The type of barrel used has a pronounced effect on the flavor of the matured product. The matured products are usually blended before bottling.

The following are examples of other distilled beverage spirits:

- Brandy is a distillate from the fermented juice, mash, or wine of fruit.
- Rum is a distillate from the fermented juice of sugar cane, sugar cane syrup, sugar cane molasses, or other sugar cane products.
- Gin is obtained by distillation from mash or by redistillation of distilled spirits, or by mixing neutral spirits, with or over juniper berries and other aromatics, or extracts from such materials. It derives its main characteristic flavor from juniper berries.
- Tequila is a distillate produced in Mexico from the fermented juice of the heads of *Agave tequilana* Weber, with or without other fermentable substances [47–61].

9.5.3.2 **Manufacture of Neutral Spirits**

A multicolumn distillation plant is used for producing neutral spirits from fermented mash. A typical system would be comprised of five columns: a whisky-separating column, an aldehyde column, a product-concentrating column, an aldehyde-concentrating column, and a fusel oils concentrating column (Figure 9.20). The whisky-separating column is fitted with sieve plates, with some bubble cap plates near the top of the column. The other four columns are fitted with bubble cap plates. The fermented mash containing 7% (v/v) of alcohol is fed to near the top of the whisky-separating column. The overhead distillate from this column is fed to the aldehyde column. The bottom product from this column is pumped to the middle of the product-concentrating column. The end product, neutral spirit, is withdrawn from near the top of this column. The top product from the aldehyde column is rich in aldehydes and esters and is fed to the aldehyde-concentrating column. The top product from this column is rich in aldehydes and is removed while the bottom product is recycled to the aldehyde column. Fusel oils concentrate

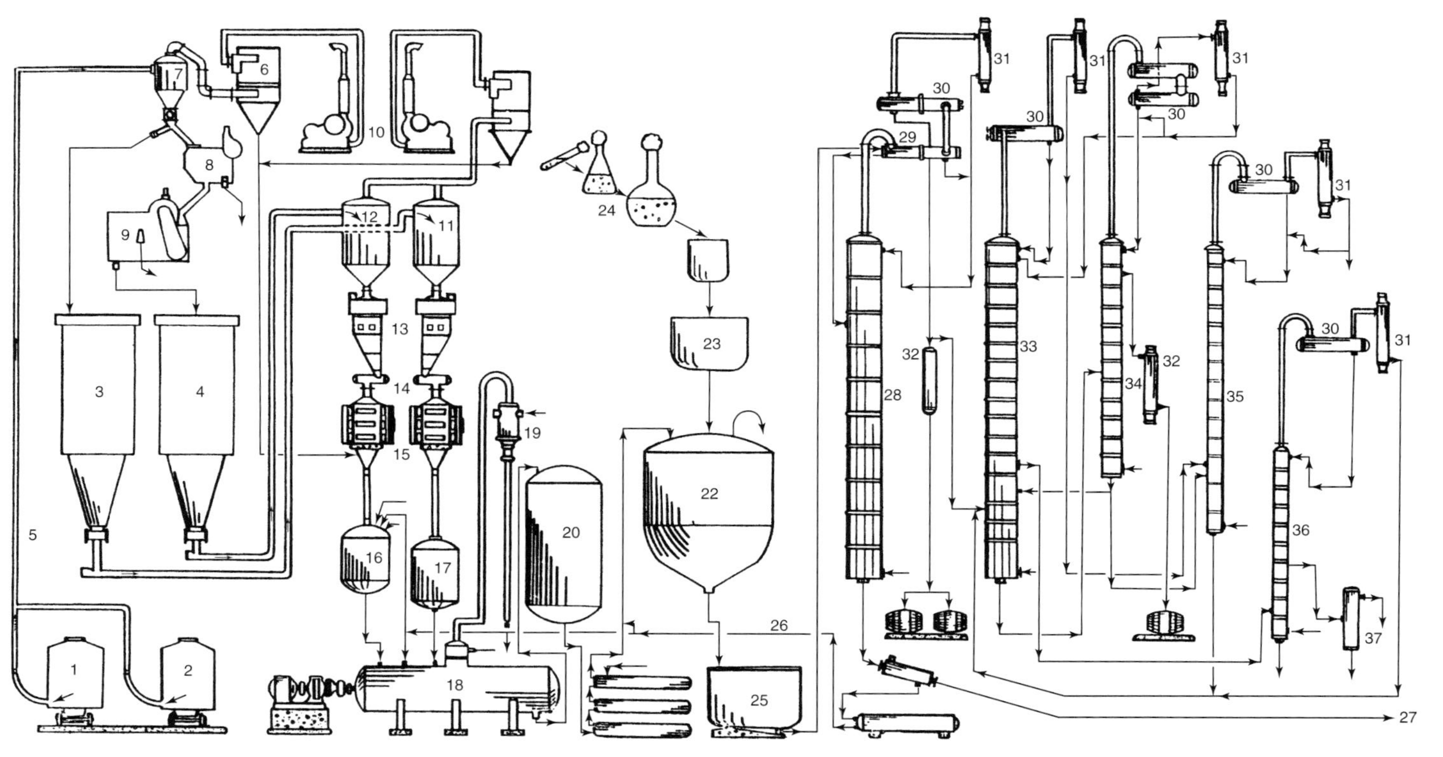

Figure 9.20 Material process flow, modern beverage spirits plant. From Ref. [47] with permission.

near the bottom of the aldehyde column and from there are fed to the fusel oil-concentrating column. The bottom product from this column is rich in fusel oils and is removed. The top product is recycled to the aldehyde column. The product from the very top of the product-concentrating column is condensed and returned as reflux to the aldehyde column [47, 50].

There are many other applications for distillation including the following examples:

Recovery of solvents from oil after extraction Most of the solvent can be recovered by evaporation using a film evaporator (see Section 3.1.2.3). However, when the solution becomes very concentrated, its temperature rises and the oil may be heat-damaged. The last traces of solvent in the oil may be removed by steam distillation or stripping with nitrogen.

Concentration of aroma compounds from juices and extracts By evaporating 10–30% of the juice in a vacuum evaporator, most of the volatile aroma compounds leave in the vapor. This vapor can be fed to a distillation column. The bottom product from the column is almost pure water and the aroma concentrate leaves from the top of the column. This is condensed and may be added back to the juice or extract prior to drying. Fruit juices and extract of coffee may be treated in this way (see Sections 3.1.5.2 and 3.1.5.3). A spinning cone evaporator (see Section 9.5.2.2) may be used for this duty.

Extraction of essential oils from leaves, seeds, etc. This may be achieved by steam distillation. The material in a suitable state of subdivision is placed on a grid or perforated plate above heated water. In some cases the material is in direct contact with the water or superheated steam may be used. If the oil is very heat sensitive, distillation may be carried out under vacuum [7].

9.6 Crystallization

9.6.1 General Principles

Many foods and food ingredients consist of or contain crystals. Crystallization has two types of purpose in food processing: (i) the separation of solid material from a liquid in order to obtain either a pure solid, for example, salt or sugar, or a purified liquid, for example, winterized salad oil and (ii) the production of crystals within a food such as in butter, chocolate, or ice cream. In either case, it is desirable to control the process such that the optimum yield of crystals of the required purity, size, and shape is obtained. It is also important to understand crystallization when considering frozen food (see Chapter 4) or where undesirable crystals are produced, for example, lactose crystals in dairy products, or precipitated fat crystals in salad oils.

9.6.1.1 Crystal Structure

Crystals are solids with a three-dimensional periodic arrangement of units into a spatial lattice. They differ from amorphous solids in having highly organized structures of flat faces and corners. A limited number of elementary crystal cells are possible with accurately defined angles; and any crystalline material has 1 of 14 possible lattice structures. Some examples of these structures are shown in Figure 9.21.

It is important to note that the final macroscopic shape of a crystal is usually not the same as its elementary lattice structure, as growth conditions can change the final "habit." The units involved in lattice structures may be metallic nuclei or atoms, but most food crystals are formed of molecular units bonded by van der Waals forces, or in a limited number of cases, ions bonded by ionic bonds. Detailed information on crystal structure and the crystallization process may be found elsewhere [52–54].

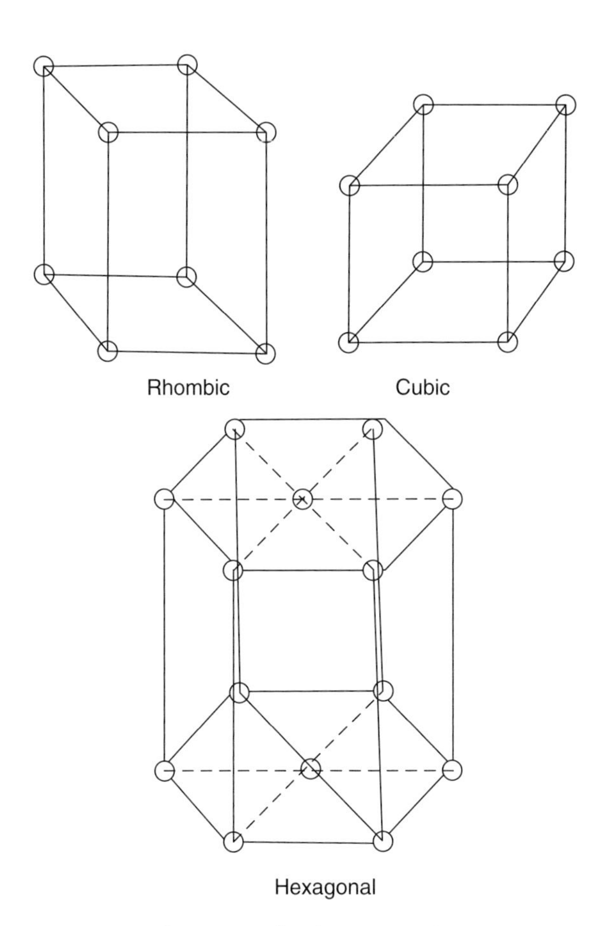

Figure 9.21 Elementary cells of some crystal systems.

9.6.1.2 The Crystallization Process

Crystallization is the conversion of one or more substances from an amorphous solid or the gaseous or liquid state to the crystalline state. In practice, we are only concerned with conversion from the liquid state. The process involves three stages: supersaturation of the liquid, followed by nucleation (formation of new crystal structures), and crystal growth. A distinction that is commonly used is "crystallization from a solution," which is the case where a purified substance is produced from a less pure mixture, as opposed to "melt crystallization" in which both or all components of a mixture crystallize into a single solid phase.

Saturation of a solution is the equilibrium concentration which would be achieved by a solution in contact with solute after a long period of time. In most solute/solvent systems, the saturation concentration increases with temperature, although this is not always the case, for example, with some calcium salts. Supersaturation occurs when the concentration of solute exceeds the saturation point ($S = 1$). The saturation coefficient (S) at any temperature is defined as: (Concentration of solute in solution)/(Concentration of solute in saturated solution).

Solutes differ in their ability to withstand supersaturation without crystallization – for example, sucrose can remain in solution when $S = 1.5–2.0$, whereas sodium chloride solutions crystallize with only a very small degree of supersaturation.

Crystals cannot form or grow in a solution at or below its saturation concentration ($S \leq 1$) at any given temperature; and thus supersaturation must be achieved in two main ways: cooling or evaporation. Cooling moves the system along the solubility curve such that a saturated solution becomes supersaturated at the same solute concentration, while evaporation increases the concentration into the supersaturated zone. It is also possible to produce supersaturation by chemical reaction or by addition of a third substance which reduces solubility, for example, addition of ethanol to an aqueous solution, but these are not commercially important in food processing.

Supersaturation does not necessarily result in spontaneous crystallization because, although the crystalline state is more thermodynamically stable than the supersaturated solution and there is a net gain in free energy on crystallization, the activation energy required to form a surface in a bulk solution may be quite high. In other words, the probability of aligning the units correctly to form a viable crystal nucleus is low and is dependent on a number of factors. Viscosity is one such factor; and crystallization occurs less readily as viscosity increases.

This can readily be seen in sugar confectionery which frequently consists of supercooled viscous liquids. Miers' theory (Figure 9.22) defines three regions for a solute/solvent mixture. Below the saturation curve (in the *undersaturated zone*) there is no nucleation or crystal growth; and in fact crystals dissolve. In the *metastable zone*, crystal growth occurs, but nucleation does not occur spontaneously. In the *labile zone*, both nucleation and crystal growth occur and, the higher the degree of supersaturation, the more rapidly these occur. While the solubility curve is fixed, the supersolubility curve is not only a property of the system, but also depends on other factors such as presence of impurities, cooling rate, or agitation of the

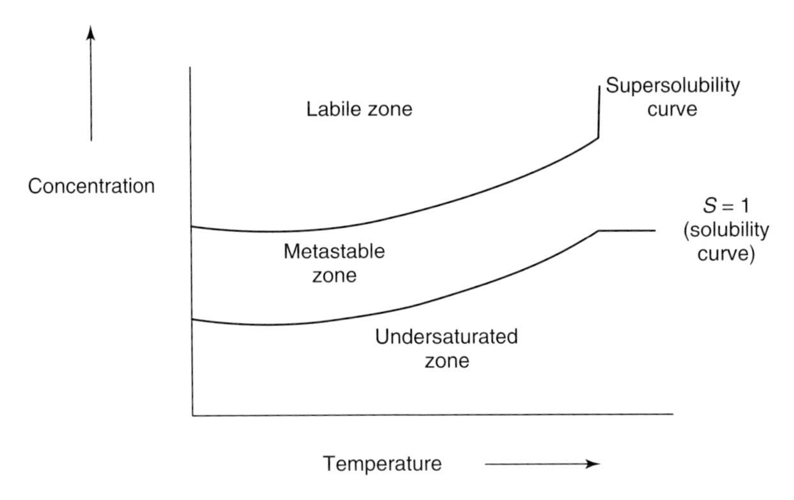

Figure 9.22 Crystallization equilibria.

system, in other words factors that affect the activation energy of the system as mentioned above. In practice, crystals can be produced by several methods:

- Homogeneous nucleation is the spontaneous production of crystals in the labile zone.
- Heterogeneous nucleation occurs in the presence of other surfaces, such as foreign particles, gas bubbles, stirrers, which form sites for crystallization in the metastable zone.
- Secondary nucleation is the presence of crystals of the crystallizing species itself and also occurs in the upper regions of the metastable zone. The reason for this phenomenon is not clear but may be due to fragments breaking off existing crystals by agitation or viscous drag and forming new nuclei.
- Nucleation is also stimulated by outside effects such as agitation or ultrasound.
- Alternatively, crystalline materials can be produced by "seeding" solutions in the metastable zone. In this case, very finely divided crystals are added to a supersaturated solution and allowed to grow to the finished size without further nucleation.

Control of crystal growth following nucleation or seeding is essential to obtain the correct size and shape of crystals. The rate of growth depends on the rate of transport of material to the surface and the mechanism of deposition. The rate of deposition is approximately proportional to S, while diffusion to the crystal surface can be accelerated by stirring. Impurities generally reduce the rate.

The final shapes or "habits" of crystals are distinguished into classes such as platelike, prismatic, dendritic, and acicular. Habits are determined by conditions of growth. Invariant crystals maintain the same shape during growth, as deposition in all directions is the same (Figure 9.23a). Much more commonly, the shape changes and overlapping of smaller faces occurs (Figure 9.23b).

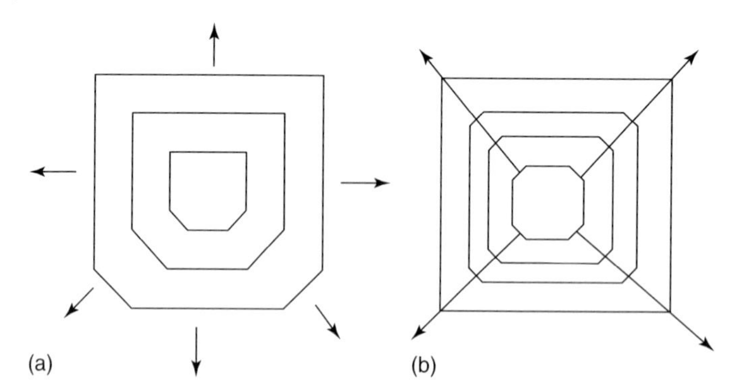

Figure 9.23 Crystal growth: (a) invariant crystal growth and (b) overlapping crystal growth.

Crystals grown rapidly from highly supersaturated solutions tend to develop extreme habits, such as needle-shaped or dendritic, with a high specific surface, due to the need to dissipate heat rapidly. The final shape is also affected by impurities which act as habit modifiers, interacting with growing crystals and causing selective growth of some surfaces, or dislocations in the structure.

Some substances can crystallize out into chemically identical, but structurally different forms. This is known as *polymorphism*; and perhaps the best recognized examples are diamond and graphite. Polymorphism is particularly important in fats, which can often crystallize out into different polymorphs with different melting points [54, 55]. In some cases, polymorphs can be converted to other, higher melting point crystal forms, which is the basis of tempering some fat-based foods.

9.6.2
Equipment Used in Crystallization Operations

Industrial crystallizers are classified according to the method of achieving supersaturation, that is, by cooling, evaporation, or mixed operations [56], as well as factors such as mode of operation (continuous or batch), desired crystal size distribution and purity. Detailed descriptions of design and operation are given by Mersmann and Rennie [57]. The crystallizing suspension is often known as "*magma*," while the liquid remaining after crystallization is the "mother liquor." Fluidized beds are common with either cooling or evaporative crystallization, in which the solution is desupersaturated as it flows through a bed of growing crystals (Figures 9.24 and 9.25).

Yield from a crystallizer (Y) can be calculated from:

$$Y = W\left[C_1 - C_2\left(1 - V\right)\right] \tag{9.9}$$

where W is the initial mass of water, C_1 and C_2 are the concentrations of solute before and after crystallization, and V is the fraction of water evaporated.

Cooling crystallizers may incorporate continuous circulation through a cooling heat exchanger (e.g., Figure 9.24) or for viscous materials, may incorporate a scraped surface heat exchanger (see Chapter 2).

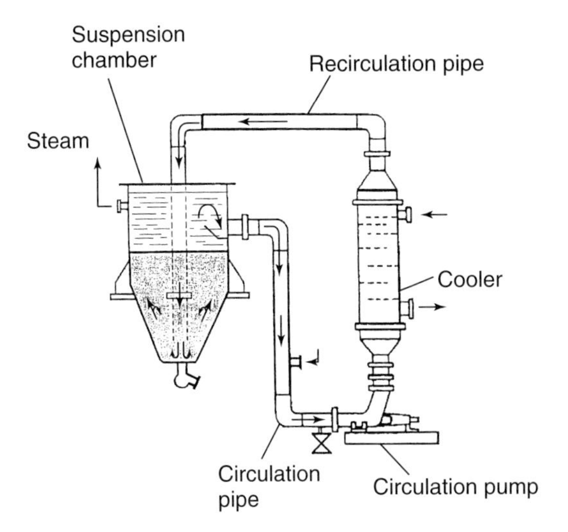

Figure 9.24 Fluidized bed cooling crystallizer. From Ref. [57] with permission.

Evaporative crystallizers are similar to forced circulation evaporators with a crystallization vessel below the vapor/liquid separator (Figure 9.25). Simultaneous evaporation and cooling can be carried out without heat exchangers, the cooling effect being produced by vacuum evaporation of the saturated solution.

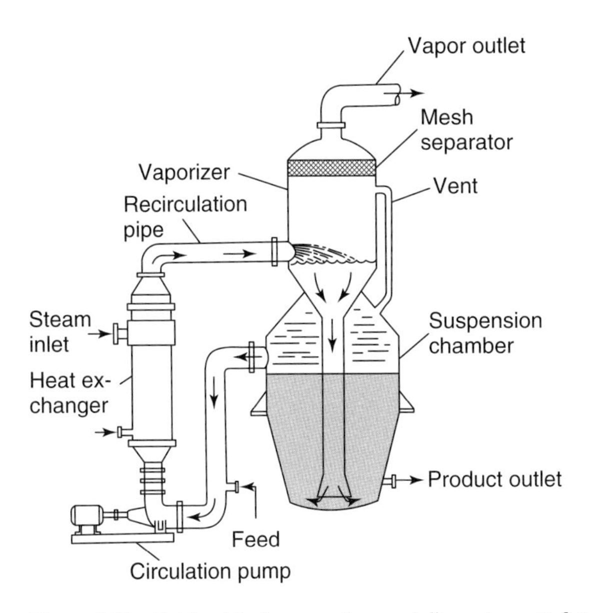

Figure 9.25 Fluidized bed evaporative crystallizer. From Ref. [57] with permission.

9.6.3
Food Industry Applications

9.6.3.1 Production of Sugar

Crystallization is a major operation in sugar manufacture. Beet or cane sugar consist essentially of sucrose; and different grades of product require uniform crystals of different sizes. Supersaturation is effected by evaporation, but the temperature range is limited by the fact that sugar solutions are sensitive to caramelization above 85 °C, while high viscosities limit the rate of crystallization below 55 °C, so all operations are carried out within a fairly narrow temperature range. The supersaturated sugar solution is seeded with very fine sugar crystals and the "massecuite," that is, syrup/crystal mixture, is evaporated with further syrup addition. When the correct crystal size has been achieved, the crystals are removed and the syrup is passed to a second evaporator. Up to four evaporation stages may be used, with the syrup becoming less pure each time.

9.6.3.2 Production of Salt

Salt is much less of a problem then sugar, in that it is not temperature-sensitive and forms crystals more easily than sugar. Evaporation of seawater in lagoons is still carried out, but factory methods are more common. Continuous systems based on multiple effect evaporators (see Chapter 3) are widespread. Sodium chloride normally forms cubic crystals, but it is common to add potassium ferrocyanide as a habit-modifier, producing dendritic crystals which have better flow properties.

A number of other food materials are produced in pure form by crystallization, including lactose, citric acid, monosodium glutamate, and aspartame.

9.6.3.3 Salad Dressings and Mayonnaise

In some cases, crystallization is carried out to remove unwanted components. Salad oils, such as cottonseed and soybean oil, are widely used in salad dressings and mayonnaise. They contain high melting point triglycerides which crystallize out on storage. This is not especially serious in pure oil, but would break an emulsion and hence lead to product deterioration. Winterization is fractional crystallization to remove higher melting point triglycerides. It is essential to remove this fraction while retaining the maximum yield of oil, which requires extremely slow cooling to produce a small number of large crystals, that is, crystal growth with little nucleation. Crystallization is carried out over 2–3 days with slow cooling to approximately 7 °C, followed by very gentle filtration. Agitation is avoided after the initial nucleation to prevent the formation of further nuclei. The winterized oil should then pass the standard test of remaining clear for 5.5 h at 0 °C.

9.6.3.4 Margarine and Pastry Fats

Fractional crystallization of fats into high (stearin) and low (olein) melting point products is used to improve the quality of fats for specific purposes, for example, margarine and pastry fats. Crystallization may be accelerated by reducing the viscosity in the presence of solvents such as hexane. Adding detergents also improves the recovery of high melting point crystals from the olein phase.

9.6.3.5 Freeze Concentration

Ice crystallization from liquid foods is a method of concentrating liquids such as fruit juices or vinegar without heating, or adjusting the alcohol content of beverages. Its use is limited as only a modest level of concentration is possible; and there is an inevitable loss of yield with the ice phase.

There are many applications of crystallization in the manufacture of foods where a separation is not involved and will therefore not be dealt with here. These include ice cream, butter, margarine, chocolate, and sugar confectionery.

References

1. Lewis, M.J. (1996) Solids separation processes, in *Separation Processes in the Food and Biotechnology Industries* (eds A.S. Grandison and M.J. Lewis), Woodhead Publishing, Cambridge, pp. 245–286.
2. Brennan, J.G., Butters, J.R., Cowell, N.D., and Lilly, A.E.V. (1990) *Food Engineering Operations*, 3rd edn, Elsevier Applied Science, London.
3. Fellows, P. (2009) *Food Processing Technology*, 3rd edn, Woodhead Publishing, Cambridge.
4. Toledo, R.T.B. (2007) *Fundamentals of Food Process Engineering*, 3rd edn, Springer, New York.
5. McCabe, W.L., Smith, J.C., and Harriott, P. (2005) *Unit Operations of Chemical Engineering*, 7th edn, McGraw-Hill, New York.
6. Berk, Z. (2009) *Food Process Engineering and Technology*, Elsevier Inc., Amsterdam.
7. Leniger, H.A. and Beverloo, W.A. (1975) *Food Process Engineering*, Reidel Publishing, Dordrecht.
8. Purchas, D.B. (1971) *Industrial Filtration of Liquids*, 2nd edn, Leonard Hill Books, London.
9. Akers, R.J. and Ward, A.S. (1977) Liquid filtration theory and filtration pre-treatment, in *Filtration Principles and Practice, Part 1* (ed. C. Orr), Marcel Dekker, New York, pp. 159–250.
10. Cheremisinoff, N.P. and Azbel, D.S. (1983) *Liquid Filtration*, Ann Arbor Science Publishers, Woburn.
11. Cheremisinoff, P.N. (1995) *Solids/Liquids Separation*, Technomic Publishing Company, Lancaster, PA.
12. Rushton, A. and Griffiths, P.V.R. (1977) Filter media, in *Filtration Principles and Practice, Part 1* (ed. C. Orr), Marcel Dekker, New York, pp. 251–308.
13. Lawson, H. (1994) *Food Oils and Fats*, Chapman and Hall, Inc., New York.
14. De Greyt, W. and Kellens, M. (2000) Refining practice, in *Edible Oil Processsing* (eds W. Hamm and R.J. Hamilton), Sheffield Academic Press, Sheffield, pp. 79–128.
15. Kellens, M. (2000) Oil modification processes, in *Edible Oil Processing* (eds W. Hamm and R.J. Hamilton), Sheffield Academic Press, Sheffield, pp. 129–173.
16. O'Brien, R.D. (2004) *Fats and Oils, Formulating and Processing for Applications*, 2nd edn, CRC Press, New York.
17. Gunstone, F.D. (2008) *Oils and Fats in the Food Industry*, John Wiley & Sons, Ltd, Chichester.
18. Hugot, E. (1986) *Handbook of Cane Sugar Engineering*, 3rd edn, Elsevier Science, Amsterdam.
19. McGinnis, R.A. (1971) Juice preparation III, in *Beet-Sugar Technology*, 2nd edn (ed. R.A. McGinnis), Beet Sugar Development Foundation, Fort Collins, CO, pp. 259–295.
20. Hough, J.S., Briggs, D.E., Stevens, R., and Yound, T.W. (1982) *Malting and Brewing Science*, vol. II, 2nd edn, Chapman and Hall, London.

21. Posada, J. (1987) Filtration of beer, in *Brewing Science*, vol. 3 (ed. J.R.A. Pollock), Academic Press, London, pp. 379–439.
22. Levy, R.V. and Jornitz, M.W. (2006) Types of filtration, in *Sterile Filtration* (ed. M.W. Jornitz), Springer, New York, pp. 1–26.
23. Farkas, J. (1988) *Technology and Biochemistry of Wine*, vol. 2, Gordon and Breach, New York.
24. Amerine, M.A., Kunkee, R.E., Ough, C.S., Singleton, V.L., and Webb, A.D. (1980) *The Technology of Winemaking*, 4th edn, AVI Publishing Co., Westport.
25. Ribereau-Gayon, P., Glories, Y., Maujean, A., and Dubourdieu, D. (2000) *Handbook of Eonology: The Chemistry of Wine Stabilization and Treatments*, vol. 2, John Wiley & Sons, Ltd, Chichester.
26. Ambler, C.M. (1952) The evaluation of centrifugal performance. *Chem. Eng. Prog.*, **48**, 150–158.
27. Walstra, P., Geirts, T.J., Noomen, A., Jellema, A., and van Boekel, M.A.J.S. (1999) *Dairy Technology, Principles of Milk Properties and Processes*, Marcel Dekker, NewYork.
28. Braddock, R.J. (1999) *Handbook of Citrus By-Products and Processing Technology*, John Wiley & Sons, Inc., New York.
29. Nelson, P.E. and Tressler, D.K. (1980) *Fruit and Vegetable Juice Processing Technology*, 3rd edn, AVI Publishing Co., Westport.
30. Schwartzberg, H.G. (1987) Leaching organic materials, in *Handbook of SeparationProcess Technology* (ed. R.W. Rousseau), John Wiley & Sons, Inc., New York, pp. 540–577.
31. Ramaswamy, H. and Marcotte, M. (2006) *Food Processing Principles and Applications*, Taylor and Francis, Boca Raton, FL.
32. Williams, M.A. (1997) Extraction of lipids from natural sources, in *Lipid Technologies and Applications* (ed. F.D. Gunstone and F.B. Padley), Marcel Dekker, New York, pp. 113–135.
33. Ebell, A. and Storz, M. (1971) Diffusion, in *Beet-Sugar Technology*, 2nd edn (ed. R.A. McGinnis), Beet Sugar Development Foundation, Fort Collins, pp. 125–160.
34. Masters, K. (1991) *Spray Drying Handbook*, 5th edn, Longman Scientific and Technical, Harlow.
35. Clarke, R.J. (1987) in *Coffee*, vol. 2 (eds R.J. Clarke and R. Macrae), Elsevier Applied Science, London, pp. 109–145.
36. Sinja, V.R., Mistra, H.N., and Bal, S. (2007) Process technology for the production of soluble tea powder. *J. Food Eng.*, **82**, 276–283.
37. McPherson, A. (1987) It was squeeze or G, now it's CCE. *Food Technol. Aust.*, **39**, 56–60.
38. Gardner, D.D. (1982) Industrial scale hop extraction with liquid CO_2. *Chem. Ind.*, **12**, 402–405.
39. Rizvi, S.S.H., Daniels, J.A., Benado, E.L., and Zollweg, J.A. (1986) Supercritical fluid extraction: operating principles and food applications. *Food Technol.*, **40**, 56–64.
40. Fair, J.R. (1987) Distillation, in *Handbook of Separation Processes* (ed. R.W. Rousseau), John Wiley & Sons, Inc., New York, pp. 229–339.
41. Foust, A.S., Wenzel, L.A., Clump, C.W., Maus, L., and Andersen, L.B. (1980) *Principles of Unit Operations*, 2nd edn, John Wiley & Sons, Inc., New York.
42. Anonymous (2000) Distillation: technology and engineering, in *Encyclopedia of Food Science and Technology*, 2nd edn (ed. F.J. Francis), John Wiley & Sons, Inc., New York, pp. 509–518.
43. Schofield, T. (1995) Natural aroma improvement by means of the spinning cone, in *Food Technology International Europe* (ed. A. Turner), Sterling Publications, Ltd, London, pp. 137–139.
44. Anonymous (2008) Production of low alcohol beverages, *Drink Technol. Mark.*, **11**, 29–30.
45. Moreira da Silva, P. and de Wit, B. (2008) Spinning cone column distillation – innovative technology for beer dealcoholisation. *Cervesia*, **33**, 91–95.
46. Belisario-Sanchez, Y.Y., Taboada-Rodriguez, A., Marin-Iniesta, F., and Lopez-Gomez, A. (2009) Dealcoholized wines by spinning cone column distillation: phenolic compounds and antioxidant activity measured by the 1,1-diphenyl-2-picrylhydrazyl method. *J. Agric. Food Chem.*, **57**, 6770–6778.

47. Owades, J.L. (2000) Distilled beverage spirits, in *Encyclopedia of Food Science and Technology*, 2nd edn (ed. F.J. Francis), John Wiley & Sons, Inc., New York, pp. 519–540.
48. Piggott, J.R. and Connor, J.M. (2003) Whisky, whiskey and bourbon, products and manufacture, in *Encyclopedia of Food Science and Nutrition*, 2nd edn (ed. B. Caballero, L.C. Trugo, and P.M. Finglas), Academic Press, London, pp. 6171–6177.
49. Nicol, D. (1989) Batch distillation, in *The Science and Technologies of Whiskies* (ed. J.R. Piggott, R. Sharp, and R.E. Duncan), Longman Scientific and Technical, Harlow, pp. 118–149.
50. Panek, R.J. and Boucher, A.R. (1989) Continuous distillation, in *The Science and Technologies of Whiskies* (eds J.R. Piggott, R. Sharp, and R.E. Duncan), Longman Scientific and Technical, Harlow, pp. 150–181.
51. Bryce, J.H. and Stewart, G.G. (eds) (2004) *Distilled Spirits: Tradition and Innovation*, Nottingham University Press, Nottingham.
52. Mersmann, A. (ed.) (1994) *Crystallisation Technology Handbook*, Marcel Dekker, New York.
53. Hartel, R.W. (2001) *Crystallisation in Foods*, Aspen Technology, Inc., Gaithersburg.
54. Singh, G. (1988) Crystallisation from solutions, in *Separation Techniques for Chemical Engineers*, 2nd edn (ed. P.A. Schweitzer), McGraw-Hill, London, pp. 151–182.
55. Rajah, K.K. (1996) Fractionation of fat, in *Separation Processes in the Food and Biotechnology Industries* (eds A.S. Grandison and M.J. Lewis), Woodhead Publishing, Cambridge, pp. 207–241.
56. Saravacos, G.D. and Kostaropoulos, A.E. (2002) *Handbook of Food Processing Equipment*, Kluwer Academic, London.
57. Mersmann, A. and Rennie, F.W. (1994) Design of crystallizers and crystallization processes, in *Crystallisation Technology Handbook* (ed. A. Mersmann), Marcel Dekker, New York, pp. 215–325.
58. Brunner, G. (2005) Supercritical fluids: technology and application to food processing. *J. Food Eng.*, **67**, 21–33.
59. Temelli, F., Chen, C.S., and Braddock, R.J. (1988) Supercritical fluid extraction in citrus oil processing. *Food Technol.*, **46**, 145–150.
60. Reverchon, E. (2003) Supercritical fluid extraction, in *Encyclopedia of Food Science and Nutrition*, 2nd edn (eds B. Caballero, L.C. Trugo, and P.M. Finglas), Academic Press, London, pp. 5680–5687.
61. Pronyk, C. and Mazza, G. (2009) Design and scale-up of pressurized fluid extractors for food and bioproducts. *J. Food Eng.*, **95**, 215–226.

Further Reading

Sahena, A., Zaidul, I.S.M., Jinap, S., Karim, A.A., Abbas, K.A., Norulaini, N.A.N., and Omar, A.K.M. (2009) Application of supercritical CO_2 in lipid extraction – A review. *J. Food Eng.*, **95**, 240–251.

10
Separations in Food Processing: Part 2 – Membrane Processing, Ion Exchange, and Electrodialysis

Michael J. Lewis and Alistair S. Grandison

10.1
Membrane Processes

10.1.1
Introduction

Over the last 50 years, several membrane processes have evolved that make use of a pressure driving force and a semi-permeable membrane in order to effect a separation of components in a solution or colloidal dispersion. The separation is based mainly on molecular size, but to a lesser extent on shape and charge. The four main processes are reverse osmosis (RO) (hyperfiltration), nanofiltration (NF), ultrafiltration (UF), and microfiltration (MF). These processes can be considered to provide a continuous spectrum, often with no clear-cut boundaries between them. Most suppliers of membranes now offer a selection of membranes covering this entire spectrum. The dimensions of the components involved typically range from less than 1 nm to over 1000 nm. Figure 10.1 illustrates these processes and how they also relate to traditional particle filtration.

10.1.2
Terminology

The feed material is applied to one side of a membrane and subjected to a pressure. In most cases the feed flows in a direction parallel to the membrane surface and the term *cross-flow filtration* is used to describe such applications. Dead-end systems are also used, but mainly for bench-scale separations. The stream which passes through the membrane under the influence of this pressure is termed the *permeate* (filtrate). The stream left after the required amount of permeate is removed is termed the *concentrate* or *retentate.*

If the membrane has a very small pore diameter (tight pores), the permeate will be almost pure water and the process is known as *reverse osmosis*; it is similar in its effects to evaporation or freeze-concentration. The permeate may be the required material; for example, in the production of "potable" water from sea-water or

Food Processing Handbook, Second Edition. Edited by James G. Brennan and Alistair S. Grandison.

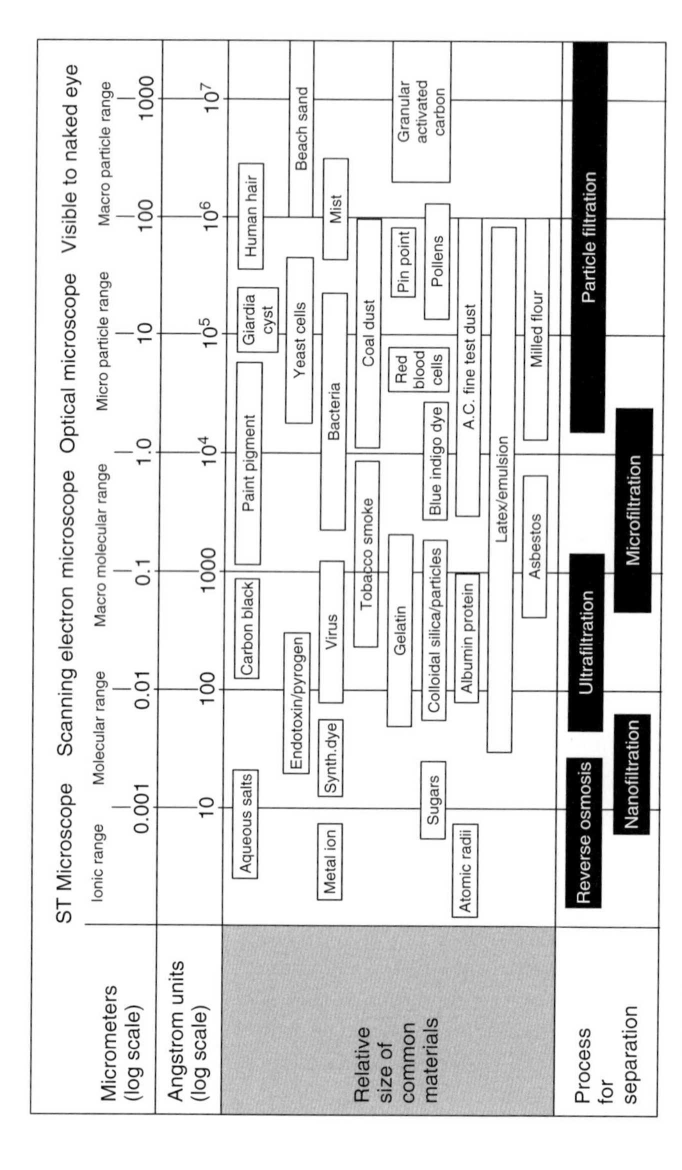

Figure 10.1 Comparison of the different membrane processes with particle filtration. Courtesy of Memtech.

polluted waters or in the production of ultrapure water. In reverse osmosis, the pressure applied needs to be in excess of the osmotic pressure of the solution. Osmotic pressures are highest for low molecular weight solutes, such as salt and sugar solutions, and large increases in their osmotic pressure occur during reverse osmosis. The osmotic pressure, Π, of a dilute solution can be predicted from the Van t'Hoff equation:

$$\Pi = iRT\left(\frac{c}{M}\right) \tag{10.1}$$

where i is degree ionization, R is the ideal gas constant, T is absolute temperature, M is molecular weight, and c is concentration (kg m^{-3}).

Nanofiltration is similar to reverse osmosis, except that the membrane will allow small molecules to permeate, for example, monovalent ions such as Na^+, K^+, and Cl^-, and to a lesser extent divalent ions. As the membrane pore size further increases, the membrane becomes permeable to other low molecular weight solutes in the feed. Lower pressure driving forces are required. However, larger molecular weight molecules (e.g., proteins) are still rejected by the membrane. It is this fractionation which makes ultrafiltration more interesting than reverse osmosis.

Microfiltration involves membranes with even larger pore sizes, which allow small macromolecules to pass through, but retain particulate matter and fat globules.

Concentration factor and rejection are two important processing parameters for all pressure-activated processes. The concentration factor (f) is defined as follows:

$$f = \frac{V_F}{V_C} \tag{10.2}$$

where V_F is feed volume and V_C is final concentrate volume.

As soon as the concentration factor exceeds 2.0, the volume of permeate will exceed that of the concentrate. Concentration factors may be as low as 1.5 for some viscous materials, to between 5 and 50 for some dilute protein solutions. Generally, higher concentration factors are used for ultrafiltration than for reverse osmosis; over 50 can be achieved for ultrafiltration treatment of cheese whey, for example, compared with about to 5 for reverse osmosis treatment of cheese whey.

The rejection or retention factor (R) of any component is defined as:

$$R = \frac{c_F - c_P}{c_F} \tag{10.3}$$

where c_F is the concentration of component in the feed and c_p is the concentration in the permeate. It can easily be measured and is very important as it will influence the extent (quality) of the separation that can be achieved.

Rejection values normally range between 0 and 1; sometimes they are expressed as percentages (0–100%). Occasionally negative rejections are found for some charged ions (Donnan effect).

- When $c_p = 0$; $R = 1$; all the component is retained in the feed.
- When $c_p = c_F$; $R = 0$; the component is freely permeating.

If the concentration factor and rejection value are known, the yield of any component can be estimated. This is defined as the fraction of that component present in the feed, which is recovered in the concentrate.

The yield (Y) can be calculated from:

$$Y = f^{R-1} \tag{10.4}$$

The derivation of this equation is provided in Lewis [1]. Thus, for a component where $R = 0.95$, at a concentration factor of 20, the yield is 0.86 (i.e., 86% is retained in the concentrate and 14% is lost in the permeate). This is an important consideration that might affect the economics of the process.

10.1.3
Membrane Characteristics

The membrane itself is crucial to the process. The earliest membranes were made from cellulose acetate and these are termed *first-generation membranes*. Temperatures had to be maintained below 30 °C and the pH range was 3–6, which limited their use, as they could not be disinfected by heat or cleaned with acid or alkali detergents. These were followed in the mid-1970s by other polymeric membranes (second-generation membranes), with polyamides (with a low tolerance to chlorine) and, in particular, polysulfones being widely used for foods. It is estimated that over 150 organic polymers have now been investigated for membrane applications. Inorganic membranes based on sintered and ceramic materials are also now available and these are much more resistant to heat and cleaning and disinfecting fluids.

Terms used to describe membranes are "*microporous*" or "*asymmetric*." Microporous membranes have a uniform porous structure throughout, although the pore size may not be uniform across the thickness of the membrane. They are usually characterized by a nominal pore size and no particle larger than this will pass through the membrane. In contrast to this, most membranes used for ultrafiltration are of asymmetric type, having a dense active layer or skin of 0.5–1 μm in thickness, and a further support layer which is more porous and thicker. The membrane also has a chemical nature. It may be hydrophilic or hydrophobic in nature. The hydrophobicity can be characterized by measuring its contact angle, θ; the higher the contact angle the more hydrophobic is the surface. Polysulfones are generally much more hydrophobic than cellulosic membranes. The surface may also be charged. These factors will result in interactions between the membrane and feed components and will influence which components will pass through the membrane, as also which will foul the membrane.

10.1.4
Permeation Rate (Flux)

Permeation rate and *power consumption* are important operating characteristics. Permeation rate is usually expressed in terms of $\text{l}\,\text{m}^{-2}\,\text{h}^{-1}$. This permits direct comparison of different membrane configurations of different surface areas. Flux values may be from less than $5\,l\,m^{-2}\,h^{-1}$ to higher than $500\,l\,m^{-2}\,h^{-1}$. Factors affecting the flux are the applied pressure, volumetric flow rate of feed across the membrane surface, feed temperature, and feed viscosity. The flux is also influenced by concentration polarization and fouling, which in turn are influenced by the flow conditions across the membrane. Inducing turbulence will increase the wall shear stress and promote higher flux rates [1].

The main energy consumption for membrane techniques is the power utilization of the pumps. The power used, W, is related to the pressure (head) developed and the mass flow rate as follows:

$$W = m'hg \tag{10.5}$$

where m is mass flow rate ($kg\,s^{-1}$), h is head developed by the pump (m), and g is acceleration due to gravity ($9.81\,m\,s^{-2}$).

This energy is dissipated within the fluid and will result in a temperature rise. Cooling may be necessary if a constant processing temperature is required.

10.1.5
Transport Phenomena and Concentration Polarization

A very important consideration for pressure-driven membrane processes is that the separation takes place not in the bulk of solution, but in a very small region close to the membrane, known as the *boundary layer*, as well as over the membrane itself. This gives rise to the phenomenon of concentration polarization over the boundary layer. It is manifested by a quick and significant reduction (2- to 10-fold) in flux when water is replaced by the feed solution, for example, in a dynamic start.

Concentration polarization occurs whenever a component is rejected by the membrane. As a result, there is an increase in the concentration of that component at the membrane surface, together with a concentration gradient over the boundary layer (Figure 10.2). Eventually a dynamic equilibrium is established, where the convective flow of the component to the membrane surface equals the flow of

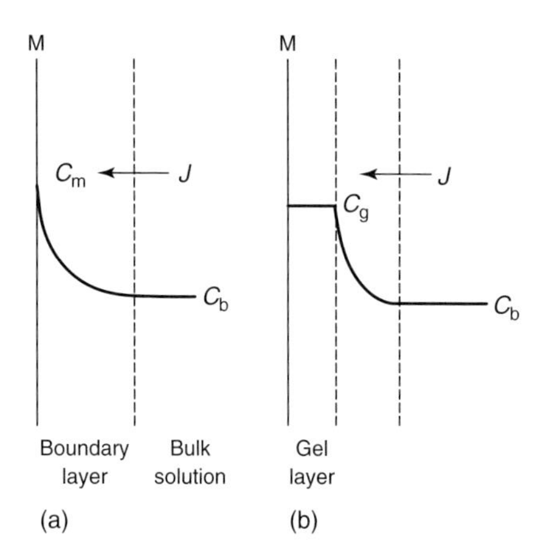

Figure 10.2 (a) Concentration polarization in the boundary layer and (b) concentration polarization with a gel layer from [1] with permission.

material away from the surface, either in the permeate or back into the bulk of the solution by diffusion, due to the concentration gradient established. This increase in concentration, especially of large molecular weight components, offers a very significant additional resistance. It may also give rise to the formation of a gelled or fouling layer on the surface of the membrane. Whether this occurs will depend upon the initial concentration of the component and the physical properties of the solution; it could be very important as it may affect the subsequent separation performance. Concentration polarization itself is a reversible phenomenon; that is, if the feed is then replaced by water, the original water flux should be restored. However, this rarely occurs in practice due to the occurrence of fouling, which is detected by a decline of flux rate at constant composition. Fouling is caused by the deposition of material on the surface of the membrane or within the pores of the membrane. Fouling is irreversible and the flux needs to be restored by cleaning. Therefore, during any membrane process, flux declines due to a combination of these two phenomena. More recently, it has been recognized that fouling can be minimized by operating at conditions at or below a "critical flux." When operating below this value, fouling is minimized, but when operating above it, fouling deposits accumulate. This can be determined experimentally for different practical situations [2].

10.1.6 **Membrane Equipment**

Membrane suppliers now provide a range of membranes, each with different rejection characteristics. For ultrafiltration, different molecular weight cut-offs in the range 1000–500 000 Da are available. Tight ultrafiltration membranes have a molecular weight cut-off value of around 1000–5000, whereas the more "open" or "loose" membranes will have a value in excess of 100 000. However, because there are many other factors that affect rejection, molecular weight cut-off should only be regarded as giving a relative guide to its pore size and true rejection behavior. Experimental determinations should always be made on the system to be validated, at the operating conditions to be used.

Other desirable features of membranes to ensure commercial viability are:

- reproducible pore size, offering uniformity both in terms of their permeate rate and their rejection characteristics;
- high flux rates and sharp rejection characteristics;
- compatible with processing, cleaning, and sanitizing fluids;
- resistance to fouling;
- an ability to withstand temperatures required for disinfecting and sterilizing surfaces, which is an important part of the safety and hygiene considerations.

Extra demands placed upon membranes used for food processing include the ability to withstand hot acid and alkali detergents (low and high pH), temperatures of 90 °C for disinfecting or 120 °C for sterilizing, and/or widely used chemical disinfectants, such as sodium hypochlorite, hydrogen peroxide, or sodium metabisulfite.

The membrane should be designed to allow cleaning both on the feed/concentrate side and the permeate side.

Membrane processing operations can range in their scale of operation from laboratory bench-top units, with samples less than 10 ml through to large-scale commercial operations, processing at rates greater than $50\,m^3\,h^{-1}$. Furthermore, the process can be performed at ambient temperatures, which allows concentration without any thermal damage to the feed components.

10.1.7 Membrane Configuration

The membranes themselves are thin and require a porous support against the high pressure. The membrane and its support together are normally known as the *module*. This should provide a large surface area in a compact volume and must allow suitable conditions to be established with respect to turbulence, high wall shear stresses, pressure losses, volumetric flow rates, and energy requirements, thereby minimizing concentration polarization. Hygienic considerations are also important; there should be no dead spaces and the module should be capable of being in-place cleaned on both the concentrate and the permeate side. The membranes should be readily accessible, both for cleaning and replacement. It may be an advantage to be able to collect permeate from individual membranes in the module to be able to assess their individual performance.

The three major designs are the tubular, flat-plate, and spiral-wound configurations (Figures 10.3 and 10.4). Sintered or ceramic membranes can also be configured in the form of tubes. Tubular membranes come in range of diameters. In general, tubes offer no dead spaces, do not block easily, and are easy to clean. However, as the tube diameter increases, they occupy a larger space, have a higher hold-up volume, and incur higher pumping costs. The two major types are the hollow fiber, with a fiber diameter of 0.001–1.2 mm and the wider tube with diameters up to 25 mm, although about 12 mm is a popular size.

For the hollow fiber system, the membrane wall thickness is about 250 μm and the tubes are self-supporting. The number of fibers in a module can be as few as 50 but there are sometimes more than 1000. The fibers are attached at each end to a tube sheet, to ensure that the feed is properly distributed to all the tubes. This may give rise to pore plugging at the tube entry point. Prefiltration is recommended to reduce this. Hollow fiber systems are widely used for desalination and in these reverse osmosis applications they are capable of withstanding high pressures. It is the ratio of the external to internal diameter, rather than the membrane wall thickness which determines the pressure that can be tolerated. Hollow fiber systems usually operate in the streamline flow regime. However the wall shear rates are high. They tend to be expensive, because if one or several fibers burst, the whole cartridge needs to be replaced.

For wider tubes, the feed is normally pumped through the tube, which may be up to 25 mm in diameter, although a popular size is about 12 mm diameter. There may be up to 20 tubes in one module; tube lengths may be between 1.2 and

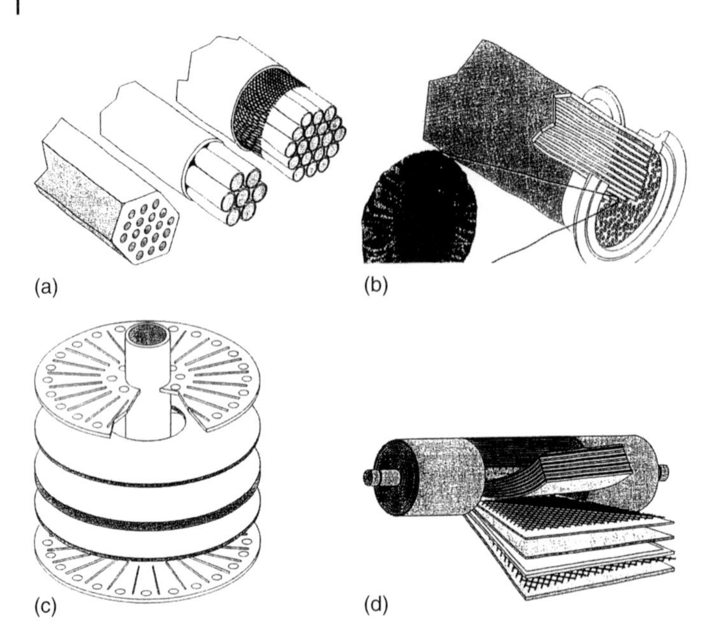

Figure 10.3 (a) tubular, (b) hollow fiber, (c) plate-and-frame, and (d) spiral-wound configurations. Courtesy of Memtech.

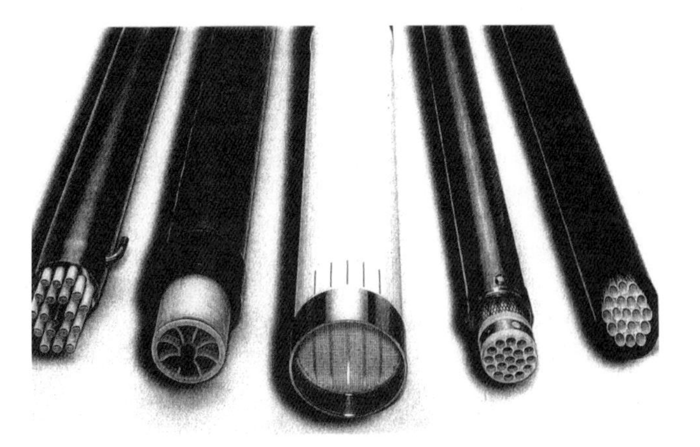

Figure 10.4 Tubular, hollow fiber, and spiral-wound configurations. Courtesy of PCI.

3.8 m and tubes within the module may be connected in series or parallel. The membrane is cast or inserted into a porous tube, which provides support against the applied pressure. Therefore they are capable of handling higher viscosity fluids and even materials with small suspended particles, up to 1/10th the tube diameter. They normally operate under turbulent flow conditions with flow velocities greater than $2\,m\,s^{-1}$. The corresponding flux rates are high, but pumping costs are also high, in order to generate the high volumetric flow rate required and the operating pressure.

The flat-plate module can take the form of a plate-and-frame-type geometry or a spirally wound geometry. The plate-and-frame system employs membranes stacked together, with appropriate spacers and collection plates for permeate removal, somewhat analogous to plate heat exchangers. The channel height can be between 0.4 and 2.5 mm. Flow may be either streamline or turbulent and the feed may be directed over the plates in a parallel or series configuration. This design permits a large surface area to be incorporated into a compact unit. Membranes are easily replaced and it is easy to isolate any damaged membrane sandwich. Considerable attention has been devoted to the design of the plate to improve performance. This has been achieved by ensuring a more uniform distribution of fluid over the plate, by increasing the width of the longer channels and reducing the ratio of the longest to the shortest channel length.

The spiral-wound system is now widely used and costs for membranes are relatively low. In this case a sandwich is made from two sheet membranes, which enclose a permeate spacer mesh. This is attached at one end to a permeate removal tube and the other three sides of the sandwich are sealed. Next to this is placed a feed spacer mesh and the two together are rolled round the permeate collection tube like a Swiss roll. The channel height is dictated by the thickness of the feed spacer. Wider channel heights will reduce the surface area-to-volume ratio, but reduce the pressure drop.

The typical dimensions of one spiral membrane unit would be about 12 cm in diameter and about 1 m in length. Up to three units may be placed in one housing, with appropriate spacers to prevent telescoping, which may occur in the direction of flow and could damage the sandwich. This configuration is becoming very popular and relatively cheap. Again the flow may be streamline or turbulent. Pressure drop/flow rate relationships suggest that flow conditions are usually turbulent.

Each system does and will continue to have its devotees. An alternative, much used unit for simple laboratory separations is the agitated stirred cell. In contrast to the systems described earlier, this is a dead-end rather than a flow-through system.

As well as the membrane module, there will be pumps, pipeline, valves and fittings, gauges, tanks, heat exchangers, instrumentation and controls, and perhaps in-place cleaning facilities. For small installations, the cost of the membrane modules may only be a relatively small component of the total cost of the finished plant, once these other items have been accounted for. This may also apply to some large installations, such as water treatment plants, where other separation processes are numerous and the civil engineering costs may also be high.

The simplest system is a batch process. The feed is usually recycled, as sufficient concentration is rarely achieved in one pass. Flux rates are initially high but decrease with time. Energy costs are high because the pressure is released after each pass. Residence times are long. Batch processes are usually restricted to small-scale operations. Batch processing with top-up is used in situations when the entire feed volume will not fit into the feed tank. Continuous processes may be single-stage

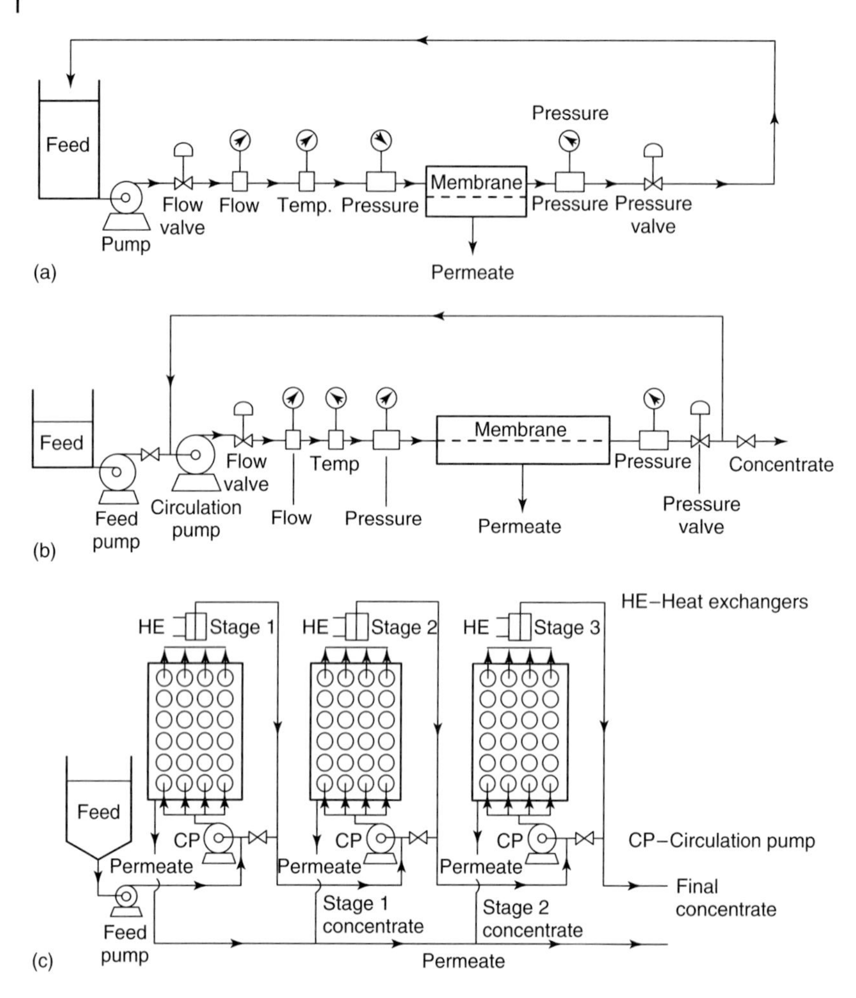

Figure 10.5 Different ultrafiltration plant layouts: (a) batch operation; (b) continuous with internal recycle (feed and bleed); (c) commercial scale with three stages. Courtesy of PCI.

(feed and bleed), or multi-stage processes, depending upon the processing capacity required.

Figure 10.5 illustrates these different systems.

10.1.8
Safety and Hygiene Considerations

It is important that safety and hygiene are considered at any early stage when developing membrane processes. These revolve around cleaning and disinfecting procedures for the membranes and ancillary equipment, as well as the monitoring and controlling of the microbial quality of the feed material. For many processes, thermization or pasteurization are recommended for feed pretreatment (see Chapter 2). Microfiltration may also be considered for heat-labile components. Relatively little has been reported about the microbiological hazards associated with membrane processing. Microorganisms will be rejected by the membrane and will increase in the concentrate by the concentration factor. There may also be some microbial growth during the process, so the residence time and residence time distribution should be known, as well as the operating temperature. If residence times are long, it may be advisable to operate either below 5 °C or above 50 °C to prevent further microbial growth.

10.1.9 Applications for Reverse Osmosis (RO)

The main applications of reverse osmosis are the concentration of fluids by removal of water. Reverse osmosis permits the use of relatively low temperatures, even lower than vacuum evaporation. It reduces volatile loss caused by the phase change in evaporation and it is very competitive from an energy viewpoint. Reverse osmosis uses much higher pressures than other membrane processes, in the range 20–80 bar, and will incur greater energy costs and require more expensive pumps. Products of reverse osmosis may be subtly different to those produced by evaporation, particularly with respect to low molecular weight solutes, which might not be completely rejected, and to volatile components, which are not completely lost.

Thin-film composite membranes based on combinations of polymers have now largely replaced cellulose acetate, allowing higher temperatures, up to 80 °C, and greater extremes of pH (3–11) to be used, thereby facilitating cleaning and disinfection. Therefore, the main applications of reverse osmosis are for concentrating liquids, recovering solids, and treatment of water.

10.1.9.1 Milk Processing

Reverse osmosis can be used for concentrating full cream milk up to a factor of 2–3 times. Flux rates for skimmed milk are only marginally higher than those for full cream milk. The product concentration attainable is nowhere near as high as that for evaporation, due to increasing osmotic pressure and fouling, due mainly to the increase in calcium phosphate, which precipitates out in the pores of the membrane. Therefore, most of the commercial applications have been for increasing the capacity of evaporation plant. Other possible applications that have been investigated and discussed include the concentration of milk on the farm for reducing transportation costs; for yoghurt production at a concentration factor of about 1.5, to avoid addition of skim milk powder; for ice cream making, also to reduce the use of expensive skim milk powder; for cheese making to increase the capacity of the cheese vats; and for recovering rinse water. Cheese whey can also be concentrated to reduce transportation costs or prior to drying. Flux values for sweet whey are higher than for acid whey, which in turn are higher than for milk, for all systems tested [3]. Reviews of the use of reverse osmosis and ultrafiltration in dairying applications include those by El-Gazzar and Marth [4] and Renner and Abd El-Salam [5].

10.1.9.2 **Other Foods**

Reverse osmosis has been used in the processing of fruit and vegetable juices, sometimes in combination with ultrafiltration and microfiltration. The osmotic pressure of juices is considerably higher than that of milk. It is advantageous to minimize thermal reactions, such as browning, and to reduce loss of volatiles. From a practical viewpoint, the flux rate and rejection of volatiles is important. Reverse osmosis modules can cope with single strength clear or cloudy juices and also fruit pulp. It can be used to produce a final product, as in the case of tomato paste and fruit purees, or to partially concentrate, prior to evaporation. Reverse osmosis is a well-established process for concentrating tomato juice from about 4.5 to 8–12°Brix. Other fruit juices which have been successfully concentrated are apple, pear, peach, and apricot. Where juices have been clarified, osmotic pressure limits the extent of concentration and up to 25°Brix can be achieved. Unclarified juices may be susceptible to fouling. With purees and pulps, the viscosity may be the limiting factor and these can be concentrated to a maximum of 1.5 times. It is possible to concentrate coffee extract from about 13 to 36% total solids at 70 °C, with little loss of solids. Tea extracts can also be similarly concentrated.

Reverse osmosis is also used for waste recovery and more efficient use of processing water in corn wet-milling processes. Commercial plant is available for concentrating egg white, to about 20% solids. In one particular application egg white is concentrated and dried, after lysozyme has been extracted.

Dealcoholization is an interesting application, using membranes which are permeable to alcohol and water. In a process akin to diafiltration, water is added back to the concentrated product, to replace the water and alcohol removed in the permeate. Such technology has been used for the production of low or reduced alcohol, beers, ciders, and wine. Leeper [6] reported that ethanol rejections for cellulose acetate ranged between 1.5 and 40%; for polyamides were between 32.8 and 60.9%, and for other hybrid membranes, as high as 91.8%.

Reverse osmosis is used in many areas world-wide for water treatment, where there are shortages of fresh water, although it is still well exceeded by multi-stage fractional distillation. Potable water should contain less than 500 ppm of dissolved solids. Brackish water (e.g., bore-hole or river water) typically contains from 1000 up to about 10 000 ppm of dissolved solids, whereas sea-water contains upward of 35 000 ppm of dissolved solids. If lower total solids are required, permeate can be subjected to a second process, known as *double reverse osmosis*.

10.1.10 **Applications for Nanofiltration**

Nanofiltration has been used for partially reducing calcium and other salts in milk and whey, with typical retention values of 95% for lactose and less than 50% for salts. Guu and Zall [7] have reported that permeate subject to nanofiltration gave improved lactose crystallization. Nanofiltration provides potential for improving the heat stability of the milk by removing some of the calcium. It has also been investigated for removing pesticides and components responsible for the color from ground water, as well as for purifying water for carbonation and soft drinks.

Nanofiltration is currently being investigated for fractionating oligosaccharides with prebiotic potential produced by the enzymatic breakdown of different complex carbohydrates.

10.1.11 Applications for Ultrafiltration

10.1.11.1 Milk Products

Milk will be taken as an example to show the potential of ultrafiltration. Milk is chemically complex, containing components of a wide range of molecular weights, such as protein, fat, lactose, minerals, and vitamins. It also contains microorganisms, enzymes, and perhaps antibiotics and other contaminants. Whole milk contains about 30–35% protein and about the same amount of fat (dry weight basis). Therefore it is an ideal fluid for membrane separation processes, in order to manipulate its composition, thereby providing a variety of products or improving the stability of the colloidal system. The same principles apply to skim milk, standardized milk, and some of its by-products, such as cheese whey. Skim milk can be concentrated up to seven times and full cream milk up to about five times Kosikowski [8]. An International Dairy Federation publication [9] gives a summary of the rejection values obtained during the ultrafiltration of sweet whey, acid whey, skim milk, and whole milk, using a series of industrial membranes. Milk protein concentrates (MPC) are now widely available in powder form.

Bastian *et al.* [10] compared the rejection values during ultrafiltration and diafiltration of whole milk. Diafiltration involves diluting the retentate with water, either continuously or intermittently. They found that the rejection of lactose, riboflavin, calcium, sodium, and phosphorus was higher during diafiltration than ultrafiltration. Diafiltration of acidified milk gave rise to lower rejections of calcium, phosphorus, and sodium. Premaratne and Cousin [11] reported a detailed study on the rejection of vitamins and minerals during ultrafiltration of skimmed milk. During a fivefold concentration the following minerals were concentrated by the following factors: Zn (4.9), Fe (4.9), Cu (4.7), Ca (4.3), Mg (4.0), and Mn (3.0), indicating high rejection values. On the other hand, most of the B vitamins examined were almost freely permeating.

Ultrafiltration has been used to concentrate cheese whey (~6.5% TS), which contains about 10–12% protein on a dry weight basis, to produce concentrates, which could then be dried to produce high-protein powders (concentrates and isolates) which retain the functional properties of the proteins. Some typical concentration factors, f, used are as follows:

- $f = 5$; protein content (dry weight basis) about 35% (similar to skim milk),
- $f = 20$; protein content about 65%,
- $f = 20$; plus diafiltration protein content about 80%.

The product starts to become very viscous at a concentration factor of about 20, so diafiltration is required to further increase the protein in the final product.

The permeate from ultrafiltration of whey contains about 5% total solids, the predominant component being lactose. Since this is produced in substantial quantities, the economics of the process are dependent upon its utilization. It can be concentrated by reverse osmosis and hydrolyzed to glucose and galactose to produce sweeteners or fermented to produce alcohol or microbial protein. Skim milk has been investigated also and protein concentrates based on skim milk are now receiving considerable commercial interest.

Yoghurt and other fermented products have been made from skim milk and whole milk concentrated by ultrafiltration [4]. Whey protein concentrates have also been incorporated [12]. Production of *labneh*, which is a strained or concentrated yoghurt at about 21% total solids, has been described by Tamime *et al.* [13], by preconcentrating milk to 21% TS. Inorganic membranes have also been used for skim milk, and Daufin *et al.* [14] have investigated the cleanability of these membranes using different detergents and sequestering agents.

As well as exploiting the functional properties of whey proteins, full cream milk has been concentrated by ultrafiltration prior to cheese making. The ultrafiltration concentrate has been incorporated directly into the cheese vats. Some advantages of this process include increased yield, particularly of whey protein; lower rennet and starter utilization; smaller vats, or even complete elimination of vats; little or no whey drainage; and better control of cheese weights. Lawrence [15] suggests that concentration below a factor of 2 gives protein standardization, reduced rennet, and vat space, but no increased yield. At concentration factors greater than 2, an increased yield is found.

Some problems result from considerable differences in the way the cheese matures and hence its final texture and flavor. The types of cheese that can be made in this way include Camembert type cheese, mozzarella, feta, and many soft cheeses. Those which are difficult include the hard cheeses such as Cheddar and also cottage cheese; the problems are mainly concerned with poor texture. More discussion is given by Kosikowski [8]. Further reviews on the technological problems arising during the conversion of retentate into cheeses are discussed by Lelievre and Lawrence [16]. Quarg is also produced successfully from ultrafiltered milk and milk is fermented prior to ultrafiltration to prevent off-flavor development.

Ultrafiltration is an extremely valuable method of concentrating and recovering many of the minor components, particularly enzymes, from raw milk, many of which would be inactivated by pasteurization. Such enzymes are discussed in more detail by Kosikowski [17]. Further reviews on membrane processing of milk are provided by Glover [3], El-Gazzar and Marth [4], Renner and El-Salam [5], and the International Dairy Federation [18].

Ultrafiltration is a useful analytical tool for measuring how minerals in milk partition between the casein micelle and soluble phase, both when pH is reduced and when temperature increases. It can be seen that Ca and P increase in its aqueous phase when pH is decreased. It has been shown that pH and ionic calcium decrease considerably in milk permeates when ultrafiltration temperature increases [19].

10.1.11.2 **Oilseed and Vegetable Proteins**

There have been many laboratory investigations into the use of ultrafiltration for extracting proteins from oilseed residues, or for removing any toxic components. Lewis [20] and Cheryan [21] have reviewed the more important of these. For soya, the separation of low molecular weight peptides from soy hydrolysates, with the aim of improving quality; the dissociation of phytate from protein, followed by its removal by ultrafiltration; the removal of oligosaccharides; the removal of trypsin inhibitor; and comparisons of performance of different membrane configurations. For cottonseed, the use of different extraction conditions has been evaluated, as have the functional properties of the isolates produced by ultrafiltration. Investigations were performed with sunflower and alfalfa to remove the phenolic compounds responsible for the color and bitter flavor and glucosinolates from rapeseed.

Many have been successful in producing good-quality concentrates and isolates, particularly with soyabean. However, few have come to commercial fruition, mainly because of the economics of the process, dictated by the relatively low value of products and the fact that acceptable food products can be obtained by simpler technology, such as isoelectric precipitation. A further problem arises from the fact that the starting residue is in the solid form, thereby imparting an additional extraction procedure. Extraction conditions may need to be optimized, with respect to time, temperature, pH, and antinutritional factors. A further problem arises from the complexity of oilseed and vegetable proteins, compared to milk products, evidenced by electrophoretic measurement. It is likely that many of these proteins are near their solubility limits after extraction, and further concentration will cause them to come out of solution and promote further fouling. Fouling and cleaning of membranes was found to be a serious problem during ultrafiltration of rapeseed meal extracts [22].

A further important area is the use of enzyme reactors. The earliest examples were used to break down polysaccharides to simpler sugars in a continuous reactor, using a membrane to continuously remove the breakdown products.

10.1.11.3 **Animal Products**

Slaughterhouse wastes contain substantial amounts of protein. Two important streams that could be concentrated by ultrafiltration are blood and wastewater.

Blood contains about 17% protein. It can be easily separated by centrifugation into plasma (70%) and the heavier erythrocytes (red blood corpuscles or cells, about 30%). Plasma contains about 7% protein, whereas the blood corpuscles contain between 28 and 38% protein. The proteins in plasma possess useful functional properties, particularly gelation, emulsification, and foaming. They have been incorporated into meat products and have shown potential for bakery products, as replacers for egg white.

Whole blood, plasma, and erythrocytes have all been subjected to ultrafiltration processes [21, 23]. The process is concentration polarization controlled and flux rates are low. High flow rate and low pressure regimes are best. Gel concentrations were approximately 45% for plasma protein and 35% for red blood cells. The fouling characteristics of different blood fractions have been investigated by Wong *et al.* [24]. Whole blood was found to be the worst foulant, when compared with lysed blood and blood plasma.

Another important material is gelatin, which can be concentrated from very dilute solutions by ultrafiltration. As well as concentration, ash is removed which improves its gelling characteristics. This is one example where there have been some high negative rejections recorded for calcium, when ultrafiltered at low pH.

Eggs have also been processed by ultrafiltration. Egg white contains 11–13% total solids (about 10% protein, 0.5% salts, and 0.5% glucose). Large amounts of egg white are used in the baking industry. The glucose can cause problems during storage and causes excessive browning during baking. Whole egg contains about 25% solids and about 11% fat, whereas egg yolk contains about 50% solids. It is unusual to evaporate eggs prior to drying, because of the damage caused. Compared to reverse osmosis, ultrafiltration also results in the partial removal of glucose; further removal can be achieved by diafiltration. Flux values during ultrafiltration are much lower than for many other food materials, most probably due to the very high initial protein concentration; rates are also highly velocity-dependent and temperature-dependent [21].

Membrane-based bioreactors appear to be a very promising application for the production of ethanol, lactic acid, acetone and butanol, starch hydrolysates, and protein hydrolysates.

10.1.12
Applications for Microfiltration

Microfiltration is generally used to separate particles suspended in liquid media, and may frequently be considered as an alternative to conventional filtration or centrifugation. For industrial use, the aim is usually to obtain either a clear permeate or the concentrate. Therefore most applications are either clarification, or the recovery of suspended particles such as cells or colloids, or the concentration of slurries. Recently, microfiltration has been used to separate the whey protein fraction from casein micelles and considerable work has been undertaken on the properties of both these fractions.

One application in the food industry has been in the treatment of juices and beverages. As microfiltration is a purely physical process, it can have advantages over traditional methods, which may involve chemical additives, in terms of the quality of the product as well as the costs of processing.

Finnigan and Skudder [25] report that very good-quality, clear permeate was found when processing cider and beer, with high flux rates and no rejection of essential components.

Clarification and biological stabilization of wine musts and unprocessed wine have also been described. This avoids the requirement for fining and, possibly, pasteurization. Another section of the industry with several applications is dairy processing. Piot *et al.* [26] and Merin [27] have clarified sweet cheese whey using cross-flow microfiltration (CMF). This has the dual benefit of removing fat and reducing the bacterial population and could eliminate the need for fat separation and heat treatment in the production of whey protein powders prior to ultrafiltration. The former authors reported that a 5-log reduction of microorganisms could be obtained in the microfiltrate compared to the whey, although some loss of whey protein was observed. Hanemaaijer [28] described a scheme for whey treatment incorporating microfiltration and ultrafiltration to produce "tailor-made" whey products with specific properties for specific applications. The products include whey protein concentrates that are rich in whey lipids, as well as highly purified protein.

Bacterial removal from whole milk by microfiltration is a problem because the size range of the bacteria overlaps with the fat globules, and to a lesser extent with the casein micelles. However, some success has been achieved with skim milk. The "Bactocatch" system can remove 99.6% of the bacteria from skim milk using ceramic membranes on commercial scale [29]. The retentate (approximately 10% of the feed) can then be sterilized by a UHT process, mixed with the permeate and the mixture pasteurized, to give a product with 50% longer shelf-life but no deterioration in organoleptic properties compared to milk that has only been pasteurized. One such product on the market in the UK is Cravendale milk (http://www.arlafoods.com). The combination of microfiltration and heat reduces the bacterial numbers by 99.99%.

Alternatively, the permeate could be used for cheese making, or the production of low-heat milk powder [30]. Piot *et al.* [31] described the use of membranes of pore diameter 1.8 μm to produce skim milk of low bacterial content. Recovery of fat from buttermilk has also been described [32].

Membranes have been used to concentrate milk prior to the manufacture of many cheese types. This results in improved yields and other associated benefits such as reduced requirement for rennet and starter, and the ability to produce much more cheese per vat [33]. However, the use of microfiltration is an attractive alternative. Rios *et al.* [32] have carried out extensive trials on this application and concluded that the use of 0.2 μm pore diameter membranes gave a product with better texture and yield than with centrifugation. The choice of ceramic membranes allowed the curd to be contacted directly with the membrane.

Other food applications have been reported with meat and vegetable products, including the following. Devereux and Hoare [34] described the use of microfiltration to recover precipitated soya protein. This could have advantages over recovery of the dissolved protein using ultrafiltration. Gelatin is a proteinaceous material derived by hydrolysis of collagen. This is purified by filtration incorporating diatomaceous earth. The latter process can be replaced by CMF, which effectively removes dirt, coagulated proteins, fats, and other particulate materials from the feed. Again, CMF gives higher yields of high-quality product on a continuous basis. Short [35] calculated that incorporating CMF plants for gelatin would have a payback time of 3 years for a capacity of $30\,t\,h^{-1}$.

Overall, microfiltration has made significant advances in new applications in the food and biotechnology industries. However, it has not yet realized its full potential, largely due to the severe problems of flux decline due to fouling. It is believed that further developments in membrane design and a greater knowledge of fouling mechanisms will result in greater application in the future, especially in the field of downstream processing.

A useful recent review article on membranes in dairy applications has been provided by Pouliot [36].

10.2 Ion Exchange

10.2.1 General Principles

Ion exchange can be used for separations of many types of molecules, such as metal ions, proteins, amino acids, or sugars. The technology is utilized in many sensitive analytical chromatography and laboratory separation procedures, frequently on a very small scale. On the other hand, industrial-scale production operations, such as demineralization or protein recovery, are also possible. More detailed information on the theory of ion exchange can be found elsewhere [37–39].

Ion exchange is the selective removal of a single, or group of, charged species from one liquid phase followed by transfer to a second liquid phase by means of a solid ion-exchange material. This involves the process of adsorption – the transfer of specific solute(s) from a feed solution on to a solid ion exchanger. The mechanism of adsorption is electrostatic, involving opposite charges on the solute(s) and the ion exchanger. After washing off the feed solution, the solute(s) is desorbed back into solution in a much purified form.

Ion-exchange solids have fixed ions covalently attached to a solid matrix. There are two basic types of ion exchanger:

- **Cation exchangers**: These bear fixed negative charges (e.g., $-SO_3^-H^+$; $-PO_3^{2-}(H^+)_2$; –COOH) and therefore retain cations.
- **Anion exchangers**: These bear fixed positive charges (amines or imines such as quaternary amine, or diethylaminoethyl groups) and thus retain anions.

Ion exchangers can be used to retain simple ionized species such as metal ions, but may also be used in the separation of polyelectrolytes, such as proteins, which carry both positive and negative charges, as long as the overall charge on the polyelectrolyte is opposite to the fixed charges on the ion exchanger. This overall charge depends on the isoelectric point (IEP) of the polyelectrolyte and the pH

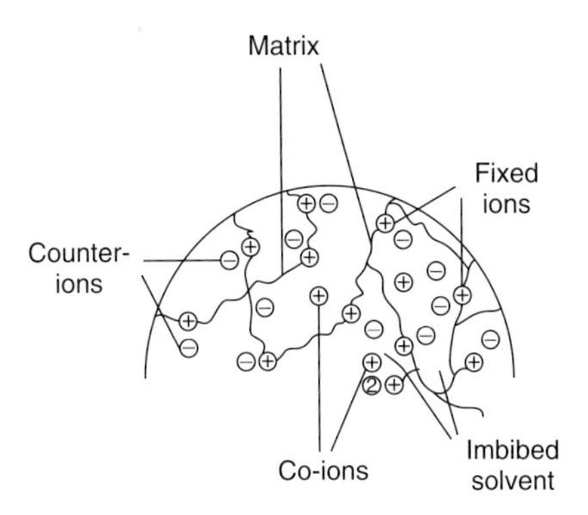

Figure 10.6 Schematic diagram of a generalized anion exchanger.

of the solution. At pH values lower than the IEP, the net overall charge will be positive and vice versa. The main interaction is via electrostatic forces, and in the case of polyelectrolytes the affinity is governed by the number of electrostatic bonds between the solute molecule and the ion exchanger. However, with large molecules, such as proteins, size and shape, and the degree of hydration of the ions may affect these interactions, and hence the selectivity of the ion exchanger for different solutes.

Figure 10.6 shows a generalized anion exchanger – that is, bearing fixed positive charges. To maintain electrical neutrality these fixed ions must be balanced by an equal number of mobile ions of the opposite charge, that is, anions, held by electrostatic forces. These mobile ions can move into and out of the porous molecular framework of the solid matrix and may be exchanged stoichiometrically with other dissolved ions of the same charge, and are termed *counterions*. As the distribution of ions between the internal phase of the ion exchanger and the external phase is determined by the Donnan equilibrium, some co-ions (mobile ions having the same sign – positive in this example – as the fixed ions) will be present, even in the internal phase. Therefore, if an anion exchanger, as in Figure 10.6, is in equilibrium with a solution of NaCl, the internal phase contains some Na^+ ions, although the concentration is less than in the external phase because the internal concentration of Cl^- ions is much larger.

When an ion exchanger is put in contact with an ionized solution, equilibration between the two phases rapidly occurs. Water moves into or out of the internal phase so that osmotic balance is achieved. Counterions also move in and out between the phases on an equivalent basis. If two or more species of counterion are present in the solution, they will be distributed between the phases according to the proportions of the different ions present and the relative selectivity of the ion exchanger for the different ions. It is this differential distribution of different counterions which forms the basis of separation by ion exchange. The relative selectivity for different ionized species results from a range of factors. The overall charge on the ion and the molecular or ionic mass are the primary determining factors, but selectivity is also related to degree of hydration, steric effects, and environmental factors such as pH or salt content.

In the adsorption stage, a negatively charged solute molecule (e.g., a protein P^-) is attracted to a charged site on the ion exchanger (R^+) displacing a counterion (X^-):

$$R^+X^-P^- \rightarrow R^+P^- + X^-$$

This is shown schematically in Figure 10.7a.

In the desorption stage, the anion is displaced from the ion exchanger by a competing salt ion (S^-), and hence is eluted:

$$R^+P^- + S^- \rightarrow R^+S^- + P^-$$

This is shown schematically in Figure 10.7b.

Alternatively, desorption can be achieved by addition of H^+ or OH^- ions. Ion exchangers are further classified in terms of how their charges vary with changes in pH, into weak and strong exchangers. Strong ion exchangers are ionized over a wide range of pH, and have a constant capacity within the range, whereas weak exchangers are only ionized over a limited pH range, for example, weak cation exchangers may lose their charge below pH 6 and weak anion exchangers above pH 9. Thus weak exchangers may be preferable to strong ones in some situations, for example, where desorption may be achieved by a relatively small change in pH of the buffer in the region of the pK_a of the exchange group. Regeneration of weak ion-exchange groups is easier than with strong groups, and therefore has a lower requirement of costly chemicals.

10.2.2
Ion-Exchange Equipment

All ion exchangers consist of a solid insoluble matrix (termed the *resin, adsorbent, medium*, or just *ion exchanger*) to which the active, charged groups are attached covalently. The solid support must have an open molecular framework which allows the mobile ions to move freely in and out, and must be completely insoluble throughout the process. Most commercial ion exchangers are based on an organic polymer network, for example, polystyrene and dextran, although inorganic materials, such as porous silica, may be used. The latter are much more rigid and incompressible. The support material does not directly determine the ionic distribution between the two phases, but it does influence the capacity, the flow rate through a column, the diffusion rate of counterions into and out of the matrix, the degree of swelling, and the durability of the material.

As the adsorption is a surface effect, the available surface area is a key parameter. For industrial processing the maximum surface area to volume should be used to minimize plant size and product dilution. It is possible for a 1 ml bed of ion exchanger to have a total surface area $>100\,m^2$. The ion-exchange material is normally deployed in packed beds, and involves a compromise between large particles (to minimize pressure drop) and small particles to maximize mass transfer rates. Porous particles are employed to increase surface area/volume.

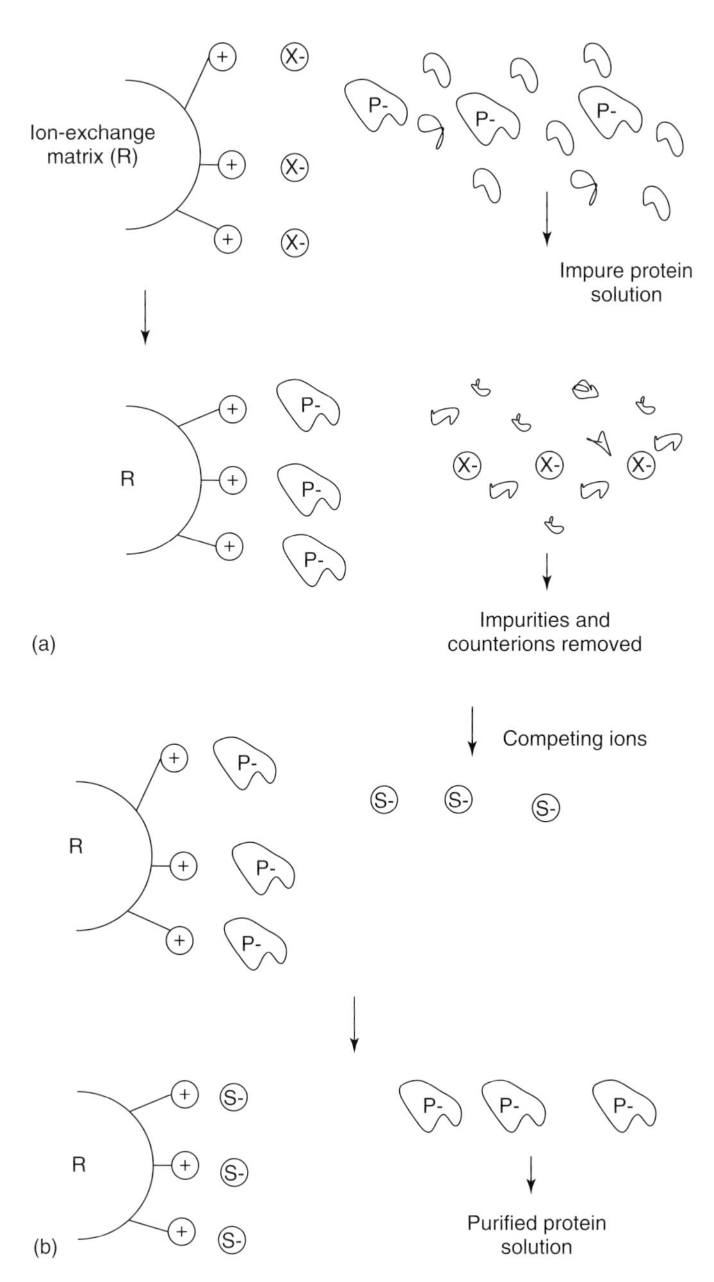

Figure 10.7 Process of anion exchange: (a) adsorption stage and (b) desorption stage.

The *capacity of an ion exchanger* is defined as the number of equivalents of exchangeable ions per kilogram of exchanger, but is frequently expressed in milliequivalents per gram (usually in the dry form). Most commercially available materials have capacities in the range 1–10 equiv./kg of dry material, but this may decline with age due to blinding or fouling – that is, nonspecific adsorption of unwanted materials, such as lipids, onto the surface, or within the pores.

The choice of method of elution depends on the specific separation required. In some cases the process is used to remove impurities from a feedstock, while the required compound(s) remains unadsorbed. No specific elution method is required in such cases, although it is necessary to regenerate the ion exchanger with strong acid or alkali. In other cases the material of interest is adsorbed by the ion exchanger while impurities are washed out of the bed. This is followed by elution and recovery of the desired solute(s). In the latter case the method of elution is much more critical – for example, care must be taken to avoid denaturation of adsorbed proteins.

The adsorbed solute is eluted from the ion exchanger by changing the pH or the ionic strength of the buffer, followed by washing away the desorbed solute with a flow of buffer. Increasing the ionic strength of the buffer increases the competition for the charged sites on the ion exchanger. Small buffer ions with a high charge density will displace polyelectrolytes, which can subsequently be eluted. Altering the buffer pH, so that the charge on an adsorbed polyelectrolyte is neutralized or made the same as the charges on the ion exchanger, will result in desorption.

Fixed-bed operations consisting of one, or two columns connected in series (depending on the type of ions which are to be adsorbed), are used in most ion-exchange separations. Liquids should penetrate the bed in plug flow, in either downward or upward direction. The major problems with columns arise from clogging of flow and the formation of channels within the bed. Problems may also arise from swelling of organic matrices when the pH changes. These problems may be minimized by the use of stirred tanks, however these batch systems are less efficient and expose the ion exchangers to mechanical damage as there is a need for mechanical agitation. The system involves mixing the feed solution with the ion exchanger and stirring until equilibration has been achieved (typically 30–90 min in the case of proteins). After draining and washing the ion exchanger, the eluant solution is then contacted with the bed for a similar equilibration time before draining and further processing. Commercial stirred-tank reactors for recovering protein from whey are shown in Figure 10.8.

Mixed-bed systems, containing both anion and cation exchangers, may be used to avoid prolonged exposure of the solutions to both high and low pH environments, as is frequently encountered when using anion- and cation-exchange columns in series, for example, during demineralization of sugar cane juice to prevent hydrolysis of sucrose as described below.

10.2.3
Applications of Ion Exchange in the Food Industry

The main areas of the food industry where the process is currently used or is being developed are sugar, dairy, and water purification. Ion exchange is also widely employed in the recovery, separation, and purification of biochemicals and enzymes.

Figure 10.8 Commercial stirred-tank reactors. Courtesy of Bio-isolates plc.

The main functions of ion exchange are:

- removal of minor components, for example, de-ashing or decolorizing;
- enrichment of fractions, for example, recovery of proteins from whey or blood;
- isolating valuable compounds, for example, production of purified enzymes.

10.2.3.1 **Softening and Demineralization**

Softening of water and other liquids involves the exchange of calcium and magnesium ions for sodium ions attached to a cation exchange resin, for example:

$$R\text{-}(Na^+)_2 + Ca(HCO_3)_2 \rightarrow R\text{-}Ca^{++} + 2NaHCO_3$$

The sodium form of the cation exchanger is produced by regenerating with NaCl solution. Apart from the production of softened water for boiler feeds and cleaning of food processing equipment, softening may be employed to remove calcium from sucrose solutions prior to evaporation (which reduces scaling of heat exchanger surfaces in sugar manufacture), and from wine (which improves stability).

Demineralization using ion exchange is an established process for water treatment, but over the last 20 years it has been applied to other food streams. Typically the process employs a strong acid cation exchanger followed by a weak or strong base anion exchanger. The cations are exchanged with H^+ ions – for example:

$$2R^-H^+ + CaSO_4 \rightarrow (R^-)_2Ca^{++} + H_2SO_4$$
$$R\text{-}H^+ + NaCl \rightarrow R\text{-}Na^+ + HCl$$

and the acids thus produced are fixed with an anion exchanger – for example:

$$R^+OH^- + HCl \rightarrow R^+Cl^- + H_2O.$$

Demineralized cheese whey is desirable for use mainly in infant formulations, but also in many other products such as ice cream, bakery products, confectionery, animal feeds, and so on. The major ions removed from whey are Na^+, K^+, Ca^{2+}, Mg^{2+}, Cl^-, HPO_4^-, citrate, and lactate. Ion-exchange demineralization of cheese whey generally employs a strong cation exchanger followed by a weak anion exchanger. This can produce more than 90% reduction in salt content, which is necessary for infant formulas.

Demineralization by ion-exchange resins is used at various stages during the manufacture of sugar from either beet or cane, as well as for sugar solutions produced by hydrolysis of starch. In the production of sugar from beet, the beet juice is purified by liming and carbonatation and then may be demineralized by ion exchange. The carbonated juice is then evaporated to a thick juice prior to sugar crystallization. Demineralization may, alternatively, be carried out on the thick juice which has the advantage that the quantities handled are much smaller. To produce high-quality sugar the juice should have a purity of about 95%. Ash removal or complete demineralization of cane sugar liquors is carried out on liquors that have already been clarified and decolorized, so the ash load is at minimum. The use of a mixed bed of weak cation and strong anion exchangers in the hydrogen and hydroxide forms, respectively, reduces the prolonged exposure of the sugar to strongly acid or alkali conditions which would be necessary if two separate columns were used. Destruction of sucrose is thus minimized.

The cation and anion resins are sometimes used in their own right for dealkalization or deacidification, respectively. Weak cation exchangers may be used to reduce the alkalinity of water used in the manufacture of soft drinks and beer, while anion exchangers can be used for deacidification of fruit and vegetable juices. In addition to deacidification, anion exchangers may also be used to remove bitter flavor compounds (such as naringin or limonin) from citrus juices. Anion- or cation-exchange resins are used in some countries to control the pH or titratable acidity of wine although this process is not permitted by other traditional wine-producing countries. Acidification of milk to pH 2.2 using ion exchange during casein manufacture by the Bridel process, has also been described.

Ion-exchange processes can be used to remove specific metals or anions from drinking water and food fluids, which has potential application for detoxification or radioactive decontamination. Procedures have been described for the retention of lead, barium, radium, aluminum, uranium, arsenic, and nitrates from drinking water. Removal of a variety of radionuclides from milk has been demonstrated. Radiostrontium and radiocaesium can be removed using a strongly acidic cation exchanger, while ^{131}I can be adsorbed on to a variety of anion exchangers. The production of low sodium milk, with potential dietetic application, has been demonstrated.

10.2.3.2 **Decolorization**

Sugar liquors from either cane or beet contain colorants such as caramels, melanoidins, melanins, or polyphenols combined with iron. Many of these are formed during the earlier refining stages, and it is necessary to remove them in the production of a marketable white sugar. The use of ion exchangers just before the crystallization stage results in a significant improvement in product quality. It is necessary to use materials with an open, porous structure to allow the large colorant molecules access to the adsorption sites. A new approach to the use of ion exchange for decolorization of sugar solutions is the application of powdered resin technology. Finely powdered resins (0.005–0.2 mm diameter) have a very high capacity for sugar colorants due to the ready availability of adsorption sites. The use of such materials on a disposable basis eliminates the need for chemical regenerants, but is quite expensive.

Color reduction of fermentation products such as wine uses a strongly basic anion exchanger to remove coloring matter, followed by a strong cation exchanger to restore the pH. It is claimed that color reduction can be achieved without substantially deleteriously affecting the other wine qualities.

10.2.3.3 **Protein Purification**

High-purity protein isolates can be produced in a single step from dilute solutions containing other contaminating materials by ion exchange. The amphoteric nature of protein molecules permits the use of either anion or cation exchangers, depending on the pH of the environment. Elution takes place either by altering the pH or increasing the ionic strength. The eluate can be a single bulk, or a series of fractions produced by stepwise or linear gradients, although fractionation may be too complex for large-scale industrial production. Separation of a single protein may take place on the basis that it has a higher affinity to the charged sites on the ion exchanger compared to other contaminating species, including other proteins present in the feed. In such cases, if excess quantities of the feed are used, the protein of interest can be adsorbed exclusively, despite initial adsorption of all the proteins in the feed. Alternatively it may be possible to purify a protein on the basis that it has a much lower affinity for the ion exchanger than other proteins present in the feed, and thus the other proteins are removed leaving the desired protein in solution. One limitation of the process for protein treatment is that extreme conditions of pH, ionic strength, and temperature must be avoided to prevent denaturation of the protein.

An area of great potential is the recovery of proteins from whey, which is a by-product of the manufacture of cheese and related products such as casein. Typically whey contains 0.6–0.8% protein, which is both highly nutritious and displays excellent physical properties, yet the vast majority of this is wasted or underutilized. Anion-exchange materials can produce high-purity functional protein from cheese whey, using a stirred tank reactor into which the whey is introduced at low pH. Following rinsing of nonadsorbed material, the protein fraction is eluted at high pH, and further purified by ultrafiltration so that the final protein content is approximately 97% (on a dry matter basis). It is further possible to fractionate the whey proteins into their separate components or groups of components. This approach has the potential of producing protein fractions with a range of functional properties which could be extremely valuable for use in the food industry. Lactoperoxidase and lactoferrin are valuable proteins with potential pharmaceutical applications which are present in small quantities in cheese whey. They may be purified from whey on the basis that these proteins are positively charged at neutral pH, whereas the major whey proteins are negatively charged. Another application of adsorption of whey protein by ion exchangers could be to improve the heat stability of milk. The use of ion exchange to recover or separate the caseins in milk is not carried out commercially, although it has been shown to be feasible.

This system has also been demonstrated for recovery of food proteins from waste streams resulting from the processing of soya, fish, vegetables, and gelatin production, plus abattoir waste streams. Such protein fractions could be used as functional proteins in the food industry or for animal feeds. A variety of other food proteins have been purified or fractionated by ion exchangers, including pea globulins, gliadin from wheat flour, egg, groundnut, and soya protein.

Purification of proteins from fermentation broths usually involves a series of separation steps and frequently includes ion exchange. Large-scale purification of a variety of enzymes with applications in the food industry has been described, for example, α-amylase and β-galactosidase.

10.2.3.4 **Other Separations**

Ion exchange has been used for various other separations in the food industry which do not fit into the above categories.

Fructose is sweeter than sucrose and glucose, and can be used as a natural sweetener at reduced caloric intake. Although present in many natural sources, it is produced commercially from corn starch by hydrolysis to dextrose which is then partially converted to fructose using the enzyme isomerase. The resulting high-fructose corn syrup may be deionized by ion exchange and then a pure fructose fraction can be recovered with a sulfonic cation exchanger. Another application is the production of lactose-free milk. A process using sulfonated cation exchangers has been used to reduce the lactose level of skim milk to $<10\%$ of that in the feed, while retaining $>90\%$ of protein, minerals, and citrate.

The purification of phenylalanine, which may be used in aspartame sweetener production, from fermentation broths using cationic zeolite material has been demonstrated. Ion exchange may also be used to purify enzymic reaction products such as flavor constituents from the enzymic degradation of fruit wastes.

10.2.4 Conclusion

There are many potential applications for ion exchange in the food industry, but few have been fully exploited in commercial practice. This is because of the complexity of the process and problems of scale-up. New applications are most likely to be developed in the food-related aspects of biotechnology, and in the production of high-value protein fractions.

10.3 Electrodialysis

10.3.1 General Principles and Equipment

Electrodialysis is a separations process in which membranes are used to separate ionic species from nonionic species. More detailed information on the theory and applications can be found elsewhere [38, 40, 41]. The process permits the separation of electrolytes from nonelectrolytes, concentration or depletion of electrolytes in solutions, and the exchange of ions between solutions. Separation depends on ion-selective membranes, which are essentially ion-exchange resins cast in sheet form, and electromigration of ions through ion-selective membranes depends on the electrical charge on the molecules, combined with their relative permeability through membranes. The membranes are composed of polymer chains which are cross-linked and intertwined into a network, and bear either fixed positive or fixed negative charges. These may be heterogeneous membranes which consist of ion-exchange resins dispersed in a polymer film, or, more commonly, homogeneous membranes in which the ionic groups ($-NH_3^+$ or $-SO_3^-$) are attached directly to the polymer. Counterions (see Section 10.2) are freely exchanged by the fixed charges on the membranes and thus carry the electric current through the membranes, while co-ions are repelled by the fixed charges and cannot pass through the membrane. Therefore cation membranes (with $-SO_3^-$ groups) allow the passage of positively charged ions, while anion membranes (with $-NH_3^+$ groups) allow the passage of negatively charged ions.

In practice, the cation and anion membranes are usually arranged alternately with plastic spacers to form thin solution compartments as shown in Figure 10.9. In commercial practice 100–200 membranes may be assembled to form a membrane stack (Figure 10.10) and an electrodialysis system may be composed of one or more stacks. Commercial electrodialysis membranes may be as large as $1–2\,m^2$. The basic unit of a membrane stack is called a *cell pair* and comprises a pair of membranes and spacers as illustrated in Figures 10.9 and 10.11. A positive electrode at one end and a negative electrode at the other permit the passage of a DC current. The electrical potential causes the anions to move toward the

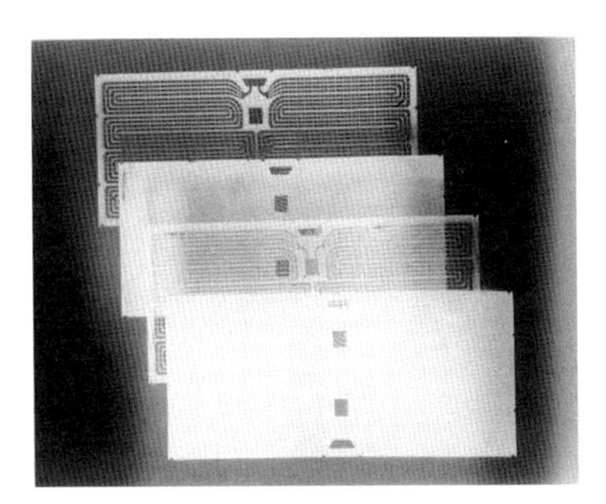

Figure 10.9 Electrodialysis membranes and spacers. With permission of Ionics Inc.

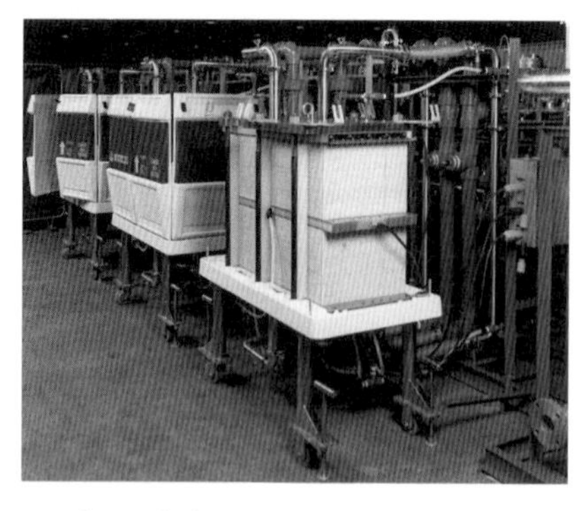

Figure 10.10 Electrodialysis membrane stacks. With permission of Ionics Inc.

anode and the cations to move toward the cathode. However, the ion-selective membranes act as barriers to either anions or cations. Hence, anions migrating toward the anode will pass through anion membranes, but will be rejected by cation membranes, and vice versa. The membranes, therefore, form alternating compartments of ion-diluting (even numbered compartments in Figure 10.11) and ion-concentrating (odd numbered) cells. If a feed stream containing dissolved salts, for example, cheese whey, is circulated through the ion-diluting cells and a brine solution through the concentrating cells, free mineral ions will leave the feed and be concentrated in the brine solution. Demineralization of the feed is, therefore,

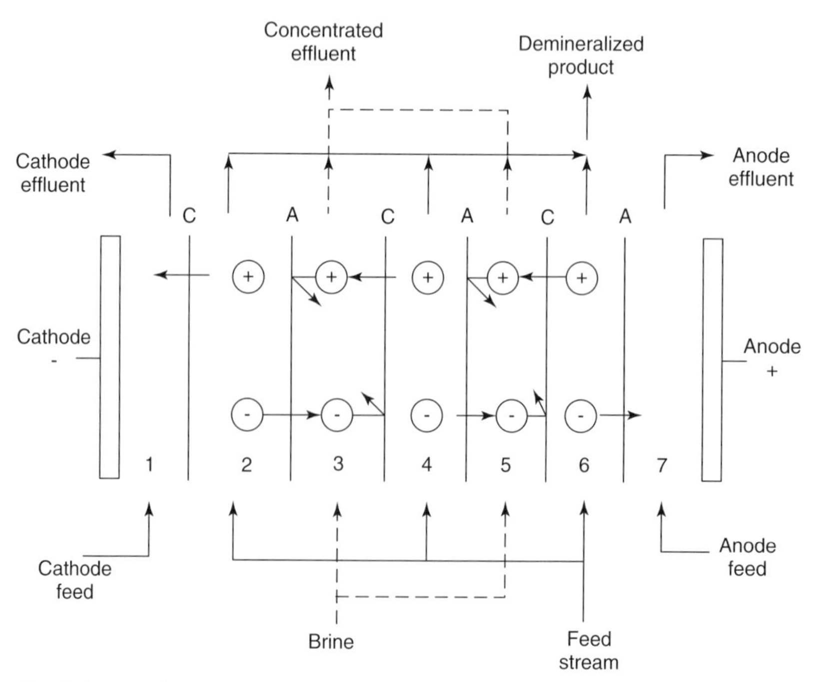

Figure 10.11 Schematic diagram of electrodialysis process.

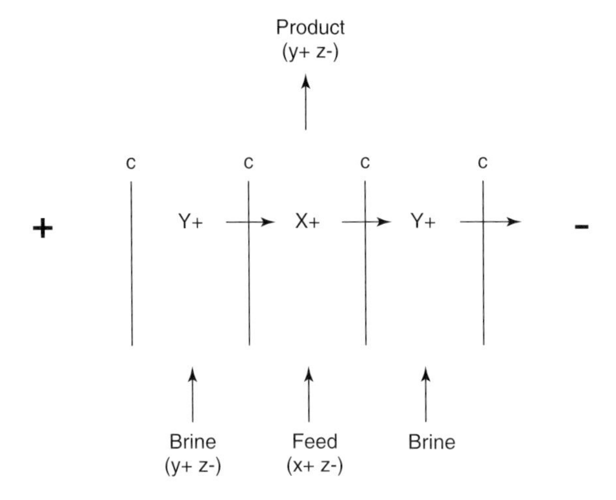

Figure 10.12 Ion replacement using cation membranes (C).

achieved. Note that any charged macromolecules in the feed, such as proteins, will attempt to migrate in the electrical field, but will not pass through either anion or cation membranes due to their molecular size. The efficiency of electrolyte transfer is determined by the current density and the residence time of the solutions within the membrane cells, and in practice efficiency is limited to about 90% removal of minerals. The membranes are subject to concentration polarization and fouling as described in Section 10.1.

Alternative membrane configurations are possible, such as the use of cation membranes only for ion replacement. In Figure 10.12, X^+ ions are replaced with Y^+ ions.

10.3.2
Applications for Electrodialysis

The largest application of electrodialysis has been in the desalination of brackish water to produce potable water. In Japan, all the table salt consumed is produced by electrodialysis of sea-water. The major application of electrodialysis in the food industry is probably for desalting of cheese whey. Following electrodialysis the demineralized whey is usually concentrated further and spray dried. Applications of electrodialysis in the dairy industry are reviewed by Bazinet [42]. Electrodialysis could potentially be employed in the refining of sugar from either cane or beet, but in fact commercial applications in these industries are limited by severe membrane fouling problems.

Other potential applications of electrodialysis in food processing include:

- demineralization of ultrafiltration permeate to improve lactose crystallization;
- separation of lactic acid from whey or soybean stock;
- removal of Ca from milk, either to improve protein stability during freezing, or to simulate human milk;

- removal of radioactive metal ions from milk;
- demineralization of fermented milk products to improve flavor and textural quality;
- extraction of salts from grape musts and wine to improve their stability;
- controlling the sugar/acid ratio in wine either by deacidification of the grape musts by ion substitution electrodialysis using anionic membranes, or acidification using cationic membranes;
- deacidification of fruit juices, either to reduce the sourness of the natural juices, or for the health food market;
- desalination of spent pickling brine;
- purification of bioactive peptides;
- precipitation of lipids from biological solutions.

In addition, the process can be integrated into continuous fermentation or reactor designs.

References

1. Lewis, M.J. (1996) Ultrafiltration, in *Separation Processes in the Food and Biotechnology Industries* (eds A.S Grandison. and M.J. Lewis), Woodhead Publishing, Cambridge. pp. 97–154.
2. Youravong, W., Lewis, M.J., and Grandison, A.S. (2003) Critical flux in ultrafiltration of skim milk. *Trans. Inst. Chem. Eng.*, **81** (Part C), 303–308.
3. Glover, F. (1985) Ultrafiltration and reverse osmosis for the dairy industry. Technical Bulletin No. 5, NIRD, National Institute for Research in Dairying, Reading.
4. ElGazzar, F.E. and Marth, E.H. (1991) Ultrafiltration and reverse osmosis in dairy technology, a review. *J. Food Prot.*, **54**, 801–809.
5. Renner, E. and Abd El-Salam, M.H. (1991) *Application of Ultrafiltration in the Dairy Industry*, Elsevier Applied Science, London.
6. Leeper, S.A. (1987) Membrane separations in the production of alcohol fuels by fermentation, in *Membrane Separations in Biotechnology* (ed. W.C. McGregor), Marcel Dekker, New York, pp. 161–200.
7. Guu, Y.K. and Zall, R.R. (1992) Nanofiltration concentration on the efficiency of lactose crystallisation. *J. Food Sci.*, **57**, 735–739.
8. Kosikowski, F.V. (1986) Membrane separations in food processing, in *Membrane Separations in Biotechnology* (ed. W.C. MeGregor), Marcel Dekker, New York, pp. 201–253.
9. International Dairy Federation (IDF) (1979) Equipment available for membrane processing. Bulletin No. 115, International Dairy Federation, Brussels.
10. Bastian, E.D., Collinge, S.K., and Ernstrom, C.A. (1991) Ultrafiltraton: partitioning of milk constituents into permeate and retentate. *J. Dairy Sci.*, **74**, 2423–2434.
11. Premaratne, R.J. and Cousin, M.A. (1991) Changes in the chemical composition during ultrafiltration of skim milk. *J. Dairy Sci.*, **74**, 788–795.
12. de Boer, R. and Koenraads, J.P.J.M. (1992) Incorporation of liquid ultrafiltration – whey retentates in dairy desserts and yoghurts in *New Applications in Membrane Processes*, Special Issue, No. 9201, International Dairy Federation (IDF), Brussels pp. 109–117.
13. Tamime, A.Y., Davies, G., Chekade, A.S., and Mahdi, H.A. (1991) The effect of processing temperatures on the quality of labneh made by ultrafiltration. *J. Soc. Dairy Technol.*, **44**, 99–103.

14. Daufin, G., Merin, U., Kerherve, F.L., Labbe, J.P., Quemerais, A., and Bousser, C. (1992) Efficiency of cleaning agents for an inorganic membrane after milk ultrafiltration. *J. Dairy Res.*, **59**, 29–38.

15. Lawrence, R.C. (1989) The use of ultrafiltration technology in cheese making. Bulletin No. 240, International Dairy Federation (IDF), Brussels.

16. Lelievre, J. and Lawrence, R.C. (1988) Manufacture of cheese from milk concentrated by ultrafiltration. *J. Dairy Sci.*, **55**, 465–447.

17. Kosikowski, F.V. (1988) Enzyme behaviour and utilisation in dairy technology. *J. Dairy Sci.*, **71**, 557–573.

18. International Dairy Federation (IDF) (1991) New applications of membrane processes, Special Issue No. 9201, International Dairy Federation, Brussels.

19. On-Nom, M., Grandison, A.S., and Lewis, M.J. (2010) Measurement of ionic calcium, pH and soluble divalent cations in milk at high temperature. *J. Dairy Sci.*, **93** (2), 515–523.

20. Lewis, M.J. (1982) Ultrafiltration of proteins, in *Developments in Food Proteins*, vol. 1 (ed. B.J.F. Hudson), Applied Science Publishers, London, pp. 91–130.

21. Cheryan, M. (1986) *Ultrafiltration Handbook*, Technomic Publishing Co., Lancaster.

22. Lewis, M.J. and Finnigan, T.J.A. (1989) in *Process Engineering in the Food Industry* (eds R.W. Field and J.A. Howell), Elsevier Applied Science, London, pp. 291–306.

23. Ockerman, H.W. and Hansen, C.L. (1988) *Animal By-product Processing*, Ellis Horwood, Chichester.

24. Wong, W., Jelen, P., and Chang, R. (1984) Ultrafiltration of bovine blood, in *Engineering and Food*, vol. 1, (ed. B., McKenna), Elsevier Applied Science, London, pp. 551–558.

25. Finnigan, T.J.A. and Skudder, P.J. (1989) The application of ceramic microfiltration in the brewing industry, in *Processing Engineering in the Food Industry* (eds R.W Field and J.A Howell), Elsevier Applied Science, London, pp. 259–272.

26. Piot, M., Maubois, J.L., Schaegis, P., Veyre, R. *et al.* (1984) Microfiltration en flux tangential des lactoserums de fromagerie. *Le Lait*, **64**, 102–120.

27. Merin, U. (1986) Bacteriological aspects of microfiltration of cheese whey. *J. Dairy Sci.*, **69**, 326–328.

28. Hanemaaijer, J.H. (1985) Microfiltration in whey processing. *Desalination*, **53**, 143–155.

29. Malmbert, R. and Holm, S. (1988) Producing low-bacteria milk by ultrafiltration. *North Eur. Food Dairy J.*, **1**, 1–4.

30. Hansen, R. (1988) Better market milk, better cheese milk and better low-heat milk powder with Bactocatch treated milk. *North Eur. Food Dairy J.*, **1**, 5–7.

31. Piot, M., Vachot, J.C., Veaux, M., Maubois, J.L., and Brinkman, G.E. (1987) Ecremage et epuration bacterienne du lait entire cru par microfiltration sur membrane en flux tangential. *Tech. Laitiere Mark.*, **1016**, 42–46.

32. Rios, G.M., Taraodo de la Fuente, B., Bennasar M., and Guidard C. (1989) Cross-flow filtration of biological fluids in inorganic membranes: a first state of the art, in *Developments in Food Preservation*, vol. 5 (ed. S. Thorne), Elsevier Applied Science, London, pp. 131–175.

33. Grandison, A.S. and Glover, F.A. (1994) Membrane processing of milk, in *Modern Dairy Technology*, vol. 1 (ed. R.K. Robinson), Elsevier Applied Science, London, pp. 273–311.

34. Devereux, N. and Hoare, M. (1986) Membrane separtion of proteins and precipitates: Studies with cross flow in hollow fibres. *Biotechnol. Bioeng.*, **28**, 422–431.

35. Short, J.L. (1988) Newer applications for crossflow membrane filtration. *Desalination*, **70**, 341–352.

36. Pouliot, Y. (2008) Membrane processes in dairy technology – From a simple idea to worldwide panacea. *Int. Dairy J.*, **18**, 735–740.

37. Anderson, R.E. (1988) Ion-exchange separations, in *Handbook of Separation Techniques for Chemical Engineers* (ed. P.A. Schweitzer), 2nd edn, McGraw-Hill, London, pp. 1–387, 1–444.

38. Grandison, A.S. (1996) Ion-exchange and electrodialysis, in *Separation Processes in the Food and Biotechnology*

Industries (eds A.S. Grandison and M.J. Lewis), Woodhead Publishing, Cambridge, pp. 155–177.

39. Walton, H.F. (1983) Ion-exchange chromatography, in *Chromatography, Fundamentals and Applications of Chromatographic Methods, Part A Fundamentals and Techniques*, Journal of Chromatography Library, vol. 22A (ed. E. Heftmann), Elsevier Scientific, Amsterdam, pp. A225–A255.

40. Lopez Leiva, M.H. (1988) The use of electrodialysis in food processing part 1: some theoretical concepts. *Lebens. Wiss. Technol.*, **21**, 119–125.

41. Lopez Leiva, M.H. (1988) The use of electrodialysis in food processing part 2: review of practical applications. *Lebens. Wiss. Technol.*, **21**, 177–182.

42. Bazinet, L. (2005) Electrodialytic phenomena and their applications in the dairy industry: a review. *Crit. Rev. Food Sci. Nutr.*, **44**, 525–544.

11
Mixing, Emulsification, and Size Reduction

James G. Brennan

11.1
Mixing (Agitation, Blending)

11.1.1
Introduction

Mixing is a unit operation widely used in food processing. Many definitions of this term have been proposed. One of the simplest is: "an operation in which a uniform combination of two or more components is formed." In addition to blending components together, mixing operations may bring about other desirable changes in the materials being mixed, such as mechanical working (as in dough mixing), promotion of heat transfer (as in freezing ice cream), and facilitating chemical or biological reactions (as in fermentation). The components in a mixing operation may be liquids, pastes, dry solids, or gases. The degree of uniformity attainable in a mixing operation varies, depending on the nature of the components. In the case of low-viscosity miscible liquids or highly soluble solids in liquids a high degree of uniformity is attainable. Less intimate mixing is likely to occur in the case of viscous liquids, pastes, and dry solids. Combining immiscible materials together usually requires specialized equipment, which is covered under emulsification in Section 11.2. Efficient utilization of energy is another criterion of mixing. This is more easily attainable in the case of low-viscosity liquids as compared with pastes and dry solids.

11.1.2
Mixing of Low- and Moderate-Viscosity Liquids

The impeller mixer is the most commonly used type of mixer for low-viscosity liquids (viscosity less than 100 P; 10 N s m^{-2}). Such a mixer consists of one or more impellers, fixed to a rotating shaft and immersed in the liquid. As the impellers rotate, they create currents within the liquid, which travel throughout the mixing vessel. If turbulent conditions are created within the moving streams of liquid, mixing will occur. Turbulence is usually most vigorous near the impeller and the

Food Processing Handbook, Second Edition. Edited by James G. Brennan and Alistair S. Grandison.

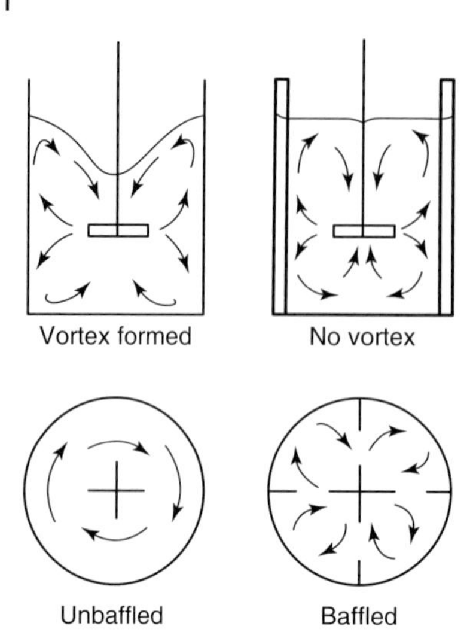

Figure 11.1 Flow patterns in baffled and unbaffled vessels with paddle or turbine agitators. From Ref. [1] with permission of the authors.

liquid should pass through this region as often as possible. The fluid velocity in the moving streams has three components: (i) a radial component acting in a direction at right angles to the shaft, (ii) a longitudinal component acting parallel to the shaft, and (iii) a rotational component acting in a direction tangential to the circle of rotation of the shaft. The radial and longitudinal components usually promote mixing but the rotational component may not.

If an impeller agitator is mounted on a vertical shaft located centrally in a mixing vessel, the liquid will flow in a circular path around the shaft. If laminar conditions prevail, then layers of liquid may form, the contents of the vessel rotate, and mixing will be inefficient. Under these conditions a vortex may form at the surface of the liquid. As the speed of rotation of the impeller increases this vortex deepens. When the vortex gets close to the impeller, the power imparted to the liquid drops and air is sucked into the liquid. This will greatly impair the mixing capability of the mixer. Rotational flow may cause any suspended particles in the liquid to separate out under the influence of centrifugal force. Rotational flow, and hence vortexing, may be reduced by locating the mixer off-center in the mixing vessel and/or by the use of baffles. Baffles usually consist of vertical strips fixed at right angles to the inner wall of the mixing vessel. These break up the rotational flow pattern and promote better mixing (Figure 11.1).

Usually four baffles are used, with widths corresponding to 1/18th (5.55%) to 1/12th (8.33%) of the vessel diameter.

Three main types of impeller mixers are used for liquid mixing: paddle mixers, turbine mixers, and propeller mixers.

11.1.2.1 **Paddle Mixer**

This type of mixer consists of a flat blade attached to a rotating shaft, which is usually located centrally in the mixing vessel (Figure 11.2). The speed of rotation is relatively low, in the range 20–150 rpm. The blade promotes rotational and radial

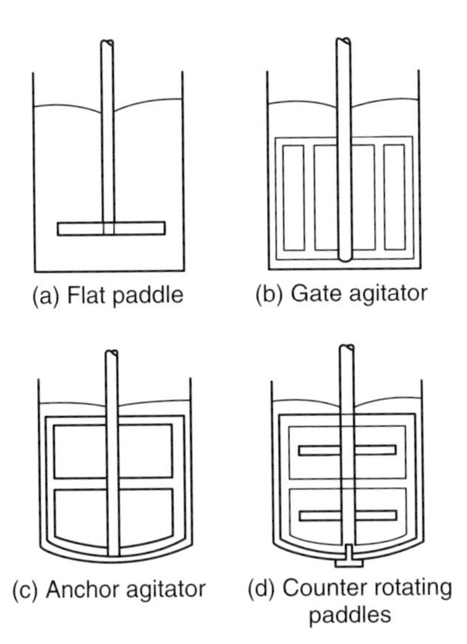

Figure 11.2 Some typical paddle impellers. From Ref. [1] with permission of the authors.

flow but very little vertical flow. It is usually necessary to fit baffles to the mixing vessel. Two or four blades may be fitted to the shaft.

Other forms of paddle mixer include: (i) the gate agitator (Figure 11.2b), which is used for more viscous liquids, (ii) the anchor agitator (Figure 11.2c), which rotates close to the wall of the vessel and helps to promote heat transfer and prevent fouling in jacketed vessels, and (iii) counter-rotating agitators (Figure 11.2d), which develop relatively high shear rates near the impeller. Simple paddle agitators are used mainly to mix miscible liquids and to dissolve soluble solids in liquids.

11.1.2.2 **Turbine Mixer**

A turbine mixer has four or more blades attached to the same shaft, which is usually located centrally in the mixing vessel. The blades are smaller than paddles and rotate at higher speeds, in the range 30–500 rpm. Simple vertical blades (Figure 11.3a) promote rotational and radial flow. Some vertical flow develops when the radial currents are deflected from the vessel walls (Figure 11.1).

Swirling and vortexing are minimized with the use of baffles. Liquid circulation is generally more vigorous than that produced by paddles and shear and turbulence is high near the impeller itself.

Pitched blades (Figure 11.3b) increase vertical flow. Curved blades (Figure 11.3c) are used when less shear is desirable, for example, when mixing friable solids. Vaned or shrouded disks (Figure 11.3d) control the suction and discharge pattern of the impeller and are often used when mixing gases into liquids. Turbine mixers are used for low and moderate viscosity liquids, up to 600 P, for preparing solutions and incorporating gases into liquids.

11.1.2.3 **Propeller Mixer**

This type of mixer consists of a relatively small impeller, similar in design to a marine propeller, which rotates at high speed, up to several thousand revolutions

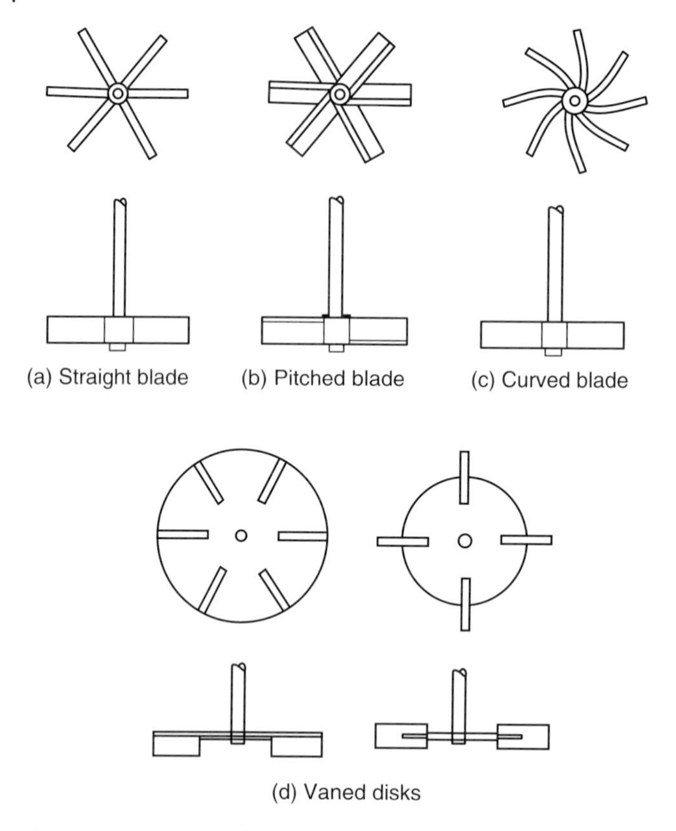

Figure 11.3 Some typical turbine impellers. From Ref. [1] with permission of the authors.

per minute. It develops strong longitudinal and rotational flow patterns. If mounted on a vertical shaft and located centrally in the mixing vessel, baffling is essential (Figure 11.4a,b).

Alternatively, the shaft may be located off center in the vessel and/or at an angle to the vertical (Figure 11.4c). When mixing low-viscosity liquids, up to 20 P, the currents developed by propeller agitators can travel throughout large vessels. In such cases the shaft may enter through the side wall of the tank (Figure 11.4d). Special propeller designs are available which promote shear, for emulsion premixing. Others have serrated edges for cutting through fibrous solids.

Many other types of impeller mixers are used for low-viscosity liquids, including disks and cones attached to shafts. These promote gentle mixing. More specialized mixing systems are available for emulsion premixing, dispersion of solids, and similar duties. One such system is shown in (Figure 11.5).

Other methods of mixing low-viscosity liquids include: (i) pumping them through pipes containing bends, baffles, and/or orifice plates, (ii) injecting one liquid into a tank containing the other components, and (iii) recirculating liquids through a holding tank using a centrifugal pump [1–11].

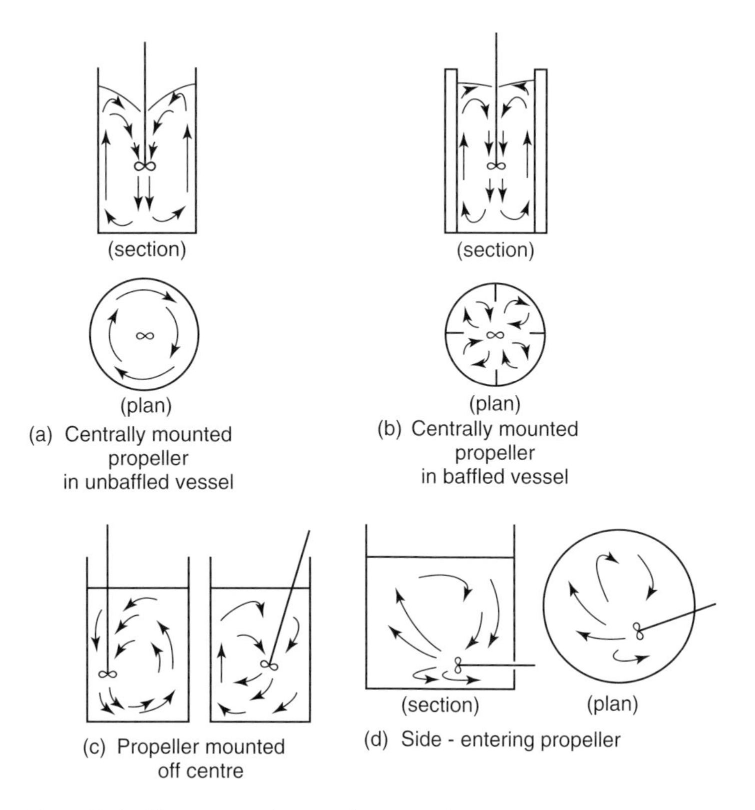

Figure 11.4 Flow patterns in propeller agitated systems. From Ref. [1] with permission of the authors.

11.1.3 Mixing of High-Viscosity Liquids, Pastes, and Plastic Solids

When mixing highly viscous and pastelike materials, it is not possible to create currents which will travel to all parts of the mixing vessel, as happens when mixing low-viscosity liquids. Consequently, there must be direct contact between the mixing elements and the material being mixed. The mixing elements must travel to all parts of the mixing vessel or the material must be brought to the mixing elements. Mixers for such viscous materials generally need to be more robust, have a smaller working capacity and have a higher power consumption than those used for liquid mixing. The speed of rotation of the mixing elements is relatively low and the mixing times long compared to those involved in mixing liquids.

11.1.3.1 Paddle Mixers

Some designs of paddle mixer, of heavy construction, may be used for mixing viscous materials. These include gate, anchor, and counter-rotating paddles, as shown in Figure 11.2a–c, respectively.

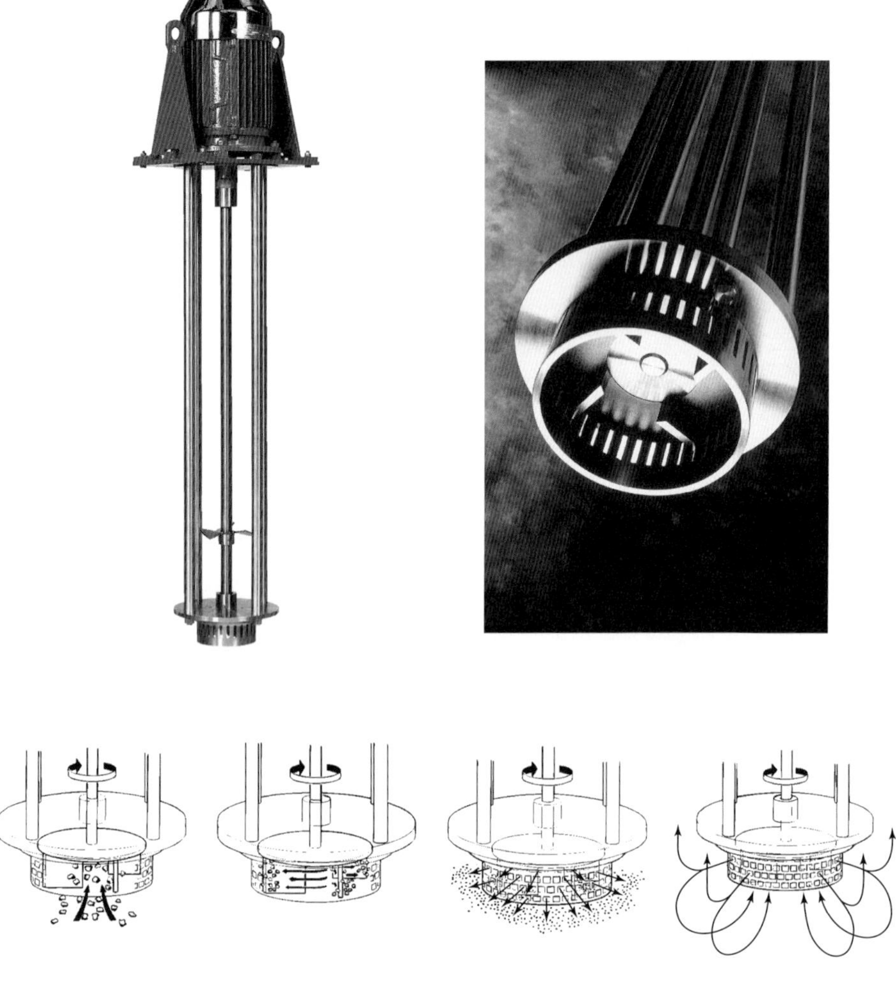

Figure 11.5 A high shear mixer: (a) the complete mixer, (b) the mixing head, and (c) high shear mixer in operation. By courtesy of Silverson Machines Ltd.

11.1.3.2 Pan (Bowl, Can) Mixers

In one type of pan mixer the bowl rests on a turntable, which rotates. One or more mixing elements are held in a rotating head and located near the bowl wall. They rotate in the opposite direction to the pan. As the pan rotates it brings the material into contact with the mixing elements. In another design of pan mixer, the bowl is stationary. The mixing elements rotate and move in a planetary pattern, thus repeatedly visiting all parts of the bowl. The mixing elements are shaped to pass with a small clearance between the wall and bottom of the mixing vessel (Figure 11.6). Various designs of mixing elements are used, including gates, forks, hooks, and whisks, for different applications.

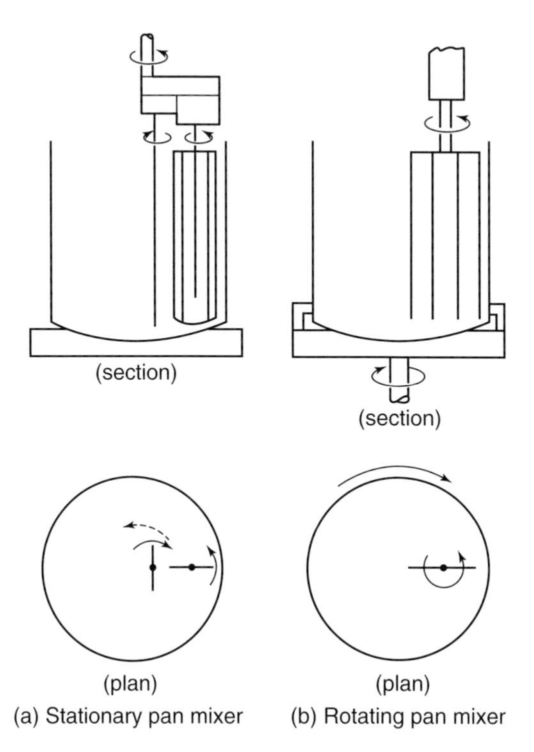

Figure 11.6 Pan mixers. From Ref. [1] with permission of the authors.

11.1.3.3 Kneaders (Dispersers, Masticators)

A common design of kneader consists of a horizontal trough with a saddleshaped bottom. Two heavy blades mounted on parallel, horizontal shafts rotate toward each other at the top of their cycle. The blades draw the mass of material down over the point of the saddle and then shear it between the blades and the wall and bottom of the trough. The blades may move tangentially to each other and often at different speeds, with a ratio of 1.5 : 1.0. In some such mixers the blades may overlap and turn at the same or different speeds. Mixing times are generally in the range 2–20 min. One type of kneader, featuring Z- or Σ-blade mixing elements is shown in Figure 11.7.

11.1.3.4 Continuous Mixers for Pastelike Materials

Many different devices are used to mix viscous materials on a continuous basis. Screw conveyors, rotating inside barrels, may force such materials through perforated plates or wire meshes. Passing them between rollers can effect mixing of pastelike materials. Colloid mills may be used for a similar purpose, see Section 11.2.3.7.

Figure 11.7 Z-blade (Σ-blade) mixer. By courtesy of Winkworth Machinery Ltd.

11.1.3.5 Static Inline Mixers

When viscous liquids are pumped over specially shaped stationary mixing elements located in pipes, mixing may occur. The liquids are split and made to flow in various different directions, depending on the design of the mixing elements. Many different configurations are available. The energy required to pump the materials through these mixing elements is usually less than that required to drive the more conventional types of mixers discussed above [1, 2, 6, 8–11, 13–15].

11.1.4 Mixing Dry, Particulate Solids

In most practical mixing operations involving dry particulate solids, unmixing or segregation is likely to occur. Unmixing occurs when particles within a group are free to change their positions. This results in a change in the packing characteristics of the solid particles. Unmixing occurs mainly when particles of different sizes are being mixed. The smaller particles can move through the gaps between the larger particles, leading to segregation. Differences in particle shape and density may also contribute to segregation. Materials with particle sizes of 75 μm and above are more prone to segregation than those made up of smaller particles. Small cohesive particles, which bind together under the influence of surface forces, do not readily segregate. In mixing operations where segregation occurs, an equilibrium between mixing and unmixing will be established after a certain mixing time. Further mixing is not likely to improve the uniformity of the mix.

There are two basic mechanisms involved in mixing particulate solids: *convection*, which involves the transfer of masses or groups of particles from one location in the mixer to another, and *diffusion*, which involves the transfer of individual particles from one location to another arising from the distribution of particles over a freshly developed surface. Usually both mechanisms contribute to any mixing operation. However, one mechanism may predominate in a particular type of mixer. Segregation is more likely to occur in mixers in which diffusion predominates.

11.1.4.1 Horizontal Screw and Ribbon Mixers

These consist of horizontal troughs, usually semi-cylindrical in shape, containing one or two rotating mixing elements. The elements may take the form of single or twin screw conveyors. Alternatively, mixing ribbons may be employed. Two such ribbons may be mounted on a single rotating shaft. They may be continuous or interrupted. The design is such that one ribbon tends to move the solids in one direction while the other moves them in the opposite direction. If the rate of movement of the particles is the same in both directions, the mixer is operated on a batch principle. If there is a net flow in one direction, the mixer may be operated continuously. The mixing vessel may be jacketed for temperature control. If enclosed, it may be operated under pressure or vacuum. Convection is the predominant mechanism of mixing in this type of mixer. Some segregation may occur but not to a serious extent.

11.1.4.2 Vertical Screw Mixers

These consist of tall, cylindrical, or cone-shaped vessels containing a single rotating screw, which elevates and circulates the particles. The screw may be located vertically at the center of the vessel. Alternatively, it may be set at an angle to the vertical and made to rotate, passing close to the wall of the vessel (Figure 11.8). The convective mechanism of mixing predominates in such mixers and segregation should not be a serious problem.

11.1.4.3 Tumbling Mixers

These consist of hollow vessels, which rotate about horizontal axes. They are partly filled, up to 50–60% of their volume, with the materials being mixed and then rotated, typically for 5–20 min, to bring about mixing of the contents. Because the main mechanism of mixing is diffusion, segregation can be a problem if the

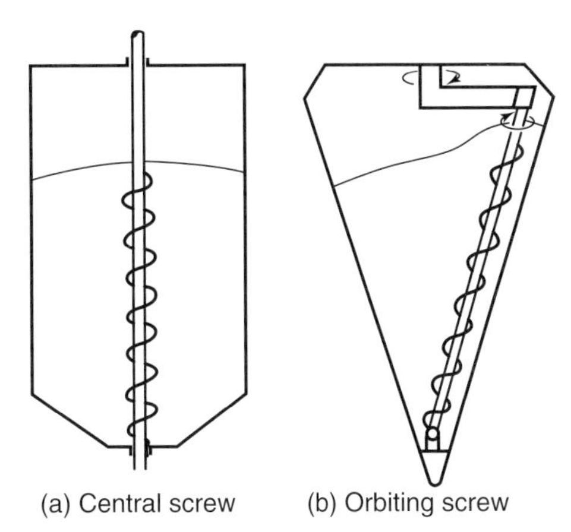

Figure 11.8 Vertical screw mixers. From Ref. [1] with permission of the authors.

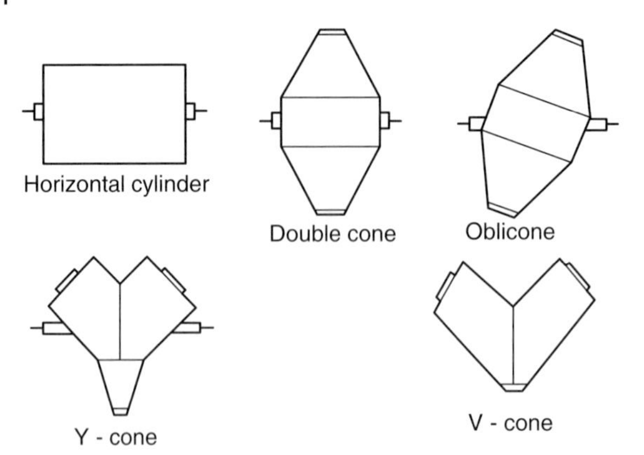

Figure 11.9 Some typical tumbling mixer shapes. From Ref. [1] with permission of the authors.

particles vary in size. Various designs are available, some of which are shown in Figure 11.9. They may contain baffles or stays to enhance the mixing effect and break up agglomerates of particles.

11.1.4.4 **Fluidized Bed Mixers**

Fluidized beds, similar to those used in dehydration (see Section 3.2.3.5), may be used for mixing particulate solids with similar sizes, shapes, and densities. The inclusion of spouting jets of air, in addition to the main supply of fluidizing air, can enhance the mixing [1–3, 6–8, 16–22].

11.1.5 Mixing of Gases and Liquids

Gases may be mixed with low-viscosity liquids using impeller agitators in mixing vessels. Turbine agitators, as shown in Figure 11.3, are generally used for this purpose. In unbaffled vessels, vortexing can draw gas into the liquid. However, the gas may not be well distributed throughout the liquid. It is more usual to use baffled vessels with relatively high-speed turbine impellers. Impellers with 6, 12, and even 18 blades have been used. Pitched blades and vaned disks (Figure 11.3b,d respectively) are particularly suited to this duty. Some special designs of vaned disks, featuring concave rather than flat blades, have been used for this purpose. Gas may be introduced into liquids using some designs of static inline mixers.

More heavy-duty equipment, such as pan and Z-blade mixers, may be used to introduce gas into more viscous materials. For example, whisk-like elements in pan mixers may be used to whip creams. Other types of elements, such as forks, hooks, and gates, are used to introduce air into doughs and batters. Dynamic, inline systems, such as scraped surface heat exchangers (see Section 2.4.3.2) may be used to heat or cool viscous materials while at the same time introducing gas to aerate them, such as in the manufacture of ice cream (see Section 11.2.4.2) [23, 24].

11.1.6 Sonic Mixer

This device was initially designed as a marine propulsion system but has been applied to mixing of food liquids and solid particles. It consists of a metal cylinder, approximately 60 cm long with an internal diameter ranging from 13 to 47 mm. Steam is injected into the cylinder, creating a vacuum which draws the ingredients into the cylinder. As the steam contracts, energy is transferred to the product and a low-density shock wave develops across the bore of the cylinder. This brings about very rapid and efficient mixing of the ingredients, which may be two or more liquids or solid particle and liquids. Sonic mixing has been used in the production of sauces, baby foods, and soft drinks. It can also be used for emulsifying, pumping, and heating food materials [25, 26].

11.1.7 Applications for Mixing in Food Processing

11.1.7.1 Low-Viscosity Liquids

Examples of applications for impeller mixers include: preparing brines and syrups, preparing liquid sugar mixtures for sweet manufacture, making up fruit squashes, blending oils in the manufacture of margarines and spreads, premixing emulsion ingredients.

11.1.7.2 Viscous Materials

Examples of applications for pan mixers and kneaders include: dough and batter mixing in bread, cake, and biscuit making, blending of butters, margarines, and cooking fats, preparation of processed cheeses and cheese spreads, manufacture of meat and fish pastes.

11.1.7.3 Particulate Solids

Examples of applications for screw, ribbon, and tumbling mixers include: preparing cake and soup mixes, blending of grains prior to milling, blending of flours, and incorporation of additives into them.

11.1.7.4 Gases into Liquids

Examples of applications of this include: carbonation of alcoholic and soft drinks, whipping of dairy and artificial creams, aerating ice cream mix during freezing, and supplying gas to fermenters.

11.2
Emulsification

11.2.1
Introduction

Emulsification may be regarded as a mixing operation whereby two or more normally immiscible materials are intimately mixed. Most food emulsions consist of two phases: (i) an aqueous phase consisting of water which may contain salts and sugars in solution and/or other organic and colloidal substances, known as *hydrophilic materials* and (ii) an oil phase which may consist of oils, fats, hydrocarbons, waxes, and resins, so-called *hydrophobic materials.* Emulsification is achieved by dispersing one of the phases in the form of droplets or globules throughout the second phase. The material which is broken up in this way is known as the *dispersed, discontinuous,* or *internal phase,* while the other is referred to as the *dispersing, continuous,* or *external phase.* In addition to the two main phases other substances, known as *emulsifying agents,* are usually included, in small quantities, to produce stable emulsions.

When the water and oil phases are combined by emulsification, two emulsion structures are possible. The oil phase may become dispersed throughout the aqueous phase to produce an oil-in-water (o/w) emulsion. Alternatively, the aqueous phase may become dispersed throughout the oil phase to produce a water-in-oil (w/o) emulsion. In general, an emulsion tends to exhibit the characteristics of the external phase. Two emulsions of similar composition can have different properties, depending on their structure. For example, an o/w emulsion can be diluted with water, colored with water-soluble dye, and has a relatively high electrical conductivity compared to a w/o system. The latter is best diluted with oil and colored with oil-soluble dye. Many factors can influence the type of emulsion formed when two phases are mixed, including the type of emulsifying agent used, the relative proportions of the phases and the method of preparation employed.

Emulsions with more complex structures, known as *multiple emulsions,* may also be produced. These may have oil-in-water-in-oil (o/w/o) or water-in-oil-in-water (w/o/w) structures. The latter structure has the most applications. w/o/w emulsions may be produced in two stages. First a w/o emulsion is produced by homogenization using a hydrophobic agent. This is then incorporated into an aqueous phase using a hydrophilic agent. In such a multiple emulsion, the oil layer is thin and water can permeate through it due to an osmotic pressure gradient between the two aqueous phases. This property is used in certain applications including prolonged drug delivery systems, drug overdose treatments, and nutrient administration for special dietary purposes [27–29].

Emulsions with very small internal phase droplets, in the range 0.0015–0.15 μm in diameter, are termed *microemulsions.* They are clear in appearance, the droplets have a very large contact area and good penetration properties. They have found uses in the application of herbicides and pesticides and the application of drugs both orally and by intravenous injection [29].

In a two-phase system free energy exists at the interface between the two immiscible liquids, due to an imbalance in the cohesive forces of the two materials. Because of this *interfacial tension*, there is a tendency for the interface to contract. Thus, if a crude emulsion is formed by mixing two immiscible liquids, the internal phase will take the form of spherical droplets, representing the smallest surface area per unit volume. If the mixing is stopped, the droplets coalesce to form larger ones and eventually the two phases will completely separate. To form an emulsion, this interfacial tension has to be overcome. Thus, energy must be introduced into the system. This is normally achieved by agitating the liquids. The greater the interfacial tension between the two liquids the more energy is required to disperse the internal phase. Reducing this interfacial tension will facilitate the formation of an emulsion. This is one important role of emulsifying agents. Another function of these agents is to form a protective coating around the droplets of the internal phase and thus prevent them from coalescing and destabilizing the emulsion [1, 2, 29–32].

11.2.2 Emulsifying Agents

Substances with good emulsifying properties have molecular structures which contain both polar and nonpolar groups. Polar groups have an affinity for water, that is, are hydrophilic, while nonpolar groups are hydrophobic. If there is a small imbalance between the polar and nonpolar groups in the molecules of a substance, it will be adsorbed at the interface between the phases of an emulsion.

These molecules will become aligned at the interface so that the polar groups will point toward the water phase while the nonpolar groups will point toward the oil phase. By bridging the interface in this way, they reduce the interfacial tension between the phases and form a film around the droplets of the internal phase, thus stabilizing them. A substance in which the polar group is stronger will be more soluble in water than one in which the nonpolar group predominates. Such a substance will tend to promote the formation of an o/w emulsion. However, if the nonpolar group is dominant the substance will tend to produce a w/o emulsion. If the polar and nonpolar groups are grossly out of balance the substance will be highly soluble in one or other of the phases and will not accumulate at the interface. Such materials do not make good emulsifying agents.

The emulsifying capability of an agent may be classified according to the hydrophile–lipophile balance (HLB) in its molecules. This is defined as the ratio of the weight percentage of hydrophilic groups to the weight percentage of hydrophobic groups in the molecule. HLB values for emulsifying agents range from 1 to 20. Agents with low values, 3–6, promote the formation of w/o emulsions, while those with high values, 8–16, favor the formation of o/w types. Agents with even higher HLB values are used in detergents and solubilizers.

Protein, phospholipids, and sterols, which occur naturally in foods, act as emulsifying agents. A wide range of manufactured agents are available, both ionic and nonionic. Ionic emulsifiers often react with other oppositely charged

Table 11.1 Examples of emulsifying agents and their HLB values.

Ionic	HLB	Nonionic	HLB
Proteins (e.g., gelatin, egg albumin)	–	Glycerol esters	2.8
Phospholipids (e.g., lecithin)	–	Polyglycerol esters	–
Potassium and sodium salts of oleic acid	18.0–20.0	Propylene glycol fatty acid esters	3.4
Sodium stearoyl-2-lactate	–	Sorbitol fatty acid esters	4.7
		Polyoxyethylene fatty acid esters	14.9–15.9
Hydrocolloids			
Agar	–	Carboxymethyl cellulose	–
Pectin	–	Hydroxypropyl cellulose	–
Gum tragacanth	11.9	Methyl cellulose	10.5
Alginates	–	Guar gum	–
Carrageen	–	Locust bean gum	–

particles, such as hydrogen and metallic ions, to form complexes. This may result in a reduction in solubility and emulsifying capacity. Nonionic emulsifiers are more widely used in food emulsions. Hydrocolloids such as pectin and gums are often used to increase the viscosity of emulsions to reduce separation of the phases under gravity (creaming). Table 11.1 lists some commonly used emulsifying agents, together with their HLB values. The treatment of the principles of emulsification and the role of emulsifying agents given above is a very simplified one. More detailed accounts will be found in the references cited [1, 33–36].

11.2.3
Emulsifying Equipment

The general principle on which all emulsifying equipment is based is to introduce energy into the system by subjecting the phases to vigorous agitation. (An exception to this principle is membrane emulsification, see Section 11.2.3.5.) The type of agitation which is most effective in this context is that which subjects the large droplets of the internal phase to shear. In this way, these droplets are deformed from their stable spherical shapes and break up into smaller units. If the conditions are suitable and the right type and quantity of emulsifying agent(s) is present, a stable emulsion will be formed. One important condition which influences emulsion formation is temperature. Interfacial tension and viscosity are temperature-dependent, both decreasing with increase in temperature.

Thus, raising the temperature of the liquids usually facilitates emulsion formation. However, for any system there will be an upper limit of temperature depending on the heat sensitivity of the components.

11.2.3.1 Mixers

In the case of low-viscosity liquids, turbine and propeller mixers may be used to premix the phases prior to emulsification. In some simple systems, a stable emulsion may result from such mixing and no further treatment may be required. Special impeller designs are available which promote emulsification, such as the Silverson system shown in Figure 11.5. In the case of viscous liquids and pastes, pan mixers, kneaders, and some types of continuous mixers (see Section 11.1.3) may be used to disperse one phase throughout another. Tumbling mixers, such as those used for mixing powders (see Section 11.1.4.3) may also be used for this purpose.

11.2.3.2 Pressure Homogenizers

The principle of operation of all pressure homogenizers is that the premixed phases are pumped through a narrow opening at high velocity. The opening is usually provided between a valve and its seat. Therefore, a pressure homogenizer consists of one or two valves and a high pressure pump. As the liquids pass through the gap, 15–300 μm wide, between the valve and seat they are accelerated to speeds of 50–300 $m\,s^{-1}$. The droplets of the internal phase shear against each other, are distorted and break up into smaller units. As the liquids exit from the gap, there is a sudden drop in pressure. Some cavitation may occur. In many valve designs, the droplets impinge on a hard surface (breaker ring) set at 90° to the direction of flow of the liquids after they emerge from the gap. All of these mechanisms stress the droplets and contribute to their disruption.

Droplets diameters of 0.1–0.2 μm are attainable in pressure homogenizers. There are many different designs of valve available. Three examples are shown in Figure 11.10. As the liquids travels between the valve and its seat, the valve lifts against a heavy duty spring or torsion bar. By adjusting the tension on this

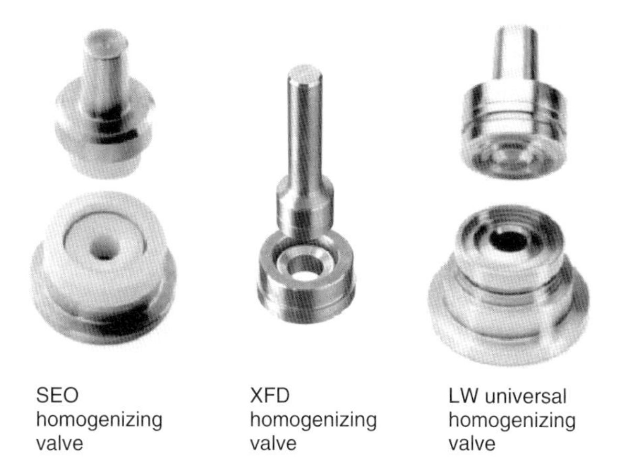

Figure 11.10 Three different types of homogenising valve, by courtesy of APV-Rannie & Gaulin Homogenisers.

spring or bar, the homogenizing pressure may be set. This may range from 3.5 to 70.0 MN m^{-2}.

The literature suggests that there is an approximately inverse linear relationship between the logarithm of the homogenizing pressure and the logarithm of the droplet diameter produced by a pressure homogenizer. Homogenizer valves are usually made of stainless steel or alloys such as stellite. More erosion-resistant materials such as tungsten carbide may be used, but not usually for food applications. It is important that a good fit is maintained between the valves and their seats. Even small amounts of damage to the surfaces can lead to poor performance. Consequently, they should be examined regularly and reground or replaced when necessary.

Valves made of compressed wire are also available. The liquids are pumped through the myriad of channels in the body of these valves. Such valves are difficult to clean and are discarded at the end of a day's run. They are known as *single-service valves.*

One passage through a homogenizing valve may not produce a well-dispersed emulsion. The small droplets of the internal phase may cluster together. These can be dispersed by passing them through a second valve. This is known as *two-stage homogenization.* The first valve is set at a high pressure, 14–70 MPa, the second at a relatively low pressure, 2.5–7.0 MPa. A representation of a two-stage homogenizing system is shown in Figure 11.11.

The liquids are pumped through the homogenizing valve(s) by means of a positive displacement pump, usually of the piston and cylinder type. In order to achieve a reasonably uniform flow rate, three, five, or seven cylinders with pistons are employed, operating consecutively and driven via a crankshaft. The mixture is discharged from the chamber of each piston into a high-pressure manifold and exits from there via the homogenizing valve. A pressure gauge is fitted to the manifold to monitor the homogenizing pressure.

11.2.3.3 **Hydroshear Homogenizers**

In this type of homogenizer, the premixed liquids are pumped into a cylindrical chamber at relatively low pressure, up to 2000 kPa. They enter the chamber through a tangential port at its center and exit via two cone-shaped discharge nozzles at the ends of the chamber. The liquids accelerate to a high velocity as they enter the chamber, spread out to cover the full width of the chamber wall and flow toward the center, rotating in ever decreasing circles. High shear develops between the adjacent layers of liquid, destabilizing the large droplets of the internal phase. In the center of the cylinder a zone of low pressure develops and cavitation, ultrahigh-frequency vibration and shock waves occur which all contribute to the break up of the droplets and the formation of an emulsion. Droplets sizes in the range 2–8 μm are produced by this equipment.

11.2.3.4 **Microfluidizers**

This type of homogenizer is capable of producing emulsions with very small droplet sizes directly from individual aqueous and oil phases. Separate streams of

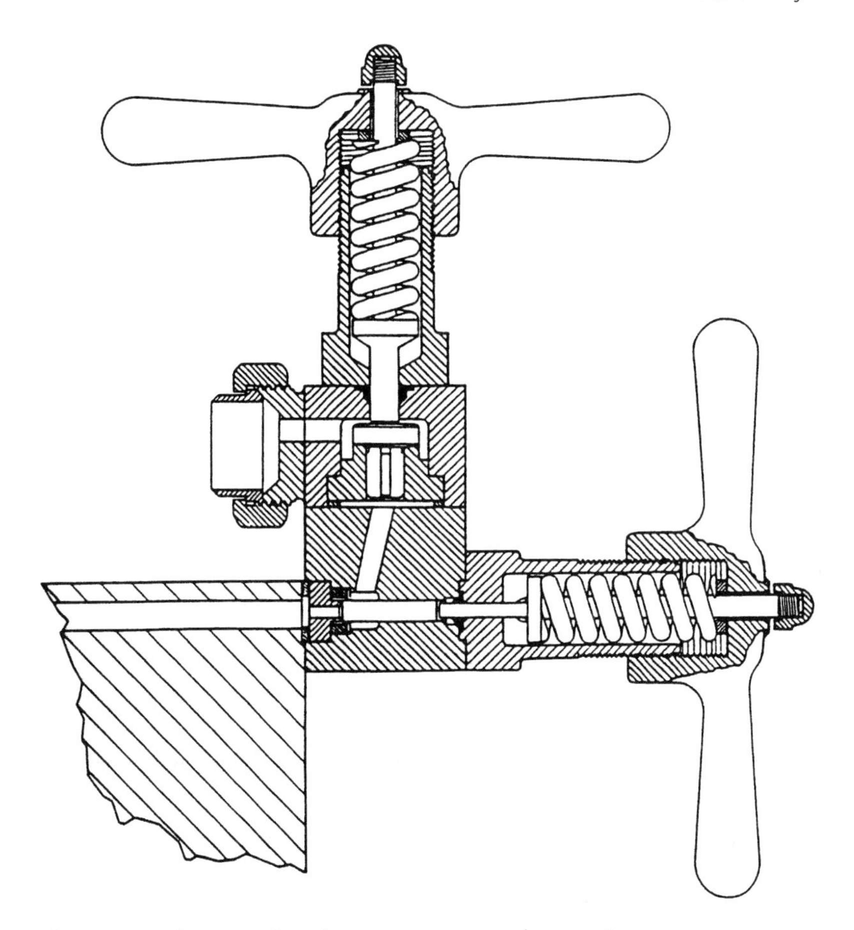

Figure 11.11 Representation of a two-stage pressure homogenizer.

the aqueous and oil phases are pumped into a chamber under high pressure, up to 110 MPa. The liquids are accelerated to high velocity, impinge on a hard surface and interact with each other. Intense shear and turbulence develop which lead to a break up of the droplets of the internal phase and the formation of an emulsion. Very small emulsion droplets can be produced by recirculating the emulsion a number of times through the microfluidizer.

11.2.3.5 **Membrane Homogenizers**

If the internal phase liquid is forced, under low pressure, to flow through pores in a glass membrane into the external phase liquid an emulsion can be formed.

The droplet size of the dispersed phase is mainly dependent on the size of the pores in the membrane and not violent agitation as in other types of homogenizers.

Glass membranes can be manufactured with pores of different diameters to produce emulsions with different droplet sizes, in the range 0.5–10 μm. Such membranes can produce o/w, w/o, and multiple emulsions with very narrow droplet size distributions. Many other materials have been used experimentally as membranes, including silicon and silicon nitride microsieves, ceramic aluminum oxide, and microporous polypropylene hollow fibers. In a batch version of this equipment the internal phase liquid is forced through a cylindrical membrane partly immersed in a vessel containing the external phase.

In a continuous version, a cylindrical membrane through which the external phase flows is located within a tube, through which the internal phase flows.

The internal phase is put under pressure, forcing it through the membrane wall into the external phase. To date, membrane homogenizers are used mainly for the production of emulsions on a laboratory scale.

11.2.3.6 **Ultrasonic Homogenizers**

When a liquid is subjected to ultrasonic irradiation, alternate cycles of compression and tension develop. This can cause cavitation in any gas bubbles present in the liquid, resulting in the release of energy. This energy can be used to disperse one liquid phase in another to produce an emulsion. For laboratory-scale applications, piezoelectric crystal oscillators may be used. An ultrasonic transducer consists of a piezoelectric crystal encased in a metal tube. When a high-intensity electrical wave is applied to such a transducer, the crystal oscillates and generates ultrasonic waves. If a transducer of this type is partly immersed in a vessel containing two liquid phases, together with an appropriate emulsifying agent(s), one phase may be dispersed in the other to produce an emulsion.

For the continuous production of emulsions on an industrial scale, mechanical ultrasonic generators are used. The principle of a wedge resonator (liquid whistle) is shown in Figure 11.12. A blade with wedge-shaped edges is clamped at one or more nodal points and positioned in front of a nozzle through which the premixed emulsion is pumped. The jet of liquid emerging from the nozzle impinges on the leading edge of the blade, causing it to vibrate at its natural frequency, usually in the range 18–30 kHz. This generates ultrasonic waves in the liquid which cause one phase to become dispersed in the other and the formation of an emulsion.

The pumping pressure required is relatively low, usually in the range 350–1500 kPa, and droplet diameters of the order of 1–2 μm are produced.

11.2.3.7 **Colloid Mills**

In a colloid mill, the premixed emulsion ingredients pass through a narrow gap between a stationary surface (stator) and a rotating surface (rotor). In doing so the liquid is subjected to shear and turbulence which brings about further disruption of

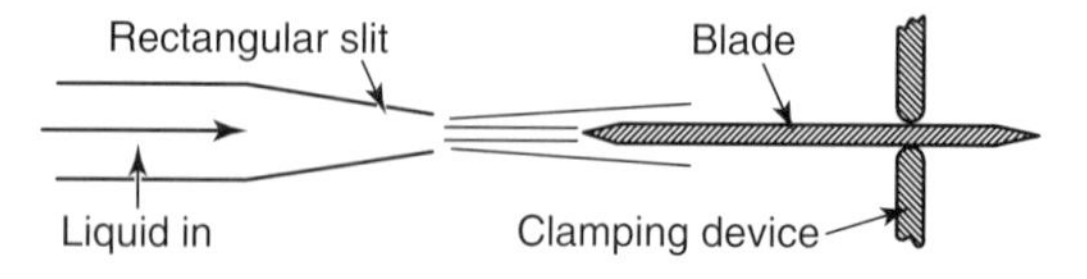

Figure 11.12 Principle of the mechanical (wedge resonator) ultrasonic generator. From Ref. [1] with permission of the authors.

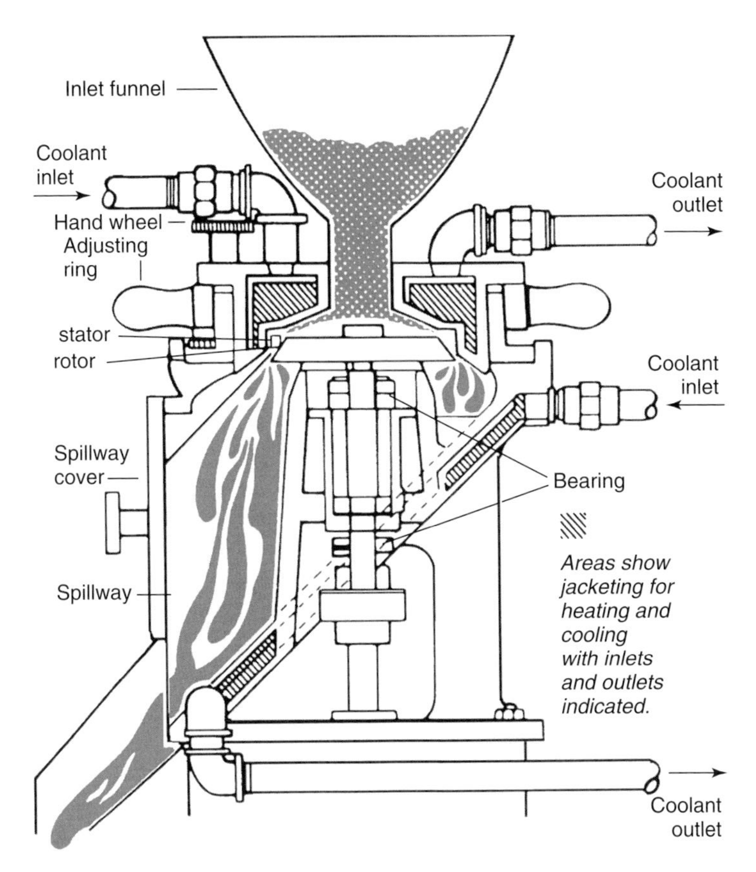

Figure 11.13 Top-feed "paste" colloid mill. By courtesy of Premier Mill Operation, SPX Process Equipment.

the droplets of the internal phase and disperses them throughout the external phase. The gap between the stator and rotor is adjustable within the range 50–150 μm. One type of colloid mill is depicted in Figure 11.13. The rotor turns on a vertical axis in close proximity to the stator. The clearance between them is altered by raising or lowering the stator by means of the adjusting ring. Rotor speed ranges from 3000 rpm for a rotor 25 cm in diameter to 10 000 rpm for a smaller rotor 5 cm in diameter. Rotors and stators usually have smooth, stainless steel surfaces.

Carborundum surfaces are used when milling fibrous materials. Colloid mills are usually jacketed for temperature control. This type of mill, also known as a *paste mill*, is suitable for emulsifying viscous materials.

For lower viscosity materials the rotor is mounted on a horizontal axis and turns at higher speeds, up to 15 000 rpm. Mills fitted with rotors and stators with matching corrugated surfaces are also available. The clearance between the surfaces decreases outwardly from the center. The product may be discharged under pressure, up to 700 kPa. Incorporation of air into the product is limited and foaming problems reduced in this type of mill [1, 2, 10, 31, 32, 37–39].

11.2.4
Examples of Emulsification in Food Processing

11.2.4.1 Oil-in-water Emulsions

Oil-in-water emulsions of importance in food processing include milk, ice cream, cream liqueurs, coffee whiteners. salad dressings, meat products, and cake products.

Milk This naturally occurring emulsion typically consists of 3.0–4.5% fat dispersed in the form of droplets throughout an aqueous phase which contains sugars and mineral salts in solution and proteins in colloidal suspension. The fat droplets or globules, range in size from less than 1 μm to more than 20 μm in diameter.

These are stabilized by a complex, multilayer coating made up of phospholipids, proteins, enzymes, vitamins, and mineral salts, known as the *milk fat globule membrane*. Under the influence of gravity, these globules tend to rise to the surface to form a cream layer when milk is standing in vats or bottles. To prevent such separation milk may be subjected to two-stage homogenization, reducing the fat globule size to not more than 2 μm in diameter. Pasteurized, homogenized milk is a widely used liquid milk product. Milk and cream which are to be UHT-treated are also homogenized to improve their stability. So also are evaporated milk and cream which are to be heat-sterilized in containers.

Ice cream mix This is an o/w emulsion typically containing 10–12% fat dispersed in an aqueous phase containing sugars and organic salts in solution and proteins and some organic salts in colloidal suspension. The stability of the emulsion is important as it has to withstand the rigors of freezing and the incorporation of air to achieve an appropriate overrun. In addition to fat, which may be milk fat, vegetable oil, or a combination of the two, ice cream mix contains about 10.5% milk solids/nonfat, usually in the form of skim milk powder, and 13% sucrose, dextrose, or invert sugar. The milk solids/nonfat acts as an emulsifying agent but additional agents such as esters of mono- and diglycerides are usually included. Stabilizers such as alginates, carrageenan, gums, and gelatin are also added. These increase the viscosity of the mix and also have an influence on the proportion of the aqueous phase which crystallizes on freezing and the growth of the ice crystals. This in turn affects the texture and melting characteristics of the frozen product. The mix is pasteurized, usually subjected to two-stage pressure homogenization and aged at 2–5 °C prior to freezing in scraped surface freezers. Some 50% or less of the aqueous phase freezes at this stage and air is incorporated to give an overrun of 60–100%. The product may then be packaged and hardened at temperatures of −20 to −40 °C.

Cream liqueurs These are further examples of o/w dairy emulsions. They need to have long-term stability in an alcoholic environment. Soluble sodium caseinate can be used to stabilize the finely dispersed emulsion.

Coffee/tea whiteners These substitutes for cream or milk are also o/w emulsions typically containing vegetable oil, sodium caseinate, corn syrup, high HLB emulsifying agents, and potassium phosphate. They are usually prepared by pressure homogenization and are available in UHT-treated liquid form or in spray dried powder form.

Salad dressings Many "French" dressings consist of mixtures of vinegar, oil, and various dry ingredients. They are not emulsified as such and the liquid phase separates after mixing. They need to be mixed or shaken thoroughly before use. Other dressings are true o/w emulsions. Salad cream, for example, contains typically 30–40% oil, sugar, salt, egg (either yolk or whole egg in liquid or dried form), mustard, herbs, spices, coloring, and stabilizer(s). The cream is acidified with vinegar and/or lemon juice. The lecithin present in the egg usually is the main emulsifying agent, but some additional o/w agents may be added. Gum tragacanth is the stabilizer most commonly used. This increases the viscosity of the emulsion. Being thixotropic, it thins on shaking and facilitates dispensing of the cream. The gum is dispersed in part of the vinegar and water, allowed to stand for up to 4 days until it is fully hydrated and then beaten and sieved. The rest of the aqueous ingredients are premixed, heated to about 80 °C, cooled to about 40 °C and sieved. An emulsion premix is prepared by adding the oil gradually to the aqueous phase with agitation. This premix is then further emulsified by means of a pressure homogenizer, colloid mill, or ultrasonic device. It is then vacuum-filled into jars or tubes. Mayonnaise is also an o/w emulsion with similar ingredients to salad cream but containing 70–85% oil. The high oil content imparts a high viscosity to the product and stabilizer(s) are usually not required. The premixing is usually carried out at relatively low temperature, 15–20 °C, and a colloid mill used to refine the emulsion.

Meat products Emulsification of the fat is important in the production of many meat products such as sausages, pastes, and pates. Efficient emulsification can prevent fat separation, influence the texture of the product, and its behavior on cooking. Meat emulsions are relatively complex systems. They are usually classed as o/w emulsions but differ in many ways from those discussed above. They are two-phase systems consisting of fairly coarse dispersions of solid fat in an aqueous phase containing gelatin, other proteins, minerals, and vitamins in solution or colloidal suspension and insoluble matter, including meat fiber, filling materials, and seasonings.

The emulsifying agents are soluble proteins. Emulsification is brought about simultaneously with size reduction of the insoluble matter in a variety of equipment typified by the mincer and bowl chopper. In the mincer the material is forced by means of a worm through a perforated plate with knives rotating in contact with its surface. It is assumed that some shearing occurs which contributes to the emulsification of the fat. The bowl chopper consists of a hemispherical bowl which rotates slowly about a vertical axis. Curved knives rotate rapidly on a horizontal axis within the bowl. As the bowl rotates is brings the contents into contact with the rotating knives which simultaneously reduce the size of the solid particles and mix the ingredients. Soluble protein is released and emulsification of the fat takes place.

Cake products *Cake* has been defined as a protein foam stabilized by gelatinized starch and containing fat, sugar, salt, emulsifiers, and flavoring materials. It is aerated mainly by gases evolved by chemical reaction involving raising agents. The fat comes from milk, eggs, chocolate, and/or added shortenings. The protein comes in the flour, eggs, and milk. Emulsifying agents are available in the milk, eggs, and flour. Additional agents may be added separately and/or included in the shortenings. Cake batters have an o/w structure. Efficient emulsification of the fat in such batters is essential. Free liquid fat adversely affects the stability of the foam formed on beating and aeration and, consequently, is detrimental to the crumb structure, volume, and shape of the baked product. Simultaneous mixing and emulsification is attained in various types of mixers such as pan mixers, operating at relatively high speed, and some continuous mixers. Colloid mills and even pressure homogenizers have been used to ensure good emulsification in low-viscosity batters.

11.2.4.2 Water-in-oil Emulsions

Water-in-oil emulsions of importance in food processing include butter, margarines, and spreads.

Butter This is usually described as a w/o emulsion but, in fact, it has quite a complex structure. The continuous phase of free fat in liquid form contains fat crystals, globular fat, curd granules, crystals of salt, water droplets, and gas bubbles. The water droplets remain dispersed due to the semi-solid nature of the continuous phase rather than being stabilized by a layer of emulsifying agent.

Pasteurized milk is separated by centrifugation to give cream containing 30–40% fat. This cream is aged by holding at a low temperature for several hours.

The purpose of aging is to achieve the optimum liquid : solid fat ratio in the butter. Butter may also be made from cultured cream which has been inoculated with lactic acid-producing microorganisms and ripened, typically for 20 h at 14 °C, to develop the flavor. In batch churning, the aged cream is tumbled inside a hollow vessel known as a *churn*. Air is incorporated into the cream and the fat globules concentrate in the surface of the air bubbles. As these break and reform, some of the fat globules break open releasing free fat. The remaining globules form clumps, known as *butter grains,* held together by some of the free fat. As churning proceeds these grains grow in size. When they reach an optimum size, about 1 cm in diameter, churning is stopped, the aqueous phase, known as *buttermilk,* is drained off and the grains are washed free of curd with chilled water. The moisture content is adjusted, salt is then added if required and the mass of butter grains is tumbled in the churn for a further period. During this stage of the process, known as *working*, more fat globules break open, releasing more free fat, water droplets, and salt crystals are dispersed throughout the bulk of the fat phase and the texture

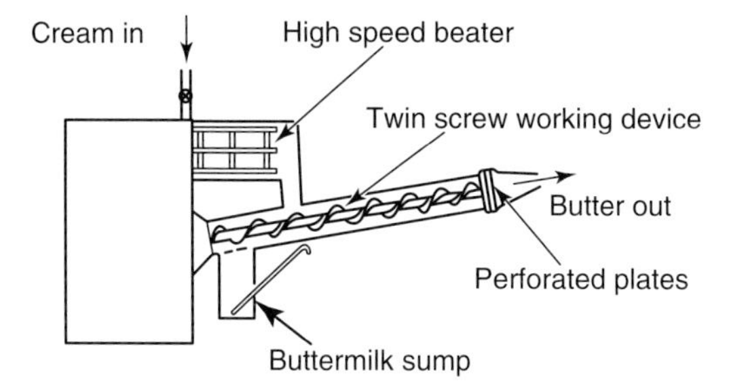

Figure 11.14 General principle of accelerated a churning and working device. From Ref. [1] with permission of the authors.

of the product develops. Working may be carried out under a partial vacuum to reduce the air content of the butter. When working is complete, about 40% of the fat remains in globular form. The butter is discharged from the churn, packaged, and transferred to chilled or frozen storage.

Most continuous methods of buttermaking work on the principles represented in Figure 11.14. Aged cream containing 30–40% fat is metered into the churning section of the equipment where it is acted on by high-speed beaters.

These bring about the rupture of some of the fat globules and the formation of small butter grains. As the grains and buttermilk exit the churning section, the buttermilk drains into a sump and the grains are carried by twin screws up the sloping barrel and extruded through a series of perforated plates. The action of the screws combined with the perforated plates brings about further disruption of fat globules and disperses the remaining water as droplets throughout the fatty phase. The butter is extruded in the form of a continuous ribbon from the working section. Salt may be introduced into the working section, if required.

Other continuous buttermaking methods involve concentrating the fat in the cream to about 80% by a second centrifugation step. This unstable, concentrated, o/w emulsion is then converted into a semi-solid, w/o system by simultaneous agitation and cooling in a variety of equipment.

Margarine and spreads These w/o products are made from a blend of fats and oils together with cultured milk, emulsifying agents, salt, flavoring compounds, and other additives.

The blend of fats and oils is selected according to the texture required in the final product. The emulsifying agents used have relatively low HLB values. A typical combination is a mixture of mono- and diglycerides and lecithin. The flavor, originally derived from the cultured milk, is now usually supplemented by the addition of flavoring materials such as acetyl methyl carbinols or aliphatic lactones.

In a typical manufacturing process, the fats and oils are measured into balance tanks. Other ingredients are added and an emulsion premix formed by high-speed agitation. This premix is then pumped through a series of refrigerated, scraped surface heat exchangers where it is simultaneously emulsified and cooled.

A three-dimensional network of long, thin, fat crystals is formed. It finally passes through a working/holding device, similar to the working section in Figure 11.14, where the final texture develops.

Margarine contains 15% water, similar to butter. Low-calorie spreads are also made from a blend of oils and contain up to 50% water. They may be based on milk fat or combinations of milk fat and vegetable oils [1, 12, 40–48].

11.3 Size Reduction (Crushing, Comminution, Grinding, Milling) of Solids

11.3.1 Introduction

Size reduction of solids involves creating smaller mass units from larger mass units of the same material. To bring this about, the larger mass units need to be subjected to stress by the application of force. Three types of force may be applied: compression, impact, and shear. Compressive forces are generally used for the coarse crushing of hard materials. Careful application of compressive forces enables control to be exercised over the breakdown of the material, for example, to crack open grains of wheat to facilitate separation of the endosperm from the bran (see Section 11.3.3.1). Impact forces are used to mill a wide variety of materials, including fibrous foods. Shear forces are best applied to relatively soft materials, again including fibrous foods. All three types of force are generated in most types of mill, but generally one predominates. For example, in most roller mills compression is the dominant force, impact forces feature strongly in hammer mills and shear forces are dominant in disk attrition mills (see Section 11.3.2).

The extent of the breakdown of a material may be expressed by the *reduction ratio*, which is the average size of the feed particles divided by the average size of the products particles. In this context, the term "average size" depends on the method of measurement. In the food industry, screening or sieving is widely used to determine particle size distribution in granular materials and powders. In this case, the average diameter of the particles is related to the aperture sizes of the screens used. Size reduction ratios vary from below 8 : 1 in coarse crushing to more than 100 : 1 in fine grinding.

The objective in many size reduction operations is to produce particles within a specified size range. Consequently, it is common practice to classify the particles coming from a mill into different size ranges. Again, screening is the technique most widely used for this purpose. To achieve a specified reduction ratio, it may be necessary to carry out the size reduction in a number of stages. A different type of mill may be used in each stage and screens employed between stages. An example of a multistage operation is depicted in Figure 11.15.

When a solid material is subjected to a force, its behavior may be represented by a plot of stress versus strain, as shown in Figure 11.16. Some materials exhibit elastic deformation when the force is first applied. The strain is linearly related to

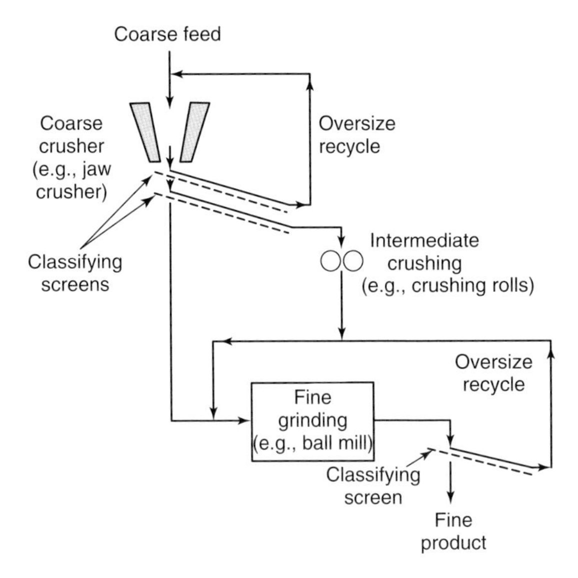

Figure 11.15 A typical size reduction flow sheet. From Ref. [1] with permission of the authors.

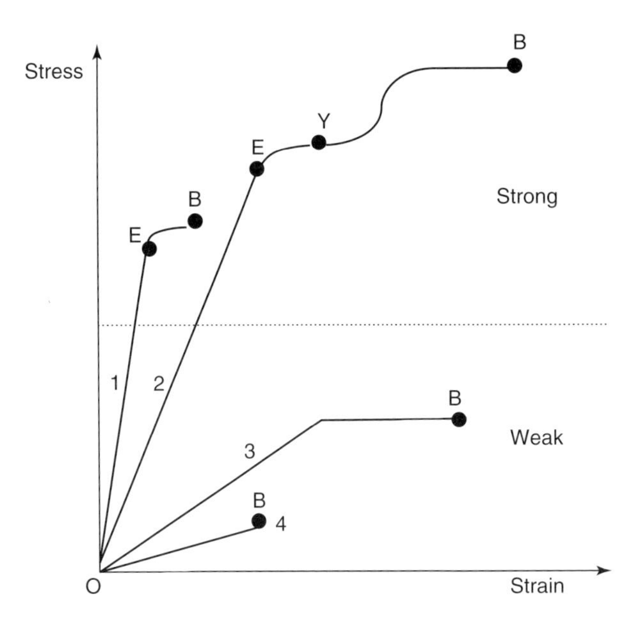

Figure 11.16 Stress-strain diagram for various foods: E, elastic limit; Y, yield point; B, breaking point; O–E, elastic region; E–Y, inelastic deformation; Y–B, region of ductility. (1) Hard, strong, brittle material. (2) Hard strong ductile material. (3) Soft, weak, ductile material. (4) Soft, weak, brittle material. Adapted from Ref. [12] with permission.

stress (see curve 2 in Figure 11.16). If the force is removed the solid object returns to its original shape. Elastic deformations are valueless in size reduction. Energy is used up but no breakdown occurs.

Point E is known as the *elastic limit*. Beyond this point, the material undergoes permanent deformation until it reaches its yield point Y. Brittle materials will rupture at this point. Ductile materials will continue to deform, or flow, beyond point Y until they reach the break point B, when they rupture. The behavior of different types of material is depicted by the four curves in Figure 11.16 and explained in the caption to that figure.

The breakdown of friable materials may occur in two stages. Initial fracture may occur along existing fissures or cleavage planes in the body of the material. In the second stage new fissures or crack tips are formed and fracture occurs along these fissures. Larger particles will contain more fissures than smaller ones and hence will fracture more easily. In the case of small particles, new crack tips may need to be created during the milling operation. Thus, the breaking strength of smaller particles is higher than the larger ones. The energy required for particle breakdown increases with decrease in the size of the particles. In the limit of very fine particles, only intermolecular forces must be overcome and further size reduction is very difficult to achieve. However, such very fine grinding is seldom required in food applications.

Only a very small proportion of the energy supplied to a size reduction plant is used in creating new surfaces. Literature values range from 2.0% down to less than 0.1%. Most of the energy is used up by elastic and inelastic deformation of the particles, elastic distortion of the equipment, friction between particles and between particles and the equipment, friction losses in the equipment and the heat, noise, and vibration generated by the equipment.

Mathematical models are available to estimate the energy required to bring about a specified reduction in particle size. These are based on the assumption that the energy, $\mathrm{d}E$, required to produce a small change, $\mathrm{d}x$, in the size of a unit mass of material can be expressed as a power function of the size of the material. Thus:

$$\frac{\mathrm{d}E}{\mathrm{d}x} = -\frac{K}{x^n} \tag{11.1}$$

Rittinger's law Rittinger's law is based on the assumption that the energy required should be proportional to the new surface area produced, that is, $n = 2$. So:

$$\frac{\mathrm{d}E}{\mathrm{d}x} = -\frac{K}{x^2} \tag{11.2}$$

or, integrating:

$$E = K\frac{1}{x_2} - \frac{1}{x_1} \tag{11.3}$$

where x_1 is the average initial size of the feed particles, x_2 is the average size of the product particles, E is the energy per unit mass required to produce this increase in surface area, and K is a constant, known as *Rittinger's constant*. Rittinger's law has been found to hold better for fine grinding.

Kick's law Kick's law is based on the assumption that the energy required should be proportional to the size reduction ratio, that is, $n = 1$. So:

$$\frac{\mathrm{d}E}{\mathrm{d}x} = -\frac{K}{x} \tag{11.4}$$

or, integrating:

$$E = -\ln\frac{x_1}{x_2} \tag{11.5}$$

Where x_1/x_2 is the size reduction ratio (see above). Kick's law has been found to apply best to coarse crushing.

Bond's law In Bond's law, n is given the value 3/2. So:

$$\frac{\mathrm{d}E}{\mathrm{d}x} = -\frac{K}{x^{3/2}} \tag{11.6}$$

or, integrating:

$$E = 2K\left[\frac{1}{(x_2)^{1/2}} - \frac{1}{(x_1)^{1/2}}\right] \tag{11.7}$$

Bond's law has been found to apply well to a variety of materials undergoing coarse, intermediate, and fine grinding [1, 2, 7, 8, 10, 49, 50].

11.3.2
Size Reduction Equipment

11.3.2.1 Some Factors to Consider When Selecting Size Reduction Equipment

Mechanical properties of the feed Friable and crystalline materials may fracture easily along cleavage planes. Larger particles will break down more readily than smaller ones. Roller mills are usually employed for such materials. Hard materials, with high moduli of elasticity, may be brittle and fracture rapidly above the elastic limit. Alternatively, they may be ductile and deform extensively before breakdown. Generally, the harder the material, the more difficult it is to break down and the more energy is required. For very hard materials, the dwell time in the action zone must be extended, which may mean a lower throughput or the use of a relatively large mill. Hard materials are usually abrasive and so the working surfaces should be made of hard wearing material, such as manganese steel, and should be easy to remove and replace. Such mills are relatively slow moving and need to be of robust construction. Tough materials have the ability to resist the propagation of cracks and are difficult to breakdown. Fibers tend to increase toughness by relieving stress concentrations at the ends of the cracks. Disk mills, pin-disk mills, or cutting devices are used to break down fibrous materials.

Moisture Content of the Feed The moisture content of the feed can be of importance in milling. If it is too high, the efficiency and throughput of a mill and the free-flowing characteristics of the product may be adversely affected. In some cases, if the feed material is too dry, it may not breakdown in an appropriate way. For example, if the moisture content of wheat grains are too high, they may deform rather than crack open to release the endosperm. Or, if they are too dry, the bran may break up into fine particles which may not be separated by the screens and may contaminate the white flour. Each type of grain will have an optimum moisture content for milling. Wheat is usually "conditioned" to the optimum moisture content before milling (see Section 11.3.3.1). Another problem in milling very dry materials is the formation of dust, which can cause respiratory problems in operatives and is a fire and explosion hazard. In wet milling, the feed materials is carried through the action zone of the mill in a stream of water.

Temperature Sensitivity of the Feed A considerable amount of heat may be generated in a mill, particularly if it operates at high speed. This arises from friction and particles being stressed within their elastic limits. This heat can cause the temperature of the feed to rise significantly and a loss in quality could result. If the softening or melting temperatures of the materials are exceeded the performance of the mill may be impaired. Some mills are equipped with cooling jackets to reduce these effects.

Cryogenic milling involves mixing solid carbon dioxide or liquid nitrogen with the feed. This reduces undesirable heating effects. It can also facilitate the milling of fibrous materials, such as meats, into fine particles.

11.3.2.2 Roller Mills (Crushing Rolls)

A common type of roller mill consists of two cylindrical steel rolls, mounted on horizontal axes and rotating toward each other. The particles of feed are directed between the rollers from above. They are nipped and pulled through the rolls where they are subjected to compressive forces, which bring about their breakdown. If the rolls turn at different speeds shear forces may be generated which will also contribute to the breakdown of the feed particles. The roll surfaces may be smooth, corrugated, grooved, fluted or they may have intermeshing teeth or lugs. In food applications smooth, grooved, or fluted rolls are most often used. Large rolls, with diameters greater than 500 mm, rotate at speeds in the range 50–300 rpm. Smaller rolls may turn at higher speeds. Usually the clearance between the rolls, the "nip," is adjustable. An overload compression spring is usually fitted to protect the roll surfaces from damage should a hard object try to pass between them (Figure 11.17).

The surface of the rolls may be cooled or heated by circulating water or some other thermal fluid within their interior. Large, smooth-surfaced rolls are used for relatively coarse crushing, usually achieving a reduction ratio of 4 or lower.

Smaller rolls, with different surface configurations, operating at higher speeds can achieve higher ratios. Smooth or fluted rolls, operating at the same or slightly different speeds, are used to crack open grains and seeds. For finer milling, shallow grooves and larger differential speeds are employed. For very fine milling, smooth surfaced rolls, operating at high differential speeds are used. To achieve high reduction ratios, the material being milled may be made to pass between two or more pairs of rolls in sequence, with the clearance decreasing from one pair to

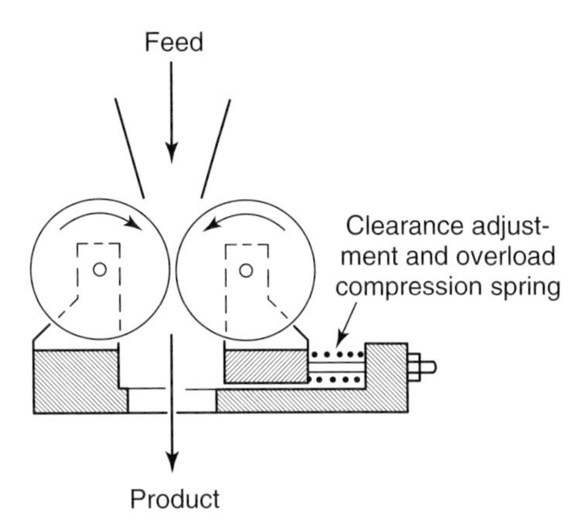

Figure 11.17 A roller mill. From Ref. [1] with permission of the authors.

the next. Some separation of the particles into different size ranges may take place between each pair of rolls. This principle is employed in the milling of wheat grains (see Section 11.3.3.1). Machines consisting of two, three, or more smooth-surfaced rolls, arranged in either a horizontal or vertical sequence, are used to mill liquid products. The product passes between the rolls in a zigzag flow pattern. The clearance between the rolls decreases in the direction of flow of the product and the speeds of consecutive rolls differ, generating shear forces.

Instead of two rolls operating against each other, one roll may operate against a flat or curved hard surface. An example is the edge mill (see Figure 11.18). The heavy rolls roll and slip over a table. Usually the rolls are driven, but in some machines the table is driven instead. The rolls and table surfaces may be smooth or grooved.

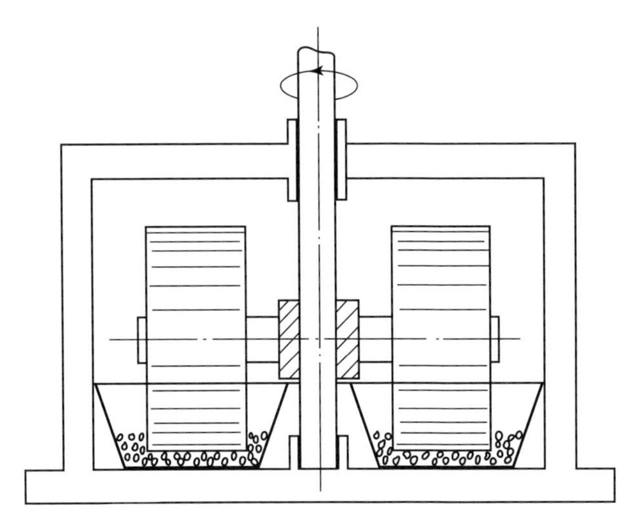

Figure 11.18 Edge mill. From Ref. [8] with kind permission of Springer Science and Business Media.

In the case of a mill consisting of two rolls rotating toward each other, the *angle of nip*, α, is the term used to describe the angle formed by the tangents to the roll faces at the point of contact between the particle and the rolls. It can be shown that:

$$\cos\frac{\alpha}{2} = \frac{D_r + D_p}{D_r + D_f} \quad (11.8)$$

where D_f is the average diameter of the feed particles, D_p is the average diameter of the product particles, corresponding to the clearance between the rolls, and D_r is the diameter of the rolls.

For the particle to be "nipped" down through the rolls by friction:

$$\tan\frac{\alpha}{2} = \mu \quad (11.9)$$

where μ is the coefficient of friction between the particle and the rolls. Equations 11.8 and 11.9 can be used to estimate the largest size of feed particle that a pair of rolls will accommodate.

The theoretical mass flow rate M (kg s^{-1}) of product from a pair of rolls of diameter D_r (m) and length l (m) when the clearance between the rolls is D_p (m), the roll speed is N (rpm) and the bulk density of the product is ρ (kg m^{-3}) is given by:

$$M = \frac{\pi D_r N D_p l \rho}{60} 60 \quad (11.10)$$

The literature suggests that the actual mass flow rate is usually between 0.1 and 0.3 of the theoretical value.

11.3.2.3 Impact (Percussion) Mills

When two bodies collide, that is, impact, they compress each other until they have the same velocity and remain in this state until restitution of the compression begins. Then the bodies push each other apart and separate. If one of the bodies is held in position, the other body has to conform with this position for a short interval of time. During the very short time it takes for restitution of compression to occur, a body possesses strain energy which can lead to fracture. The faster the bodies move away from each other the more energy is available to bring about fracture. The faster the rate at which the force is applied the more quickly fracture is likely to occur [40].

Hammer mill In this type of mill, a rotor mounted on an horizontal shaft turns at high speed inside a casing. The rotor carries hammers which pass within a small clearance of the casing. The hammers may be hinged to the shaft so that heads swing out as it rotates. In some designs, the hammers are attached to the shaft by rigid connections. A toughened plate, known as the *breaker plate*, may be fitted inside the casing (Figure 11.19). The hammers and breaker plate are made of hard-wearing materials such as manganese steel. The hammers drive the feed particles against the breaker plate. Fracture of the particles is brought about mainly by impact forces, but shear forces also play a part.

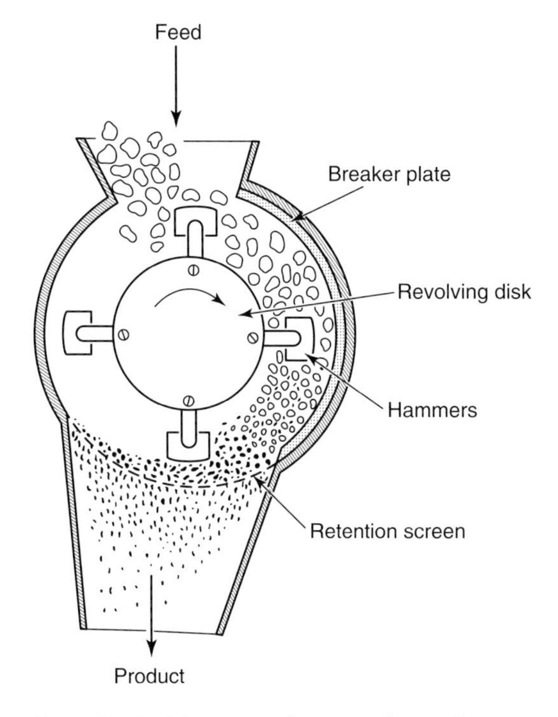

Figure 11.19 A hammer mill. From Ref. [1] with permission of the authors.

The casing may be fitted with a screen through which the product is discharged. The size of the screen aperture determines the upper limit particle size in the product. This way of operating a mill is known as *choke feeding*. When milling friable materials, choke feeding may result in a high proportion of very small particles in the product. When milling fibrous or sticky materials, the screen may become blocked. In some such cases the screen may be removed. The mill casing may be equipped with a cooling jacket. The hammer mill is a general purpose mill used for hard, friable, fibrous, and sticky materials.

Beater bar mill In this type of mill, the hammers are replaced by bars in the form of a cross. The tips of the bars pass within a small clearance of the casing. Beater bars are mainly used in small machines.

Comminuting mill Knives replace the hammers or bars in this type of mill. They may be hinged to the shaft so that the swing out as it rotates. Alternatively, they may be rigidly fixed to the shaft. Such mills are used for comminuting relatively soft materials, such as fruit and vegetable matter. In some designs, the knives are sharp on one edge and blunt on the other. When the shaft rotates in one direction the machine has a cutting action. When the direction of rotation of the shaft is reversed, the blunt edges of the knives act as beater bars.

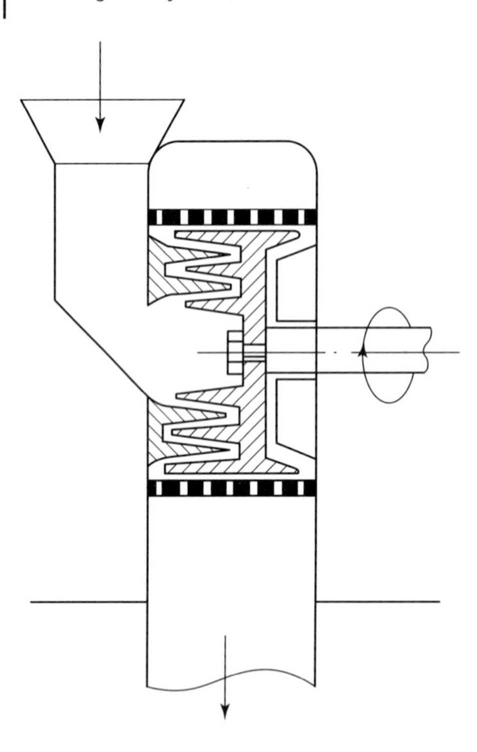

Figure 11.20 A pin mill fitted with screen over the periphery. From Ref. [8] with kind permission of Springer Science and Business Media.

Pin (pin-disk) mill In one type of pin mill a stationary disk and rotating disk are located facing each other, separated by a small clearance. Both disks have concentric rows of pins, pegs, or teeth. The rows of one disk fit alternately into the rows of the other disk. The pins may be of different shapes; round, square, or in the form of blades. The feed is introduced through the center of the stationary disk and passes radially outwards through the mill where it is subjected to impact and shear forces between the stationary and rotating pins. The mill may be operated in a choke feed mode by having a screen fitted over the whole or part of the periphery (Figure 11.20). Alternatively, it may not have a screen and the material is carried through the mill by an air stream (Figure 11.21). In another type of mill, both disks rotate either in the same direction at different speeds or in opposite directions. In some mills the clearance between the disks is adjustable. Disk speeds may be up to 10 000 rpm. Pin mills may be fitted with jackets for temperature control. Such mills are suitable for fine grinding friable materials and for breaking down fibrous substances.

Fluid energy (jet) mill In this type of mill, the solid particles to be comminuted are suspended in a gas stream traveling at high velocity into a grinding chamber. Breakdown occurs through the impact between individual particles and with the wall of the chamber. The gases used are compressed air or superheated steam, which are admitted to the chamber at a pressure of the order of 700 kPa. An air–solids separation system, usually a cyclone, is used to recover the product.

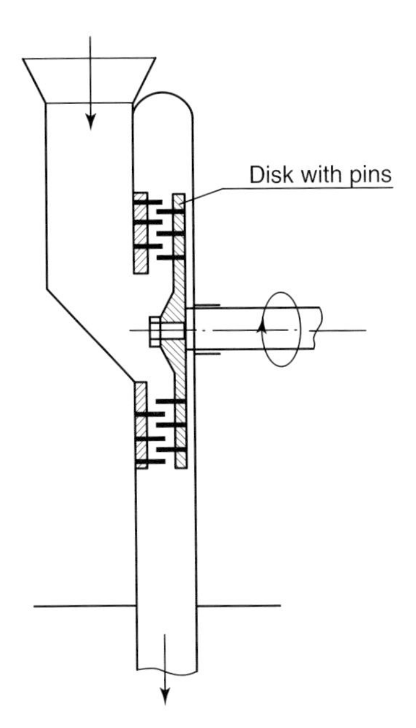

Figure 11.21 A pin mill without screen. From Ref. [8] with kind permission of Springer Science and Business Media.

Particles up to 10 mm can be handled in these mills but usually the feed consists of particles less than 150 μm. The product has a relatively narrow size range. Since there are no moving parts or grinding media involved, product contamination and maintenance costs are relatively low. However, the energy efficiency of such mills is relatively low.

11.3.2.4 **Attrition Mills**

The principle of attrition mills is that the material is rubbed between two surfaces. Both pressure and frictional forces are generated. The extent to which either of these forces predominates depends on the pressure with which both surface are held together and the difference in the speed of rotation of the surfaces.

Buhrstone mill This is the oldest form of attrition mill. It consists of two stones, one located above the other. The upper stone is usually stationary while the bottom one rotates. Matching grooves are cut in the stones and these make a scissor action as the lower one rotates. The feed is introduced through a hole in the center of the upper stone and gradually moves outwards between the stones and is discharged over the edge of the lower stone. In some such mills both stones rotate, in opposite directions. Siliceous stone is used. In more recent times toughened steel "stones" have replaced the natural stone for some applications.

Such stone mills were used for milling of wheat for centuries and are still used to produce wholegrain flour today. They are often driven by windmills. They are also used for wet milling of corn.

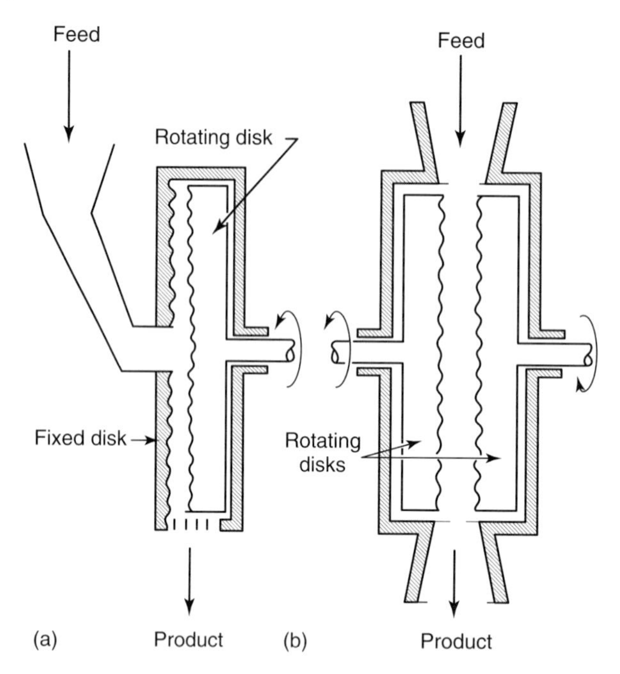

Figure 11.22 Disk mills. (a) A single disk mill. (b) A double disk mill. From Ref. [1] with permission of the authors.

Single-disk attrition mill In this type of mill a grooved disk rotates in close proximity to a stationary disk with matching grooves. The feed is introduced through the center of the stationary disk and makes its way outwards between the disks and is discharges from the mill via a screen (Figure 11.22a). Shear forces are mainly responsible for the breakdown of the material, but pressure may also play a part. The clearance between the two disks is adjustable.

Double-disk attrition mill This attrition mill consists of two counter-rotating disks, with matching grooves, located close to each other in a casing. The feed is introduced from the top and passes between the disks before being discharged, usually through a screen at the bottom of the casing (Figure 11.22b). Shear forces, again, predominantly cause the breakdown of the material.

Both types of disk attrition mills are used for milling fibrous materials such as corn and rice.

The Foos mill This design of mill has studs fitted to the disks instead of grooves. It could be regarded also as a modified pin mill (see Section 11.3.2.3). It is used for similar applications as the other disk mills.

Colloid mill This is another example of an attrition mill which is used for emulsification and in the preparation of pastes and purees (see Section 11.2.3.7).

11.3.2.5 Tumbling Mills

A typical tumbling mill consists of a cylindrical shell, sometimes with conical ends, which rotates slowly about a horizontal axis and is filled to about 50% of its volume with a solid grinding medium. As the shell rotates the loose units of the grinding medium are lifted up on the rising side of the shell to a certain height. They then cascade and cataract down the surface of the other units. The material being comminuted fills the void spaces between the units of the grinding medium. Size reduction takes place between these units in the jostling as they are lifted up and in the rolling action and impact as they fall down. The most commonly used grinding media are balls and rods. Contamination of the feed material due to wear of the grinding medium is a problem with tumbling mills and needs to be monitored.

Ball mills In ball mills the grinding medium consists of spherical balls, 25–150 mm in diameter, and usually made of steel. Alternatively, flint pebbles or porcelain or zircon spheres may be used. Mills employing such grinding materials are known as *pebble mills*. At low rotation speeds the balls are not lifted very far up the wall of the cylindrical and tumble over each other as they roll down.

In such circumstances, shear forces predominate. At higher speeds, the balls are lifted up higher and some fall back down generating impact forces which contribute to the size reduction of the feed material. Above a certain critical speed, the balls will be carried round against the cylinder wall under the influence of centrifugal force and comminution ceases. This critical speed can be calculated from the expression:

$$N = \frac{42.3}{(D)^{1/2}} \tag{11.11}$$

where N is the critical speed (rpm) and D is the diameter of the cylinder (m). Ball mills are run at 65–80% of their critical speeds.

Ball mills may be batch or continuous. In the latter case, the feed material flows steadily through the revolving shell. It enters at one end through a hollow trunnion and leaves at the other end through the trunnion or through peripheral openings in the shell. Ball mills are used for fine grinding and may be operated under dry or wet conditions.

In the vibration ball mill, the shell containing the balls is made to vibrate by means of out of balance weights attached to each end of the shaft of a doubleended electric motor. In such mills, impact forces predominate and very fine grinding is attainable.

Another variation of the ball mill, known as an *attritor*, consists of a stationary cylinder filled with balls. A stirrer keeps the balls and feed material in slow motion generating shear and some impact forces. It is best suited to wet milling and may be operated batchwise or continuously. This type of mill is used in chocolate manufacture (see Section 11.3.3.2).

Rod mills Grinding rods, usually made of high carbon steel, are used instead of balls in rod mills. They are 25–125 mm in diameter and may be circular, square, or hexagonal in cross-section. They extend to almost the full length of the shell and occupy about 35% of the shell volume. In such mills, attrition forces predominate but impacts also play a part in size reduction. They are classed as intermediate grinders and are more useful than ball mills for milling sticky materials [1, 7, 8, 10, 49–51].

11.3.3 Examples of Size Reduction of Solids in Food Processing

11.3.3.1 Cereals

Wheat The structure of the wheat grain is complex but, in the context of this section, it can be assumed to consist of three parts: the outer layer or bran, the white starchy endosperm, and the embryo or germ. According to Meuser [41]: "The objectives of milling white flour are: 1. to separate the endosperm, which is required for the flour, from the bran and embryo, which are rejected, so that the flour shall be free from bran specks, and of good color, and so that the palatability and digestibility of the product shall be improved and its storage life lengthened. 2. To reduce the maximum amount of endosperm to flour fineness, thereby obtaining the maximum extraction of white flour from the wheat."

The number of parts of flour by weight produced per 100 parts of wheat is known as the *percentage extraction rate*. The wheat grain contains about 82% of endosperm, but it is not possible to separate all of it from the bran and embryo. In practice, the extraction rate is in the range 70–80%.

Prior to milling, the wheat grains are cleaned to remove metal fragments, stones, animal matter, and unwanted vegetable matter and conditioned to the optimum moisture content for milling. A mill consists of two sections: a break section and a reduction section. The clean grain is fed to the break section, which usually consists of four or five pairs of fluted rolls. The rolls rotate toward each other at different speeds. As the grains pass through the first pair of break rolls they are split open. Some large fragments of endosperm (semolina) are released together with a small amount of small particles, less than 150 μm in size, which is collected as flour. The fragments of bran coming from these rolls will have endosperm attached to them. The fractions from the first break rolls are separated by sieving (Figure 11.23). The bran passes through a second set of fluted rolls where more semolina and flour is released. This is repeated two or three more times, until relatively clean bran is collected as a by-product from the last set of sieves. The clearance between the pairs of break rolls and the depth of the fluting decrease in the direction of flow of the bran. The semolina, which contains the germ and some bran particles, goes to the reduction section. This section consists of up to 16 pairs of smooth surfaced rolls, rotating toward each other at different speeds. The speed differential is less than that between the break rolls. As the semolina passes through the first set of reduction rolls, some of the large fragments are broken down into flour; and the germ, which is relatively soft, is pressed into flakes. These fractions are separated by sieving. The germ

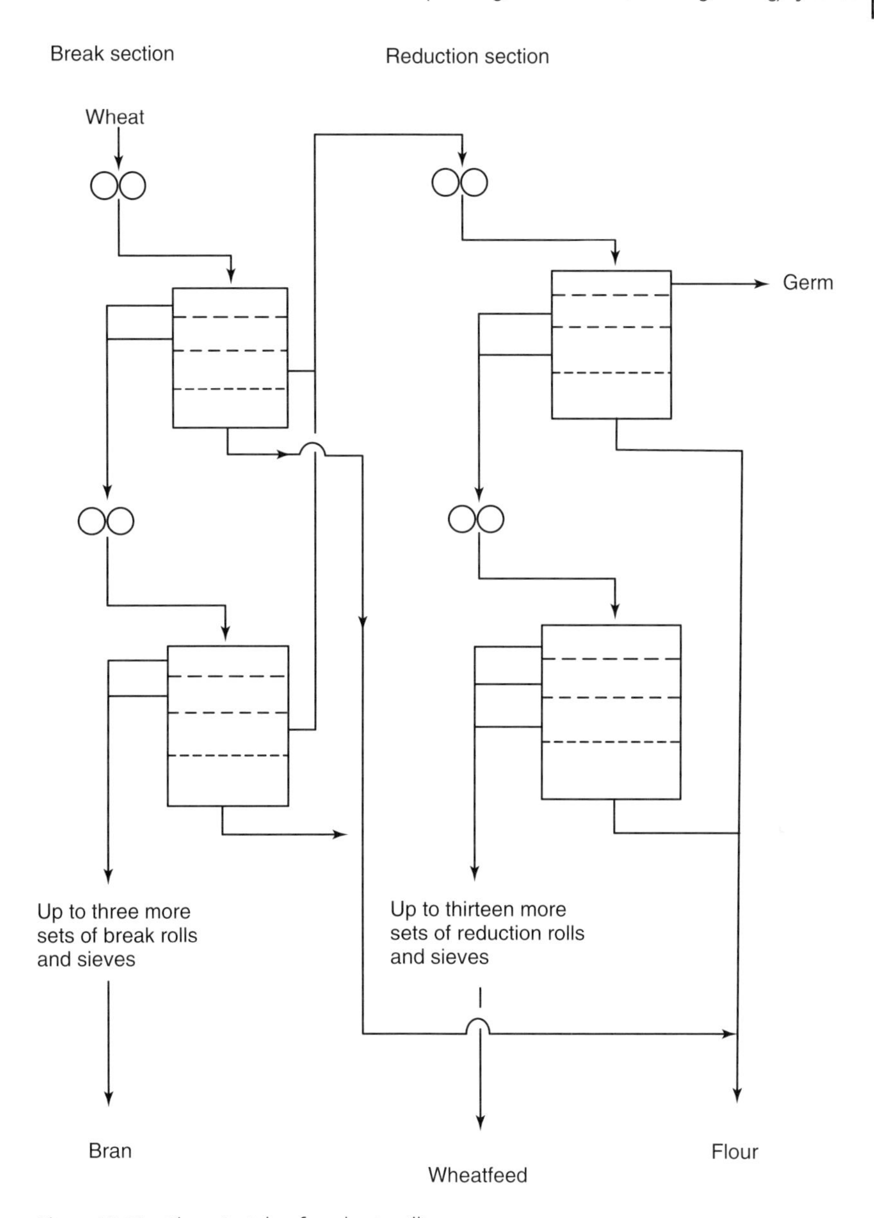

Figure 11.23 The principle of a wheat milling process.

is usually discharged from this set of sieves as another by-product. The large particles of semolina pass through another set of reduction rolls where more flour is produced. This is repeated up to 14 more times. In addition to some flour, another by-product, consisting mainly of fine bran particles and known as *wheat feed*, is discharged from the last set of sieves. The clearance between the pairs of reduction rolls decreases in the direction of flow of the semolina. The greatest proportion of flour is collected from the early reduction rolls. The flour coming from the later reduction rolls contains more fine bran particles and is darker in color than that collected from the early rolls. This is a rather simplified description of white flour milling. More detailed accounts are available in the literature [52–57].

Wholemeal flour contains all the products of milling cleaned wheat. It is produced using millstones (see Section 11.3.2.4), by roller mills using a shortened system or by attrition mills. Brown flours, which have extraction rates of 85–98%, are produced by a modified white flour milling system. The wheat may be conditioned to a lower moisture content than that used for white flour so that the bran breaks into smaller particles. Durum wheat is milled to produce coarsely ground endosperm, known as *semolina*, which is used in the manufacture of pasta and as couscous.

Sorghum, millets, maize, and rye may be dry milled using modified flour milling systems.

Rice Unlike most milled cereal products, in the production of milled rice the endosperm is kept as intact as possible. Brown rice is dehulled using rubber-covered rolls or by means of abrasive disks. The bran and embryo are then removed, a process known as *whitening*. This may be achieved by means of abrasive cones or rolls. A further step, known as *polishing*, removes the aleurone layer, which is rich in oil. This extends the storage life of the rice by reducing the tendency for oxidative rancidity to occur. Polishing is achieved by means of a rotating cone covered with leather strips.

Maize This cereal is wet milled to produce a range of products including starch, oil, and various types of cattle feed. Cleaned maize is steeped in water, containing 0.1–0.2% sulfur dioxide, at 50 °C for 28–48 h. The steeping softens the kernel and facilitates separation of the hull, germ, and fiber from each other. The SO_2 may disrupt the -SS- bonds in the protein, enabling starch/protein separation. After steeping, the steep water is concentrated by vacuum evaporation and the protein it contains is recovered by settling. The steeped maize is coarsely milled in a Foos mill (see Section 11.3.2.4). The grain is cracked open and the germ released. The germ is recovered by settling or by means of hydrocyclones. Oil may be extracted from the germ by pressing. The degermed material is strained off from the liquid and milled in an impact or attrition mill. The hulls and fiber are separated from the protein and starch by screening. The suspension of starch and protein coming from the screen is fed to high-speed centrifuges where they are separated from each other. The starch is purified in hydrocyclones, filtered, and dried. The protein is also filtered and dried. The products from the wet milling of maize include about 66% starch, 4% oil, and 30% animal feed. Most of the starch is converted into modified starches, sweeteners, alcohol, and other useful products [52–57].

11.3.3.2 Chocolate

Size reduction equipment is employed at several stages in the manufacture of chocolate. The first stage is breaking the shell of the cocoa bean to release the nib. The beans are predried in heated air, or steam puffed, or exposed to infrared radiation to weaken the shell and loosen the nib. The bean shells are generally broken by a type of impact mill. The beans are fed onto plate or disk, rotating at high speed. They are flung against breaker plates by centrifugal force. Sometimes they fall onto a second rotating disk, which flings them back against the plate again. The size distribution of the product from the mill should be as narrow as possible, ideally with the size range 7.0–1.5 mm.

The nibs are separated from the fragments of shell by a combination of sieving and air classification (winnowing). It is important that this separation is as complete as possible, as the presence of shell can cause excessive wear on the equipment used subsequently to grind the nibs and refine the chocolate. The next application for size reduction equipment is the grinding of cocoa nibs. These are ground to cocoa liquor for the removal of cocoa butter from within the cellular structure. Cocoa nib contains about 55% butter, contained within cells which are 20–30 μm in size. These cells must be broken to release the butter. The cell wall material and any shell and germ present are fibrous and tough. The particle size range after grinding should be in the range 15–50 μm. The grinding of the nibs is usually carried out in two steps. They are first preground in a hammer, disk, or pin mill. The second step is carried out in an agitated ball mill or attrition mill. The first of these mills consists of a cylinder filled to as much as 90% of its volume with balls made of steel or ceramic material. For large throughputs, a series of ball mills may be used, with the size of the balls decreasing in the direction of flow of the liquor, from 15 mm down to 2 mm. Some contamination of the liquor is to be expected due to wear of the equipment. Three single-disk attrition mills, working in series, may replace the ball mill. The three pairs of disks may be housed in the one machine. These can be fed directly with dry nibs, but pregrinding is more usual.

A further stage in chocolate manufacture is refining. The purpose of refining is to ensure that the particle size of the dispersed phase is sufficiently small so as not to impart a gritty texture to the chocolate when it is eaten. Usually this means that the largest particles should be in the range 15–30 μm. If the cocoa liquor has been properly ground, the purpose of refining is to reduce the size of the added sugar crystals and, in the case of milk chocolate, the particles of milk powder. Refining is usually carried out using a five-roll machine. The gap between the rolls decreases from bottom to top while the speed of rotation increases. The cocoa liquor is introduced between the bottom two rolls. The film of chocolate leaving each gap is transferred to the faster roll and so moves upwards. The product is scraped off the top roll.

Uncompressed rolls are barrel-shaped. Hydraulic pressure is applied to the roll stack to compress the camber on the rolls and obtain an even coating across the length of each roll. The reduction ratio attainable is 5–10, which should result in the particles in the product being in the correct size range. The temperature of the rolls is controlled by circulation of water through them. It is important that the feed material to the rolls is well mixed. In some systems, a two-roll refiner known as a *prerefiner* is employed to prepare the feed for the main roll stack.

Cocoa powder is produced by partially defatting ground cocoa nibs in a press. The broken cake from the press is blended prior to milling to standardize the color. The blend is fed to a pin mill were it is broken down to a powder. The powder is cooled to about 18 °C by being conveyed through jacketed tubes by compressed air. At the end of the cooling line, the powder is recovered from the air by means of a cyclone. It is transported to a stabilizer, where the latent heat is removed to solidify the remaining fat [58–60].

11.3.3.3 **Coffee Beans**

After roasting, coffee beans are cooled and then ground. The extent of the grinding depends on the how the coffee is to be brewed. Generally, coffee which is to be percolated will have a larger particles size than filter coffee. The actual particle sizes vary from country to country. In Europe, coarse ground coffee for percolators has an average particle size of about 850 μm, while that of a medium grind, intended to be filtered, is about 450 μm. Finely ground coffee, about 50 μm, is required for some types of Turkish coffee and for espresso machines. In the latter case, the beans are ground in small amounts just prior to use.

Ground beans for large-scale percolation (extraction) for the production of instant coffee may be somewhat coarser than the normal coarse ground product.

The original method of grinding coffee beans was by means of a buhrstone-type mill (see Section 11.3.2.4). These are still in use in some developing countries, particularly in markets where customers grind their roast beans at the time of purchase. The stones have been replaced by serrated, cast iron disks. For large-scale grinding of coffee beans, multiroll mills are mostly used. Three or four pairs of rolls are located one above the other. In each pair, the rolls rotate at different speeds. The fast roll has U-shaped corrugations running lengthwise while the other has peripheral corrugations. As the beans pass down between successive pairs of rolls, the clearance between them decreases and the roll speed increases. A three-stand mill is used to produce coarse or medium ground coffee. A fourth pair of rolls is used for fine grinding. Chaff remaining after roasting is released during grinding. This may be removed or incorporated with the ground coffee in a type of ribbon mixer. The latter procedure is known as *normalizing*. Various types of attrition and hammer mills are used to grind roasted beans on a smaller scale.

A considerable amount of carbon dioxide is released during grinding of roasted beans. Quite high concentrations of the gas can develop in the vicinity of the milling equipment. If ventilation is poor, operatives may be affected physiologically. The main effect is that the person becomes silly and acts oddly, without realizing why. Water vapor and volatile flavor compounds are also released during grinding.

Cryogenic grinding, using solid carbon dioxide or liquid nitrogen, may be employed to maximize the retention of the flavor volatiles [61–63].

11.3.3.4 Oil Seeds and Nuts

It is common practice to flake or grind oil seeds prior to extraction, either by pressing or solvent extraction, in order to rupture the oil cells. Hammer mills and attrition mills are sometimes used from the preliminary breakdown of large oil seeds such as copra or palm. However, flaking rolls are usually used to prepare seeds for extraction. A common arrangement used for cottonseed, flaxseed, and peanuts consists of five rolls located one above the other. The seed is fed in between the top two rolls and then passes back and forth between the rolls down to the bottom of the stack. Each roll supports the weight of the rolls above it, so the seeds are subjected to increasing pressure as they move down the stack. The top roll is grooved, but the others have smooth surfaces [64].

11.3.3.5 Sugar Cane

Large, heavy duty, two- or three-roll crushers are used to break and tear up the cane to prepare it for subsequent milling for the extraction of the sugar. The surfaces of the rolls are grooved to grip the cane and very high pressures are applied to the cane, 8–12 $t\,dm^{-2}$. Following this initial crushing, the cane may be further broken down in shredding devices, which tear open the cane cells and release some of the juice. These devices are essentially large-capacity hammer mills, with rotors turning at 1000–1200 rpm, carrying hammers pivoted to disks or plates. The hammers pass very close to an anvil plate made up of rectangular bars. The material leaving the shredders is made up of cell material and long thread-like fibers which hold it together when it is subjected to pressure during extraction of the juice by milling.

To extract the juice, the shredded cane is passed through a series, usually five, of triple-roll mills. These are heavy-duty rolls, with grooved surfaces which exert high pressure, 15–30 $t\,dm^{-2}$ on the cane, expressing the juice. This is not a size reduction operation as such, but is an example of another use of roller mills. Water is added to the material after each set of rolls. Alternatively, the dilute juice from the last mill is sent back before the preceding mill. This procedure, known as *imbibition*, increases the amount of sugar extracted, compared with dry crushing [65, 66].

Numerous other food materials are size-reduced in mills. Mustard seeds are milled in a similar fashion to wheat grains, passing through grooved break rolls and then through smooth-surfaced reduction rolls. Spices are milled using a variety of impact and attrition mills. Considerable heat may be generated and result in a loss of volatile oils and development of undesirable aromas and flavors. Mills may be jacketed and cooled. Cryogenic milling, using liquid nitrogen, can result in higher quality products. Sugar crystals may be ground in impact mills to produce icing sugar. Particles of milk powder, lactose, and dry whey may be reduced in size in impact mills.

References

1. Brennan, J.G., Butters, J.R., Cowell, N.D., and Lilly, A.E.V. (1990) *Food Engineering Operations*, 3rd edn, Elsevier Applied Science, London.
2. Fellows, F.P. (2009) *Food Processing Technology*, 3rd edn, Woodhead Publishing, Cambridge.
3. Nienow, A.W., Edwards, M.F., and Harnby, N. (1992) Introduction to mixing problems, in *Mixing in the Process Industries*, 2nd edn (eds N. Harnby, M.F. Edwards, and A.W. Nienow), Butterworth-Heinemann, Oxford, pp. 1–24.
4. Edwards, M.F. and Baker, M.R. (1992) A review of liquid mixing equipment, in *Mixing in the Process Industries*, 2nd edn (eds N. Harnby, M.F. Edwards, and A.W. Nienow), Butterworth-Heinemann, Oxford, pp. 118–136.
5. Edwards, M.F. and Baker, M.R. (1992) Mixing of liquids in stirred tanks, in *Mixing in the Process Industries*, 2nd edn (eds N. Harnby, M.F. Edwards, and A.W. Nienow), Butterworth-Heinemann, Oxford, pp. 137–158.
6. Lindley, J.A. (1991) Mixing processes for agricultural and food materials 1: fundamentals of mixing. *J. Agric. Eng. Res.*, **48**, 153–170.
7. McCabe, W.L., Smith, J.C., and Harriott, P. (1985) *Unit Operations of Chemical Engineering*, 4th edn, McGraw-Hill, New York.
8. Leniger, H.A. and Beverloo, W.A. (1975) *Food Process Engineering*, Reidel Publishing, Dordrecht.
9. Dickey, D.S. (2009) Equipment design, in *Food Mixing: Principles and Applications* (ed. P.J. Cullen), John Wiley & Sons Ltd, Chichester, pp. 73–89.
10. Berk, Z. (2009) *Food Process Engineering and Technology*, Academic Press, Amsterdam.
11. Edwards, M.F. (1992) Laminar flow and distributive mixing, in *Mixing in the Process Industries*, 2nd edn (eds N. Harnby, M.F. Edwards, and A.W. Nienow), Butterworth-Heinemann, Oxford, pp. 200–224.
12. Yang, S.C. and Lai, L.S. (2003) Dressing and mayonnaise, in *Encyclopedia of Food Sciences and Nutrition*, 2nd edn (eds B. Cabellero, L.C. Trugo, and P.M. Finglas), Academic Press, London, pp. 1892–1896.
13. Lindley, J.A. (1991) Mixing processes for agricultural and food materials: part 2, highly viscous liquids and cohesive materials. *J. Agric. Eng. Res.*, **48**, 229–247.
14. Richards, G. (1997) Motionless mixing: efficiency with economy. *Food Process.*, **66**, 29–30.
15. Godfrey, J.C. (1992) Static mixers, in *Mixing in the Process Industries*, 2nd edn (eds N. Harnby, M.F. Edwards, and A.W. Nienow), Butterworth-Heinemann, Oxford, pp. 225–249.
16. Harnby, N. (2003) Mixing of powders, in *Encyclopedia of Food Science and Nutrition*, 2nd edn (eds B. Caballero, L.C. Trugo, and P.M. Finglas), Academic Press, London, pp. 4028–4033.
17. Harnby, N. (1992) Characterization of powder mixtures, in *Mixing in the Process Industries*, 2nd edn (eds N. Harnby, M.F. Edwards, and A.W. Nienow), Butterworth-Heinemann, Oxford, pp. 25–41.
18. Harnby, N. (1992) The selection of powder mixers, in *Mixing in the Process Industries*, 2nd edn (eds N. Harnby, M.F. Edwards, and A.W. Nienow), Butterworth-Heinemann, Oxford, pp. 42–61.
19. Lindley, J.A. (1991c) Mixing processes for agricultural and food materials: 3 powders and particulates. *J. Agric. Eng. Res.*, **49**, 1–19.
20. Geldart, D. (1992) Mixing in fluidized beds, in *Mixing in the Process Industries*, 2nd edn (eds N. Harnby, M.F. Edwards, and A.W. Nienow), Butterworth-Heinemann, Oxford, pp. 62–78.
21. Harnby, N. (1992) The mixing of cohesive powders, in *Mixing in the Process Industries*, 2nd edn (eds N. Harnby, M.F. Edwards, and A.W. Nienow), Butterworth-Heinemann, Oxford, pp. 79–98.

22. Fitzpatrick, J.J. (2009) Particulate and powder mixing, in *Food Mixing: Principles and Applications* (ed. P.J. Cullen), John Wiley & Sons Ltd, Chichester, pp. 269–286.
23. Middleton, J.C. (1992) Gas–liquid dispersion and mixing, in *Mixing in the Process Industries*, 2nd edn (eds N. Harnby, M.F. Edwards, and A.W. Nienow), Butterworth-Heinemann, Oxford, pp. 322–363.
24. Sahu, J.K. and Niranjan, K. (2009) Gas–liquid mixing, in *Food Mixing: Principles and Applications*, John Wiley & Sons Ltd, Chichester, pp. 230–252.
25. Higgins, K.T. (2006) Sonic mixing and other oddities of the new age. *Food Eng.*, **78**, 81–86.
26. Dunn, J. (2007) Reliability is a turn on. *Food Manuf.*, **82**, 63–64.
27. Matsumo, S. (1986) Review paper – W/O/W type emulsions with a view to possible food applications. *J. Texture Stud.*, **17**, 141–159.
28. Garti, N. and Benichou, A. (2004) Recent developments in double emulsions for food applications, in *Food Emulsions*, 4th edn (eds S.E. Friberg, K. Larsson, and J. Sjoblom), Marcel Dekker, New York, pp. 353–412.
29. Friberg, S.E. and Kayali, I. (1991) Surfactant association structures, microemulsions, and emulsions in foods: an overview, in *Microemulsions and Emulsions in Foods* (eds M. El-Nokaly and D. Cornell), American Chemical Society, Washington, DC, pp. 7–24.
30. Robins, M.M. and Wilde, P.J. (2003) Colloids and emulsions, in *Encyclopedia of Food Sciences and Nutrition*, 2nd edn (eds B. Caballero, L.C. Trugo, and P.M. Finglas), Academic Press, London, pp. 1517–1524.
31. McClements, D.J. (2004) *Food Emulsions*, 2nd edn, CRC Press, New York.
32. Spyropoulos, F., Cox, P.W., and Norton, I.T. (2009) Immiscible liquid–liquid mixing, in *Food Mixing: Principles and Applications* (ed. R.J. Cullen), John Wiley & Sons Ltd, Chichester, pp. 175–198.
33. Kinyanjul, T., Artz, W.E., and Mahungu, S. (2003) Emulsifiers, in *Encyclopedia of Food Sciences and Nutrition*, 2nd edn (eds B. Caballero, L.C. Trugo, and P.M. Finglas), Academic Press, London, pp. 2070–2077.
34. Grindsted Products (1988) *Emulsifiers and Stabilisers for the Food Industry*, Grindsted Products, Brabrand.
35. Lewis, M.J. (1987) *Physical Properties of Foods and Food Processing Systems*, Ellis Horwood, Chichester.
36. Whitehurst, R.J. (2004) *Emulsifiers in Food Technology*, Blackwell Publishing Ltd, Oxford.
37. Wilbey, R.A. (2003) Homogenization, in *Encyclopedia of Food Sciences and Nutrition*, 2nd edn (eds B. Caballero, L.C. Trugo, and P.M. Finglas), Academic Press, London, pp. 3119–3125.
38. Walstra, P. (1983) Formation of emulsions, in *Encyclopedia of Emulsion Technology*, vol. **1**, (ed. P. Becher), Marcel Dekker, New York, pp. 57–127.
39. Charcosset, C. (2009) Preparation of emulsions and particles by membrane emulsification for the food processing industry. *J. Food. Eng.*, **92**, 341–249.
40. Walstra, P., Geurts, T.J., Noomen, A., Jellema, A., and van Boekel, M.A.J.S. (1999) *Dairy Technology*, Marcel Dekker, New York.
41. Lane, R. (1998) Butter and mixed fat spreads, in *The Technology of Dairy Products*, 2nd edn (ed. R. Early), Blackie Academic and Professional, London, pp. 158–107.
42. Bucheim, W. and Dejmek, P. (1997) Milk and dairy-type emulsions, in *Food Emulsions*, 3rd edn (eds S.E. Friberg, K. Larsson, and J. Sjoblom), Marcel Dekker, New York, pp. 235–278.
43. Ford, L.D., Borwanker, R.P., Pechak, D., and Schwimmer, B. (2004) Dressings and sauces, in *Food Emulsions*, 4th edn (eds S.E. Friberg, K. Larsson, and J. Sjoblom), Marcel Dekker, New York, pp. 525–572.
44. Berger, K.G. (1997) Ice cream, in *Food Emulsions*, 3rd edn (eds S.E. Friberg and K., Larsson), Marcel Dekker, New York, p. 490.
45. Eliasson, A.-C. and Silvero, J. (1997) Fat in baking, in *Food Emulsions*, 3rd edn (eds S.E. Friberg, K. Larsson, and J. Sjoblom), Marcel Dekker, New York, pp. 525–548.

46. Mizukoshi, M. (1997) Baking mechanism in cake production, in *Food Emulsion*, 3rd edn (eds S.E. Friberg, K. Larsson, and J. Sjoblom), Marcel Dekker, New York, pp. 549–575.
47. Krog, N.J., Riison, T.H., and Larsson, K. (1985) Applications in the food industry: I, in *Encyclopedia of Emulsion Technology*, vol. **2** (ed. E. Becher), Marcel Dekker, New York, pp. 321–365.
48. Jaynes, E. (1985) Applications in the food industry: II, in *Encyclopedia of Emulsion Technology*, vol. **2** (ed. E. Becher), Marcel Dekker, New York, pp. 367–384.
49. Loncin, M. and Merson, R.L. (1979) *Food Engineering – Principles and Selected Applications*, Academic Press, London.
50. Lowrison, G.C. (1974) *Crushing and Grinding – the Size Reduction of Solid Materials*, Butterworths, London.
51. Meuser, F. (2003) Types of mill and their uses, in *Encyclopedia of Food Science and Nutrition*, 2nd edn (eds B. Caballero, L.C. Trugo, and P.M. Finglas), Academic Press, London, pp. 3987–3997.
52. Kent, N.L. and Evers, A.D. (1993) *Technology of Cereals*, 4th edn, Elsevier Science, Oxford.
53. Barnes, P.J. (1989) Wheat milling and baking, in *Cereal Science and Technology* (ed. G.H. Palmer), Aberdeen University Press, Aberdeen, pp. 367–412.
54. Posner, E.S. (2009) Wheat flour milling, in *Wheat Chemistry and Technology*, 4th edn (eds K. Khan and P.R. Shewry), AACC International, St. Paul, MN, pp. 119–152.
55. Owens, W.G. (2001) Wheat, corn and coarse grains milling, in *Cereals Processing Technology* (ed. G. Owens), Woodhead Publishing, Cambridge, pp. 27–52.
56. Posner, E.S. (2003) Principles of milling, in *Encyclopedia of Food Science and Nutrition*, 2nd edn (eds B. Caballero, L.C. Trugo, and P.M. Finglas), Academic Press, London, pp. 3980–3986.
57. Morrison, W.R. and Barnes, P.J. (1983) Distribution of wheat acyl lipids and tocols into flour millstreams, in *Lipids in Cereal Technology* (ed. P.J. Barnes), Academic Press, London, pp. 149–163.
58. Heemskerk, R.F.M. (1999) Cleaning, roasting and winnowing, in *Industrial Chocolate Manufacture and Use*, 3rd edn (ed. S.T. Beckett), Blackwell Science Ltd., Oxford, pp. 79–100.
59. Meursing, E.H. and Zijderveld, J.A. (1999) Cocoa mass, cocoa butter and cocoa powder, in *Industrial Chocolate Manufacture and Use*, 3rd edn (ed. S.T. Beckett), Blackwell Science, Oxford, pp. 101–114.
60. Ziegler, G. and Hogg, R. (1999) Particle size reduction, in *Industrial Chocolate Manufacture and Use*, 3rd edn (ed. S.T. Beckett), Blackwell Science, Oxford, pp. 115–136.
61. Sivetz, M. and Desrosier, N.W. (1979) *Coffee Technology*, AVI Publishing, Westport.
62. Clarke, R.J. (1987) Roasting and grinding, in *Coffee Technology*, vol. **2** (eds R.J. Clarke and R. Macrae), Elsevier Applied Science, London, pp. 73–107.
63. Petracco, M. (2005) Grinding, in *Espresso Coffee, The Science of Quality* (eds A. Illy and R. Viani), Elsevier Academic Press, Amsterdam, pp. 215–229.
64. Williams, M.A. and Hron, R.J. Sr. (1996) Obtaining oils and fats from source materials, in *Bailey's Industrial Oil and Fat Products*, 5th edn, vol. 4 (ed. Y.H. Hui), John Wiley & Sons, Inc., New York, pp. 61–155.
65. Hugot, E. (1986) *Handbook of Cane Sugar Engineering*, 3rd edn, Elsevier Science, Amsterdam.
66. Rein, P.W. (2007) Developments in sugarcane processing over the last 25 years. *Sugar Ind./Zuckerindustrie*, **132**, 435–444.

12
Baking

Stanley P. Cauvain

12.1
Introduction

Baking is a term commonly applied to the practice of making a wide range of food products, including breads, cakes, pastries, cookies, and crackers (Figure 12.1). The common threads which link the manufacture of such a diverse group of food products is that they all use recipes which are based on wheat flour and the products undergo a heat processing step. It is the latter which is the more precise definition of baking and it represents a very important difference between baked products and other foods, because during the baking stage baked products undergo dramatic changes in both form and structure. The transition from raw materials to baked product is most often referred as being a change from a foam to a sponge. In the foam, gas bubbles which have been incorporated during mixing are separate and discrete from one another, while in the sponge, the gas cells are open and interconnected. Bread and other fermented products and cakes are the ones which most closely fit the "foam-to-sponge" model, while in biscuits, cookies, and pastries the model is less well defined.

Common English dictionary definitions for groups of baked products include:

- **Bread** – *n.* food made of flour or meal (and) baked.
- **Cake** – *n.* baked, sweetened bread.
- **Biscuit** – *n.* dry, small, thin variety of cake.
- **Pastry** – *n.* article of food made chiefly of flour, fat, and water.

These definitions confirm the central role that flour (usually wheat) plays in defining baked products but make no reference to the unique role that wheat proteins play in the formation of baked product structures. Wheat proteins are almost unique among cereals in their ability to form a rubbery mass when mixed with water [1]. This rubbery mass is referred to by bakers as "gluten" and it is responsible for the ability of a wheat flour–water mix to expand under the inflation of carbon dioxide gas from fermentation; this is a fundamental process in bread making and will be discussed in more detail below. In cakes, cookies, and pastries the formation of a gluten network is less critical or even undesired

Food Processing Handbook, Second Edition. Edited by James G. Brennan and Alistair S. Grandison.

Figure 12.1 Variety of baked products. Reproduced with permission from BakeTran.

and so in the manufacture of such products different steps are taken to limit gluten formation at mixing. The relationships between gluten formation and the different baked product groups are summarized in Table 12.1. Cauvain and Young [2] used a two-dimensional diagram in which they plotted the ratio of sugar to flour in typical baked products recipes against the ratio of fat to flour in the same recipe to illustrate the relationship between the different groups of baked products and the potential for gluten formation. The authors chose the ingredients fat and sugar because of their potential in limiting gluten formation; the former by disrupting the gluten network and the latter through its impact on water activity.

12.2 The Key Characteristics of Existing Bakery Product Groups

In order to understand the processes involved in the manufacture of the different groups of baked products it is necessary to appreciate their key characteristics. This is because these characteristics are now such an integral part of product quality that modern processing technologies have evolved to ensure that historical product characteristics are maintained.

12.2.1 Bread, Rolls, and Buns

This group of products is generally characterized by having a "crust," a dry, thin layer enclosing a soft, sponge-like cellular structure. The crust is commonly a light golden brown color, though darker colors are seen when wholemeal (wholewheat),

Table 12.1 Relationship between gluten formation and different groups of baked products.

Product group	Desirability and level of gluten formation	Factors which contribute to gluten formation or its inhibition
Bread	Essential and significant	Encouraged by Choice of flour (protein content and quality) Energy input during mixing Oxidation processes
Rolls and buns	Essential and significant	Encouraged by Choice of flour (protein content and quality) Energy input during mixing Oxidation processes Limited by Recipe fat and sugar levels
Laminated products	Desirable but moderate	Encouraged by Choice of flour (protein content and quality) Limited by Energy input during mixing Recipe fat and sugar levels
Crackers	Desirable but modest	Encouraged by Choice of flour (protein content and quality) Limited by Energy input during mixing
Semi-sweet biscuits	Required but modest	Limited by Choice of flour Energy input during mixing Recipe fat and sugar levels
Short-dough biscuits and cookies	Not required and limited	Limited by Choice of flour Energy input during mixing Recipe fat and sugar levels
Cakes	Not desirable and none	Limited by Choice of flour Recipe fat, sugar, and water levels
Wafers	Not desirable and none	Limited by Choice of flour Recipe water sugar levels

brown, or rye flour are used in bread production. The bread crust has a considerably lower moisture content than that of the crumb; typically crust moisture contents are in the range 12–17%, while for bread crumb they will range from 35 to 42%, depending on bread type. The low moisture content yields a crust with a hard and brittle eating character which may be accentuated by the thickness of the crust. All fermented bread products have an open, cellular crumb structure. A key contribution to the cellular structure of breads comes from the release of carbon dioxide gas from bakers' yeast fermentation. A fundamental requirement of bread crumb is that it should be relatively soft combined with a degree of resilience or springiness and a degree of "chewiness." An important contributor to the character of bread crumb is the nature of the cellular structure as determined by the size of the individual cells, their distribution throughout the product, and the thickness of the cell wall material. Bread products are not highly flavored by comparison with other baked products and many other foods.

12.2.2
Cakes and Sponges

These terms encompass a diverse group of products, though there are some unifying characteristics which distinguish them from other baked products. They may be classified as intermediate moisture foods with moisture contents in the range 18–30% of the product mass. Cake products do have a thin crust but it does not usually have a significantly lower moisture than the crumb. The cellular structure of cakes tends to be less well defined than that of bread. However, there is considerable variation, with sponge cakes having a comparatively well-defined cell structure. Cake and sponge products usually have a soft and friable eating character with resilience in the crumb. The flavors found in sponges and cakes are totally dominated by the choice of ingredients and the recipe used. A key attribute of cakes is the relatively longer shelf life which they enjoy (often many weeks) compared with that of bread (a few days).

12.2.3
Biscuits, Crackers, and Cookies

These product have many significant differences from other classes of baked products. The most obvious is their smaller unit size and weight. The moisture content of biscuits products is very low, typically under 5%. The low moisture content coupled with the thinness of the products gives them a crisp, hard eating character though this may be moderated in biscuit types with higher recipe fat levels. The low moisture content and low water activity of products in this group means that they have long mold-free shelf-lives, typically many months. As is the case with cakes, the flavor of biscuits is dominated by the ingredients and the recipe used.

12.2.4
Pastries

Few pastry products are eaten "alone," that is, without some filling or topping, or both. They are a versatile medium which can be considered as an "edible packaging." While the fillings used in the manufacture of pastry products may have a wide range of textures, moisture contents, and water activities, the pastry itself tends to have a moisture content above that of biscuits but below that of cake. Typically the moisture content of pastries confers a firm and relatively crisp eating character to the product when freshly baked. Since the water activities of pastries are commonly below that of the fillings used in pies, and so on, water readily moves from the filling to the pastry with the result that the pastry softens and loses its crispness [3]. The "shelf life" of the pastry can be quite long but the migration of moisture from filling to paste reduces this life considerably so that typical shelf lives will range from a few days for meat-containing pastries to a few weeks for pastries with sweet fillings.

12.3 Bread Making

Bread is a staple foodstuff made and eaten in most countries around the world. Over the centuries craft bakers have developed traditional bread varieties making the best use of their available raw materials and this has led to a proliferation of bread making methods. However, all bread types are based on three underlying principles: the development of a gluten network in the dough, the expansion of the dough through yeast fermentation, and the conversion of the foam structure trapped in the gluten network into the sponge structure formed in the oven.

The few basic steps which form the basis of all modern bread making processes can be listed as follows [4]:

1) The mixing of wheat flour and water, together with yeast and salt, and other specified ingredients in appropriate ratios.
2) The development of a gluten structure in the dough through the application of energy during mixing.
3) The incorporation of air bubbles within the dough during mixing.
4) The continued "development" of the gluten structure initially created in order to modify its rheological properties and to improve its ability to expand when gas pressures increase during fermentation.
5) The creation and modification of particular flavor compounds in the dough.
6) The sub-division of the dough mass into unit pieces for processing.
7) A preliminary modification of the shape of the divided piece.
8) A short delay in processing to further modify physical and rheological properties of the dough pieces.
9) The forming of the dough pieces to their required shape.
10) The fermentation and expansion of the shaped dough pieces during "proof."
11) Further expansion of the dough pieces and fixation of the final bread structure during baking.
12) Cooling and storage of the final product before consumption.

12.3.1
Dough Development

Dough development is a relatively undefined term which covers a number of complex changes that begin when the ingredients are first mixed. These changes are associated with the formation of gluten, which requires both the hydration of the proteins in the flour and the application of energy through the process of kneading. The role of energy in the formation of gluten is not always fully appreciated but it is a significant contributor to the bread making process. The development process brings about changes in the physical properties of the dough and in particular improvement in its ability to retain the carbon dioxide gas which will later be generated by yeast fermentation. This improvement in gas retention ability is particularly important when the dough pieces reach the oven. In the early stages of baking before the dough has set, yeast activity is at its greatest and large quantities of carbon dioxide gas are being generated and released from solution in the aqueous phase of the dough. If the dough pieces are to continue to expand at this time then the dough must be able to retain a large quantity of that gas being generated and it can only do this if we have created a gluten structure with the appropriate physical properties.

12.3.2
Dough Mixing and Bread Making Methods

The different methods used to manufacture bread dough may be placed into four broad categories: bulk fermentation, mechanical dough development (e.g., the Chorleywood Bread Process (CBP)), sponge and dough, and no-time dough making.

12.3.2.1 Bulk Fermentation

In these types of process the bulk dough is rested for a period of time after mixing under defined temperature condition before being sub-divided into unit pieces for processing. A key controlling factor in optimizing consistent bread quality with bulk fermentation is the skill of the baker to judge when sufficient change had taken place in the rheological properties of the dough to yield the desired characters in the final bread. The process is often referred to as "*ripening*." Key issues with fermentation processes are:

- the quantity of protein in the flour, with lower protein flours requiring shorter fermentation times to achieve maturity for further processing, and vice versa;
- the modification of bread flavor because of the extended fermentation processes. The flavor formation reactions are complex and commonly not from the added yeast but from a range of different microorganisms naturally present in the flour. The choice of bulk fermentation temperature can have a profound impact on the nature of the flavor development.

12.3.2.2 The Chorleywood Bread Process (Mechanical Dough Development)

The historical development of the CBP and its uses are discussed in detail by Cauvain and Young [5]. They may be summarized as follows:

- Mixing and dough development in a single operation lasting between 2 and 5 min to a fixed energy expenditure during mixing.
- The addition of an oxidizing improver, usually ascorbic acid [6].
- The inclusion of a high melting point fat, emulsifier, or fat and emulsifier combination [6].
- The addition of extra water to adjust dough consistency to be comparable with those from bulk fermentation.
- The addition of extra yeast to maintain final proof times comparable with that seen with bulk fermentation.
- The control of mixer headspace atmosphere to achieve given bread cell structures.

As the level of energy per kilogram dough in the mixer increases, bread volume increases, and with the increase in bread volume comes a reduction in crumb cell size, increased cell uniformity, and improved crumb softness [4]. The creation of bubble structures in CBP dough and indeed for many other no-time processes, depends on the occlusion and, to some extent, the sub-division of air during mixing. The numbers, sizes, and regularity of the gas bubbles depend in part on the mixing action, energy inputs, and control of mixer headspace atmospheric conditions.

12.3.2.3 Sponge and Dough

The sponge and dough process is most common in the United States [4]. Elements of the processes are similar to those for bulk fermentation in that a prolonged period of fermentation is required to effect physical and chemical changes in the dough. Only part of the total ingredient mass is fermented – the sponge – and fermentation times may vary considerably, as may their compositions. The key features may be summarized as follows:

- A two-stage process in which part of the total quantity of flour, water, and other ingredients from the formulation (commonly 25–40% of the total) are mixed to form an homogeneous soft dough – the sponge.
- The resting of the sponge in bulk for a prescribed time at a prescribed temperature.
- Mixing of the sponge with the remainder of the recipe ingredients to form an homogeneous dough.
- Immediate processing of the final dough after mixing.

The sponge contributes to flavor modification and the development of required rheological properties in the final dough. Additions of improvers are commonly made in the dough rather than the sponge making stage. Flours used in typical sponge and dough production will be at least as strong as those used in bulk fermented dough, with protein contents not less than 12% (on a 14% moisture basis).

12.3.2.4 **No-Time Dough Processes**

In many smaller bakeries no-time dough making processes are used with mixers running with a single, vertical S-shaped mixing tool. Typically the bowl will rotate and often there is a single, vertical, fixed bar to increase energy input to the dough. Mixing speeds and energy input levels are somewhat lower than those seen with the CBP or sponge and dough bread making methods. The bulk dough may be given a short period of fermentation after mixing and before dividing but this is becoming less common. In the case of the Dutch green dough process [4] more than one rounding stage may be employed with extended resting periods between (about 40 min).

12.3.2.5 **The Contribution of Flour to Bread Quality**

Bread quality is determined by the complex interactions of the raw materials, their qualities and quantities used in the recipe, and the dough processing method. It is therefore not possible to point to a single aspect of bread making and identify with clarity the factor which will predict bread quality. However, it is well understood that the contribution of the wheat flour proteins probably outweighs all of the other individual ingredient contributions, not least because the formation of gluten is an essential component of every bread making process and the many variations which exist.

The level and quality of the gluten-forming proteins depends on the wheat variety, agricultural practices, and environmental effects. In general, the higher the protein content in the wheat the higher the protein content of the flours produced from it and the better is its ability to trap and retain carbon dioxide gas. The nature of the polymers which comprise gluten is complex [1]. The main functional polymers are the gliadin and the glutenin and it is these that are largely responsible for the visco-elastic properties of gluten. In the three-dimensional network of gluten in dough, disulfide and hydrogen bonding and protein entanglements form the network structure which is so important for trapping air during dough mixing and retaining carbon dioxide from fermentation. It is now clear that the molecular size, structure, and molecular weight distribution of the gluten-forming polymers are all major factors in determining their performance in baking and various models for the polymer structures have been proposed (for example, see contributions in Cauvain [7]).

12.3.2.6 **Dough Mixing**

In essence, mixing is the homogenization and hydration of the ingredients, whereas kneading is the development of the dough (gluten) structure by "work done" after the initial mixing. The changes which take place in bread dough during mixing can be followed by measuring the resistance of a flour–water mixture with time during mixing. The overall shape of the curve which is obtained is shown in Figure 12.2. Initially there is little resistance as the ingredients are first blended; gradually as the proteins and damaged starch in the flour are hydrated the resistance of the mixture begins to increase. Resistance to the movement of the mixer arms increases dramatically as the gluten network is developed in the dough, and reaches

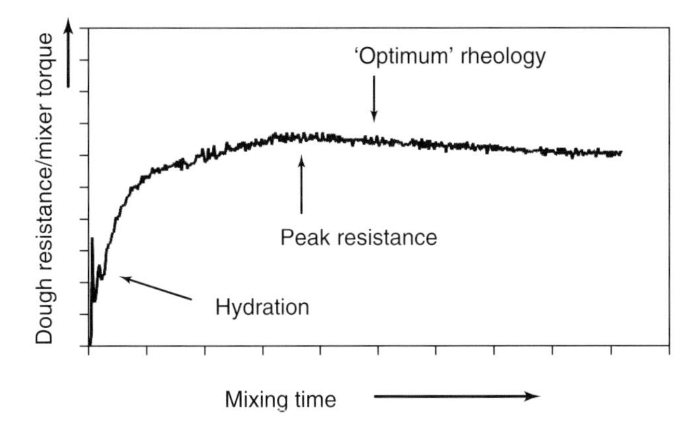

Figure 12.2 Stylized mixer torque-time trace obtained when mixing wheat flour and water.

a maximum before gradually falling. The precise shape of the curve depends heavily on the level and quality of the proteins in the wheat flour and analytical tests have been developed utilizing this principle to assess the suitability of wheat flours for bread making [8].

The determination as to when a dough is "fully" developed by mixing is not precisely defined, not least because the particular definition used will be related to the combination of dough rheology and bread characteristics considered to be desirable or optimal. In many cases optimum dough development is not considered to be the point (mixing energy or time) at which peak resistance is reached, instead many empirical definitions consider a point beyond peak resistance as being beyond the optimum, typically 10–20% of the time taken to reach peak resistance [9]. Applying near-infrared (NIR) technology and principal components analysis to wavelength data gathered during dough mixing Millar [10] identified a "turning point" on the derived NIR curve which approximated to peak dough resistance. Maximum bread volume was achieved a short time prior to the turning point while the finest bread crumb cell structure was achieved shortly afterwards. Such findings illustrate the difficulties in defining when optimum dough development has been achieved.

While the development of a gluten network in the dough is critical there are other processes taking place which are also worthy of note. Marsh and Cauvain [11] summarized the various sub-processes taking place during mixing as follows:

- the uniform dispersion of the recipe ingredients;
- dissolution and hydration of those ingredients, in particular the flour proteins and the damaged starch;
- the development of a gluten (hydrated flour protein) structure in the dough arising from the input of mechanical energy by the mixing action;
- the incorporation of air bubbles within the dough to provide the gas bubble nuclei for the carbon dioxide which will be generated by yeast fermentation and oxygen for dough oxidation and yeast activity;

- the formation of a dough with suitable rheological properties for subsequent processing.

In both the CBP and sponge and dough mixing processes the velocity of the dough being moved around within the mixing chamber is used to incorporate the full volume of ingredients into the mix and impart energy to the dough from the mixing tool. In the United Kingdom energy levels of around 11 Wh/kg of dough in the mixer are common, while in other parts of the world or with other products, such as breads in the United States, this may rise to as much as 20 Wh/kg of dough [12, 13]. Horizontal bar mixers are capable of mixing large quantities of dough in one batch and mixing speeds typically range up to a maximum of 150 rpm. The horizontal mixer is most often used with the sponge and dough process [14]. The lower mixing speed in horizontal bar mixers means that a longer mixing time is required than with the CBP-type.

12.3.2.7 Cell Creation and Modification during Mixing

The production of a defined cellular structure in the baked bread depends entirely on the creation of gas bubbles in the dough during mixing, their retention and the degree of coalescence during subsequent processing [15]. After mixing has been completed the only "new" gas which becomes available is the carbon dioxide gas generated by the yeast fermentation. Carbon dioxide gas has high solubility relative to other gases and in bread dough cannot form gas bubbles [16]. As the yeast produces carbon dioxide gas, the latter goes into solution in the aqueous phase within the dough until saturation is achieved. Thereafter continued fermentation causes dough expansion if the carbon dioxide gas is retained within its structure. Air already present in the flour and subsequently entrained during mixing provides oxygen and nitrogen gases. The residence time for oxygen is relatively short since it is quickly used up by the yeast cells within the dough and by the end of, or soon after mixing the only gas which remains entrapped is nitrogen. These nitrogen gas bubbles provide bubble nuclei into which the carbon dioxide gas can diffuse. The numbers and sizes of gas bubbles in the dough at the end of mixing are strongly influenced by the mechanism of dough formation, mixer design, and mixing conditions in a particular machine [15].

The modification of gas bubble populations through the control of mixer headspace atmospheric conditions has been known for many years. In the CBP it was commonly controlled through the application of partial vacuum to CBP-compatible mixers [17] to create a fine and uniform cell structure in the baked product. In more recently developed CBP-compatible pressure–vacuum mixers (Figure 12.3), which are able to work sequentially at pressures above and below atmospheric, it has become possible to obtain a wider range of baked product cell structures. Pressure settings above atmospheric may be used enhance dough oxidation by ascorbic acid and deliver an open cell structure (e.g., for French sticks, see Figure 12.4). In modern CBP-compatible mixers it is possible to move sequentially from positive to negative pressures without stopping the mixer [18] and thereby obtain greater control of dough bubble

Figure 12.3 Commercial-scale CBP-type mixer. Reproduced with permission from BakeTran.

Figure 12.4 Open cell structure of French sticks made by the Chorleywood Bread Process.

populations in the dough to yield a wide range of product structure options [19, 20].

12.3.3 Dough Processing

The stages required to process the bulk dough into bread are as follows:

1) dividing;
2) rounding and first molding;
3) intermediate or first proving;
4) final molding or shaping;
5) structure control during dough processing;
6) proving;
7) baking.

12.3.3.1 Dividing

The bulk dough produced by mixing is divided into unit pieces of the required shape and size. This sub-division is usually achieved volumetrically with portions of a given size cut either by filling a chamber with dough and cutting off the excess (*piston dividing*) or by pushing the dough through an orifice at a fixed rate and cutting billets from the end at regular intervals (*extrusion dividing*). Different dividers need to be matched to different dough types in order to give optimum dividing accuracy with minimal compression damage. For example, "strong" North American bread doughs can withstand high compression loads whereas more delicate French baguette doughs are readily damaged.

12.3.3.2 Rounding and First Molding

Modification of the shape of the divided dough piece is common. Mechanical molding subjects the dough to stresses and strains and the rheological properties of the dough are important if damage to the relatively delicate gas bubble structure present in the dough is to be avoided.

12.3.3.3 Intermediate or First Proving

A period of rest between the work carried out by dividing and rounding and before final shaping may be used. The length of time chosen for this process is related to the dough rheology required for final molding. Changes occur in dough rheology during this time which make the final shaping of the dough piece easier to achieve.

12.3.3.4 Final Molding or Shaping

The most common practices involve:

- passage of the round dough piece through sets of parallel rolls moving at high speed to reduce its thickness;
- curling of the ellipse which has been created by sheeting by trapping the leading edge underneath a static chain (or some other device) which creates a "Swiss roll" of dough;
- compression and shaping of the Swiss roll to give a uniform cylinder of dough. This is achieved by compressing the dough piece underneath a molding or pressure board while it is still being moved along the length of the molder by the action of a moving belt;
- in some cases the dough pieces may be re-oriented to further modify the final bread cell structure; for example, by cutting the cylinder coming from the molding board into four equal pieces and turning these through 90° before placing them in the pan [21].

12.3.3.5 Structure Control during Dough Processing

It is in the final molding stages that one of two significant changes may occur in the gas bubble populations in the dough piece. Both changes are related to the final structure of the baked product; they are the potential elongation of gas bubbles and a slight, though potentially important degassing of dough pieces during sheeting and shaping. Commonly, as round dough pieces pass through pairs of sheeting rolls in the final molder some elongation of gas bubbles in the direction of sheeting occurs and this orientation of the bubbles is retained during subsequent curling. Elongation is most likely to occur with the larger gas bubbles located nearer to the surface of the dough during sheeting and it is unlikely that the pressures applied during sheeting will affect the smaller gas bubbles located in the center of the dough. The elongation of gas bubbles affects product quality because when baked into the bread and sliced the elongated form tends to be shallower than surrounding gas bubbles and they cast less shadow in the cut surface which makes the crumb appear whiter. Sheeting and bubble elongation also contribute to the physical strength of the breadcrumb during slicing and its eating qualities.

The degree to which a dough piece may be degassed during the sheeting stages of final molding depends on its rheology and interactions with the equipment. Whitworth and Alava [22] have shown that the de-gassing of no-time doughs is small but examination of computer tomography X-ray scans of CBP doughs shows that it does occur. The rupture of the gas-stabilizing films in the dough leads to the coalescence of small gas bubbles to form ones of larger size which can contribute to the formation of unwanted holes and dark colored streaks in the crumb structure of the finished loaf. Unwanted bubble coalescence arising from "molder damage" also contributes to firmer eating characteristics because of the thickening of the adjacent cell wall material (gluten).

12.3.3.6 **Proving**

Proving is the name given to the dough fermentation period which occurs after the molded pieces have been put into tins or placed in trays and before they enter the oven. The atmosphere in which this fermentation is conducted is controlled to encourage bakers' yeast activity and limit dehydration of the dough. Typically the conditions are 35–45 °C and 70–85% relative humidity. During proof the damaged starch from the flour is progressively converted into dextrins and sugars by the action of the enzymes which are present in the dough. Yeast feeds on the sugars naturally present in the flour and those resulting from the enzymic action to produce carbon dioxide and alcohol.

The carbon dioxide which is produced by bakers' yeast diffuses out of the liquid phase in the dough and into the entrapped gas bubbles, causing them to grow and the dough to expand. The growth and stability of the gas bubbles during proof have a significant impact on the ultimate bread volume and the crumb cell structure in the product [23]. The dough expansion process in proof (and in the early stages of baking – see below) is based on a balance of gas production and gas retention. Total gas production is driven by the level of recipe yeast, the dough temperature and the fermentation time allotted to the proving process. On the other hand, gas retention is driven by a large number of different ingredient, recipe, and processing factors, including the protein content of the flour, the use of oxidizing agents, fats, emulsifiers, and different enzymes and the degree of development achieved during mixing (see above).

A key process in the proving stage is the growth and subsequent coalescence of the small gas bubbles. At this time the rheological properties of the bubble cell walls are very important in limiting bubble coalescence. In breads baked in pans much of the dough expansion is vertical and so the elongation strain hardening properties of the gluten polymers are important [24, 25]. In free-standing breads the forces of extension are more uniformly distributed in the dough. These differences account in part for the finer, more uniform crumb cell structure of pan breads compared with free-standing (oven-bottom) type product.

12.3.3.7 **Baking**

After proof the dough must be heat-set, that is, baked, so that the conversion of the foam to a sponge can take place to yield the light and aerated palatable product that we call bread. Typical oven temperatures lie in the region of 200–250 °C and a key parameter of loaf quality is to achieve a core temperature of about 92–96 °C by the end of baking to ensure that the product structure is set sufficiently to be handled. For the center of a dough piece, the move from prover to the oven has little impact because it is so well insulated by surrounding dough and this means that in effect the center of the dough gets additional proof. The driving force for heat transfer is the temperature gradient from regions near the crusts, where the temperature is limited to the boiling point of water, to the center. The heat transfer mechanism is conduction along the cell walls and the center temperature will rise independently of the oven temperature and approach boiling point asymptotically. There is no significant movement of moisture and the moisture content will be the same at the end of baking as at the beginning.

As dough warms there are a series of complex physical, chemical, and biochemical changes. These are summarized in Table 12.2. Yeast activity decreases from 43 °C and ceases by 55 °C; even so there is significant release of carbon dioxide in the first third or so of the time that the dough piece spends in the oven. If the dough has sufficient gas retention capability then there will be an increase in product volume referred to by bakers as "oven spring." Bread products which lack oven spring not only have low volume but also a firmer eating quality and tend to stale faster.

The stability of the dough structure is maintained by the expansion of the trapped gases. Since the gas bubbles are rapidly expanding there will be increased coalescence of the small bubbles to form what will eventually be referred to as the *product crumb structure*. Coalescence sees an increase in bubble sizes from (typically) about 200 μm in the dough to 1–3 mm in the bread crumb. Not only does the size of the gas cells increase because of coalescence but their numbers fall by about 80–90%. During baking, strain hardening of the gluten polymers as they stretch under the influence of gas expansion is one of the mechanisms which determine product volume and crumb structure [26].

The formation of a crust provides strength of the finished loaf and the greater part of its flavor. Condensation on the surface of the loaf at the start of baking is essential for the formation of gloss, but quite soon the temperature of the surface rises above the local dew point temperature and evaporation starts. After the surface reaches the boiling point of the free liquid the rate of moisture loss

Table 12.2 Some changes taking place in dough, batter, and pastes during baking; the nature and extent of the changes depends on the product type.

Activity	Approximate temperature range (°C) of changes
Starch swelling leading to gelatinization	Swelling begins at about 40 and gelatinization around 65. The addition of sugar (as in cakes) raises the gelatinization temperature up to 90 depending on the concentration
Protein coagulation	Typically 65–75, less affected by recipe than starch
Evolution of carbon dioxide	Up to 55 in the case of yeast raised products, baking powder reactions may continue up to around 60
Thermal expansion of gases	Continues throughout the baking stage as long as the foam is intact in the product. Gas may be evolved after the sponge has been set but cannot contribute to product expansion
Evolution of water vapor	Achieved once the boiling point is reached (note may be higher than 100 because of presence of soluble materials). Continues throughout the whole baking process but cannot contribute to expansion once the sponge is set
Amylase activity	Commonly amylase activity on the starch as it swells and gelatinizes. The extent of the amylase activity depends on the source of the alpha-amylase
Enzyme activity	May come from a number of different sources through additions in the product recipe
Maillard browning	Achieved as the temperature approaches and exceeds 115

accelerates. There is conduction within the cell walls and water evaporates at the hot end of the cell [27]. Some is lost to the outside but the rest moves across the cell toward its center and condenses at the cold end of the cell. In doing so it transfers its latent heat before diffusing along the cell wall to evaporate again at the hot end. The evaporation front will develop at different rates depending on the bread types. The crust is outside the evaporation front and here the temperature rises toward the air temperature in the oven. As water is driven off and the crust acquires its characteristic crispness and color, flavor and aroma develop from the Maillard reactions, which start at temperatures above 115 °C.

Some breads are characterized by the crispness of their crust, for example, baguette. The first few moments in the oven are vital for the formation of a glossy crust and it is essential that vapor condenses on the surface to form a starch paste that will gelatinize, yield dextrins and eventually caramelize to give both color and shine. If there is excess water, paste-type gelation takes place, while with insufficient water crumb-type gelation occurs. To deliver the necessary water, steam is introduced into the oven.

Gelatinization of the starch in the dough starts at about 60 °C. Alpha-amylase activity converts the starch into dextrins and then sugars and reaches its maximum activity between 60 and 70 °C. As the amylase enzymes begin to break down the gelatinizing starch there is a release of water from the granule which appears to be taken up by the expanding gluten network. As shown by Cauvain and Chamberlain [28] using loaf height measurements during baking, the addition of fungal alpha-amylase (a common component of bread "improvers") allows the dough piece to expand to greater height and slightly delays the point at which the structure finally sets. The authors also showed that the foam-to-sponge conversion in bread dough (and cakes) is characterized by a small loss in product height during baking (see also Section 12.4).

Much of the initial firming effect that we see when bread cools comes not from the coagulated gluten but from the retrograding amylose portion of the starch [29]. In the oven the starch is gelatinized as it absorbs water and is heated but on cooling it begins to retrograde and this contributes to the firming of the bread crumb. The short-term firming effect allows the product to be sliced while the longer term firming effect is associated with the process of amylopectin retrogradation and is commonly described as "staling." This intrinsic firming of the bread crumb with time occurs even in the absence of moisture losses [30].

12.4 The Manufacture of Cakes

Cake batters are complex emulsions and foam systems [31] but, unlike bread dough, the foam is not stabilized by the formation of a gluten network. In their simplest form, the batters comprise wheat flour, sugar, and whole egg. If the egg and the sugar are whisked together the sugar goes into solution in the water present in the egg and large numbers of minute air bubbles are trapped in the batter by the surface active proteins in the egg. These proteins form a protective film around the air bubbles, preventing them from coalescing and escaping from the batter. In basic cake recipes it is important to avoid destabilizing the egg foam through the addition of other ingredients. However, many cake recipes contain a proportion of oil or fat to deliver a softer eating product and their addition changes the batter to an oil- (fat-)in-water emulsion where the aqueous, continuous phase contains the dissolved sugars, hydrated proteins and suspended flour, and other ingredient particles. Oil or solid fat has a considerable de-foaming effect on the egg properties and so it is necessary to change the foam creation and stabilization mechanism. This involves the addition of an emulsifier, usually glycerol monostearate, for the manufacture of sponge-type products and for more dense cake products the use of a composite fat with the necessary level of crystalline solids to take over the role of foam stabilization. In practice, a number of complex mixing procedures have evolved to form cake batters [32]. In many cases the need for elaborate multi-stage mixing processes was based on the use of ingredients in their "traditional" form, for example, milk, or to compensate for significant variations in ingredient character, for example, butter composition. Today, provided that sufficient water is available to dissolve and hydrate the necessary ingredients, many cake batters can be based on a single-stage, all-in mixing method.

Cake batters are low-viscosity systems by comparison with bread and biscuit doughs, and there is limited opportunity for a gluten structure to develop during mixing, in part because of the low resistance of the batter to the action of the mixer and in part because of the gluten-inhibitory effects of the ingredients in the recipe. The low viscosity of the batter after mixing makes it easy to deposit individual portions of batter into a container for subsequent baking. A few cake products are baked based on highly aerated sponge recipes and are deposited directly onto the oven band or trays in thin sheets for rapid baking (e.g., Swiss roll and sponge drops, Figure 12.5).

The characteristic eating qualities of cakes are significantly influenced by the level of batter aeration during mixing, the air retention, and the thermal gas expansion and generation of steam during baking. Chemical aeration through the addition of a suitable baking powder (combination of food acid and sodium bicarbonate) in the recipe usually augments the mechanical aeration which comes from mixing. The progression from foam in the cake batter to sponge (in the generic sense) follows a similar pattern to that of bread dough, though in the case of cakes the bubble stabilizing role is assumed by the egg protein, recipe fat, and recipe emulsifier rather than the gluten-forming proteins in the flour.

This is not to say that flour assumes a passive role in cake making. In cake systems the gelatinization of the starch plays a highly significant role in controlling batter viscosity and bubble coalescence and in establishing the final product structure. The temperature at which starch gelatinization takes place depends to a significant degree on the levels of recipe water (including that from water-containing ingredients, such as liquid egg) and sugar. The presence of a sucrose solution in the batter will delay the temperature at which the starch gelatinizes and thus the moment when the foam-to-sponge conversion takes place in the oven. The stronger the sucrose solution the higher the temperature at which the starch will gelatinize and so the later the foam-to-sponge conversion will occur in the oven.

Figure 12.5 Cake product (sponge drop) baked directly on the oven band. Reproduced with permission from BakeTran.

Figure 12.6 Fruit buns showing a wrinkled surface. Reproduced with permission from BakeTran.

The advantage of delaying the conversion is that the baked product can achieve a greater volume and softness of eat.

However, it is important that conversion does occur in the oven otherwise as the cake product leaves the oven it will collapse. This is because as the cake leaves the oven and its temperature falls the internal pressure of any gas cells trapped in elements of any unconverted foam will fall below that of the surrounding atmosphere and as the cake structure is still relatively unstable it collapses. A similar effect is seen with fermented products which contain significant quantities of sugar, for example, fruit buns, most often manifest as a "wrinkled" top crust (Figure 12.6). The pragmatic solution to this phenomenon is for the baker to bang the tray or pan of products and the mechanical shock that the product experiences will rupture the intact gas cells and cause an instantaneous equalization of internal and external pressures [21], otherwise reformulation of the recipe is required.

12.5 Biscuit and Cookie Making

Biscuits and cookies may be separated into five broad categories: hard-dough semi-sweet, rotary-molded short-dough, wire-cut cookie, crackers, and wafers, see Table 12.3. The individual groups may be distinguished from one another according to the degree to which gluten development occurs, or is desirable, as well as on the basis of the type of equipment used in their production. There is no significant foam creation in biscuit and cookie dough, so that expansion of these products in the oven is of a much lower order of magnitude than with bread and cakes.

For short-dough and cookie dough products, gluten development needs to be limited so that the forming processes for individual pieces can be easily accomplished and to avoid shrinkage of the pieces after forming and during baking.

Table 12.3 The characteristics of the main types of biscuits and cookies.

Product	Main characteristics	Gluten formation	Processing technology
Crackers	Flaky eating	Significant	Sheeted dough, laminated with fat-flour dust, sheeted, and cut
Water biscuits	Flinty eating	Modest	Sheet and cut
Semi-sweet biscuits	Hard eating	Limited	Sheet and cut
Short-dough biscuits	Short, crumbly texture	Very little	Rotary molded or rout press
Cookies	Crumbly texture	Very little	Rotary molded or rout press
Wafers	Crisp, powdery texture	None	Deposited batter

Sugar and fat levels also contribute to limiting gluten formation. Short-dough biscuits are usually shaped by pressing the soft dough into a mold cut into the metal roll of a rotary molder. After extraction from the mold the pieces move quickly to the oven for baking. Wire-cut cookies are based on similar dough consistency. In this case the individual pieces are formed by forcing the soft dough through a tube and as it emerges from the end a wire or knife passes through the dough to cut off a unit piece of relevant size. This technique is particularly useful in the manufacture of cookies that contain particulate materials, like nuts and chocolate chips.

Hard-dough, semi-sweet biscuits require a greater degree of gluten formation and so added water levels tend to be a little higher (typically 20–25% flour weight), and fat and sugar levels lower. Modification of the dough rheological character may also be undertaken through the addition of a reducing agent, commonly sodium metabisulfite, or a suitable proteolytic enzyme or through the adjustment of the recipe water level. These biscuits are made by sheeting the dough and then passing the sheet under a cutter, or series of cutters to deliver the final biscuit shape for baking.

The control of biscuit and cookie dough temperature is as important as that for any baked product in order to obtain consistent processing and final product quality. There can be significant temperature rise during dough mixing from the frictional heat which occurs as the stiff dough moves around the mixer. One of the more unusual features of biscuit and cookie manufacture is that the doughs may be mixed to a fixed final temperature rather than for a fixed time or to a fixed energy level [33]. In short-biscuit doughs the choice of final dough temperature is often related to the type of fat being used since in low water dough the consistency of the fat makes a significant contribution to effective processing of the dough on the plant.

The degree of gluten formation in the manufacture of crackers needs to be greater than that with other biscuits in order to maintain the integrity of the dough and contribute to product lift. This means that water levels tend to be higher than with the biscuit types discussed above and fat levels lower. The mixing methods used deliberately set out to encourage gluten formation though not to the same degree as achieved with bread dough. Dough temperature control is important, especially as some forms of crackers may contain yeast and a fermentation step may be employed [33]. Crackers and some other biscuit forms are made by sheeting the dough through pairs of smooth rolls, folding the sheet to create one or more layers (laminating) with further sheeting to reduce the thickness of the paste by passing through more rolls. Fat or a fat–flour dust may be incorporated between layers in order to increase the "flakiness" of the product and there may be more than one laminating step. After the final sheeting reduction the dough sheet passes under a cutter and the individual dough pieces are removed for baking.

Wafers are low-fat biscuits derived from a batter. Recipe water levels are high and the viscosity of the mix is sufficiently low to allow the batter to be deposited onto hot plates for baking. After depositing onto one plate a second is placed over the top and the steam pressure which is created from the heat of the oven and the restricting effect of the gap between the plates forces the batter deposit into a sheet. Some deposited forms of biscuit are baked directly on a hot plate and because of their high sugar content may remain flexible enough immediately after baking to be folded and shaped. In the case of all types of deposited biscuits the formation of a gluten network is a distinct disadvantage and can lead to processing problems [21].

12.6 The Manufacture of Pastry Products

Short-dough pastes are used in a variety of bakery applications and sweet and savoury products. The other main form of pastry is commonly referred to as *laminated pastry* and includes puff pastry and yeasted examples, such as croissant and Danish. Gluten formation is not normally required in short-pastry products and if it occurs may lead to problems during processing and baking. On the other hand, a reasonable degree of gluten formation is required in laminated pastries in order for the paste to withstand the considerable processing that is required to make laminated products. As with biscuits, added water levels in short paste formulations are kept to a minimum because much of the water is baked out in the oven to give a crisp eating character to the pastry. Mixing methods for short pastes may be all-in or multi-stage and in all cases the aim is to limit gluten formation. After mixing, the short paste may be rested for a short period of time to modify the rheological properties of the paste and limit the risk of shrinkage. Short paste products are usually made by cutting shapes from a sheet of the paste and then folding and forming or by a process known as "*blocking*," in which a portion of paste (the "billet") is placed in a foil or pan held in a shaped die and subjected to pressure from a second die moving downward. The force of the downward moving die squeezes the paste into the narrow gap which is formed between the moving and static dies, the moving die is withdrawn upward and a paste shell remains behind ready for removal, filling with suitable sweet or savoury filling, and then baking, sometimes a sheeted (or rotary molded) paste lid may be placed on the top of the filled product.

Laminated products have a distinctive flaky eating character which is achieved by creating alternate layers of paste and fat. Little fat is added to the base paste formulation and so gluten development is more likely to occur during mixing. Usually the gluten structure in the base dough is less well developed than that in bread dough because there is significant energy transfer to the paste during subsequent processing and this adds to the gluten development that would have occurred during mixing. The key rheological character of the base dough is that it should be easily formed into a continuous sheet onto which the laminating fat is placed. A series of sheeting (thickness reduction) and folding (laminating) operations follows and this progressively builds up alternate and discrete layers of dough and fat. Resting stages may be used to modify the rheological properties of the paste depending on the flour qualities and the product requirements. After sheeting and forming, unyeasted laminated products (e.g., puff pastry) usually pass quickly to the oven while yeasted laminated products (e.g., croissant) require a period of proof before they are ready for baking.

There is no significant foam creation in any pastry product. The expansion of short pastry on baking is relatively limited. However, in the case of laminated products the degree of expansion during baking can be significant. This expansion comes from the steam generated from the water in the dough during baking. The movement of the steam out of the laminated dough is impeded by the melting fat layers and pressure builds up between the dough layers forcing them part. Expansion in such products reaches its maximum about halfway through the short baking cycle and thereafter the structure becomes set as the gluten coagulates.

References

1. Stauffer, C.E. (2007) Principles of dough formation, in *Technology of Breadmaking*, 2nd edn (eds S.P. Cauvain and L.S. Young), Springer Science + Business Media, New York, pp. 299–332.
2. Cauvain, S.P. and Young, L.S. (2006) *Baked Products; Science, Technology and Practice*, Blackwell Publishing, Oxford.
3. Cauvain, S.P. and Young, L.S. (2008) *Bakery Food Manufacture and Quality: Water Control and Effects*, 2nd edn, Wiley-Blackwell, Oxford.
4. Cauvain, S.P. (2008) Breadmaking processes, in *Technology of Breadmaking*, 2nd edn (eds S.P. Cauvain and L.S. Young), Springer Science + Business Media, New York, NY, pp. 21–50.
5. Cauvain, S.P. and Young, L.S. (2006) *The Chorleywood Bread Process*, Woodhead Publishing Ltd, Cambridge.
6. Williams, A. and Pullen, G. (2008) Functional ingredients, in *Technology of Breadmaking*, 2nd edn (eds S.P. Cauvain and L.S. Young), Springer Science + Business Media, New York, pp. 51–92.
7. Cauvain, S.P. (2003) *Bread Making: Improving Quality*, Woodhead Publishing Ltd, Cambridge.
8. Cauvain, S.P. and Young, L.S. (2009) *The ICC Handbook of Cereals, Flour, Dough and Product Testing: Methods and Applications*, DEStech Publications Inc., Lancaster, PA.
9. Millar, S. (2003) Controlling dough development, in *Bread Making: Improving Quality* (ed. S.P. Cauvain),

Woodhead Publishing Ltd, Cambridge, pp. 401–423.

10. Millar, S. (2000) Mixed fractions. *Eur. Baker*, **5/6**, 18–23.
11. Marsh, D. and Cauvain, S.P. (2008) Mixing and dough processing, in *Technology of Breadmaking*, 2nd edn (eds S.P. Cauvain and L.S. Young), Springer Science + Business Media, New York, pp. 93–140.
12. Tweedy of Burnley Ltd (1982) Dough mixing for farinaceous foodstuffs. UK Patent GB 2,030,883B, HMSO, London.
13. Gould, J.T. (2008) Breadmaking around the world, in *Technology of Breadmaking*, 2nd edn (eds S.P. Cauvain and L.S. Young), Springer Science + Business Media, New York, pp. 223–244.
14. Stear, C.A. (1990) *Handbook of Breadmaking Technology*, Elsevier Applied Science, London.
15. Cauvain, S.P., Whitworth, M.B., and Alava, J.M. (1999) The evolution of bubble structure in bread doughs and its effects on bread cell structure, in *Bubbles in Food* (eds G.M. Campbell, C. Webb, S.S. Pandiella, and K. Niranjan), Eagen Press, Saint Paul, pp. 85–88.
16. Baker, J.C. and Mize, M.D. (1941) The origin of the gas cell in bread dough. *Cereal Chem.*, **18**, 19–34.
17. Pickles, K. (1968) Improvements in or relating to dough production. UK Patent 1, 133, 472, Tweedy (Chipping) Ltd., HMSO, London.
18. APV Corporation Ltd (1992) Dough mixing. UK Patent GB 2, 264, 623A, HMSO, London.
19. Cauvain, S.P. (1994) New mixer for variety bread production. *Eur. Food Drink Rev.*, **51**, 53.
20. Cauvain, S.P. (1995) Creating the structure: the key to quality. *S. Afr. Food Rev.*, **22**, 33, 35, 37.
21. Cauvain, S.P. and Young, L.S. (2001) *Baking Problems Solved*, Woodhead Publishing, Cambridge.
22. Whitworth, M.B. and Alava, J.M. (1999) The imaging and measurement of bubbles in bread doughs, in *Bubbles in Food* (eds G.M. Campbell, C. Webb, S.S. Pandiella, and K. Niranjan), Eagen Press, Saint Paul, pp. 221–231.
23. Chiotellis, E. and Campbell, G.M. (2003) Proving of bread dough II. Measurement of gas production and retention. *Trans. Inst. Chem. Eng.*, **81**, 207–216.
24. Kokelaar, J.J., van Vliet, T., and Pins, A. (1996) Strain hardening and extensibility of flour and gluten doughs in relation to breadmaking performance. *J. Cereal Sci.*, **24**, 199–214.
25. Dobraszczyk, B.J. and Morgenstern, M.P. (2003) Review: rheology and the breadmaking process. *J. Cereal Sci.*, **38**, 229–245.
26. van Vliet, T., Janssen, A.M., Bloksma, A.H., and Walstra, P. (1992) Strain hardening of dough as a requirement for gas retention. *J. Texture Stud.*, **23**, 439–460.
27. Wiggins, C. and Cauvain, S.P. (2008) Proving, baking and cooling, in *Technology of Breadmaking*, 2nd edn (eds S.P. Cauvain and L.S. Young), Springer Science + Business Media, New York, pp. 141–174.
28. Cauvain, S.P. and Chamberlain, N. (1988) The bread improving effect of fungal alpha-amylase. *J. Cereal Sci.*, **8**, 239–248.
29. Schoch, T.J. and French, D. (1947) Studies on bread staling. 1. Role of starch. *Cereal Chem.*, **24**, 231–249.
30. Pateras, I.M.C. (2008) Bread spoilage and staling, in *Technology of Breadmaking*, 2nd edn (eds S.P. Cauvain and L.S. Young), Springer Science + Business Media, New York, NY, pp. 275–298.
31. Cauvain, S.P. (2003) Nature of cakes, in *Encyclopaedia of Food Science and Nutrition*, 2nd edn (eds B. Caballero, L. Trogo, and P.M. Finglas), Academic Press, Saint Louis, MO, pp. 751–756.
32. Cauvain, S.P. (2003c) Methods of manufacture, in *Encyclopaedia of Food Science and Nutrition*, 2nd edn (eds B. Caballero, L. Trogo, and P.M. Finglas), Academic Press, Saint Louis, MO, pp. 756–759.
33. Manley, D. (2000) *Technology of Biscuits, Crackers and Cookies*, 3rd edn, Woodhead Publishing Ltd, Cambridge.

13
Extrusion

Paul Ainsworth

13.1
General Principles

Extrusion can be defined as the process of forcing a pumpable material through a restricted opening. It involves compressing and working a material to form a semi-solid mass under a variety of controlled conditions and then forcing it, at a predetermined rate, to pass through a hole.

The origins of extrusion are in the metallurgical industry, where in 1797 a piston-driven device was used to produce seamless lead pipes [1]. The current understanding of extrusion technology and the developments in machine design are largely a result of research carried out by the plastics industry.

Extrusion technology was first applied to food materials in the mid 1800s, when chopped meat was stuffed into casings using a piston type extruder. In the1930s, a single-screw extruder was introduced to the pasta industry, to both mix the ingredients (semolina and water) and to shape the resulting dough into macaroni in one continuous operation.

Today, a wide variety of intermediate or food products are produced by extrusion.

13.1.1
The Extrusion Process

Extrusion is predominantly a thermomechanical processing operation that combines several unit operations, including mixing, kneading, shearing, conveying, heating, cooling, forming, partial drying, or puffing, depending on the material and equipment used. During extrusion processing, food materials are generally subjected to a combination of high temperature, high pressure, and high shear. This can lead to a variety of reactions with corresponding changes in the functional properties of the extruded material.

In the extrusion process, there are generally two main energy inputs to the system. First, there is the energy transferred from the rotation of the screws and

Food Processing Handbook, Second Edition. Edited by James G. Brennan and Alistair S. Grandison.

second, the energy transferred from the heaters through the barrel walls. The thermal energy generated by viscous dissipation and/or transferred through the barrel wall results in an increase in the temperature of the material being extruded. As a result of this, there may be phase changes, such as melting of solid material, and/or the evaporation of moisture.

The ingredients used in extrusion are predominantly dry powdered materials, the most commonly used being wheat, maize, and rice flours. The conditions in the extruder transform the dry powdered materials into fluids and so characteristics such as surface friction, hardness, and cohesiveness of particles become important. In the high solids concentration of the extruder melt, the presence of other ingredients, such as lipids and sugars, can cause significant changes in the final product characteristics.

In addition to starch-based products, a range of protein-rich products can be manufactured by extrusion, using raw materials such as soya or sunflower, fava beans, field beans, and isolated cereal proteins.

Figure 13.1 shows a schematic diagram of an extruder.

In order to convey the dry raw material to the extruder barrel, volumetric and gravimetric feeders tend to be used. Volumetric devices include single- and twin-screw feeders, rotary airlock feeders, disk feeders, vibratory feeders, and volumetric belt feeders. In all of these feed mechanisms, it is assumed that the density of the feed material remains constant over time and hence a constant volume of feed will result in a constant mass flow rate [2]. Gravimetric feeders are more expensive and more complex than volumetric feeders. They are usually microprocessor controlled to monitor the mass flow rate, and adjust the feeder speed as required. The most commonly used gravimetric feeders are the weight-belt and the loss-in-weight feeders [3].

Addition of liquid feed ingredients to the extruder can be achieved using a variety of devices including rotameters, fluid-displacement meters, differential-pressure meters, mass flow meters, velocity flow meters, and positive displacement pumps [2].

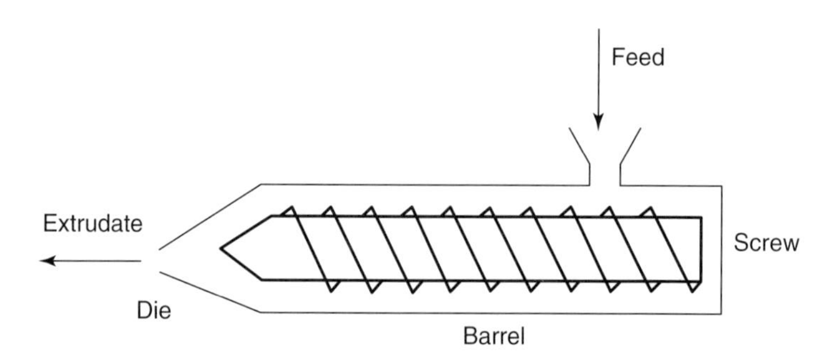

Figure 13.1 Schematic diagram of an extruder.

13.1.2
Advantages of the Extrusion Process

There are many benefits to using extruders to process food materials. They are capable of producing a wide variety of different product types and shapes, often with only small changes to the extruder, its operational settings or the raw materials used. From an engineering perspective, extruders can be described as a combination of a pump and a scraped surface heat exchanger for which operating conditions are relatively insensitive to material viscosity [4]. Thus extrusion systems are able to process highly viscous materials that are difficult or impossible to handle using conventional methods.

The ability of extruders to process biopolymers and ingredient mixes at relatively high temperatures (250 °C) and pressures (e.g., 25 MPa) with high shear forces and low moisture contents (10–40%), leads to a variety of fast and comparatively efficient chemical reactions and functional changes of the extruded material [5].

The ability of extrusion systems to carry out a series of unit operations simultaneously and continuously gives rise to savings in labor costs, floor space costs, and energy costs while increasing productivity [6]. These production efficiencies, combined with the ability to produce shapes not easily formed with other production methods, have led to extensive use of extrusion in the food industry. An indication of the range of applications is given in Table 13.1 [5].

Table 13.1 Extrusion cooking applications [5].

Applications of extrusion cooking	
Bread crumbs	Degermination of spices
Precooked starches	Flavor encapsulation
Anhydrous decrystallization of sugars to make confectioneries	Enzymic liquefaction of starch for fermentation into ethanol
Chocolate conching	Quick-cooking pasta products
Pretreated malt and starch for brewing	Oilseed treatment for subsequent oil extraction
Stabilization of rice bran	Preparation of specific doughs
Gelatin gel confectioneries	Destruction of aflatoxins or gossypol in peanut meal
Caramel, licorice, chewing gum	Precooked soy flours
Corn and potato snack	Gelation of vegetable proteins
Coextruded snacks with internal fillings	Restructuring of minced meat
Flat crispbread, biscuits, crackers, cookies	Preparation of sterile baby foods
Pre-cooked flours, instant rice puddings	Oilseed meals
Cereal-based instant dried soup mixes or drink bases	Sterile cheese processes
Transformation of casein into caseinate	Animal feeds
Precooked instant weaning foods or gruels	Texturized vegetable proteins

13.2
Extrusion Equipment

Extruders come in a wide variety of shapes, sizes, and methods of operation, but can be categorized into three main types: piston, roller, and screw extruders [4, 7]. The simplest of these is the piston extruder, which consists of a single or a battery of pistons that force the material through a nozzle onto a wide conveyor. The pistons can deliver very precise quantities of materials and are often used in the confectionery industry to deposit the center fillings of chocolates. Roller extruders consist of two counter-rotating drums placed close together. The material is fed into the gap between the rollers that rotate at similar or different speeds and have smooth or profiled surfaces. A variety of product characteristics can be obtained by altering the rotation speeds of the rollers and the gap between them. This process is used primarily with sticky materials that do not require high pressure forming. Screw extruders are the most complex of the three categories of extruder, and employ a single, twin, or multiple screws rotating in a stationary barrel to convey material forward through a specially designed die. Among the screw-type extruders, it is common to classify machines on the basis of the amount of mechanical energy they can generate. A low-shear extruder is designed to minimize the mechanical energy produced and is used primarily to mix and form products. Conversely, high-shear extruders aim to maximize mechanical energy input, and are used in applications where heating is required.

In the food industry, single- and twin-screw extruders predominate and hence are discussed in more detail. Figure 13.2 shows a co-rotating twin-screw extruder (Continua 37, Werner and Pfleiderer, Stuttgart, Germany).

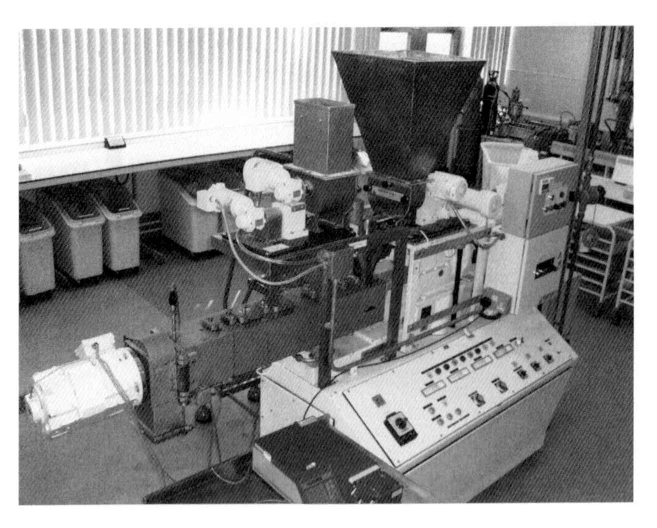

Figure 13.2 Co-rotating twin-screw extruder (Continua 37, Werner and Pfleiderer, Stuttgart, Germany).

13.2.1 Single-Screw Extruders

Single-screw extruders were first used in the 1940s to produce puffed snacks from cereal flours and grits. During transport through the extruder, mechanical energy from the rotation of the screw is converted to heat, raising the temperature of the mixture to over 150 °C. The resulting plasticized mixture is then forced through the die. The sudden reduction in pressure at the die results in a rapid flash off of the moisture as steam, puffing the product. From the late 1950s, extrusion processes were developed to manufacture products such as dry expanded pet food, dry expanded ready-to-eat breakfast cereals, and textured vegetable protein.

Because single-screw extruders have relatively poor mixing ability, they are often used with materials that have been either premixed or preconditioned. Preconditioning is used to increase residence time, to reduce mechanical power consumption, and/or to increase capacity. The preconditioner is an atmospheric or pressurized chamber in which raw granular food ingredients are uniformly moistened or heated or both by contact with live steam or water before entering the extruder.

The single-screw extruder relies upon *drag flow* to convey the feed material through the barrel of the extruder and to develop pressure at the die. In order for the product to advance along the barrel it must not rotate with the screw. The frictional force between the material and the barrel wall is the only force that can keep the material from turning with the screw and hence many single-screw machines have grooves cut in the barrel to promote adhesion to the barrel wall.

The rotation of the screw in the barrel gives rise to a second flow, called the *cross-channel flow*. This flow does not contribute to the net movement of material along the barrel but instead recirculates it within the screw flights, and hence is responsible for some of the mixing action of the extruder.

In forcing the feed material along the barrel of the extruder and through the restricted opening of the die, a third flow known as the *pressure flow* is generated. The pressure flow causes movement backwards down the extruder barrel, causing further mixing of the product. The three flows combine to give the net flow of material out of the die.

Single-screw extrusion operation depends upon the pressure requirements of the die, the slip at the barrel wall (controlled by the barrel wall temperature, the presence of barrel wall grooves, or both), and degree to which the screw is filled. Screw fill is dictated by feed rate, screw speed, melt characteristics, and viscosity of the material extruded. The coupling of these variables limits the operating range and flexibility of single-screw extruders [8].

In the compression section of the screw, the compression ratio increases rapidly, causing most of the mechanical energy used to turn the screw to be dissipated leading to an increase in the temperature of the material. This results in the plasticization of the dry feed ingredients. Energy input to the extrusion system may also arise from heat transfer through the jacket, and latent heat from steam injected into the ingredients in the preconditioner.

13.2.2 Twin-Screw Extruders

Twin-screw extruders were introduced to the food industry in the 1970s and are now extensively used in food production. In addition to manufacturing foods similar to those produced by single-screw extruders, twin-screw extruders have found a wide application in the food industry because of their better process control and versatility, their flexible design permitting easy cleaning and rapid product changeover, and their ability to handle a wide variety of formulations.

Twin-screw extruders differ from the single-screw extruder in terms of their processing capability and mechanical characteristics, and are largely responsible for the increasing popularity of extrusion processing. The screws in a twin-screw extruder are positioned adjacent to each other and are retained in position by a profiled barrel housing, having a horizontal "figure of eight" appearance. The position of the screws in relation to one another and their direction of rotation, can be used to categorize twin-screw machines.

Twin-screw extruders can have intermeshing screws in which the flights of one screw engage the other or they can have non-intermeshing screws in which the threads of the screws do not engage one another, allowing one screw to turn without interfering with the other. Non-intermeshing screw extruders function like single-screw extruders but have a higher capacity.

Twin-screw extruders may have co-rotating or counter-rotating screws. Both co-rotating and counter-rotating extruders can have fully, partially, or non-intermeshing screws. Co- and counter-rotating screws differ in their transport characteristics and are therefore suited to different technological applications [7, 9].

Intermeshing twin-screw extruders generally act like positive displacement pumps, forcing the material within the flights to move toward the die by rotation of the screws. The movement of the material is dependent upon the screw geometry and occurs independent of the operating conditions. Twin-screw intermeshing extruders have found wide applications due to their positive pumping action, efficient mixing, and self-cleaning characteristics.

Intermeshing counter-rotating machines are particularly suited to the processing of relatively low-viscosity materials that require low screw speeds and a long residence time. Examples of products suited to this type of extruder include chewing gum, jelly, and licorice confections [7]. These extruders exhibit poor mixing characteristics as each half of the chamber housing the screws acts independently, thus two streams of material that have little interaction are generated. Hence the only mixing that is done is due to the recirculation within the chamber itself.

Although high pressures can be achieved in counter-rotating extruders if the screw speed is increased, large separating forces are generated at the interface between the screws, giving a calendaring effect which can cause excessive wear. Hence, the production of expanded products with counter-rotating extruders is considered uneconomic [2, 7].

Intermeshing co-rotating extruders are particularly suited to applications where a high degree of heat transfer is required but not forced conveyance and thus are widely used for the production of expanded products. In this type of extruder, the material being extruded is transferred from one screw to the other. The flow mechanism is a combination of both drag and positive displacement flow [10]. The self-wiping style of co-rotating extruders is most commonly used due to their high capacity and enhanced mixing ability [2]. Co-rotating extruders can be operated at higher speeds than counter-rotating extruders because the radial forces generated are more evenly distributed. The conveying ability of twin-screw extruders allows them to handle sticky and other difficult to handle ingredients [7].

In order to improve the mixing, heat transfer and viscous dissipation of mechanical energy, sections of the extruder are completely filled with material. To create these filled sections, some type of restriction is placed on the screw configuration. The addition of forward- or reverse-conveying disks into the screw configuration alters the pressure profile within the barrel. The forward-conveying disks push the material toward the die, increasing the pressure in the barrel. Reverse-conveying disks reduce the pressure by delaying the passage of material through the extruder, increasing the barrel fill prior to the restriction, allowing additional processing and improved efficiency of heat transfer through the barrel wall. Restrictions in the screw configuration are placed under greater stresses, hence they tend to wear, requiring more frequent replacement than other elements of the screw.

Non-intermeshing twin-screw extruders can be described as two single-screw extruders sitting side by side with only a small portion of the barrels in common [11]. Like the single-screw extruder, these extruders rely on friction for extrusion.

13.2.3 Comparison of Single- and Twin-Screw Extruders

A comparison of single- and twin-screw extruders is shown in Table 13.2 [9]. For a given throughput, twin-screw extruders are 1.5–2.0 times as expensive as single-screw extruders, primarily due to the complexity of the screw, drive, and heat transfer jackets [12].

The preconditioning of feed ingredients with live steam is widely used in conjunction with single-screw extrusion processes, and provides around half of the heat necessary for cooking/processing, the remainder of the heat being derived from the mechanical energy inputs. The rate of heat transfer to a material using the direct injection of live steam is very high, and thus represents the lowest cost method of heating the product. For this reason, single-screw extruders usually have lower energy costs than twin-screw extruders.

The cost of twin-screw extruders can, however, be offset by their ability to process at lower moisture levels, thus reducing or eliminating the need for additional post-process drying [12].

The geometry and characteristics of the screws used in twin-screw extrusion present some advantages over single-screw extruders and enables the twin-screw extruder to process a wide range of ingredients that cause feeding problems for single-screw machines [11]. The conveying angle combined with the self-wiping

Table 13.2 Relative comparison of single- and twin-screw extruders [9].

Item	Single-screw	Twin-screw
Relative cost/unit capacity		
Capital		
Extruder	1.0	1.5–2.5
System	1.0	0.9–1.3
Relative maintenance	1.0	1.0–2.0
Energy		
With preconditioner	Half from steam	Generally not used
Without preconditioner	Mechanical energy	Mix of mechanical and heat exchange
Screw		
Conveying angle	10°	30°
Wear	Highest at discharge and transition sections	Highest at restrictions and kneading disks
Positive displacement	No	No, but approached by fully intermeshing screws
Self cleaning	No	Self-wiping
Variable flight height	Yes	No
L/D	4–25	10–25
Mixing	Poor	Good
Uniformity of shear	Poor	Good
Relative RTD spread	1.2	1.0
Venting	Requires two extruders	Yes
Drive		
Relative screw speed	1.0–3.0	1.0
Relative thrust bearing capacity	Up to 2.5	1.0
Relative torque/pressure	Up to 5	1.0
Heat transfer	Poor. Jackets control barrel wall temperature and slip at wall	Good in filled sections
Operations		
Moisture (%)	12–35	6 to very high
Ingredients	Flowing granular materials	Wide range
Flexibility	Narrow operating range	Greater operating range

feature results in an extruder that is less prone to surging. Increased uniformity of processing also occurs in twin-screw extruders due to the consistency of shear rate across the channel depth, leading to a narrower residence time distribution and increased mixing within the screw channel [2].

A single-screw extruder is relatively ineffective in transferring heat from barrel jackets because convective heat transfer is limited by poor mixing within the channel. Instead, the jackets in single-screw extruders control barrel wall temperature to regulate slip between the materials and the wall. Twin-screw extruders have considerably more heat exchange capability that expands their application to heating and cooling viscous pastes, solutions, and slurries [12].

Single- and twin-screw extruders have found many different applications in the food industry. For example, single-screw extruders are considered an economical and effective method for the thermal processing and forming of pet foods, while twin-screw extruders have been widely used in snack food production, where better control and flexibility is required [9].

13.3 Effects of Extrusion on the Properties of Foods

13.3.1 Extrusion of Starch-Based Products

Cereal flours and other starchy materials are widely used as raw materials in the production of many extruded products. The physical characteristics of cereal fluids developed within an extruder and their extrudates are predominantly due to the starch component present, which usually represents between 50 and 80% of the dry solids in the mixture. Accordingly, many of the studies relating to cereal extrusion focus on changes occurring in the starch component of the product under varying extrusion conditions and the resulting effects on its physical, chemical, and organoleptic properties. The type of extruder, feed moisture, feed rate, barrel temperature, screw speed, screw profile, and die size are all important in developing the characteristics of the extruded product.

Cereals and starch-based products have been processed by extrusion since the initial introduction of the process to the food industry.

A study of the effect of feed moisture and barrel temperature on the extrusion of commercial yellow corn grits found that an increase in feed moisture or barrel temperature, up to 177 °C, causes an increase in water absorption index (WAI) [13]. At temperatures above 177 °C, WAI was found to decrease. Similarly, water solubility index (WSI) was found to gradually increase with an increase in barrel temperature up to a value of 177 °C, with a more pronounced increase above this temperature. For any given set of operating conditions, a decrease in feed moisture content resulted in reduced WAI and increased WSI. Determination of the final cooked-paste viscosities showed maximum viscosity at 25% feed moisture.

A study carried out by Mercier and Feillet [14] looked at the modification of carbohydrate components in a range of cereal products, namely, corn grits, corn, waxy corn, Amylon 5 and 7 (52 and 61% amylose), wheat, and rice. The barrel temperature was varied from 70 to 225 °C with a moisture content of 22%. Their study showed a consistent increase in WSI with increasing barrel temperature, agreeing with earlier work [13]. Analysis of WAI for waxy corn showed a decrease with increasing barrel temperature from 70 to 225 °C, while the Amylon 5 and 7 samples showed little change until 200 °C, after which there was a sharp increase.

WAI data for corn, wheat, and rice products showed a gradual increase with barrel temperature reaching a maximum at around 180 °C. At an extrusion temperature of 135 °C, cooked viscosity of corn and rice starch was similar, while wheat starch gave a higher final cooked viscosity. Data obtained from the extrusion of starch with different amylose contents at different temperatures showed that the WSI and water-soluble carbohydrate increased less with increasing amylose content.

Research on the effect of several extrusion variables such as moisture, barrel temperature, screw geometry, and screw speed on the gelatinization of corn starch indicated that barrel temperature and moisture had the greatest effect on the gelatinization of starch [15]. The results showed that the maximum gelatinization occurred at high moisture and low barrel temperatures or vice versa. Higher screw speed reduced gelatinization and was related to lower residence times. During the extrusion of cereal starches, it was found that starch was solubilized in a macromolecular form and no small oligo- or monosaccharides were formed [14]. Research on potato starch [16] showed that extrusion preferentially broke the α-1–4 linkages of amylose and not the outer chains of amylopectin. Linear oligosaccharides having no α-1–6 linkages were found.

The effect of feed moisture, barrel temperature, screw speed, and die size on the gelatinization of wheat flour during extrusion showed that the degree of starch gelatinization increased sharply with increasing temperature, when the feed moisture contents were 24–27%, but increased more gradually when moisture contents were 18–21% [17]. At lower temperatures (65–80 °C) an increase in feed moisture content was found to cause a slight decrease in gelatinization whilst at higher temperatures (95–110 °C), an increase in moisture resulted in a significantly increased degree of gelatinization. Increase in the screw speed was found to decrease the degree of starch gelatinization despite the increased shear. This was explained by the decrease in retention time of the sample in the extruder. Increasing the die size was found to reduce the degree of starch gelatinization. This was explained by the possible decrease in the retention time of the sample in the extruder due to lower pressures and decreased surface shear.

At a constant moisture content, the effect of barrel temperature, feed rate, and screw speed on starch gelatinization during the extrusion of a yoghurt–wheat mix, showed that barrel temperature had the most pronounced effect, followed by feed rate and screw speed [18].

A comparison of the appearance of starch granules from wheat semolina prior to and after extrusion at 60 °C, showed little difference in the shape of the starch granules [19]. Increasing the barrel temperature resulted in a flattening of the granule although the original shape (unextruded) was still recognizable. A complete destruction of the granule was not seen until the barrel temperature had reached 125 °C.

The physicochemical properties of several blends of raw, gelatinized, and dextrinized commercial yellow corn flour were evaluated [20] and results compared to those of raw commercial yellow corn flour extruded at a range of moisture contents. The extruded samples had properties similar to blends containing gelatinized and dextrinized corn. Reduction of the extrusion moisture level resulted in an increase in the relative proportion of dextrinized corn from 10 to 60%. It was suggested that dextrinization is the dominant mechanism for starch degradation during extrusion, especially at low moisture contents.

Colonna *et al.* [21] extruded modified wheat starch and found this led to a macromolecular degradation of amylose and amylopectin, by random chain splitting. The water-soluble fractions were composed of partly depolymerized amylose and amylopectin. It was also concluded that shear in the extruder completely disperses starch components by decreasing molecular entanglement.

A study [22] to evaluate the structural modifications occurring during extrusion cooking of wheat starch revealed that the amylopectin fraction of the starch was significantly degraded during extrusion and the degradation products were also macromolecules. It was felt that the structural modifications occurring during the extrusion was limited debranching attributed to mechanical rupture of covalent bonds.

A study [23] of the effect of moisture content, screw speed, and barrel temperature on starch fragmentation of corn starch found that carbohydrate from the extruded samples dissolved at a significantly faster rate than that from native corn starch, with a pronounced difference in the amount of material solubilized within the first 2 h. It was suggested that this was indicative of an increase in the amount of linear polysaccharide and thus an increase in the degree of fragmentation. The amount of large molecular weight material decreased from 68% to a range of 24–58% depending upon the extrusion conditions applied. Reductions in moisture content or temperature resulted in an increase in fragmentation while decreasing the screw speed resulted in a decrease in fragmentation.

Bhattacharya and Hanna [24] studied the effect of moisture content, barrel temperature, and screw speed on the textural properties of extrusion-cooked corn starch. Their results showed that the feed moisture content and barrel temperature affected the expansion of the extruded products predominantly. Increasing the moisture content reduced expansion in both waxy and non-waxy corn starch samples. They suggest that this is due to a reduction in the dough/melt temperature that in turn reduces the degree of gelatinization. An increase in the barrel temperature was found to increase the degree of expansion, reflecting an increase in gelatinization. Similar results were reported on the extrusion of potato flakes [25]. Screw speed was shown to be insignificant in its effect on expansion of the product. Although an increase in the screw speed would increase the rate of shear and hence the degree of starch modification, this was accompanied by a reduction of residence time which canceled out the effect of the additional shear. The shear strength of the products was found to increase with increasing moisture content, reflecting the decrease in expansion and increase in density.

The effect of screw speed on the extrusion of mixtures of yellow corn meal with wheat and oat fibers has been reported [26]. An increase in the screw speed was found to reduce the torque and die pressure while increasing specific mechanical energy, resulting in a decrease in radial expansion with increases in axial expansion, bulk density, and breaking force. It is suggested that the reduction in torque and die pressure is the result of a decrease in the screw fill. The reduction in radial expansion and the increases in axial expansion, bulk density, and breaking force are reported to be due to the reduced die pressure and resistance to the flow of extrudate at the die. These results are contrary to those of Fletcher *et al.* [27] who observed an increase in die pressure and radial expansion with increasing screw speed. The conflicting results in these studies suggest that the textural characteristics may or may not be affected by screw speed, depending upon the feed material and the geometry and design of the screw used.

Although the extrusion characteristics of cereals are dominated by the physical and chemical changes occurring in the starch component, cereals typically contain between 6 and 16% protein and between 0.8 and 7% lipid, which can significantly affect the properties of the extruded product [28]. In addition to the components of the cereal itself, a wide variety of materials are incorporated into cereal extrusion mixes to modify the characteristics of the final product.

13.3.1.1 **Protein**

Typically protein acts as a "filler" in cereal extrudates and is dispersed in the continuous phase of the extrusion melt, modifying the flow behavior and characteristics of the cooled extrudates. Proteinaceous materials hydrate in the mixing stage of the process and become soft viscoelastic doughs during formation of the extrusion melt. The shearing forces generated in the extruder cause breakage of the protein into small particles of roughly cylindrical and globular shapes. At levels of around 5–15%, they tend to reduce the extensibility of the starch polymer foam during its expansion at the die exit, reducing the degree of expansion [28].

Extrudates show a reduction in the cell size with addition of protein proportional to the amount of protein added. At higher levels of protein, severely torn regions in the cell walls of the extrudate are noted, indicating a loss of elasticity in the extrusion melt.

Martinez-Serna *et al.* [29] investigated the effects of whey protein isolates on the extrusion of corn starch. The isolates were blended with corn starch at concentrations of between 0 and 20%, and then extruded at varying barrel temperatures, screw speeds, and moisture contents. Their results showed that an increase in barrel temperature increased the level of starch–protein interaction. Extrusion of the blend showed a 30% reduction in expansion when compared to extruded corn starch alone. They suggest that this is due to modification in the viscoelastic properties of the dough, as a result of competition for the available water between the starch and protein fractions leading to a delay in starch gelatinization. Increasing the screw speed and hence shear rate, led to an increase in viscosity due to unfolding of the protein, involving rupture of covalent bonds, or interaction with the starch. A similar trend was also observed when protein concentration was increased.

13.3.1.2 **Fat**

Fats and oils have two functions in starch extrusion processes. They act as a lubricant in the extrusion melt and modify the eating qualities of the final extruded product. The action of the extruder screws causes the oils to be either dispersed into small droplets or smeared on the polymers [28].

Extrusion of starches with low lipid contents, such as potato or pea starch, at low moisture contents (<25%) is extremely difficult due to degradative dehydration of the starch polymers. This results in the formation of a very sticky melt which tends to cause blockages. Addition of 0.5–1.0% of an oil to the starch reduces the degradation of the starch, and enables extrusion without blockages [28].

The macromolecular modifications occurring in manioc starch extruded without and with a range of lipids (oleic acid, dimodan, copra, and soya lecithin) have been investigated [30]. The results showed that an increase in extrusion temperature or screw speed increased the degree of macromolecular degradation of the native manioc starch. Addition of all lipids at 2% was shown to reduce the degree of macromolecular degradation, with all samples having higher intrinsic viscosities than the native starch extruded under the same conditions. While all of the added lipids gave rise to an increase in the intrinsic viscosity, differences were apparent between each of the lipid–starch extrudates, suggesting different modes of action for each of the lipids studied.

A highly expanded oat cereal product is difficult to achieve due to its high fat content. The effects of the process conditions on the physical and sensory properties of an extruded oat–corn puff showed that increasing the level of oat flour (and hence fat content) caused an increase in the extrudate bulk density and a reduction in specific length and expansion [31].

An investigation into the improvement of extruded rice products showed that defatting of the flour resulted in improved expansion and lower bulk densities [32].

13.3.1.3 **Sugars**

Sucrose and other sugars are commonly added to extruded products, particularly breakfast cereals. The level of sugar added to a product varies but is typically within the range from 6 to 25 wt% on a final dry product basis. Sugar contributes to binding, flavor, and browning characteristics and is important in controlling texture and mouthfeel. In addition, it can act as a carrier and potentiator of other flavors.

The effects of sucrose on the structure and texture of extrudates has been studied extensively. When extrusion is carried out at moisture contents above 16%, addition of sucrose progressively reduces extrudate expansion with an accompanying increase in the product density. This effect was noted at sucrose concentrations as low as 2% when extruding with a feed moisture of 20% [33]. In addition to the reduction in expansion and increase in density, an increase in mechanical strength and the number of cells formed per unit area have been observed [34]. The structural changes in the extrudates brought about by the addition of sucrose have been attributed to competition for moisture, inhibition of gelatinization, and plasticization of starch-based systems by low molecular weight constituents during extrusion.

The effect of sucrose and fructose on the extrusion of maize grits indicated inhibition of starch conversion due to a reduction in specific mechanical energy input [35]. In addition, a change in the packing of amylose–lipid complexes in the extrudate was noted. It is suggested that this rearrangement process is accelerated by the addition of the sugars due to the enhanced molecular mobility.

The effect of sucrose on both maize and wheat flour extrudates showed that, in contrast to maize extrudates, sucrose addition had little effect on the degree of starch conversion and sectional expansion of wheat flour extrudates [36]. It was postulated that the observed differences between the wheat–sucrose and maize–sucrose extrudates may be a result of the particle size and the presence of gluten.

13.3.1.4 Dietary Fiber

There is a growing interest in increasing the dietary fiber content of foods from a health standpoint. A number of researchers have increased dietary fiber levels in extrudates by supplementing with oat bran [37], wheat bran [38], barley bran [39], brewer's spent grain [40, 41], and cauliflower by-products [42].

It was found that the addition of dietary fiber affected the textural characteristics of the extrudates, increasing the hardness of the products [38, 40] as a result of its effect on cell wall thickness.

The level of fiber has also been found to affect the expansion of extrudates. In a study by Grenus *et al.* [43], both radial and axial expansion of extrudates increased with the addition of 10% rice bran but decreased at higher levels. The addition of brewers' spent grain [40] resulted in reduced expansion of extrudates at a fixed screw speed but expansion increased with increasing screw speed. Stojceska *et al.* [41] found, in high-fiber extrudates, a high correlation between mean cell area to the sectional expansion index (SEI), indicating that the higher level of fiber lowered the expansion of the products, giving a structure containing more small cells.

13.3.2 Nutritional Changes

13.3.2.1 Protein

Extrusion cooking, like other food processing, may have both beneficial and undesirable effects on the nutritional value of proteins. During extrusion, chemical constituents of the feed material are exposed to high temperature, high shear, and/or high pressure that may improve or damage the nutritional quality of proteins in the extruded materials by various mechanisms. These changes depend on temperature, moisture, pH, shear rate, residence time and their interactions, the nature of the proteins themselves, and presence of materials such as carbohydrates and lipids.

The proteins present in the feed material may undergo structural unfolding and/or aggregation when subjected to heat or shear during extrusion. Intact protein structures represent a significant barrier to digestive enzymes and the combination of heat and shear are a very efficient way of disrupting such structures.

In general, denaturation of protein to random configurations improves nutritional quality by making the molecules more accessible to proteases and, thus, more digestible. In addition, protein digestibility will also be increased due to extrusion processing reducing the level of antinutritional factors, such as trypsin inhibitors, phytates, tannins, and lectins. The destruction of antinutritional factors are favored by high barrel temperatures, increase in residence time, and high moisture content [44–46]. This is especially important in legume-based foods that contain active enzyme inhibitors in the raw state.

Disulfide bonds are involved in stabilizing the native tertiary configurations of most proteins. Their disruption aids in protein unfolding and thus digestibility. Mild shearing can contribute to the breaking of these bonds. Partial hydrolysis of proteins during extrusion increases their digestibility by producing more open configurations and increasing the number of exopeptidase-susceptible sites. Conversely, production of an extensively isopeptide cross-linked network could interfere with protease action, reducing the digestibility [47].

The Maillard reaction may take place during extrusion cooking of protein foods containing reducing sugars. The chemical reaction between the reducing sugars and a free amino group on an amino acid has important nutritional and functional consequences for extruded products. Maillard reactions can result in a decrease in protein quality, by lowering digestibility and producing nonutilizable products. During extrusion, the Maillard reaction is favored by conditions of high temperatures ($>180\,^\circ C$) and shear (>100 rpm) in combination with low moisture ($<15\%$) [48].

Loss of total lysine and changes in *in vitro* availability of amino acids in protein-enriched biscuits when extruded at different mass temperatures and moisture contents have been investigated [44, 49]. It was found that digestibility of the product extruded at 170 °C was not different from that of the raw material. However, increasing the mass temperature to 210 °C decreased the *in vitro* digestibility.

In a similar study [50], it was reported that screw speed was not a major factor in the retention of available lysine, but 40% of the available lysine present in the unextruded mix was lost during extrusion conditions above 170 °C mass temperature at a moisture content of 13%. The loss of lysine decreased when the water was increased to 18%, despite an increase in barrel temperature to produce an equivalent mass temperature to the 13% moisture mixture. Higher losses in lysine were observed with increased sucrose content and reduced pH.

Extrusion of corn gluten meal and blends of corn gluten meal and whey at various screw speeds and barrel temperatures showed an increase in the *in vitro* digestibility of the extruded product when compared to the raw material [51]. It was also found that the addition of whey had no significant effect on digestibility.

The protein nutritional value of extruded wheat flours was studied by Bjorck *et al.* [52]. Amino acid analysis showed that lysine retention was between 63 and 100%. It was found that the retention was positively affected by an increase in feed rate, and negatively by an increase in screw speed. The authors felt that the prominent lysine damage under severe conditions was probably due to formation of reducing carbohydrates through hydrolysis of starch. The loss of other amino acids was found to be small.

The effects of a range of extrusion process variables on *in vitro* protein digestibility of minced fish and wheat flour mixes have been studied [53]. The extrudates showed slight increases in *in vitro* protein digestibility values. The authors found that of the process variables studied, only the effect of feed ratio and temperature of extrusion were significant.

The effect of extrusion on *in vitro* protein digestibility of sorghum showed that varying screw speed and moisture content did not have a significant effect on the digestibility of sorghum, but temperature was significant in improving the digestibility of sorghum [54].

Dahlin and Lorenz [55] investigated the effect of extrusion on the protein digestibility of various cereal grains (sorghum, millet, quinoa, wheat, rye, and corn). When the extrusion feed moisture was decreased from 25 to 15% an increase in protein digestibility was observed. It was found that a screw speed of 100 rpm and a product temperature of 150 °C improved *in vitro* protein digestibility of all cereals studied.

Extrusion of a yoghurt–wheat mixture at a constant moisture content of 43% showed no decrease in *in vitro* protein digestibility, up to barrel temperatures of 120 °C and screw speeds of 300 rpm [56].

In addition to lysine loss and decrease in protein digestibility, the color of the extruded product is another indication of the extent of the Maillard reactions. A correlation was found between the degree of browning of extruded wheat flour and the total lysine content of the samples [52]. It was also found that there was a positive correlation between Hunter L values for wheat-based breakfast cereals and *in vitro* protein digestibility and available lysine [57].

13.3.2.2 **Dietary Fiber**

A variety of changes take place in plant cell wall polysaccharides during processing and cooking which affect the physicochemical properties of dietary fiber and possibly its nutritional effects. A number of papers evaluating the changes in dietary fiber during extrusion have been published but the findings are often conflicting primarily due to the source of fiber used and the category of fiber measured.

It has been reported that extrusion cooking increases the total dietary fiber level of cereals due to the formation of resistant starch [58], while in vegetables, heat and moisture solubilizes and degrades pectic substances [59] leading to a decrease in fiber content.

Sandberg *et al.* [60] found that extrusion cooking did not change the content of dietary fiber in extruded bran products compared with non-extruded products. However, orange pulp during extrusion was found to lose a considerable amount of total dietary fiber [61]. The total dietary fiber content in the orange pulp decreased with higher barrel temperatures and lower moisture contents with the screw speed fixed at 160 rpm. The authors found that extrusion modified the properties of the fiber components resulting in a decrease in insoluble dietary fiber and an increase in soluble dietary fiber.

Extrusion processing is expected to mechanically rupture glycosidic bonds in the total dietary fiber polysaccharides, leading to the release of oligosaccharides and therefore to the increase of soluble dietary fiber [62]. Also if small molecular mass fragments are formed then there will be a falsely low value in total dietary fiber content [63]. However, in some extrusion processing an increase of insoluble fiber has been observed [64]. It is suggested that this is due to the gelatinization and the retro degradation of the starch occurring during the process changing part of it into not-degradable polysaccharides such as resistant starch.

The effect of extrusion cooking on the total dietary fiber level of snacks produced from food by-products, such as brewers' spent grain and red cabbage trimmings, showed that a significant increase in total dietary fiber occurred [65]. This was more evident as the feed moisture level increased up to 15%, probably as a result of the formation of resistant starch. This increase was more evident at feed moisture between 12 and 15%. Similar results were reported by Østergård *et al.* [66] where total dietary fiber content of barley increased upon extrusion cooking, accompanied by a decrease in total starch content, probably due to formation of indigestible starch fragments. Later, Vasanthan *et al.* [67] reported that extrusion cooking increased total dietary fiber of barley flours, which may be attributed, primarily, to a shift from insoluble to soluble dietary fiber, as well as the formation of resistant starch and enzyme-resistant indigestible glucans. Esposito *et al.* [68] tested the extrusion cooking process to evaluate the possibility of increasing the amount of soluble fiber in extruded products made with durum wheat bran fractions. It was found that extrusion cooking did not affect the amount of soluble fiber but increased the insoluble fiber. Ainsworth *et al.* [40] found changes in screw speed between 100 and 300 rpm had no effect on total dietary fiber. However, it was noted that increasing screw speed reduced resistant starch content which may be due to modification of the starch molecules by the increased shear effect at higher screw speeds.

13.3.2.3 **Vitamins**

The retention of vitamins in extrusion cooking generally decreases with increasing temperature, increasing screw speed, decreasing moisture, decreasing throughput, decreasing die diameter, and increasing specific energy input [69].

The stability of thiamin, riboflavin, and niacin during extrusion cooking of full-fat soy flour has been studied [70, 71]. Different barrel temperatures and water contents of feed material with a residence time of 1 min did not affect riboflavin and niacin. However, some loss of thiamin was observed at high moisture ($>$15%), high barrel temperature ($>$153 °C), and long residence times ($>$1 min).

The retention of thiamin and riboflavin in maize grits during extrusion has been studied [72]. The average thiamin retention was 54 and 92% for riboflavin. These workers found that moisture content (13–16%) did not have any effect on the stability of thiamin. An increased degradation was seen for thiamin with increasing temperature and screw speed. Riboflavin degradation increased with increasing moisture content and screw speed.

Thiamin losses in extruded potato flakes at a range of barrel temperatures and screw speeds, different screw compression ratios and die diameters, and a moisture content of 20% have been studied [25]. It was reported that thiamin loss did not exceed 15% in all runs.

The importance of moisture content on the retention of thiamin in potato flakes during extrusion indicated that at moisture contents of 25–59%, retention ranged from 22 to 97% [73]. The retention was poor at low moisture contents.

Thiamin losses in extruded legume products with increase in process temperature (93–165 °C), pH (6.2–7.4), and screw speed (10–200 rpm) have been reported [74]. The retention of thiamin increased with higher moisture contents (30–45%).

The effect of throughput and moisture content on the stability of thiamin, riboflavin, B6, B12, and folic acid when extruding flat bread showed that vitamin stability improved with increasing throughput and increased water content [75].

Guzman-Tello and Cheftel [76] studied the stability of thiamin as an indicator of the intensity of thermal processing during extrusion cooking. They found that retention of thiamin decreased from 88.5 to 57.5% when the product temperature increased from 131 to 176 °C. Other extrusion parameters were kept constant. More thiamin was retained by increasing the moisture content (14–28.5% wet basis), while higher screw speeds (125 and 150 rpm) caused higher losses.

At screw speeds up to 300 rpm at a barrel temperature of 120 °C, no loss of thiamin and riboflavin was observed when extruding a yoghurt–wheat mixture at a moisture content of 43% [56].

Although extrusion systems can be successfully used with higher vitamin retentions, food manufacturers often carry out vitamin fortification post-extrusion by dusting, enrobing, spraying, or coating.

Studies on vitamin retention in extruded products largely center on the B vitamins with little work on other vitamins. Lorenz and Jansen [77] studied the extrusion of maize/soy blends and found it was possible to retain up to 80% of the vitamin C with most of the vitamin being lost during subsequent storage of the extrudate. Losses of ascorbic acid during the extrusion of potato flakes have been reported [73] to be between 14 and 68%, with greatest retention occurring at high moisture contents (59%) and low barrel temperatures (70 °C). However, at high barrel temperatures and lower moisture contents (10%), ascorbic acid decreased in extruded wheat flour [78]. The influence of barrel temperature (75–150 °C) and screw speed (100–300 rpm) on the retention of ascorbic acid in a rice-based extrudate have been studied [79]. At all screw speeds greater losses occurred as barrel temperature increased. Retention of ascorbic acid varied between 56 and 79%, with the highest retentions being associated with low screw speed and low barrel temperature.

13.3.2.4 **Minerals**

Very little research has been carried out on the changes in mineral levels during extrusion. Minerals are stable to heat processing and levels are unlikely to change during extrusion. An increase in iron content with increasing screw speed and temperature has been noted during the extrusion of potato flakes [73] and this has been attributed to the wear of the extruder screws [80]. The extrusion of rice flour showed an increase in iron content compared to unextruded flour [79]. An increase in screw speed for any given temperature showed an increase in iron content of the extrudate. However, as temperature was increased for any given screw speed a fall in iron content was noted which may be a result of the extruded material becoming more fluid and less abrasive in nature.

13.3.3 Flavor Formation and Retention during Extrusion

Much of the flavor, or the volatile components of flavor are either lost to the atmosphere as the extrudate exits the die or become bound to starch or proteins during extrusion. Because of the high losses, flavors are generally added after extrusion. Post-extrusion flavoring processes have many disadvantages, including difficulties in obtaining even application, possibility of contamination, and rubbing off of the coating in the package or consumer's hands. A limited range of encapsulated and extrusion stable flavors have been developed to address these problems, however, these are often restricted in application due to their cost. An alternative approach to the problem of flavoring extruded products involves the addition of reactive flavor precursors to the extrusion mix. The conditions applied during the extrusion process cause reaction of the precursors to produce the desired flavor compounds. These precursors are usually compounds known to participate in the browning and flavor reactions that occur normally during the extrusion process.

In the extrusion of cereal-based products, flavors are predominantly generated by non-enzymic browning reactions typified by caramelization, the Maillard reaction, and oxidative decomposition. The temperature and shear conditions generated in the extruder provide the physical and chemical means whereby starch and protein can be partially degraded to provide the reactants that can then participate in non-enzymic browning reactions. In addition, lipids (especially unsaturated lipids) may undergo thermal degradation thereby providing additional flavor compounds.

An evaluation of the volatile compounds and color formation in a whey protein concentrate–corn meal extruded product showed 71 volatile compounds in the headspace of the samples, 68 of which were identifiable [81]. The compounds present included 12 aldehydes, 10 ketones, 6 alcohols, 2 esters, 6 aromatics, and 10 hydrocarbons. In addition, 11 pyrazines, 4 furans, 5 heterocyclics, and 2 sulfur compounds were isolated. They suggest that these compounds were products of the Maillard reaction and are possibly the most important in contributing to the corn flavor of the product. The concentrations of pyrazines, furans, and other heterocyclics were found to increase with increasing levels of whey protein in the extrudates.

Nair *et al.* [82] identified 91 compounds in the die flash-off condensate and 56 compounds from the headspace of the extrudate of corn flour. They suggest that the difference in the composition of the two samples is a reflection of the volatility with water of the compounds. Whilst their study identified a significant number of compounds formed during extrusion were retained in the extrudate, it was concluded that flash-off at the die is still a major hindrance to flavoring of extruded products.

The effects of product temperature, moisture content, and residence time on the aroma volatiles generated during extrusion of maize flour have been investigated by Bredie *et al.* [83]. Their results showed that low temperatures and high moisture contents favored production of volatiles associated with lipid degradation. Increasing both the temperature and residence time, along with a reduction in moisture content, was reported to increase the production of compounds derived from the Maillard reaction while reducing the levels of those compounds associated with lipid degradation.

The effect of pH on the volatiles formed in a model system containing wheat starch, lysine, and glucose showed that both the total yield and number of compounds formed are greater at pH 7.7 than at pH 4.0 and 5.0, where the total yields are similar [84]. At pH 7.7, the volatile compounds from the extrudate were dominated by pyrazines that gave the extrudates a nutty, toasted, and roasted character. Modification of the pH to 4.0 or 5.0 reduced production of pyrazines and increased production of 2-furfural and 5-methylfurfural, which accounted for over 80% of the total volatiles. It is suggested that the presence of these compounds is an indication of starch degradation, leading to an additional source of carbohydrate precursor during extrusion. They conclude that it may be possible to control flavor and color development by careful control of the processing conditions.

Using a trained sensory panel to study the aroma profile of extrudates produced from wheat flour and wheat starch fortified with mixtures of cysteine, glucose, and xylose [85], a vocabulary of 24 odor attributes was developed with 17 of the attributes significantly differentiating the samples. The results indicated that extrusion of the wheat flour alone resulted in an extrudate aroma that was characterized by the "biscuity," "cornflakes," "sweet," and "cooked milk" terms. Addition of cysteine and glucose resulted in major differences in aroma, with the cereal terms "popcorn," "nutty/roasted," and "puffed wheat" predominating. Cysteine/xylose mixtures gave rise to sulfur odor notes, such as "garlic-like," "onion-like," "rubbery," and "sulfury" with the less desirable "acrid/burnt," "sharp/acidic," and "stale cooking oil" terms becoming apparent.

A more extensive study using wheat starch, cysteine, xylose, and glucose extruded at a range of pH and target die temperatures has been carried out [86]. Extrudates prepared using glucose were more frequently described as "biscuity" and "nutty," whereas those prepared using xylose were commonly described using the terms "meaty" or "onion-like." Analysis of the extrudates identified 80 compounds. Yields of the compounds formed were generally higher in extrudates prepared using xylose than those using glucose, under the same processing conditions. Increases in temperature and pH were reported to increase the yields of the compounds formed. Pyrazines and thiophenes were among the most abundant classes of compounds identified. Results from the analysis of the extrudates prepared using glucose showed aliphatic sulfur compounds and thiazoles to feature strongly in the aroma, whereas use of xylose gave rise to both nonsulfur-containing and sulfur-containing furans.

References

1. Hsieh, F. (1992) Extrusion and extrusion cooking, in *Encyclopedia of Food Science and Technology* (ed. Y.H. Hui), John Wiley & Sons, Inc., New York, pp. 795–800.
2. Harper, J.M. (1989) Food extruders and their applications, in *Extrusion Cooking* (eds C. Mercier, P. Linko, and J.M. Harper), AACC, St Paul, MN, pp. 1–16.
3. Lanz, R. (1983) Successful multi-component, continuous-extruder feeding, in *Progress in Food Engineering* (eds C. Canterelli and C. Peri), Forster, Kisnacht, pp. 625–629.
4. Janssen, L.P.B.M. (1993) Extrusion cooking – principles and practice, in *Encyclopedia of Food Science Food Technology and Nutrition* (eds R. Macrae, R.K. Robinson, and M.J. Sadler), Academic Press, London, pp. 1700–1705.
5. Cheftel, J.C. (1986) Nutritional effects of extrusion cooking. *Food Chem.*, **20**, 263–283.
6. Heldman, D.R. and Hartel, R.W. (1997) *Principles of Food Processing*, Chapman and Hall, New York.
7. Frame, N.D. (1994) Operational characteristics of the co-rotating twin-screw extruder, in *The Technology of Extrusion Cooking* (ed. N.D. Frame), Blackie Academic and Professional, Glasgow, pp. 1–51.
8. Harper, J.M. (1990) Extrusion of foods, in *IFT Symposium Series: Biotechnology and Food Process Engineering* (eds H.G. Schwartzberg and M.A. Rao), Marcel Dekker, New York, pp. 295–308.
9. Harper, J.M. (1986) Processing characteristics of food extruders, in *Food Engineering and Process Applications, Unit Operations*, vol. 2 (eds M.L. Maguer and P. Jelem), Elsevier Applied Science, London, pp. 101–114.
10. Jager, T., van Zuilichem, D.J., and Stolp, W. (1992) Residence time distribution, mass flow, and mixing in a co-rotating twin-screw extruder, in *Food Extrusion Science and Technology* (eds J.L. Kokini, C. Ho, and M.V. Karwe), Marcel Dekker, New York, pp. 71–78.
11. Dziezak, J.D. (1989) Single- and twin-screw extruders in food processing. *Food Technol.*, **43**, 163–174.
12. Harper, J.M. (1992) A comparative analysis of single and twin screw extruders, in *Food Extrusion Science and Technology* (eds J.L. Kokini, C. Ho, and N.V. Karwe), Marcel Dekker, New York, pp. 139–148.
13. Anderson, R.A., Conway, H.F., Pfeifer, V.F., and Griffin, E.L. (1969) Gelatinization of corn grits by roll- and extrusion-cooking. *Cereal Sci. Today*, **14**, 1–12.
14. Mercier, C. and Feillet, P. (1975) Modification of carbohydrate components by extrusion cooking of cereal products. *Cereal Chem.*, **52**, 283–297.
15. Lawton, B.T., Henderson, G.A., and Derlatka, E.J. (1972) The effects of extruder variables on the gelatinization of corn starch. *Can. J. Chem. Eng.*, **50**, 168–173.
16. Mercier, C. (1997) Effect of extrusion-cooking on potato starch using a twin screw french extruder. *Staerke*, **29**, 48–52.
17. Chiang, B.Y. and Johnson, J.A. (1977) Gelatinization of starch in extruded products. *Cereal Chem.*, **54**, 436–443.
18. Ibanoglu, S., Ainsworth, P., and Hayes, G.D. (1996) Extrusion of tarhana: effect of operating variables on starch gelatinization. *Food Chem.*, **57**, 541–544.
19. Kim, J.C. and Rottier, W. (1980) Modification of aestivum wheat semolina by extrusion. *Cereal Foods World*, **24**, 62–66.
20. Gomez, M.H. and Aguilera, J.M. (1983) Changes in the starch fraction during extrusion cooking of corn. *J. Food Sci.*, **48**, 378–381.
21. Colonna, P., Doublier, J.L., Melcion, J.P., Monredon, F., and Mercier, C. (1984) Physical and functional properties of wheat starch after extrusion cooking and drum drying, in *Thermal Processing and Quality of Foods* (eds P. Zeuthen, J.C. Cheftel, C. Eriksson, M. Jul, H. Leniger, P. Linko, G. Varela, and G. Vos), Elsevier Applied Science, London, pp. 96–112.

22. Davidson, V.J., Paton, D., Diosady, L.L., and Larocque, G. (1984) Degradation of wheat starch in a single-screw extruder: characterization of extruded starch polymers. *J. Food Sci.*, **49**, 453–459.
23. Wen, L.F., Rodis, P., and Wasserman, B.P. (1990) Starch fragmentation and protein insolubilization during twin-screw extrusion of corn meal. *Cereal Chem.*, **67**, 268–275.
24. Bhattacharya, M. and Hanna, M.A. (1987) Textural properties of extrusion-cooked corn starch. *Lebensm. Wiss. Technol.*, **20**, 195–201.
25. Maga, J.A. and Cohen, M.R. (1978) Effect of extrusion parameters on certain sensory, physical and nutritional properties of potato flakes. *Lebensm. Wiss. Technol.*, **11**, 195–197.
26. Hsieh, F., Mulvaney, S.J., Huff, H.E., Lue, L., and Brent, J. (1989) Effect of extrusion parameters on certain sensory, physical and nutritional properties of potato flakes. *Lebensm. Wiss. Technol.*, **22**, 204–207.
27. Fletcher, S.I., Richmond, P., and Smith, A.C. (1985) An experimental study of twin-screw extrusion-cooking of maize grits. *J. Food Eng.*, **4**, 291–312.
28. Guy, R.C.E. (1994) Raw materials for extrusion cooking processing, in *The Technology of Extrusion Cooking* (ed. N.D. Frame), Blackie Academic and Professional, Glasgow, pp. 52–72.
29. Martinez-Serna, M., Hawkes, J., and Villota, R. (1990) Extrusion of modified and natural whey proteins in starch-based systems, in *Engineering and Food: Advanced Processes*, vol. 3 (eds W.E.L. Spiess and H. Schubert), Elsevier Applied Science, London, pp. 346–365.
30. Colonna, P. and Mercier, C. (1983) Macromolecular modifications of manioc starch components by extrusion cooking with and without lipids. *Carbohydr. Polym.*, **3**, 87–108.
31. Liu, Y., Hsieh, F., Heymann, H., and Huff, H.E. (2000) Effect of process conditions on the physical and sensory properties of extruded oat-corn puff. *J. Food Sci.*, **65**, 1253–1259.
32. Kumagai, H., Lee, B.H., and Yano, T. (1987) Flour treatment to improve the quality of extrusion-cooked rice flour products. *J. Agric. Biol. Chem.*, **51**, 2067–2071.
33. Jin, Z., Hsieh, F., and Huff, H.E. (1994) Extrusion cooking of corn meal with soya fiber, salt and sugar. *Cereal Chem.*, **71**, 227–234.
34. Ryu, G.H., Neumann, P.E., and Walker, C.E. (1993) Effects of some baking ingredients on physical and structural properties of wheat flour extrudates. *Cereal Chem.*, **70**, 291–297.
35. Fan, J., Mitchell, J.R., and Blanshard, J.M.V. (1996) The effect of sugars on the extrusion of maize grits II. Starch conversion. *Int. J. Food Sci. Technol.*, **31**, 67–76.
36. Carvalho, C.W.P. and Mitchell, J.R. (2000) Effect of sugar on the extrusion of maize grits and wheat flour. *Int. J. Food Sci. Technol.*, **35**, 569–576.
37. Martianez-Tomea, M., Murcia, A., Frega, N., Ruggieri, S., Jimea, A., Roses, F., and Parras, P. (2004) Evaluation of antioxidant capacity of cereal brans. *J. Agric. Food Chem.*, **52**, 4690–4699.
38. Yanniotis, S., Petraki, A., and Soumpasi, E. (2007) Effect of pectin and wheat fibers on quality attributes of extruded cornstarch. *J. Food Eng.*, **80**, 594–599.
39. Baik, B.K., Powers, J., and Nguyen, L.T. (2004) Extrusion of regular and waxy barley for production of expanded cereals. *Cereal Chem.*, **81**, 94–99.
40. Ainsworth, P., Ibanoglu, S., Plunkett, A., Ibanoglu, E., and Stojceska, V. (2007) Effect of brewers spent grain addition and screw speed on the selected physical and nutritional properties of an extruded snack. *J. Food Eng.*, **81**, 702–709.
41. Stojceska, V., Ainsworth, P., Plunkett, A., and Ibanoglu, S. (2008) The recycling of brewer's processing by-product into ready-to-eat snacks using extrusion technology. *J. Cereal Sci.*, **47**, 469–479.
42. Stojceska, V., Ainsworth, P., Plunkett, P., Ibanoglu, E., and Ibanoglu, S. (2008) Cauliflower by-products as a new source of dietary fibre, antioxidants and proteins in cereal based ready-to-eat expanded snacks. *J. Food Eng.*, **87**, 554–563.

43. Grenus, K.M., Hsieh, F., and Huff, H.E. (1993) Extrusion and extrudate properties of rice flour. *J. Food Eng.*, **18**, 229–245.
44. Bjorck, I. and Asp, N.G. (1983) The effects of extrusion cooking on nutritional value- a literature review. *J. Food Eng.*, **2**, 281–308.
45. Asp, N.G. and Bjorck, I. (1989) Nutritional properties of extruded foods, in *Extrusion Cooking* (eds C. Mercier, P. Linko, and J.M. Harper), American Association of Cereal Chemists, St Paul, MN, pp. 399–434.
46. Alonso, R., Aguirre, A., and Marzo, F. (2000) Effects of extrusion and traditional processing methods on antinutrients and in vitro digestibility of protein and starch in feba and kidney beans. *Food Chem.*, **68**, 159–165.
47. Phillips, R.D. (1989) Effect of extrusion cooking on the nutritional quality of plant proteins, in *Protein Quality and the Effect of Processing* (eds R.D. Phillips and J.W. Finley), Marcel Dekker, New York, pp. 219–246.
48. Camire, M.E., Camire, A., and Krumhar, K. (1990) Chemical and nutritional changes in foods during extrusion. *Crit. Rev. Food Sci. Nutr.*, **29**, 35–57.
49. Bjorck, I., Asp, N.G., and Dahlqvist, A. (1984) Protein nutritional value of extrusion-cooked wheat flours. *Food Chem.*, **15**, 203–214.
50. Noguchi, A., Mosso, K., Aymard, C., Jeunink, J., and Cheftel, J.C. (1982) Protein nutritional value of extrusion-cooked wheat flours. *Lebensm. Wiss. Technol.*, **15**, 105–110.
51. Bhattacharya, M. and Hanna, M.A. (1988) Extrusion processing to improve nutritional and functional properties of corn gluten. *Lebensm. Wiss. Technol.*, **21**, 20–24.
52. Bjorck, I., Matoba, T., and Nair, B.M. (1985) In-vitro enzymatic determination of the protein nutritional value and the amount of available lysine in extruded cereal-based products. *Agric. Biol. Chem.*, **49**, 945–951.
53. Bhattacharya, S., Das, H., and Bose, A.N. (1988) Effect of extrusion process variables on in vitro protein digestibility of fish-wheat flour blends. *Food Chem.*, **28**, 225–231.
54. Fapojuwo, O.O., Maga, J.A., and Jansen, G.R. (1987) Effect of extrusion cooking on in vitro protein digestibility of sorghum. *J. Food Sci.*, **52**, 218–219.
55. Dahlin, K. and Lorenz, K. (1993) Protein digestibility of extruded cereal grains. *Food Chem.*, **48**, 13–18.
56. Ibanoglu, S., Ainsworth, P., and Hayes, G.H. (1997) In vitro protein digestibility and content of thiamine and riboflavin in extruded tarhana, a traditional turkish cereal food. *Food Chem.*, **58**, 141–144.
57. McAuley, J.A., Kunkel, M.E., and Acton, J.C. (1987) Relationships of available lysine to lignin, color and protein digestibility of selected wheat-based breakfast cereals. *J. Food Sci.*, **52**, 1580–1582.
58. Englyst, H.N., Bingham, S.A., Runswick, S.A., Collinson, E., and Cummings, J.H. (1989) Dietary fibre (non-starch polysaccharides) in cereal products. *J. Hum. Nutr. Diet.*, **2**, 253–271.
59. Anderson, N.E. and Clydesdale, F.M. (1986) Effects of processing on the dietary fibre content of wheat bran, pureed green beans and carrots. *J. Food Sci.*, **45**, 1533–1537.
60. Sandberg, A.S., Andersson, H., Kivisto, B., and Sandstrom, B. (1986) Extrusion cooking of high-fibre cereal product. *Br. J. Nutr.*, **55**, 245–254.
61. Larrea, M.A., Chang, Y.K., and Bustos, F.M. (2005) Effect of some operational extrusion parameters on the constituents of orange pulp. *Food Chem.*, **89**, 301–308.
62. Lue, S., Hsieh, F., and Huff, H.E. (1991) Extrusion cooking of corn meal and sugar beet fiber: effects on expansion properties, starch gelatinization, and dietary fiber content. *Cereal Chem.*, **68**, 227–234.
63. Nyman, M. (1995) Effects of processing on dietary fibre in vegetables. *Eur. J. Clin. Nutr.*, **49**, S215–S218.
64. Unlu, E. and Faller, F. (1998) Formation of resistant starch by a twin-screw extruder. *Cereal Chem.*, **75**, 346–350.
65. Stojceska, V., Ainsworth, P., Plunkett, A., and Ibanoglu, S. (2009) The effect of extrusion cooking using

different water feed rates on the quality of ready-to-eat snacks made from food by-products. *Food Chem.*, **114**, 226–232.
66. Østergård, K., Björck, I., and Vainionpää, J. (1989) Effects of extrusion cooking on starch and dietary fibre in barley. *Food Chem.*, **34**, 215–227.
67. Vasanthan, T., Gaosong, J., Yeung, J., and Li, J. (2002) Dietary fibre profile of barley flour as affected by extrusion cooking. *Food Chem.*, **77**, 35–40.
68. Esposito, F., Arlotti, G., Bonifati, A.M., Napolitano, A., Vitale, D., and Fogliano, V. (2005) Antioxidant activity and dietary fibre in durum wheat bran by-products. *Food Res. Int.*, **38**, 1167–1173.
69. Killeit, U. (1994) Vitamin retention in extrusion cooking. *Food Chem.*, **49**, 149–155.
70. Mustakas, G.C., Griffin, E.L., Alien, L.E., and Smith, O.B. (1964) Production and nutritional evaluation of extrusion cooked full fat soybean flour. *J. Am. Oil Chem. Soc.*, **41**, 607–615.
71. Mustakas, G.C., Albreeth, W.J., Bookwalter, G.N., McGhee, J.E., Kwolek, F., and Griffin, E.L. (1970) Extruder processing to improve nutritional quality, flavour and keeping quality of full fat soy flour. *Food Technol.*, **24**, 1290–1296.
72. Beetner, G., Tsao, T., Frey, A., and Harper, J.M. (1974) Degradation of thiamine and riboflavin during extrusion processing. *J. Food Sci.*, **39**, 207–208.
73. Maga, J.A. and Sizer, C.E. (1978) Ascorbic acid and thiamine retention during extrusion of potato flakes. *Lebensm. Wiss. Technol.*, **11**, 192–194.
74. Pham, C.B. and Rosario, R.R. (1986) Studies on the development of texturized vegetable products by the extrusion process. III. Effects of processing variables on thiamine retention. *J. Food Technol.*, **21**, 569–576.
75. Millauer, C., Wiedmann, W.M., and Killeit, U. (1984) Influence of extrusion parameters on the vitamin stability, in *Thermal Processing and Quality of Foods* (eds P. Zeuthen, J.C. Cheftel, C. Eriksson, M. Jul, H. Leniger, P. Linko, G. Varela, and G. Vos), Elsevier Applied Science, London, pp. 208–213.
76. Guzman-Tello, R. and Cheftel, J.C. (1987) Thiamine destruction during extrusion cooking as an indication of the intensity of thermal processing. *Int. J. Food Sci. Technol.*, **22**, 549–562.
77. Lorenz, K. and Jansen, G.R. (1980) Nutrient stability of full-fat soy flour and corn-soy blends produced by low-cost extrusion. *Cereal Foods World*, **25**, 161–172.
78. Andersson, Y. and Hedlund, B. (1990) Extruded wheat flour: correlation between processing and product quality parameters. *Food Qual. Prefer.*, **2**, 201–216.
79. Plunkett, A. and Ainsworth, P. (2007) The influence of barrel temperature and screw speed on the retention of L-ascorbic acid in an extruded rice based snack product. *J. Food Eng.*, **78**, 1127–1133.
80. Alonso, R., Rubio, L.A., Muzquiz, M., and Marzo, F. (2001) The effect of extrusion cooking on mineral bioavailability in pea and kidney bean seed meals. *Anim. Feed Sci. Technol.*, **94**, 1–13.
81. Bailey, M.E., Gutheil, R.A., Hsieh, F., Cheng, C., and Gerhardt, K.O. (1994) Maillard reaction volatile compounds and color quality of a whey protein concentrate-corn meal extruded product, in *Thermally Generated Flavours: Maillard, Microwave and Extrusion Processes*, ACS Symposium Series, vol. 543 (eds T.H. Parliment, M.J. Morello, and R.J. Mc.Gorrin), American Chemical Society, Washington, DC, pp. 315–327.
82. Nair, M., Shi, Z., Karwe, M., Ho, C.T., and Daun, H. (1994) Collection and characterisation of volatile compounds released at the die during twin screw extrusion of corn flour, in *Thermally Generated Flavours: Maillard, Microwave and Extrusion Processes*, ACS Symposium Series, vol. 543 (eds T.H. Parliment, M.J. Morello, and R.J. McGorrin), American Chemical Society, Washington, DC, pp. 334–347.
83. Bredie, W.L.P., Mottram, D.S., and Guy, R.C.E. (1998) Aroma volatiles generated during extrusion cooking of maize flour. *J. Agric. Food Chem.*, **46**, 1479–1487.

84. Ames, J.M., Defaye, A.B., and Bates, L. (1997) The effect of pH on the volatiles formed in an extruded starch-glucose-lysine model system. *Food Chem.*, **58**, 323–327.
85. Bredie, W.L.P., Hassell, G.M., Guy, R.C.E., and Mottram, D.S. (1997) Aroma characteristics of extruded wheat flour and wheat starch containing added cysteine and reducing sugars. *J. Cereal Sci.*, **25**, 57–63.
86. Ames, J.M., Guy, R.C.E., and Kipping, G.L. (2001) The effect of ph and temperature on the formation of volatile compounds in cysteine/reducing sugar/starch mixtures during extrusion cooking. *J. Agric. Food Chem.*, **49**, 1885–1894.

14
Food Deep-Fat Frying

Pedro Bouchon

14.1
General Principles

Deep-fat or immersion frying is an old and popular process, which originated in and was developed around the Mediterranean area, because of the availability of olive oil [1]. Today, numerous processed foods are prepared by deep-fat frying all over the world, since in addition to cooking, frying provides unique flavors and textures that improve the overall palatability. This chapter briefly describes the frying process from industrial and scientific perspectives. First, it introduces the process, and presents some fried food quality characteristics along with the most important microstructural changes that occur during the process. Thereafter, it describes most commonly used oils, most relevant pathways of oil degradation, as well as the effect of fried food consumption on human health. It then describes most important aspects of oil absorption kinetics and discusses most important factors affecting oil absorption. Finally, this chapter describes the most important characteristics of frying equipment and some features of French fries, potato chips, fabricated chips, and third-generation snacks production.

14.1.1
The Deep-Fat Frying Process

Deep-fat frying can be defined as a process for the cooking of foods, by immersing them in an edible fluid (fat), at a temperature above the boiling point of water [2]. Frying temperatures can range from 130 to 190 °C, but the most common temperatures are 170–190 °C. Immersion frying is a complex process that involves simultaneous heat and mass transfer resulting in counterflow of water vapor (bubbles) and oil at the surface of the piece. The high temperatures of the oil bath lead to the evaporation of water at the surface of the food. Due to evaporation, water in the external layers of the product leaves the food to the surrounding oil and surface drying occurs, leading to crust formation. In addition, oil is absorbed by the food, replacing part of the water.

Food Processing Handbook, Second Edition. Edited by James G. Brennan and Alistair S. Grandison.

Frying induces physicochemical alterations of major food components and significant microstructural changes [3]. One of major aims of deep-fat frying is to seal the food surface while immersing the food into the oil bath so that flavors and juices can be successfully retained within the food. In fact, most of the desirable characteristics of fried foods are derived from the formation of a composite structure: a dry, porous, crisp, and oily outer layer or crust, and a moist cooked interior or core. The crust is the result of several alterations that mainly occur at the cellular and subcellular level, and are located in the outermost layers of the product. These chemical and physical changes include physical damage produced when the product is cut and a rough surface is formed with release of intracellular material, starch gelatinization and consequent dehydration, protein denaturation, breakdown of the cellular adhesion, water evaporation, and rapid dehydration of the tissue, and finally, oil uptake itself [4].

Dehydration, high temperatures, and oil absorption distinguish frying from simmering, which occurs in a moist medium and where the temperature does not exceed the boiling point of water. During baking, heat transfer coefficients are much lower than during frying, and although there is surface dehydration and crust formation, there is no oil uptake. In addition, the high temperatures achieved during frying (usually more than 150 °C) allow enzyme inactivation, intercellular air reduction and destruction of microorganisms, including pathogens [5].

Certainly, frying technology is important to many sectors of the food industry, including suppliers of oils and ingredients, fast-food shop and restaurant operators, industrial producers of fully fried, par-fried, and snack foods, and manufacturers of frying equipment [6].

14.1.2
Heat and Mass Transfer during Deep-Fat Frying

Deep-fat frying is a thermal process, in which heat and mass transfer occur simultaneously. A schematic diagram of the process is shown in Figure 14.1, where it can be observed that convective heat is transferred from the frying media to the surface of the product and, thereafter, conductive heat transfer occurs inside

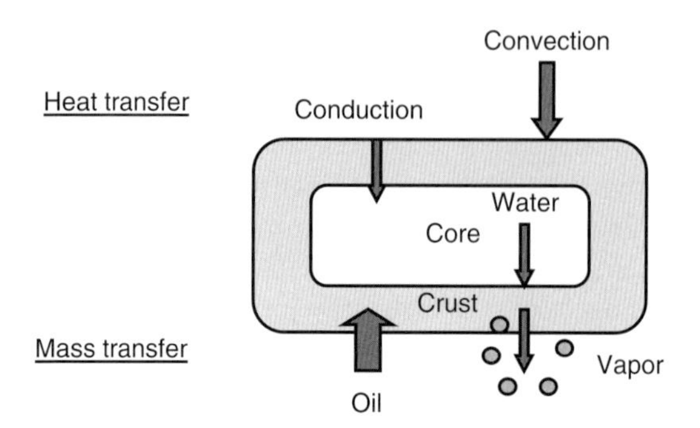

Figure 14.1 Schematic diagram of simultaneous heat and mass transfer during frying.

the food. Mass transfer is characterized by the loss of water from the food as water vapor and the movement of oil into the food [3].

Farkas *et al.* [7] observed that the temperature at any location in the core region is limited to values below the boiling point of the interstitial liquid (~105 °C). When all the liquid is evaporated from the region, the moving front propagates toward the interior and the temperature begins to rise, approaching the oil temperature. On the basis of visual observations and analysis of temperature profiles and moisture data, they suggested that the frying process is composed of four distinct stages:

1) **initial heating**, which lasts a few seconds and corresponds to the period of time while the surface temperature reaches the boiling point of the liquid; heat transfer is by natural convection and no vaporization of water occurs,
2) **surface boiling**, which is characterized by the sudden loss of water, the beginning of the crust formation and a forced convection regime due to high turbulence, associated with nucleate boiling,
3) **falling rate**, which is the longest, in which the internal moisture leaves the food, the core temperature rises to the boiling point, the crust layer increases in thickness, and finally the vapor transfer at the surface decreases, and
4) **bubble end-point**.

Several authors have attempted to measure natural convective heat transfer coefficients. Using the lumped capacity method with a spherical aluminum transducer, Miller *et al.* [8] estimated natural convective heat transfer coefficients for canola oil, palm oil, corn oil, and soybean oil at 170, 180, and 190 °C. They obtained values ranging from 250 to 280 W m^{-2} K^{-1}. Certainly, the absence of water vapor escape around the metal transducer yielded a natural convective heat transfer coefficient that is different from that when a food is undergoing frying and that is only meaningful during the early stages of frying when few bubbles are present. Using a similar approach, Moreira *et al.* [9] estimated the natural convective heat transfer coefficient for soybean oil, which was 280 W m^{-2} K^{-1} when heating at 190 °C. Bouchon and Pyle [10] obtained values in the same range when heating palm olein at 155, 170, and 185 °C, determining natural convective heat transfer coefficients of 262, 267, and 282 W m^{-2} K^{-1}, respectively. In relation to boiling convective heat transfer coefficient estimation, Hubbard and Farkas [11] obtained maximum averages values of 610, 650, and 890 W m^{-2} K^{-1}, when frying potato cylinders at 120, 150, and 180 °C, respectively, which gradually decreased to ~300 W m^{-2} K^{-1} toward the end of the process. Similar maximum average values (443 and 650 W m^{-2} K^{-1}) and conclusions were reported by Costa *et al.* [12]. Interestingly, Sahin *et al.* [13] found differences when determining the boiling convective heat transfer coefficient at the top and bottom surfaces of potato slices during frying (150–190 °C). They determined higher coefficients at the bottom surface (450–480 W m^{-2} K^{-1}) compared to the top surface (300–335 W m^{-2} K^{-1}) until the crust was formed, in contrast to what might be expected. The authors attributed this behavior to the strong insulating effect that is produced by the vigorous escape of bubbles at the top surface. Bouchon and Pyle [10] estimated boiling convective heat transfer coefficients for increasing frying times at different temperatures, which ranged approximately from 260 to 600 $W m^{-2} K^{-1}$, similar to those reported by previous authors, and adjusted a first-order kinetic model to experimental data to describe the change with frying time. Overall, all studies have found that convective heat transfer coefficients can achieve maximum values up to two or three times higher that those measured in the absence of bubbling. This value gradually decreases over the duration of the process.

14.1.3
Fried Products

A wide variety of food materials are used to make fried products, including meats, dairy products, grains, and vegetables. Different shapes and forms, such as French fries, chips, fabricated snacks, doughnuts, battered and breaded food, among others, can be found in the market. Therefore, frying technology is important to many sectors of the food industry.

Huge quantities of fried food and oils are used at industrial and commercial levels. For example, the United States produced back in the 1990s, on average, over 2.3 million tons of sliced frozen potato and potato products every year, the majority of which was fried or partially fried [6]. Commercial deep-fat frying was estimated to be worth £45 billion in the United States and at least twice this amount for the rest of the world [14]. In the United Kingdom, the potato chip market was estimated to be worth £693 million, while all other snack products together were estimated to be worth £751 million [15]. Consumption of snack food in the United States is estimated to be higher than 6.5 kg of snack food per capita annually [16].

A critical quality parameter of fried food is the amount of fat absorbed during the process, which undermines recent consumer trends toward healthier food and low-fat products. Per capita consumption of oils and fats was estimated to be 62.7 lb per year in the United States, far exceeding the recommendations found in the Surgeon General's Report on Nutrition and Health [17]. Several studies have revealed that excess consumption of fat is a key dietary contributor to coronary heart disease and perhaps cancer of the breast, colon, and prostate, imposing an alert to human consumption [18, 19]. Despite this, consumption of oils and fats is still high, and salty snacks account for slightly over half of total snack sales [16].

The snack categories that are showing the greatest growth are those that offer a wide range of alternatives and coincide with convenience, sensory, and health trends. In relation to health, salty snack products that are low in fat, calories, carbohydrates, and sodium, and rich in fiber and vitamins are becoming of great interest as are organic items and those that offer some health-promoting benefit. Nevertheless, even health-conscious consumers are not willing to sacrifice organoleptic properties, and intense full-flavor snacks remain an important trend in the salty snack market [20]. In fact, consumption of snack food is increasing in developed and developing countries, and fried products still contain large amounts of fat. An example of the total oil content of selected snack and fast foods is presented in Table 14.1, from [21]. Most of these products have an oil content varying from 5% (frozen French fries) to 40% (potato chips).

Table 14.1 Oil absorption in fried foods.

Food item	Fat (g absorbed/ 100 g edible portion)
Frozen French fries	≈5
Fresh French fries	≈10
Battered food (fish/chicken)	≈15
Low-fat chips	≈20
Breaded food (fish/chicken)	15–20
Potato chips	35–40
Doughnuts[a]	15–20

[a] Doughnuts also contain about 10% fat used in preparation of the dough.

As mentioned, there is an increased demand for low-fat products. However, reduced-fat and no-fat chips varieties represent only 11% of potato chip market sales in the United States [22]. Sales of fat-free snacks are increasing, but, as these products are baked rather than fried, they have different flavor and textural characteristics to fried chips, and therefore, consumer acceptance is low. On the other hand low-fat snacks, such as chips or tortilla chips, are acquiring greater acceptance. These products are usually dried prior to frying and research is focused on developing a product with enough fat to impart the desired organoleptic properties.

14.1.4 Microstructural Changes during Deep-Fat Frying

The importance of microstructural changes during deep-fat frying has been recognized in modeling heat and mass transport and unraveling their mechanisms. Three of five papers by leading scientists in the field published in an overview on frying of foods (*Food Technology*, October 1995) included scanning electron microscope (SEM) photomicrographs of fried products, suggesting the importance of microstructure. In fact, Baumann and Escher [23] recommended that the explanation of some factors in oil absorption needs to be validated by structural analysis in relation to the location of oil deposition and to the mechanisms of oil adhesion.

Since the first histological studies of deep-fat fried potatoes by Reeve and Neel [24], using light microscopy, evidence has accumulated that except for the outermost layers damaged by cutting, the majority of the inner cells retain their individuality after frying and contain in their interior dehydrated but gelatinized starch granules. The microstructural aspect of the core tissue is similar to that of cooked potatoes. In the case of potato chips (or outer layers in French fries), cells shrink during frying but do not rupture while cell walls become wrinkled and convoluted around dehydrated gelled starch [4, 24]. It is thought that the rapid dehydration occurring during deep-fat frying, reduces the starch swelling process, and therefore cell walls do not break as sometimes occurs during ordinary cooking. Similar observations were determined by Costa *et al.* [25], when studying structural changes of potato during frying and by McDonough *et al.* [26], when evaluating the physical changes during deep-fat frying of tortilla chips.

Keller *et al.* [27] and Lamberg *et al.* [28] determined the extent of oil penetration in fried French fries using Sudan Red B, a heat-resistant and oil soluble dye, which was added to the frying medium before frying. They concluded that oil uptake during deep-fat frying was localized on the surface of the fried product and restricted to a depth of a few cells. In fact, Bouchon *et al.* [29], using high spatial resolution infrared microspectroscopy, determined oil distribution profiles within fried potato cylinders. Results confirmed that oil was confined to the outer region and that oil distribution reflected the anisotropic nature of the porous network developed during the process. Lisinska and Golubowska [30] used electron scanning microscopy to follow structural transformation during the production of French fries and confirmed that oil was mainly located in the outer layers where cells suffered maximum deformation.

Blistering in potato chips takes place after separation of neighboring intact cells alongside cell walls, similar to fracture observed in steam-cooked potatoes [31], and oil was found to be mainly distributed in the cell walls, intercellular spaces, and blister area [24]. Aguilera and Gloria [32] demonstrated that three distinct microstructures exist in finished fried commercial French fries: (i) a thin outer layer (~250 μm thick) formed by remnants of cell walls of broken or damaged cells by cutting; (ii) an intermediate layer of shrunken intact cells which extends to the evaporation front; and (iii) the core with fully hydrated intact cells containing gelatinized starch.

It is thought that standard microscopic techniques may produce artefacts in samples like swelling of the interiors by solvents and smearing of oil during sectioning. In an attempt to reduce invasion and destruction of samples, Farkas *et al.* [33], used magnetic resonance imaging (MRI) to determine water location and oil penetration depth in immersion fried potato cylinders. They confirmed that oil was mainly located on the surface of the product and penetrated only slightly inside the structure. Confocal laser scanning microscopy (CLSM) has been introduced recently as a new methodology for studying oil location directly in fried potato French fries with minimal intrusion [4, 34], as shown in Figure 14.2. This was achieved by frying in oil containing a heat-stable fluorochrome (Nile Red) and observing the fried crust, without further preparation, under a CLSM. This technique allows optical sectioning to be carried out using a laser beam, avoiding any physical damage on the specimen. It was shown that cells seem to be quite preserved and surrounded by oil, which is not uniformly located at each depth, suggesting that at least some of the oil penetrates into the interior of a potato strip by moving between cells. Using noninvasive X-ray tomography, Miri *et al.* [35] observed the porous structure of potato cylinders during frying at 170 °C. They were able to reconstitute and characterize some crust structural parameters, such as the thickness, the porosity, and the interconnectivity between pores. They concluded

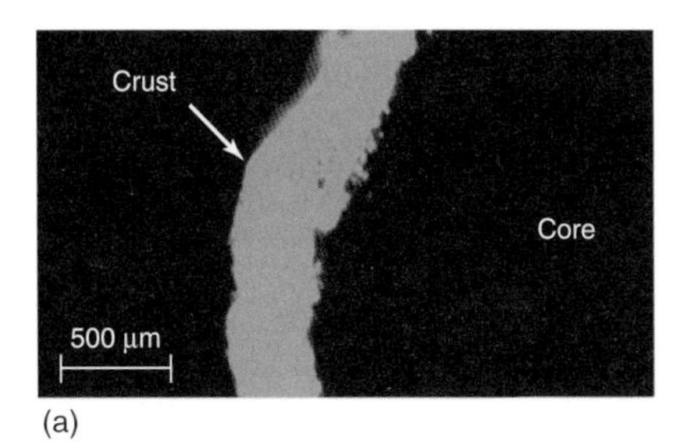

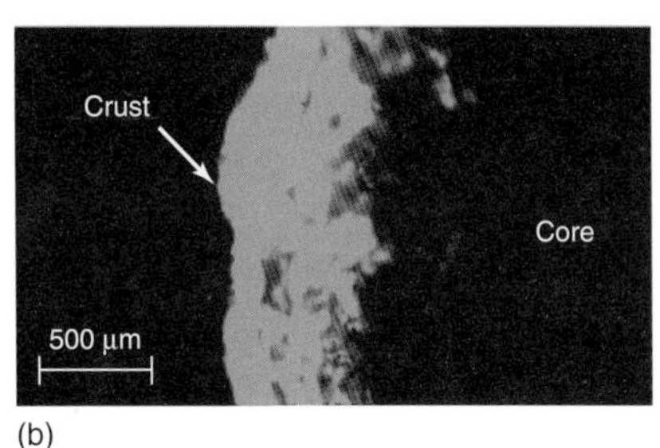

Figure 14.2 Cross-sections of a potato cylinder fried at 185 °C using confocal laser scanning microscopy (CLSM). (a) Fried for 1 min and (b) fried for 5 min. From Bouchon *et al.* [29].

that the crust seemed to be mainly formed during the last minutes of frying and that product porosity was mainly constituted by few large pores rather than small ones.

14.2
Effect of Deep-Fat Fried Food on Human Health

Chronic multifactorial diet-related diseases are the major causes of death and illness worldwide. The amount and composition of fat in the diet are important determinants of the pathobiology of many of these conditions [36]. Accordingly, the World Health Organization/Food and Agriculture Organization (WHO/FAO Expert Consultation 2003) have defined clear recommendations with respect to fat intake, which include consumption of less than 10% of total energy derived from saturated fat (or <7% in high-risk individuals), less than 1% of energy from trans fatty acids, and 6–10% total energy from polyunsaturated fatty acids (PUFAs) maintaining a balance around 5–8% between *n*-6 and *n*-3 PUFAs. Even though they do not make any specific recommendation in terms of total fat consumption, they suggest up to 35% in highly active groups, with diets rich in fruit, vegetables, legumes, and wholegrain cereal. Otherwise considerably lower fat intakes are recommended. Certainly, one of the most important quality parameters of fried food is the amount of fat absorbed during the process, which is incompatible with consumer trends toward healthier food and low-fat products. In addition, important nutritional compounds degrade during the process, and toxic molecules may generate either in the foodstuff or in the frying oil itself, whose intake should be at least limited [37].

14.2.1 Frying Oils

Food can be fried in a wide range of oils and fats, which include vegetable oils, shortenings, animal fats, or a mixture thereof. The main criteria used to select frying oils are long frying stability, fluidity, bland flavor, low tendency to foam or smoke formation, low tendency to gum (polymerize), oxidative stability of the oil in the fried food during storage, and good flavor stability of the product and price [21, 38]. Saturated fatty acids provide a greater stability in frying applications, but they are undesirable from a nutritional standpoint [39]. Conversely, oils high in PUFAs show lower thermo-oxidative stability than rich monoenoic unsaturated fatty acids or saturated fatty acids oils [40]. The fat melting point is a very important parameter as it affects the temperature of heating in tanks and pipes during storage and handling, and also it may affect the sensory attributes of fried products that are eaten at colder temperatures. Conversely, high melting point fats frequently have lower tendency to oxidation.

Frying oils are principally from vegetable sources. Some traditional oils used for frying are corn, cottonseed, and groundnut oils, which are used as a stable source of polyunsaturated fatty acids due to their low linolenic acid content [41]. However, over the last decades, the use of groundnut oil has diminished due to its cost and also due to production problems related to naturally occurring aflatoxins [21, 38].

Nowadays, palm oils are increasingly used in industrial frying. Palm oil is a semi-solid fat, which is fractionated at low temperature to give a liquid fraction called *palm olein* with a melting point that ranges between 19 and 24 °C and a high melting point fraction called *palm stearin* which melts at not less than 44 °C. Also, a double fractioned oil called *super olein* can be obtained, which has a melting point ranging from 13 to 16 °C. Palm olein is liquid enough for frying use, however, super olein is the grade used to produce fully liquid frying oil blends together with sunflower and groundnut oils. Palm oil and palm olein have a very good frying performance due to their high resistance to oxidation and to flavor reversion because of their low unsaturation and are now becoming virtually the standard oils for industrial frying in West Europe [42]. However, the high level of saturated (palmitic) fatty acids (38.2–42.9% in palm olein and 40.1–47.5% in palm oil) may be criticized from a nutritional view point.

Rapeseed (canola) and soya bean oils have a high level of linolenic acid (8–10%), making them vulnerable to oxidation and off-flavor development, and therefore, can be slightly hydrogenated for industrial frying. This procedure may also be applied to sunflower oil and may be attractive where an oil with a high polyunsaturated to saturated ratio is needed for dietary purposes [21]. In addition, new oils, such as high oleic sunflower oil are also a clear option [37].

Olive oil has excellent attributes that make it suitable as frying oil. It has a low level of PUFAs and a mixture of phenolic antioxidants that make it resistant to oxidation. Extra-virgin and virgin olive oils are expensive for industrial use, however, refined solvent-extracted olive oil can be satisfactory for industrial frying.

Animal fats are also used for frying in some regions due to the characteristic flavors that they impart to the fried food, despite their high level of saturated fatty acids. Fish oils are rarely used for frying, as their high level of long-chain PUFAs makes them prone to oxidation [21].

Fat substitutes, such as sucrose polyesters (olestra) have been extensively studied. In fact, Olean, which is Procter & Gamble's brand name for olestra has been introduced in the Unites States and several snacks fried in olestra are already on the market (fat-free Pringles, Frito-Lay Wow). Olestra is synthesized from sucrose and fatty acid methyl esters and has no calories because its structure prevents digestive enzymes from breaking it down. Two main negative aspects are currently discussed in relation to olestra that impair its acceptance in other countries. First, olestra can cause gastrointestinal discomfort in some people and second, its ability to reduce the absorption of fat-soluble nutrients is often discussed [22].

Antioxidants such as tertiary butyl hydroquinone (TBHQ), butylated hydroxyanisole (BHA), and butylated hydroxytoluene (BHT) are usually added to improve oil stability. TBHQ is regarded as the best antioxidant for protecting frying oils against oxidation and, like BHA and BHT, it provides carry through protection to the finished fried product.

14.2.2
Oil Degradation

Oil selection is of great importance, since frying oils may undergo thermal, oxidative, and hydrolytic degradation due to their exposure to elevated temperatures in the presence of air and moisture [40]. Figure 14.3 summarizes most important reactions occurring to a triacylglycerol molecule subjected to normal frying conditions, identifying the effect of temperature, oxygen, and moisture release from the foodstuff in oil degradation [43]. In this figure, linoleic acid was selected as an example because it is the precursor for the formation of the toxic aldehyde hydroxy-2-*trans*-nonenal (HNE), a secondary lipid peroxidation product that has shown cytotoxic and mutagenic properties, as well as some HNE-related compounds such as 4-hydroxy-2-*trans*-hexenal (HHE) and 4-hydroxy-2-*trans*-octenal (HOE), derived from the degradation of linolenic and linoleic acids, respectively [44]. The toxicity of these compounds seems to be due to their high reactivity with proteins, nucleic acids, DNA, and RNA, and reports have related them to several diseases, including atherosclerosis, Alzheimer's disease, and liver disease.

In addition, water release gives rise to di- and mono-acylglycerols, glycerols and free fatty acids, free molecules which are more susceptible to oxidative and thermal degradation than when esterified to the glycerol molecule [45]. Hydroperoxides, which occur by loss of hydrogen in the presence of trace metals, heat and light, are not stable under deep-fat frying conditions and may undergo fission to produce a wide variety of secondary lipid peroxidation products, such as aldehydes, ketones, and other carbonyl-containing compounds. These compounds contribute to the volatile fraction of the degraded frying oil and therefore determine the development of off-flavors in the fried products [46, 47]. As a summary, the oil changes from a

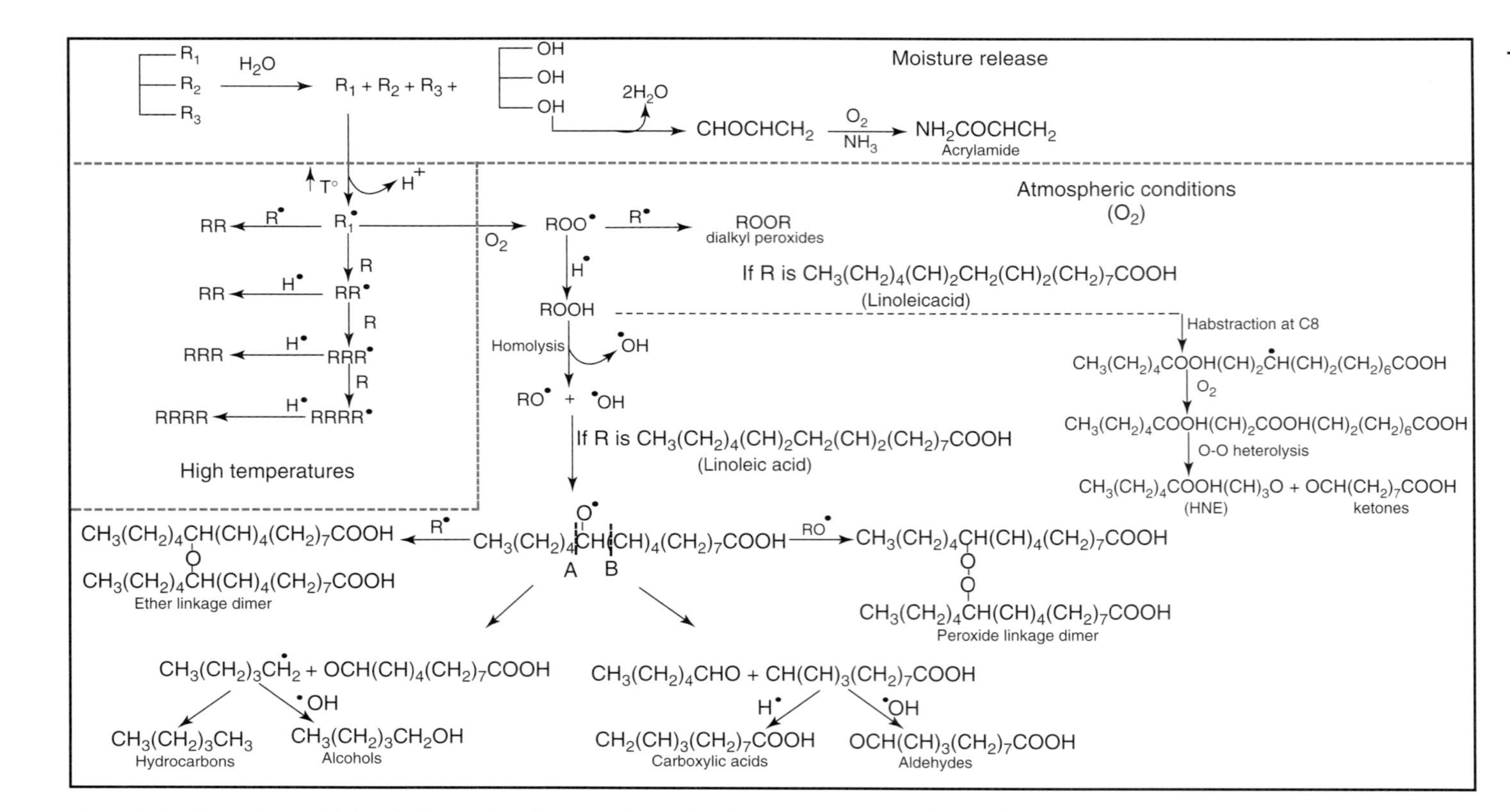

Figure 14.3 Mechanism of frying oil degradation, from Dueik and Bouchon [43]. Linoleic acid was selected as an example since it is the precursor of hydroxy-2-*trans*-nonenal (HNE) formation. Adapted from Choe and Min [45], Mestdagh *et al.* [48], Uchida [49], and Dobarganes [50].

medium that is almost pure triglyceride to a mixture of hundreds of compounds, which are absorbed together with the fried food [6].

14.2.3
Harmful Compounds Generated during Frying

In April 2002, Swedish scientists sound an alarm when they discovered that certain cooked foods, particularly potato chips and French fries, contained high levels of acrylamide (30–2300 $\mu g\,kg^{-1}$) [51], a chemical compound that is listed by the WHO as a probable human carcinogen [52]. There is no consensus about whether acrylamide in food is a danger and the WHO has not called yet for any reduction in foods containing high levels of this compound. The Maillard reaction between amino acids (e.g., asparagine) and reducing sugars (e.g., fructose and glucose) at temperatures above 120 °C has been suggested to be the main route for the formation of acrylamide, a substance that is often used to produce plastics and dyes [53]. The abundance of free asparagine in potatoes has been suggested to be responsible for chips and French fries being the food category with probably the highest concentration of acrylamide recorded so far [54]. Certainly, the Maillard reaction plays a relevant role during the deep-fat frying of carbohydrate-rich foods, since this highly temperature-dependent reaction is instrumental in the development of color, flavor, and aroma compounds. However, despite its desired effects, this reaction has been stated to be the principal mechanism for harmful compounds generation in fried food. It is important to note that even though results signal Maillard reaction as an important route for acrylamide formation, it is not possible to exclude any other possible mechanisms.

Several strategies have been followed to reduced acrylamide content. Among successful attempts to reduce acrylamide formation we can find oil temperature reduction [55], blanching or soaking pretreatments [55, 56], pH reduction [57], and use of an asparaginase solution prior to frying [56].

It is important to note that living organisms may convert acrylamide into glycinamide, a compound that is thought to be considerably more toxic than acrylamide, however, little research is available in this topic [58, 59]. Other heat-induced harmful compounds may be found in certain foods. Among them, we can find hydroxymethylfurfural in carbohydrate-rich foods [60] and heterocyclic amines in protein-rich foods [61]. An in-depth discussion about harmful compounds generation in food during frying may be found in Dueik and Bouchon [43].

14.3
Oil Absorption in Deep-Fat Fried Food

As previously mentioned, consumer trends are slowly moving toward healthier foods and low-fat products, creating the need to reduce the amount of oil in end products. In order to obtain products low in fat, it is essential to understand the mechanisms involved during the frying process, so that oil migration into the structure can be effectively minimized. Deep-fat frying is a complex unit operation involving high temperatures, significant microstructural changes both to the surface and the body of the product, and simultaneous heat and mass transfer resulting in flows in opposite directions of water vapor (bubbles) and certainly oil, a quality factor of great concern.

14.3.1 Kinetics of Oil Uptake

The selection of a model needs to be in accordance with experimental observations that reveal how oil absorption takes place. It is not clearly understood yet when and how the oil penetrates into the structure, however, it has been shown that most of the oil is confined to the surface region of the fried product [4, 26, 28, 29, 33, 34, 62] and there is strong evidence that it is mostly absorbed during the cooling period [15, 22, 32, 63].

Gamble *et al.* [64] gave a reasonable initial explanation of the deep-fat frying mechanism. They proposed that most of the oil is pulled into the product when it is removed from the fryer due to condensation of steam producing a vacuum effect. They suggested that oil absorption depends on the amount of water removed and on the way this moisture is lost. As a matter of fact, as dehydration proceeds, water vapor finds selective weaknesses in the cellular adhesion that leads to the formation of capillary pathways, increasing surface porosity.

Ufheil and Escher [63] suggested that oil uptake is primarily a surface phenomenon, involving equilibrium between adhesion and drainage of oil upon retrieval of the slice from the oil. The authors carried out successive frying experiments adding dyed oil, using a fat-soluble and heat-stable dye (Sudan Red B), to the oil bath at different moments before ending the frying process and quantifying the amount of dyed oil in the fried potato French fries with an extraction–refractometric method. They found that even if the dyed oil was added just before ending the immersion, the proportion of dyed oil in the fried product was very high, concluding that oil does not seems to penetrate within the product during the deep-fat frying process. Rather, it seems to be absorbed upon cooling, from the oil layer that remains on the surface of the product, after it is removed from the fryer. In a similar way, Matz [65] commented that if potato chips are promptly removed from the fryer while their temperature is still rising, only 15% of the oil is absorbed into the food, while the remainder is held on the surface. These studies show that, while the counterflows of water vapor and oil are related to each other, they are not necessarily synchronized.

Moreira *et al.* [66] found that the largest amount of oil penetrates into the structure of tortilla chips during the cooling period and not during frying. They determined that only 20% of the total oil content is absorbed during frying and ~80% remains on the product surface. In addition, they found that almost 64% of the total oil content was absorbed during the cooling (post-frying) period. Interestingly, they determined that oil uptake takes place during the first 20 s of cooling, that is, when the temperature in their experiments was still above the condensation temperature

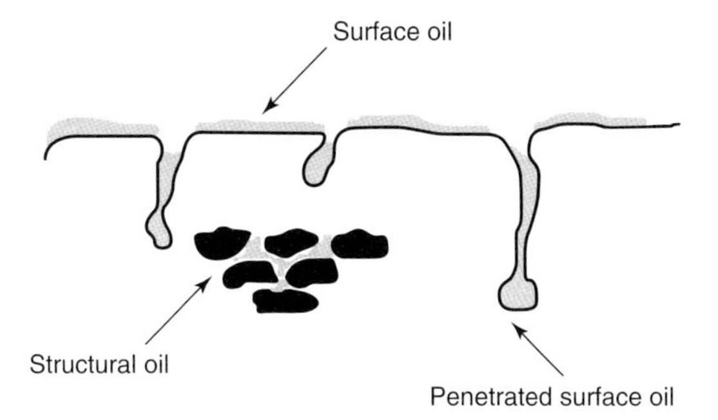

Figure 14.4 Diagram showing the three locations of oil in the product microstructure after frying process. From Bouchon *et al.* [29].

(around 100 °C), suggesting that the effect of water condensation seemed to be negligible. Accordingly, Moreira and Barrufet [67] explained the mechanism of oil absorption in tortilla chips during cooling solely in terms of capillary forces.

Using an approach similar to that of Ufheil and Escher [63], when frying potato cylinders, Bouchon *et al.* [29] determined that three different oil fractions could be identified as a consequence of the different absorption mechanisms involved, that is: (i) structural oil (STO), which represents the oil absorbed during frying, (ii) penetrated surface oil (PSO), which represents the oil suctioned into the food during cooling after removal from the fryer, and (iii) surface oil (SO), that is, the oil that remains on the surface. A schematic diagram showing these oil fractions is presented in Figure 14.4.

The authors determined that only a small amount of oil penetrated during frying as most of the oil was picked up at the end of the process, corresponding mainly to a surface phenomenon. After cooling, oil was located either on the surface of the chip (SO) or suctioned into the porous crust microstructure (PSO), with an inverse relationship between them for increasing frying times. On the basis of their results they suggested that the microstructure (that is, the mean pore size, connectedness, and permeability) of the crust region, which is formed while the potato is being fried, is the single most important product-related determinant of the final oil uptake into the chip.

14.3.2
Modeling the Deep-Fat Frying Process

Many of the studies related to frying have been limited to observations on the frying process. Mathematical models have been centered on the prediction of temperature profiles inside the fried food and the prediction of the kinetics of moisture loss, and little has been carried out in relation to oil uptake. Models with different levels of complexity have been reported, which vary considerably in terms of the assumptions made. Most of the models for water evaporation during frying consider the fried product as a single phase, with no differentiation between the crust and the core regions and without the existence of a moving crust/core interface, where either energy or mass diffusion are considered to be the rate-controlling mechanisms [13, 22, 68–75].

However, some authors have recognized the existence of two separate regions, the crust and the core, and have proposed a model considering the presence of a moving boundary [7, 10, 76, 77]. In fact, Farkas *et al.* [7] were the pioneers in developing a comprehensive model of thermal and moisture transport during frying. They formulated their model by analogy to freezing [78] and to the solution of the uniform retreating ice front during freeze drying [79]. They provided different sets of equations for the two regions, separated by a moving boundary where the evaporation occurred. In their model, they did not include the oil phase; they described the heat transfer in both regions using the unsteady heat transfer conduction equation, they considered water diffusional flow within the core region and they assumed that water vapor movement was pressure-driven. The final set of equations consisted of four nonlinear partial differential equations, which were solved using finite differences. The results were compared with experimental data, and they obtained a reasonable prediction for temperature profiles, water content, and thickness of the crust region [80]. However, as Farid and Chen [76] pointed out, simulations were time consuming, making it difficult to extrapolate the model to a multidimensional geometry; they therefore simplified the model.

Ni and Datta [77] developed a multiphase porous media model to predict moisture loss, oil absorption, and energy transport in a potato slab, also considering a moving front. They assumed that vapor and air transport were considered to be driven by convective and diffusive flows, while liquid water and oil were supposed to be driven by convective and capillary flows. Model predictions were compared with experimental data from the literature, mainly from Farkas *et al.* [80]. They centered their validation on temperature profiles, moisture content, and crust thickness predictions, however, they did not include oil uptake absorption as part of their model validation. In relation to oil uptake, their model allowed oil absorption to occur during the immersion period as water moved out from the food, and did not take into account the effect of cooling time on oil absorption. They explained that the rate of oil uptake was more important in the early stage of frying when there is a larger difference between surrounding and absorbed oil concentrations. As a summary, this model gave valuable insights about how to model transport phenomena in a multiphase porous media, but it was not really close to what has been observed experimentally.

Moreira and Barrufet [67] explained the mechanism of oil absorption in tortilla chips during cooling in terms of capillary forces, since they found that most of the absorption took place when the temperature was still above the boiling point of water. They modeled heat transfer during cooling using the heat conduction equation including an effective thermal diffusivity and considered that mass transfer from the SO into the structure was controlled by capillary pressure. The mathematical model gave good agreement between theoretical and experimental data, showing that oil absorption during cooling could be described in terms of capillary flow mechanisms.

As explained in the previous section, some studies have shown that cooling is also critical to initiate oil infiltration. This so-called *condensation mechanism* refers to the oil post-cooling suction due to water vapor condensation, a mechanism that may well be mediated by capillary forces. Using a modified form of the Washburn equation, Bouchon and Pyle [10] focused on the pressure barrier that needs to be overcome by the oil film to penetrate within the structure upon cooling. The model was developed for two different geometries, an infinite slab and an infinite cylinder, and was divided into two main submodels, one describing the immersion frying period itself and the other describing the post-frying cooling period. The immersion frying period was described by a transient moving-front model that considered the movement of the crust/core interface, which was controlled by heat transfer, based on the one proposed by Farid and Chen [76]. The cooling problem was also formulated as an unsteady heat conduction process. The amount of water that evaporated during the cooling period was considered to be negligible, therefore, the boundary could be considered to be stationary. The final temperature profile obtained after a fixed immersion time gave the initial temperature distribution within the product, and thus the initial condition. Therefore, the final state of the product at the end of frying was realistically the initial state during cooling. Results were successfully compared with those found in the literature (infinite slab) or obtained experimentally (infinite cylinder) [81].

A key element in this model is the hypothesis that oil suction would only begin once a positive pressure driving force had developed. Temperature prediction during cooling is therefore a critical element since it determines the moment when water condensation occurs. Post-frying cooling oil absorption was considered to be a pressure-driven flow mediated by capillary forces. A laminar, steady, and fully developed oil flow through a uniform pore (perfect cylinder of radius r) was considered, which was directly related to the piezometric pressure drop within the pore (cylinder) according to Equation 14.1:

$$Q = \frac{\pi r^4}{8\mu l} \times \Delta P^* \tag{14.1}$$

where Q is the oil volumetric flow ($m^3\,s^{-1}$), r is the pore radius (m), μ is the oil viscosity (Pa s), l is the length (m), and ΔP^* is piezometric pressure difference (Pa).

In capillary flow, due to the existence of a meniscus at the moving front, the total driving pressure is made up of three different components, as Washburn [82] pointed out: the unbalanced atmospheric pressure, the static pressure of the fluid, and the capillary pressure, which relates the pressure difference across the curved surface of the meniscus to the surface tension of the fluid and the contact angle between the fluid and the tube wall of the capillary (Equation 14.2).

$$\text{Capillary pressure} = P_a - P_b = \frac{2 \times \sigma \times \cos\theta}{r} \tag{14.2}$$

where σ is the oil surface tension ($N\,m^{-1}$) and θ is the contact angle between the oil and the capillary (rad).

As shown in Figure 14.5, the resulting net driving force depends on the different configurations that can be found. If a reference datum plane ($h = 0$) is set at the

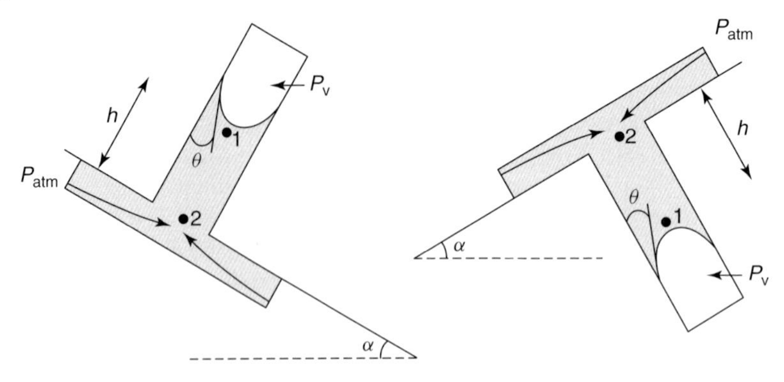

Figure 14.5 Schematic diagram showing the capillary penetration phenomena when having different arrangements. (a) Upward configuration, the action of gravity restricts capillary penetration. (b) Downward configuration, the action of gravity enhances capillary penetration. From Bouchon and Pyle [10].

bottom of each of the capillaries, the expression for the total driving force, that is the piezometric pressure difference along the penetration length h, for a capillary with an upward ($+\rho gh\cos\alpha$) or a downward ($-\rho gh\cos\alpha$) orientation, is given by the following equation:

$$\Delta P^* = P_2^* - P_1^* = P_{\mathrm{atm}} - P_{\mathrm{pore}} = P_{\mathrm{atm}} - \left(P_{\mathrm{V}} - \frac{2\sigma\cos\theta}{r} \pm \rho gh\cos\alpha\right) \tag{14.3}$$

where P_i^* are piezometric pressures (Pa), P_v is the water vapor pressure at the crust/moving-front (Pa), ρ is the density of the oil (kg m^{-3}), g is the acceleration due to gravity (m s^{-2}), h is the oil penetration distance (m), and α is the angle between the normal axis of the capillary and the vertical axis (rad).

As can be seen from Equation 14.3, the oil will penetrate as long as the pressure difference is positive, and hence until the unbalanced atmospheric pressure at the bottom of the capillary persists (that is as far as $P_{\mathrm{atm}} > P_{\mathrm{pore}}$). Since P_{pore} depends heavily on vapor pressure, this is expected to begin after some cooling takes place. As explained by Mellema [83], the condensation mechanism should predominate in large samples and short frying times, since in smaller samples or longer frying times moisture loss rate may diminish considerably (and therefore P_{pore}), allowing oil absorption to take place at an earlier stage.

14.3.3
Factors Affecting Oil Absorption

There has been much research to examine the different factors affecting oil absorption during frying and many empirical studies have correlated oil absorption measurements with process and/or product characteristics.

The effect of frying time and oil temperature on oil uptake has been extensively studied by different authors. These two process parameters are closely related since products must be fried until reaching certain final moisture content, so a lower oil temperature implies a longer frying time. In order to understand adequately results reported in the literature, it is important to understand the different outcomes when reporting results on a wet and a dry basis. When results are expressed on a wet-weight basis (w.b.), which is of industrial interest, there is a systematic reduction in the basis as the water content diminishes. Also, results are biased by oil uptake itself, which affects the basis too. When oil uptake results are measured as a percentage on a dry-weight basis (d.b.) and the solids remain constant throughout the whole process, it may provide a consistent basis for comparison [84].

Gamble *et al.* [85] concluded that a lower oil temperature resulted in lower oil content in potato chips in the early stages of frying, with a greater difference between 145 and 165 °C than between 165 and 185 °C. Similarly, Moreira *et al.* [66] determined higher differences in oil absorption in tortilla chips (wet basis) between 130 and 160 °C than between 160 and 190 °C. However, results were expressed as oil uptake on a wet basis. Consequently, there was a systematic reduction in the basis, as the water content was constantly reduced due to the higher dehydration that resulted when the product was fried at a higher temperature for the same period of time. On the other hand, Bouchon *et al.* [86] determined that the rate of oil absorption (dry basis) by potato cylinders was not significantly affected by oil temperature (150, 170, and 185 °C). In addition, no significant differences were found in oil absorption between the two higher temperatures. However, when frying at 155 °C, a significantly lower absorption compared to the previous ones was found. Similarly, Moreira *et al.* [87] determined that the oil absorption rate was unaffected by the oil temperature when frying tortilla chips and that a frying temperature of 190 °C gave a higher oil content (3–5%) than a frying temperature of 155 °C. Nonaka *et al.* [88] also found that oil content increased with increasing frying temperature when frying potato French fries. Krokida *et al.* [89] found that potato strips absorb less oil when fried at lower oil temperatures for long frying times (over 3 min), the difference getting greater as frying proceeded. They also determined that oil content increased for increasing frying times, especially for thinner products. However, other researchers have found a different behavior. For instance, Kita *et al.* [40] found that chips absorbed less fat (wet basis) when increasing the frying temperature (150, 170, and 190 °C). With every increase in frying temperature by 20 °C, fat absorption was reduced by 3%, on average, with all the oils under investigation. As can be seen, results may be contradictory. Certainly, the way results are reported (wet or dry basis), the statistical analysis that is carried out, as well as some "unrevealed" details, such as whether the samples were shaken after frying or whether they were blotted with paper towel, and so on, are critical to really assessing the effect of these parameters.

The influence of oil type and quality on oil absorption and residues absorbed by fried foods is widely documented (e.g., Blumenthal and Stier [6], Nonaka *et al.* [88], Krokida *et al.* [89], Blumenthal [90, 91], Pokorny [92]). In relation to oil type, most of the research had shown no relationship between oil type and oil absorption. However, Kita *et al.* [93], when studying the effect of eight kinds of vegetable oils (sunflower, rapeseed, soybean, olive, peanut, palm, partially hydrogenated rapeseed oil, and a mixture of hydrogenated rapeseed oil with palm oil) on oil absorption, determined that the amount of fat absorbed by the chips, as well as their texture, depended on the kind of oil used for frying. They found that chips fried in rapeseed oil absorbed the smallest amount of fat (36.8%), irrespective of frying temperature, whereas the chips fried in olive oil and a mixture of hydrogenated rapeseed oil with palm oil absorbed the highest quantity of fat (41%).

Pinthus and Sanguy [94] studied the relationship between the initial interfacial tension of a restructured potato product and various frying media and the medium uptake during deep-fat frying. They found that total oil uptake was higher for lower initial interfacial tension, reflecting the importance of wetting phenomena. Blumenthal [90, 91], also noticed the importance of surface wetting on oil absorption and developed the so-called "surfactant theory of frying" in which he explained that several classes of surfactants are formed during frying, which act as wetting agents, reducing the interfacial tension between the food and the frying oil, producing excessive oil pick-up by the fried product. However, the formation of active agents has lately been shown to provide only a limited explanation for the increased oil uptake during prolonged frying. Dana and Saguy [95] have shown that new data present contradicting values that do not support the theory that during extended frying surfactant formation would reduce the contact angle and/or the interfacial tension of the oil, and consequently, would influence oil uptake significantly. Higher oil uptake is thought to be linked to higher oil viscosity caused by polymerization reactions.

An increase in additional factors such as initial solids content [96–98], slice thickness [23, 99, 100], and gel strength [101] have been shown to reduce oil uptake. Additionally, it has been found that an increase in the initial porosity of the food increases the absorption [102]. However, as explained by Saguy and Pinthus [17], crust formation plays an additional and fundamental role as soon as frying commences. Accordingly, the microstructure of the crust has been defined as the single most important product-related determinant of the final oil uptake into the product [29]. In fact, pore development [103] and pore size distribution [104] have been found to directly influence oil absorption during frying.

Some pre-frying treatments have been shown to significantly reduce oil absorption during frying. Lowering the moisture content of the food prior to frying using microwave and hot-air drying results in a reduction in the final oil content [22, 28, 84, 85, 104], whereas freeze drying increases the absorption [84, 85]. Because of the empirical relationship between moisture loss and oil absorption many studies aim at reducing initial water content in order to decrease the uptake. The effectiveness of these pretreatments though, which is usually achieved through drying, is not due to a reduction of the moisture content on its own, as commonly believed, but due to the structural changes occurring at the surface of the food, which reduce surface permeability [84]. On the other hand, posttreatments such as hot-air drying [88] have been shown to reduce oil uptake.

Osmotic dehydration has also been reported as an effective pretreatment in oil absorption reduction. Its effectiveness greatly depends on the solution employed [105–107]. However, as revealed in a recent study, osmotic dehydration has not only been shown to be ineffective in oil reduction, but it has even been reported as a means of increasing oil absorption. In fact, Moreno and Bouchon [84] found that potato cylinders absorbed more oil when they were immersed either in sucrose or NaCl solutions prior to frying. The difference between results found in their study and those reported in the literature was explained to be caused by the different ways in which results are reported. Normally oil-uptake results are expressed on a dry basis, a basis that, as revealed in their study, can be noticeably increased during osmotic dehydration due to solid impregnation. Consequently, even if samples absorb more oil, the increase in the basis can lead to an apparent oil uptake reduction. In fact, they reported an apparent oil-uptake reduction in osmotic dehydrated samples compared to a control sample, up to 27% (d.b.) when using a sucrose solution. However, this large decrease in oil absorption was attributed to the increase in solids content occurring during the osmotic dehydration process rather than to a reduction in the amount of oil taken up. In fact, when the amount of oil absorbed per cylinder was determined, they verified that oil uptake of osmotically dehydrated samples was even higher than the control (>45%), in contrast to what has been previously reported in the literature. They concluded that in order to carry out adequate comparisons between treatments in absolute terms, it is necessary to compute the net amount of oil absorbed by each sample. The increase in oil uptake of osmotically dehydrated samples was mainly attributed to the greater space available due to the considerable water lost during both the pretreatment and the frying process.

Different ingredients may be applied to reduce oil absorption. Their potential is mainly based on their film-forming capability and/or because they can reduce the porosity of the external layers. In this respect, much attention has been given to the use of hydrocolloids such as methylcellulose (MC), hydroxypropyl methylcellulose (HPMC), long fiber cellulose, corn zein, alginates, among others, to inhibit oil uptake [70, 108–113]. The hydrocolloid mixture can be added to the product in several ways: (i) directly in the formula, such as in doughnuts and restructured potato products (also known as *fabricated products*), (ii) incorporating the ingredient into the batter or breading, or (iii) by spraying the mixture on the surface of the product as a solution [108].

Batters and breadings are gaining more importance as they may contribute significantly to product added value [104]. Most of research has been focused on coatings and batters, and little research can be found in relation to direct incorporation into the dough formula in restructured sheeted products. In formulated products, the permeability of the outer layer of the product depends on the thickness of the sheeted dough since it determines the structural resistance to vapor escape. A stronger and more elastic network can result in a less permeable outer layer that may act as an effective barrier against oil absorption [114]. In fact, the authors examined the oil absorption capacity of different restructured potato chips during deep-fat frying, using low leach potato flake as the major ingredient and native and pregelatinized potato starch as complementary ingredients, and found that oil absorption increased significantly when reducing product thickness in all products. Interestingly, they found that the product containing native potato starch as ingredient picked up the lowest amount of oil when sheeted into a thick chip, whereas it absorbed the largest amount of oil when sheeted into a thin one. Those findings were mainly attributed to crust microstructure development as revealed by electron microscopy and confocal microscopy.

In a complementary study, Gazmuri and Bouchon [115] studied the oil absorption capacity of a restructured matrix made with native wheat starch and vital wheat gluten. They analyzed four different product formulations, using two levels of gluten content (8 and 12% d.b.) and two levels of water content (38 and 44% w.b.). Dough was sheeted into two thicknesses (1 and 2 mm) and cut into disks that were either directly fried or fried after predrying with dry air (2 min at 150 °C). Results showed that gluten film-forming capacity was the most relevant factor influencing oil absorption. In fact, dough with higher gluten content presented a significant reduction in oil uptake in almost every case under study (dried and undried) with the exception of products with high moisture content. This was probably because high water content led to explosive vaporization, creating tissue disruption, and therefore, increasing surface permeability to oil absorption. Interestingly, they found that even though predried disks with 8% gluten (d.b.) lost more water during the drying step than products with 12% gluten (d.b.), they absorbed a significantly higher amount of oil, showing that oil uptake is not so clearly related to the amount of moisture lost but to product microstructure and external layers permeability. Predried disks absorbed on average half the oil when compared to undried samples, reflecting the effect of shrinkage and case hardening on oil surface permeability. Results did not show any relationship between product expansion and oil absorption, supporting the hypothesis that oil absorption is a surface phenomenon.

It has been suggested that surface roughness increases surface area, enhancing oil absorption [104]. In an effort to quantify the irregular conformation of the surface Pedreschi *et al.* [116] and Rubnov and Saguy [117] have used fractal geometry, confirming the significant role of crust roughness in oil absorption. Certainly, oil adherence and drainage must also play an important role in oil absorption since they define the oil layer to be sucked during post-frying cooling. Moreno *et al.* [118] examined the relationship between surface roughness and oil uptake in fried formulated products, which were divided into two main categories: (i) formulations based on potato flakes and (ii) formulations based on wheat gluten. The surface of fried products was measured using a scanning laser microscope and characterized by area-scale fractal analysis. The authors determined that in each product category products with higher surface roughness absorbed more oil. However, this relationship was restricted to products of similar nature (gluten or potato flake-based products categories) and could not be extended when comparing different product categories, explaining that other food-related properties may explain differences among product categories.

14.4 Deep-Fat Frying Equipment

Frying equipment can be divided into two broad categories: (i) batch frying equipment, which is used in small plants and catering restaurants and (ii) continuous frying equipment, which is used on the industrial scale and facilitates processing larger amounts of product. Normally, deep-fat fryers operate under atmospheric conditions, however, new developments may consider operation under lower or higher pressures.

14.4.1 Batch Frying Equipment

A batch fryer consists in one or more chambers with an oil capacity that ranges from 5 to 25 l. The oil can be heated directly by means of an electrical resistance heater that may be installed a few inches above the bottom of the fryer. This allows arrangement for a cool zone at the bottom of the vessel where debris can fall, remaining there, minimizing oil damage. The fryer can also be heated by direct gas flames underneath the bottom of the vessel; however, this arrangement makes difficult the provision of a cold zone under the heaters [21]. High-efficiency fryers include turbo-jet infrared burners which use less than 30–40% energy than standard gas-fired fryers with the same capacity [22]. Modern batch fryers are constructed with high-grade stainless steel and no copper or brass is used in any valve fitting or heating element to avoid oxidation catalysis. Usually, the operators immerse and remove the baskets manually from the oil but new equipment may include an automatic basket lift system that rises automatically when the frying time is finished. Removal of food scraps and oil filtration is an essential practice that needs to be carried out on a daily basis to increase oil shelf life and avoid smoking, charring, and off-flavor development. New equipment can also have a built-in pump filtration unit for the removal of sediments [38].

14.4.2 Continuous Frying Equipment

Large-scale processing plants use continuous fryers. These are automated machines that consist of a frying vessel where oil is maintained at the desired temperature, a conveyor that carries the product through the unit and an extraction system that eliminates the fumes, primarily made up of moisture and a fine mist of fatty acids.

The oil can be heated either directly by means of an electric heater or a battery of gas burners, or indirectly, by pumping a heated thermal fluid into pipes immersed in the oil bath. Some fryers are equipped with external heat exchangers. In these systems, the oil is continuously removed at the discharge end of the tank, pumped through a filter unit and then through an external heat exchanger, before it is returned to the receiving end of the vessel (Figure 14.6). Some continuous fryers

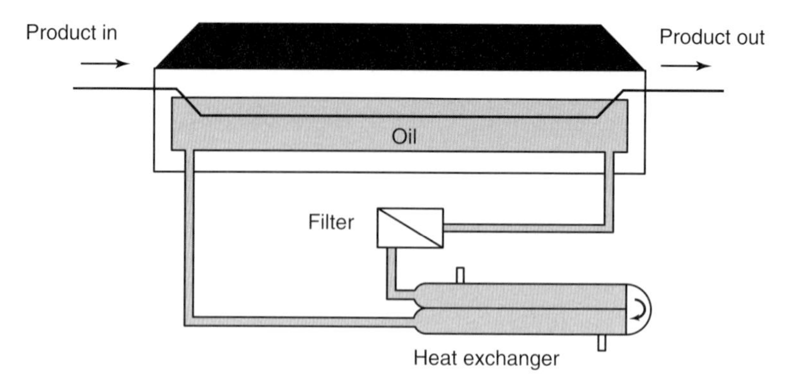

Figure 14.6 Scheme of a fryer with an external heating system and a fat filter unit.

are designed with multiple heating zones along the fryer that can be adjusted separately, providing optimal temperature control to improve product quality.

As the oil is absorbed by the fried product, it has to be made up with fresh oil continuously. The amount of fresh oil added to the vessel defines the oil turnover [119]:

$$\text{Oil turnover} = \frac{\text{weight of oil in fryer}}{\text{weight of oil added per hour}} \tag{14.4}$$

which represents the time needed to replace all the oil contained in the fryer. A fast oil turnover is desired as it maintains satisfactorily the level of free fatty acids, preserving oil quality for longer periods of time. Normally, the oil turnover is kept between 3 and 8 h [34]. Industrial fryers have oil capacities ranging from 200 to 1000 kg and can have a throughput that varies from 250 to 25 000 kg product per hour [120]. Some of the main frying equipment manufacturers are Florigo B.V. in The Netherlands, Heat and Control, and Stein in the United States.

Given the current concern with lowering fat contents of diets, an oil-reducing unit has been developed. The equipment is mounted at the discharge end of the fryer and removes the fat excess from the recently fried food using steaming and drying technology. Super-heated steam at 150–160 °C, which is circulated by a fan through heat exchangers, flows through the product bed, removing nonabsorbed SO from the hot surface of the food. Consecutively, the oil–vapor mixture is filtered, and the oil is pumped back to the fryer. The low-fat stripping system can reduce oil content in chips by 25% [121]. Units range from batch strippers for pilot plants or product development to continuous production units.

14.4.3
Pressure Fryers

Deep-fat fryers may also operate at a higher pressure, from 60 to 220 Pa (18–65 in.Hg), still lower than ordinary pressure cookers. These have been developed to meet particular needs, primarily in certain catering outlets, especially in the preparation of fried chicken in some restaurants, because of the uniform color and improved texture (higher moisture content) conferred to food. Pressure fryers may reduce frying time considerably, but they can also increase frying oil deterioration rate since steam retention within the fryer increases free fatty acids content [22].

14.4.4
Vacuum Fryers

Through vacuum it is possible to lower substantially the boiling point of product moisture in a low-oxygen environment. This is one reason why vacuum technology is a recognized route to protect heat-sensitive food during dehydration [53]. This is the main reason why a vacuum frying system developed by Florigo B.V. during the 1960s has been reintroduced. The equipment was first created to produce high-quality French fries, however, due to the improvement in blanching technology and in raw material quality, the use of this technology almost disappeared [22]. Nowadays, vacuum frying technology is being used to maintain natural colors, flavors, and nutrients in high added-value products, such as vegetables and fruits, due to the fact that much lower temperatures can be used during frying. The low temperatures employed and minimal exposure to oxygen account for most of its benefits, which include natural color, flavor, and nutrient preservation [20, 122–125], as well as oil quality protection [126] and reduction of acrylamide content [127]. In terms of oil uptake, this technology has been shown to decrease oil absorption compared to atmospheric frying [20, 119, 125, 128] and several vacuum fried foods contain low amounts of oil [129, 130].

Vacuum should be preferably lower than 6.8 Pa (2 in.Hg), making it possible to reduce substantially the frying temperature due to water boiling-point depression [129]. In the market, only a few continuous (e.g., H&H Industry Systems B.V., The Netherlands) and batch (e.g., Qinhuangdao TongHai Science & Technology Development Co., Ltd., China; Archigama technic, Indonesia) vacuum fryers are available, since this technology is still under development. A diagram of a vacuum batch deep-fat fryer together with its main components is shown in Figure 14.7. The food is normally placed inside the basket once the oil reaches the target temperature and the chamber depressurized. The product remains in the oil for the required amount of time, the basket is lifted and the vessel is pressurized. This translates into a sudden increase in the surrounding pressure at a constant temperature, which may force SO within the food before even cooling begins (i.e., $P_{atm} - P_{pore} > 0$) and therefore, it is a critical step. An in-depth discussion about vacuum frying technology can be found by Dueik and Bouchon [43].

14.5
French Fries, Potato Chips, and Fabricated Chips Production

Potato is the primary raw material used in the frying industry. The storage organ and hence the food part of potato is the tuber, essentially a thickened underground

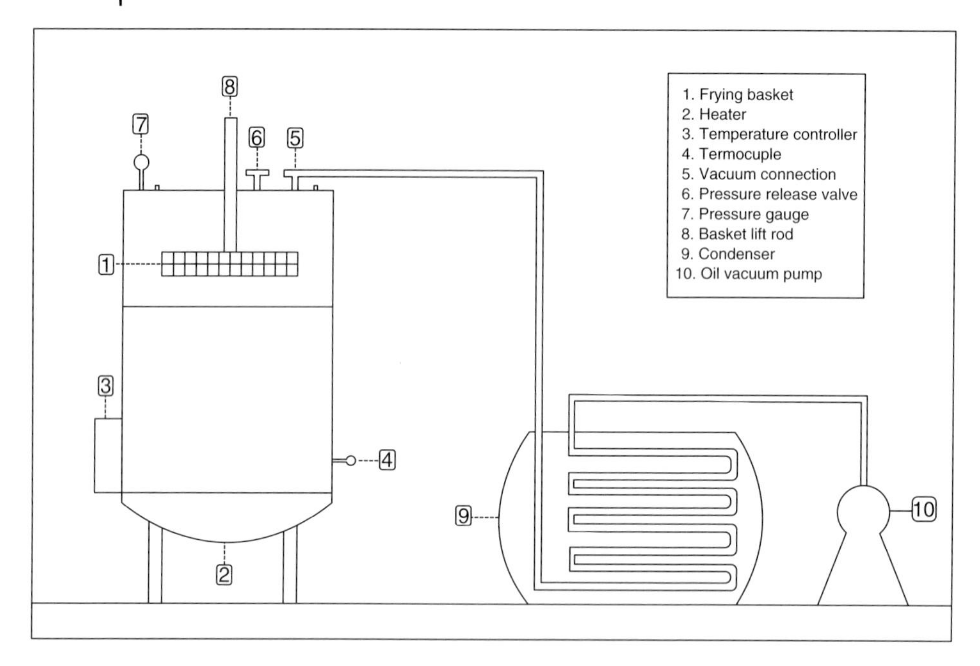

Figure 14.7 Diagram of a batch vacuum frying system. From Dueik and Bouchon [43].

stem, which is made of 2% skin, 75–85% parenchyma, and 14–20% pith [131]. The anatomy of the potato plant has been described in detail by Artschwager [132, 133], from Talburt *et al.* [134]. In terms of their composition, potatoes are mainly made up of water, having an average water content of 77.5%, ranging between 63.2 and 86.9% [134]. The chemical composition of the remaining solid part can vary greatly depending on a wide range of factors, including variety, maturity, cultural practice, environmental differences, chemical application, and storage conditions. Starch represents 65–89% of the dry matter weight, and amylose and amylopectin are usually present in a 1 : 3 ratio [135]. Starch granules are ellipsoidal in shape, about 100 by 60 μm, much larger than the average starch granules of cereal grains.

For potato fried products, potatoes with high solids content (20–22%) are preferred, as they result in better finished product texture, higher yields, and a lower oil absorption [136]. Also, low reducing sugar contents are required to minimize color development during processing, which is generated by the Maillard, non-enzymatic browning reaction [137].

14.5.1
French Fries Production

French fries are traditionally produced by cutting potato strips from fresh potatoes (parallelepiped of $1 \times 1\ \text{cm}^2$ cross-section by 4–7 cm in length), which are then deep-fat fried. Three major kinds of French fries are produced at a commercial scale: (i) deep-frozen completely fried fries, which just require oven heating, (ii) deep-frozen partially fried fries, which require additional frying before eating, and (iii) refrigerated partially fried fries, which have a short shelf life and need additional frying [97]. A summary of the production process is described below.

The technology of French fries production progressively improves as the industry develops modern equipment and entire technological lines for its manufacture [97]. Processors of frozen French fries wash and peel potatoes with lye or steam, as abrasion peeling results in higher losses. Peeled potatoes are conveyed over trimmers and cut. Strip cutters are aimed to orient potatoes along the long axis, in order to obtain the greatest yield of long cuts. Subsequently, French fries are blanched in hot water prior to frying. The usual range of water temperatures is from 60 to 85 °C. The positive effects of blanching include a more uniform color of the fried product, reduction of the frying time, since the potato is partially cooked, and improvement of texture of the fried product [138]. After blanching, excessive water is removed, in order to minimize the frying time and lower the oil content of the product. This is carried out by means of dewatering screens and, subsequently, blowing warm air in continuous dryers. Afterwards, potato strips are par-fried in a continuous fryer. The frying time is controlled by the rate of the conveyor's movement, temperature of the oil, dry matter content in potato tubers, size of strips, and type of processed French fries (par-fried or finish fried). The most commonly used temperature range falls within 160 and 180 °C. Temperatures above 190 °C are not used due to the possibility of more rapid oil breakdown. The recommended frying parameters are, for finish-fried French fries, 5 min at 180 °C, and for par-fried, 3 min at 180 °C [97].

The excess fat is removed by passing the product over a vibrating screen immediately after emerging from the fryer, allowing the fat to drain off and, thereafter, the product is air-cooled for about 20 min while it is conveyed to the freezing tunnel. Finally, the product is packed in polyethylene/polypropylene bags or in cartons for the retail trade. A flow sheet for frozen French fries is presented in Figure 14.8.

14.5.2 Potato Chip Production

Processing lines for potato chips include similar steps as in French fries production. Potatoes are washed, peeled, and sliced, commonly using a rotary slicer. Shape and thickness can be varied to meet marketing needs, but thickness is usually in the range 1.3–1.5 mm (50–60 thousandths of an inch). To remove excess starch, the potato slices are washed and dried on a flat wire conveyor to remove as much surface starch and water as possible [139]. Some potato processing plants use blanching prior to frying to improve the color of the chip. The blanching solution may be heated to 65–95 °C and blanching may take around 1 min. Excess water is removed. Thereafter, potatoes are usually fried in a continuous fryer, where they

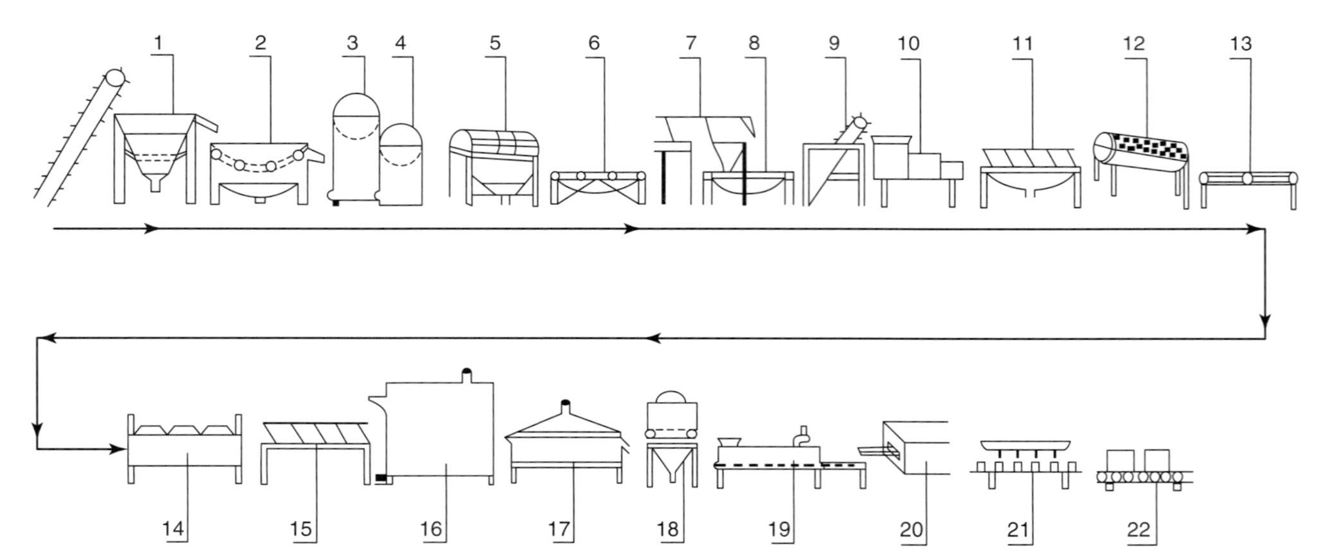

Figure 14.8 Flow sheet for frozen chips. 1, Trash remover; 2, washer; 3, preheater; 4, lye peeler; 5, washer; 6, trimming belt; 7 and 8, size grader; 9, surge hopper; 10, strip cutter; 11, sliver eliminator; 12, nubbin eliminator; 13, inspection belt; 14, blancher; 15, dewatering screen; 16, dryer; 17, fryer; 18, deffater; 19, cooler; 20, freezer; 21, package filler; and 22, freezing storage. From Lisinska and Leszczynski [97].

remain from $1\frac{1}{2}$ to 3 min at 170–190 °C, until the moisture level is less than 2% of the total weight. The frying time depends on the flow of slices to the fryer, the initial moisture level of the potato, and the desired browning. Subsequently, chips are conveyed, allowing excess oil to drain off, to be salted, or flavored. Finally, chips are cooled on a conveyor and sorted by size before packaging.

14.5.3
Fabricated Chips Production

Development of new food (micro)structures that respond to new consumers demands is the target of many food product developers. In fact, food manufacturing is facing new challenges, which include providing products that contribute to the health and well-being of consumers. In response, the food industry is shifting from an industry whose task was, during the twentieth century, the scaling-up of small production processes into highly efficient fabrication lines, into an industry that is asked to produce, in addition to microbiologically safe and high-quality foods, products that fulfill new demands of very well-informed consumers [140]. Consumers ask for products that contribute to their wellness and health, but also they require specific textures, flavors, colors, among other parameters, and certainly, a consistent product. That is, consumers expect minimal variations in food products from batch to batch. These demands are redirecting the focus of the food industry and food research ("fork to farm" approach), creating a challenge to the food industry to develop new products or modify existing ones to meet their needs. This product-driven process engineering era, as coined by the author, requires the construction of controlled right structures and, therefore, an understanding of the functionality of the structural elements prior to or formed during processing.

In the light of this approach, product formulation appears to be a good alternative to developing new products with controlled attributes. In this context, fabricated products (also known as *formulated products*) are gaining importance in the snack industry as a good alternative to the use of raw materials, because of the advantages of reproducibility, uniformity, and lack of defects [139], in contrast, for example, to raw vegetables, whose heterogeneity can cause major variations in final products [23].

Most common fabricated snacks include expanded ones, which are mainly produced through one-step extrusion cooking, and sheeted snacks, which are mainly fried, and to a lesser extent baked (or baked followed by frying). In a normal operation, sheeted snacks are made from dough, which is obtained by first mixing dry ingredients followed by the addition of liquid ingredients. Thereafter, the mixture is introduced to the sheeting line, where they are formed into discrete pieces. Subsequently, the product is fried in a continuous fryer, following a similar procedure to the one described above for potato chips manufacture. Production lines can vary depending on the final requirements and uses of the final product.

Formulated fried snacks are mainly made from potato dough, corn dough, and wheat flour dough. Potato dough fried snacks are usually made with a dehydrated potato flake base, where low leach potato flake is the standard. Restructured potato chips may not have the same flavor and textural characteristics as fresh potato chips, but they have the advantages of uniformity and absence of defects. Corn-based snack products, such as tortilla chips, are also extremely popular, and are mainly made from alkaline-cooked corn (nixtamal). An important difference between tortilla chips and potato chips is that in this situation the food is first baked, prior to frying, and most of the starch gelatinizes during that stage. Traditionally, tortilla chips were baked in three-tier (triple-pass) ovens. A new trend is to use single-pass ovens, which have easier maintenance and longer belt life, among other advantages [141].

Many formulated products are also based on wheat flour among other components. Wheat popularity is largely determined by the ability of wheat flour to be processed into different foods, mainly because of the unique properties of wheat flour gluten proteins. Products based on wheat flour dough are often used in frying operations to produce products such as doughnuts, battered food, and fritters among others, but they may be also sheeted and cut into small pieces to be fried. The development of new products requires specific knowledge about ingredient functionality, assemblage, effects of processing conditions, and microstructural aspects to design the right structures [115]. To do so, the integration of basic science, traditional processing/laboratory techniques, and new approaches, such as advance microstructural analysis, are of paramount importance.

14.5.4 Third-Generation Snack Products

Third-generation snack products are dense pellets, which are obtained through extrusion cooking after drying to a stable moisture content to assure stability during storage. Pellets available on the market range from potato to wheat based, sheeted, or extruded, 2D or 3D shapes. These products are then sold to food processing companies, where they are puffed or expanded by immersion in hot oil or hot air puffing. This is why they are sometimes referred as *semi-products* or *half products*. They may be sold to restaurants or for home puffing or frying, followed by the addition of salt or flavor. Stability during storage together with the high bulk density of the product account for most of its benefits and enhance its marketing potential. Newer variants can be expanded or puffed by infrared or microwave heating, which together with hot air puffing are gaining increasing importance because of the lower caloric content of the final product.

Pellets are mostly starch-based products and may be produced either through cooking extrusion or cold forming extrusion, which requires the use of pregelatinized ingredients. In a normal frying operation, pellets are completely immersed in hot oil at 150–200 °C for just 15–30 s [142]. These particular frying conditions and product characteristics led to the development of a new set of frying equipment, referred to as *pellet frying systems*, which normally include a submerger unit and a take-out and de-oiling conveyor, which allows accurate control over the frying process.

References

1. Varela, G. (1988) Current facts about the frying of food, in *Frying of Food: Principles, Changes, New Approaches* (eds G. Varela, A.E. Bender, and I.D. Morton), Ellis Horwood, Chichester, pp. 9–25.
2. Farkas, B.E. (1994) Modeling immersion frying as a moving boundary problem. PhD dissertation. University of California, Davis.
3. Singh, R.P. (1995) Heat and mass transfer in foods during deep-fat frying. *Food Technol.*, **49**, 134–137.
4. Bouchon, P. and Aguilera, J.M. (2001) Microstructural analysis of frying of potatoes. *Int. J. Food Sci. Technol.*, **36**, 669–676.
5. Aguilera, J.M. (1997) Fritura de alimentos, in *Temas en Tecnología de Alimentos* (ed. J.M. Aguilera), Instituto Politécnico Nacional, México, DF, pp. 187–214.
6. Blumenthal, M.M. and Stier, R.F. (1991) Optimization of deep-fat frying operations. *Trends Food Sci. Technol.*, **2**, 144–148.
7. Farkas, B.E., Singh, R.P., and Rumsey, T.R. (1996) Modeling heat and mass transfer in immersion frying, I. Model development. *J. Food Eng.*, **29**, 211–226.
8. Miller, K.S., Singh, R.P., and Farkas, B.E. (1994) Viscosity and heat transfer coefficients for canola, corn, palm, and soybean oil. *J. Food Process. Preserv.*, **18**, 461–472.
9. Moreira, R.G., Palau, J., Sweat, V., and Sun, X. (1995) Thermal and physical properties of tortilla chips as a function of frying time. *J. Food Process. Preserv.*, **19**, 175–189.
10. Bouchon, P. and Pyle, D.L. (2005) Modelling oil absorption during post-frying cooling. II Solution of the mathematical model, model testing and simulations. *Trans. Inst. Chem. Eng. Part C, Food Bioprod. Process.*, **83**, 253–260.
11. Hubbard, L.J. and Farkas, B.E. (1999) A method for determining the convective heat transfer coefficient during immersion frying. *J. Food Process. Preserv.*, **22**, 201–214.
12. Costa, R.M., Oliveira, F.A.R., Delaney, O., and Gekas, V. (1999) Analysis of the heat transfer coefficient during potato frying. *J. Food Eng.*, **39**, 293–299.
13. Sahin, S., Sastry, S.K., and Bayindirli, B. (1999) Heat transfer during frying of potato slices. *Lebensm. Wiss. Technol.*, **32**, 19–24.
14. Blumenthal, M.M. (1996) Frying technology, in *Bailey's Industrial Oil and Fat Products*, vol. 3, 5th edn (ed. Y.I. Hui), Wiley-VCH Verlag GmbH, New York, Chichester, pp. 429–482.
15. Bouchon, P. (2002) Modelling oil uptake during frying. PhD dissertation. The University of Reading, Reading.
16. Mintel International Group Ltd (2006) Salty Snacks – US, *http://www.marketresearch.com* (accessed August 21, 2006).
17. Saguy, I.S. and Pinthus, E.J. (1995) Oil uptake during deep-fat frying: Factors and mechanism. *Food Technol.*, **49**, 142–145, 152.
18. Browner, W.S., Westenhouse, J., and Tice, J.A. (1991) What if American ate less fat? A quantitative estimate of the effect on mortality. *J. Am. Med. Assoc.*, **265**, 3285–3291.
19. Saguy, S. and Dana, D. (2003) Integrated approach to deep fat frying: engineering, nutrition, health and consumer aspects. *J. Food Eng.*, **56**, 143–152.
20. Mariscal, M. and Bouchon, P. (2008) Comparison between atmospheric and vacuum frying of apple slices. *Food Chem.*, **107**, 1561–1569.
21. Rossell, J.B. (1998) Industrial frying process. *Grasas Aceites*, **49**, 282–295.
22. Moreira, R.G., Castell-Perez, M.E., and Barrufet, M.A. (1999) *Deep-Fat Frying: Fundamental and Applications*, Aspen Publication Inc., Gaithersburg.
23. Baumann, B. and Escher, E. (1995) Mass and heat transfer during deep fat frying of potato slices – I. Rate of drying and oil uptake. *Lebensm. Wiss. Technol.*, **28**, 395–403.

24. Reeve, R.M. and Neel, E.M. (1960) Microscopy structure of potato chips. *Am. Potato J.*, **37**, 45–57.
25. Costa, R.M., Oliveira, F.A.R., and Boutcheva, G. (2000) Structural changes and shrinkage of potato during frying. *Int. J. Food Sci. Technol.*, **36**, 11–24.
26. McDonough, C., Gomez, M.H., Lee, J.K., Waniska, R.D., and Rooney, L.W. (1993) Environmental scanning electron microscopy evaluation of tortilla chip microstructure during deep-fat frying. *J. Food Sci.*, **58**, 199–203.
27. Keller, Ch., Escher, F., and Solms, J.A. (1986) Method of localizing fat distribution in deep-fat fried potato products. *Lebensm. Wiss. Technol.*, **19**, 346–348.
28. Lamberg, I., Hallstrom, B., and Olsson, H. (1990) Fat uptake in a potato drying/frying process. *Lebensm. Wiss. Technol.*, **23**, 295–300.
29. Bouchon, P., Hollins, P., Pearson, M., Pyle, D.L., and Tobin, M.J. (2001) Oil distribution in fried potatoes monitored by infrared microspectroscopy. *J. Food Sci.*, **66**, 918–923.
30. Lisinska, G. and Golubowska, G. (2005) Structural changes of potato tissue during French fries production. *Food Chem.*, **93**, 681–687.
31. van Marle, J.T., Clerkx, A.C.M., and Boekstein, A. (1992) Cryo-scanning electron microscopy investigation of the texture of cooked potatoes. *Am. Potato J.*, **11**, 209–216.
32. Aguilera, J.M. and Gloria-Hernández, H. (2000) Oil absorption during frying of frozen parfried potatoes. *J. Food Sci.*, **65**, 476–479.
33. Farkas, B.E., Singh, R.P., and McCarthy, M.J. (1992) Measurement of oil/water interface in foods during frying, in *Advances in Food Engineering* (eds R.P. Singh and A. Wirakartakusumah), CRC Press, Inc., Boca Raton, pp. 237–245.
34. Pedreschi, F., Aguilera, J.M., and Arbildua, J.J. (1999) CLSM study of oil location in fried potato slices. *Micosc. Anal.*, **37**, 21–22.
35. Miri, T., Bakalis, S., Bhima, S.D., and Fryer, P.J. (2006) Use of x-ray micro-CT to characterize structure phenomena during frying. Paper presented at IUFoST, 13th World Congress of Food Science and Technology, Nantes, France.
36. Minihane, A.M. and Harland, J.I. (2007) Impact of oil used by the frying industry on population fat intake. *Crit. Rev. Food Sci. Nutr.*, **47**, 287–297.
37. Bouchon, P. (2009) Understanding oil absorption during deep-fat frying, in *Advances in Food and Nutrition Research*, vol. 57 (ed. S. Taylor), Academic Press, Burlington, MA, pp. 209–234.
38. Kochhar, S.P. (1998) Security in industrial frying processes. *Grasas Aceites*, **49**, 282–296, 302.
39. Sanibal, E.A.A. and Mancini-Filho, J. (2004) Frying oil and fat quality measured by chemical, physical, and test kit analyses. *J. Am. Oils Chem. Soc.*, **81**, 847–852.
40. Kita, A., Lisinska, G., and Powolny, M. (2005) The influence of frying medium degradation on fat uptake and texture of French fries. *J. Sci. Food Agric.*, **85**, 1113–1118.
41. Pavel, J. (1995) *Introduction to Food Processing*, Reston Publishing Company, Inc., Reston.
42. Pantzaris, T.P. (1999) Palm oil in frying, in *Frying of Food* (eds D. Boskou and I. Elmadfa), Technomic Publishing, Lancaster, pp. 223–252.
43. Dueik, V. and Bouchon, P. (2011) Development of healthy low-fat snacks: understanding the mechanisms of quality changes during atmospheric and vacuum frying. *Food Rev. Int.*, **27**, 408–432.
44. Seppanen, C.M. and Csallany, A. (2004) Incorporation of the toxic aldehyde 4-hydroxy-2-trans-nonenal into food fried in thermally oxidized soybean oil. *J. Am. Oils Chem. Soc.*, **81**, 1137–1141.
45. Choe, E. and Min, D. (2007) Chemistry of deep-fat frying oils. *J. Food Sci.*, **72**, R77–R86.
46. Melton, S.L., Jafar, S., Sykes, D., and Trigiano, M.K. (1994) Review of stability measurements for frying oils and fried food flavor. *J. Am. Oils Chem. Soc.*, **71**, 1301–1308.

47. Subramanian, R., Nandini, K.E., Sheila, P.M., Gopalakrishna, A.G., Raghavarao, K.S.M.S., Nakajima, M., Kimura, T., and Maekawa, T. (2000) Membrane processing of used frying oils. *J. Am. Oils Chem. Soc.*, **77**, 323–328.
48. Mestdagh, F., Castelen, P., Van Peteghem, C., and De Meulenaer, B. (2008) Importance of oil degradation on the formation of acrylamide in fried foodstuffs. *J. Agric. Food Chem.*, **56**, 6141–6144.
49. Uchida, K. (2003) 4-Hydroxy-2-nonenal: a product and mediator of oxidative stress. *Progr. Lipid Res.*, **42**, 318–343.
50. Dobarganes, M.C. (2009) Frying oils – chemistry: formation of volatiles and short-chain bound compounds during the frying process, *http://lipidlibrary.aocs.org/frying/c-volatile/index.htm* (accessed July 21, 2009).
51. Coughlin, J.R. (2006) Acrylamide: what we have learned so far. *Food Technol.*, **57**, 100.
52. Mitka, M. (2002) Fear of frying: Is acrylamide in foods a cancer risk? *J. Am. Med. Assoc.*, **288**, 2105–2106.
53. Mottram, D.S., Wedzicha, B.L., and Dodoson, A.T. (2002) Food chemistry: acrylamide is formed in the Maillard reaction. *Nature*, **419**, 448–449.
54. Zyzak, D., Sanders, R., Stojanovic, M., Tallmadge, D., Loye, B., Ewald, D., Gruber, D., Morsch, T., Strothers, M., Rizzi, G., and Villagran, M. (2003) Acrylamide. Formation mechanism in heated foods. *J. Agric. Food Chem.*, **51**, 4782–4787.
55. Haase, N.U., Matthäus, B., and Vosmann, K. (2008) Acrylamide formation in foodstuffs: minimising strategies for potato crisps. *Dtsch. Lebensmitt. Rundsch.*, **99**, 87–90.
56. Pedreschi, F., Kaack, K., and Granby, K. (2008) The effect of asparaginase on acrylamide formation in French fries. *Food Chem.*, **109**, 386–392.
57. Jung, M.Y., Choi, D.S., and Ju, J.W. (2003) A novel technique for limitation of acrylamide formation in fried and baked corn chips and in french fries. *J. Food Sci.*, **68**, 1287–1290.
58. Besaratinia, A. and Pfeifer, G. (2004) Genotoxicity of acrylamide and glycidamide. *J. Natl Cancer Inst.*, **96**, 1023–1029.
59. Koyama, N., Sakamoto, H., Sakuraba, M., Koizumi, T., Takashima, Y., Hayashi, M., Matsufuji, H., Yamagata, K., Masuda, S., Kinae, N., and Honma, M. (2006) Genotoxicity of acrylamide and glycidamide in human lymphoblastoid TK6 cells. *Mutat. Res.*, **603**, 151–158.
60. Teixidó, E., Santos, F.J., Puignou, L., and Galceran, M.T. (2006) Analysis of 5-hydroxymethylfurfural in foods by gas chromatography–mass spectrometry. *J. Chromatogr. A.*, **1135**, 85–90.
61. Ngadi, M.O. and Hwang, D.K. (2007) Modeling heat transfer and heterocyclic amines formation in meat patties during frying, Agricultural Engineering International: the CIGR Ejournal, Manuscript BC 04 004, Vol. IX, August 2007.
62. Saguy, I.S., Gremaud, E., Gloria, H., and Turesky, R.J. (1997) Distribution and quantification of oil uptake in french fries utilizing a radiolabeled 14C palmitic acid. *J. Agric. Food Chem.*, **45**, 4286–4289.
63. Ufheil, G. and Escher, F. (1996) Dynamics of oil uptake during deep-fat frying of potato slices. *Lebensm. Wiss. Technol.*, **29**, 640–644.
64. Gamble, M.H., Rice, P., and Selman, J.D. (1987) Relationship between oil uptake and moisture loss during frying of potato slices from c.v. Record U.K. tubers. *Int. J. Food Sci. Technol.*, **22**, 233–241.
65. Matz, S.A. (1993) *Snack Food Technology*, Van Nostrand Reinhold/AVT, New York.
66. Moreira, R.G., Sun, X., and Chen, Y. (1997) Factors affecting oil uptake in tortilla chips in deep-fat frying. *J. Food Eng.*, **31**, 485–498.
67. Moreira, R.G. and Barrufet, M.A. (1998) A new approach to describe oil absorption in fried foods: a simulation study. *J. Food Eng.*, **35**, 1–22.

68. Ashkenazi, N., Mizrahi, S., and Berk, Z. (1984) Heat and mass transfer in frying, in *Engineering and Food* (ed. B.M. McKenna), Elsevier Applied Science Publishers, London, pp. 109–116.

69. Rice, P. and Gamble, M.H. (1989) Technical note: modelling moisture loss during potato slice frying. *Int. J. Food Sci. Technol.*, **24**, 183–187.

70. Kozempel, M.F., Tomasula, P.M., and Craig, J.C. Jr. (1991) Correlation of moisture and oil concentration in french fries. *Lebensm. Wiss. Technol.*, **24**, 445–448.

71. Ateba, P. and Mittal, G.S. (1994) Modelling the deep-fat frying of beef meatballs. *Int. J. Food Sci. Technol.*, **29**, 429–440.

72. Rao, V.N.M. and Delaney, R.A.M. (1995) An engineering perspective on deep-fat frying of breaded chicken pieces. *Food Technol.*, **49**, 138–141.

73. Dincer, I. (1996) Modelling for heat and mass transfer parameters in deep frying of products. *Heat Mass Transfer*, **32**, 109–113.

74. Chen, Y. and Moreira, R.G. (1997) Modelling of a batch deep-fat frying process for tortilla chips. *Trans. Inst. Chem. Eng., Part C (Food Bioprod. Process.)*, **75**, 181–190.

75. Ngadi, M.O., Watts, K.C., and Correia, L.R. (1997) Finite element method modelling of moisture transfer in chicken drum during deep-fat frying. *J. Food Eng.*, **32**, 11–20.

76. Farid, M.M. and Chen, X.D. (1998) The analysis of heat and mass transfer during frying of food using a moving boundary solution. *Heat Mass Transfer*, **34**, 69–77.

77. Ni, H. and Datta, A.K. (1999) Moisture, oil and energy transport during deep-fat frying of food materials. *Trans. Inst. Chem. Eng., Part C (Food Bioprod. Process.)*, **77**, 194–204.

78. Carslaw, H.S. and Jaeger, J.C. (1959) *Conduction of Heat in Solids*, 2nd edn, Oxford University Press, New York.

79. King, C.J. (1970) Freeze drying of foodstuffs. *Crit. Rev. Food Technol.*, **1**, 379–451.

80. Farkas, B.E., Singh, R.P., and Rumsey, T.R. (1996) Modeling heat and mass transfer in immersion frying, II. Model solution and verification. *J. Food Eng.*, **29**, 227–248.

81. Bouchon, P. and Pyle, D.L. (2005) Modelling oil absorption during post-frying cooling. I model development. *Trans. Inst. Chem. Eng. Part C, Food Bioprod. Process.*, **83**, 261–272.

82. Washburn, E.W. (1921) The dynamics of capillary flow. *Phys. Rev.*, **17**, 273–283.

83. Mellema, M. (2003) Mechanism and reduction of fat uptake in deep-fat fried foods. *Trends Food Sci. Technol.*, **14**, 364–373.

84. Moreno, M.C. and Bouchon, P. (2008) A different perspective to study the effect of freeze, air, and osmotic drying on oil absorption during potato frying. *J. Food Sci.*, **73**, E122–E128.

85. Gamble, M.H. and Rice, P. (1987) Effect of pre-fry drying on oil uptake and distribution in potato crisp manufacture. *Int. J. Food Sci. Technol.*, **22**, 535–548.

86. Bouchon, P., Aguilera, J.M., and Pyle, D.L. (2003) Structure oil-absorption relationships during deep-fat frying. *J. Food Sci.*, **68**, 2711–2716.

87. Moreira, R.G., Palau, J., and Sun, X. (1995) Simultaneous heat and mass transfer during the deep fat frying of tortilla chips. *J. Food Process Eng.*, **18**, 307–320.

88. Nonaka, M., Sayre, R.N., and Weaver, M.L. (1977) Oil content of French fries as affected by blanch temperatures, fry temperatures and melting point of frying oils. *Am. Potato J.*, **54**, 151–159.

89. Krokida, M.K., Oreopoulou, V., and Maroulis, Z.B. (2000) Water loss and oil uptake as a function of frying time. *J. Food Eng.*, **44**, 39–46.

90. Blumenthal, M.M. (1991) A new look at the chemistry and physics of deep-fat frying. *Food Technol.*, **45**, 68–71.

91. Blumenthal, M.M. (2001) A new look at frying science. *Cereal Foods World*, **46**, 352–354.

92. Pokorny, J. (1980) Effect of substrates on changes of ftas and oils during frying. *Riv. Ital. Sost. Grasse*, **57**, 222–225.

93. Kita, A., Lisinska, G., and Golubowska, G. (2007) The effects of oils and frying temperatures on the texture and fat content of potato crisps. *Food Chem.*, **102**, 1–5.
94. Pinthus, E.J. and Saguy, I.S. (1994) Initial interfacial tension and oil uptake by deep-fat fried foods. *J. Food Sci.*, **59**, 804–807, 823.
95. Dana, D. and Saguy, S. (2006) Review: mechanism of oil uptake during deep-fat frying and the surfactant effect-theory and myth. *Adv. Colloid Interface Sci.*, **128–130**, 267–272.
96. Gamble, M.H. and Rice, P. (1988) Effect of initial tuber solids content on final oil content of potato chips. *Lebensm. Wiss. Technol.*, **21**, 62–65.
97. Lisinska, G. and Leszczynski, W. (1991) *Potato Science and Technology*, Elsevier Applied Science Publishers, London.
98. Lulai, E.G. and Orr, P.H. (1979) Influence of potato specific gravity on yield and oil content of chips. *Am. Potato J.*, **56**, 379–390.
99. Gamble, M.H. and Rice, P. (1988) The effect of slice thickness on potato crisp yield and composition. *J. Food Eng.*, **8**, 31–46.
100. Selman, J.D. and Hopkins, M. (1989) Factors affecting oil uptake during the production of fried potato products, Technical Memorandum 475, Campden Food and Drink Research Association, Chipping Campden, Gloucestershire, UK.
101. Pinthus, E.J., Weinberg, P., and Saguy, I.S. (1992) Gel-strength in restructured potato products affects oil uptake during deep-fat frying. *J. Food Sci.*, **57**, 1359–1360.
102. Pinthus, E.J., Weinberg, P., and Saguy, I.S. (1995) Oil uptake in deep-fat frying as affected by porosity. *J. Food Sci.*, **60**, 767–769.
103. Thanatuksorn, P., Kajiwara, K., and Suzuki, T. (2007) Characterization of deep-fat frying in a wheat flour–water mixture model using a state diagram. *J. Sci. Food Agric.*, **87**, 2648–2656.
104. Saguy, I.S., Ufheil, G., and Livings, S. (1998) Oil uptake in deep-fat frying: review. *Oleag., Corps Gras, Lipids*, **5**, 30–35.
105. Krokida, M.K., Oreopoulou, V., Maroulis, Z.B., and Marinos-Kouris, D. (2001) Effect of osmotic dehydration pretreatment on quality of French fries. *J. Food Eng.*, **49** (4), 339–345.
106. Moyano, P. and Berna, A. (2002) Modeling water loss during frying of potato strips: effect of solute impregnation. *Drying Tech.*, **20** (7), 1303–1318.
107. Bunger, A., Moyano, P., and Rioseco, V. (2003) NaCL soaking treatment for improving the quality of french-fried potatoes. *Food Res. Int.*, **36** (2), 161–166.
108. Pinthus, E.J., Weinberg, P., and Saguy, I.S. (1993) Criterion for oil uptake during deep-fat frying. *J. Food Sci.*, **58**, 204–205, 222.
109. Balasubramaniam, V.M., Chinnan, M.S., Mallikarjunan, P., and Phillips, R.D. (1997) The effect of edible film on oil uptake and moisture retention of a deep-fat fried poultry product. *J. Food Process. Eng.*, **20**, 17–29.
110. Williams, R. and Mittal, G.S. (1999) Low-fat fried foods with edible coatings: modelling and simulation. *J. Food Sci.*, **64**, 317–322.
111. Mallikarjunan, P., Chinnan, M.S., Balasubramaniam, V.M., and Phillips, R.D. (1997) Edible coatings for deep-fat frying of starchy products. *Lebensm. Wiss. Technol.*, **30**, 709–714.
112. Albert, S. and Mittal, G.S. (2002) Comparative evaluation of edible coatings to reduce fat uptake in a deep-fat fried cereal product. *Food Res. Int.*, **35**, 445–458.
113. García, M.A., Ferrero, C., Bertola, N., Martino, M., and Zaritzky, N. (2002) Edible coatings from cellulose derivatives to reduce oil uptake in fried products. *Innov. Food Sci. Emerg. Technol.*, **3**, 391–397.
114. Bouchon, P. and Pyle, D.L. (2004) Studying oil absorption in restructured potato chips. *J. Food Sci.*, **69**, E115–E122.
115. Gazmuri, A.M. and Bouchon, P. (2009) Analysis of wheat gluten and starch matrixes during deep-fat frying. *Food Chem.*, **115**, 999–1005.

116. Pedreschi, F., Aguilera, J.M., and Brown, C. (2000) Characterization of food surfaces using scale-sensitive fractal analysis. *J. Food Process. Preserv.*, **23**, 127–143.

117. Rubnov, M. and Saguy, I.S. (1997) Fractal analysis and crust water diffusivity of a restructured potato product during deep-fat frying. *J. Food Sci.*, **62**, 135–137, 154.

118. Moreno, M.C., Brown, A.B., and Bouchon, P. (2010) Effect of food surface roughness on oil uptake of deep-fat fried food. *J. Food Eng.*, **101**, 179–186.

119. Banks, D. (1996) Food service frying, in *Deep Frying* (eds E.G. Perkins and M.D. Erickson), AOCS Press, Champaign, IL, pp. 245–257.

120. Moreira, R.G. (2006) The engineering aspects of deep-fat frying, in *Handbook of Food Science, Technology, and Engineering*, vol. 3 (ed. Y.H. Hui), CRC Press Inc., Boca Raton, FL, pp. 111-1–111-5.

121. Kochhar, S.P. (1999) Safety and reliability during frying operations-effects of detrimental components and fryer design features, in *Frying of Food* (ed. D. Boskou and I. Elmadfa), Technomic Publishing, Lancaster, PA, pp. 253–269.

122. Shyu, S.L. and Hwang, L.S. (2001) Effects of processing conditions on the quality of vacuum fried apple chips. *Food Res. Int.*, **34**, 133–142.

123. Da Silva, P. and Moreira, R. (2008) Vacuum frying of high-quality fruit and vegetable-based snacks. *Lebensm. Wiss. Technol.*, **41**, 1758–1767.

124. Dueik, V., Robert, P., and Bouchon, P. (2010) Vacuum frying reduces oil uptake and improves the quality parameters of carrot chips. *Food Chem.*, **119**, 1143–1149.

125. Dueik, V. and Bouchon, P. (2011) Quality parameters of vacuum and atmospheric fried apple, potato and carrot chips. *J. Food Sci.*, **76**, E188–E195.

126. Shyu, S., Hau, L., and Hwang, L. (1998) Effect of vacuum frying on the oxidative stability of oils. *J. Am. Oils Chem. Soc.*, **75**, 1393–1398.

127. Granda, C., Moreira, R.G., and Tichy, S.E. (2004) Reduction of acrylamide formation in potato chips by low-temperature vacuum frying. *J. Food Sci.*, **69**, 405–411.

128. Garayo, J. and Moreira, R. (2002) Vacuum frying of potato chips. *J. Food Eng.*, **55**, 181–191.

129. Perez-Tinoco, M., Perez, A., Salgado-Cervantes, M., Reynes, M., and Vaillant, F. (2008) Effect of vacuum frying on main physicochemical and nutritional quality of parameters of pineapple chips. *J. Sci. Food Agric.*, **88**, 945–953.

130. Mir-Bel, J., Oria, R., and Salvador, M.L. (2009) Influence of the vacuum break conditions on oil uptake during potato. *J. Food Eng.*, **95**, 416–422.

131. Montaldo, A. (1984) *Cultivo y Mejoramiento de la Papa*, Instituto Interamericano de Cooperacion para la Agricultura, San José, Costa Rica.

132. Artschwager, E. (1918) Anatomy of the potato plant, with special reference to the ontogeny of the vascular system. *J. Agric. Res.*, **14**, 221–252.

133. Artschwager, E. (1924) Studies on the potato tuber. *J. Agric. Res.*, **27**, 809–835.

134. Talburt, W.F., Schwimmer, S., and Burr, H.K. (1975) in *Potato Processing*, 3rd edn (eds W. Talburt and O. Smith), AVI Publishing Company, Westport, CT, pp. 11–42.

135. Banks, W. and Greenwood, C.T. (1959) The starch of the tuber and shoots of the sprouting potato. *Biochem. J.*, **73**, 237–241.

136. True, R.H., Work, T.M., Bushway, R.J., and Bushway, A.A. (1983) Sensory quality of French fries prepared from Belrus and Russet Burbank potatoes. *Am. Potato J.*, **60**, 933–937.

137. Mottur, G.P. (1989) A scientific look at potato chips – the original savory snack. *Am. Assoc. Cereal Chem.*, **34**, 620–626.

138. Weaver, M.L., Reeve, R.M., and Kueneman, R.W. (1975) Frozen french fries and other frozen potato products, in *Potato Processing*, 3rd edn (eds W. Talburt and O. Smith), AVI Publishing Company, Westport, pp. 403–442.

139. Gebhardt, B. (1996) Oils and fats in snack foods, in *Bailey's Industrial Oil and Fat Products*, vol. 3, 5th edn (ed. Y.I. Hui), Wiley-VCH Verlag GmbH, Weinheim, pp. 409–428.

140. Aguiera, J.M. (2006) Food product engineering: building the right structures. *J. Sci. Food Agric.*, **86**, 1147–1155.

141. Mehta, U. and Swinburn, B. (2001) A review of factors affecting fat absorption in hot chips. *Crit. Rev. Food Sci. Nutr.*, **41** (2), 133–154.

142. Huber, G. (2001) Snack foods from cooking extruders, in *Snack Foods Processing* (eds E.W. Lusas and L.W. Rooney), CRC Press, Boca Raton, FL, pp. 315–367.

Further Reading

WHO/FAO Expert Consultation (2003) Diet, Nutrition and the Prevention of Chronic Diseases. WHO Technical Report Series 916, WHO, Geneva.

15
Safety in Food Processing

Carol A. Wallace

15.1
Introduction

It is a fundamental requirement of any food process that the food produced should be safe for consumption. Food safety is a basic need but there is a danger that it may be overlooked in the development of effective and efficient processes.

There are three key elements to ensuring food safety is achieved in food manufacture:

1) safe design of the process, recipe, and packaging format;
2) prerequisite programs (PRPs) or good manufacturing practice (GMP) to control the manufacturing environment; and
3) use of the HACCP (hazard analysis and critical control point) system of food safety management.

This chapter will outline these current approaches to effective food safety management and consider how they fit with the design and use of different food processing technologies.

15.2
Safe Design

"When designing a new food product it is important to ask if it is possible to manufacture it safely. Effective HACCP systems (and PRPs) will manage and control food safety but what they cannot do is make safe a fundamentally unsafe product" [1]. It is important, therefore, to understand the criteria involved in designing and manufacturing a safe product. These include:

- an understanding of the likely food safety hazards that may be presented through the ingredients, processing and handling methods;

Food Processing Handbook, Second Edition. Edited by James G. Brennan and Alistair S. Grandison.

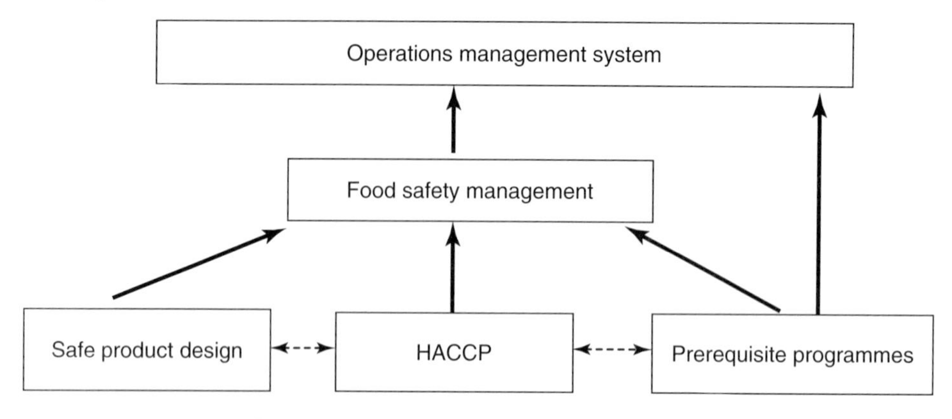

Figure 15.1 Safe food processing achievement model. Adapted from Ref. [4].

- the intrinsic factors involved in developing a safe recipe;
- a thorough knowledge of the chosen food processing and packaging technologies;
- manufacturing in a facility operating to PRPs of GMP; and
- management of production within the framework of a validated HACCP program.

PRPs and HACCP will be covered in Sections 15.3 and 15.4, respectively. Before further considering the design of safe products, we can look at how these different criteria fit together to ensure safe food processing. Figure 15.1 shows a model for the achievement of safe food processing. The safety management criteria, that is, safe product design, HACCP, and prerequisite GMP programs are all managed within the framework of the operational management system, which could be a formal quality management system such as ISO 9000:2000 [2] or ISO 22000:2005 [3].

15.2.1 Food Safety Hazards

Food safety hazards are contaminants that may cause a food product to be unsafe for consumption. Hazards are defined by Codex 2009 [5] as follows:

> "Hazard: a biological, chemical or physical agent in, or condition of, food with the potential to cause an adverse health effect."

Hazards may enter a food product from its ingredients or may contaminate during processing or handling. Table 15.1 shows examples of common hazard types for consideration. At the product design stage, it is important to understand the likely hazards that might be encountered in the chosen ingredient types, or that might be present in the processing environment. This allows the development team to identify the best ways to control these hazards, either by

Table 15.1 Examples of food safety hazards.

	Type of hazard		
	Biological	**Chemical**	**Physical**
Considerations	Organisms that can cause harm through infection or intoxication	Chemicals that can cause harm through toxic effects, either immediate or long term	Items that can cause harm through direct injury or choking
Examples	Pathogenic bacteria, for example, *Escherichia coli*, *Bacillus cereus*, *Campylobacter jejuni*, *Clostridium botulinum*, *Cl. botulinum* (nonproteolitic), *Cl. perfringens*, *Salmonella* spp., *Shigella* spp., *Staphylococcus aureus*, *Vibrio parahaemolyticus*; viruses; protozoan parasites, e.g., *Cryptosporidium parvum*, *Giardia intestinalis*, *Cyclospora cayetanensis*	Mycotoxins, e.g., aflatoxins, patulin, vomitoxin, fumonisin; pesticides; allergenic materials; heavy metals; polychlorinated biphenyls (PCBs); dioxins; cleaning chemicals	Glass, metal, stones, wood, plastic, pests, intrinsic natural material, for example, bone, nut shell

This table provides examples only and is not intended to be an exhaustive list of food safety hazards.

preventing their entry to the process, destroying them, or reducing the contamination to a level where it no longer poses a food safety risk. This information on likely hazards and proposed control options should link with the PRPs and HACCP systems to ensure everyday control is established in the manufacturing operation.

Consideration of likely hazards at an early stage in the development process can also, in some cases, help to design these hazards out of the product, either through careful choice and sourcing of ingredients or through identification of appropriate processing technologies and/or equipment. For example, if there is a concern about physical hazards gaining entry to a product during manufacture due to the use of open vessels, the redesign of the equipment to use enclosed vessels would prevent this hazard from ever occurring at that processing step. Similarly, if there is concern about pathogen contamination in a raw ingredient, for example, *Salmonella* spp. contamination in coconut that is to be used as a topping ingredient after heat processing, it may be possible to replace this ingredient with a preprocessed ingredient, in this example pasteurized coconut.

15.2.2
Intrinsic Factors

Intrinsic factors are the formulation criteria that control the ability of microorganisms to survive and grow in foods. These factors have been used traditionally to prevent problems with spoilage organisms and pathogens in a wide variety of foodstuffs. The most commonly used intrinsic factors in food processing are water activity, pH, organic acids, and preservatives.

Water activity (a_w) is a measure of the amount of water available in a foodstuff for microbial growth. Pure water has an a_w of 1.0 and, as solutes such as salt and sugar are added to make a more concentrated solution, the a_w decreases. There is also a characteristic pH range across which microorganisms can grow; and the limiting pH for growth varies widely between species. The use of pH to control the growth of microorganisms is very common in food processing, finding uses in pickled foods, such as pickled vegetables, and fermented foods, such as cheese and yoghurt. Table 15.2 shows the a_w and pH limits for growth of a number of key microbial pathogens.

Organic acids, such as acetic, citric, lactic, and sorbic acids, are widely used as preserving factors in food processing. The antimicrobial effect of organic acids is due to undissociated molecules of the acid and, since the dissociation of the molecules is pH-dependent, the effectiveness is related to pH. Chemical preservatives may be added to food products to prevent the growth of pathogens and spoilage organisms. The use of preservatives is normally controlled by legislation, with different levels of various preservatives allowed for use in different groups of foodstuffs. Further detailed information on the effects of intrinsic factors on a wide

Table 15.2 Control of key microbiological hazards through intrinsic factors.

Organism	Minimum pH for growth	Minimum water activity (a_w) for growth
Bacillus cereus	5.0	0.93
Campylobacter jejuni	4.9	0.99
Clostridium botulinum	4.7	0.94
Clostridium botulinum (nonproteolytic)	5.0	0.97
Clostridium perfringens	5.5	0.93
Escherichia coli	4.4	0.95
Listeria monocytogenes	4.4	0.92
Salmonella spp.	3.8	0.94
Shigella spp.	4.9	0.97
Staphylococcus aureus	4.0	0.85
Vibrio parahaemolyticus	4.8	0.94

Adapted from Refs [6, 7].

range of microorganisms can be found in other publications, such as those of the International Commission on Microbiological Specifications of Foods (ICMSF) [7].

15.2.3 Food Processing Technologies

A wide variety of food processing technologies are available, as highlighted in the other chapters of this book. It is important for food safety that the chosen food process is thoroughly understood so that any potential food safety hazards can be effectively controlled. Table 15.3 shows the effects of various food processing techniques on food safety hazards.

It can be seen that most types of food processing illustrated in Table 15.3 are designed to control microbiological hazards and involve either destruction, reduction of numbers, or prevention from growth of various foodborne pathogens. To a lesser extent, a number food processing techniques, for example, cleaning and separation, involve the removal of physical hazards. Very few food processing techniques are designed to control chemical hazards in foods; and therefore it is important to source high-quality ingredients that are free from chemical hazards and to prevent contamination during processing.

15.2.4 Food Packaging Issues

The chosen packaging type should also be evaluated as part of the "safe design" process. Food packaging systems have evolved to prevent contamination and ensure achievement of desired shelf life; however, there may be hazard considerations if they are inappropriate to the type of food or proposed storage conditions. Table 15.4 lists a number of considerations for choosing a safe packaging system.

15.3 Prerequisite Programs

Prerequisite programs (PRPs) or "GMP" provide the hygienic foundations for any food operation. The terms "*prerequisite programs*" and "*good manufacturing practice*" have been used interchangeably in different parts of the world but have the same general meaning. For simplicity, the term PRPs will be used in this chapter.

Several groups have suggested definitions for the terms and the most commonly used are reproduced here. PRPs are:

- practices and conditions needed prior to and during the implementation of HACCP and which are essential to food safety (World Health Organization (WHO) [8]);
- universal steps or procedures that control the operating conditions within a food establishment, allowing for environmental conditions that are favorable for the production of safe food (Canadian Food Inspection Agency [9]); and

Table 15.3 Effects of food processing on food safety hazards.

Processing operation	Intended effect on food safety hazards	Example food types
Cleaning		
Dry	Removal of foreign material and dust	Grain crops
Wet	Reduction in level of microorganisms and foreign material	Raw foods, for example, vegetables, fruit, dried fruit
Antimicrobial dipping/spraying	Reduction in levels of microorganisms	Fruit and vegetables
Fumigation	Destruction of certain microorganisms and pests	Nuts, dried fruit, cocoa beans
Thermal processing		
Pasteurization/cooking	Destroys vegetative pathogens, for example, *Salmonella* spp., *Listeria monocytogenes*	Milk products, meat, fish, ready-meals.
Sterilization – UHT/aseptic	Destroys pathogens and prevents recontamination in packaging system	UHT milk, fruit juices
Sterilization – cans/pouches	Destroys pathogens	Canned meats, soups, pet food, and so on
Evaporation/ dehydration	Halts growth of pathogenic bacteria at a_w 0.84 all microorganisms at a_w 0.60	Various foodstuffs, for example, dried fruit, milk powder, cake mixes, and so on
Salt preserving	Halts growth of pathogenic bacteria at a_w 0.84 all microorganisms at a_w 0.60. Growth of many microorganisms halted at about 10% salt	Fish, meats, vegetables
Sugar preserving	Halts growth of pathogenic bacteria at a_w 0.84 all microorganisms at a_w 0.60.	Jam, fruits, syrups, jellies, confectionery.
Chilling (<5 °C)	Slows or prevents growth of most pathogens	Cooked meats, dairy products, fruit juices
Freezing (< – 10 °C)	Prevents growth of all microorganisms. Destroys some parasites	Many foodstuffs, for example, fruit, vegetables, meat, fish, icecream, and so on
Irradiation	Destroys microorganisms	Can be used for various products, e.g., fruit, shellfish, however, consumer pressure has limited its application
High pressure processing	Destroys/inactivates microorganisms; also affects functional and organoleptic qualities	Fruit juice, guacamole, yoghurt, oysters
Electric field processing	Destroys microorganisms, inactivates some enzymes	Potential applications include fruit juice, milk
Fermentation/ acidification	Halts growth of pathogens. Destroys some organisms depending on pH/acid used	Cheese, yoghurt, vegetables, fruit, and so on
Separation (for example, filtration)	Removes physical hazards and/or pathogens (depending on filter pore size), adjusts chemical concentration (e.g., reverse osmosis)	Various foodstuffs, e.g., sugar, grains, water, and so on

Table 15.4 Food packaging considerations.

Packaging type	Considerations
Retortable containers – cans	Hygienic container suitable for a wide range of foods. Suitable for retort sterilization/pasteurization and ambient storage, dependent on formulation suitability (pasteurization) Careful handling required after sealing and retorting Type of can and inner lacquer needs to be matched to food type to prevent degradation and leaching of metal into the product
Retortable containers – pouches	Hygienic container suitable for a wide range of foods Suitable for retort sterilization/pasteurization and ambient storage, dependant on formulation suitability (pasteurization) Careful handling required after sealing and retorting Need to check that film constituents (e.g., plasticizers and additives) cannot transfer to food during packaging use
Glass	Hygienic container suitable for a wide range of foods Suitable for hot and cold fill For ambient storage, need to ensure that the recipe intrinsic factors keep the product safe over the shelf life High-quality glass required and container design for strength necessary Careful handling required to prevent breakage and glass hazards Glass breakage procedures needed
Gas-flushed containers	Intended to extend life of product by preventing growth of spoilage organisms Need to ensure that any pathogens present (e.g., anaerobic spore formers) cannot grow in the chosen gas mix
Vacuum packaging	Intended to extend life of product by preventing growth of spoilage organisms Anaerobic conditions provided can allow growth of some pathogens (e.g., *Clostridium botulinum*) May need to use vacuum packaging in conjunction with additional control measures such as chilling
Gas permeable packaging	Is it possible for other materials (e.g. moisture) to pass through into the product and cause contamination?
Product contact films and plastics	Need to check that constituents (e.g., plasticizers and additives) cannot transfer to food during packaging use
Paper/cardboard	Most suitable for secondary/tertiary packaging Need to ensure that inks and adhesives cannot transfer to foodstuff
Wood	May introduce hazards, for example, splinters In most cases wood is best kept for secondary/tertiary packaging rather than direct product contact

Detailed legislation exists in many countries for materials coming into contact with foods, such as packaging.

- procedures, including GMP, that address operational conditions, providing the foundation for the HACCP system (US National Advisory Committee for Microbiological Criteria for Foods [10]).

The internationally accepted requirements for prerequisites are defined in the Codex general principles of food hygiene [11].

15.3.1 Prerequisite Programs – the Essentials

Using the headings given in the Codex General Principles of Food Hygiene [11], the following notes describe the general requirements for PRPs in each area. Further, more detailed information can be found in other publications such as the Codex document itself [11], and those produced by the Canadian Food Inspection Agency [9], the Institute of Food Science and Technology [12], and Sprenger [13].

15.3.1.1 Establishment: Design and Facilities

The location of food premises is important and care should be taken to identify and consider the risks of potential sources of contamination in the surrounding environment. Suitable controls to prevent contamination should be developed and implemented.

The design and layout of the premises and rooms should permit good hygiene and protect the products from cross-contamination during operation. Internal structures and equipment should be built of materials able to be easily cleaned/disinfected and maintained. Surfaces should be smooth, impervious, and able to withstand the normal conditions of the operation, for example, moisture and temperature ranges.

Facilities should be provided to include adequate potable water supplies, suitable drainage and waste disposal, appropriate cleaning facilities, storage areas, lighting, ventilation, and temperature control. Suitable facilities should also be provided to promote personal hygiene for the workforce, including adequate changing areas, lavatories, and hand washing and drying facilities.

15.3.1.2 Control of Operation

The rationale for operational control listed in [11] is "to reduce the risk of unsafe food by taking preventive measures to assure the safety and suitability of food at an appropriate stage in the operation by controlling food hazards." This includes the need to control potential food hazards by using a system such as HACCP.

Codex also describes key aspects of hygiene control systems, including:

- time and temperature control;
- microbiological and other specifications;
- microbiological cross-contamination risks; and
- physical and chemical contamination.

Incoming material requirements and systems to ensure the safety of materials and ingredients at the start of processing are necessary, along with a suitable packaging design (see also Section 15.2.4).

Codex [11] also lists the importance of hygienic control of water, ice, and steam, appropriate management and supervision, the need to keep adequate documentation and records, and the need to develop and test suitable recall procedures so that product can be effectively withdrawn and recalled in the event of a food safety problem.

15.3.1.3 **Establishment: Maintenance and Sanitation**

Maintenance and cleaning are important both to keep the processing environment, facilities, and equipment in a good state of repair so they function as intended and to prevent cross-contamination with food residues and hazards, such as microorganisms or allergens, that might otherwise build up. Facilities should operate preventative maintenance programs as well as attending to breakdowns and faults without delay.

Cleaning programs should be developed to encompass all equipment and facilities as well as general environmental cleaning. Cleaning methods need to be developed that are both suitable for the item to be cleaned, including the use of appropriate chemical cleaning agents, disinfectants, hot/cold water, and cleaning tools, and are validated as effective in cleaning the relevant items/areas. Methods should describe how the item is to be cleaned and personnel should be trained to apply the methods correctly. A cleaning schedule should also be developed to identify the frequency of cleaning needed in each case and records of cleaning and monitoring should be kept.

Cleaning in place (CIP) systems may be used in certain types of equipment, for example, tanks and lines. Here it is important that the CIP program is properly designed for the equipment to be cleaned, taking into account the flow rates, coverage, and the need for rinsing and disinfection cycles. Hygienic design of CIP routes is also essential to prevent cross-contamination during cleaning, for example, from equipment used to handle raw ingredients to that used with processed products.

Pest control systems are important to prevent the access of pests that might cause contamination to the product. Pest management is often contracted out to a professional pest control contractor and it is important that this contractor is suitably qualified to perform the work. Buildings need to be made pestproof and regularly inspected for potential ingress points. Interior and exterior areas need to be kept clean and tidy to minimize potential food and harborage sources. Suitable interior traps and monitoring devices should also be considered and any pest infestations need to be dealt with promptly, without adversely affecting food safety.

Waste management should ensure that waste materials can be removed and stored safely so that they do not provide a cross-contamination risk or become a food or harborage source for pests.

All maintenance and sanitation systems should be monitored for effectiveness, verified and reviewed, with changes made to reflect operational changes.

15.3.1.4 Establishment: Personal Hygiene

The objectives for personal hygiene stated in [11] are as follows: "To ensure that those who come directly or indirectly into contact with food are not likely to contaminate food by:

- maintaining an appropriate degree of personal cleanliness and
- behaving and operating in an appropriate manner."

Food companies should, therefore, have standards and procedures in place to define the requirements for personal hygiene and staff responsibility; and staff should be appropriately trained. This should include the establishment of health status where individuals may be carrying disease that can be transmitted through food, a consideration of illness and injuries where affected staff members may need to be excluded or wear appropriate dressings, the need for good personal cleanliness and effective hand washing, the wearing of adequate protective clothing, and the prevention of inappropriate behavior such as smoking, eating, or chewing in food handling areas. Visitors to processing and product handling areas should be adequately supervised and required to follow the same standards of personal hygiene as employees.

15.3.1.5 Transportation

To ensure continuation of food safety throughout transportation, transport facilities need to be designed and managed to protect food products from potential contamination and damage and to prevent the growth of pathogens. This includes the need for cleaning and maintenance of vehicles and containers and the use of temperature control devices where appropriate.

15.3.1.6 Product Information and Consumer Awareness

It is important that sufficient information is easily identifiable on the products so that the lot or batch can be identified for recall purposes and that the product can be handled correctly, for example, stored at $<5\,^{\circ}C$. Product information and labeling should be clear such that it facilitates consumer choice and correct storage/use. Codex [11] also highlights the importance of consumer education, particularly the importance of following handling instructions and the link between time/temperature and foodborne illness.

15.3.1.7 Training

Food hygiene training is essential to make personnel aware of their roles and responsibilities for food control. Companies should develop and implement appropriate training programs and should include adequate supervision and monitoring of food hygiene behavior. Training should be evaluated and reviewed with refresher or update training implemented as necessary.

15.3.2
Validation and Verification of Prerequisite Programs

PRPs are the basic standards for the food facility, in which the safely designed product can be manufactured. They form the hygiene foundations on which the HACCP system is built to control food safety every day of operation. As such, it is essential that PRPs are working effectively at all times and it is therefore necessary that each prerequisite element is validated to establish that it will be effective and that an ongoing program of monitoring and verification is developed and implemented.

15.3.3
Operational Prerequisites

Operational prerequisites is a new term that has been introduced under ISO 22000:2005 [3]. It is defined as:

> Operational PRP: a PRP identified by the hazard analysis as essential in order to control the likelihood of introducing food safety hazards to and/or the contamination or proliferation of food safety hazards in the product(s) or in the processing environment [3].

This operational PRP definition is quite different from the definitions listed in Section 15.3 above and refers specifically to the control of food safety hazards. This has been quite controversial since in HACCP [5] significant food safety hazards are expected to be controlled by CCPs (critical control points). Although PRPs have always been considered important as a foundation for HACCP and therefore need to be established as formal programs that are validated, monitored, and verified, ISO 22000 now seems to be saying that operational PRPs are as critical as CCPs for the control of food safety hazards.

The idea for operational PRPs seems to have come from the fact that the manifestation of some hazards, in particular, cross-contamination risks, might require elements of control that are generally thought to be part of PRPs. For example, the potential allergen contamination of a shared production line that is used to make both products containing and products not containing certain allergens is likely to need specific targeted cleaning procedures as part of the management of this issue. A traditional way of dealing with this was to elevate that specific cleaning procedure to the status of a CCP, ensuring that it was validated in the same way as all other CCPs and then monitored and verified as effective. In this example, the general cleaning of all other aspects of the operation would have remained part of PRPs. Using the same example but taking the ISO 22000 definition [3] into account, it is likely that the specific targeted cleaning of the line to remove the likelihood of allergen contamination would now be thought of as an operational PRP.

At this point of time there is still debate over whether the concept of operational PRPs is useful. Some practitioners like them whilst others are worried that this is a move away from Codex HACCP. Only time will tell whether they become fully adopted into the HACCP system, and this is only likely to come as more companies get exposed to the concept through the adoption of ISO 22000 [3].

15.4
HACCP, the Hazard Analysis and Critical Control Point System

HACCP stands for the "hazard analysis and critical control point" system, a method of food control based on the prevention of food safety problems. The HACCP story began in the early 1960s, when the Pillsbury Company was working with the National Aeronautics and Space Administration (NASA) and the US Army Natick laboratories to provide food for the American manned space program. Up until this time, most food safety control systems had been based on end product testing but it was realized that this would not give enough assurance of food safety for such an important mission. Taking the failure mode and effect analysis (FMEA) approach as a starting point, the team adapted this into the basis of the HACCP system that we know today: a system that looks at what can go wrong at each step in the process and builds in control to prevent the problem from occurring.

The HACCP system has become the internationally accepted approach to food safety management [5, 14]. It is based on the application of seven principles that show how to develop, implement, and maintain a HACCP system:

- Principle 1: Conduct a hazard analysis.
- Principle 2: Determine the critical control points (CCPs).
- Principle 3: Establish critical limit(s).
- Principle 4: Establish a system to monitor control of the CCP.
- Principle 5: Establish the corrective action to be takern when monitoring indicates that a particular CCP is not under control.
- Principle 6: Establish procedures for verification to confirm that the HACCP system is working effectively.
- Principle 7: Establish documentation concerning all procedures and records appropriate to these principles and their application.

The use of HACCP is promoted by the WHO [15] and is increasingly being seen by government groups worldwide as a cornerstone of food safety legislation.

15.4.1
Developing a HACCP System

In order to develop a HACCP system, a food company applies the Codex HACCP principles to its operations. This is most easily achieved using the Codex logic sequence [5]:

Step 1: Assemble HACCP team
Step 2: Describe product

Step 3: Identify intended use
Step 4: Construct flow diagram
Step 5: On-site confirmation of flow diagram
Step 6: List all potential hazards, conduct a hazard analysis and consider control measures
Step 7: Determine CCPs
Step 8: Establish critical limits for each CCP
Step 9: Establish a monitoring system for each CCP
Step 10: Establish corrective actions
Step 11: Establish verification procedures
Step 12: Establish documentation and record keeping.

However, before applying the logic sequence it is necessary to consider the overall structure of the HACCP system. HACCP systems can be linear, where the principles are applied to the whole operation from ingredients to end product, or modular, where the operation is split into process stages or modules and HACCP principles are applied to each module. Modular systems are common in complex manufacturing operations and are practical to develop; however, a key point is to ensure that the modules add up to the entire operation and that no process stages are missed out. Figure 15.2 shows the linear and modular approaches to HACCP.

The outcome of HACCP principle application is the HACCP plan, a document that shows how significant hazards will be controlled, and this needs to be implemented into every day control within the operation for food safety to be achieved. The HACCP plan is defined by Codex [5] as follows:

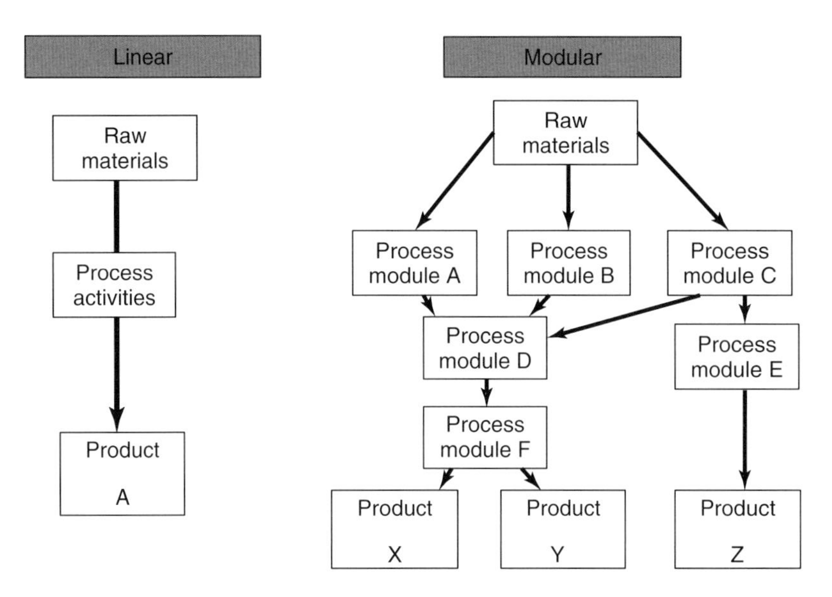

Figure 15.2 Linear and modular HACCP system example layouts.

"HACCP plan: a document prepared in accordance with the principles of HACCP to ensure control of hazards that are significant for food safety in the segment of the food chain under consideration [5]."

Step 1. Assemble HACCP team HACCP is normally applied by a multidisciplinary team, so that the system is the output of a group with the necessary combined experience and knowledge to take decisions about product safety. This approach works well in manufacturing operations and normally includes, as a minimum, the following disciplines:

- manufacturing or operations personnel who understand the process operations on site;
- quality or technical personnel who understand the product's technical characteristics regarding hazard control and have up to date information on likely hazards in that sector of the food industry; and
- engineering personnel who have knowledge and experience of the equipment and process operations in use on site.

In addition to the above disciplines, it can be helpful to include personnel from the following areas; however, the total size of a HACCP team is best kept to four to six personnel for ease of management:

- microbiology;
- supplier/vendor assurance;
- storage and distribution; and
- product development.

Step 2. Describe product It is important for all members of the HACCP team to understand the background to the product/process that they are about to study. This is achieved by constructing a product description (also known as a *process description*). The product description is not simply a specification for the product, but rather contains information important to making safety judgments. The following criteria are normally included:

- hazard types to be considered;
- main ingredient groups to be used in the product/process line;
- main processing technologies;
- key control measures;
- intrinsic (recipe) factors;
- packaging system; and
- start and end points of the study.

The task of constructing a product description helps to familiarize all HACCP team members with the product/process under study. It is normal practice to document the product description and include it with the HACCP plan paperwork. The document is also useful at later stages as a familiarization tool for HACCP system auditors or any personnel who need to gain an understanding of the HACCP plan.

Step 3. Identify intended use It is necessary to identify the intended use of the product, including the intended consumer target group, because different uses may involve different hazard considerations and different consumer groups may have varying susceptibilities to the potential hazards. This information is usually included as part of the product description (Step 2).

Step 4. Construct flow diagram A process flow diagram, outlining all the process activities in the operation being studied, needs to be constructed. This should list all the individual activities in a stepwise manner and should show the interactions of the different activities. The purpose of the process flow diagram is to document the process and provide a foundation for the hazard analysis (Step 5). A simple example of a process flow diagram is shown in Figure 15.3. This shows a process module taken from a modular HACCP system at a milk processing plant. Notice that the steps are shown as activities rather than as a list of equipment, and storage functions are also included.

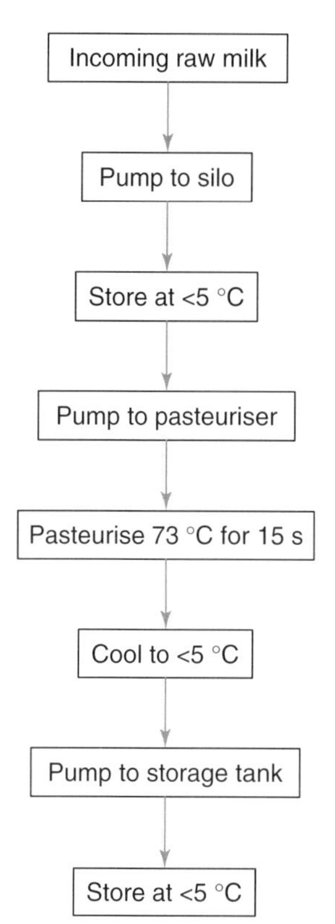

Figure 15.3 Example of a process flow diagram.

Step 5. On site confirmation of flow diagram Since the process flow diagram is used as a tool to structure the hazard analysis, it is important to check and confirm that it is correct. This is done by walking the line and comparing the documented diagram with the actual process activities, noting any changes necessary. This exercise is normally done by members of the HACCP team but could also be done by process line operators. The completed process flow diagram should be signed off as valid by a responsible member of staff, for example, the HACCP team leader.

Step 6. List all potential hazards, conduct a hazard analysis and consider control measures Using the process flow diagram, the HACCP team now needs to consider each step in turn and list any potential hazards that might occur. They should then carry out an analysis to identify the significant hazards and identify suitable control measures. These terms are defined by Codex [4] as follows:

- "**hazard**: a biological, chemical, or physical agent in, or condition of, food with the potential to cause an adverse health effect";
- "**hazard analysis**: the process of collecting and evaluating information on hazards and conditions leading to their presence to decide which are significant for food safety and therefore should be addressed in the HACCP plan"; and
- "**control measure**: an action or activity that can be used to prevent, eliminate, or reduce a hazard to an acceptable level."

An example of hazard analysis for two steps from the milk process flow diagram (Figure 15.3) is given in Table 15.5. Note, only one potential hazard has been detailed for each process step – there may be others.

The process of hazard analysis requires the team to transcribe each process activity to a table such as the example given, consider any potential hazards, along with their sources or causes, and then evaluate their significance. To identify the significant hazards, it is necessary to consider the likelihood of occurrence

Table 15.5 Example of hazard analysis process.

Process step	Hazard and source/cause	Significance evaluation		Significant hazard?	Control measure
		Likelihood	Severity	(yes or no)	
Incoming raw milk	Presence of vegetative pathogens (e.g., *Salmonella*) due to contamination from animal	Likely	High	Yes	Control by pasteurization step in process
Pasteurization	Survival of vegetative pathogens (e.g., *Salmonella*) due to incorrect heat process	Likely	High	Yes	Effective heat process (correct time/ temperature combination)

of the hazard in the type of operation being studied as well as the severity of the potential adverse effect. This may be done using judgment and experience or using a structured "risk evaluation" method, where different degrees of likelihood and severity are weighted to help with the significance decision. Effective control measures then need to be identified for each significant hazard.

Step 7. Determine CCPs CCPs are the points in the process where the hazards must be controlled in order to ensure product safety. They are defined by Codex [5] as follows:

> "Critical control point (CCP): a step at which control can be applied and is essential to prevent or eliminate a food safety hazard or reduce it to an acceptable level."

It is important to identify the correct points as CCPs so that resources can be focused on their management during processing. CCPs can be identified using HACCP team knowledge and experience or by using tools such as the Codex CCP decision tree (Figure 15.4). More detailed explanations on the identification of CCPs and use of decision trees can be found in other publications, for example [1, 4, 10].

Step 8. Establish critical limits for each CCP Critical limits are the safety limits that must be achieved for each CCP to ensure that the products are safe. As long as the process operates within the critical limits, the products will be safe but if it

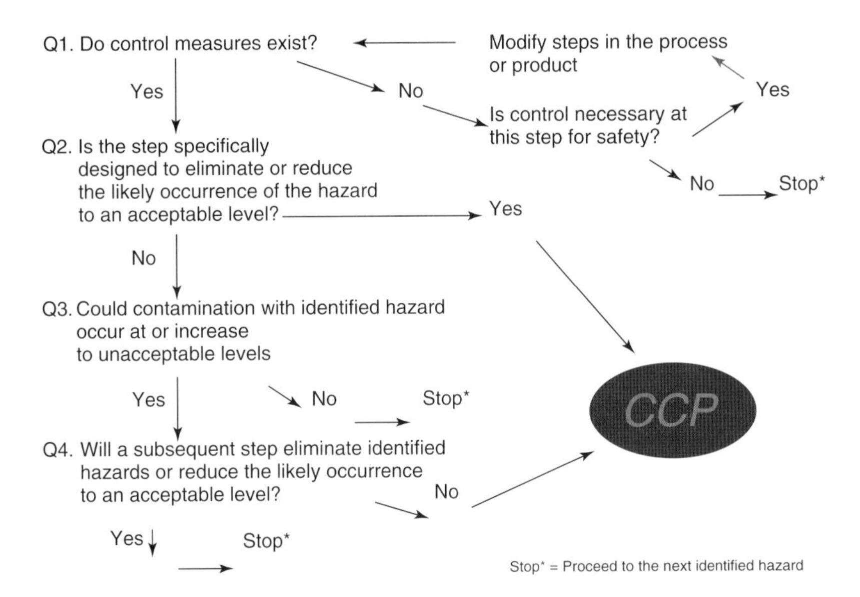

Figure 15.4 CCP decision tree. From Ref. [5].

goes beyond the critical limits then the products made will be potentially unsafe. Critical limits are defined by Codex [5] as follows:

> "Critical limit: a criterion that separates acceptability from unacceptability."

Critical limits are expressed as absolute values (never a range) and often involve criteria such as temperature and time, pH and acidity, moisture, and so on. It is important that the chosen critical limits are valid for food safety control and can always be achieved by the control measures. This might require process capability studies for some CCPs.

Step 9. Establish a monitoring system for each CCP Monitoring is necessary to demonstrate that the CCPs are being controlled within the appropriate critical limits. Monitoring requirements are specified by the HACCP team during the HACCP study but will usually be done by the process operators when the HACCP plan is implemented in the operation.

> "Monitoring: the act of conducting a planned sequence of observations or measurements of control parameters to assess whether a CCP is under control" [5].

Monitoring should be defined in terms of the monitoring activity itself, along with the frequency and responsibility for doing the task. It is important to check that the monitoring procedure is capable of detecting loss of control at the CCP.

Step 10. Establish corrective actions Corrective action needs to be taken where monitoring shows that there is a deviation from a defined critical limit. Corrective actions will deal with the material produced while the process is out of control and will also bring the process back under control.

> "Corrective action: any action to be taken when the results of monitoring at the CCP indicate a loss of control" [5].

As for monitoring, the corrective action procedures and responsibility need to be identified by the HACCP team during the HACCP study, but will be implemented by the appropriate operations personnel if deviation occurs.

A completed table demonstrating control of CCPs using critical limits, monitoring and corrective action is shown as Table 15.6.

Step 11. Establish verification procedures The HACCP team needs to consider how to determine if the HACCP system is working effectively over time. Verification procedures are the methods that will be used to demonstrate compliance and verification is defined by Codex [5] as:

> "Verification: the application of methods, procedures, tests, and other evaluations, in addition to monitoring, to determine compliance with the HACCP plan."

Table 15.6 Example of critical control point (CCP) control.

Process step	Hazard	Control measure	Critical limit	Monitoring			Corrective action	
				Procedure	Frequency	Responsibility	Activity	Responsibility
Pasteurization	Survival of vegetative pathogens (e.g., *Salmonella*)	Correct temperature and time regime: effective heat process	71.7| °C for 15 s	Chart recorder: visual check and sign off	Each batch	Pasteurizer operator	Report to supervisor; ensure divert working correctly; if not, dump/reprocess	Pasteurizer operator, production supervisor, plant engineer
				Check auto-divert function	Daily at start-up and shut-down	Pasteurizer operator	Hold product until correct heat process verified; dump/ re-process if not	Pasteurizer operator, production supervisor, plant engineer, QA manager

Commonly used verification procedures include:

- HACCP audits;
- review of CCP monitoring records; and
- review of deviations, including product disposition and customer complaints.

In addition to demonstrating compliance with the HACCP plan in practice, the verification HACCP principle (principle 6) also requires determination of the adequacy of the HACCP plan. This is known as *validation* and is defined by Codex [5] as:

> "Validation: obtaining evidence that the elements of the HACCP plan are effective."

This check that the HACCP plan will control all relevant significant hazards to the operation is normally done before implementing the HACCP plan into everyday practice and then both periodically, to ensure that remains effective, and whenever there are changes to the operation or new information on hazards indicates a need to update/strengthen the HACCP plan.

Step 12. Establish documentation and record keeping It is important to document the HACCP system and to keep adequate records. The HACCP plan will form a key part of the documentation, outlining the CCPs and their management procedures (critical limits, monitoring, corrective action). It is also necessary to keep documentation describing how the HACCP plan was developed, that is, the hazard analysis, CCP determination, and critical limit identification processes.

When the HACCP plan is implemented in the operation, records will be kept on an ongoing basis. Essential records include:

- CCP monitoring records;
- records of corrective actions associated with critical limit deviation;
- records of verification activities; and
- records of modifications to processes and the HACCP plans.

15.4.2
Implementing and Maintaining a HACCP System

The 12 steps of the HACCP logic sequence outlined above describe how to develop HACCP plans and their associated verification and documentation requirements. However, they do not describe how to implement the HACCP plans into everyday practice. Implementing HACCP requires careful preparation and training of the workforce and is, perhaps, best managed as a change management process. Depending on the maturity of the operation, this may be a straightforward implementation of the HACCP requirements or may require a culture change.

The implementation stage is where the HACCP plans are handed over from the HACCP team(s) that worked on the development process to the operations personnel who will manage the CCPs on a day-to-day basis. Training for the personnel who will monitor CCPs and take corrective action is essential and HACCP awareness training for the operations workforce is advisable. HACCP monitoring personnel need to understand the monitoring procedures and frequency, as well as how to record results and when corrective action must be taken.

After implementation, the HACCP verification procedures identified in Step 11 of the HACCP logic sequence need to commence. Results of verification should be reviewed regularly and actions should be taken where necessary to strengthen the HACCP system.

15.5 Ongoing Control of Food Safety in Processing

In order to ensure ongoing control of food safety, the PRPs, HACCP, and safe design processes need to work together as a cohesive system. The key points to ongoing control of food safety are as follows:

- verification of food safety system elements effectiveness;
- review of system elements and their suitability for food safety, including periodic validation of control procedures;
- change control procedures that require safety assessment and approval for all proposed changes to ingredients, process activities, and products;
- ongoing management and update of system elements; and
- training of staff.

As shown at the start of this chapter (Figure 15.1), the management of food safety system elements is often done using an overall operations management system, for example, the quality management framework ISO 9001:2000 [2] or the food safety management standard ISO 22000:2005 [3]. This document includes HACCP, quality management, and PRP requirements and is being used by increasing numbers of food companies in parts of the world to gain confidence through external certification of their food safety systems. Other external Standards also require food companies to manage food safety through PRPs, management practices, and HACCP. These include retail-driven Standards such as the widely used British Retail Consortium (BRC) Global Standard – Food [16], manufacturing Standards such as the American Grocery Manufacturers Association, GMA-Safe Program (*www.gma-safe.org*), and national expert Standards such as the Netherlands National Board of HACCP Experts HACCP Code [17].

External standards for food safety management can be helpful in giving an external perspective as well as keeping the requirements for food safety in the forefront of people's minds. Many of these external standards are coming together under the framework of the Global Food Safety Initiative (GFSI) [18], which allows benchmarking of audit standards against the GFSI requirements. This helps to facilitated acceptance of standards around the world and these schemes are now a requirement for doing business in many areas, required by manufacturers and retailers alike.

The essential requirement for any food processor is that they can manage their facility, ingredients, processes, and products to ensure that only safe products reach the customer. The food safety system elements – safe design, PRPs, and HACCP – described in this chapter will allow the requirement for safe food to be achieved. The use of an external audit standard to assess the operation of system elements may be a business requirement for some companies or may be regarded as an optional extra by others. Either way, it is important to assess regularly whether the systems are working and therefore that there is ongoing control of food safety.

References

1. Mortimore, S.E. and Wallace, C.A. (1998) *HACCP: A Practical Approach*, 2nd edn, Aspen Publishers, Gaithersburg, MD.
2. International Organization for Standardization (2000) ISO 9000: 2000. *Quality Management Systems – Requirements*, International Organization for Standardization, London.
3. International Organization for Standardization (ISO) (2005) BS EN ISO 22000:2005. *Food Safety Management Systems – Requirements for Any Organization in the Food Chain*, International Organization for Standardization.
4. Mortimore, S.E. and Wallace, C.A. (2001) *HACCP – Food Industry Briefing*, Blackwell Science, Oxford.
5. Codex (Joint FAO/WHO Food Standards Programme, Codex Alimentarius Commission) (2009) Hazard analysis and critical control point (HACCP) system and guidelines for its application, in *Food Hygiene Basic Texts*, 4th edn, Joint FAO/WHO Food Standards Programme, Food and Agriculture Organization of the United Nations, Rome, *http://www.fao.org/docrep/012/a1552e/a1552e00.htm* (accessed Sept 2, 2011).
6. Kyriakides, A.L. (Forthcoming) The principles of food safety, in *Training and Education for Food Safety* (ed. C.A. Wallace), Blackwell Science, Oxford.
7. International Commission on Microbiological Specifications of Foods (1996) *Microbiological Specifications of Food Pathogens*, Microorganisms in Foods, vol. 5, Blackie Academic and Professional, London.
8. World Health Organization (1999) Strategies for Implementing HACCP in Small and/or Less Developed Businesses. Report of a WHO Consultation, WHO/SDE/PHE/FOS/99.7, WHO, Geneva.
9. Canadian Food Inspection Agency (2000) Prerequisite programs, in *Guidelines and Principles for the Development of HACCP Generic Models, Food Safety Enhancement Program Implementation Manual*, vol. 2, 2nd edn, Canadian Food Inspection Agency, Ottawa
10. Wallace, C.A., Sperber, W.H., and Mortimore, S.E.I.P. (2010) *Food Safety in the 21st Century*, Wiley-Blackwell, Oxford.
11. Codex Committee on Food Hygiene (1999) *Recommended International Code of Practice – General Principles of Food Hygiene*, CAC/RCP 1-1969, rev. 3-1997, amended 1999, Food and Agriculture Organization of the United Nations, Rome.
12. Institute of Food Science and Technology (2007) *Food and Drink – Good Manufacturing Practice: A Guide to its Responsible Management*, 5th edn, Institute of Food Science and Technology, London.
13. Sprenger, R. (2003) *Hygiene for Management*, Highfield Publications, Doncaster.
14. NACMCF (1997) Hazard analysis and critical control point principles and applications guidelines, *http://www.fda.gov/Food/FoodSafety/HazardAnalysisCriticalControlPointsHACCP/ucm114868.htm* (accessed 2009).

15. World Health Organization (WHO) (2007) Hazard Analysis Critical Control Point System (HACCP), *http://www.who.int/foodsafety/fs_management/haccp/en/* (accessed May 14, 2009).
16. British Retail Consortium (2008) *BRC Global Standard for Food Safety*, Issue 5, The Stationery Office, London.
17. Netherlands National Board of HACCP Experts (2007) *Dutch HACCP Code (Requirements for a HACCP Based on Food Safety System, Version 4)*, Netherlands National Board of HACCP Experts, The Hague, www.foodsafetymanagement.info (accessed 2009).
18. Global Food Safety Initiative (GFSI) (2007) GFSI Guidance Document, 6th edn, *www.mygfsi.com* (accessed Sept 2, 2011).

16
Traceability in Food Processing and Distribution

Christopher Knight

16.1
What Is Traceability?

Traceability is a widely used term and is one of those broad concepts, like quality, for which there are many definitions and applications. In reality, the term "*traceability*" is quite difficult to define, there being different ideas about what is required or important. In practical terms it is about meeting external marketplace requirements, including legal demands and customer expectations, as well as implementing internal quality management objectives and improving business performance.

There is no single universally acceptable definition or system of traceability; it depends on many factors, including the nature of the product and type of production operation. Traceability may also have different objectives, such as food safety, product identity, and reliability of information provided to the next business in the food supply chain. Although legal requirements, international standards, and private standards require traceability systems, none is prescriptive in the way traceability is achieved. This is perhaps not surprising as there are many options available and it is up to the business to define the scope of the traceability system and how it is to be achieved based on their particular needs. These issues highlight the practical difficulties in establishing and implementing a traceability system in a food business operation including food processing and distribution.

Traceability identifies the path from where a product originated to where it has been supplied, and consists of an interlinking chain of records between steps in a process operation and/or between different stages in a food supply chain. Traceability systems have three basic components:

- **supplier traceability**, which enables the source of materials used or handled to be identified;
- **process traceability**, which enables the identity of raw materials and process or handling records for each lot; and
- **customer traceability**, which enables the customer to whom the product has been supplied to be identified.

Food Processing Handbook, Second Edition. Edited by James G. Brennan and Alistair S. Grandison.

Linked to these basic components of traceability are efficient record keeping and the ability to provide relevant information on demand. There are two categories of information relating to traceability:

- internal traceability, which relates to the processing history within an operation and
- external traceability, which relates to product information that an operation receives from suppliers or provides to customers.

Most operations within the food supply chain cannot readily create traceability throughout the whole food supply chain, but each has a role to play in collecting and storing information about raw materials, products, and processes under their control.

Traceability features the procedures for identity, production history, and source of a product. Traceability is a management tool; it does not make food safe or identify product. Traceability does, however, give the assurance of food safety or identity. This chapter outlines the general principles and basic system requirements for the design and implementation of a traceability system, with special reference to food processing and transport operations.

16.2 Traceability and Legislation

Traceability has a legal framework in many countries. In European Union (EU) food law, for example, food business operators must be able:

- to identify from whom and to whom product has been supplied and
- to have systems and procedures in place that allow for this information to be made available to competent authorities upon their request.

This requirement relies on the "one step back" and "one step forward" approach, that is, external traceability (Figure 16.1), and implies that food business operators must be able to:

- identify from whom product has been received, including suppliers of raw materials,
- identify the businesses to whom they have supplied products, and
- provide the information to the Competent Authorities in a timely manner.

The EU legal requirements do not include internal traceability (see Figure 16.1), that is, the matching up of all inputs to outputs, which is an additional feature of international and private standards. Nor is there any requirement for records to be kept identifying how lots are split and combined within an operation to create new products or lots. The principal purpose of the legal requirements is to assure food safety and assist in accurate withdrawal or recall of product from the food chain.

The EU legal requirements apply to any food business operation that trades in food or feed at all stages of the food chain from farm to retail, including food

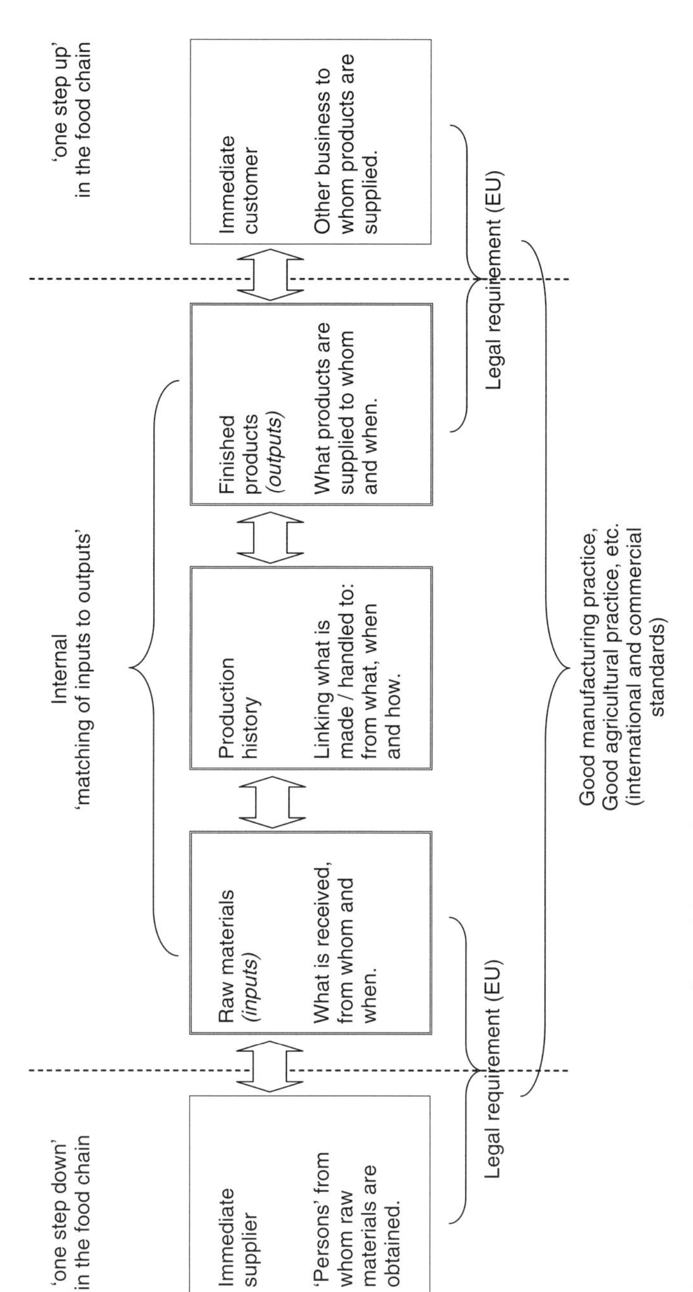

Figure 16.1 Basic components of traceability in a food business operation.

processing, storage, and distribution, even if their supplier is not in the EU. The requirements do not, however, extend to food business operators outside the EU.

As a general rule the "one step back/one step forward" (i.e., external traceability) approach is a general principle that is applied in food legislation in many countries, not just the EU. It is also a recommendation of Codex. For example, where traceability/product tracing is a tool within a food inspection and certification system, this should be able to identify at any specified stage of the food chain from where the food came (the one step back) and to where the food went (the one step forward). Codex defines traceability/product tracing in this context as the ability to follow the movement of food through specified stages of production, processing, and distribution.

16.3 Traceability and International/Private Standards

Traceability may be managed as part of a formal business management system, for example, the international series of standards for Quality System Management (the ISO 9000 series). ISO 90001: 2008 specifies management system attributes including in the section on production and service provisions requirements for identification and traceability and preservation of product. This states that where appropriate, the organization shall identify product by suitable means throughout product realization and, where traceability is a requirement, the organization shall control the unique identification of the product and maintain records.

Traceability is also a constituent part of the ISO 22000 standard (food safety management systems – requirements for any organization in the food chain). ISO 22000: 2005 specifies requirements for a food safety management system where an organization in the food chain needs to demonstrate its ability to control food safety hazards in order to ensure that food is safe at the time of human consumption. It is applicable to all organizations which are involved in any aspect of the food chain. There is a specific requirement for traceability, which covers the establishment and application of a traceability system that enables the identification of product lots and their relation to batches of raw materials, processing, and delivery records. This implies a combination of the one up/one down approach (external traceability) plus internal traceability (Figure 16.1).

Traceability systems attributes are also specified in ISO 22005: 2007, which outlines the principles and specifies the basic requirements for the design and implementation of a food and feed traceability system. As such it covers the management system attributes of a traceability system. As with ISO 9001 and ISO 22000 it does not specify how traceability is to be achieved. It does, however, imply that traceability should include external and internal traceability, plus appropriate record keeping.

16.4 Traceability and Private Standards

There are many private standards which apply to specific sectors in the food chain or product types. Most, if not all, have provisions relating to product identity and/or traceability. The British Retail Consortium (BRC) Global Standard for Food, for example, requires that the company shall have a system in place to identify and trace product lots and follow through all raw materials, all stages of processing, and the distribution of the finished product to the customer in a timely manner. Similarly, the BRC Global Standard for Storage and Distribution specifies that the site shall have a traceability system with the ability to trace products through receipt, storage, dispatch and, where applicable, distribution, and vice versa.

The Global Food Safety Initiative (GFSI) benchmarks existing food standards against food safety criteria. Benchmarking is a procedure by which a food safety-related scheme is compared to the GFSI Guidance Document such that benchmarked schemes have a common foundation of requirements, including traceability. In respect of traceability the guidance specifies that the standard shall ensure:

- identification of outsourced product, ingredient, or service;
- complete records of batches in process or final product and packaging throughout the production process;
- record of purchaser and delivery destination for all products supplied.

This implies a requirement for internal and external traceability plus record keeping. Manufacturing schemes recognized by GFSI scheme include BRC Global Standard, International Featured Standard (IFS) and SQF 2000.

The relevance of these private standards is that they help organizations that adopt the specified standard meet legal requirements (the one up/one down approach) and the expectations of the food supply chain for internal traceability, that is follow the movement of food through the specified stage from receipt of supplied materials to distribution of the product.

16.5 Traceability in the Food Supply Chain

Taking all the requirements described above into account, that is as required by legislation and adopted international or private standards, then the basis of a traceability system in the food supply chain is:

- identify and trace what is received (one step down external traceability),
- identify and trace what is made or handled, from what, when and how (internal traceability), and
- identify and trace the delivery destination of supplied product (one step up external traceability).

This is, in turn, linked to efficient record keeping. Figure 16.1 provides a summary of the basic elements of a traceability system as should be applied in the food supply chain. Each operator is responsible only for the sector of the food chain operation under their control, including food processors and distributors. By linking each stage in the chain, whole chain traceability can be achieved.

The food supply chain is a series of separate operations in sequence (farm to market), each linked by the products supplied to them from the preceding operation (the one step back or down) and the products they supply to the next operation (the one step up or forward). Each operator in the chain records information which links the separate operations with their own traceability system and provides whole chain traceability.

16.6 Product Identification

To achieve traceability of food and drink raw materials within a food operation or supply chain it is essential to identify the food item concerned and to provide some form of data carrier facility for maintaining identification of these items.

Two types of identification may be recognized; primary and secondary. Primary identification generally relates to techniques that identify an entity by measuring a unique component or characteristic. These are primarily chemical or molecular-based analytical methods (such as DNA) that can be used to identify an individual entity.

Secondary identification is based on attributing an identifier to a discrete unit (e.g., a separate batch or lot) of materials in a form that can be attached to, or conveniently accompany, that unit through, or partly through, an operation or supply chain. Where such identifiers are used partly within the chain, other identifiers may be introduced to accompany the entities, part-entities, or combined entities as they are processed or handled.

Secondary identification relies upon assigning a unique mark or code based on numbers or alphanumeric strings, including exploiting available standards for numbering and identification where applicable. In the food supply chain these are invariably open systems where the identification codes and any additional information relating to the item, such as a batch number, weight, other identification numbers, production or use/sell-by date, adhere to a particular identification convention or standard.

In order to use a secondary identifier for traceability purposes a data carrier is required. This is the physical thing which is attached to, directly marked on, or accompanies the item, and carries the identification code. Identifiers can either be in human or machine-readable form, or both. For automated system approaches to traceability, machine-readable data carriers are required.

A range of data carrier technologies and an even wider range of commercial products and systems are available to support identification at various levels of sophistication, including, for example, linear (1D) and 2D barcodes (Figure 16.2),

Linear (1D) bar code

2D data matrix bar code

Figure 16.2 Barcode examples.

radiofrequency identification (RFID) tags (see Chapter 8, Section 8.7.4), and smart labels (passive and active devices). The technology that is perhaps most widely used and recognized is the linear barcode.

A barcode is a machine-readable representation of data. Linear (or one-dimensional) barcode data carriers are probably the most prominent and well-established data carrier technology. They are used widely in a variety of situations including asset management, manufacturing, retail, warehouse management, and distribution.

The data-carrying part of a linear barcode comprises a number of alternating dark (bar) and light (space) parallel lines of variable width. The bars and spaces are structured to carry data in digital form. The rules by which they are structured determine the type of barcode and the attributes they exhibit.

The differences between the different barcode types is mainly in the range of characters that they can encode, and the way in which these characters are represented by patterns of bars and spaces. Different barcode types have been developed to accommodate numeric or alphanumeric data and the use of different printing technologies for higher security of data and for reliability in reading.

Barcodes can be read by optical scanners called *barcode readers*. Reading barcodes typically involves directing a beam of red light and detecting the light reflected. This is converted into a digital signal from which to determine, through a decoding process, the characters the signals represent.

16.7
Management of Traceability Information

There are several types of data management in a typical food operation, for example, transfer, joining, and splitting. Transfer is the simplest of operations, where product identification is transferred with the product through one or more steps in a process. That is where the traceability information is retained and the identification is transferred between the process steps. Joining is where one process step combines several traceability units, each with a unique identification code and a new identification code is established. Splitting, on the other hand, is where a traceability unit is split and used in the production of new traceability units, each with a new identity code, for example, in different processes, products, or customer destinations.

In addition to the data operations described, the method of recording will depend on the operation being carried out, including whether it is a batch or continuous operation. Batch operations are where the production operation is carried out on one batch at a time. The batch operation records can provide a direct linkage between the batch process and product produced. In a continuous operation, on the other hand (e.g., processing of products in a continuously operated line for a defined period), the linkage with the product is based on the particular production run start and finish time, date, and so on.

Efficient and accurate record keeping is essential to the successful application of traceability. Traceability records need to be kept in a way appropriate to the nature and size of the operation and need to be organized and retained in good condition to enable easy retrieval. A working traceability system can generate many records. The method of recording may be paper-based or an electronic computerized system (including commercial record keeping system) or a combination of both. The procedures used will depend on the needs of the operation.

Retention and target retrieval time will be another important consideration. The length of time records are retained will depend on the nature of the product, and any legal or commercial requirements. The EU general food law does not foresee a minimum retention period. It is considered that commercial documents are usually retained for a period of 5 years for taxation purposes. This 5-year period would, therefore probably meet the requirements of the regulations. However, this common rule would need to be adapted in some cases, for example, for perishable, shorter shelf life products.

A traceability system should allow necessary actions to be taken, such as isolation of a nonconforming batch, within an appropriate time frame. There is therefore a need to determine a target time of reaction for traceability data availability. In the EU, food operators are required to have in place systems and procedures to ensure traceability of their products, which implies structured mechanisms to deliver required information upon request. In the United States this has been defined as 4 h during normal working periods and 8 h outside these periods.

The target time for retrieval of traceability data will depend on a number of factors, such as any regulatory requirements and customer expectations, and industry norms taking into account the nature of the product and production operation or section of the food chain. This is likely to be measured in hours rather than days, for example, between 2 and 24 h depending on circumstance and data to be retrieved.

16.8 The Traceability System

Traceability systems, where applicable, should be underpinned by adherence to appropriate legislation and adopted international or private standards. The system approach outlined in this chapter is based on these general requirements but adapted to demonstrate good practice for establishing and implementing a traceability system, with special reference to food processing and storage and distribution operations.

There are four basic components of a traceability system.

- organize and plan traceability;
- implement traceability;
- ensure effective operation of traceability; and
- document and record traceability.

When planning a traceability system it is helpful to conduct a traceability study. A typical traceability study comprises seven stages (Table 16.1). These stages include essential preparation activities (Stages 1–4) and the application and maintenance of traceability (Stages 5–7).

The traceability analysis (Stage 5) depicts the traceability system and indicates the traceability control points in an operation. The analysis is systematically applied to all the steps in the process in sequence as defined in the flow diagram (Stage 4). In practice this involves asking three questions at each process step:

- What identification details (codes or identifiers) are read?
- What records relevant to traceability are kept?
- What identification codes or identifiers are transferred to the next step (new or retained)?

The purpose is to identify the traceability identification information that is read at the beginning of the process step and applied at the end, together with the records taken. If, in the analysis, it is determined that traceability is compromised at a process step and if control of traceability is necessary then the step, process, or procedures can be modified to ensure that the appropriate level of traceability is implemented.

Performing test and review activities of the traceability system is akin in many ways to verification of hazard analysis and critical control point (HACCP) systems. The aim is to establish procedures for verification to confirm that the system

Table 16.1 Stages in conducting a traceability system.

Stage 1	Define the scope of the study	The terms of reference of the traceability system should be defined clearly to enable the personnel involved with traceability to focus on the key issues and ensure the traceability system is effective
Stage 2	Define authority and responsibility	A traceability study will require the collation and evaluation of technical data, and is best carried out by suitably qualified persons with appropriate knowledge and experience of the product and process operation
Stage 3	Describe the product	A full description of the product(s) under study should be prepared, including defining key identity parameters which relate to traceability where applicable
Stage 4	Define the process	Prior to the traceability study beginning it is necessary to carefully examine the product/process operations under study and produce a flow diagram around which the study can be based
Stage 5	Conduct a traceability analysis	Identify and list the traceability attributes; conduct a traceability analysis to determine where identity needs to be read and applied
Stage 6	Perform test and review activities	The traceability personnel should put into place procedures that can be used to demonstrate compliance with the traceability system and to determine its effectiveness in use
Stage 7	Establish documentation and record keeping	Efficient and accurate record keeping is essential to the successful application of traceability. It is important for the operation to be able to demonstrate that the traceability system has been implemented and maintained, and that documentation and records have been kept in a way appropriate to the nature and size of the business

From Campden BRI Guideline No. 60 [1].

is working effectively. Verification demonstrates conformance (e.g., with stated procedures) and that the traceability system is effective (i.e., traceability objectives are being met). In traceability systems there are two key questions:

- Does traceability work in practice? That is, is there compliance with the traceability system as implemented and is it working in practice?
- Is the traceability system up to date? That is, has there been any change that affects traceability? This will involve a periodic review of the traceability system.

Typical examples of compliance testing are audits or other inspections of procedures and associated records, and testing the system in some way. Typically, this involves selecting a product lot and following production backward from despatch to receipt of raw materials or a raw material to finished product and collating associated records. It may also include a quantity check or mass balance, that is a reconciliation of the amount of supplied materials against the amount used in the resulting product, taking into account waste and rework. Review, on the other hand, is the mechanism that drives the vital maintenance of the traceability system, that is, keeps it up to date and relevant. There should be a formal scheduled review of the traceability system; typically this should be performed annually. In addition, there should also be a mechanism in place that will automatically "trigger" a review prior to significant changes due to internal or external factors, for example, changes to product or process (internal) or legislation or customer requirements (external).

16.9 Examples of Traceability Systems

To demonstrate the application of traceability in food business operations, two case studies are presented, namely food processing and storage and distribution. These examples are used to illustrate a generic approach to the application of traceability, and represent typical traceability scenarios. The details are given for illustrative purposes only and are not exhaustive, nor are they recommendations for similar operations.

The examples are not complete traceability plans as outlined in Table 16.1, they are extracts and focus on the terms of reference and traceability analysis. The details such as identification read, records taken, and identification applied are shown for indication only and will in practice always depend on specific circumstances.

Example 1. Food Processing

Terms of Reference

Product	Frozen potato chips
Process	The production of frozen potato chips (preparation, blanching, frying, freezing, packing, cold storage and distribution) Start point: Intake of potatoes Finish point: Transport of finished product
Process flow	The sequence of unit operations (process steps) is shown in the traceability analysis
Objectives	To support food safety and quality objectives To fulfill legal requirements and customer expectations To document the history and origin of the product To facilitate the search for the cause of nonconformity and the ability to withdraw and/or recall products if necessary

Traceability Analysis

Process step	Identification read	Recorded information	Identification applied
1. Potato receiving			
Receipt of purchased potatoes, intake checks, grading, and transfer to on-site storage bins	Delivery documents	Supplier details, potato variety, intake checks, and storage bin destination (Step 2) entered on intake record	Intake records are retained
2. Storage			
Temporary holding of received potatoes	Intake records	Potato variety, supplier details, and so on recorded on intake record (Step 1)	Storage bin identification code
3. Potato preparation			
Transfer from storage bin, and prepare the raw potatoes (peel, etc.)	Storage bin identification code	Potato variety, potato source (storage bin identification code), and time of use entered on process records	No specific identity applied at this stage as it is a continuous inline process
4. Cut			
Cut prepared potatoes into chips	No specific identify read at this stage it is as a continuous inline process	None	No specific identity applied at this stage as it is a continuous inline process
5. Defect removal			
Automatic removal of visual defects and offcuts	No specific identify read at this stage as it is a continuous inline process	None	No specific identity applied at this stage as it is a continuous inline process
6. Processing aids			
Receipt and storage of purchased processing aids	Delivery documents and manufacturer's lot identification codes	Supplier details and manufacturer's lot number of received materials recorded on intake record	Manufacturer's lot number is retained with materials

Traceability Analysis (continued)

Process step	Identification read	Recorded information	Identification applied
7. Blanch			
Blanching, addition of processing aids, and drying	Lot numbers of processing aids	Processing aids used and lot number of materials recorded on process records. Blanch time and temperature is recorded continuously on process control system	No specific identity applied at this stage as it is a continuous inline process
8. Oil			
Receipt of purchased oil and transfer to bulk storage tank	Deliver documents	Supplier details, tanker information, delivery date, and time are recorded on intake record	No specific identity is applied
9. Fry			
Frying in oil and defat	No specific identify read at this stage as it is a continuous inline process	Fry time and temperature is recorded continuously by process control system	No specific identity applied at this stage as it is a continuous inline process
10. Freezing			
Blast freezing and transfer product direct to packing line and/or tip to bulk tote bins and transfer of bins to cold store	No specific identify read at this stage as it is a continuous inline process	Freezing time and temperature is recorded continuously by process control system. Product details in tote bins is recorded on cold store management system	Unique identification code is applied to each tote bin
11. Packaging			
Receipt and storage of purchased packing materials (polythene film and cases)	Deliver documents and manufacturer's lot identification codes	Supplier details and manufacturer's lot numbers of materials received are recorded on intake record	Manufacturer's lot number is retained with packaging materials

(continued overleaf)

Traceability Analysis (continued)

Process step	Identification read	Recorded information	Identification applied
12. Packing			
Weighing and filling of bags with product. Direct feed from freezer and/or tipping of tote bins to packing line	Lot number of polythene film used to make bags. Tote bin identification code	Lot number of polythene used is recorded on line packing record. Tote bin used are recorded on packing line records. QC records include pack and product quality checks	A unique product lot number[a], packing time, and best before date are printed on bags
13. Metal detection			
Detection and rejection of metal in packs	No specific identify is read at this stage as it is a continuous inline process	Metal detector checks are recorded on QC records	No additional identity is applied as it is a packed product
14. Case packing			
Packing of product bags into outer cases and case sealing	Lot number of case materials used to pack product	Lot numbers of case materials used is recorded on packing records	Product lot number[a], packing time, product code, and best before date is printed on case
15. Palletizing			
Automatic stacking and transfer to cold store	Product lot number[a], packing time, product code, and best before date	Product details for each pallet are recorded on cold store management system	Unique pallet identification code is applied to each pallet
16. Cold storage			
Storage of packed and palletized finished product	Pallet identification code	Product and pallet details and location in cold store is recorded on cold store management system	Pallet identification code is retained
17. Distribution			
Despatch of palletized product to customer	Pallet identification codes	Consignment inventory and customer destination are recorded on delivery records	Pallet and product identity codes are retained. Delivery identification code

[a]Product lot number is an alphanumeric code to identify product, packing date, packing line, and factory.

Traceability in this example may be summarized as follows:

- **Suppliers:** information relating to suppliers of inputs is linked to intake records.
- **Process inputs:** information relating to inputs used is linked to production records.
- **Finished product:** information from production records is linked to a finished product identity codes (product packs – bags and cases – and pallets), that is a production run date and time.
- **Customers:** information relating to whom product is supplied is linked to despatch records.
- **Consumer:** information from product name and codes (on product bags) links consumer with process records and inputs.

Example 2. Distribution

Terms of Reference

Product	Frozen foods
Process	The storage and distribution of frozen foods Start point: collection of products from producer End point: transport to consignee
Process flow	The sequence of unit operations (process steps) is shown in the traceability analysis
Objectives	To support food safety and quality objectives To fulfill legal requirements and customer expectations To document the history and origin of the product To facilitate the search for the cause of nonconformity and the ability to withdraw and/or recall products if necessary

Traceability Analysis

Process step	Identification read	Recorded information	Identification applied
1. Collection			
Pick-up of product from producer	Delivery documents and inventory (product identification codes)	Delivery documents	Delivery identification code
2. Transport			
Transport of goods to Regional Distribution Centre (RDC)/ warehouse	Delivery identification code	Delivery documents	Delivery identification code is retained
3. Intake			
Intake at RDC/warehouse	Delivery identification code (delivery documents and inventory)	Intake records	Product identification codes

(continued overleaf)

Traceability Analysis (continued)

Process step	Identification read	Recorded information	Identification applied
3. Intake			
Intake at RDC/warehouse	Delivery identification code (delivery documents and inventory)	Intake records	Product identification codes
4. Storage			
Storage of goods at RDC/warehouse	Product identification codes	Cold store management system (product details and location record)	Cold store identification codes
5. Picking			
Picking, collation, and labeling of consignment at RDC/warehouse	Product identification codes (pallets and or cases)	Consignment inventory (product and consignee destination record)	Consignment identification code/label
6. Loading			
Loading of vehicle at RDC/warehouse	Consignment identification code	Delivery documents (consignment inventory and destination)	Consignment identification code is retained
7. Transport			
Transport to consignee destination	Consignment identification code	Delivery documents (consignment inventory and destination)	Consignment identification code and product identification codes are retained

References

1. Campden BRI (2009) Traceability in the Food and Feed Chain: General Principles and Basic System Requirements, Guideline No. 60. *http://www.campden.co.uk/publ/pubDetails.asp?pubsID=2489* (accessed May 2011).
2. British Retail Consortium (2008) *Best Practice Guideline Traceability*, The Stationery Office, London.
3. British Retail Consortium (2008) *Global Standard for Food Safety*, The Stationery Office, London.
4. British Retail Consortium (2010) *Global Standard for Storage and Distribution*, The Stationery Office, London.
5. Codex Alimentarius Commission (2006) Principles for Traceability/Product Tracing as a Tool Within a Food Inspection and

Certification System, CAC/GL 60-2006. *http://www.codexalimentarius.net/download/standards/10603/CXG_060e.pdf* (accessed June 2011).

6. Derek, S. and Dillon, M. (2004) *A Guide to Traceability in the Fish Industry*, SIPPO/Eurofish, Zurich.
7. European Commission (2004) Guidance on the Implementation of Articles 11, 12, 16, 17, 18, 19 and 20 of Regulation (EC) No. 178/2002 on General Food Law, *http://www.ec.europa.eu/food/food/foodlaw/guidance/guidance_rev_7_en.pdf.*
8. Heap, R., Kierstan, M., and Ford, G. (eds) (1998) *Food Transportation*, Blackie Academic and Professional, London.
9. International Featured Standard (2007) *IFS Food 5 Standard for auditing retailer and wholesaler branded food product*, HDE Trade Services GmbH, Germany, *http://www.ifs-certification.com.*
10. International Organization for Standardization (2008) ISO 9001 : 2008. *Quality Management Systems – Requirements*, International Organization for Standardization, London.
11. International Organization for Standardization (2005) ISO 22000 : 2005. *Food Safety Management Systems – Requirements for Any Organization in the Food Chain*, International Organization for Standardization, London.
12. International Organization for Standardization (2007) ISO 22005 : 2007. *Traceability in the Feed and Food Chain – General Principles and Basic Requirements for System Design and Implementation*, International Organization for Standardization, London.
13. Lees, M (ed.) (2003) *Food Authenticity and Traceability*, Woodhead Publishing Ltd, Cambridge.
14. Smith, I. and Furness, A. (2006) *Improving Traceability in Food Processing and Distribution*, Woodhead Publishing Ltd, Cambridge.
15. Safe Quality Food (SQF) Institute (2005) SQF 2000 Code, Washington, DC, www.sqfi.com.

17
The Hygienic Design of Food Processing Plant

Tony Hasting

17.1
Introduction

In comparison to many other industrial sectors, the food processing industry is exceptionally diverse and complex, as can be seen from the size and nature of companies, the wide range of raw materials, products, and processes, and the numerous combinations of each. Products may be manufactured on a large scale for a global market as well as numerous specialist or traditional products at national and even regional levels. The sector is also subject to diverse local economic, social, and environmental conditions, as well as varying national legislation.

Food manufacturing is a dynamic, rapidly changing sector, which is strongly influenced by trends in consumer behavior, the industry itself, and political and regulatory pressures. Changes in social and economic conditions have an immediate impact on the food industry, for example, increased life expectancy, more single person households, increased time pressure, more disposable income, increased health awareness and changing taste, fashion and culture. While consumer safety is always an absolute priority for food manufacturers, consumer and industry trends can potentially increase the risk of contamination as shown in Table 17.1.

The industry also continues to face increasing regulation from both national and international authorities, which is unlikely to decrease in the future. In addition, developments in measurement and assessment techniques enable the authorities to be more effective at detecting the source and cause of food poisoning. The potential consequences of failure to prevent contamination may therefore become even more serious for both the individuals and companies involved. Contamination may result in product recalls, loss of confidence in the brand and the company as well as the economic implications and potential legal action.

Environmental pressures to reduce greenhouse gases, energy, and water use can also have an impact on food safety, particularly where recycling or reuse of out-of-specification product or utilities is implemented.

Food Processing Handbook, Second Edition. Edited by James G. Brennan and Alistair S. Grandison.

Table 17.1 Implications of consumer and industry trends on food safety.

Consumer trends	Implications for food safety
Reduce/eliminate chemical preservatives	Products may be more susceptible to microbial growth
Milder processing, for example, reduced heat treatment to improve quality	Reduced safety margin in ensuring pathogens inactivated
Increasing population susceptible to allergens	Greater risk to consumers
Industry trends	
Increased outsourcing of manufacture to reduce costs	Less control over manufacturing operations
Increased flexibility, that is, wider product range, shorter production runs, rapid changeover between batches, reduced cleaning frequencies and times	Potential risks from allergen cross contamination, increased challenges for cleaning systems
Loss of technical expertise from both food manufacturers and their suppliers due to cost pressures	Reduced level of process understanding

17.2
Engineering Factors Influencing Hygiene

The design of a piece of equipment or a complete process line starts with a specified objective and the designer evaluates a number of potential designs before arriving at what is believed to be the optimum way of achieving the specified objective. Chaddock [1] defined design as, "the conversion of an ill-defined requirement into a satisfied customer." The designer will be constrained by many factors that limit the number of possible designs, but a number of alternatives will usually be technically feasible.

Design constraints arise in a number of forms [2] and can be characterized in various ways. There are those that the designer has no control over, such as:

- physical laws;
- government regulations;
- standards or codes; and
- economic and resource constraints.

These define the overall boundary of possible designs and within this there are further constraints over which the designer may have some degree of control, such as choice of process, process conditions, materials, and equipment. Time will invariably be a major constraint and will usually limit the number of alternative designs that can be assessed in detail before the final selection is made.

Food hygiene is a key design constraint for both equipment and food manufacturers and may be defined as:

- all measures necessary to ensure the safety and wholesomeness of foodstuffs (EU General Food Hygiene Directive 93/43/EEC, 1993);
- all conditions and measures necessary to ensure the safety and suitability of food at all stages of the food chain (Codex Alimentarius Commission 2003).

The hygienic design of equipment used within the food manufacturing sector is a critical element in ensuring that all products are consistently safe from allergens, physical, chemical, or microbiological contaminants. The equipment that converts the raw materials into a final product results in a direct interaction between the food materials and the equipment, and poor design of such equipment can compromise the safety of the final product. Food safety is usually directly associated with the consumer response to a contaminant, such as food poisoning or an allergenic reaction. However, it should be noted that there are situations where microbiological contamination may not pose any significant risk to the consumer, for example, mold spoilage of a soft drink. In this case, spoilage may have an adverse impact if it results in a product recall and thus poses a risk to the credibility and hence security of the product, brand, and company as well as the associated costs of such events. This may be particularly relevant when the product has an extended shelf life, such as 12–18 months, is distributed to a number of different countries, and the contamination is only discovered after a number of months.

From an engineering perspective, equipment design, while important, is only one of a number of factors where an engineering understanding is essential and an emphasis on equipment design alone can increase rather than decrease the risk to final product quality. Other key factors are [3]:

- **Process design:** Even if the equipment meets the highest standards of hygienic design, an inadequate process design will result in the product safety being compromised. For example, the intrinsic safety of product from a pasteurization process designed with an insufficient time–temperature combination will not be influenced by the hygienic design of the equipment but by the inadequate process design.
- **Installation and layout:** Even if the individual items of equipment are acceptable from a hygienic design perspective, their integration into a complete process line can create many hygiene issues that can negate the benefits of using equipment designed to rigorous hygiene standards.
- **Process operation and control:** The control of the process plant during start up, production, shutdown, and cleaning are again critical to the safety of the final product. There can be a risk that the main focus is on the production phase and key areas such as cleaning and plant start up are not given the attention they require. Measurement is also a key element of operation and control as without the ability to measure process parameters reliably, it is not possible to monitor, control, validate, or optimize the process.

- **Factory environment:** The factory environment within which the process equipment is installed can have a significant impact on product quality, particularly where the product is exposed for all or part of the process. Management of material and personnel flows, generation and distribution of services, segregation of different areas of the plant, and the fabric of the building, such as floors, walls, and ceilings, are all key aspects of the overall design.
- **Storage and distribution:** The conditions the final product is subjected to during storage and distribution, which may range from a few days to many months, are factors where engineering understanding is important but equipment design may not be a critical factor.

A crucial requirement is the need to include hygienic design as early as possible in any project, as the ability to influence a project is usually greatest at the beginning when little detail has been fully specified, reducing as the design becomes more definite and finalized. Equally the cost of putting things right is least at the beginning of the project but rapidly increases, particularly at the commissioning and commercial production stages. An overemphasis on hygienic equipment design can result in hygiene only being considered at the stage when the process has been fully defined and the detailed equipment specifications are being prepared. By this time it is likely to be too late in the project to have any significant influence in making changes to the process.

17.3 Hygienic Equipment Design

Hygienic equipment design has been the subject of considerable discussion over many years by both national and international working groups, usually with participants from equipment suppliers, food manufacturers, and research institutes [4]. This work has resulted in a widely accepted set of principles related to equipment design, which can be characterized under three main headings: materials of construction, equipment geometry and fabrication, and equipment cleanability.

17.3.1 Materials of Construction

A wide range of materials are used in the construction of food process plant, including metals, plastics and composites, elastomers, and lubricants [5]. Whatever the material, when it is in direct contact with food it must meet a number of minimum requirements. It must be:

- inert to the product as well as any chemicals used, such as detergents or biocides, under all in use conditions in particular temperature and concentration;
- corrosion resistant;
- nontoxic;

- nonporous; and
- mechanically stable, smooth, free from surface imperfections and easily cleanable.

Nonproduct contact materials, although posing less of a risk to the product, should also be mechanically stable, smooth, and easily cleanable.

In practice, the main metal of choice in the food industry is stainless steel, with the most widely used grades being the austenitic 18%Cr/10%Ni AISI-304 (or DIN Werkstoff No. 1.4301) and 17%Cr/12%Ni/2.5%Mo AISI-316 (or Werkstoff No. 1.4401-4) forms. Stainless steel has a good overall corrosion resistance, is mechanically robust, as well as being easy to clean and sterilize. It is therefore possible to expect a long service life and hence is considered a cost-effective choice. All such steels are susceptible, however, to localized corrosion, almost always associated with chlorides. Hence in applications where chlorides are present at high levels it may be necessary to use a more corrosion-resistant alloy or alternative materials such as titanium. Aluminum and its alloys can be used for dry materials where only dry cleaning is used but is not corrosion resistant where wet cleaning with alkaline detergents is required and has limited mechanical strength.

Plastics are increasingly used in many applications within food manufacturing, both as components of process equipment such as conveyors and molds, as well as packaging materials in the form of bottles, trays, and pots. Temperature- and chemical-resistance requirements are similar to those for metals. There are a wide range of plastics available and it is essential that only those that comply with the appropriate national standards are used (e.g., the Food and Drug Administration in the United States [6]). The supplier of such materials must provide evidence of such compliance.

Cleanability is also an important feature in selecting the most appropriate material. Some plastics, such as polytetrafluoroethylene (PTFE), have excellent temperature and corrosion resistance but have been considered to be difficult to clean as well as being porous [5]. A number of plastics are easy to clean, nontoxic, and widely used in hygienic applications including:

- polypropylene,
- polyvinyl chloride (unplasticized),
- acetal copolymer,
- polycarbonate,
- high density polyethylene, and
- polyethylene tetraphalate.

Other key issues related to plastics are dimensional stability, tainting, and porosity. Good dimensional stability is important as changes in dimension or shape during operation may allow food access to areas where they will be difficult to clean and pose a contamination risk. It is also essential that the plastic does not taint the product and thus render it unacceptable to the consumer. If the plastic is porous, either due to its normal structure or degradation during operation, it will pose a significant risk as organisms will be able to access the internal structure of the plastic and will be extremely difficult to clean and disinfect. Composites such as glass reinforced plastic can be susceptible to delamination, which can lead to the glass fiber breaking up and creating a physical hazard.

Rubber elastomers are widely used, particularly for sealing applications, a key property being elasticity, the ability to return to its original shape once a stress has been removed. Recommended elastomer choices are [5]:

- nitrile rubber,
- nitrile/butyl rubber (NBR),
- ethylene propylene diene monomer (EPDM),
- silicon rubber, and
- fluoroelastomer (Viton).

The choice will depend on application; for example, silicon and fluoroelastomer can be used for high-temperature environments of up to around 180 °C but incur a higher cost. Other materials may be cheaper but less robust and may be unsuitable for contact with certain food materials; for example, EPDM is unsuitable for applications involving oils and fats.

Routine replacement of elastomers is a normal part of plant maintenance, the frequency being application dependent as well as a function of the physical and chemical stresses imposed on the material. Symptoms of elastomer deterioration are usually observed as a hardening of the material, leading to a loss of elasticity and eventually a loss of sealing function. This deterioration is frequently associated with the cleaning process with the combination of aggressive chemicals and elevated temperatures. Failure to replace elastomers in such circumstances can result in more rapid corrosion due to chemicals being able to access the area between the elastomer and the construction material.

Mechanical stresses can also create hygiene issues when the elastomer deteriorates to such an extent that the material can be abraded and physically break up. This can result in physical contamination of the product with elastomer particles as well as creating a roughened surface with cavities that is much more difficult to clean and susceptible to build-up of microorganisms.

Many items of food processing equipment require lubrication to prevent contact between moving and static surfaces and thus avoid excessive wear and potential overheating. Where contact between lubricants and food cannot be assured, food grade lubricants must be used as defined in a number of documents, for example, NSF (National Sanitation Foundation) H1 registered food grade lubricants may only contain ingredients listed as acceptable on the US Code of Federal Regulations Title 21, Food and Drugs, Section 178.3570. These lubricants are not intended for human consumption but must be acceptable for incidental food contact. The major hazards associated with lubricants are [7]:

- incorrect labeling on packaging,
- leakage from bearings,
- drips from open lubrication points,

- leakage from circulation systems, and
- spillage of oil during maintenance.

17.3.2 Equipment Geometry and Fabrication

Key principles for equipment geometry and fabrication are as follows:

- Equipment should be self emptying and draining.
- Stagnant areas or dead spaces where product may accumulate and be retained for extended periods should be avoided.
- Avoid sharp corners and metal-to-metal contact points.
- Equipment should protect product from external contamination where practical.

Stagnant zones are defined as areas outside the main product flow where material can accumulate and remain for an extended period of time (Figure 17.1). These can have important implications for process hygiene:

1) During production, if the local environmental conditions such as temperature are favorable, microbial growth may occur within the product held up in the dead space. After a period of time the microorganisms may be able to transfer back into the main product flow, for example, by diffusion, and contaminate the product.
2) It may also be difficult to clean and disinfect or sterilize such an area after production has finished due to the lack of fluid movement within the dead space. This would result in a potentially contaminated zone at the beginning of production, which could have a detrimental impact on product quality immediately on start up.
3) If a highly preserved product, such as one with a very high salt content, is retained within the dead space, it may prevent microbiological growth occurring during production. However, during rinsing and cleaning it is possible that the residual product may become diluted to such an extent that the preservative effect of the salt may be insufficient to inhibit growth. This can result in growth followed by recontamination of the product during the subsequent production.

The boundary between the bulk flow and the totally static fluid in Figure 17.1 will not be a clearly defined interface; rather there will be a volume of fluid between

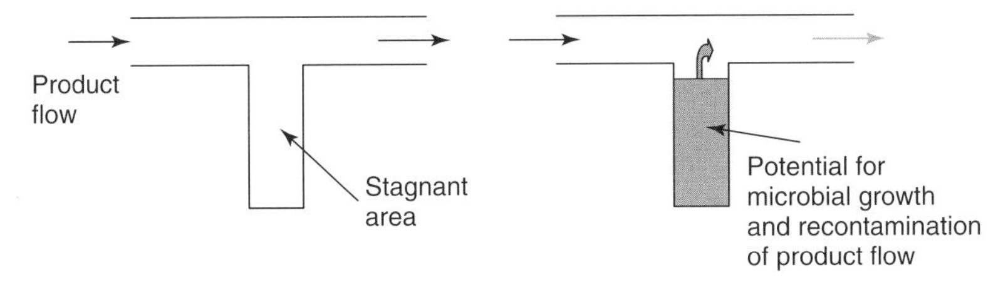

Figure 17.1 A stagnant zone in process pipework.

the two extremes across which the fluid motion steadily reduces in intensity from the maximum at the center of the tube to zero in the dead space. The length of this interface, L_i, will be a complex function of fluid velocity and rheology as well as the geometry of the equipment. The cleaning of a range of pipework geometries was studied by Tamplin [8]. Different orientations of dead space were studied under bulk flow velocities of 0.3–1.5 m s^{-1} and highlighted the importance of fluid velocity but showed that in some cases even the highest flow velocity used was unable to fully clean the dead space. Other workers [9, 10] have carried out both theoretical and experimental studies which have confirmed these earlier findings.

It is a common misconception that a dead space will still be capable of being disinfected or sterilized using either chemical or thermal means, provided additional contact time is used. For chemical biocides it is clearly going to be difficult to ensure that the chemical can be delivered to the whole of the dead zone as this will take place by diffusion, which is likely to take substantially longer than the circulation time of the biocide. In addition there will also be chemical reaction occurring as the biocide reacts with any residual soil in the dead zone, which will slow the process down even further. Heat is often considered to be more effective than chemical for such geometries with penetration into a dead zone due to conduction. However, recent studies [10] have shown that even with turbulent flow within a horizontal pipeline, the temperature within a dead zone orientated vertically downwards will fall rapidly with distance from the pipe centerline. This is to be expected since the natural inclination is for heat to rise and even if heat starts to penetrate the dead space, there will be a cooling effect from the walls of the pipe in contact with the external ambient environment. This will potentially compromise the thermal process due to a failure to reach the required minimum temperature conditions and ensure inactivation of the target microorganisms.

Frequently such problems occur during the physical assembly, installation, and integration of individual equipment components with connecting pipework, examples being T junctions in pipework, incorrect orientation of equipment and incorrect installation of sensors into pipework. Figure 17.2 shows an installation of a positive gear pump with product inlet and outlet mounted horizontally [3]. As installed, the pump cannot be completely drained, a problem that could be avoided by mounting the inlet and outlet vertically. The figure also highlights further poor design features of such an installation, which are often seen in practice. The drain valve on the inlet line to the pump will not help in draining the pump and the valve itself creates a dead zone between the valve and the product pipework. There would also be a safety hazard if the valve were a manual type as an operator opening the valve during cleaning in place (CIP) to drain or clean the line through this valve would risk being splashed with hot CIP solution.

While the practical concern about stagnant areas tends to concentrate on product being retained for extended periods, potentially resulting in microbial growth and recontamination, failure to fully drain nonproduct contact areas can also pose significant risks. Figure 17.3 shows a jacketed aseptic tank, frequently used to provide buffer storage between a product sterilizer and aseptic filler. Prior to production, the tank is sterilized with steam typically at 130 °C, followed by cooling

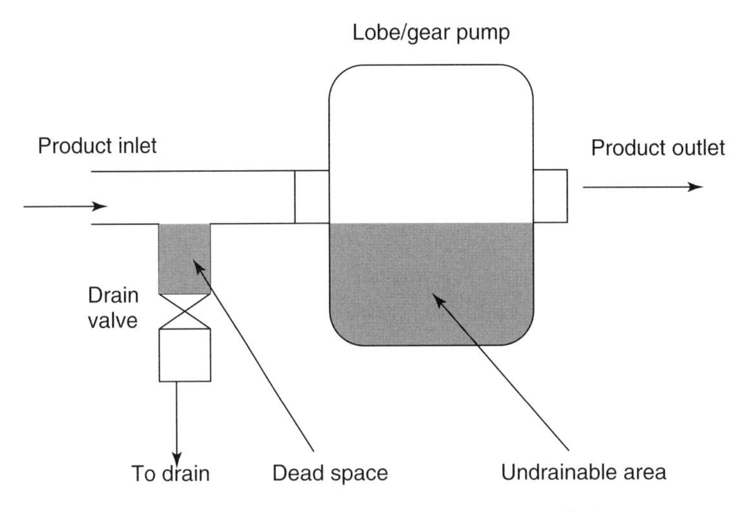

Figure 17.2 Positive pump installation with hygienic design faults [3].

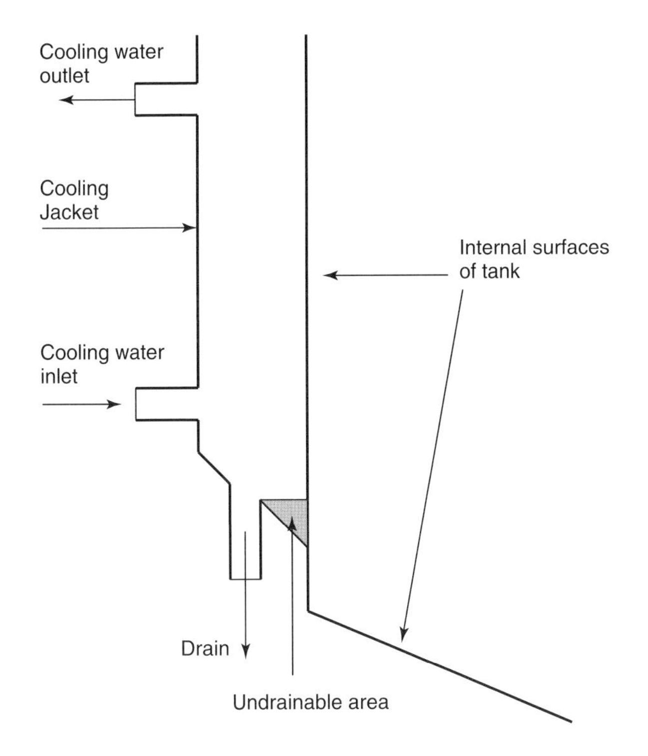

Cooling water inlet and outlet closed during steam sterilization; drain open
Cooling water inlet and outlet open during cooling phase; drain closed

Figure 17.3 Design of a section of a jacketed vessel.

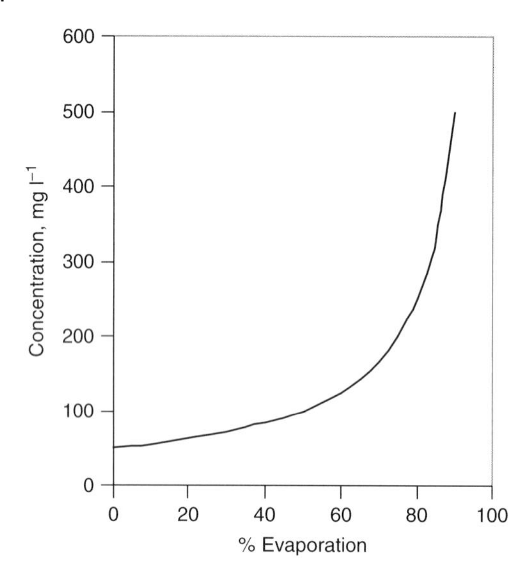

Figure 17.4 Chloride concentration as a function of percentage water evaporation.

to ambient temperature by circulating cooling water through the jacket. In order to minimize the time taken to heat the tank up, the jacket is drained of water before steam sterilization. The cooling water has a chloride content of less than 50 mg l^{-1}, which is desirable to minimize the risk of corrosion of stainless steel.

The design of jacket is such that the water cannot be fully drained, leaving a small volume in the bottom. During steam sterilization, the residual water will be heated up and will rapidly start to boil and evaporate. The chloride concentration in the water will correspondingly increase, as shown in Figure 17.4. This provides a combination of elevated temperature and concentration that will be extremely aggressive to stainless steel even though the contact times will be relatively short, approximately 1 h for each sterilization process. This may lead to corrosion of the tank wall and potential product contamination from the cooling water. The detailed design of the jacket must ensure that it can be fully drained before sterilization and visual inspection carried out, where practical, to confirm the integrity of the surface.

17.3.3 Cleanability

Cleaning of process plant may be required for microbiological or process reasons but it is essential that the equipment can be effectively cleaned and, where necessary, disinfected or sterilized when required. Key requirements for cleanability are that:

- equipment should be easily dismantled for cleaning;
- equipment should be easily visible for inspection; and
- CIP results, if used, should be satisfactory without dismantling.

Cleaning is frequently given insufficient attention during the design process and is considered to be mainly associated with the detergent used. In reality other factors such as flowrate, temperature, and equipment design have a significant impact which may, in some cases, be more significant than the detergent. If the equipment design is such that the cleaning chemicals cannot be effectively delivered to all product contact surfaces, cleaning will be ineffective regardless of the quality of the detergent. In practice it is often the case that with equipment becoming larger and more complex it is impractical to dismantle and visually inspect and thus there is greater reliance on CIP.

17.3.4
Limitations of Hygienic Design Principles

Although the principles of hygienic design are widely accepted as being technically desirable, logical, and simple to understand, they do pose a number of challenges when being applied in commercial practice [3]:

- They are difficult to apply in absolute terms.
- They are defined in isolation from the product being processed.
- They are related to specific items of equipment rather than the process itself.
- They usually refer to new equipment and do not take into account the possible wear and deterioration the material may be subjected to during operation.

For example, materials should ideally have the following desirable attributes:

- smooth, nonporous, and crevice free;
- resistant to the product and any chemical solutions such as detergents or biocides under in use conditions such as temperature and concentration.

The attachment and removal of deposits from surfaces is believed to be a function of the surface roughness such that the rougher the surface the longer the cleaning time. The roughness average, R_a value, is the most commonly used measure to define or specify a surface, being defined as the "average departure of the surface profile from a calculated center line" [11]. However this is an average value measured over a limited area of the material and does not take account of the surface topography, which can vary widely depending on the method of finishing the material (e.g., mechanical polishing, casting, milling, or turning). The surface finish also cannot take account of any interactions between the fluid and material that may promote adhesion, for example, van der Waals and electrostatic forces.

There are accepted techniques for measuring the surface finish of a material. such as surface or laser profilometry and atomic force microscopy (AFM). The recommendation is that for food contact surfaces the roughness should be less than 0.8 μm. While achieving the desired surface finish should be feasible provided good manufacturing and finishing techniques are applied, the surface quality may change significantly during operation due to factors such as wear, abrasion, corrosion, and impact.

Table 17.2 Equipment definitions [4].

Class of equipment	General definition
Aseptic equipment	Hygienic equipment that is in addition, impermeable to microorganisms
Hygienic equipment Class 1	Equipment that can be cleaned in place and freed from relevant microorganisms without dismantling
Hygienic equipment Class 2	Equipment that is cleanable after dismantling and that can be freed from relevant microorganisms by steam sterilization or pasteurization after reassembly

The requirement for materials to be nonporous is to avoid the ingress of fluid into the interior of the material, from where it would be difficult to clean and thus prevent microbial growth and avoid recontamination. While metals may be nonporous, other material such as elastomers and plastics are unlikely to be completely nonporous, particularly if the material degrades during its life.

The thermo-physical properties of the product, in particular its rheology, will have a major impact on its interaction with the equipment geometry. A low-viscosity newtonian fluid such as a soft drink will interact with equipment in a very different way to a viscous non-newtonian material such as ice cream mix or tomato paste. The low-viscosity fluid will achieve far greater degree of fluid transfer between bulk and stagnant zone than the viscous fluid. In addition, the low-viscosity fluid will be capable of being pumped at a significantly higher flow rate than the viscous fluid, which will again enhance this interaction.

In many practical applications, different items of equipment may require different levels of hygienic design depending on the risk to the final product. In general, the closer the equipment is to the end of the line, the more likely it is that the hygiene requirements will increase. For example, the hygienic design of a mixer for raw meat need not achieve the same standards as a slicer for cooked meats [12]. With this in mind the European Hygienic Engineering and Design Group (EHEDG) have developed definitions associated with different hygiene levels, ranging from unhygienic to fully aseptic (Table 17.2). It may, however, still be possible to produce microbiologically acceptable product using unhygienic equipment but there will almost certainly be significant risk implications in such an approach as well as increased cost. Use of unhygienic equipment is likely to result in [3]:

- shorter run lengths between cleaning;
- longer cleaning times;
- more aggressive cleaning regimes;
- a less consistent and robust process;
- increased product and equipment testing, thus moving from quality assurance toward quality control.

17.4 Process Design

The design of process lines often has to reflect a balance between costs, both capital and operational, as well as product quality. A typical example is the design of thermal processing equipment such as pasteurizers and sterilizers. In such plant the product has to be heated to a specified process temperature and held for a defined minimum residence time in order to achieve the desired level of microbial kill, followed by cooling to the final temperature required.

If a continuous sterilization process involves heating product from 5 to 145 °C, a certain overall energy input will be required depending on the capacity and product properties. The simplest way of reducing the energy required is to use the product at 145 °C to preheat the incoming product to an intermediate temperature, following which the additional energy required to heat to 145 °C would be reduced. The higher the preheat temperature achieved, the less energy is required; for example, if the product can be preheated to 126 °C, the energy required and hence cost would be reduced by 90%. This saving would, however, come at the expense of requiring a larger heat exchanger to achieve the overall heating process, resulting in a higher capital cost. The larger heat exchanger would result in the product having an increased residence time within the unit and this may have a negative effect on product quality.

As product flows through a process line, its interaction with the equipment will have an impact on the residence time within the line and hence potentially the quality attributes that the final product is expected to deliver. In an ideal world the flow of material would be plug flow, whereby every element of the product has the same residence time and is thus subjected to identical process conditions. In practice, particularly with liquid products, there will be a distribution of residence times; for example, in a pipe, the fluid at the center will be moving more rapidly than fluid near the wall and will consequently have a lower residence time. The fluid at the wall will have a significantly higher residence time and can in some cases give rise to concerns.

Consider the case of a simple tube in a tube heat exchanger shown in Figure 17.5. A chilled product at 4 °C is to be heated to 30 °C prior to further processing using warm water at 35 °C. The mean residence time of product within the exchanger is 2 min. Due to the viscosity of the product, the flow is laminar and hence there will be a temperature gradient between the wall and the center of the tube and heat transfer will be primarily via conduction. As the fluid moves through the exchanger, this gradient will become smaller until the center reaches the desired temperature. There will be a range of product residence times within the system with the maximum at the wall and it is possible to predict the fluid velocity and hence residence time close to the wall. This indicates that fluid close to the wall can theoretically have a residence time of hours rather than minutes. The temperature at or close to the wall will tend toward that of the water temperature, 35 °C, and whereas a residence time of a few minutes at 30 °C, would not be a cause for concern, this would not be the case if this is potentially increased to several hours.

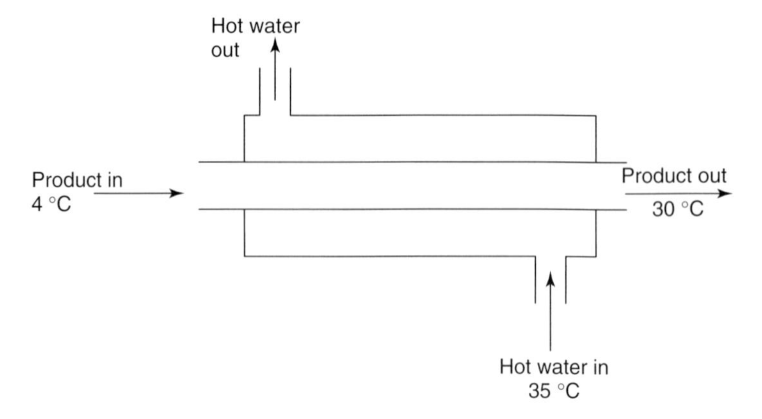

Figure 17.5 Tube in tube heat exchanger.

One solution would be to use an alternative heat exchanger geometry but in practice it was simpler to modify the process design in order to manage the risk.

The solution proposed was to reduce the size of the heat exchanger and to use a higher water temperature and hence driving force. If a water temperature of 60–65 °C was to be used, the temperature at or near the wall would be close to these temperatures, which would not pose a microbiological risk. The portion of fluid at the temperatures of concern, such as 30–35 °C, would be further away from the wall and thus have a lower residence time within the system.

17.4.1 Case Study for Frozen Food

A frozen food manufacturer identified an intermittent microbiological contamination problem with a production line producing a shepherd's pie product consisting of a meat base with a topping of mashed potato. The main steps in the process are shown in Figure 17.6, with the meat base being cooked and cooled in a jacketed batch vessel, which had the capability of using steam in the jacket and direct steam injection into the vessel for heating. For cooling, chilled water could be fed to the jacket and there was a vacuum system to provide evaporative cooling within the vessel itself. The tank included a number of agitators to mix the product and enhance heat transfer. The product was cooled in the vessel until the temperature was reduced below 10 °C, at which point it was discharged into an open batch tank of approximately 1 m^3 capacity. The tank was of cylindrical cross-section with a hemispherical base and after filling with product was transferred into a chill store maintained at an ambient temperature of 5 °C. The tank could be held for up to 48 h before being transferred to the dosing unit where a fixed volume was dosed into a foil tray, followed by the potato topping. The final product was then packed into a cardboard carton and then frozen prior to storage and distribution.

The factory personnel believed that the problem was associated with the hygienic design of the dosing unit, which was very poor with stagnant areas present and

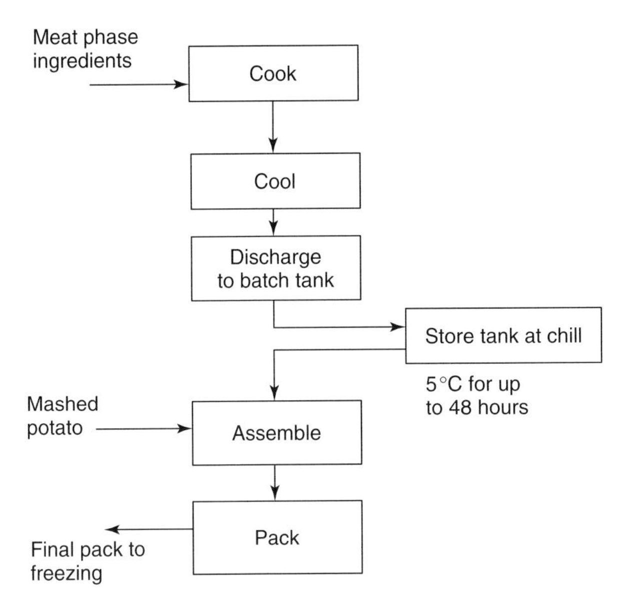

Figure 17.6 Process flow diagram for frozen foods line.

leaking seals resulting in a build-up of product. The cleaning of the piston type doser was manual, usually carried out during the night shift and therefore relied on the operator to ensure effective cleaning. The decision was therefore taken by the factory to reduce the temperature of the chill store to 2–3 °C, in the belief that this would further cool the product and hence the dosing unit. This reduced temperature would help to slow down the growth rate of any product retained within the stagnant areas. However, this was found to have no positive effect on the microbiological quality of the final product and intermittent contamination was still experienced.

An assessment was therefore carried out on the complete process line, starting at the cooker-cooler. The data from the cooking process was found to meet the specified temperature and residence time, which in itself was considered sufficient to achieve the required degree of microbial inactivation. However, it was noted that the cooling process was sometimes inconsistent and the temperature of the product being discharged to the batch vessel was occasionally as high as 15 °C. Although the data from microbiological sampling indicated no problems at this stage, this increased temperature was considered potentially significant and indicated the need to evaluate the batch cooling process in more detail.

The cooled product in the batch tank was very viscous and hence heat transfer could be considered to be solely by conduction. The chilled air cooling the external surfaces of the vessel was delivered to the chill store via fans in the ceiling. The room had a high ceiling, ~8 m high, and the flow of air at the lower part of the room was limited. In addition the batch tanks were often stored close to each other, which would further restrict air circulation. The critical area of the vessel from a microbiological perspective was the thermal center, which would be the slowest element of the product to cool. The heat transfer mechanisms involved in the cooling process were:

- conduction through product to the inner surface of the tank;
- conduction through the metal wall of the tank; and
- convection from the air flow at the outer surface, which could either be natural or forced convection.

Prior to carrying out experimental work, simple heat transfer models were developed to enable the temperature profiles at different parts of the vessel to be predicted. These predictions could then be compared against the experimental data. Since the geometry of the vessel was a cylinder with a hemispherical bottom section, existing solutions to the standard heat transfer equations based on the shape, that is, cylinder or sphere could be used to provide a relatively rapid solution [13]. Comparison of the solutions for a finite cylinder and a sphere indicated that the difference in the cooling temperature profiles was sufficiently small for both to be usable. The thermal center of the cylinder was defined as the mid point along the length of the cylinder and the center of the cylinder. The cooling profile for the surface or close to the surface of the product exposed to the chilled air could be calculated from the solutions for an infinite slab.

If the temperature profile could be predicted or measured directly, it would be possible to couple this with a microbiological growth model and thus assess the potential growth occurring as a function of time. The microbiological model was based on the target organism of concern and incorporated both lag and growth phases.

The heat transfer model was relatively simple to set up; the thermophysical properties of the product were known and the main unknown was the heat transfer coefficient between the air and the external surface of the vessel. Literature suggests coefficients of between 2 and $40\,\mathrm{W\,m^{-2}\,K^{-1}}$ for natural and forced air convection [14].

Initial predictions of the temperature profile assumed a coefficient of $5\,\mathrm{W\,m^{-2}\,K^{-1}}$ on the basis that there was a flow of air within the chill store but this was relatively limited around the tank. The predicted temperature profiles are shown in Figure 17.7 for an initial product temperature within the tank of 15 °C. These highlight the major difference between the two profiles with the product close to the surface cooling down relatively rapidly to below 10 °C. The thermal center, however, has an extensive time lag of nearly 20 h before any change in temperature is noted and even after 48 h the predicted temperature had only fallen by around 1.5 °C. These predictions were then confirmed by experimental measurements taken over a 48 h period, which showed good agreement between predicted and measured temperatures. The effect of these cooling profiles on the predicted microbiological growth is shown in Figure 17.8, indicating that at the surface the relatively rapid reduction in temperature has the effect of both extending the lag time as well as reducing the growth rate. At the center, the slow cooling results in a significantly shorter lag time as well as more rapid growth

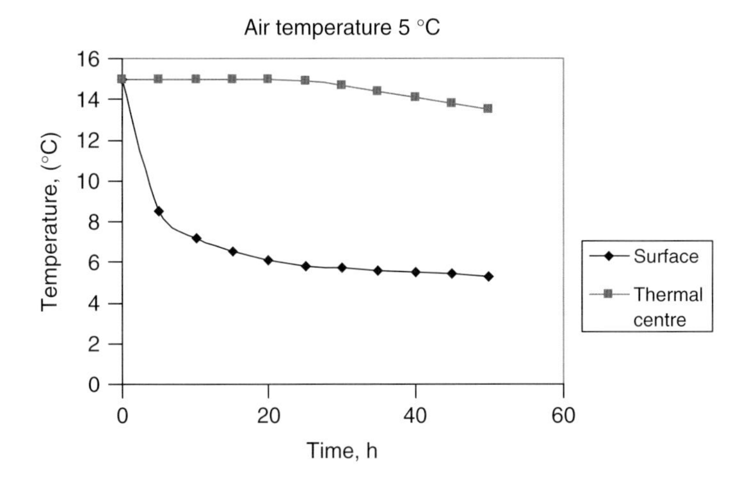

Figure 17.7 Predicted temperature profiles within batch tank.

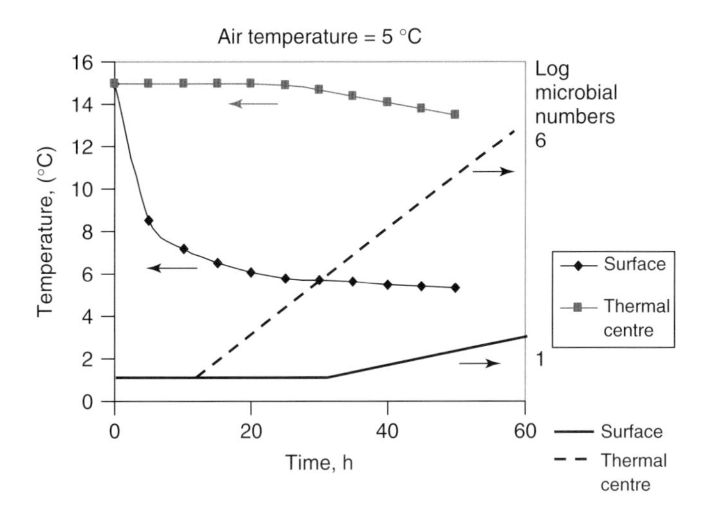

Figure 17.8 Predicted microbial growth within batch tank can you distinguish symbols/lines for temperature from those for micro growth? I know it is obvious but it is not clear!

rates, such that the microbial numbers at the center could in theory be almost 5 logs higher after 48 h.

It was interesting to note the apparent inconsistency between these findings and those from the factory, where it was stated that samples taken from the batch tank had never shown any indications of microbial growth or excessively slow cooling. On further investigation it was found that the sampling procedure involved taking a sample only from the top of the tank and measuring the temperature in the same location. These are areas which as can be seen from Figure 17.7, cool rapidly and are not a concern with regard to microbiological growth. Hence the sampling regime was insufficient to identify the problem, leading to the initial conclusion by the factory that the problem lay in the poor hygienic design of the dosing system.

The value of having even a simple model of the heat transfer process is that it enables changes in process conditions to be rapidly evaluated. Hence the impact of reducing the chill store temperature to 2 °C could be predicted as well as the effect of forced convection achieving a higher heat transfer coefficient of 40 W m^{-2} K^{-1}. Neither of these changes had any significant impact, the difference in center temperature after 48 h only reducing by an additional 0.5 °C. The reason for this was that the controlling parameter in the cooling process was the tank diameter as this provided the main resistance to heat transfer and only a reduction in this could enhance the rate of cooling.

The simplest way of overcoming the problem was to ensure that the cooled product could not be discharged from the agitated jacketed tank until it was below 10 °C, as once it was in the batch tank effective cooling could not take place. The analysis also highlighted that these potential problems could and should have been identified at the design stage. A worst case scenario could have been evaluated with the microbial growth as a function of product temperature calculated from the microbiological growth model, assuming no cooling took place in the tank and the temperature remained constant throughout. This would very rapidly identify the critical temperature below which the product could be held for 48 h without creating growth issues.

The analysis also raised a number of other issues:

- Since the cooling of the product in the tank was negligible, there was no benefit in running the chill store at 5 °C. A temperature of 10 °C would have been sufficient to prevent any temperature rise at the product surface. The cost of running a chill store at 5 °C rather than 10 °C is significant in terms of energy consumption.
- The chill store was a large volume all of which had to be cooled to the required temperature. The actual tanks only took up a very small fraction of this volume and hence significant reductions could be made by modifying the design to have a much smaller chill store and aiming to achieve localized cooling.
- The requirement for a 48 h buffer storage capability was challenged and found to be based more on convenience than necessity and could in principle be reduced to a maximum of 24 h. The buffer capacity was required due to the difference in production capacities of the processing and filling/packaging parts of the line such that there was a build-up of product between the two stages.

This study demonstrated a number of limitations in the design of the line and the way the initial investigation was carried out:

- The hygienic design of the dosing equipment was not the root cause of the problems.
- There was a lack of understanding of the process implications both at the line design stage and commercial production.

- Greater emphasis on the process design could have resulted in a simpler, more energy efficient line with improved quality assurance.

17.5 Process Operation and Control

Process lines are designed to deliver a defined set of conditions that will consistently produce a product or range of products of the desired quality at a commercially acceptable cost. There will usually be a set of individual sequences that will be carried out to move from start up of the line, which may include disinfection or sterilization, through to production, cleaning, and shutdown, and each of the individual sequences may contain a number of steps.

Automation is increasingly used within the industry, although manual operation is still preferred in situations where short production runs and rapid changeover are required. Automation has two main functions:

- control of the routing of product through the plant during the various stages of the production process;
- control of specific process parameters within the system such as temperature, flow, and pressure.

The control of the process plant during start-up, production, shutdown, and cleaning is critical to the safety of the final product. There can be a risk that the main emphasis at the design stage is on the production phase, and key areas such as cleaning and plant start-up are not given the attention they require. The whole production cycle including start-up, shutdown, and cleaning must be considered as a complete process, as problems can often be traced back to the changeover between sequences.

Measurement is a critical element of operation and control, as without the ability to measure process parameters reliably, it is not possible to monitor, control, validate, or optimize the process. The availability of reliable sensors provides the opportunity to move from a situation where the process data are limited, retrospective, and often of doubtful reliability, to one where the data are real and immediate. This can be used proactively to highlight potential concerns as to whether the process is under control. Real time data may not always correspond with what would be expected but further investigation to resolve this disparity can lead to an improved understanding of the design and operation of the line.

Peracetic acid (PAA) is now widely used as a biocide in the food industry for both disinfection and sterilization applications and its performance is a function of the concentration, temperature, and contact time. Prior to the development of PAA sensors, the concentration could only be determined by titration and results were therefore retrospective and often of limited value.

The application of an early design of PAA sensor in a factory environment was described by Hasting *et al.* [15]. The sensor was installed in a PAA recovery system and initially used purely for monitoring rather than control. The normal operation of the PAA system in the factory was that at the beginning of each day, a sample was titrated and concentrated PAA added to the recovery tank until the target value of 300 $mg\,l^{-1}$ was reached. There was no assessment made of the implications of how much PAA had been used during the previous day on the likely microbiological performance. The real time monitoring using the sensor showed that for a considerable part of the day the concentration was lower than the 150 $mg\,l^{-1}$ minimum required for effective performance. Further investigation showed that due to the way the recovery system operated, there was both dilution and losses of PAA every time it was used to disinfect a section of process plant. The fall-off in concentration during the day was therefore what would be expected and there was nothing fundamentally wrong with the equipment, but rather with the control procedures. A simple control system was subsequently installed to monitor and control concentration in real time, resulting in consistent control of PAA concentration and allowing the set point to be reduced to 250 $mg\,l^{-1}$. This also enabled the factory to eliminate a number of microbiological problems that had been encountered previously but never fully understood.

The continuing pressure on the industry to reduce costs has resulted in an increase in subcontracting analyzes to external laboratories and a corresponding reduction of the in house laboratory capability. Some routine analyses have therefore been transferred to the line operator with either a simple laboratory station in the production area or the operator carrying out the analysis in the main laboratory. Any such approach requires methods to be as simple, robust, and reliable as possible so that the data obtained are meaningful and thus can genuinely add value. Unreliable data are of less value than no data at all.

Food safety assurance has been significantly enhanced over the last decades by the widespread use of the HACCP system to identify the critical control points (CCPs), within a process and determine how these can be managed. Many of the CCPs involve real time measurements of process parameters such as temperature, flowrate, and pressure, which enable the control system to take immediate action if the critical limits are exceeded. The measurement sensors in such cases are therefore important to ensure that the information provided is timely, reliable and of the required accuracy. However, it is apparent that such instruments are often taken for granted and insufficient attention is paid to installation and the need for regular calibration. On a cost-for-cost basis, sensors may have more impact on product safety and quality than most other items of equipment. There is often an expectation with sensors such as temperature and flowrate that the output value displayed either on the sensor itself or the control panel is accurate to the number of decimal points displayed.

Critical instruments, such as those associated with CCPs, must be regularly calibrated to confirm that they are still operating within the stated instrument specifications. Typical calibration frequencies will vary widely but the usual action is to adjust the instrument if required, and then recheck the calibration at the same frequency. It is considered, however, that this is an insufficiently robust approach. A more rigorous approach would be to determine what constitutes an

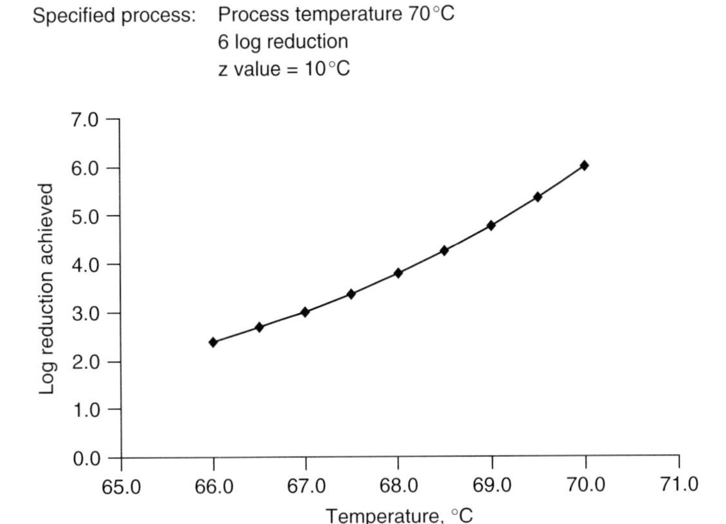

Figure 17.9 Effect of sensor inaccuracies on a pasteurization process. Specified process: Process temperature 70 °C, 6 log reduction, z value = 10 °C.

acceptable difference between the measured and actual temperature, for example, a temperature difference of 1 °C. If the calibration shows a difference greater than this, there should be a clear policy as to the action required. It may be sufficient to adjust the instrument but require the calibration to be carried out more frequently. However, where the differences are above a certain level, consideration should be given to replacing the instrument. Temperature is such a critical parameter in processes such as pasteurization and sterilization that even small differences can have a significant impact on the level of microbial inactivation.

Figure 17.9 shows the impact of sensor inaccuracies on the log reduction of target organisms in a pasteurization process. The standard process at 70 °C is equivalent to delivering a 6 log reduction of the target organism. However, if the actual temperature is lower than this due to an inaccurate sensor the effect on the log reduction achieved is significant, with even an inaccuracy of 1 °C, resulting in a log reduction of only 4.8, a 20% reduction. A 2 °C inaccuracy would result in a log reduction of 3.8, a 37% fall.

Cleaning, equipment disinfection, and sterilization are increasingly being identified as essential prerequisites that must be in place before a HACCP study can be undertaken. Failure to achieve the required standards may have adverse consequences both operationally and on product quality. An inability to clean plant effectively can result in fouling building up more rapidly, while failure to adequately disinfect or sterilize may result in microbiological contamination of the product.

The view is sometimes taken that even if the equipment is not fully cleaned, the subsequent disinfection or sterilization sequence will ensure that the product can be safely produced. However, this is an approach that is fundamentally unsound.

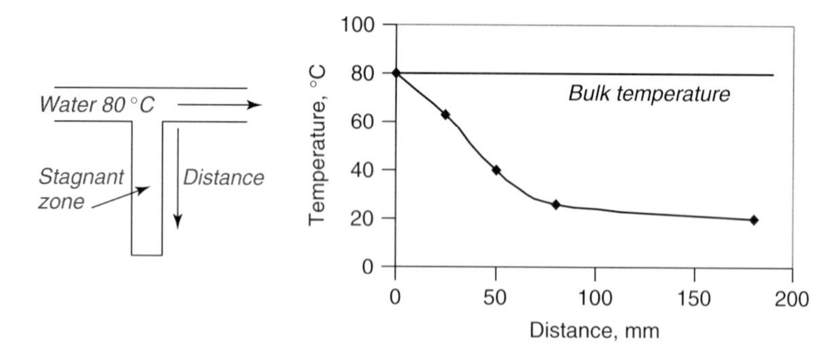

Figure 17.10 Thermal disinfection of a stagnant zone.

Disinfection and sterilization generally use either a thermal or chemical process to achieve the required microbial inactivation. Heat is often preferred on the basis that it is believed to be more effective than chemicals in penetrating into difficult to access areas. In the stagnant zone shown in Figure 17.1, the vertical orientation means that the hot fluid in the bulk flow has to interact sufficiently with the geometry so there is transfer of hot fluid throughout the zone or sufficient thermal diffusion to heat the stagnant fluid. Figure 17.10 shows that the fluid temperature within the zone falls rapidly with distance away from the main bulk flow, due to the cooling effect from the wall and the natural tendency of heat to rise [10]. At a point 25 mm into the stagnant zone the fluid temperature has already fallen below 60 °C, compared to a bulk temperature of 80 °C.

A similar issue arises when considering chemical disinfection, and in this case the concern occurs if the soil has not been effectively cleaned from the surface and a biocide is then used for disinfection. Figure 17.11 shows a 1 mm thick biofilm on a surface, which has not been removed by the cleaning process. During disinfection the chemical biocide is circulated past the biofilm at a constant concentration of 200 mg l^{-1}. In terms of disinfecting the plant, the area of greatest concern is the interface between the residual biofilm and the surface as it will take a finite time for

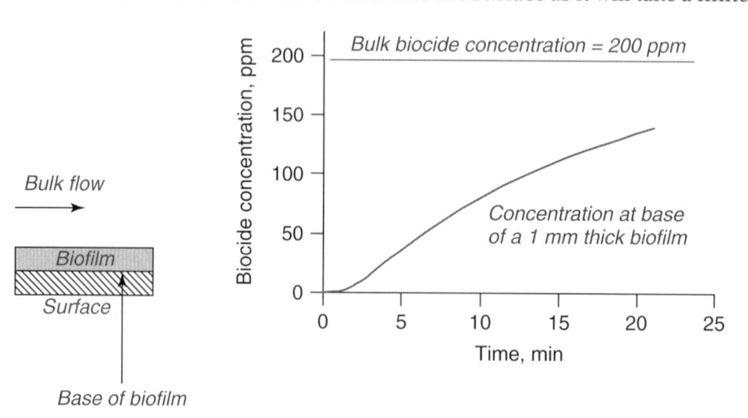

Figure 17.11 Chemical disinfection of a residual biofilm after cleaning.

the biocide to diffuse through the biofilm and act on the surface. Figure 17.11 shows the predicted concentration at the interface as a function of time and assuming diffusion as the mass transfer mechanism. This indicates that even after a typical disinfection process of 10 min, the interfacial concentration is less than 100 mg l^{-1}. In reality the concentration would be expected to be even lower as no account has been taken of the reaction of the biocide species with the biofilm as it diffuses toward the surface and hence the true process is a combination of mass transfer and chemical reaction. In both these examples, in line measurements within the bulk flow would not indicate any potential problem as the required conditions would appear to be consistently achieved.

The ability of modern supervisory control and data acquisition (SCADA) systems to monitor and record large quantities of data provides an opportunity to obtain valuable information on the consistency of the production and cleaning processes. However, in order to achieve this it is necessary to target the key parameters influencing performance. A practical example of this is the fouling of a plate heat exchanger being used for the pasteurization of a range of food materials. Typically, fouling of heat exchangers can be monitored by temperature difference and pressure drop, both of which tend to increase with amount of fouling on the plate surfaces. In the particular factory application, there was insufficient temperature instrumentation to determine the temperature differences and unwillingness to install what was required. The system was designed to provide a constant feed pressure and flowrate and hence the pressure drop could be monitored. Since the exchanger operated at a constant feed pressure, the reduction in outlet pressure could also be used as a single direct measure of the fouling process. Classical fouling studies often refer to the existence of two stages in the fouling process; an initial induction or conditioning phase where there is no apparent change in process conditions, followed by a fouling period where conditions do change [16]. In this application the fouling period would be indicated as a reduction in outlet pressure (Figure 17.12). During the trials, the fouling of a single product was followed, this being one where there were inconsistent and relatively short run lengths before the exchanger had to be cleaned. It was found that the outlet pressure followed the classical fouling example with an induction stage of constant pressure followed by a fouling period where the outlet pressure fell. It was interesting to note that the fouling period followed a straight line and it was therefore possible to characterize each production run in terms of the length of the induction period and the gradient of the fouling period (Table 17.3).

The exact point at which the induction stage changed into the fouling period was subject to some uncertainty but the gradients were considered to be remarkably consistent given the limited factory instrumentation available. This work highlighted that the inconsistent run lengths were almost exclusively due to variations in the length of the induction period of up to 8 h. It was also considered that the variations in induction period were likely to be due to fouling residues resulting from the cleaning process. The immediate action implemented, therefore, was to improve the efficiency and consistency of the cleaning process to maximize the induction period and enable longer runs to be achieved.

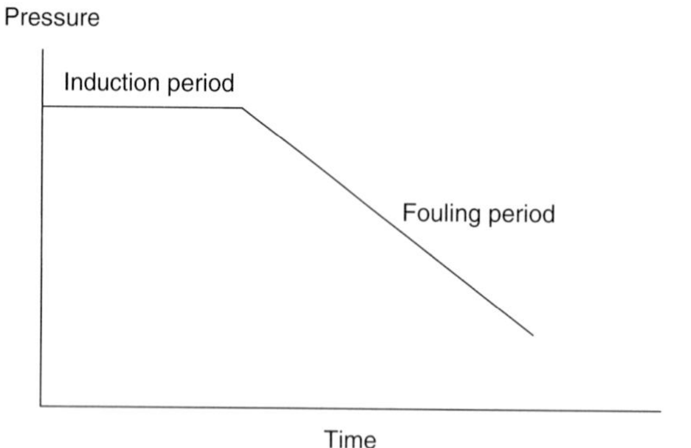

Figure 17.12 Typical fouling curve based on outlet pressure from heat exchanger.

Table 17.3 Fouling characteristics of a plate heat exchanger.

Trial	Induction period (h)	Fouling period gradient (bar per h)	Run length before cleaning required (h)
1	2.0	0.21	14.0
2	3.0	0.20	15.8
3	2.5	0.19	19.8
4	8.0	0.22	20.3
5	0	0.20	10.7

17.6 Future Trends

Consumer, economic, and legislative pressures will continue to place increasing demands on the industry to maximize production efficiency, flexibility, and cost effectiveness. Product safety will still remain an absolute requirement and improved levels of hygiene assurance will be expected by the food retailers and their customers. Environmental pressures will continue to increase and greater emphasis will be required on minimizing the use of resources and more effective conversion of raw materials into final product by, for example, reducing product losses.

Increased flexibility with an ever wider range of products processed on each process line is likely to result in shorter runs and more frequent cleaning. This will have an increasing impact on production downtime and hence operational efficiency. Further developments in modifying equipment surfaces to minimize fouling or enhance cleaning performance will continue, particularly for equipment such as heat exchangers, where the elevated temperatures will often cause fouling that will restrict the production times that can be achieved.

Greater use should be made of the process data available from the plant to confirm that the process is genuinely under control, as well as trend analysis to proactively identify changes to process operation that indicate the system performance is not optimal. Computer-based statistical models will become available to provide an improved predictive capability to the user. For example, when a process line is handling a wide range of products it is unlikely that a single cleaning process will be optimal in all cases. A predictive model built up from process data from a number of previous runs would enable the optimum cleaning conditions for each product to be determined, taking a number of factors into account, such as the length of production run before cleaning. What constitutes an optimum clean may not necessarily always be the same and may depend on a number of factors, such as minimizing downtime, such that a more aggressive, though more costly, cleaning could, taken overall, be the most cost-effective approach.

The continuing cost pressures are likely to result in an ever increasing loss of technical experience and capability within commercial operations. It is essential that greater effort is made to ensure that as much of the knowledge and expertise of experienced people within the factory is captured and not lost when they leave. Such expertise is a precious resource, often having been gained when things go wrong and need to be resolved, and usually incurring considerable cost.

17.7 Conclusions

The hygienic design and operation of process plant is central to the production and distribution of safe food. From an engineering perspective, it is essential to appreciate that the hygienic design of equipment is only one element in the engineering chain and other key factors must be taken into account. It is essential that the hygienic processing issues are considered as early as possible, whether considering a new investment, a new product formulation or a modification to an existing process or its operation. One of the most underutilized capabilities of modern automated plant is the process data that can provide valuable information in such areas as quality of control or trend analysis to identify potential problems before they arise.

References

1. Chaddock, D.H. (1975) Thought structure or what makes a designer tick, Paper read to South Wales Branch, Institution of Mechanical Engineers, 27 February 1975.
2. Sinnott, R.K. (2005) *Chemical Engineering Design*, Elsevier, pp. 481–482.
3. Hasting, A.P.M. (2008) Designing for cleanability, in *Cleaning-in-Place: Dairy, Food and Beverage Operations* (ed. A.Y. Tamine), Blackwell Publishing, Oxford, pp. 81–106.
4. Anonymous (1993) Hygienic equipment design criteria. *Trends Food Sci. Technol.*, 4, 225–229.
5. Lewan, M. (2003) Equipment construction materials and lubricants, in *Hygiene in Food Processing* (eds H.L.M. Lelieveld, M.A. Mostert, J. Holah, and B. White), Woodhead Publishing, Cambridge, pp. 167–178.
6. Anonymous (1996) *Title 21 The Code of Federal Regulations Parts 170-199*, US

Government Printing Office, Washington, DC.
7. Moon, M. (2007) How clean are your lubricants. *Trends Food Sci. Technol.*, **18**, S74–S88.
8. Tamplin, T.C. (1990) Designing for cleanability, in *CIP: Cleaning in Place* (ed. AJ.D. Romney), Society of Dairy Technology, Huntingdon, pp. 41–103.
9. Grasshof, A. (1980) Hygienic design: the basis for computer controlled automation. Proceedings of Conference Food Engineering in a Computer Climate, St John's College, Cambridge, 30 March–1 April 1992, pp. 89–109.
10. Asteriadou, K., Hasting, A.P.M., Bird, M.R., and Melrose, J. (2006) Computational fluid dynamics for the prediction of temperature profiles and hygienic design in the food industry. *Food Bioprod. Process.*, **84** (C4), 157–163.
11. Verran, J. (2005) Testing surface cleanability in food processing, in *Handbook of Hygiene Control in the Food Industry* (eds H.L.M. Lelieveld, M.A. Mostert, and J. Holah), Woodhead Publishing, Abingdon, pp. 556–572.
12. Timperley, D.A. and Timperley, A.W. (1993) Hygienic design of meat slicing machines, Technical Memorandum No. 679, Campden and Chorleywood Food Research Association, Chipping Campden.
13. Earle R.L. (1983) *Unit Operations in Ffood Processing*, Pergamon Press, Oxford, pp. 24–38.
14. Perry, J.H. (1963) *Chemical Engineers Handbook*, 4th edn, McGraw Hill, New York.
15. Hasting, A.P.M., Burns, I.W., de Goederen, G., and Luijendijk, P. (1992) The monitoring and control of disinfection processes, Food Engineering in a Computer Climate, Cambridge, St John's College, 30 March–1 April 1992, pp. 111–115.
16. Lewis, M.J. and Heppell, N.J. (2000) *Continuous Thermal Processing of Foods*, Aspen Publishers Inc., New York.

18
Process Control in Food Processing

Keshavan Niranjan, Araya Ahromrit, and Ashok S. Khare

18.1
Introduction

Process control is an integral part of modern processing industries, and the food processing industry is no exception. The fundamental justification for adopting process control is to improve the economics of the process by achieving, among others, the following objectives: (i) reduced variation in the product quality, achieving more consistent production and maximizing yield, (ii) ensuring process and product safety, (iii) reduced manpower and enhanced operator productivity, (iv) reduced waste, and (v) optimized energy efficiency [1, 2].

Processes are operated under either steady state (i.e., process conditions do not change) or unsteady state conditions, and conditions vary depending on time. The latter occurs in most real situations and requires control action in order to keep the product within specifications. Although there are many types of control actions and many different reasons for controlling a process, the following two steps form the basis of any control action:

1) accurate measurement of process parameters;
2) manipulation of one or more process parameters using control systems in order to alter or correct the process behavior.

It is essential to note that a well-designed process ought to be easy to control. More importantly, it is best to consider the controllability of a process at the very outset, rather than attempt to design a control system after the process plant has been developed [1].

18.2
Measurement of Process Parameters

As mentioned earlier, accurate measurement of the process parameters is absolutely critical for controlling any process. There are three main classes of sensors used for the measurement of key processing parameters, such as temperature, pressure,

Food Processing Handbook, Second Edition. Edited by James G. Brennan and Alistair S. Grandison.

mass, material level in containers, flowrate, density, viscosity, moisture, fat content, protein content, pH, size, color, turbidity, and so on [3]:

- **Penetrating sensors:** These sensors penetrate inside the processing equipment and come into contact with the material being processed.
- **Sampling sensors:** These sensors operate on samples that are continuously withdrawn from the processing equipment.
- **Nonpenetrating sensors:** These sensors do not penetrate into the processing equipment and, as a consequence, do not come into contact with the materials being processed.

Sensors can also be characterized in relation to their application for process control as follows [3]:

- **Inline sensors:** These form an integral part of the processing equipment, and the values measured by them are used directly for process control.
- **Online sensors:** These too form an integral part of the processing equipment, but the measured values can only be used for process control after an operator has entered these values into the control system.
- **Offline sensors:** These sensors are not part of the processing equipment, nor can the measured values be used directly for process control. An operator has to measure the variable and enter the values into a control system to achieve process control.

Regardless of the type of the sensor selected, the following basic characteristics have to be evaluated before using it for measurement and control: (i) response time, gain, sensitivity, ease, and speed of calibration; (ii) accuracy, stability, and reliability; (iii) material of construction and robustness; and (iv) availability, purchase cost, and ease of maintenance. Detailed information on sensors, instrumentation, and automatic control for the food industry can be obtained from Refs [4, 5].

18.3 Control Systems

Control systems can be of two types: manual control and automatic control.

18.3.1 Manual Control

In manual control, an operator periodically reads the process parameter which is required to be controlled and, when its value changes from the set value, initiates the control action necessary to drive the parameter toward the set value. Figure 18.1 shows a simple example of manual control where a steam valve is adjusted to regulate the temperature of water flowing through the pipe [3].

An operator is constantly monitoring the temperature in the pipe. As the temperature changes from the set point on the thermometer, the operator adjusts

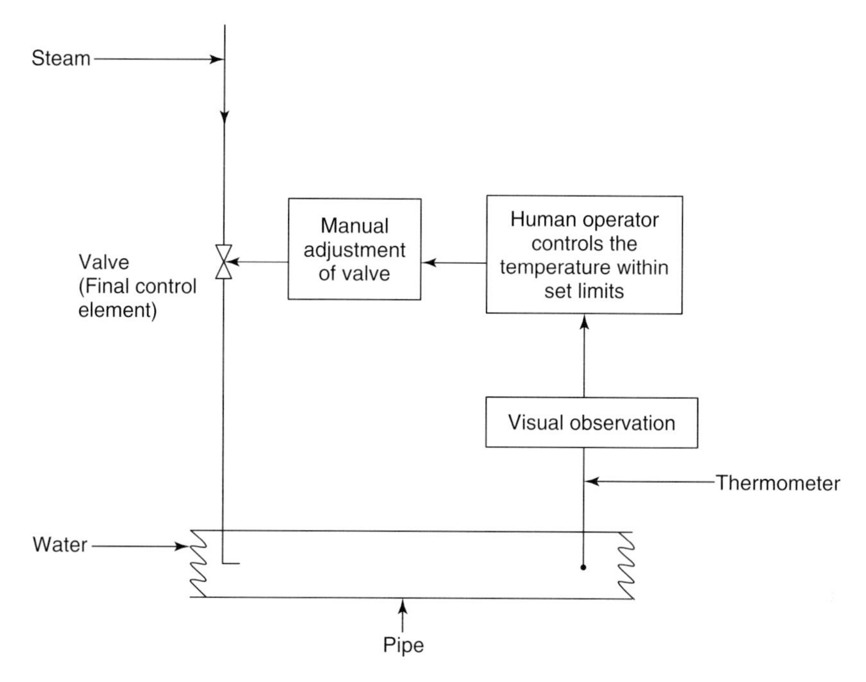

Figure 18.1 A simple example of manual control. From Ref. [3].

the steam valve, that is, either increases or decreases its flow in order to get the temperature in the pipe back to the set point. Further action may be required if the temperature does not return to the set point within a reasonable time.

It is clear that the success of manual control operation depends on the skill of individual operators in knowing when and how much adjustment to make. Therefore, manual control may be used in those applications where changes in the manipulated parameter cause the process to change slowly and by a small amount. This is possible in plants where there are few processing steps with infrequent process upsets and the operator has sufficient time to correct before the process parameter overshoots acceptable tolerance. Otherwise, this approach can prove to be very costly in terms of labor, product inconsistencies, and product loss.

18.3.2 Automatic Control

In automatic control, the process parameters measured by various sensors and instrumentation may be controlled by using control loops. A typical control loop consists of three basic components [3]:

- **Sensor:** The sensor senses or measures process parameters and generates a measurement signal acceptable to the controller.

- **Controller:** The controller compares the measurement signal with the set value and produces a control signal to counteract any difference between the two signals.
- **Final control element:** The final control element receives the control signal produced by the controller and adjusts or alters the process by bringing the measured process property to return to the set point, for example, liquid flow can be controlled by changing the valve setting or the pump speed.

An automatic control system can be classified into four main types:

- on/off (two position) controller;
- proportional controller (P-controller);
- proportional integral (PI) controller;
- proportional integral derivative (PID) controller.

18.3.2.1 On/Off (Two Position) Controller

This is the simplest automatic controller for which the final control element (e.g., a valve) is either completely open or at maximum, or completely closed or at minimum. There are no intermediate values or positions for the final control element.

Thus, final control elements often experience significant wear, as they are continually and rapidly switched from open to closed positions and back again. To protect the final control element from such wear, on/off controllers are provided with a *dead band*. The dead band is a zone bounded by an upper and a lower set point. As long as the measured process parameter remains between these set points, no changes in the control action are made. On/off controllers with a dead band are found in many instances in our daily lives: home heating systems, ovens, refrigerators, and air conditioners.

All these appliances oscillate periodically between an upper and lower limit around a set point. Figure 18.2 illustrates the action of the control system. It is interesting to note from the figure that the use of dead band reduces the wear and tear on the final control element, but amplifies the oscillations in the measured process parameter. Such controllers have three main advantages: (i) low cost, (ii)

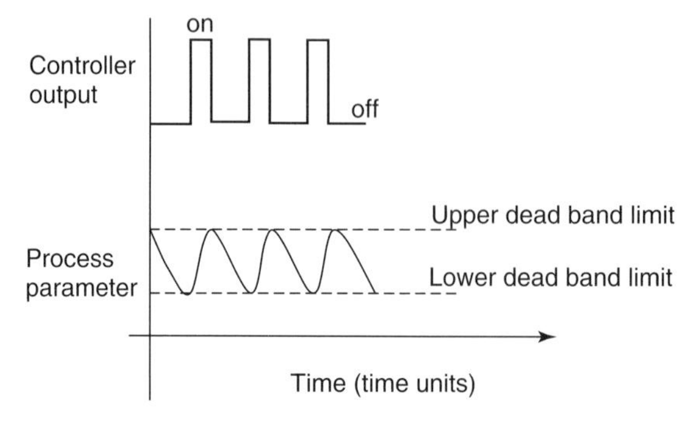

Figure 18.2 Action of an on/off controller with dead bands. From Ref. [6].

instant response, and (iii) ease of operation. However, it is important to ascertain that the upper and lower limit values are acceptable for a specific process. The main disadvantages of this type of control action are: (i) it is not suitable for controlling any process parameter likely to suffer large sudden deviations from the set point and (ii) the quality of control is inferior to the continuous controller.

18.3.2.2 **Proportional Controller**

The P-controller is one of the most commonly used controllers. It produces an output signal to the final control element that is proportional to the difference between the set point and the value of the measured process parameter given by the sensor (this difference is also known as *controller error* or *offset*). Mathematically, it can be expressed as:

$$\mathrm{COS}_{(t)} = \mathrm{COS}_{(\mathrm{NE})} + K_\mathrm{C} E_\mathrm{t} \tag{18.1}$$

where COS is the controller output signal at any time t, $\mathrm{COS}_{(\mathrm{NE})}$ is the controller output signal when there is no error, K_C is known as the *controller gain* or *sensitivity* (controller tuning parameter) and E_t is the controller error or offset. The proportional controller gain or sensitivity (K_C) can also be expressed as:

$$K_\mathrm{C} = \frac{100}{\mathrm{PB}} \tag{18.2}$$

where PB is known as the proportional band, which expresses the value necessary for 100% controller output. The proportional controller gain (K_C), through the value of PB, describes how aggressively the P-controller output will move in response to changes in offset or controller error (E_t). When PB is very small, K_C is high and the amount added to $\mathrm{COS}_{(\mathrm{NE})}$ in Equation 18.1 is large. The P-controller will therefore respond aggressively like any simple on/off controller, with no offset, but a high degree of oscillations. In contrast, when PB is very high, K_C is small and the controller will respond sluggishly, with reduced oscillations, but increased offset. Thus, K_C through PB can be adjusted for each process to make the P-controller more or less active by achieving a compromise between degree of oscillations and offset. The main disadvantage of the P-controller is the occurrence of the offset, while its key advantage is that there is only one controller tuning parameter: K_C. Hence, it is relatively easy to achieve a best final tuning. Figure 18.3 shows a step set point change under a P-controller for two different K_C values [6].

18.3.2.3 **Proportional Integral Controller**

The PI controller produces an output signal to the final control element which can be expressed mathematically as:

$$\mathrm{COS}_{(t)} = \mathrm{COS}_{(\mathrm{NE})} + K_\mathrm{C} E_\mathrm{t} + K_\mathrm{C} E_\mathrm{t} + \frac{K_\mathrm{C}}{\tau_\mathrm{I}} \int E_\mathrm{t} \mathrm{d}t \tag{18.3}$$

where τ_I is a tuning parameter called the *reset time* and the remaining notations are explained under Equation 18.1. The integral term continually sums the controller error and its history over time to reflect how long and how far the measured process parameter has deviated from the set point. Thus, even if a small error

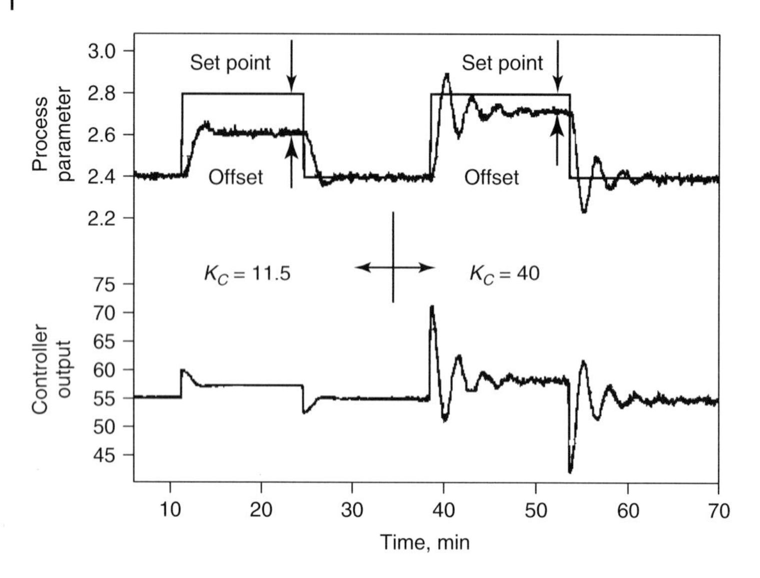

Figure 18.3 P-controller with two different K_C values. From Ref. [6].

persists over a long duration of time, the effects will add up. However, according to Equation 18.3, the contribution of this integral term depends on the values of the tuning parameters K_C and τ_I. It is evident from Equation 18.3 that higher values of K_C and lower values of τ_I will increase the contribution of the integral term.

Figure 18.4 shows a typical PI controller response. It is clear from the case considered in the figure that, from 80 min onward, E_t is constant at zero, yet the integral of the complete transient has a final residual value (obtained by subtracting A_2 from the sum $(A_1 + A_3)$). This residual value, when added to $COS_{(NE)}$, effectively creates a new overall $COS_{(NE)}$ value, which corresponds to the new set point value.

The consequence of this is that integral action enables the PI controller to eliminate the offset, which is the key advantage of the PI controller over a P controller. However, it is important to note that in a PI controller, two tuning parameters interact and it is difficult to find the "best" tuning values once the controller is placed in automatic mode.

Moreover, a PI controller increases the oscillatory behavior, as shown in Figure 18.4.

18.3.2.4 **Proportional Integral Derivative Controller**

In line with the PI controller, the PID controller produces an output signal to the final control element which can be expressed as:

$$COS_{(t)} = COS_{(NE)} + K_C E_t + \frac{K_C}{E_t}\int E_t dt + K_C \tau_D \frac{dE_t}{dt} \tag{18.4}$$

where τ_D is a new tuning parameter called the *derivative time* and the remaining notations are already explained above. Higher values of τ_D provide a higher weighting to the fourth (i.e., derivative) term which determines the rate of change

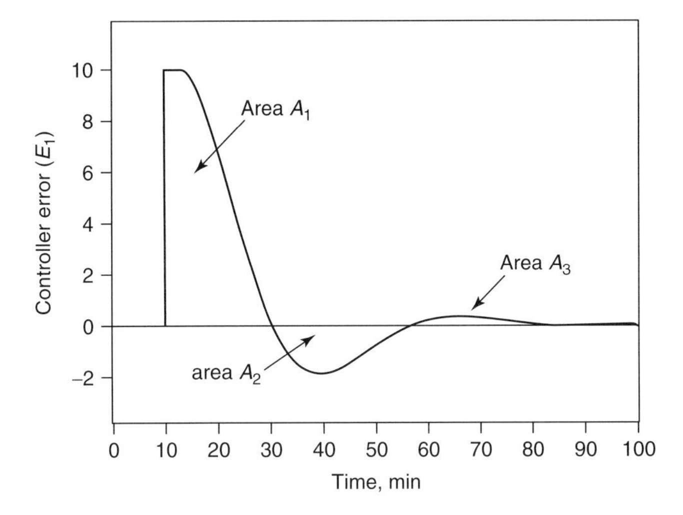

Figure 18.4 Proportional integral controller – integral of error continually increases and decreases with time. From Ref. [6].

of the controller error (E_t), regardless of whether the measurement is heading toward or away from the set point (i.e., whether E_t is positive or negative).

This implies that an error which is changing rapidly will yield a larger derivative. This will cause the derivative term to dominate in determining, provided K_C is positive. Figure 18.5 shows a situation where the derivative values are positive, negative, and zero (when the derivative term momentarily makes no contribution to

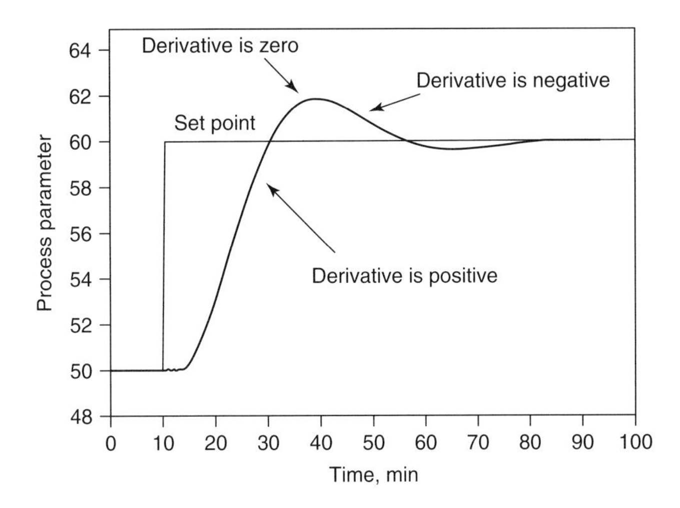

Figure 18.5 Proportional integral derivative controller – positive, negative, and zero derivative values. From Ref. [6].

the control action; note that the proportional and integral terms definitely influence at that point in time).

The major advantage of the PID controller (i.e., the introduction of the derivative term) is that it modifies the drawback of the PI controller: it works to decrease the oscillating behavior of the measured process parameter. A properly tuned PID controller action can achieve a rapid response to error (proportional term), offset elimination (integral term), and minimize oscillations (derivative term). The key disadvantage of the PID controller is that it has three tuning parameters, which interact and must be balanced to achieve the desired controller performance. Just as in the case of the PI controller, the tuning of a PID controller can be quite challenging, as it is often hard to determine which of the three tuning parameters is dominantly responsible for an undesirable performance. To summarize, in all the above control actions, there are one or more parameters to be set when the controller is installed. However, it is most likely that the process information will be insufficient to give best values for these variables.

Most control loops are therefore capable of becoming unstable and potentially result in serious consequences. Formal procedures are therefore necessary to arrive at the right controller settings. A number of procedures and techniques, each with their own advantages and disadvantages, have been developed over the years for single and multivariable control [7, 8]. A range of mathematical concepts have been applied in order to seek improvements in control quality. These are generally known as *advanced control* and include parameter estimation, fuzzy logic [9], and neural networks [10].

Without going into any further details, we will now consider how process control is implemented in modern food processes.

18.4
Process Control in Modern Food Processing

Control applications in food processing, according to McFarlane [1], can be discussed in the context of three categories of products: (i) bulk commodity processing (e.g., grain milling, milk, edible oil, sugar, and starch production), where control is arguably most advanced; (ii) manufactured products (e.g., pasta, cheese, in-container and aseptically processed products); and (iii) products which have been subjected to processing methods essentially designed to retain their original structure (e.g., meat, fish, fruits, and vegetables). Regardless of the nature of the products, process control in food processing has moved on from just attempting to control single variables (e.g., level, temperature, flow, and so on), to systems which ensure smooth plant operation with timely signaling of alarms. The systems are also geared to provide vital data at shopfloor level right through to vertically structured systems which encompass supervisory control and data acquisition (SCADA), manufacturing execution systems (MESs) and interfacing with complex enterprise resource planning systems (ERP), which may be connected across multiple production sites.

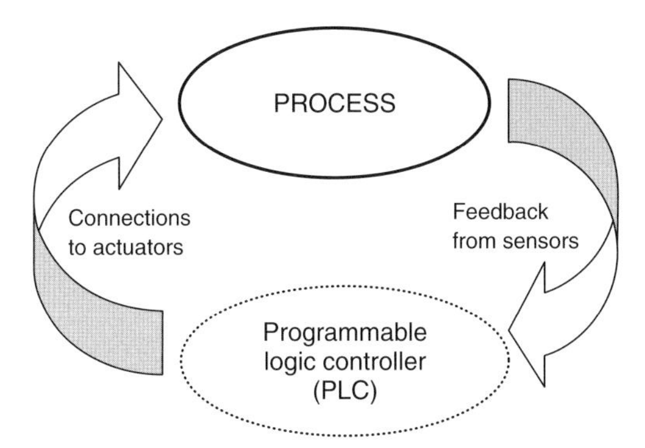

Figure 18.6 Schematic illustration of the working of a programmable logic controller (PLC).

18.4.1 Programmable Logic Controller

The programmable logic controller (PLC) is the most common choice in modern control [11]. It is a microprocessor-based system which can communicate with other process control components through data links. PLCs commonly use the so-called *ladder logic* which was originally developed for electrical controls using relay switches. Programs can be written using a variety of languages.

Once the program sequence has been entered into the PLC, the keyboard may be locked or removed altogether for security. When a process is controlled by a PLC, it uses inputs from sensors to take decisions and update outputs to drive actuators, as shown in Figure 18.6. Thus, a control loop is a continuous cycle of the PLC reading inputs, solving the ladder logic and then changing the outputs. A real process will inevitably change over time and the actuators will drive the system to new states (or modes of operation). This implies that the control performance relies on the sensors available and its performance is limited by their accuracy.

18.4.2 Supervisory Control and Data Acquisition

The SCADA system is not a full control system, but is a software package that is positioned at a supervisory level on top of hardware to which it is interfaced, generally via PLCs, or other hardware modules [12]. SCADA systems are designed to run on common operating systems. Two basic layers can be distinguished in a SCADA system: (i) the client layer which serves as the human–machine interface and (ii) the data server layer which handles process data and control activities, by communicating with devices such as PLCs and other data servers. Such communication may be established by using common computing networks. Modern data servers and client stations often run on Windows NT or Linux platforms. SCADA-based control systems also lend themselves to being scalable by adding more process variables, more specialized servers, for example, for alarm handling or more clients. This is normally done by providing multiple data servers connected to multiple controllers.

Each data server has its own configuration database and real time database (RTDB) and is responsible for handling a section of the process variables, for example, data acquisition, alarm handling, or archiving. Reports can also be produced when needed, or be automatically generated, printed, and archived. SCADA systems are generally reliable and robust and, more often than not, technical support and maintenance are provided by the vendor.

18.4.3 Manufacturing Execution Systems

MESs [13] are software packages which have been used for a number of years in process industries to support key operations and management functions ranging from data acquisition to maintenance management, quality control, and performance analysis. However, it is only in recent years that there has been a concerted attempt to integrate factory floor information with ERP systems. Modern MESs include supply chain management, combine it with information from the factory floor and deliver results to plant managers in real time, thus integrating supply chain and production systems with the rest of the enterprise. This gives a holistic view of the business that is needed to support a "manufacture to order" model. MESs can manage production orders and can track material use and material status information. The software collects data and puts it into context, so that the data can be used for both real time decision making and performance measurement and for historical analysis.

A typical control system currently available for dairy plants is shown in Figure 18.7 [14]. The system combines supervision and control into a single concept, but its architecture is essentially open and modular, thereby delivering the operational flexibility needed. Operator control is the link between the plant operator and the process control modules, which enables actions such as routing and storage tank selection to be initiated at the click of a mouse. Using a variety of software tools, it is possible to incorporate a graphical presentation of the plant and other written information. Using the zoom facility, an operator can quickly access more detail on any given section of the plant. This feature can also be used to train new operators. Process data are also stored in this section, which can be easily accessed, for instance, via "pop up" control windows. The process control model contains information on the process and the parameters which have to be controlled. It can also include flow routing and control, storage information, and sequences of the cleaning cycles. These modules can provide real time control capability. The batch and recipe section contains information on recipe, ingredients, and product specifications. This ensures use of ingredients in the right proportions and consistent product quality. All critical parameters can be monitored in real time graphs and displayed onscreen as well as logged for reports.

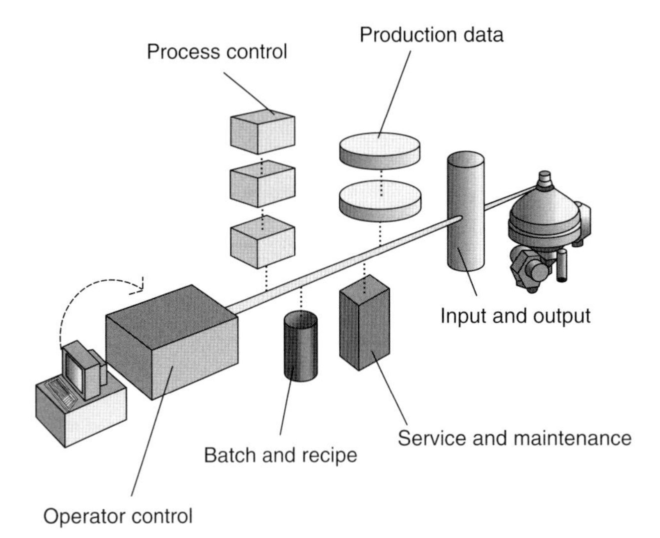

Figure 18.7 A typical control system for dairy plant: FOOCOS system from ABB Automation [14]. Reproduced with permission.

The production data modules form a logbook of the activities of each processing unit – giving the origin of the product, how it has been treated and its final destination. The logbook can also be used for process evaluation and establishing traceability. The service and maintenance modules log information on equipment run times, number of valve strokes, alarm limits on equipment, and so on, which identify poor functioning or worn units, well ahead of a possible breakdown.

This enables the implementation of a preventative maintenance programme. Finally, the input and output modules handle the actual connections from the process control and supervision to the physical parts (like valves, pumps, etc.) to give a complete and detailed inventory.

18.5 Concluding Remarks

Most food businesses are currently facing constraints on capital budgets. There is a perpetual need to improve efficiency and seek out valuable incremental improvements in manufacture. Proper implementation of process control is one way to release latent potential in existing processing facilities. Savings to be made by various sectors of the food and drink industry by implementing appropriate control of the manufacturing process are given in "Energy Wizard" (*www.actionenergy.org.uk*), which is an interactive energy efficiency guide aiming to provide companies with free, independent, and authoritative advice on many aspects of efficient energy use.

References

1. McFarlane, I. (1995) *Automatic Control of Food Manufacturing Process*, 2nd edn, Chapman and Hall, London.
2. Pyle, D.L. and Zaror, C.A. (1997) Process control, in *Chemical Engineering for the Food Industry* (eds P.J. Fryer, D.L. Pyle, and C.D. Rielly), Blackie Academic and Professional, London, pp. 250–294.
3. Stanbury, P.F., Whitaker, A., and Hall, S.J. (1995) *Principles of Fermentation Technology*, 2nd edn, Elsevier, Oxford, pp. 215–241.
4. Kress-Rogers, E. and Brimelow, C.J.B. (2001) *Instrumentation and Sensors for the Food Industry*, 2nd edn, Woodhead, Cambridge.
5. Moreira, R.G. (2001) *Automatic Control for Food Processing Systems*, Aspen, New York.
6. Cooper, D. (2002) Practical process control using control station, Course Notes, Control Station.
7. Wellstead, P.E. and Zarrop, M.B. (1995) *Self-Tuning Systems: Control and Signal Processing*, John Wiley & Sons, Ltd, Chichester.
8. Albertos, P. and Antonio, S. (2004) *Multivariable Control Systems: An Engineering Approach*, Springer, New York.
9. Zhang, Q. and Litchfield, J.B. (1991) Fuzzy mathematics for product development. *Food Technol.*, **45**, 108–115.
10. Eerikainen, T., Linko, P., Linko, S., Siimes, T., and Zhu, Y.-H. (1993) Fuzzy logic and neural network applications in food science and technology. *Trends Food Sci. Technol.*, **4**, 237–242.
11. *http://www.control-systems-principles.co.uk/whitepapers/programmable-logic-control.pdf* (accessed August 15, 2011).
12. Daneels, A. and Salter, W. (1999) Selection and evaluation of commercial SCADA systems for the controls of the CERN LHC experiments. Proceedings of the 1999 International Conference on Accelerator and Large Experimental Physics Control Systems, p. 353.
13. Saenz de Ugarte, B., Artiba, A., and Pellerin, R. (2009) Manufacturing execution system – a literature review. *Production planning and control*, **20**, 525–539. (doi: 10.1080/09537280902938613)
14. *http://www.abb.com/global/abbzh/abbzh251* (accessed July 5, 2005).

19
Environmental Aspects of Food Processing

Niharika Mishra, Ali Abd El-Aal Bakr, Keshavan Niranjan, and Gary Tucker

19.1
Introduction

Waste is inevitably produced in all human endeavors; and its volume is proportional to the resources consumed. Waste is generally thought of as something that is no longer needed by the original user and is subsequently discarded. It is defined in UK legislation as: "any substance which constitutes a scrap material or an effluent or other unwanted surplus substance arising from the application of any process" [1]. It is further defined as: "any substance or article which requires to be disposed of as being broken, worn out, contaminated or otherwise spoiled" [1], or "that the holder discards, intends to or is required to discard" [2] (based on the definition of waste in EC Directive 91/156/EEC) [3].

The increased culture of consumerism within our societies has escalated the problem of waste because of the use of disposable goods. Processed food wastes constitute one of the largest fractions of municipal waste these days. Manufacturing processes operate under strict quality control and retailing has stringent "sell by" date regulations, which has resulted in the generation of large volumes of food and packaging waste. The food industry is facing increasing pressure to reduce its environmental impact, both from consumers and regulators. Initial results from an Environment Agency study indicate that the food, drink, and tobacco sector contributes $8–11 \times 10^6$ t year^{-1} to the industrial/commercial total of $70–100 \times 10^6$ t year^{-1}. This partly reflects the importance and size of the food and drink industry within the UK [4].

Transferring food from the field to the plate involves a sophisticated production and supply chain, but for the purposes of waste production this can be simplified into three main steps: agriculture, food processors/manufacturers, and the retail/commercial sector. Each of the sectors generates waste and wash water. Given the complexity of the food chain, environmental impacts can occur at various points in the chain, even for a single food product. It is therefore necessary to take a holistic systems-based approach to tackle the problem. This, however, demands that the entire food chain be considered in the context of dealing with environmental issues. Since such an approach would become too unwieldy in the context of

Food Processing Handbook, Second Edition. Edited by James G. Brennan and Alistair S. Grandison.

this book, this chapter merely aims to identify key environmental issues relating to food processing and manufacture; and it discusses food waste characteristics, the relation between processing operations, and the types of waste generated, waste processing options, energy issues in food manufacture, the environmental impact of refrigerants and packaging wastes.

19.2 Waste Characteristics

The quality and quantity of wastes produced depend on the type of food being processed. There are big differences from sector to sector, and even site to site: generalization is not only difficult, but could also be misleading. Food wastage levels are often inferred from mass balances. It is estimated that about 21% of food product at the farm gate is lost, much due to spoilage, and only about 7%, on average, is lost during processing [5]. From the data cited in [5] (Table 19.1), it can be inferred that, although the percentage loss during food processing is low, wastage mass or volumes are very high. The wastes produced in any food industry depend mainly on the type of food being processed. Food processing operations produce many varied types of wastes that can be categorized into solid, liquid, and gaseous wastes.

Table 19.1 Solid wastes generated in selected processes [5].

Processed food waste	Total solids (g kg^{-1})	Liquid volume (m^3 kg^{-1})
Vegetables		
Kale	16	0.004
Spinach	20	–
Mustard greens	16	–
Turnip greens	15	–
Potatoes	66	0.012
Peppers (caustic peeling)	65	0.020
Tomatoes (caustic peeling)	14	0.010
Dairy		
Cheese whey	–	9.00
Skim milk	–	0.07
Ice cream	–	0.08
Meat		
Red meat	0.440	25.00
Poultry	0.270	50.00
Eggs	0.111	–

19.2.1
Solid Wastes

Solid wastes emanating from food processing plants may include: the unnecessary leftover from the preliminary processing operations, residues generated as an integral part of processing, wastes resulting from processing inefficiencies, sludge produced from the treatment of wastewater, containers for the raw materials, and finished products. Table 19.1 summarizes typical solid wastes generated from a selection of food processes [6, 7]. In general, solid wastes are poorly characterized, both in terms of quality and quantity; and estimates of solid wastes are usually inferred from mass balances [6, 7].

19.2.2
Liquid Wastes

Wastewater from the processing industry is the main stream that is produced. It includes: wastewater resulting from using water as a coolant, water produced by different processing operations like washing, trimming, blanching, and pasteurizing, and a large amount of wastewater produced from cleaning equipment [7].

19.2.3
Gaseous Wastes

The gaseous emissions from the food processing industry are mainly manifested in terms of emanating odors and, to a lesser extent, dust pollution. Other emissions include solvent vapors commonly described as volatile organic emissions and gases discharged by combustion of fuels. Even though the characteristics of food wastes can be discussed in terms of their physical states, it is necessary to note that solid wastes contain a substantial proportion of water, just as liquid wastes may contain a significant proportion of solids. It is therefore absolutely critical to note that food wastes are not only multicomponent but also multiphase in nature [7].

19.3
Wastewater Processing Technology

Treatment of the wastes produced from food industries is an important concern from the environmental point of view. As discussed earlier, the waste products from food processing facilities include bulky solids, wastewater, and airborne pollutants. All of these cause potentially severe pollution problems and are subject to increasing environmental regulations in most countries. In general, wastewater is most common, because food processing operations involve a number of unit operations, such as washing, evaporation, extraction, and filtration. The wastewaters resulting from these operations normally contain high concentrations of suspended solids and soluble organics, such as carbohydrates, proteins, and lipids, which cause disposal problems. To remove these contaminants from water, different technologies are adopted in the food industry, as described in detail in Chapter 20 and in [8, 9].

19.4 Resource Recovery from Food Processing Wastes

The wastes from food industry, after recovery, and further processing, can be used for different purposes: the recovered materials can either be recycled, or be used to recover energy by incineration or anaerobic digestion. Recycling not only reduces the environmental impact of the material, but also helps to satisfy the increasing demands for raw materials. In addition, it also reduces disposal costs, a key driver of recycling technologies. For instance, fruit, vegetables, and meat processors generate large quantities of solid wastes. Table 19.2 lists examples of useful materials which can be recovered from fruit and vegetable wastes. Recovered materials can be used in various ways. Solid food wastes can be used as animal feed after reducing their water content. A good example of this practice is soybean meal, a by-product of soybean oil extraction, which was simply discarded previously but is now used as animal feed on account of its high nutritive value [10]. Solid wastes can also be upgraded by fermentation. A number of fermented foods are produced this way. Composting and ensilaging are also examples of solid waste fermentation process [11]. Solid wastes rich in carbohydrate can also be converted to sugars by enzyme-assisted hydrolysis; an example is the enzymatic hydrolysis of lactose and galactose sugar using β-galactosidase [12]. Solid wastes rich in sugar can be fermented to produce carbon dioxide and ethanol. The latter is a valuable product, and has also been earmarked as an important source of fuel for the future [13].

As mentioned above, solid wastes can also be utilized as fuel directly or converted to methane by anaerobic digestion in a bioreactor. Biological hydrogen is produced by fermentation of both glucose and sucrose in food processing wastes under slightly acidic conditions in the absence of oxygen. This can be achieved by

Table 19.2 Some examples of products which can be recovered or made from fruit and vegetable wastes [7].

Source of waste	Product
Apple pomace	Pectins
Apple skin	Aromatics
Tomato pomace	Pectins, tomato seed oil, color from skin
Stalk of paparika and pumpkin seeds	Natural coloring agents
Green pea pods	Leaf proteins, chlorophyll
Stones from stoned fruits	Active carbon, kernels (after debittering)

using a variety of bacteria through the actions of well-studied anaerobic metabolic pathways and hydrogenase enzymes. Hydrogen has 2.4 times the energy content of methane (on a mass basis) and its reaction with oxygen in fuel cells produces only water, a harmless by-product. Hydrogen gas has valuable potential as a clean and economical energy source in the near future [14].

19.5 Environmental Impact of Packaging Wastes

Packaging is acknowledged to perform a number of useful functions. It acts as a physical barrier between a product and the external environment, protecting it from external contamination and maintaining hygienic conditions, it protects and preserves the product during handling and transportation, it serves to attract the attention of consumers, giving the product a good market value, and it also serves to provide information on the product and instructions on how to use it (see also Chapter 8). Despite these advantages, the environmental impact of packaging wastes is considerable and, in many cases, outweighs their benefits. Recent studies have shown that in Europe packaging comprises about 16% of municipal solid waste (MSW) and 2% of non-MSW [15]. The key environmental issues related to packaging are:

- the use of packaging materials, such as plastics and steel, which are either nonrecyclable or uneconomic to recycle (a large amount of such wastes invariably end up in landfills);
- the use of material-intensive packaging, which requires an energy-intensive process to manufacture;
- the use of substances in the packages having high chemical and biological oxygen demand (some even hazardous and toxic to the environment) which cannot be discharged safely into natural water streams.

In most countries, regulations are in place for reducing the impact of packaging and packaging wastes on the environment. This is mostly done by limiting the production of packaging wastes, enforcing the recycling of packaging material and reuse of packages where possible, and encouraging the use of minimal packaging at source.

19.5.1 Packaging Minimization

The foremost strategy in packaging waste management is to reduce the use of packaging to a bare minimum level at all stages of production, marketing, and distribution. This can be achieved by: (i) decreasing the weight of material used in each pack (known as *lightweighting* or *downgauging*), (ii) decreasing the size or volume of the package or using less material in the first place, for example, reducing the thickness of the packaging material, (iii) using consumable or edible package, and (iv) modifying the product design, for example, avoiding unnecessary multiple wrapping of a product with different materials [16].

19.5.2 Packaging Materials Recycling

The purpose of recycling is to use a material as raw material for the production of a new product after it has already been used successfully. If recycling is done properly and in conjunction with good design, many materials can be recovered after their first useful life is over. The two major objectives should be to conserve limited natural resources and to reduce and rationalize the problems of managing MSW disposal [15]. *Recycling* is defined as the reprocessing of the waste material in a production process either for the original purpose or for other purposes. The European Union (EU) definition [17] also includes organic recycling, that is, aerobic or anaerobic treatment of the biodegradable part of the packaging waste to produce stabilized organic residues or methane. In general, recycling involves physical and/or chemical processes which convert collected and sorted packaging, or scrap, into secondary raw materials or products. *Secondary raw material* is defined as the material recovered as a raw material from used products and from production scrap. Before sending packaging materials for recycling, they should be properly sorted (i.e., separated from other packaging materials) and cleaned (i.e., free from any contamination). Sorting and cleaning are two important operations before processing, since they affect the quality of the input stream which finally determines the quality and value of the secondary materials.

The materials commonly used for food packaging are: paper and board, plastic, glass, aluminum, and steel. Given the widespread use of paper and board as packaging material, their recycling is critical from the environmental point of view as well as resource recovery. Recycled paper is a major source of raw material for the paper industry. About 44.7×10^6 t of waste paper were recycled in Europe in 2003, which is substantially higher than 10 years earlier, when only about 26×10^6 t were recycled. This represents 53.2% of the total paper used in Europe [18]. This figure reached 64.5% in 2007 [19]. Packaging is the largest sector; and it uses almost two-thirds of the recycled paper in Europe to manufacture case materials, corrugated board, wrapping, and so on. The total consumption of plastics in Europe was about 48.5×10^6 t in 2008, of which 38% was through packaging materials. Plastics account for 17% of the total packaging usage in Western Europe [1]. The most widely use packaging plastics include low density polyethylene (LDPE) and high density polyethylene (HDPE), linear low density polyethylene (LLDPE), polypropylene (PP), polystyrene (PS), polyethylene tetraphthalate (PET), and polyvinyl chloride (PVC). After collection, the material is sorted, to separate the plastics from other materials such as paper, steel, aluminum, and so on. The sorting step also includes the separation of plastics by their resin type (like PET, HDPE, etc.). The sorted plastics are then recycled by different technologies, such as mechanical, feedstock, and chemical recycling [20]. Mechanical recycling involves processes like extrusion, co-extrusion, injection, blow molding, and so on (see also Chapter 8). Feedstock recycling includes pyrolysis, in which plastics are subjected to high temperature in the absence of oxygen which enables the hydrocarbon content of the polymer to be recycled. Pyrolytic processes have been studied extensively for the last two decades. However, most of this research has been undertaken using pure and clean plastics, or mixtures of pure plastics. There is a strong need to develop processes capable of dealing with wastes that have plastics attached to other contaminants, such as paper, metals or bioproducts. Microwave-induced pyrolysis of plastics is a novel process in which microwave energy is applied to carbon mixed with plastic waste [21]. Chemical recycling involves depolymerization of PET, resulting in the monomers terephthalic acid and ethylene glycol, which, after purification, can be reused to produce new polymers. Another method, called the "*super clean recycling process*" uses mechanical and nonmechanical procedures to recycle high-quality postconsumer material, producing polymers suitable for use in monolayer application, that is, use in direct contact with food. The processes are proprietary, but they are believed to involve a combination of standard mechanical recycling processes with nonmechanical procedures such as high-temperature washing, high-temperature and pressure treatments, use of pressure/catalysts, and filtration to remove polymer-entrained contaminants [22].

Recycled plastics have been used in food contact applications since 1990 in various countries around the world. To date, there have been no reported issues concerning health or off taste resulting from the use of recycled plastics in food contact applications. This is due to the fact that the criteria that have been established regarding safety and processing are based on extremely high standards that render the finished recycled material equivalent in virtually all aspects to virgin polymers [21].

Various food contact materials and constituents can be used, provided they do not pose health concerns to consumers, which may occur when some substances from the food packaging migrate into the food. To ensure the safety of such materials, food packaging regulations in Europe require that the packaging materials must not cause mass transfer (migration) of harmful substances to the food, by imposing restrictions on substances from the materials itself that could migrate into the food. Consequently, food packaging materials must comply with many chemical criteria and prescribed migration limits. The migration of substances from the materials into the foodstuffs is a possible interaction that must be minimized or even avoided, since it may affect the food or pose longer term health concerns to the consumer [23] (see also Chapter 8). With regard to recycled PET, there is strong need to have relevant analytical data on the nature and concentration of the contaminants that can be found in the recycled material in order to ascertain the safety of reusing PET for food purposes. Knowledge of the contaminants and information on practical and effective test methods would help in the formulation of future legislation [24].

With a view to make packaging from sustainable materials, a number of biodegradable alternatives have been developed. Traditionally, biobased packaging materials have been divided into three types, which illustrate their historical development. First generation materials consist of synthetic polymers and 5–20% starch fillers. These materials do not biodegrade after use, but will biofragment, that is, they break into smaller molecules. Second generation materials consist of a mixture of synthetic polymers and 40–75% starch. Some of these materials are fully biobased and biodegradable [25]. The market value of biobased food packaging materials is expected to incorporate niche products, where the unique properties of the biobased materials match the food product concept [26]. Packaging of high-quality products such as organic products, where extra material costs can be justified, may form the starting point. Biopolymer-based materials are not expected to replace conventional materials in the short term. However, due to their renewable origin, they are indeed the materials of the future [27]. According to [4], targets for recovery and recycling have been set by the EU as follows: 50–65% by weight of packaging waste to be recovered, 25–45% to be recycled, and 15% to be recovered by materials. These targets refer to packaging composed of plastics, paper, glass, wood, aluminum, and steel. Further, the combined content of lead, mercury, cadmium, and chromium (VI) has been limited to 100 ppm. The law also ensures that packaging materials are introduced in the marketplace only if they meet "essential requirements," that is, characteristics that include minimization of weight and volume, and suitability for material recycling.

19.6 Refrigerants

Refrigeration systems are essential for the production, storage, and distribution of chilled and frozen foods. The commonly used refrigerants in these systems are chlorofluorocarbons (CFCs) and hydrofluorocarbons (HFCs). Although highly efficient, these refrigerants have been shown to be responsible for severe environmental threats like global warming and depletion of the ozone layer. CFCs are organic compounds containing chlorine, fluorine, and carbon atoms and having ideal thermodynamic properties for use as refrigerants. But their chlorine content is mainly responsible for the depletion of the ozone layer in our environment. When CFCs are released into atmosphere, they dissociate in the presence of ultraviolet (UV) light to give free chlorine. This free chlorine atom decomposes ozone to oxygen and regenerates itself by interacting with a free oxygen atom, as follows [28]:

$$CF_2Cl_2 \rightarrow CF_2Cl + Cl$$

$$Cl + O_3 \rightarrow ClO + O_2$$

$$ClO + O \rightarrow Cl + O_2$$

The regeneration of chlorine sustains the process and depletes the ozone layer. This layer is known to protect life on earth from UV radiation by absorbing a large portion of it and allowing only a small fraction to reach the Earth. But its depletion will expose us to UV radiation, causing skin cancer, damage to eyes, damage to

Table 19.3 Refrigerant characteristics [29].

Refrigerant	Ozone-depleting potential	Global warming potential ($CO_2 = 1.0$)	Stratospheric lifetime
CFC 11	1.0	4100	55.0
CFC 12	1.0	7400	116.0
CFC 113	1.07	4700	110.0
CFC 114	0.8	6700	220.0
CFC 115	0.5	6200	550.0
HCFC 22	0.055	2600	15.2
HCFC 123	0.02	150	1.6
HCFC 124	0.022	760	6.6
HCFC 141b	0.11	980	7.8
HCFC 142b	0.065	2800	19.1
HFC 125	0.0	4500	28.0
HFC 134a	0.0	1900	5.5
HFC 143a	0.0	4500	41.0
HFC 152a	0.0	250	1.7

crops, global warming, climate change, and so on [28]. Besides this effect, such refrigerants are also known to contribute to global warming, along with CO_2 and other gases such as methane, nitrous oxides, CFCs, and halocarbons. The extent to which a substance can destroy the ozone layer is measured in terms of a parameter called ozone-depleting potential (ODP). The ODPs of CFCs are significantly greater than those of hydrochlorofluorocarbons (HCFCs) and hydrofluorocarbons (HFCs). Hence, CFCs are gradually being replaced by these other two. It may be noted that HFCs have zero ODP, since they do not contain any chlorine atoms. However, the F-C bonds in CFCs, HCFCs, and HFCs are very strong in absorbing infrared radiations escaping from the Earth's surface. Their absorption capacity is much more than that of CO_2 [29]. To measure the contribution of different gases to global warming, a scale called the global warming potential (GWP) has been set up. Table 19.3 lists the ODP and GWP values of different refrigerants. It is clear from the table that CFCs have high ODPs and GWPs compared to HCFCs and HFCs [30].

To control the production and consumption of substances which cause ozone depletion, the Montreal Protocol on Substances that Deplete the Ozone Layer was signed in 1987 and has been effective since 1989 [31]. The purpose of this agreement was to phase out CFCs by the year 2000 and to regularly review the use of transitional ozone-safe alternative refrigerants, which are scheduled to be replaced by 2040. Similarly, although HCFCs are used as replacements for CFCs, they are still responsible for ozone depletion and need to be phased out by 2020 as specified by the amended Montreal Protocol [15]. HCFCs are expected to be replaced by HFCs.

19.7
Energy Issues Related to the Environment

In most countries the energy consumed by the food and drink industry is a significant proportion of the total energy used in manufacturing industries. For instance, in the United Kingdom this proportion is around 1/10th [32]. Energy is consumed by the food industry to keep food fresh and safe for consumption. This is achieved by different processing operations (boiling, evaporation, pasteurization, cooking, baking, frying, and so on), safe and convenient packaging (aseptic packaging), and storage (freezing, chilling). The energy required for these processes is obtained from either electricity or burning fossil fuel. When the cost of energy consumption is considered, it has received very low priority in many organizations because it accounts for only 2–3% of the total production cost [32]. But considering the other side of the coin, that is, the environmental effects, the energy consumption cannot be ignored. The food industry will be affected by all international measures aimed at reducing industrial energy consumption. The background to some of the international measures is discussed below.

The burning of fossil fuel results in the emission of large amounts of CO_2, the most important greenhouse gas (GHG), which is responsible for about two-thirds of potential global warming. CO_2 produced from burning fossil fuel is responsible for 80% of the world's annual anthropogenic emissions of CO_2. Methane (CH_4), the second most important GHG, is responsible for about 15% of the build-up, and nitrous oxide (N_2O), which also has a high stratospheric lifetime, is responsible for 3% of the build-up [33]. Other GHGs, for example, CFCs, HCFCs, perfluorinated carbons (PFCs), sulfur hexafluoride (SF_6), and so on, are produced from various sources, which include the refrigeration systems used in food processing. Methodology for the calculation of GHGs is illustrated in Section 19.9. The average temperature rise experienced by the planet on account of greenhouse emissions has been estimated to be approximately 0.5 °C over the past 100 years. But sophisticated computer models solely based on CO_2 emissions are predicting a temperature rise of 5 °C over the next 200 years [34]. The average rate of warming due to emission of these gases would probably be greater than that ever seen in the last 10 000 years. This increasing temperature may cause many catastrophic events, like melting of the polar ice cap, rising of global sea levels and unbearably hot climates all over the world. The global sea level has risen by 10–25 cm in the last 100 years and it is expected to increase by between 13 and 94 cm by the year 2100, which might cause widespread flooding [34].

Burning of fossil fuels also gives rise to SO_2, which is converted to sulfate in the atmosphere, known as *sulfate aerosols*. These aerosol particles absorb and scatter solar radiation back into space and hence tend to cool the Earth. But, due to their shorter lifespans, it is difficult to assess the impact of aerosols on the global climate. However, it has been concluded that the increase in sulfate aerosols has had a cooling effect since 1850 [34]. To minimize the chances of catastrophic events occurring in the future, we must slow down the emission of GHGs. This can be achieved by limiting the combustion of fossil fuels, which ultimately leads to reduced energy demand by increasing the drive for energy efficiency and improving its use.

Energy efficiency can be achieved by the use of combined heat and power (CHP) or renewable energy. CHP is a fuel-efficient energy technology in which a major part of the heat that is being wasted to the environment is recovered and used in other heating systems. CHP can increase the overall efficiency of fuel use to more than 75%, compared with around 40% from conventional electricity generation. CHP plays an important role in the UK Government's energy policy, whose ambition is to achieve a 60% reduction in CO_2 emissions by 2050 [35]. Following good process design practices can also make a difference [36]: for instance, insulating valves, flanges, autoclaves, heated vessels, and pipes during steam production can prevent leakage of steam and hence reduce heat loss; also, using an optimum air–fuel ratio prevents unnecessary burning of fuel, and so on.

Renewable sources of energy such as solar radiation, wind, sea waves and tides, biomass, and so on, and the use of fuel containing low or no carbon (e.g., hydrogen) can reduce the emission of GHGs to a significant extent. Another option is to capture the CO_2 emitted by a burning fuel and then utilize it or store it for later use [37]. CO_2 can be captured by various methods such as adsorption onto molecular sieves, absorption into chemically reacting solvents (e.g., ethanolamines), membrane separation methods, and so on. After separation, it can be used as a feedstock for the manufacture of chemicals which enhance the production of crude oil or in the growth of plants or algae which could be used as a biofuel. Several methods of storing the CO_2 have been proposed, such as storing it inside ocean beds, in deep saline reservoirs, in depleted oil and gas reservoirs, and so on [38]. These options may not be economically viable at this stage, but technology will need to be improved so that these options could be exercised more easily.

It is evident from the above discussion that environmental problems resulting from energy consumption cannot be resolved by nations unilaterally. A number of international treaties and agreements have been formulated through the annual Conferences of Parties (COP) to protect the environment from the hazards of GHGs, such as the Kyoto Protocol. During the 1992 Framework Convention on Climate Change (FCCC), the first formal international statement of concern and agreement was formulated to take a concerted action for stabilizing atmospheric CO_2 concentrations. In this context, the 1997 Kyoto Protocol was negotiated (which includes several decisions such as reducing GHG emissions, based on 1990 levels, by 5.2% in the period 2008–2012) by the industrialized countries [39]. The United Kingdom voluntarily committed to reducing emissions by 12.5% by 2010. Other measures include enhancement of energy efficiency in different sectors, increased use of new and renewable forms of energy, advanced innovative technology for CO_2 separation, and the protection and enhancement of sinks and reservoirs of GHGs. In addition, there was a commitment to reduce fiscal incentives, tax and duty exemptions, and subsidies in all GHGs-emitting sectors that ran counter to the objective of the convention [35].

The EU aimed to control the environmental impacts of industrial activities by formulating an "integrated pollution prevention and control" (IPPC) directive (Directive 96/61/EC of 24 September 1996), which sets out measures to ensure the sensible management of natural resources. These provisions enable a move toward a sustainable balance between human activity and the environment's resources and regenerative capacity [4].

The Climate Change Levy (CCL) was introduced by the UK government as a tax on fuels or energy sources used by industry on April 1, 2001. The levy package aimed to reduce CO_2 emissions by at least 2.5×10^6 t year^{-1} by 2010. The levy did not apply to waste used as fuel. To encourage the reduction of fuel consumption, the UK government also announced that a discount of 80% from the levy be given to companies who agree to reduce the CO_2 emission by reducing their energy consumption [40]. Food processing industries are expected to work within the above parameters and there is no doubt that manufacturing practices will continue to change for the foreseeable future to comply with national and international regulations formulated to protect our environment.

19.8 Life Cycle Assessment

The life cycle assessment (LCA) is a tool standardized by the International Standardization Organization (ISO) to evaluate the environmental risks associated with a product from "cradle" to "grave." It takes into account the environmental impact associated with its production, starting from the raw materials and energy needed to produce it, to its disposal, along with processing, transportation, handling, distribution, in between [41]. LCA studies have been carried out for a variety of products, including food. The first LCA studies on food products were undertaken at the beginning of the 1990s [42]. LCA identifies the material, energy, and waste flows associated with a product during the different stages of life cycle and the resulting environmental impact. For example, if we consider the production of orange juice, the LCA analysis will involve: the weight of oranges and energy associated with transporting raw materials, the amount of waste produced (both processing and packaging wastes), the energy or power consumed during processing, the mass and energies associated with the use of utilities such as cleaning water, steam, and air, the emissions released into air, water, and land from the processing site and other relevant factors depending on the operating technology and regional location of the processing facility.

19.9 Calculating Greenhouse Gas Emissions

Global warming, also referred to as *climate change*, is the principal environmental issue of concern because of its effects on changing climates for growing

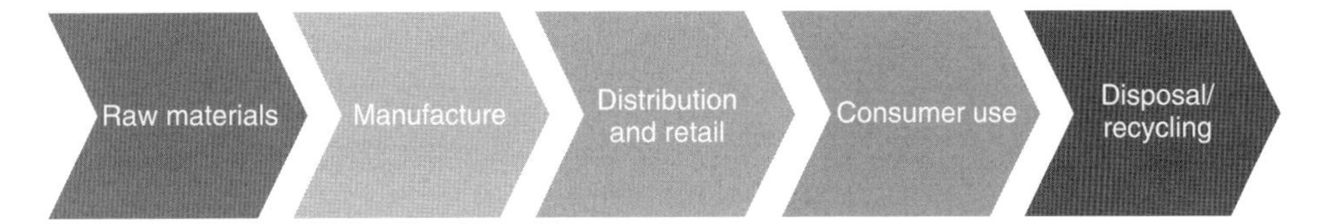

Figure 19.1 Stages required in the greenhouse gas assessment of a food product.

food globally. It is increasingly accepted that the change in climate over the last 100 years is due to an increase in atmospheric concentrations of GHGs and that anthropogenic sources are playing a major role in these increases. Implications for agricultural systems in most growing regions of the world are substantial.

Calculating GHG emissions in terms of their carbon dioxide equivalent values (kg CO_2e) assesses the effect of an activity on global warming. The terms carbon footprint (CF) and GHG emissions have similar meanings and are interchangeable. To calculate a CF value for the production of a food product or for a food business requires all of the stages involved in that activity to be considered. Figure 19.1 illustrates the linkage between raw materials production through to disposal of the waste and final product at the end of its life. Each stage must be assessed for its GHG emissions using a detailed series of calculations.

Calculation of GHG emissions is carried out relative to CO_2 that is given a value of unity (Table 19.4). The recently released Publicly Available Specification (PAS) 2050 [43] contains a more complete table that includes gases used in refrigeration, solvents, and packaging. It is apparent from Table 19.4 that a process releasing N_2O has the potential for much greater global warming impact than one releasing CO_2 [44]. For example, beef cattle produce large quantities of methane as a result of enteric fermentation. Methane has a GWP 25 times that of CO_2 and so this can lead to cattle products such as beef and milk with high CF values. In addition, the overuse of nitrogenous fertilizers when growing crops results in the release of N_2O into the atmosphere. This will result in a high CF for that crop because N_2O has a GWP 298 times higher than that of CO_2.

The next section presents a case study for a bottled apple juice, which demonstrates the methods involved when calculating a CF value.

Table 19.4 Relative global warming potential (GWP) of the main greenhouse gases of relevance to the food industry.

Greenhouse gas	Formula	GWP
Carbon dioxide	CO_2	1
Methane	CH_4	25
Nitrous oxide	N_2O	298

19.9.1
Case Study: Bottled Apple Juice

This study estimated the embodied GHG emissions during manufacture of a 75 cl bottle of Cox's apple juice. The bottled juice was in-pack pasteurized to destroy yeasts and molds, which requires a process of at least 2 min at 70 °C. In addition to the primary packaging (glass bottle), the bottles were packaged into cardboard boxes of 12 bottles per box.

The apple juice CF included all material emissions generated as a direct or indirect result of the product unit being produced. PAS 2050 allows immaterial emissions to be excluded, which are any single sources <1% of the total emissions, up to a maximum of 5%. In the apple juice example, ascorbic acid was added in very small quantities to minimize browning of the pressed juice; this was excluded from the calculations.

Figure 19.2 shows an outline flow diagram from the point at which apples were taken into the juicing room, up to when the bottled juice was ready for distribution from the farm. It shows the single raw material input (apples), the various packaging inputs, and the two waste streams (pressed pulp and wastewater). The juicing operation took place in a dedicated room, shown with a dotted line.

19.9.1.1 Raw Materials (0.407 kg CO_2e per product unit)

The variety of apple was Cox's orange pippin, although the data calculated for manufacture and packaging of apple juice was not specific to this variety. A CF for Cox's apples of 0.075 kg CO_2 kg^{-1} was used [44] based primarily on assessments of fertilizer use and machinery outputs, with apples grown in a modern orchard. Allowing for a 70% juice recovery from milled apples, each 75 cl bottle contained 1.4 kg of apples, which is 0.105 kg CO_2e per product unit.

This study used a product sold in green bottles, which have the advantage of allowing a higher proportion of recycled glass than for a clear bottle. Data for GHG emissions for glass was obtained from the British Glass Manufacturers Confederation [45], in which virgin glass and recycled glass have GHG values of 0.843 and 0.529 kg CO_2e kg^{-1}, respectively. The recycle rate for green glass was 81%, therefore the GHG value for glass was 0.589 kg CO_2e kg^{-1}. The plastic cap was polypropylene, which has a GHG value of 4.4 kg CO_2e kg^{-1} [46]. Minor components of the packaging were also considered, which were the paper label and adhesive, and the plastic wrapping over the cap (assumed to be low-density polyethylene). Table 19.5 details the mass and GHG values for each packaging component.

Bottles were packed into cardboard boxes (secondary packaging) in units of 12. A GHG value of 1.03 kg CO_2e kg^{-1} for cardboard was taken from the FEFCO LCA inventory [47]. Each box weighed 570 g and so the pro-rata weight for a one bottle product unit was 47.5 g, therefore the GHG value was 0.0489 kg CO_2e per product unit.

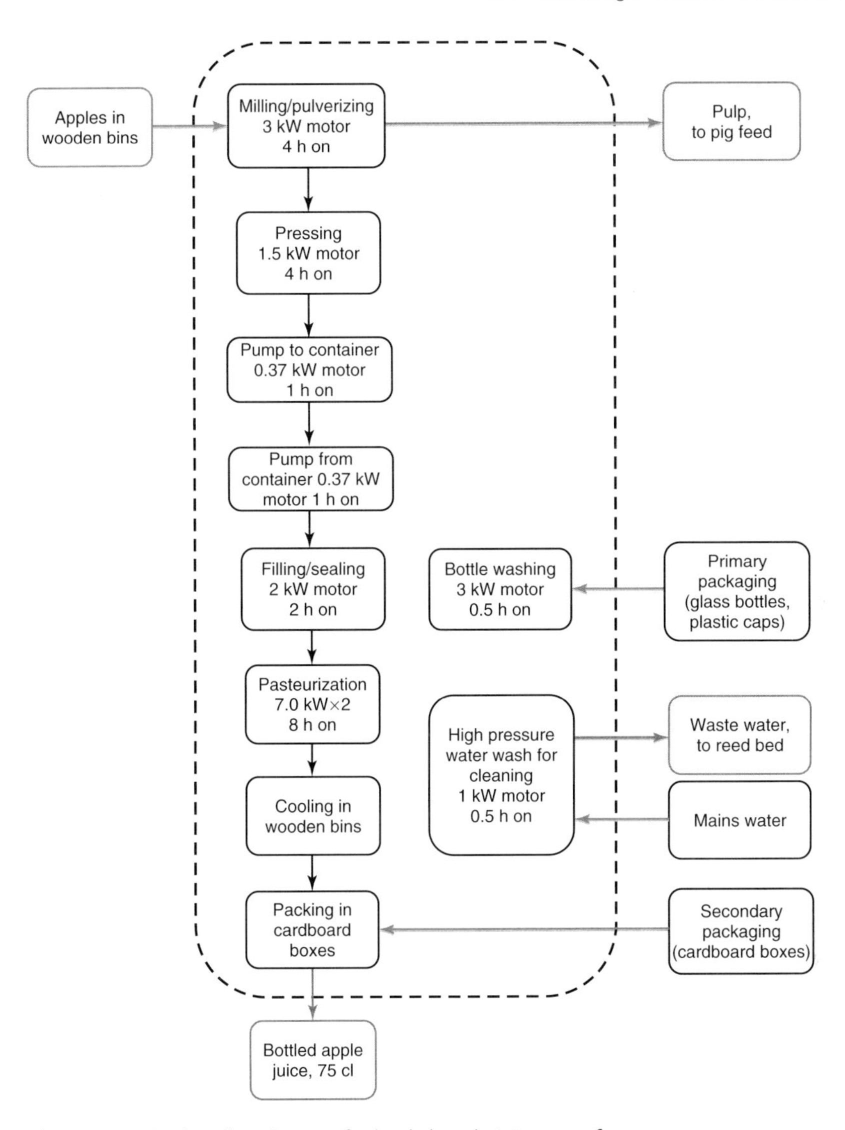

Figure 19.2 Outline flow diagram for bottled apple juice manufacture.

19.9.1.2 **Manufacture (0.061 kg CO_2e per product unit)**

Juice bottling included the process steps within the dotted line in Figure 19.2. Estimations were made for the duration of each stage that involved an electric motor, which was required because operation of these stages was intermittent. Calculated data for each process step was given in Table 19.6.

Total electrical energy required to process a batch of 1200 bottles was 136.74 or 0.114 kWh per product unit (bottle). Conversion of electrical energy to kilograms CO_2e used the emission factor from Defra [48], which resulted in the juice bottling process contributing 0.061 kg CO_2e per product unit.

Table 19.5 GHG emissions (kg CO_2e) for the packaging materials used for 75 cl bottles.

Component	Weight (g per product unit)	GHG value (kg CO_2e kg^{-1})	GHG value (kg CO_2e per product unit)
Glass bottle	400	0.589	0.2350
Plastic cap	3.0	4.4	0.0130
Paper label/adhesive	0.5	1.03	0.0005
Plastic cap wrapping	0.5	2.4	0.0012
Cardboard boxes	47.5	1.03	0.0489
Total			0.2989

Table 19.6 Electrical energy required for the milling, pressing, and pasteurizing stages of 1200 bottles of apple juice.

Unit operation	Motor power (kW)	Duration (h)	Power used (kWh)
Two pasteurizers	7.0	8.0	112.0
Two pumps	0.37	1.0	0.74
Bottle sealer	2.0	2.0	4.0
Pulverizer/mill	3.0	4.0	12.0
Press	1.5	4.0	6.0
Bottle washing machine	3.0	0.5	1.5
High pressure washer	1.0	0.5	0.5
Total			136.74

19.9.1.3 Transportation (0.057 kg CO_2e per product unit)

Transportation of apples from the trees to the juicing process was by tractor, with the apples packed in large wooden bins. These wooden bins were reused. Estimations were made for the distances traveled in delivering the raw materials to the farm, and for the number of journeys made. Table 19.7 presents the breakdown of transportation GHG emissions for the various raw materials.

19.9.1.4 Waste (0 kg CO_2e per product unit)

Two categories of waste were generated; wastewater from cleaning and washing operations, and apple pulp from milling and pressing. Wastewater was gravity fed to a reed bed, which removed much of the organic matter, and fixed the carbon and nitrogen within the plant material. It is arguable that within the 100 year life cycle suggested by PAS 2050 all of the organic carbon and nitrogen will end up as gases, and in doing so will contribute to GHG emissions. This was not considered during this study because no data were available on the quantities of organic materials in the wastewater. No caustic products or detergents were used for the cleaning operation.

Table 19.7 Greenhouse gas emissions (kg CO_2e) for transportation of the raw materials used for 75 cl bottles of apple juice.

Component	GHG value (kg CO_2e per product unit)
Wooden bins of apples	0.0092
Glass bottles	0.0243
Plastic caps	0.0122
Cardboard boxes	0.0110
Total	0.0566

Apple pulp was fed to the pigs that live on the farm, which was a carbon zero activity.

19.9.1.5 Overall Carbon Footprint (0.525 kg CO_2e per product unit)

Table 19.8 compiles the GHG emissions for each the categories detailed above. The CF for these 75 cl bottles of Cox's apple juice was 0.525 kg CO_2e per product unit. The most significant component of this footprint was the glass bottles, which made up 45%, followed by the apples at 20% and the processing at 12%.

19.9.2 GHG Emissions for Other Food Products

There are now numerous publications freely available on the internet that quote CF values for food products [49, 50]. Some of these have used the procedures defined in PAS 2050 [43], although the source of the calculation procedures is not always clear. Data presented in Table 19.9 was calculated by a group of researchers whose aim was to advise Defra on applying PAS 2050 for food products and their raw materials [44, 51, 52].

Table 19.8 Greenhouse gas emissions (kg CO_2e) for a 75 cl bottle of apple juice.

Category	GHG emissions (kg CO_2e per product unit)
Raw materials	0.407
Manufacture	0.061
Transportation	0.057
Waste	0.000
Total	0.525

Table 19.9 Summary of GHG emissions for manufactured food products, per kg product and per product unit.

Food product	kg $CO_2e\ kg^{-1}$ or l^{-1}	kg CO_2e/PU	PU description
Beef cottage pie	7.6	3.3	Single 434.9 g chilled ready meal
White loaf of bread	0.73	0.60	827 g loaf in plastic bag
Packed mild cheddar cheese	9.8	4.9	500 g plastic pack
Cox's apple juice	0.71	0.53	75 cl glass bottle
Chocolate coated cakes	2.5	0.42	165 g packet
Duck in Hoisin sauce	2.0	0.88	Single chilled ready meal (430 g inc. packaging)
Lamb shanks and roasted potatoes	19	25	Single chilled ready meal (1300 g inc. packaging)
Thai chicken pizza	3.5	1.6	1 pizza (460 g inc. packaging)

Food commodities with low emissions (<1 kg $CO_2e\ kg^{-1}$ or l^{-1}) tended to be crop commodities with high yields and low inputs, for example:

- apples (0.066–0.100) and bottled apple juice (0.71),
- potatoes (0.12–0.16),
- spring onions (0.23),
- animal feed crops (0.0043–0.7400),
- carrots (0.35),
- UK conventional tomatoes grown using "waste" heat (0.39),
- wheat (0.40–0.74),
- onions (0.42–0.59).

These GHG data were used to calculate GHG emissions for manufacture of the food products in Table 19.9. Each of these products has a heat processing step, either for preserving a raw material (e.g., milk for cheese) or as the final cook or process (e.g., bread or beef cottage pie).

Food commodities with medium emissions (1–5 kg $CO_2e\ kg^{-1}$ or l^{-1}) tended to be high yielding livestock products or manufactured products, for example:

- milk (1.2–1.4 kg $CO_2e\ l^{-1}$),
- duck in Hoisin sauce ready meal (2.0),
- chocolate-coated cakes (2.5),
- Thai chicken pizza (3.5),
- chicken meat (3.1–4.4),
- duck meat (4.1),
- tea bags (4.1).

Food commodities with high emissions (>5kg $CO_2e\ kg^{-1}$ or l^{-1}) tended to be livestock products and highly manufactured foods, for example:

- pig meat (5.5–9.9),
- beef cottage pie ready meal (7.6),
- packed mild cheddar cheese (9.8),
- beef (10–40),
- lamb shanks and roasted potatoes ready meal (19),
- lamb (27–39).

It is clear from the apple juice example, together with the data presented in Table 19.7 and the above examples, that the heat processing step does not contribute significantly to the CF. The packaging materials have an influence because of the energy-intensive methods for their manufacture, but are insignificant if the food product contains materials derived from a ruminant origin. However, glass and metal packaging, while having relatively high CFs, are environmentally friendly materials in terms of their minimal effect on abiotic resource depletion. This is not the case for plastic packaging that uses oil as its raw material.

References

1. Association of Plastics Manufacturers in Europe (2002) *An Analysis of Plastics Consumption and Recovery in Western Europe 2000*, APME, London.
2. Balch, W.E., Fox, G.E., Magrm, L.J., Woese, C.R., and Wolfe, R.S. (1979) Methanogens: reevaluation of a unique biological group. *Microbiol. Rev.*, **43**, 260–269.
3. European Commission (1991) Council directive 91/156/EEC of 18 March 1991 amending directive 75/442/EEC on waste. *Off. J. Eur. Commun.*, **L 78**, 32–27.
4. Cybulska, G. (2000) *Waste Management in the Food Industry, an Overview*, Campden and Chorleywood Food Research Association Group, Chipping Campden.
5. Gorsuch, T.T. (1986) *Food Processing Consultative Committee Report*, Ministry of Agriculture, Fisheries and Foods, London.
6. Mardikar, S.H. and Niranjan, K. (1995) Food processing and the environment. *Environ. Manage. Health*, **6**, 23–26.
7. Niranjan, K. (1994) An assessment of the characteristics of food processing wastes, in *Environmentally Responsible Food Processing*, Symposium Series, vol. 300 (eds K. Niranjan, M.R. Okos, and M. Rankowitz), American Institute of Chemical Engineers, Washington, DC, pp. 1–7.
8. Hansen, C.L. and Hwang, S. (2003) Waste treatment, in *Environmentally Friendly Food Processing* (eds B. Mattsson and U. Sonesson), Woodhead Publishing, Cambridge, pp. 218–240.
9. Eckenfelder, W.W.J.R. (1961) *Biological Treatment*, Pergamon Press, Oxford.
10. Lancaster Farming *http://www.lancasterfarming.com/18.html.* (accessed July 3, 2005).
11. Litchfield, J.H. (1987) Microbiological and enzymatic treatments for utilizing agricultural and food processing wastes. *Food Biotechnol.*, **1**, 27–29.
12. Kirsop, B.H. (1986) Food wastes. *Prog. Ind. Microbiol.*, **23**, 285–306.
13. Gong, C.S. (2001) Ethanol production from renewable resources. *Fuel Energy Abstr.*, **42**, 10.
14. Yang, S.T. (2002) Effect of pH on hydrogen production from glucose by a mixed culture. *Bioresour. Technol.*, **82**, 87–93.
15. Sturges, M. (2002) *Packaging and Environment – Arguments For and Against Packaging and Packaging Waste Legislation*, Pira International, Leatherhead.
16. De Leo, F. (2003) The environmental management of packaging: an overview, in *Environmentally-Friendly Food Processing* (eds B. Mattsson and U. Sonesson),

Woodhead Publishing, Cambridge, pp. 130–153.
17. European Commission (1994) Council directive 94/62/EC, *Off. J. Eur. Commun.*, **L365**, 10–23.
18. Confederation of European Paper Industry (2004) Special Recycling 2003 Statistics, *http://www.forestindustries.se/pdf/SpecRec2003-092022A.pdf.* (accessed July 5, 2005).
19. European Declaration on Paper Recycling (2006–2010) Monitoring Report 2007. European Recovered Paper Council, *http://www.intergraf.eu/Content/ContentFolders/PressReleases/2008-09_ERPC_AnnualReport_2007.pdf.* (accessed August 15, 2011).
20. Dainelli, D. (2003) Recycling of packaging materials, in *Environmentally Friendly Food Processing* (eds B. Mattsson and U. Sonesson), Woodhead Publishing, Cambridge, pp. 154–179.
21. Bayer, A.L. (1997) The threshold of regulation and its application to indirect food additive contaminants in recycled plastics. *Food Addit. Contam.*, **14**, 661–670.
22. Recoup (2002) Fact Sheet: Use of Recycled Plastics in Food Grade Applications, *www.recoup.org.*
23. Simoneau, C., Raffael, B., and Franz, R. (2003) Assessing the safety and quality of recycled packaging materials, in *Environmentally-Friendly Food Processing* (eds B. Mattsson and U. Sonesson), Woodhead Publishing, Cambridge, pp. 241–265.
24. Baner, A.L., Franz, R., and Piringer, O. (1994) Alternative fatty food stimulants for polymer migration testing, in *Food Packaging and Preservation* (ed. M. Mathlouthi), Chapman and Hall, Glasgow, pp. 23–47.
25. Gontard, N. and Gulbert, S. (1994) Bio-packaging: technology and properties of edible and/or biodegradable materials of agricultural origin, in *Food Packaging and Preservation* (ed. M. Mathlouthi), Blackie Academic and Professional, Glasgow, pp. 159–181.
26. Weber, C.J., Haagard, V., Festersen, R., and Bertelsen, G. (2002) Production and applications of biobased packaging materials for the food industry. *Food Addit. Contam.*, **19** (Suppl.), 172–177.
27. Weber, C.J. (ed.) (2000) *Biobased Packaging Materials for the Food Industry, Status and Perspectives* (EU Concerted Action Project Report, Contract PL98 4045), KVL, The Royal Vetarinary and Agricultural University, Fredriksberg, ISBN 87-90504-07-0, *http://www.biodeg.net/fichiers/Book%20on%20biopolymers%20(Eng).pdf.* (accessed on August 15, 2011).
28. Infoplease (2007) Chlorofluorocarbons, *http://www.infoplease.com/ce6/sci/A0812001.html.* (accessed August 15, 2011).
29. ChemCases.com (2011) Refrigerants for the 21st century, 17. Global warning, *http://chemcases.com/fluoro/fluoro17.htm.* (accessed August 15, 2011).
30. Dellino, C.V.J. and Hazle, G. (1994) Cooling and temperature controlled storage and distribution systems, in *Food Industry and the Environment* (ed. J.M. Dalzell), Blackie Academic and Professional, Glasgow, pp. 259–282.
31. AFEAS *http://www.afeas.org/montreal_protocol.html.* (accessed August 15, 2011).
32. Walshe, N.M.A. (1994) Energy conservation and the cost benefits to the food industry, in *Food Industry and the Environment* (ed. J.M. Dalzell), Blackie Academic and Professional, Glasgow, pp. 76–105.
33. *http://www.iclei.org/EFACTS/ GREENGAS.HTM.*
34. *http://www.ieagreen.org.uk/ghgs.htm.*
35. *http://www.defra.gov.uk/environment/chp/index.htm.*
36. *http://cleanerproduction.curtin.edu.au/industry/foods/energy_efficiencyfoods.pdf.*
37. *http://www.ieagreen.org.uk/doc3a.htm.*
38. *http://www.ieagreen.org.uk/removal.htm.*
39. *http://www.uic.com.au/nip24.htm.*
40. *http://www.defra.gov.uk/environment/ccl.*
41. Berlin, J. (2000) Life cycle assessment – an introduction, in *Environmentally-Friendly Food Processing* (eds B. Mattsson and U. Sonesson), Woodhead Publishing, Cambridge, pp. 5–15.

42. Mattsson, B. and Olsson, P. (2001) Environmental audits and life cycle assessment, in *Auditing in the Food Industry* (eds M. Dillon and C. Griffith), Woodhead Publishing, Cambridge, pp. 174–194.
43. PAS 2050 (2007) Publicly Available Specification for the Assessment of the Life Cycle Greenhouse Gas Emissions of Goods and Services, *www.bsigroup.com*. (accessed in 2008–2009).
44. Wiltshire, J.J., Fendler, A., Tucker, G., and Wynn, S. (2009) Scenario Building to Test and Inform the Development of a BSI Method for Assessing Greenhouse Gas Emissions from Food. Technical annex to the final report to Defra, Project Reference Number: FO0404, *http://www.defra.gov.uk*. (accessed in 2008–2009).
45. Enviros Consulting Limited (2003) British Glass Manufacturers Confederation – Public Affairs Committee, Glass recycling – life cycle carbon dioxide emissions, *http://www.britglass.org.uk/Files/Enviros_LCA.pdf*. (accessed in 2008–2009).
46. Plastics Europe (2005) Reports by I Boustead. "Eco-profiles of the European Plastics Industry LDPE Film Extrusion, Polyethylene Terephthalate (PET) Bottle grade", PET film, *www.plasticseurope.org*. (accessed in 2008–2009).
47. FEFCO (2006) European Database for Corrugated Board Life Cycle Studies, European Federation of Corrugated Board Manufacturers, *www.fefco.org*. (accessed in 2008–2009).
48. Defra (2008) Guidelines to Defra's GHG Conversion Factors – Annexes Updated April 2008, *http://www.defra.gov.uk/environment/business/envrp/pdf/ghg-cf-guidelines-annexes2008. pdf*. (accessed in 2008–2009).
49. Tate and Lyle Carbon Footprint, *http://www.tateandlyle.com/TateAndLyle/social_responsibility/environment/default.htm*. (accessed in 2008–2009).
50. British Sugar Carbon Footprint, *http://www.silverspoon.co.uk/home/about-us/carbon-footprint*. (accessed in 2008–2009).
51. Wiltshire, J.J., Fendler, A., Tucker, G., and Wynn, S. (2008) The contribution of primary food production to lifecycle greenhouse gas emissions of food products. *Aspects Appl. Biol.*, **87**, 73–76.
52. Tucker, G., Wiltshire, J.J., and Fendler, A. (2008) Carbon footprint of British food production. *Food Sci. Technol.*, **22** (4), 23–26.

Further Reading

Ludlow-Palafox, C. (2002) Microwave induced pyrolysis of plastic wastes. PhD thesis. University of Cambridge, Cambridge.

20
Water and Waste Treatment

R. Andrew Wilbey

20.1
Introduction

Most food manufacturers use much more water than the ingredients or raw materials that they are processing. While some water may be used as an ingredient, the greater use will be for cleaning of raw materials, plant cleaning, cooling water, and boiler water feed, each use potentially requiring water to a different specification.

This, in turn, creates similar quantities of used water containing variable concentrations of food components, cleaning chemicals, biocides, and boiler treatment chemicals. Small enterprises may find it preferable to discharge their waste to the municipal sewage system, though larger process plants normally need to carry out either partial or complete treatment of their trade effluent.

20.2
Fresh Water

Water quality requirements in food processing will vary from product to product. Extreme cases of product-led specifications are to be found with beer and whisky production. For English beer production, a very hard water is needed and product waters are "Burtonised," that is hardened by addition of salts to approximate the composition of ground water from Burton-on-Trent. At the other extreme, water used in Scotch malt whisky production is very soft and may contain soluble organic compounds from the peaty highland soils.

Thus the water quality in an area, which is largely determined by its geology, can be a historical determinant of the development of specific sectors within the food industry. Modern water treatment can overcome this constraint by the introduction of a range of physical and chemical treatments, which will be adjusted to the source and end use of the water. These treatments must cope with suspended matter, from trees in floodwater at one extreme to grit and microorganisms at the other;

Food Processing Handbook, Second Edition. Edited by James G. Brennan and Alistair S. Grandison.

plus dissolved minerals, gases, and organic compounds that may give rise to color, taste, and odor problems in the final product.

While most food processors draw their water from the municipal supply and need to carry out very little treatment themselves, some will be required to provide part or all of their water from an untreated source. In most countries an abstraction license or permit will be required.

20.2.1
Primary Treatment

Surface waters, whether drawn from rivers or lakes, are assumed to contain large suspended matter so intakes must be of robust construction and located away from direct flow so that collision damage may be avoided. The intakes would typically be faced with 15–25 mm vertical mild steel bars, gap-width 25–75 mm. Flow through the intake should be <0.6 m s^{-1}, ideally <0.15 m s^{-1}, which would minimise the drawing in of silt [1]. Where ice formation is expected, the intakes must be in sufficiently deep water to permit adequate flow in cold weather despite the surface being frozen over.

Incoming water should then be passed through an intermediate filter, typically with an aperture of 5–10 mm (sometimes down to 1 mm), to remove the smaller debris. Drum or traveling band screens are often used as frequent back washing is needed to prevent blockage. Flow through the screens should be at <0.15 m s^{-1}. With ground water sources there should be little suspended matter and only light duty intermediate screening is needed. The screened water should then be pumped to the treatment plant, the velocity in the pipeline being ≥ 1 m s^{-1} to avoid deposition in the pipe.

Sedimentation may be employed if there are significant quantities of suspended matter in the water, the process being described by the Stokes' law equation:

$$v = d^2 g(\rho_S - \rho)/18\,\mu \tag{20.1}$$

where v is the velocity of the particle, d is the equivalent diameter of the particle, ρ_S is the density of the suspended particle, ρ is the density of the water, and μ is the viscosity of the water at that temperature.

Some water treatment systems use a simple, up-flow sedimentation basin for pre-treatment. In this case the throughput or surface overflow rate is given by

$$Q = uA \tag{20.2}$$

where u is the upward velocity of the water and A is the cross-sectional area. Providing that $u < v$ then sedimentation will occur. Brownian motion will prevent very small particles, ≤ 1 μm diameter, from separating.

Water that has undergone this primary treatment is adequate for cooling refrigeration plant, for example, ammonia compressors, where there is no risk of contact with foodstuffs. For other uses, further treatment is required.

20.2.2
Aeration

Some ground waters may contain gases and volatile organic compounds that could give rise to taints, off-flavors, and other problems. For instance, while hydrogen sulfide may be regarded as a curiosity in spa waters, it is considered objectionable in potable water. Fresh water may be treated by aeration, mainly using either waterfall aerators or diffusion/bubble aerators. The transfer of a volatile substance to or from water is dependent upon:

1) the characteristics of that compound;
2) temperature;
3) gas transfer resistance;
4) partial pressures of the gases in the aeration atmosphere;
5) turbulence in the gas and liquid phases;
6) time of exposure.

Henry's law for sparingly soluble gases states that the weight of a gas dissolved by a definite volume of liquid at constant temperature is directly proportional to the pressure. However, there is seldom time for equilibrium to be achieved so the extent of the interchange will depend on the gas transfer that has occurred in the interfacial film, the diffusion process being described by Fick's first law. Reducing the bubble size will considerably increase aeration efficiency. With very high surface : volume ratios, such as with a spray nozzle, an exposure time of 2 s may be adequate while an air bubble in a basin aerator may need to have a contact time of at least 10 s [2]. All aeration systems must be well ventilated, not only to maximise efficiency but also to avoid safety risks, for example, explosion with methane, asphyxia with carbon dioxide, and poisoning with hydrogen sulfide. Hydrogen sulfide can be difficult to remove from water as it ionizes in solution, the anions being nonvolatile. However, removal may be accelerated by enriching the atmosphere to over 10% carbon dioxide in order to lower the pH and thus reduce ionisation. In aerating such waters, there is the risk that oxygen will also oxidise the hydrogen sulfide to give colloidal sulfur that will be difficult to remove.

20.2.3
Coagulation, Flocculation, and Clarification

Particles of $\leq 1\ \mu m$ in size, including microbes, are maintained in suspension by brownian motion. Many particles are also stabilised by their net negative charge within the normal pH range of water. In soft waters, colloidal dispersions of humic and fulvic acids may give rise to an undesirable "peat stain." This suspended matter may be destabilised by addition of salts, sometimes combined with alkali to raise the pH of the water. Trivalent are more effective than divalent cations, which are much more effective than monovalent cations [3].

Coagulant dosing may be preceded by injection with ozone or hypochlorite to oxidise organic compounds as well as to reduce the microbial loading.

Aluminum sulfate is effective as a coagulant over a pH range of 5.5–7.5 but its popularity in the United Kingdom has declined since an accident at Camelford, Cornwall, in 1988. Ferric chloride or ferric sulfate, with an effective pH range of 5.0–8.5, are now more commonly used in the United Kingdom. The salts may be used in conjunction with a cationic polymer. Where necessary, lime may be added prior to coagulant addition, allowing about 10 s for mixing, though the pH correction and coagulant dosing may be carried out at the same time if high energy mixing is employed. A wide range of mixing systems have been used; including air injection which, though aiding aeration of the water, can cause scum problems. The principle is that rapid dispersion of the chemicals enables a homogeneous dispersion so that aggregation of the colloidal particles to form flocs can then progress under low shear conditions, followed by settlement [4]. Where aeration is used, then much of the flocs may be removed from the surface as a foam.

Inorganic cations (e.g., Pb, As, Se) may also be removed from the water, the extent depending on the pH, coagulant, and oxidation state of the cation.

Clarification of the water is achieved by allowing the flocs to settle out in settling basins, which may be rectangular or circular and be of downflow or upflow contact design. Figure 20.1 illustrates the principles of a circular clarifier basin.

The pH correcting and coagulating agents may be dosed into the feedline to the clarifier or at the discharge into the basin. In the latter case, higher shear mixing would be needed. Coagulant dosing may be accompanied by the addition of water-softening agents such as calcium hydroxide and sodium carbonate at 100–200 $g\,m^{-3}$ [5, 6]. Flocculation is encouraged by gentle mixing in the central portion of the basin (shown as B in Figure 20.1), as the influent water moves downward within the central ducting. Sedimentation then takes place as the water slowly moves outward and then up to the radial collecting trays at the surface. The sediment forms a sludge, which is directed down the sloping base of the basin toward the center by the slowly rotating sweep arms and is then periodically

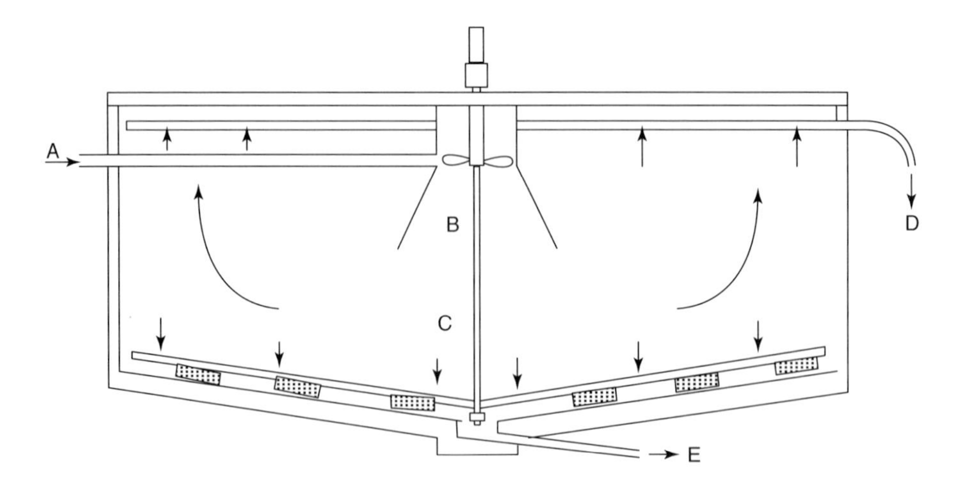

Figure 20.1 Schematic layout of circular clarifier.

pumped away [7]. This sludge is of no further use for fresh water treatment but may be used to aid flocculation of effluent in wastewater treatment.

20.2.4 Filtration

For large-scale fresh water filtration, for example, with water utility companies and where extensive land use can be justified, slow sand filters are very effective. The efficiency is due to a combination of physical separation and biological activity. Each tank is approximately 3 m deep, with gravel over a porous base to allow the filtered water to drain away. Fine silica sand, particle size typically 0.2–0.4 mm, is deposited evenly to a depth of up to 1 m with the influent water maintained at 1–2 m above the sand. Influent water must be distributed evenly across the filter to avoid disturbing the bed. The top 20–30 mm of the sand rapidly develops a complex biofilm, commonly referred to as *zoogleal slime* or *Schmutzdecke*. This biofilm is made up from polysaccharide secreted by bacteria such as *Zoogloea ramigera* and is colonised by protozoa. Bacteria and fine particles become trapped in the slime and are ingested by the protozoa. The resulting increase in the efficiency of filtration can give a two order (99%) drop in the microbial population of the water as well as reducing the levels of dissolved organic and nitrogenous matter [6, 8]. The removal of organic matter may be further improved by including a layer of granular activated carbon in the sand bed and by ozone injection prior to filtration. Throughput can be up to $0.7\ \mathrm{l\ m^{-2}\ s^{-1}}$ ($\approx 60\ \mathrm{m^3\ m^{-2}\ day^{-1}}$). The build-up of biofilm and retained debris plus algal growth slowly reduces the throughput of the filter so that the surface layer needs to be removed and cleaned at intervals of one to six months. Complete replacement of the sand is needed at longer intervals, an expensive operation.

In most industrial sites space is at a premium, so either high rate or pressure filters will be employed. High rate filters are also commonly used by water utilities, either instead of or in conjunction with slow filters. High rate filters may be up-flow or down-flow (Figure 20.2), the flow rate for the former being limited so that fluidisation of the bed does not occur. These filters are normally built as a series of modules with the pipework so arranged that one module can be taken out of service at a time for regeneration by backflushing, which is usually carried out on a daily basis [9]. Units are normally up to $\approx 200\ \mathrm{m^2}$ in surface area.

High rate filters use a coarse silica sand, for example, particle size 0.4–0.7 mm, in a thinner layer, ≈0.4–0.7 m deep, than for the trickling filters. Sometimes dual media beds are used, which can almost double the throughput to $\approx 24\ \mathrm{l\ m^{-2}\ s^{-1}}$ ($2000\ \mathrm{m^3\ m^{-2}\ day^{-1}}$). The faster throughput and more frequent cleaning prevents the build-up of a biofilm and bacterial removal is typically less than 80% (less than one order). The sand must be periodically topped up as some is lost during backflushing when the bed is fluidised.

Pressure filters are suitable for smaller scale operations where space is limited. These units are normally supplied as prefabricated mild steel pressure vessels, with diameter up to 3 m for ease of transport. The main axis may be horizontal or vertical to suit the site (Figure 20.3). Operating pressures are higher, with drops of up to

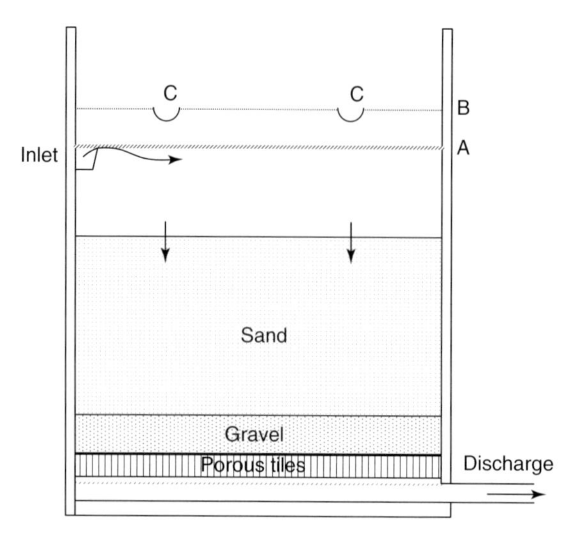

Figure 20.2 Schematic of high rate down-flow filter.

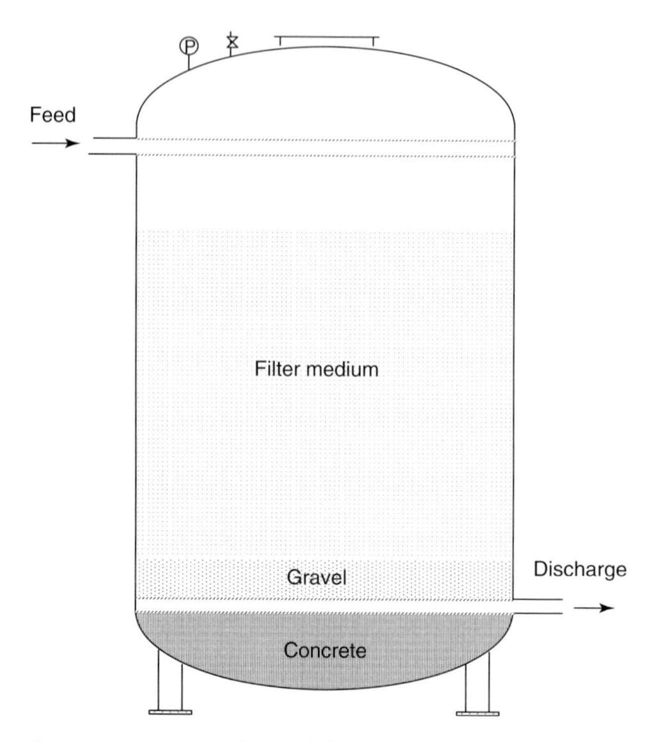

Figure 20.3 Pressurised vertical filter unit.

80 kPa across the filter medium. Filter media may range from sand or anthracite to diatomaceous earth on stainless support mesh or plastic formers wound with monofilament. Pressure filters should be installed in a duplex arrangement to allow one to be cleaned while the other remains in operation.

Cartridge filters have been used extensively for medium and small-scale water filtration, using a range of filter media from stainless meshes and sintered materials to plastics and paper filters. While paper filters are single use, others are more robust and can be cleaned by backflushing. Ceramic filters have been used for small-scale and portable filtration equipment. These can operate down to the micron level and be used for removal of microorganisms, that is, for microfiltration which has been used commercially to provide water with a high microbiological quality. In small-scale personal filters the ceramic filter may also contain silver to add a bactericidal stage.

Occasionally, where ground water has to be used and is contaminated by heavy metals, the water may be treated by nanofiltration or reverse osmosis, the latter also being used for treatment of saline waters (see Chapter 10) [3, 10–12]. In these treatments, the pore size is so small that the filtration is carried out at the molecular level, reverse osmosis being regarded as a diffusion-based process. With nanofiltration there will be appreciable leakage of small ions, for example, Na^+ through the membrane but larger ions and molecules will be retained. The retention of small molecules and ions leads to an increase in the osmotic pressure of the retentate, requiring a higher driving force (>2 MPa) than for other filtration processes [12]. This energy requirement is less than would be required for distillation processes and reverse osmosis provides an economic alternative for desalination. Ion-exchange processes may also be employed for removal of heavy metals [11, 13].

The application of membrane processes for water treatment has been covered in detail by Duranceau [14]. Nanofiltration has been applied to the treatment of contaminated river water to provide drinking water. At Méry-sur Oise, 80% of the water being treated was put through an additional treatment, using microfiltration followed by nanofiltration with 230 Da cut-off at a maximum throughput of 180 000 $m^3\ day^{-1}$. The system achieved a 90% reduction in total organic compounds from a maximum intake value of 3.5 $mg\ l^{-1}$ to typical output values of 0.1–0.25 $mg\ l^{-1}$ [15, 16].

20.2.5 Disinfection

It is assumed that any water used within a food processing plant must, at the very least, be of potable quality. This means freedom from taints, chemical contaminants, and pathogenic organisms. Potable water often contains low levels of organisms capable of causing product spoilage problems. The microbiology of water is discussed in detail by McFeters [17]. Disinfection is any process whereby pathogenic organisms are removed or inactivated so that there is no risk of infection from consuming the treated water. Both chemical and physical methods

Table 20.1 Specific lethalities of common disinfectants ($mg\ l^{-1}\ min^{-1}$, $n = 1$).

Disinfectant	Enteric bacteria	Viruses	Spores	Amoebic cysts
Ozone (O_3)	500.0	5.0	2.0	0.5
Chlorine dioxide (ClO_2)	10.0	1.5	0.6	0.1
Hypochlorous acid (HOCl)	20.0	>1.0	0.05	0.05
Hypochlorite ion (OCl^-)	0.2	0.02	0.0005	0.0005
Monochloramine (NH_2Cl)	0.1	0.005	0.001	0.002

Source: Ref. [1], by courtesy of McGraw-Hill.

may be used. Chemical methods are the most common and are normally based on chlorine or ozone, while physical methods may include microfiltration, irradiation with ultraviolet (UV), and heat. Since the disinfected water is only safe until it is recontaminated, disinfection should be regarded as the terminal treatment in the process. It is imperative that disinfected water be protected from recontamination. Both pasteurisation and UV treatment have been used for preparing washwaters and additive water where the addition of trace compounds is not desired, for example, chlorine-free washwater for cottage cheese curd preparation and for dilution of orange concentrates.

Chemical disinfection using chlorine or chlorine derivatives is the most commonly applied method for large-scale water treatment, followed by the use of ozone, the latter particularly in Canada, France, and Germany.

The Chick–Watson theory remains the principal theory to explain the kinetics of disinfection, where the lethality of the process may be described by the following equation [18]:

$$\ln \frac{N}{N_o} = kC^n t \tag{20.3}$$

where N is the number of pathogens surviving, N_0 is the number of pathogens at t_0, C is the concentration of disinfectant, t is the time, k is the coefficient of specific lethality, and n is the dilution coefficient.

Specific lethalities vary considerably between disinfectants and for the target microorganisms. These lethalities also change at different rates with pH and temperature; thus Table 20.1 should be treated as a semi-quantitative indication of the ranking of disinfectants and their effectiveness against groups of organisms.

20.2.5.1 **Chlorination**

Liquid chlorine was the most commonly used agent for large-scale chlorination. Smaller capacity plant has also used chlorine dioxide, sodium, or calcium hypochlorites.

On dissolution of chlorine in water, chloride ions and hypochlorous acid are generated:

$$Cl_2 + 2H_2O = H_3O^+ + Cl^- + HOCl \tag{20.4}$$

The equilibrium is temperature sensitive: $pK_H = 3.64$ at 10 °C, 3.42 at 25 °C. Ionization of the hypochlorous acid is pH sensitive: $pK_I \approx 7.7$ at 10 °C, 7.54 at 25 °C [1].

$$HOCl + H_2O = H_3O^+ + OCl^- \quad (20.5)$$

Thus, below pH 7, the un-ionised form will predominate and between pH 6–9 the proportion will increase rapidly with pH. Chlorine existing in solution as chlorine, hypochlorous acid, or hypochlorite ions is known as *free available chlorine.*

Safety concerns have led to the replacement of chlorine and chlorine dioxide by sodium hypochlorite solutions. Concentrated hypochlorite solutions are both corrosive and powerful oxidising agents so frequently it is generated as a dilute solution by electrolysis of brine in all but the smallest water treatment plants [5]. Care must be taken to avoid accumulation of hydrogen.

$$NaCl + H_2O \longrightarrow NaOCl + H_2 \quad (20.6)$$

Hypochlorous acid will react with ammonia, ammonium compounds, and ions to form a range of compounds; for example:

$$NH_4{}^+ + HOCl \longrightarrow NH_2Cl + H_2O + H^+ \quad (20.7)$$

$$NH_2Cl + HOCl \longrightarrow NHCl_2 + H_2O \quad (20.8)$$

$$NHCl_2 + HOCl \longrightarrow NCl_3 + H_2O \quad (20.9)$$

$$2NH_4{}^+ + 3HOCl \longrightarrow N_2 + 3Cl^- + 3H_2O + 5H^+ \quad (20.10)$$

$$NH_4{}^+ + 4HOCl \longrightarrow NO_3{}^- + H_2O + 6H^+ + 4Cl^- \quad (20.11)$$

Up to 8.4 mg of chlorine could be needed to react with 1 mg of ammonia. Doses of 9–10 mg chlorine per mg ammonia are recommended in order to guarantee that there would be free residual chlorine in the treated water. The addition level at which the added chlorine has oxidized the ammonia is referred to as the *break point.* Addition of chlorine beyond the break point will enable the disinfection process to continue after treatment and will confer some resistance to postprocess contamination. This can be critical in maintaining the safety of some food processes, for example, the cooling of canned products.

Superchlorination may be used when there are potential problems, for example, due to a polluted source or breakdown. The water is chlorinated well beyond the "break point" to ensure rapid disinfection. If the water is left in this state it will be unpalatable, thus the water then may be partially dechlorinated by adding sulfur dioxide, sodium bisulfite, ammonia, or by adsorption onto activated carbon. Superchlorinated water should be used in the food industry for cooling cans and other retorted products to avoid postprocess contamination due to seal leakage during the cooling stage.

In contrast to superchlorination, the ammonia–chlorine or combined residual chlorination methods make use of chloramines or "bound chlorine" left by incomplete oxidation. Chloramine is 40–80 times less effective as a disinfecting agent than hypochlorous acid but is more persistent and causes less problems with off-flavors and odors than chlorine or dichloramine (which is ≈200 times less effective). Thus any residual chloramine may have a bactericidal effect during distribution. The most common way to apply this principle is to first chlorinate then add ammonia to the disinfected water [18].

20.2.5.2 Ozone

Ozone treatment has been used widely, for both general supplies and for smaller scale treatment of water for breweries and mineral water plants as well as swimming pools. The advantage with ozone is its rapid bactericidal effect, good colour removal, taste improvement, and avoidance of problems with chlorophenols. Chlorophenol production can be critical in the reconstitution of orange juice from concentrates; chlorine-free water is essential if this taint is to be avoided. There is no cost advantage over chlorine.

Ozone is more soluble than oxygen in water and is a more powerful oxidizing agent than chlorine. As such it is frequently used in the pretreatment of water to break down pesticide residues and other organic compounds in the raw water. This has a secondary benefit of inactivating protozoan contaminants such as *Cryptosporidium*, providing a higher quality intermediate for filtration and final treatment. The effectiveness of ozone is related to oxygen, superoxide, and hydroperoxyl radicals formed on its auto-decomposition [19].

$$O_3 \longrightarrow O_2 + O \text{ (in air)} \tag{20.12}$$

$$O_3 + OH^- \longrightarrow HO_2 + O_2^- \text{ (initiation in water by hydroxide ion)} \tag{20.13}$$

$$HO_2 \longrightarrow H^+ + O_2^- \tag{20.14}$$

$$O_2^- + O_3 \longrightarrow O_2 + O_3^- \tag{20.15}$$

$$H^+ + O_3^- \longrightarrow HO_3 \tag{20.16}$$

$$HO_3 \longrightarrow HO + O_2 \tag{20.17}$$

$$HO + O_3 \longrightarrow HO_2 + O_2 \tag{20.18}$$

Auto-decomposition rates increase with pH, radicals, UV, and hydrogen peroxide, but are reduced by high concentrations of carbonate or bicarbonate ions. Ignoring the effect of carbonate, the auto-decomposition rate was described [1] as:

$$\frac{[O_3]_t}{[O_3]_0} = 10^{-At} \tag{20.19}$$

where $[O_3]_t$ is the concentration at time t, $[O_3]_0$ is the concentration at time 0, t is the time, and the value of A is $10^{(0.636\,\text{pH} - 6.97)}$.

Thus at pH 8, the half-life of ozone is about 23 min, depending upon water quality.

20.2.6 Boiler Waters

Boiler waters must be soft so that total solids build up only slowly in the boiler but the water should not be corrosive. Simple treatments rely on lime and ferric chloride for softening and/or carbonate reduction with sodium hydroxide addition to give a final pH of 8.5–10. Residual carbon dioxide should be removed. Further softening can be achieved by phosphate addition to sequester calcium, by ion exchange or by nanofiltration. Dissolved oxygen must be removed, either by de-aeration or by dosing in a scavenger such as sodium sulfite. Dispersants such as tannins or polyacrylates should be added to aid dispersal of the accumulating suspended matter in the boiler while antifoaming agents can reduce the carry-over of water droplets in the steam [20]. Boiler water demand should be reduced and run time extended by returning condensate to the boiler feed [21, 22]. Where relatively unpolluted evaporator condensate is available, this can be treated to provide either boiler feed or soft cleaning water.

While all boiler water additives must be compatible with food production, any steam being generated for direct injection into food must be of exceptional quality and should be supplied from a dedicated generator.

20.2.7 Refrigerant Waters

Any water used in a heat exchanger for food should be of good microbiological quality, on the assumption that, despite precautions, leakage may occur. The risk of leakage may be reduced by operating the refrigerant at a lower pressure than the product; any refrigerant must be checked regularly to ensure that it has not been contaminated by leakage of the food product.

For refrigerant waters at 3–8 °C, potable water should be sufficient. Below 3 °C antifreeze additives may be necessary, becoming essential for operating temperatures below 1 °C unless an extensive ice bank system is used. Such additives are often covered by national legislation. Glycol solutions are often used and calcium chloride ($CaCl_2$) solutions have been widely used throughout the food industry, both as a refrigerant and for freezing, for example, in ice lolly baths.

20.3 Wastewater

All processes will create waste and by-products to a greater or lesser degree, as illustrated in Table 20.2 [23]. These represent not only a loss of ingredients and hence reduced profit from their conversion but an increased fresh water cost plus the additional cost of disposal of the waste created. It is essential that any manufacturing process should be designed and managed so as to minimise both the amount of fresh water used and the quantity of waste produced.

Waste minimisation must start at raw material delivery, assuming that the method of transporting the materials has already been optimised. Bulk tankers must be adequately drained before cleaning and a burst rinse should minimise waste before cleaning [21]. Fresh fruit and leaf vegetables are particularly

Table 20.2 Examples of effluent loads from food processing.

	COD (mg l^{-1})	SS (mg l^{-1})	Water : product ratio (W : P)	Source
Beet sugar refining	1600	1015	–	[21]
Bread, biscuits, confectionery	5100 (275–9500)	3144	–	[22]
Brewing	2105 (1500–3500)	441	4	[22]
Cocoa, chocolate confectionery	9500 (up to 30 000)	500	4	[22]
Fruit and vegetables	3500 (1600–11 100)	500	15–20	[22]
Meat, meat products, poultry	2500 (500–8600)	712	10	[22]
Milk and milk products	4500 (80–9500)	820	1.5	[22]
Milk: liquid processing	About 700	–	12	[24, 25]
Milk: cheese making	≥2000	–	3	[24, 25]
Potato products	2300	656	–	[22]
Starch, cereals	1900	390	–	[22]

susceptible to damage and are likely to arrive in plastic trays that will need to be cleaned before reuse. As in the factory, cleaning solutions should be collected and reused. Wherever possible, solid waste should be collected for separate disposal or recycling and certainly not flushed into the drains to add to effluent disposal problems.

The level of contamination of wastewater is normally measured in terms of the biochemical oxygen demand (BOD_5). This is defined as the weight of oxygen (milligrams per liter, equivalent to grams per cubic metre) which is absorbed by the liquid on incubation for 5 days at 20 °C [3]. The method is time-consuming and may underestimate the potential for pollution if the sample is deficient in microflora capable of degrading the materials present. Measurement of the chemical oxygen demand (COD), based on the oxidation by potassium dichromate in boiling sulfuric acid, provides a more rapid measure of the capacity for oxygen uptake. Sometimes the milder oxidation by potassium permanganate may be used to yield a permanganate value. These values are typically lower than the total oxygen demand based on incineration and CO_2 measurement [24]. There is no direct link between BOD_5 and COD since the relationship depends on the biodegradability of the components in the waste stream. For readily biodegradable wastes there is an empirical relationship: 1 unit $BOD_5 \approx 1.6$ COD units, the latter value rising for less biodegradable materials. The difference between these values may be less if citing the difference between influent and effluent values for a treatment step or process, as these differences will be based on changes in readily metabolized components.

It can be argued that the role of a food factory is to produce food and not to become involved with a nonproductive issue such as effluent disposal, which should be left to specialists. This could certainly be true of small enterprises but for larger factories, it may prove more cost effective to undertake either partial or complete treatment of its trade effluent. In the United Kingdom, the charge for treatment of effluent is calculated by the Mogden formula, based primarily on the volume, COD, and total suspended solids (TSS) [25].

Where an enterprise is to treat its own trade waste, the treatment plant should be located as far away from the production plant as possible, downwind and yet must avoid being a nuisance to neighbors.

In most cases where effluent treatment is undertaken, this is kept separate from sewage which poses greater public health problems and is normally dealt with on a community basis. Occasionally a plant may be built as a joint operation to handle sewage as well as trade waste, in which case more rigorous isolation from the food plant is essential.

20.3.1 Types of Waste from Food Processing Operations

The types of wastewater produced by food processing operations reflect the wide variety of ingredients and processes carried out. Washing of root vegetables, including sugar beet, can give rise to high TSS levels in the effluent. Further processing of vegetables involving peeling and/or dicing will increase the dissolved solids, as will also be the case with fruit processing where sugars are likely to be the major dissolved component. Cereals processing and brewing will create a carbohydrate-rich effluent, while effluent from processing legumes will contain a higher level of protein. The processing of oilseeds will result in some loss of fats, usually as suspended matter. Milk processing creates an effluent with varying proportions of dissolved lactose and protein plus suspended fat. Meat and poultry processing will give rise to effluents rich in both protein and fat.

In most of these examples, there will be particulate waste (i.e., particles greater than 1 mm in size) in addition to the fine suspended matter. These should be removed by screening prior to disposal of the plant effluent into the drain and strainers should be fitted into each drain to collect those particles that have bypassed the screens. Both screens and strainers must be cleaned daily.

Material recovered from screens within the process plant may be suitable for further processing. If screens are not fitted into the process plant, then particles collected on screens at the entry to effluent treatment cannot be reclaimed within the processing plant and must go to solid waste disposal.

20.3.2 Physical Treatment

Sedimentation and/or flotation usually form the first stage of effluent treatment, depending on the particular effluent [3]. Both processes are applications of Stokes' law (see Section 20.2.1). Where the fat may be recovered and recycled within the process, the flotation must be carried out within the production area. Centrifugation provides a rapid and hygienic technique.

The simplest flotation technique is to use a long tank. Wastewater enters at one end over a distribution weir. Flows at ≥ 0.3 m s^{-1} will prevent sedimentation of fine suspended matter but a residence time of about an hour may be needed. Flotation can be hastened by aeration, fine gas bubbles being introduced at the base of the tank by air or oxygen injection, or by electrolysis. Fat globules will associate with the gas bubbles, forming larger, less dense particles that will rise more rapidly to the surface. Bubbles 0.2–2 mm in diameter will rise at 0.02–0.2 m s^{-1} in water. The presence of free, that is unemulsified, liquid fat will act as an antifoam and prevent excessive foaming as the fat agglomerates into a surface layer. This can be scraped off, dewatered and, for instance, sold off for fatty acid or soap production.

Production processes starting with dirty raw materials such as root vegetables will produce an effluent with high TSS, some of which may get past the primary screening within the plant. In this instance a sedimentation or grit tank is needed. A rectangular tank is often used, with flows ≥ 0.3 m s^{-1} to allow grit and mineral particles to sediment without loss of suspended organic material. The grit may be removed from the tank by a jog conveyor and dumped into a skip for disposal, either back onto the farmland, if relatively uncontaminated, or by landfill.

Following either of these pretreatments, the effluent should be collected into balance tanks. These serve to even out the fluctuations in pH, temperature, and concentration throughout the day. Some form of mixing is desirable, both to aid standardisation and to maintain an aerobic environment, thus reducing off-odor generation. This will minimise the use of acids and alkali to standardise the pH of the wastewater to render it suitable for subsequent treatments. Lime (calcium hydroxide) or sodium hydroxide has been used to raise pH while hydrochloric acid is a common acidulant. With dairy wastes the use of sodium hydroxide as the principal cleaning agent normally results in an alkaline effluent, while plants handling citrus and vine products could expect an acid waste stream.

20.3.3
Chemical Treatment

Most of the organic contaminants remaining in the wastewater will either be in solution or colloidal dispersion. At around neutral pH these colloidal particles usually have a net negative charge so addition of polyvalent cations, for instance, aluminum sulfate at pH 5.5–7.5, or ferric chloride (or sulfate) at pH 5.0–8.5, will promote the formation of denser agglomerates that can be sedimented and recovered as sludge. Chemical addition is usually by dosing a solution into the waste stream followed by rapid mixing to ensure even distribution. The treated wastewater is then allowed to stand, to permit formation of the flocs and their sedimentation. Sedimentation may be carried out in rectangular or circular basins. This process is similar to that employed for fresh water treatment (Section 20.2.3) but the quantities of sludge settling out are much higher. Effluent will leave the settling tank via an overflow weir, which should be protected in order to prevent any surface fat and scum from overflowing too. Such fat should be scraped off periodically.

Sludge from settlement vessels typically contains about 4% solids and must be pumped over to an additional settlement tank where about half of the volume can be removed as supernatant and returned to the beginning of the treatment process.

The effluent from the settlement tank may then either be discharged to the sewer as partially treated effluent, incurring a much lower disposal charge, or else taken on to biological treatment.

20.3.4 Biological Treatments

Biological treatments may be divided into aerobic and anaerobic processes. In aerobic processes oxygen acts as the electron acceptor so the primary products are water and carbon dioxide. In anaerobic treatment the primary products are methane and carbon dioxide, with sulfur being reduced to hydrogen sulfide.

While properly run aerobic treatments produce the less polluting effluents, anaerobic treatment has great potential for large-scale treatment of sludge and highly polluted wastewaters. The relative advantages are summarized in Table 20.3.

In general, smaller plants will opt for aerobic treatment while the larger plants may use a combination of aerobic and anaerobic methods.

20.3.4.1 Aerobic Treatment – Attached Films

The trickling or percolating filter provides a simple and flexible means of oxidising dilute effluents. It normally takes the form of a circular (typically 7–15 m diameter) or rectangular concrete containment wall, 2–3 m high, on a reinforced concrete base that includes effluent collecting channels (Figure 20.4). The infill is preferably a light, porous material (about 50% voidage) with a high surface area (up to $100\ m^2\ m^{-3}$), for example, coke or slag. Solid rock may also be used. The particle size varies from 30 to 50 mm, sometimes up to 75–125 mm for pretreatment prior to discharge to a sewer. The use of less dense, synthetic, filter media allows deeper beds to be constructed.

The influent of clarified wastewater is spread over the top surface of the filter by nozzles, mounted either on rotating arms for circular filters or reciprocating bars for the rectangular beds. The surface must be evenly wetted, the liquid then

Table 20.3 Comparison of aerobic and anaerobic processes.

Factor	Aerobic	Anaerobic
Capital cost	(Lower)	Higher
Energy cost	Medium–high	Net output
Influent quality	Flexible	Demanding
Sludge retention	High	Low
Effluent quality	Potentially good	Poor

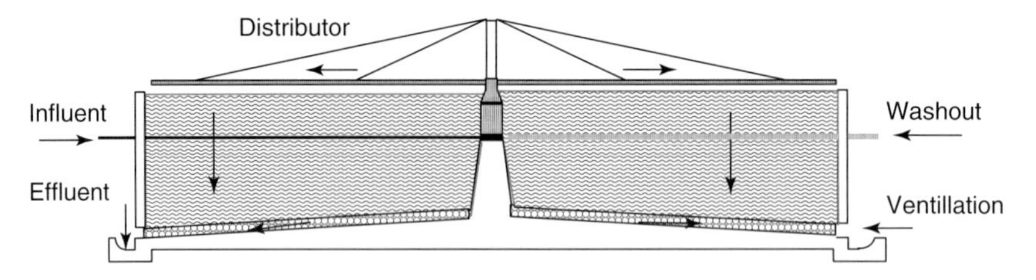

Figure 20.4 Section through a percolating filter.

trickling down through the filter. Filamentous algae often grow on the surface while, within the bed, the medium provides a physical support for a complex ecosystem. This biofilm contains bacteria, protozoa, fungi, rotifers, worms, and insect larvae. *Zoogloea ramigera* is the predominant colonising bacterial species, producing an exopolysaccharide-based support medium, the thickness of which increases with the richness of the nutrients in the influent liquor. The slime may be less than 1 mm thick with lean wastes but reach several millimetres with concentrated wastes, with a higher proportion of fungal mycelium in the latter. The biofilm also contains a range of other heterotrophic species, including *Pseudomonas, Flavobacterium,* and *Alcaligenes* spp., which will absorb soluble nutrients from the influent [24, 26]. Fungi and algae are also present. Fine suspended matter is trapped by the slime and ingested by protozoa, while the burrowing activity of the larvae helps maintain flow through the biomass on the filter. Most of the BOD_5 reduction occurs in upper layers while oxidation of ammonia to nitrate take place in the lower portion of the bed.

With such a complex ecosystem, trickling filters take time to adapt to changes of feedstock. Initiation time can be reduced by seeding the filter with material from another filter running on similar effluent. Care must be taken to avoid feeding inhibitory materials or making sudden, major changes to the feedstock.

Careful management of the filter is essential. Since the biofilm builds up when fed rich effluents, blockage of the channels can occur with ponding of feedstock on the top of the filter. Algae growth on the surface can become excessive and sometimes weeds can grow too, requiring the surface layers to be periodically turned over with a fork. The biofilm can be most easily managed by using two filters in series, alternating their position at intervals of 10–20 days, as illustrated in Figure 20.5. The filter with a rich biofilm is then supplied with a much leaner nutrient stream and loses biofilm, which is sloughed off into the effluent at a higher rate than before. This system is referred to as alternating double filtration (ADF). In some small effluent plants a pseudo-ADF system is used, where the effluent from the trickling filter is collected then passed back through the filter a second time while the primary effluent is held back. The ADF system could be run on a daily cycle basis.

The hydraulic loading on the filter should ideally be less than $1\ m^3\ m^{-3}\ day^{-1}$ (being doubled with an ADF system), with a BOD_5 loading of less than $300\ g\ m^{-3}\ day^{-1}$, normally $\approx 60\ g\ m^{-3}\ day^{-1}$. Filter effluent will contain flocs of biofilm, which must be removed by sedimentation. Clarification is achieved by

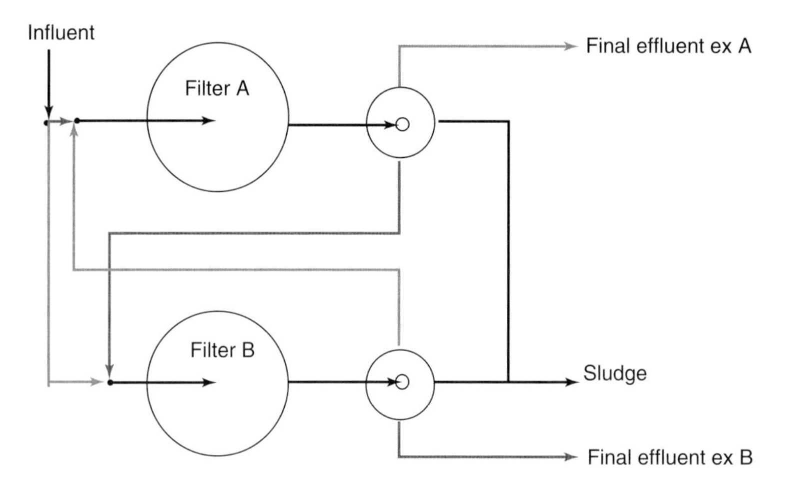

Figure 20.5 Illustration of flows with alternating double filtration.

passing the effluent through a settling tank, for example, as shown in Figure 20.6. The sludge from the settling tank may be combined with that from earlier settling processes. The settled effluent BOD_5 from the primary filter should be less than 30 mg l^{-1}, while that from the final filter should be less than 15, typically ≈4 mg l^{-1}.

While trickling filters are relatively tolerant of varying effluent quality, their oxidative capacity is fairly low. In all but the smallest effluent plants there can be an advantage in using higher rate aerobic systems, if only for pretreatment.

High rate aerobic filters use a very low density medium, for example, plastic tube section, with high voidage (up to 90%) and specific surface areas up to 300 m^2 m^{-3}.

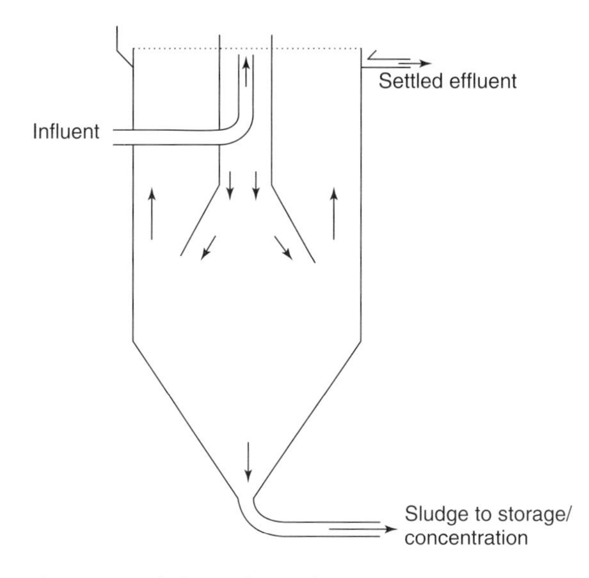

Figure 20.6 Sludge settling tank.

Table 20.4 Examples of aerobic treatments used for food effluents.

Sector	Organization	Location	Reactor	Capacity (t d^{-1} BOD_5)	COD[a]
Brewery	San Miguel	Philippines	Trickling filter + activated sludge	14.0	22.4
Dairy	Entrement	Malestroit, France	Extended aeration	3.0	4.8
Vegetables	Findus	Beauvais, France	Activated sludge	7.5	12.0

[a]COD converted from BOD, assuming 1.6 COD ≈ 1 BOD_5, to aid comparison with data in Table 20.5.
Source: Ref. [19].

Being very light, high rate filters can be mounted above ground level, sometimes being used as modifications mounted above pre-existing treatment plants. The standardised, pretreated effluent is pumped over the filter at a relatively high rate. The hydraulic load is typically 5–10 m^3 m^{-2} day^{-1} with high recirculation rates to give more than 50% removal of BOD_5 [26].

Various types of disk and other rotating contactors have been used for effluent treatment though initially these were more successful for general sewage than for food industry wastes. Rotating contactors use slowly rotating surfaces, which are less than half immersed in the effluent. As the device rotates the damp film is taken up into the air and oxygenated, enabling the surface biofilm to metabolise the effluent. The build-up of biomass must be removed by periodic flushing. Greater success has been achieved with submerged filters, operating with forced aeration in either upflow or downflow modes. The method has found use in smaller plants where activated sludge treatment may be difficult to use.

Some examples of aerobic treatments are given in Table 20.4.

20.3.4.2 Aerobic Treatment – Suspended Biomass

Activated sludge processes have been adapted successfully to the treatment of food wastes, and are attractive for larger plants (>8 t COD day^{-1}) despite the higher energy costs, as they take up less land than trickling filters [27, 28].

The biomass, consisting primarily of bacteria, is suspended in the medium as flocs. Protozoa are present in the flocs as well as free-swimming ciliated species [24]. Oxygen is introduced via compressed air injection, oxygen injection, or by rigorous stirring using surface impellers. Surface mixing is less efficient but simpler and is used in smaller plants. The suspended biomass requires a constant influent composition for optimal operation. The efficiency in removing BOD_5 will depend on the rate of oxygenation, which can be reduced as the medium flows through the tank. Where phosphate reduction is also desired, part of the tank is run anaerobically to encourage additional phosphate uptake into the biomass once aerobic conditions are reintroduced. Partial denitrification is achieved by recycling biomass, a 1 : 1 recycling ratio being associated with 50% denitrification as a result

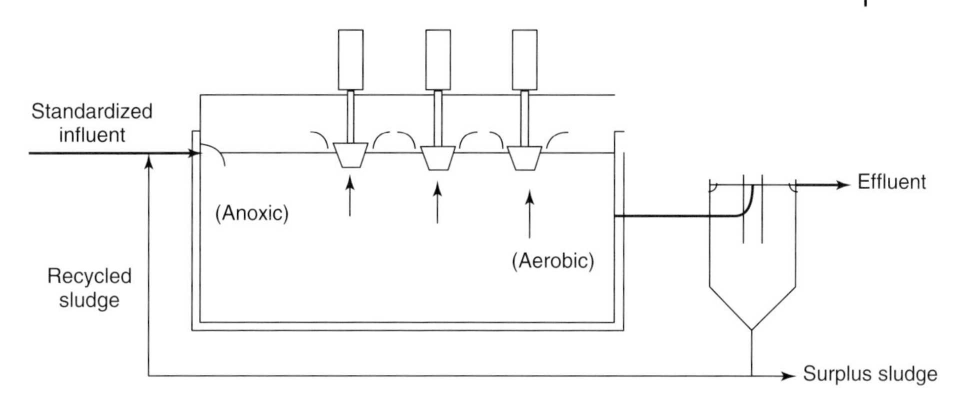

Figure 20.7 Activated sludge fermentation plant.

of anoxic conditions being set up in the first stage of the digestion. Figure 20.7 illustrates the basic principles of an activated sludge plant. Where air or oxygen injection is employed, a tall cylindrical shape may also be used.

Though activated sludge plants are effective in reducing BOD_5, energy costs and sludge production are relatively high. A typical BOD_5 reduction of 95% can be achieved with hydraulic retention times of 10–20 h and biomass retention times of 5–10 days [27]. Higher BOD_5 removal can be achieved by reducing the substrate concentration and throughput, the longer residence time also resulting in lower sludge production. Increasing the substrate loading and/or throughput will increase sludge production while giving a lower percent BOD_5 removal. Both microfiltration and ultrafiltration have now been used as an alternative to settling for removal of a highly clarified effluent from the activated sludge reactor [3, 28].

With dilute process effluents, similar results can be achieved using an oxidation ditch where, again, the biomass is largely suspended. Wastewater is circulated under turbulent conditions ($\geq 0.3\ m\ s^{-1}$) around a channel, 1–2 m deep. The propulsion and aeration is carried out by a series of rotating brushes, as illustrated in Figure 20.8.

Wastewater is constantly fed to the ditch, the overflow passing through a settling tank. Retention times vary between 1 and 4 days with sludge retention times of 20–30 days. Part of the sludge can be fed back to the ditch, giving a relatively low net sludge production. While the oxidation ditch is simpler to operate than the activated sludge system, more land is required.

20.3.4.3 **Aerobic Treatment – Low Technology**

In some areas, where land is inexpensive and rainfall moderate or low, it may be possible to simply use a shallow lagoon, typically 0.9–1.2 m deep, for slow oxidation and settlement of effluents prior to irrigation [9]. Some lagoons may be up to 2 m deep, in which case anaerobic conditions will occur in the lower levels [29]. The lagoon should be lined with clay or other impermeable material to minimise the loss of polluted water into the surrounding soil. This approach has been used for effluent from seasonal canning operations and some dairy wastes. The ecosystem

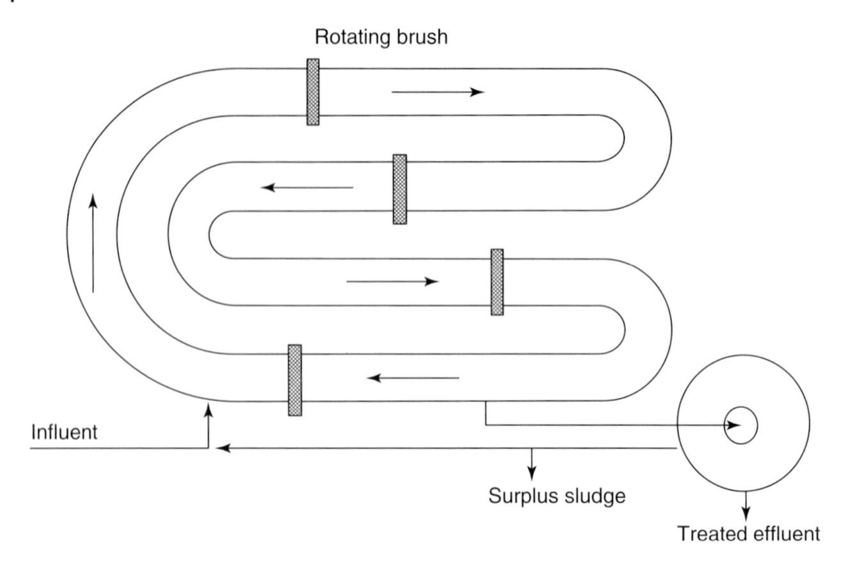

Figure 20.8 Plan view of an oxidation ditch.

in such lagoons is extremely complex with bacteria, protozoa, and invertebrates plus algae and aquatic plants. Though these systems tolerate high organic shock loadings, care must be taken to site such a lagoon well away from the factory, to avoid overloading its oxidation capacity and to prevent leakage into any watercourse. Such partially treated effluent can be used for irrigation using low pressure sprays to minimize drift. Up to 25 mm ($1\ l\ m^{-2}\ day^{-1}$, including rainfall) have been used per 25 day irrigation cycle on grassland.

The lagoon approach can be improved by providing a number of ponds operating in series, providing a residence time of up to three weeks for stabilisation of BOD_5 levels [24].

Treatment of effluents has been achieved by trickling through beds of reeds and/or other semi-aquatic plants, where the root structure supports a complex aerobic ecosystem, similar to that found in lagoons. Soil-based wetland has been found to be more stable than gravel-based systems which can block up. Findlater *et al.* [30] reported 70–80% BOD_5 removal at loadings of 4–20 $g\ m^2\ day^{-1}$, while Halberl and Perfler [31] reported 80–90% BOD_5 removal at slightly lower loadings of 1.5–15 $g\ m^{-2}\ day^{-1}$. Extending the loading to 2–25 $g\ m^{-2}\ day^{-1}$ gave a still wider range of BOD_5 removal (56–93%), the higher loading giving a reduced percentage removal [32]. A reduced reduction in BOD_5 was also noted when feeding the reed beds with treated wastewater where the BOD_5 was $\approx$50 $mg\ l^{-1}$. In experiments with meat processing effluent in gravel-bed wetland trenches ($18\ m^2$) and soil-based surface-flow beds ($250\ m^2$), van Oostrom and Cooper [33] found that COD removal rates increased with loading up to $\approx$20 $g\ m^{-2}\ day^{-1}$. BOD_5 reduction was 79% with untreated effluent from balance tanks at a loading of 117 $g\ m^{-2}\ day^{-1}$, rising to 84% for partially treated, anaerobic, effluent fed at 24 $g\ m^{-2}\ day^{-1}$. These data suggest a bed requirement of 50 $m^2\ kg^{-1}$ BOD day^{-1}

with feed rates of ≈50 l m^{-2} day^{-1}, depending upon influent strength. A complex of beds both in parallel and in series would be desirable to consistently produce a low BOD effluent, with an annual harvest of the above-ground biomass.

20.3.4.4 Anaerobic Treatments

Anaerobic treatments have been applied both to the sludges from aerobic treatment and to treat highly polluted waste streams from food plants. Aerobic treatments can produce 0.5–1.5 kg of biomass per kilogram of BOD_5 removed, so potentially large quantities of sludge may need to be treated. Much of the sludge is low in solids and should first be concentrated, by settlement or centrifugation, to at least 8–10% dry matter.

Most anaerobic reactors are run at 35 ± 5 °C, to ensure methanogenesis. The catabolism of the food components is summarised in Figure 20.9. While the main gaseous products will be methane and carbon dioxide, there will also be small quantities of hydrogen sulfide and other noxious compounds. The gases are normally collected and used on site, for example, to drive a combined heat and power (CHP) plant, with scrubbing of the waste gases where necessary. As with aerobic treatments, membrane processes are now being introduced for sludge concentration and water recovery [28, 33].

There are five main types of anaerobic reactor:

- stirred tank reactor,
- upflow sludge blanket,
- upflow filter,
- downflow filter,
- fluid bed.

The stirred tank reactor is similar to the sludge fermenters used for domestic sewage but with part of the effluent sludge recycled to the influent (Figure 20.10). Sludge from domestic sewage treatment is increasingly being heat-treated before

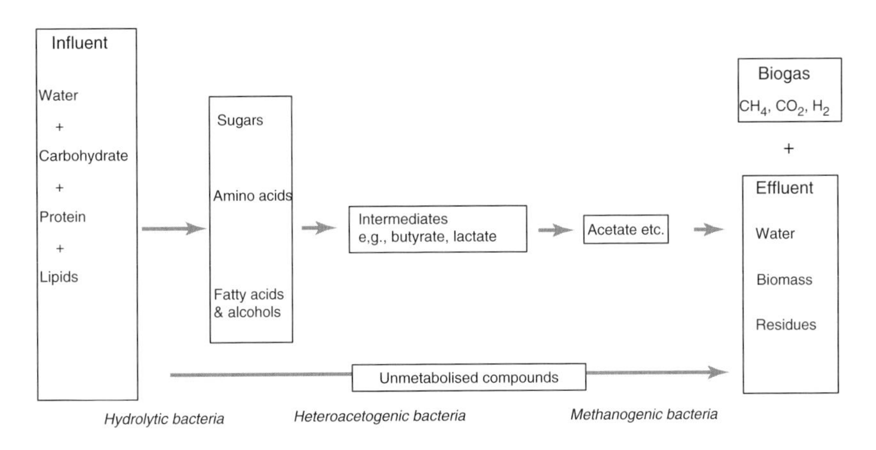

Figure 20.9 Anaerobic catabolism of food components.

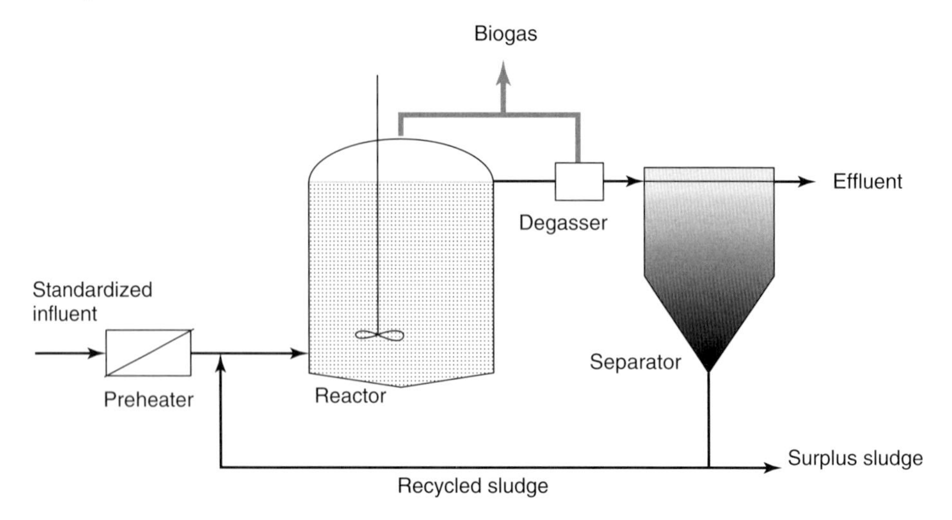

Figure 20.10 Stirred tank reactor.

fermentation to remove pathogens; similar treatment may be required for some food wastes such as slaughterhouse effluent. Mixing may also be achieved by returning biogas from the gas separator. Heat from the CHP plant or other low-grade source is used to preheat the influent and to maintain temperature at 35 °C in the reactor. Biogas may be recovered from both the reactor and separator.

The upflow sludge anaerobic blanket (USAB) reactor does not use mixers but relies on the evenly distributed upflow of the influent (Figure 20.11) plus the bubbles from gas generation during fermentation. The bulk of the biomass forms a granular floc in the lower layer of the reactor, encouraged by a high proportion of short-chain fatty acids (which may be produced by pre-fermentation), the presence of Ca^{2+} and pH >5.5, preferably ≈7.5. For many influents, dosing with $Ca(OH)_2$ will be needed. A lighter floc of biomass also covers the granular floc. The influent quality demands have restricted the application of this type of reactor to wastes from yeast, sugar beet and potato processing wastes. Problems have been encountered with abattoir, dairy, distillery, and maize processing wastes [34].

Anaerobic upflow filter reactors (Figure 20.12) contain a media fill. Crushed rock is the lowest cost fill, similar to that with the aerobic trickling filters, with 25–65 mm rock giving approximately 50% void. About half of the biomass is attached to the medium, the rest being in suspension in the voids. More expensive media can give up to 96% void but with less biomass attached to the medium. The high proportion of suspended biomass limits the throughput as excessive biomass can be lost from the reactor. Similarly, the risk of excessive biomass growth also limits the influent concentration.

Anaerobic downflow reactors tend to use random packed high void media. Though the bulk of the biomass is in suspension, the downward flow of the influent is opposed by the upward movement of the gas bubbles (Figure 20.13).

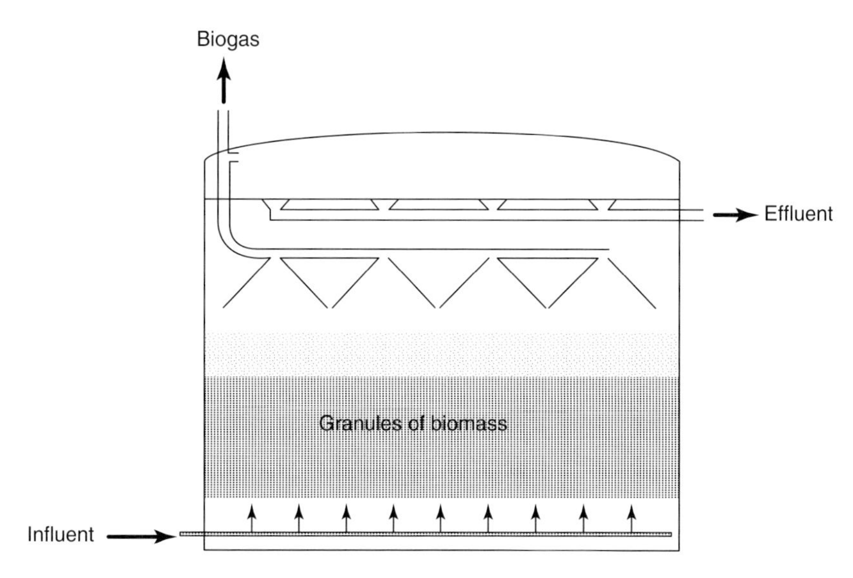

Figure 20.11 USAB reactor.

This upward movement both buoys up the biomass to reduce losses and promotes mixing within the reactor.

Fluid bed reactors provide an improvement on the USAB and upflow filter reactors, combining some of their properties. Fine particulate material, typically sand (particle size $\leq$1 mm) is fluidised by the upflow of the influent liquid plus the gas evolved, expanding the bed volume by 20–25%. Up-flow velocity is critical,

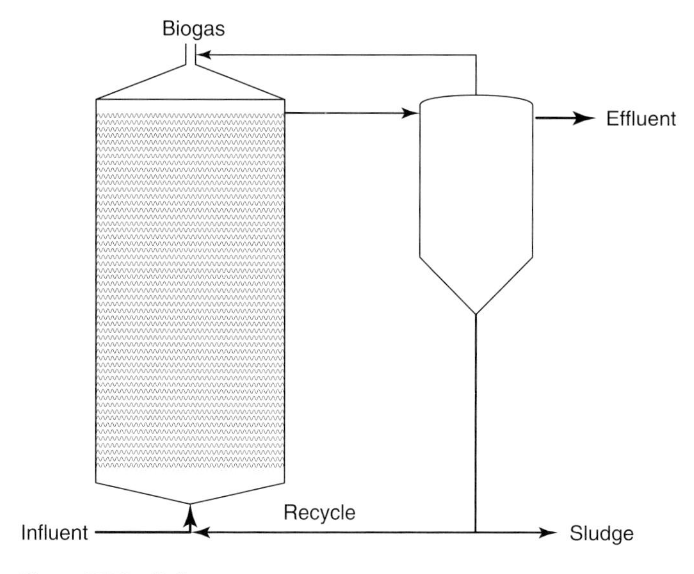

Figure 20.12 Upflow reactor.

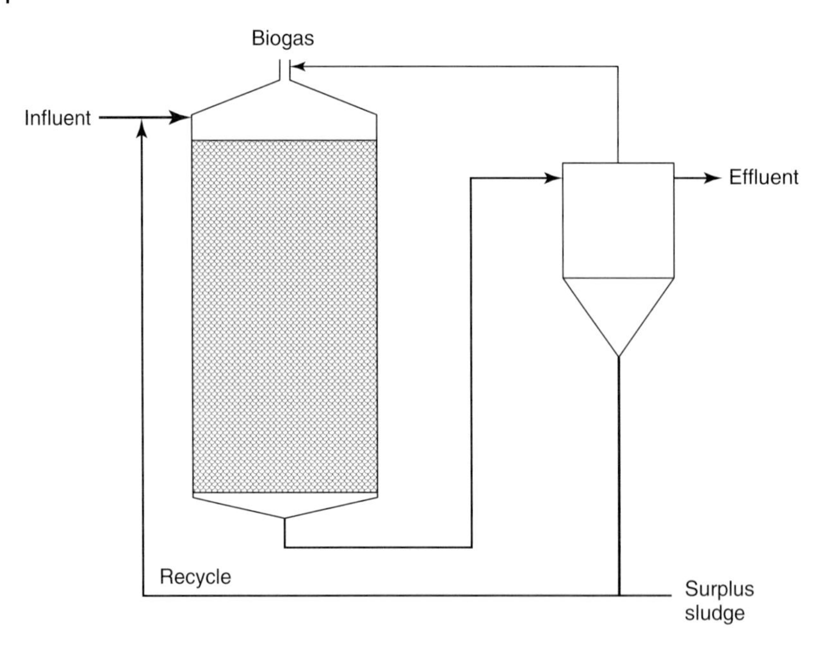

Figure 20.13 Downflow reactor.

typically being 3–8 × 10^{-3} m s^{-1}. The sand and attached biomass are retained within the reactor by reducing the upflow velocity in the wider section at the top of the reactor, suspended biomass being returned to the primary vessel, as shown in Figure 20.14.

The various types of reactor have been increasingly adopted for large-scale processing of food processing wastes, particularly where these wastes are relatively concentrated (Table 20.5). A comparison between types is given in Table 20.6. Typically, between 10 and 30% of the influent COD remains in the effluent so, irrespective of the type of reactor, further processing of the effluent is required. This would entail separation of the sludge and final aerobic treatment of the water.

20.3.4.5 **Biogas Utilisation**

Gas collected from anaerobic treatments will contain primarily methane and carbon dioxide in ratios varying from 1 : 1 to 3 : 1, with traces of hydrogen, hydrogen sulfide, and other volatiles depending on the substrate and operating conditions [35]. It is normally collected in a floating dome gas holder at relatively low pressure, 1 ± 0.5 kPa. The capacity of the gas holder will depend on the output from the digester and whether the gas will be used constantly or only during part of the day. Gas may be used in boilers to raise steam or in a CHP engine. The power produced by the CHP should exceed that used in the anaerobic process, the heat being used to maintain the fermentation temperature in the reactor(s). A flare may be used to automatically burn off surplus gas in the event of a breakdown.

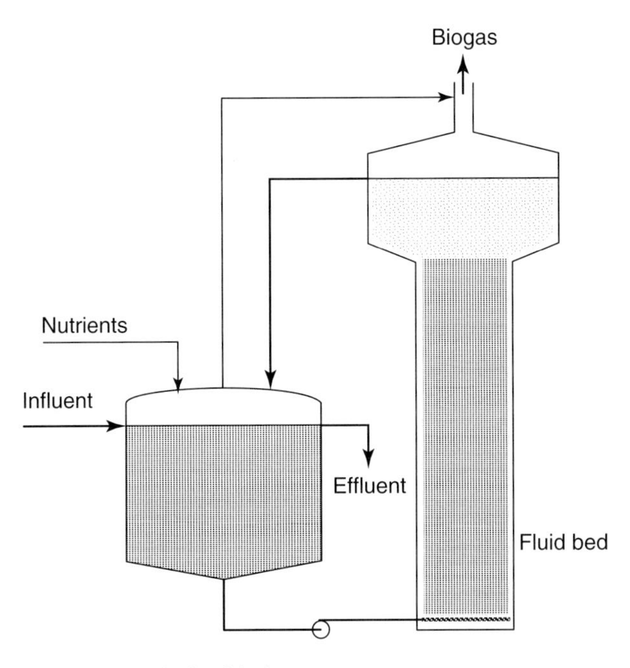

Figure 20.14 Fluidized bed reactor.

Table 20.5 Examples of anaerobic treatments used for food effluents.

Sector	Organisation	Location	Reactor	Capacity ($t\ d^{-1}$ COD)
Brewing	El Aguila	San Sebastian de Los Reyes, Spain	Fluidized bed	50.0
Brewing	Sébastien-Artois	Armentières, France	Sludge blanket	10.0
Canning	Bonduelle	Renescue, France	Digester	18.0
Dairy	St Hubert	Magnières, France	Upflow filter	2.8
Distillery	APAL	Paraguay	Digester	54.0
Distillery	DAA	Ahausen, Germany	Upflow filter	12.0
Sugar	Julich	Julich, Germany	Digester + activated sludge	30.0
Sugar	Südzucker	Platting, Germany	Digester	30.0–38.0
Winery	Canet C.C.	Canet, France	Fluidized bed	4.2

Source: Ref. [18].

Table 20.6 Comparison of anaerobic reactor designs.

Type	Loading (kg COD m^{-3} d^{-1})	Feed (g l^{-1})	Retention time (h)	COD removal (%)
Stirred tank	0.2–2.5	5–10	24–120	80–90
Filter	2–15	1	10–50	70–80
USAB	2–15	10	10–50	70–90
Fluid bed	2–60	2.5	0.5–24.0	70–80

Source: Refs [19, 25, 32].

20.4 Sludge Disposal

Sludge production is a major problem with both aerobic and anaerobic processes, in terms of both their immediate offensive nature and the potential for pollution. Sludge disposal at sea is no longer permitted so the choice is between disposal on land or incineration. Sludges, particularly those from primary settlement, can be highly putrescible but have an advantage over municipal sewage sludge in that levels of heavy metals and organic contaminants are likely to be low.

If agricultural land is nearby it may be economic to dispose of the unconcentrated sludge direct to the land, preferably by injection below the surface to minimise nuisance and avoid run-off [36]. In many cases it will be necessary to first concentrate or thicken the sludge. This may be carried out by gravity settlement for 2 days, sometimes aided by addition of polyelectrolytes. The supernatant must be fed back to the effluent treatment plant. Disposal of sludge from food processing by landfill is uncommon in the United Kingdom but has been used widely in other parts of Europe.

Sludge has been used to aid bio-remediation of contaminated land, where its nutrients and humus help raise the activity of the soil bacteria. It has also been useful in raising the productivity of the poor soils often used for forestry. With further dehydration, such as by belt drying, there is potential for mixing the sludge with straw and composting. The composting process is an aerobic batch process, during which the temperature can rise to 70 °C so that further dehydration occurs. The resulting compost is suitable for horticultural use as a soil conditioner.

Exceptionally, dehydration by belt-press could be carried out to give solids in excess of 30%, for instance, with fibrous sludges from vegetable processing. These high solids sludges could be disposed of by high temperature incineration. As with other waste incineration, the waste gases must be brought up to 800–900 °C to destroy any volatile organic compounds, particularly dioxins, that may be formed at lower temperatures in the initial stages of combustion. Within the European Union, Germany is the greatest user of incineration for sludge disposal [27].

Table 20.7 Simplified example of a river classification. BOD_5 (mg ml^{-1}) to the 90th percentile, DO (minimum percent oxygen saturation) to the tenth percentile, NH_3 (mg N l^{-1}) to the 90th percentile.

Class	Description	BOD_5	DO	NH_3	Biology
A	Very good	2.5	80	0.25	No problems
B	Good	4.0	70	0.6	No significant problems
C	Fairly good	6.0	60	1.3	Some restriction to fish species
D	Fair	8.0	50	2.5	Extraction for potable water after advanced treatment
E	Poor	15.0	20	9.0	Only low-grade abstraction, for example, cooling water
F	Bad	>15.0	<20	>9.0	Very polluted, severely restricted ecosystem and potential nuisance

Source: adapted from Ref. [35].

Incineration will still yield a solid waste, the ash making up to 30% of the original solids and which would normally be sent for landfill as a hazardous waste.

20.5 Final Disposal of Wastewater

In most cases, final disposal of treated wastewater is into a water course where it will be diluted by the existing flow. General requirements are covered by regulations, in the European Community (EC) based on the Urban Wastewater Directive (91/271/EC), usually complemented by consent limits based on avoidance of pollution. The EC approach is now complemented by a move to systems for integrated pollution prevention and control [37] under Directive 96/61/EC. While it may be desirable to recycle water within a factory [38, 39], with food processing such recycling would be constrained by aesthetic as well as cost considerations.

Measurement of river quality is complex as the river is effectively an aerobic fermenter; flowrate, and hence oxygenation playing a vital part of the natural bioremediation processes. An illustration of a simplified river classification is given in Table 20.7.

Discharge licenses may include maxima for flow, temperature, suspended solids, dissolved solids, BOD_5, nitrogen, phosphorus, and turbidity. One processor's waste stream may subsequently (certainly eventually) be another's water source [40].

References

1. American Water Works Association, American Society of Civil Engineers (1990) *Water Treatment Plant Design*, 2nd edn, McGraw-Hill, New York.
2. Dyksen, J.E., (1998) Aeration, *Water Treatment Plant Design*, 3rd edn, American Water Works Association, American Society of Civil Engineers, McGraw-Hill, New York, pp. 61–86.
3. Dégremont (2007) *Water Treatment Handbook*, 7th edn, vol. 1, Lavoisier Publishing, Rueil-Malmaison.
4. Wesner, G.M. (1998) Mixing coagulation and flocculation, *Water Treatment Plant Design*, 3rd edn, American Water Works Association and American Society of Civil Engineers, McGraw-Hill, New York, pp. 87–110.
5. Twort, A.C., Law, F.M., Crowley, F.W., and Ratnayake, D.D. (1994) *Water Supply*, 4th edn, Edward Arnold, London.
6. Benefield, L.D. and Morgan, J.M. (1998) Chemical precipitants, *Water Treatment Plant Design*, 3rd edn, American Water Works Association and American Society of Civil Engineers, McGraw-Hill, New York, pp. 87–110.
7. Willis, J.R. (1998) Clarification, *Water Treatment Plant Design*, 3rd edn, American Water Works Association and American Society of Civil Engineers, McGraw-Hill, New York, pp. 111–152.
8. Choreser, M. and Broder, M.Y. (1998) Slow sand and diatomaceous earth filtration, *Water Treatment Plant Design*, 3rd edn, American Water Works Association and American Society of Civil Engineers, McGraw-Hill, New York, pp. 193–220.
9. Degrémont (1991) *Water Treatment Handbook*, 6th edn, vol. I, Lavoisier Publishing, Paris.
10. Taylor, J.S. and Wiesner, M. (1999) Membranes, in *Water Quality and Treatment*, 5th edn (ed. R.D. Letterman), McGraw-Hill, New York, pp. 11.1–11.71.
11. Bergman, R.A. (1988) Membrane processes, *Water Treatment Plant Design*, 3rd edn, American Water Works Association and American Society of Civil Engineers, McGraw-Hill, New York, pp. 335–378.
12. Lewis, M.J. (1996) Pressure-activated membrane processes, in *Separation Processes in the Food and Biotechnology Industries* (eds A.S. Grandison and M.J. Lewis), Woodhead Publishing, Cambridge, pp. 65–96.
13. Gottlieb, M.C. and Meyer, P. (1998) Ion exchange processes, *Water Treatment Plant Design*, 3rd edn, American Water Works Association and American Society of Civil Engineers, McGraw-Hill, New York, pp. 299–334.
14. Duranceau, S.J. (ed.) (2001) *Membrane Practices for Water Treatment*, American Water Works Association, Denver, CO.
15. Peltier, S.J., Benezet, M., Gatel, D., and Cavard, J. (2001) What are the expected improvements of a distributed system by nanofiltered water? in *Membrane Practices for Water Treatment* (ed. S.J. Duranceau), American Water Works Association, Denver, CO, pp. 371–384.
16. Ventresque, C., Gisclon, V., Bablon, G., and Chagneau, G. (2001) First-year operation of the Méry-sur-Oise membrane facility, in *Membrane Practices for Water Treatment* (ed. S.J. Duranceau), American Water Works Association, Denver, CO, pp. 421–446.
17. McFeters, G.A. (ed.) (1990) *Drinking Water Microbiology*, Springer-Verlag, New York.
18. Haas, C.N. (1999) Disinfection, in *Water Quality and Treatment*, 5th edn (ed. R.D. Letterman), American Water Works Association, McGraw-Hill, New York, pp. 14.1–14.60.
19. Singer, P.C. and Reckhow, D.A. (1999) Chemical oxidation, *Water Quality and Treatment*, 5th edn, American Water Works Association, McGraw-Hill, New York, pp. 12.1–12.51.
20. Anonymous (1991) *Water Treatment Handbook*, 6th edn, vol. II, Degrémont, Lavoisier Publishing, Paris.
21. Hills, J.S. (1995) *Cutting Water and Effluent Costs*, 2nd edn, Institution of Chemical Engineers, Rugby.

22. Wang, L. (2009) *Energy Efficiency and Management in Food Processing Facilities*, CRC Press, Boca Raton, FL.
23. Anonymous (1986) *Food Processing Research Consultative Committee Report to the Priorities Board*, Ministry of Agriculture, Fisheries and Food, London.
24. Horan, N.J. (1990) *Biological Wastewater Treatment Systems: Theory and Operation*, John Wiley & Sons, Ltd, Chichester.
25. Walker, S. (2000) Water charges: the Mogden formula explained? *Int. J. Dairy Technol.*, **53** (2), 37–40.
26. Gray, N.F. (1989) *Biology of Wastewater Treatment*, Oxford University Press, Oxford.
27. Wheatley, A.S. (2000) Food and wastewater, in *Food Industry and the Environment in the European Union: Practical Issues and Cost Implications*, 2nd edn (ed. J.M. Dalzell), Aspen Publishers, Gaithersburg, MD, pp. 111–229.
28. Dégremont (2007) *Water Treatment Handbook*, 7th edn, vol. 2, Lavoisier Publishing, Rueil-Malmaison.
29. UNEP International Environmental Technology Centre (2002) *Environmentally Sound Technologies for Wastewater and Stormwater Management: An International Source Book*, IWA Publishing, London.
30. Findlater, B.C., Hobson, J.A., and Cooper, P.F. (1990) Reed bed treatment systems: performance evaluation, in *Constructed Wetlands in Water Pollution Control* (eds P.F Cooper and B.C. Findlater), Pergamon Press, Oxford, pp. 193–204.
31. Halberl, R. and Perfler, R. (1990) Seven years of research work and experience with wastewater treatment by a reed bed system, in *Constructed Wetlands in Water Pollution Control* (eds P.F. Cooper and B.C. Findlater), Pergamon Press, Oxford, pp. 205–214.
32. Coombes, C. (1990) Reed bed treatment systems in Anglian Water, in *Constructed Wetlands in Water Pollution Control* (eds P.F. Cooper and B.C. Findlater), Pergamon Press, Oxford.
33. van Oostrom, A.J. and Cooper, R.N. (1990) Meat processing effluent treatment in surface-flow and gravel bed constructed wastewater wetlands, in *Constructed Wetlands in Water Pollution Control* (eds P.F. Cooper and B.C. Findlater), Pergamon Press, Oxford, pp. 321–332.
34. Stronach, S.M., Rudd, T., and Lester, J.N. (1986) *Anaerobic Digestion Process in Industrial Waste Treatment*, Springer-Verlag, Berlin.
35. Barnes, D. and Fitzgerald, P.A. (1987) Anaerobic wastewater treatment processes, in *Environmental Biotechnology* (eds C.F. Forster and D.A.J. Wase), Horwood, Chichester, pp. 57–113.
36. Department of the Environment (1989) *Code of Practice for the Agricultural Use of Sewage Sludge*, HMSO, London.
37. DEFRA (2002) *Integrated Pollution Prevention and Control*, 2nd edn, Department for Environment, Food and Rural Affairs, London.
38. Environment Agency (2003) GQA Methodologies for the Classification of River and Estuary Quality, *http://www.environment-agency.gov.uk/science/219121/monitoring/184353/* (accessed January 25, 2003).
39. Lens, P., Pol, L.H., Wilderer, P., and Asano, T. (eds) (2002) *Water Recycling and Resource Recovery in Industry: Analysis, Technologies and Implementation*, IWA Publishing, London.
40. Asano, T., Burton, F.L., Leverenz, H.L., Tsuchlhashi, R., and Tchobanoglous, G. (2007) *Water Reuse: Issues, Technologies and Applications*, McGraw-Hill, New York.

21
Process Realisation

Kevan G. Leach

21.1
Synopsis

This chapter sets out to cover key techniques in realising a product concept into manufacturing, designing a process around the concept, and assessing the plant performance. It covers the basic steps needed to ensure that a practical and economic design for the plant can be realised for the product concept.

Starting with these steps in realising a process, this chapter builds in stages to produce a methodology that is generic to all processes. A section on process economics is included and the chapter concludes with an outline of systems thinking, an analysis of process performance through natural process run charts and other techniques, and an outline of lean principles.

In essence, the chapter leads the reader through the stages of concept, design, financial justification, and process performance. Although project management is not covered here, the reader is directed towards a useful text on the topic [1].

The chapter concludes with a description of several tools useful in addressing and improving performance.

21.2
Manufacturing Design

There is increasing pressure throughout the food industry to improve efficiency. This includes improving process yields and minimising wastes, using human and physical resources efficiently and maintaining product quality consistently. It is also apparent that improvements in instrumentation, computing software and hardware, and information technology can be deployed to great effect to help meet these objectives. However, most courses in food science and technology barely touch on these issues, so that there is a knowledge gap between the food specialists and professionals on the one hand and the technology specialists on the other. Historically, processes have (and the author sees that many still have)

Food Processing Handbook, Second Edition. Edited by James G. Brennan and Alistair S. Grandison.

been configured on existing physical constraints and time pressures to achieve manufacturing output.

The main aim of this chapter is to help remedy this situation. The chapter covers some of the key principles and concepts underlying modern techniques for operations management, which are directly applicable in the food industry. It is based on courses taught and on experience gained in building and operating a wide variety of food processes by the author over many years. It is not a catalogue of the latest software or technology. Instead it aims to explain the how and why and limitations of some of the more important techniques. Thus it is hoped that it will help students and professionals in the industry understand how a variety of important and recurrent problems can be approached.

The chapter is divided into four main sections covering:

- preliminary specifications,
- investment criteria,
- operational efficiency, measurement, and systems, and
- tools for process improvement.

21.3 Process and Plant Design

As in all design processes there is a preferred sequence of steps to go through in order to develop an optimum design. An optimum design would be one that also accommodates ease of operation, access, avoids waste of materials, and is economical to run in terms of materials conversion, energy, hygiene, and financial aspects. Practically, things are never laid out in order; briefs may be sketchy and objectives a little hazily defined. However, and while there is no substitution for asking many and varied questions, the plant designer should have the overall project delivery in mind and be focused on the end objectives. These are simply to specify a manufacturing operation that produces the required products to the right quality and required food safety parameters and which can be built and run economically.[1)]

As mentioned, a brief may be sketchy and/or the manufacturing premises may already be determined, so the following methodology will guide the practitioner through the appropriate path to a satisfactory solution.

The preferred order for approaching plant design is as follows:

1) Statement of requirements (SoR)
2) Product
3) Process
4) Services
5) Plant/building.

1) While the plant has to be safe within the requirements of legislation, coverage of such requirements is not the province of this text.

Adherence to this methodology, as much as is practical, will go a long way to mitigating subsequent limitations in capability, capacity, control, and both the economic viability and quality of the product and plant.

Clearly, a simple statement of the five base elements of approach does little in itself to suggest a solution. However, adherence to this process as closely as possible will go a long way to producing an optimal solution.

21.3.1
Statement of Requirements

At the start of a project – whether building an entire factory or just changing one production line, in whole or in part – the best place to begin is for the project leader or champion to produce a statement of requirements.

The generation of such a document may at first sight seem trivial or simply not necessary; after all, surely the product development team have produced all the documents needed? And the business, surely it knows how many items per day or per week are needed? Of course the finance and production teams know just how much this product has to cost to make, its specification, and the margin? Often, the reality is that the answers to these statements are not fully defined or agreed at the same level by all. At best, different groups can be working from different versions of criterion; at worst the groups may not have sufficient information to make complete decisions about the new product or process.

And if the business is fully communicating and aware of all the requirements, what about the suppliers of new machinery or facilities that might be needed? How will they know what is needed in sufficient detail without a common and agreed set of information? This assumes, of course, that at the tender stage all the potential suppliers were given the same criteria, including the way in which the tender documents were to be submitted, on which to base their quotations. Without these tender comparison is made much more complex. Lack of commonality and specification also leaves the door open for specification creep and uncosted and unscheduled changes which have the potential to add cost and delay into any project.

The simplest and most comprehensive solution is to pull together a full statement of requirements for the project. At the initial stage, the document is used by the business to say to itself what it intends to do and no more. This sets boundaries to the investment, and helps avoid specification creep with all the adverse financial consequences. Later, the document can be changed to form the ITB (invitation to bid), which will be sent out to potential suppliers of equipment and services. With an agreed scope on volumes rates and sizes, the business can optimise the time needed to launch.

The next question is: What should a statement of requirements contain? We will answer this by looking at a typical table of contents for such a document and expanding the elements by way of description. Of course not all statements of requirements will be the same and some information may need to be simplified (or better, separated form the whole) for the sake of confidentiality. However the principle remains the same and the list below serves well as a template.

A typical statement of requirements will contain the following as a minimum:

- **Introduction, purpose, and contents**: A simple statement of what the project is about, why it is being considered and what the document contains.
- **Schedule of data and reference document**: Reference to and copies of as much data as is to hand, such as quantities to be made (needed so that plant can be sized and ingredients sourced); production rates (needed for plant sizing and process design); recipe and ingredient specification (including photographs from new product development (NPD); process and cost and margin basis for proposed product) at various manufacturing volumes; process steps and timings (estimates of hazard analysis and critical control point (HACCP) requirements and estimates of hygiene and cleaning requirements); and packaging specifications.
- **Operation requirements**: These include any specific product handling characteristics; area size and shape availability; environmental issues; noise; fume extraction; machinery outline requirements; overall equipment effectiveness (OEE) and variation criteria (see later sections in this chapter); performance criteria and yields.
- **Full financial analysis**: Based on net present value (NPV) or internal rate of return (IRR) criteria (see later sections in this chapter).
- **Project team members**
- **Outline program**
- **Drawing of proposed factory areas**
- **Proposed tender format.**

At the start of this process, not all the information will be to hand. Indeed the whole process will be iterative as specifications, sales volumes, and pricing structures are refined. But the presences of the document will help to focus efforts and will allow the business to make a more informed and rational assessment of the likely benefits, opportunities, and potential areas of overlap or conflict.

In summary, in its statement of requirements the business:

- explains to itself just what it wants to do, which entails explaining to all what is going to be done – no more, no less;
- avoids specification creep, which is the financial graveyard of many projects due to unplanned expenditure;
- produces a starting point for an ITB, which will be needed by potential contractors to allow proper and practical comparison of received tenders.

The statement of requirements is a critical component in any project realisation.

21.3.2 Product

As early as possible the designer needs to know in some detail what is to be made and what role the product needs to fulfil. This is not simply a statement of the item to be made, that is, a ready meal or a chilled ready to eat item, but a description of what type of food, what ingredients, what conditions (storage and use) for those ingredients, what states and what spoilage or risk criteria those ingredients may pose. In conjunction with the ingredients and finished product there needs to be a HACCP analysis of the ingredients, processing steps, and finished product. In short it is a comprehensive statement of what the process has to provide and the environmental conditions required to safely handle the ingredients, process, and store the food product.

In addition, it is useful at this stage to describe the intended market and how the food product is to be packaged, distributed, and consumed. At first sight, this may seem a little excessive. However, much can change in a process and equipment to accommodate specific processing or packaging requirements.

Lastly, estimates of the anticipated volumes and the likely consumption profiles are needed so that a volume and rate analysis can be made. There is a whole world of difference between making a single meal under kitchen conditions, making several meals for a small event, or continuously making thousands of meals per day every day. In order to handle the larger volumes the equipment specifications need to be more tightly defined and the service requirements (including building flows) are greater.

21.3.3
Process

Once the product and the HACCPs (see also Chapters 15–17) have been defined in relation to each other, the next step is to consider how the product processing stages and control parameters for each step of manufacture are defined and fit together.

This is, by definition, an iterative operation as the product, manufacturing, and handling processes will need to be refined to accommodate the HACCP requirements.

The development of the manufacturing process is a two-stage paper exercise for which the preferred tool is a process flow chart. The stages are:

1) Develop the outline process
2) Develop the detailed stage-by-stage processes.

Process flow charts are a powerful graphical tool for describing how a series of events or stages combine to form a process.[2] An extension of this is the deployment flow chart, where different departments and processes (e.g., preparation or high care assembly) combine in the manufacturing process. The deployment flow chart is particularly useful in identifying physical transfers that lead to delay and excessive intermediate stock levels.

The purpose is that as layers of complexity are added to the basic flow chart, the determinations of process equipment start to emerge. With the addition of further layers, such as the service requirements (power, etc.) for the process machines, measurement requirements, waste streams, and factory room environments become evident. This process also allows the designer to calculate room sizes and

2) Flow charts produced with FlowMap4 – Management-NewStyle *www.management-newstyle.co.uk/*. See also Refs. [2, 3].

locations for the services such that those locations do not compromise the logical flow for the product or people.

An essential element in determining the process and the process flow is that of the point at which the process becomes irreversible. The action of processing food sets in motion a series of physiochemical reactions, which run at a given rate. Some of these are slow, some fast. Some only start when ingredients are brought together (i.e., mixing a cake) and some require a change of temperature. For example, the transformation of dough to a finished cake suitable to eat which takes place during baking. What is important here is that the point of irreversibility is often coincidental with a transfer of the part-processed product from low risk operations to high care operations.[3)]

Moreover, the process point at which this occurs is often (but not always) a constraint in the process. The management of constraints has a considerable effect on overall process economics and careful process design to manage these with regard to both upstream and downstream processes yields considerable operational and financial benefits if done early in the process design. See also Refs. [4] and [5] for more on the Theory of Constraints.

An example of an outline process flow chart for our intended product will have a basic form as shown in Figure 21.1. The chart needs both a start point and an end point, which are traditionally placed at the top and bottom of the chart. It is common practice to product these charts in a vertical form, starting at the top of the page. From this chart, the reader will see that there are only basic steps. These describe the basic stages of manufacture. At this stage, the diagram does not show volumes, waste, or specific process details.

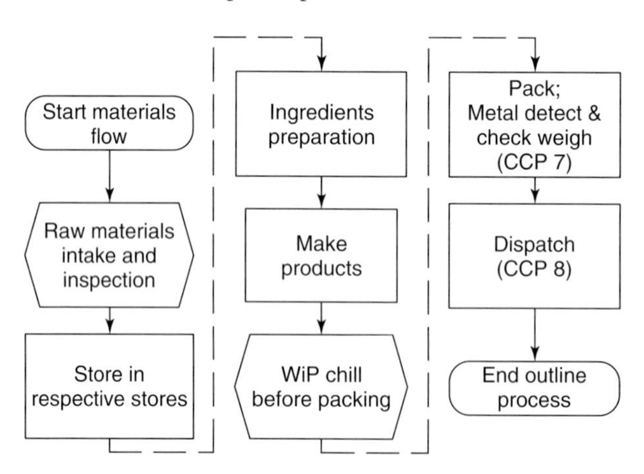

Figure 21.1 Example outline process flow chart.

3) *Low risk* is where the product is to receive further processing, say a heat treatment stage that will destroy spoilage organisms before the product is consumed. This may be by the processor or the consumer. *High care* operations are where there is to be no further treatment of the product before it is consumed and as such, the product needs to be safe to consume.

Expanding this basic diagram to show the interrelationship with other process steps and streams we obtain the flow chart shown in Figure 21.2. In addition, the high care boundary has been added to the diagram. Operations within this boundary need to be carried out in separate areas and waste from earlier and later processes cannot pass back through this area. Also, the critical control points (CCPs) have been added to show where in the process these are to be located.

At this stage, the flow charts represent the basic functional blocks of both the process and the basic functional areas of the factory. It is now possible to structure the basic process areas as blocks and to see how these might start to fit together. However, before there can be a detailed area or room design, more information is needed: specifically, product volumes, manufacturing rates, assessments of types and size of processing machines, specific environments, waste, and number of staff to allow a staff circulation analysis.

Detailing these operations allows for the stages of:

- room sizing,
- specifying air change and air-condition requirements,
- lighting, power, and services,
- estimates of plant sizing,
- estimates of plant room requirements.

The process continues in finer detail until the individual component flows, the processing requirements, the volumes and run rates, the machine sizing and services requirements, the measuring equipment, and approximate room size determinations are made. The input and output flow arrows also show where this part of the process links to predecessors and successors.

At this stage, a whole factory layout can be constructed. Subsequent iterations of the above methodologies and processes refine the solutions to a workable compromise.

21.4 Process Economics – Investment Criteria

Criteria for investment can be expressed in a number of forms. These may range from a simple return on investment (RoI), the application of a discounted cash flow (DCF) model or a model based on the internal rate of return (IRR). While these different criteria will be discussed individually, they all rely on one premise: that the enterprise or project to be undertaken will generate a return on the investment to a given principle within a given time frame. Different organisations will themselves have individual preferences for recovery, payback, and the time in which this is to be achieved. But what is common is that there needs to be a benefit from the investment.

In considering an investment, the essence is to evaluate the returns by looking at the time value of money. This is an important concept as it means that the value of a pound (£) or Euro (€) today is more than the value of a pound (£) or Euro (€) at some time in the future. This because over the period of time between today

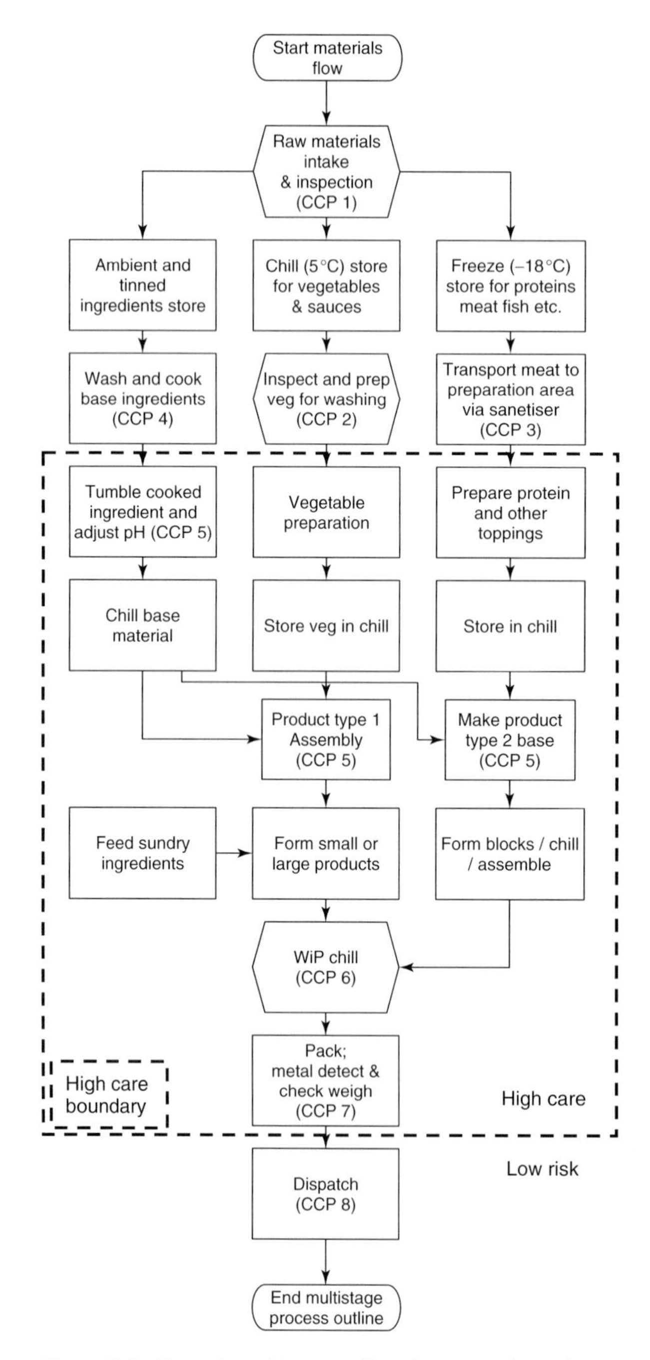

Figure 21.2 Example multi-process flow diagram with parallel processing streams, CCPs, and high care boundary.

and the point in the future there is the opportunity to invest and make a return on that investment. There is also the cost of paying interest to a lender on borrowed money. Capital purchases are noted in the company books as having a write-down value. That is, they theoretically have a resale value. The amount of interest earned will be a function of the interest rate, the amount invested (as interest rates vary with level of investment), and the period of investment. In addition, interest rates are coupled to inflation rates which also reduces the value of money over time.

This can be thought of in mortgage terms; that is, you borrow money to buy a house. Provided the realisable market value of the house is greater than the debt, then the loan is theoretically secure and the house can be sold to settle the debt. But if the value of the house (or asset) falls below the value of the debt then additional funds will be needed to settle the debt. This is effectively negative equity and in business terms a nonrealisable project. General inflation may be stated as a given number, but if a project or investment needs the input of specialist services which are (or may become) in short supply, then the premium needed to source these services may be substantially larger than the general inflation rate. In practice, when considering and investment or project opportunity, all monetary flows, both positive (return from the investment) and negative (into the investment) need to be analysed.

21.4.1 Considerations for Investment

In considering an investment there are a number of monetary flow that need to be taken into account, including:

- costs associated with acquiring or building the facility including site preparation;
- costs associated with researching the process parameters; and
- costs associated with procuring and commissioning the process plant.

Up to this point all the monetary (cash) flows are negative (i.e., expenditures). This is expenditure on gaining the assets. Not everything in this group of cash flows will be attributable to capital, as some items such as computer software are generally classed as revenue items.

In making the investment the objective is that at some point in the future, the investment pays off the debt and the project starts to generate an income stream. Even when the new process or factory is working and producing saleable product all is not necessarily secure. For not only has the project to pay back the investment, with interest, it has to produce product at an economically viable rate to secure the positive cash flows.

Excluding repayment of the capital investment, economic viability of a manufacturing operation is governed by market factors. To produce a product economically, the cost of manufacture must be less than the price at which it can be sold. This looks straightforward at first sight. The business buys in materials at a price; processes those materials to add value and then sells the resultant product for a profit. But selling price and market price are not the same thing. And the internal costs of manufacture can be a major controlling factor affecting the selling price. Of course, the process yield and efficiency are also major influencing factors on internal costs.

The main factor of a product or process that determines success is economic viability.

For a process to be economically viable, it must cost the business less to make a product than the price it receives from selling the product. This may be obvious, but the questions that need to be asked are: Which, how, and by what magnitude individual elements of the manufacturing process and the raw material supply affect the manufacturing price? The NPV procedure (see Section 21.4.2.2 and Figure 21.4) can be used to assess various material price scenarios on viability.

21.4.2 Returns on Investment and the Time Value of Money

21.4.2.1 Simple Return on Investment

Simple return on investment is the simplest criterion for evaluating whether an investment is worth undertaking. It is calculated from the value of the investment divided by the anticipated profit to be generated. This excludes the daily running costs of the plant and hence masks the financial analysis of this part of the operation. RoI is calculated as:

$$\text{RoI} = \frac{€(\text{capital})}{€_{\text{year}}(\text{annual profit})} = x_{\text{years}}$$

where annual profit is the anticipated annual income minus the annual expenditure.

RoI is commonly used where a brief overview of the investment potential is needed. Basic weaknesses are that it does not account for the cost of money and is not suitable for comparing projects of dissimilar investment or duration and generally only looks at the capital investment. It does not easily allow for the testing of various scenarios on materials/labour/energy price fluctuations (i.e., the revenue streams).

21.4.2.2 Net Present Value

The NPV is a method used to analyse, in advance, the likely profitability of an investment or project. It particularly relates to capital budgeting where the present values of monetary (cash) inflows is reduced by the present value of monetary (cash) outflows (i.e., expenditure). The process is sensitive to the reliability of future income that an investment or project will generate. Although it is often calculated over a given investment period at a fixed interest rate, it can for greater precision be decomposed into small time scale steps where the interest rates can be adjusted over time blocks. This produces a more nonlinear model. The test for NPV is: if NPV > 0 then the investment may be viable for the test conditions only. Where NPV < 0 it is best not to proceed as the cash flows are probably negative. NPV basically compares the value of a pound (£) or Euro (€) today to the value of a pound (£) or Euro (€) some time in the future, taking into account interest and inflation.

NPV is calculated as:

$$\mathrm{NPV} = (I - E)_0 + \frac{(I - E)_1}{(1 + i)} + \frac{(I - E)_2}{(1 + i)^2} + \ldots + \frac{(I + E)_n}{(1 + i)^n}$$

where i can be considered as a weighting function since: – [or discount factor: $1/(1 + i)^n$]; I_0 is income in year 0 and E_0 is expenditure (all) in year 0. or:

$$\mathrm{NPV} = (I - E)_0 + w_1(I - E)_1 + w_2(I - E)_2 + \ldots + w_n(I - E)_n \quad \{w_j = 1/(1 + \mathrm{i})^n\}$$

where w_j is weighting associated with future year Y_j. The equation is only valid for $i =$ constant.

High values of interest imply or weight for early returns. Later income/benefits being less important. Also with higher interest rates (above 10%) the future of the project beyond year 10 is less important.

21.4.2.3 Inflation

If inflation at a constant rate f is added, it further reduces the present value (PV) by:

$$X\mathrm{now} \equiv \frac{X}{(1 + f)^n}$$

If one now lets $P_j =$ net income in year j. Then:

$$\mathrm{NPV} = \sum_{j=0}^{n} \frac{p_j}{(1 + i)^j} \text{ For a project of length } n \text{ years.}$$

n is normally 5–10 years for industrial projects. But there is an increasing trend toward shorter product durations and consequently lower capital investment and shorter return times. Some projects can have returns as low as 5 years and there is a growing trend to shorten this even further where the product lifecycle is short.

21.4.2.4 Discounted Cash Flow

This is a method for capitalising income. DCF is an evaluation method for projects (or investments) based on an estimation of future cash flows and which takes into account the time value of money. It is in effect an estimate of what value of investment is worth considering (making) in order to gain the predicted income at a point in time in the future. The future cash flows are discounted so as to express their present values so that the true value of a project under consideration can be determined. In addition the time at which future monetary flows are earned must also be estimated.

DCF then reflects two things about a proposed investment:

- the time value of money, that is, how much interest needs to be paid for the delay in earning income; and
- the risk premium, that is, the additional return based on the risk that the project might not make the returns predicted.

$$\mathrm{DCF} = \frac{\mathrm{CF}_1}{(1 + r)^1} + \frac{\mathrm{CF}_2}{(1 + r)^2} + \ldots + \frac{\mathrm{CF}_n}{(1 + r)^n}$$

where CF is cash flow; r is discount rate; n is period (usually a year).

21.4.2.5 **Future Value**

The sum of the current values of one or more future monetary outflows (payments) that have been discounted at a given predicted interest rate.

$$\mathrm{FV} = I^{*}(1+i)^{n} \text{ or } \mathrm{FV} = \mathrm{PV}^{*}(1+i)^{n} \text{ or } \mathrm{PV} = \mathrm{FV}^{*}(1+i)^{-n}$$

where: FV is future value; I is the original investment (or PV); i is interest rate per period; n is number of periods (usually months or years).

The difference or relationship between present value and future value is shown in Figure 21.3.

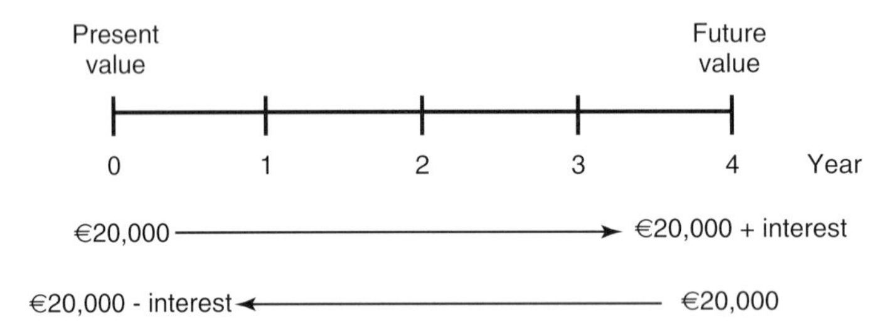

Figure 21.3 Relationship between present value and future value of money.

21.4.2.6 **Internal Rate of Return**

Internal rate of return is defined as the value of the discount rate at which the NPV becomes zero. It may require a trial and error approach. Generally, if the IRR is greater than the company target (interest rate + a factor), then the project should go ahead.

Instead of calculating NPV for a given i, one can calculate the IRR from P's (income) where IRR is the value of i for which:

$$\mathrm{NPV} = \sum_{i=0}^{n} \frac{p_j}{(1+i)^j} = 0$$

Or: if IRR > Target investment rate (i.e., the bank rate or other internal criterion) then one can proceed with the investment.

Some advantages of the IRR method are that it considers the time value of (including discounts and future returns) all the cash flows over the economic life of the project and is appropriate for comparing investments with unequal initial costs and/or unequal durations.

Calculating the IRR requires an iterative approach. It can be calculated directly on a spreadsheet using inbuilt functions. Alternatively one can adjust the discount rate in the NPV calculation until the final (point in time at which you want it to be) value is zero. The value of the interest rate then equates to the value for IRR.

21.4.2.7 **Project Appraisal – Cumulative Cash Flow – Typically**

A typical NPV curve is shown in Figure 21.4.

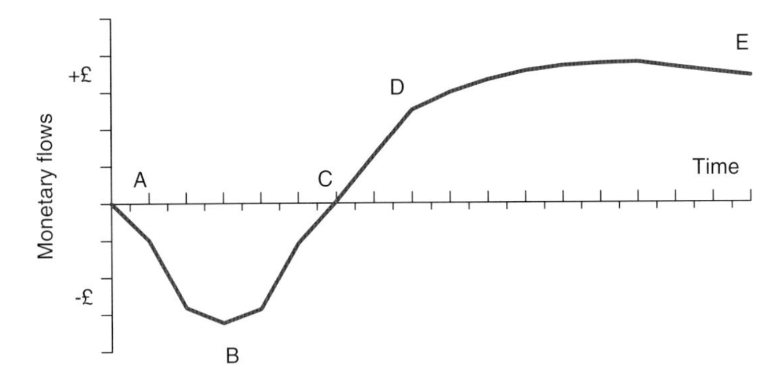

Figure 21.4 A typical net present value curve. The curve A–B represents the investment period. Point B represents the maximum depth of borrowings (or investment) and the point at which positive cash flows start. The curve B–E is the period to payback/breakeven (point C). The curve C–D–E represents the period of profit to process cessation. The fall-off at E represents lower sales/increases in maintenance costs/plant renewal, and so on, which result in lower profits.

21.5
Determining and Improving Process Performance

While this chapter so far has been directly focused on building and financially assessing a physical process for the manufacture of a food product, this next section will deal with the analysis of performance and improvement of the process. The tools and techniques described are applicable to any process in a business: purchasing procedures, finance, technical operations, and are equally applicable in the service sector. Here we shall be using examples from manufacturing. For fuller coverage of the service sector and other business functions, the reader may find the following authors of interest: Wheeler (6–11), Scholtes [12], Deming [13, 14], Neave [15], and Gaster [16].

We begin by looking at continuous improvement and the PDSA (Plan – Do – Study – Act) cycle. This is followed by reference to process improvement in addressing "The Six Big Losses," specifically through OEE. This leads to an in-depth study of the natural process behavior chart (sometimes called SPC (statistical process control) or control charts) and its use in improving process performance. There then follows a short coverage of "lean" principles and waste. And the section concludes with a description of seven further tools for process improvement. It is again worth iterating that all these techniques are applicable to social systems (i.e., purchasing, finance, human resources, etc.) found in any business whether engaged in making physical objects or providing main or support services.

21.5.1
Plan – Do – Study – Act

This is about managing the process of improvement, simply if one goes about trying to improve a process without considering the end objectives or without proper analysis of data, current state and what is a logical move then the whole initiative will run out of momentum and at best be sub-optimal. Any gains made will soon be lost. The PDSA cycle was originally devised by W. E. Deming in his book *Out of the Crisis* [14]. Some practitioners find that a more cohesive path forward is to start by studying what the system is already doing, acting on this to find the issues, planning what to change – why and how, making the changes and then studying the effect of those changes on the system a more workable alternative. This modified process is often called CPAD Check – Plan – Appraise – Do. Checking the process state, planning what to change and how, addressing the outcomes and risks and then implementing the plan.

Whatever the starting point, the essence is planning change and carefully checking / studying outcomes of those applied changes. Simply making changes to one part of a process that is observed as underperforming may have undesirable results elsewhere if the system as a whole is not fully understood. Better to understand the system, plan the approach, make known and measured changes and then study the outcome before acting on any individual small and isolated change. This whole activity is iterative and involves the whole team.

21.5.2
OEE Measures of Performance

OEE is a tool that is both simple to apply to a process and which is effective in communicating measures of performance about that process. It is often used as a key metric (or KPI – key performance indicator) in a lean program. It comprises three primary categories of measure which express how the process is performing and where improvements can be made. For many processes the required OEE is often specified at the design stage and forms part of the purchasing performance criteria.

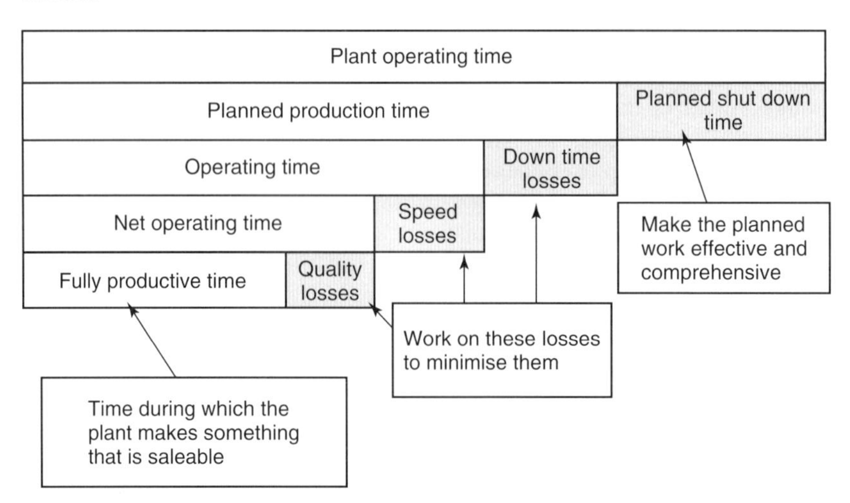

Figure 21.5 Diagrammatic representations of the overall equipment effectiveness losses.

OEE is built on five definitions of time and loss. See Figure 21.5 for a pictorial explanation of the following. It starts with the **Plant Operating Time**. This is defined as the amount of time the production line is available to run. A whole day (24 h) is generally a good starting point.

However, the production line may not be scheduled to run for the whole day due to a number of factors. These can be: scheduled (but not unscheduled) maintenance, defined breaks and periods where the plant is simply not scheduled to run (i.e., there is nothing to make). The sum of these times is called the *planned shutdown time*. Subtracting this from plant operating time yields the planned production time.

21.5.2.1 The Overall Equipment Effectiveness Factors

- **Availability:** is the measure of losses due to unplanned down time. These include all events which cause the line to stop: running out of materials, equipment failures (breakdowns), services failures, and time for changeover. All changeover times are included as it is a form of down time. Although changeovers are necessary, they can be reduced by addressing design and technique (see SMED – single minute exchange of dies). Subtracting the down time losses from the planned production time yields the **Operating Time**.
- **Performance:** is the measure that looks at the speed losses on the line. It is the total of all the factors that contribute to the line operating below its maximum possible speed, when it is running. Typical factors are machine wear (mechanical, air losses in pneumatic cylinders) inappropriate feedstock (i.e., poorly mixed dough slowing down a sheeting operation), misfeeds, operator actions, and actual slowing of the line to fill out a production window. Subtracting the speed losses from the operating time yields the **Net Operating Time.**[4)]
- **Quality:** is the measure of products produced that do not meet the appropriate standards. This includes waste (scrap) and rework. Subtracting the quality losses from the net operating time yields the **Fully Productive Time**.

The objective behind OEE is to maximise the fully productive time. This is shown diagrammatically in Figure 21.5.

And for a plant producing multiple products with change-overs the diagram can be considered in the form shown in Figure 21.6.

Examples from the food industry have show that while quality often scores well, performance and availability are often low due to multiple small stoppages which

4) The reader will find that not all the components in a production line will run at the same speed. Indeed, good line design will ensure that the upstream and downstream elements around a bottleneck run faster on the upstream side so as to keep the bottleneck fully supplied with work and also on the downstream side to remove finished (or part finished) products to avoid constraining the bottleneck.
For the OEE measures:

$$\text{OEE}\,\% = \text{Availability}\% \times \text{Performance}\% \times \text{Speed}\%$$

$$\text{Availability}\% = \frac{\text{Actual running time}}{\text{Planned production time}}$$

$$\text{Performance}\,\% = \frac{\text{Quantity produced}}{\text{Theoretical line rate}}$$

$$\text{Quality}\,\% = \frac{\text{Good pieces made}}{\text{Total pieces made}}$$

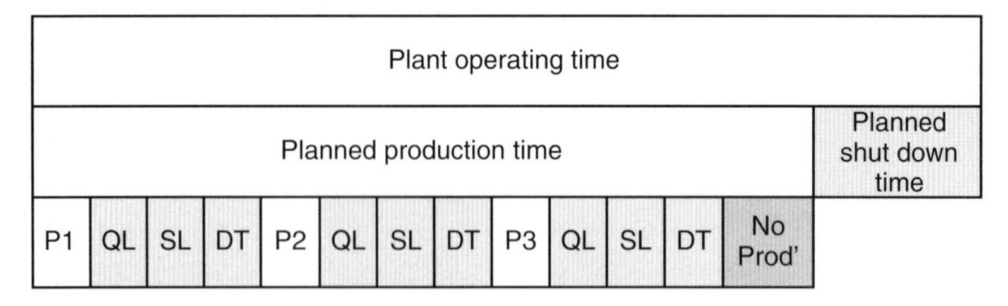

Figure 21.6 Multiple products (P1, P2, etc.) scheduled sequentially through a process have their own down time, speed, and quality losses. QL, quality loss; SL, speed loss, DT, down time.

can lead to underloading a plant. The financial rewards to a business from a high performing plant have been seen to be substantial.

21.5.2.2 The Six Big Losses

The six most common categories of loss which reduce efficiency are often referred to as the "*Six Big Losses.*" Collectively they can have a significant effect on plant performance and ironically they often go unaddressed due the seemingly small nature of an individual event. It is for this type of reason that collecting the line data and summarising it through the OEE calculations leads to the opportunity to improve the line efficiency and hence profitability (Table 21.1).

One shortcoming of OEE is that it does not address schedule attainment or that the plant could be making the wrong product. As such it should not be used independently of measures of schedule attainment. It is useful for identifying multiple small plant failures that cannot be tolerated in a lean procedure and which distort analysis of variation (see Section 21.6). When looking at multiple small failures a cause and effect investigation (see Section 21.8.2) on each failure needs to be accompany the OEE study.

21.6 Variation

While OEE is a good and simple tool for finding and quantifying losses of time and quality within a process it is only part of the story. As it addresses only time, count, and source it does not consider variation. Variation is simply the everyday changes in the value of a measured variable (i.e., deposit or filling weight of food). It can also be an attribute (i.e., count of things). What is important about variation is that it cannot be eliminated. It can only be reduced to the point that it is of little consequence to the customer. Variation also comes in two forms – common cause and special cause – and, to further add to complexity, all measured data from a process are not what at first sight they appear to be. Simply, the numbers presented may not be a real representation of the process. Untangling these aspects of data may at first sight seem complex, however there is one robust and straightforward tool for this: the natural process run chart (NPRC).

Table 21.1 Overall equipment effectiveness (OEE) categories of the "Six Big Losses."

OEE category	Loss category	Examples	Dialogue
Down time losses	Set-up and running adjustments	Set-up and running adjustments. Ingredient shortages. Operator absence. Plant preconditioning (i.e., oven warm-up or cool-down times	Minimising this type of loss is part of pre-preparing change components so the line down time for changeover is minimised. See also SMED
	Breakdowns	Machine failures. Tool breaks. Component jams leading to damage	This category should include all plant failures
Speed or running rate losses	Reduced speed running	Plant wear causing slow running. Leaking air cylinders. Operator generated. Slipping belts	Theses are the things that prevent the plant from running at its design rate
	Short stops	Product jams. Miss feeds. Air in depositor lines. Cleaning after spillage	This category is for misfeeds, and so on, which are not plant failures due to something breaking or failing to work
Quality losses	Start-up reject pieces	Scrap/waste. Rework. QA sample. Damage due to process. Inferior raw materials (i.e., a batch of poorly mixed sauce)	Although these two items share common events and examples they are separated here as start-up scrap is part of the set-up process.
	In-production scrap products	Scrap/waste. Rework. QA samples. Damage due to process. Inferior raw materials (i.e., a batch of poorly mixed sauce)	Some of this may be required as in confirming deposit weights. However, if the start-up scrap is excessive then it probably indicates that whole changeover and set-up routines need to be revised. See also SMED

SMED, single minute exchange of dies.

Before we look in detail at natural process run charts (NPRCs) and how to interpret them, a slight diversion into data acquisition is called for.

21.6.1 Collecting and Analysing Process Data

The objective of collecting and analysing data is to bring clarity to a process and its behavior. It is also to provide management with appropriate tools in order to gain insight on which to base business decisions. Business decisions based on either no data or a lack of properly collected and analysed data will lead to erroneous decisions which will always have a detrimental effect on the business. It is better to know clearly and unambiguously about the state of a business, even if the numbers are unpalatable, than proceeding in the dark with a badly drawn map.

> **In the absence of reliable and properly analysed data, it is easy for speed to become essential and direction optional. Anon.**

At first sight, many processes, both industrial and commercial, appear to be a bewildering array of factors that may be influencing the process. Indeed, some factors may be direct and some hidden or indirect. And while collected data are often in tabular form, tables of numbers do not immediately give a clear or obvious picture of what is going on. To compound this, numbers relating to the process may be presented in a noncoherent fashion or indeed in a preprocessed way, which detracts from analysis rather than aiding it.

It is essential that one looks at all data in context, original and unmodified and in a graphical form. Looking at pictures of the process output always leads to greater and more rapid insight to the underlying issues. There are two initial points to consider:

- collecting the data and
- analysing the data.

See also Shewhart [17, 18] and Wheeler [6, 10, 11] on the foundations of data.

21.6.1.1 Collecting the Numerical Data

The purpose of collecting data is to provide a solid basis for making sound judgments and taking the appropriate action(s). First, a descriptive view of data is useful: According to Wheeler "All data contain noise; some data contain signals" [6]. The objective in data analysis is to extract the signals from the noise and to analyse these signals in such a way that the context of the original data is preserved. It is only on this basis that sound managerial action can be taken on the business.

In order to achieve this satisfactorily the data must satisfy three conditions:

- **The data must be the right data:** That is, the data must relate to the process, be sampled appropriately, and be unmodified. This last point is particularly important: often data are presented in a prefiltered or processed form. Data of this nature is not useful as the prefiltering process may lose or suppress the signals in the data.

- **The data must be analysed such that the results are meaningful:** This means looking at the data in a structured and graphical way. The tools described below expand on this.
- **The results must be interpreted in the context of the original data:** That no data have any meaning outside their original context (Wheeler [6]) leads to focusing the analysis on what the data display and not what the data do not display. Indeed, focusing on reality and not simply business wishes.

21.6.2 What is Variation?

In describing variation and methods of analysis we first need to ask a question: Why does one need to consider variation? After all, there is a process specification and the (say) weights of the product are recorded. Anything outside the specified limits is removed from the line and treated as waste. Unfortunately this view of the world ignores the point that simply meeting specification does not resolve why there were products produced which were outside the specification in the first instance. And second, this approach will not lead to suitable mechanisms (methods) for analysing the process to resolve the issue of waste or describe techniques for improving the process. What is needed is an understanding of what leads to the waste products; a prediction of how likely further waste product might be in the future and what might be understood about the process that will lead to an improvement in the process. Looking at specifications alone will not lead anywhere.

The issue is one of variation. In practice all processes do over time exhibit some form of variation. That is, if looking at the weight of a component, the weight of the component will vary both up and down over time. Whether hour to hour, day to day, or even shift to shift, the weight of the product will vary. What is important is just how much the weight is varying (magnitude), when (frequency), whether it is predictable, and whether the changes are a result of the process itself (mechanical fluctuations in, for example, a blocking machine) or due to external influences (i.e., operator intervention or fluctuations in, perhaps, factory air supply pressure).

To resolve these influences on the product we need to understand two important aspects about the causes of variation in the product. The first is that variation is (and always will be) present. It cannot be removed totally. It can only be reduced to the point that any residual variation in the product is of no consequence to the customer. The second is that there are two types of variation:

- special cause, which come from outside the system, are few in number but need dealing with individually; and
- common cause, which come from within the process, may or may not be few in number, but which need to be dealt with by systematic improvement of the process to both reduce variation and to move the process mean to the desirable state.

Paradoxically, it is not possible to work on both types of variation at the same time. Indeed to do so will most likely lead to making matters worse. The right approach is to resolve the special cause (external) events first and then when the process is stable, work on the common cause (from within the process) events to reduce the variation and to drive the process mean to the desired value.

In tackling the variation in a process to make the product more consistent and to reduce waste the difference between common and special cause variation becomes important, because treating special cause variation (that from outside the process) as though it were part of the process will lead to overreaction, higher costs and will fail to deal with the underlying process characteristics. Conversely, treating common cause variation as special cause has similar adverse effects. For a fuller exploration of variation see Wheeler [6].

Given two numbers from a process, do we know that they represent a change in the process output, or are they a result of a change in the measurement system? Simply, we need to ensure that the measurement process is reliable, robust, and consistent. This is an essential element to concluding that the numbers do actually represent a change in the process and can be relied on. This is not explored here but further exploration can be found in Balestracci [19].

Data may not always be what it seems, for all data contain noise and some data may contain signals [6]. The question is what technique can be used to differentiate between the two and the two types of variation reliably and repeatably. Fortunately, work in this area was pioneered by Walter Shewhart in his work for the Bell Telephone Laboratory, later expanded by Deming and more recently by Wheeler.

Shewhart observed that the company was putting best efforts into trying to improve the quality and repeatability of components, but the harder they tried the worse things got.

What Shewhart found was that the company was making two mistakes that were making matters worse. The first was treating a special cause error as common cause. The second was treating a common cause error as special cause. Mistakes of this type are costly and lead to matters being worse. Rather than a concern for the statistics behind the method, the practitioner needs a full understanding of these two types of error and the differentiation between them. This is most important.

Deming went on to show that it is not possible to either reduce both types of cause of error simultaneously or eliminate either cause completely. After all, that would be to imply that variation can be removed completely and that is not possible. All systems contain variation, it is what systems produce.

Shewhart also developed a tool to distinguish these two causes of error quite uniquely. This tool is the Process Control Chart (PRC), often referred to as a *Statistical Process Control Chart* and now more commonly referred to as a Natural Process Run Chart (NPRC) as the original title, which is still in common use, might imply that a rigorous knowledge of statistics is needed in order to use the chart and method. Nothing is further from the truth. Use of the method does not rely on knowledge of statistics or require a prefiltering of the data. Indeed, its use on raw unmodified data is the better approach.

Below we will look at the construction and use of NPRCs, but first a note about software. Examples in this chapter are based on Minitab® with some additional examples from Winchart® and a very low cost but thoroughly rigorous application, Baseline®.

Prior to using any software it is recommended that some manual checks are carried out to ensure that the software does indeed calculate the UNPL (upper natural process limit) and the lower natural process limit (LNPL) at 3σ correctly from the moving range data. In addition, check also that the limits can be locked when the base process performance is established. Third, ensure that if the run is split, the limits are calculated from the correct subset of the moving range data. If these checks produce the correct results then the software under consideration may be useful.

An outline of the basic calculations for an XmR chart based on single measurements is given below. For a full and comprehensive treatise on SPC the reader is referred to Wheeler [6].

It could be asked "Why use a NPRC?" To answer this we need to think in terms of continuous improvement and understanding the process. The use of NPRCs guides the user to the correct course of action in response to the signals in a given set of data. It helps to differentiate between common and special causes of variation and the underlying reason or source of each within the process. Above all, it leads to a better understanding of the process itself and as such more profitable, higher quality product.

Undertaking process improvement needs to be carried out in three separate sequential stages as described by the Society of Motor Manufactures and Traders (SSMT) [20]:

- **Stage 1:** Stabilise the process by the identification and elimination of special causes of variation.
- **Stage 2**: Improvement of the process by reducing the common causes of variation.
- **Stage 3**: Monitoring the process, to ensure that the improvements are working. And incorporating additional improvements as opportunities arise.

NPRCs are an important tool in all three phases. As they show data points outside the UNPLs and LNPLs; along with other indicators, they are the principal diagnostic tool for Stage 1. In the subsequent stages, other tools can effectively be used for specific analysis of the data and investigation into the individual components of the process. But even in these stages, the NPRC has a unique role to play in observing the process and in reporting the results of action in changing the process. This leads to some new questions to ask:

- How did we expect the data (NPRC) to look?
- Are there any special causes of variation present? What can we do to prevent them happening in future?
- Is the amount of common cause variation too large?

- Are we where we should be? If not, how can we get there? Over time, are we doing better?[5]
- What can we learn about the causes of variation?

21.6.3 Some Background

21.6.3.1 Natural Process Limits

For single point measurements, the process standard deviation (PSD, σ) is calculated from the moving range information. Where the moving range is the positive difference between successive data points (Table 21.2). The relationship between the mean of the moving range $\overline{\mathrm{mR}}$ and the PSD is given by:

$$\text{Process standard deviation (PSD)} = \overline{mR}/\ 1.128^{6)}$$

Hence $3 \times$ standard deviation $= 3 \times \overline{mR}\ \ 1.128 = 2.66 \times \overline{mR}$. Therefore, the UNPL $=$ mean $+ 3 \times$ PSD and the LNPL $=$ mean $- 3 \times$ PSD. From these formulas for the above data, we obtain:

$$\text{Upper natural process limit} = 21.924$$
$$\text{Mean} = 20.284$$
$$\text{Lower natural process limit} = 18.644.$$

Figure 21.7 shows the natural process run chart for the block weight data.

21.6.3.2 Why Use the Moving Range to Estimate the Process Standard Deviation?

The mathematical expression for standard deviation is

$$\text{SD} = \sqrt{\frac{\sum (x - \overline{x})^2}{n - 1}}$$

To show why the moving range method is needed in estimating the PSD consider the following sequence of numbers:

12, 11, 10, 11, 13, 14, 15, 14, 13, 12

The mean is 12.5 and the mathematical standard deviation is 1.5811. If the natural process limits were calculated from the mathematical standard deviation then the

Table 21.2 Table of block weights

Block numbers	1	2	3	4	5	6	7	8	9	10
Weight	20.22	20.78	21.01	21.05	19.99	19.55	20.01	20.89	19.45	19.89
Moving range	–	0.56	0.23	0.04	1.06	0.44	0.46	0.88	1.44	0.44

5) If the natural process limits are recalculated from a known point at which the process was changed; do the limits get narrower (better performance) or wider (worsening performance)?

6) 1.128 is from statistical theory.

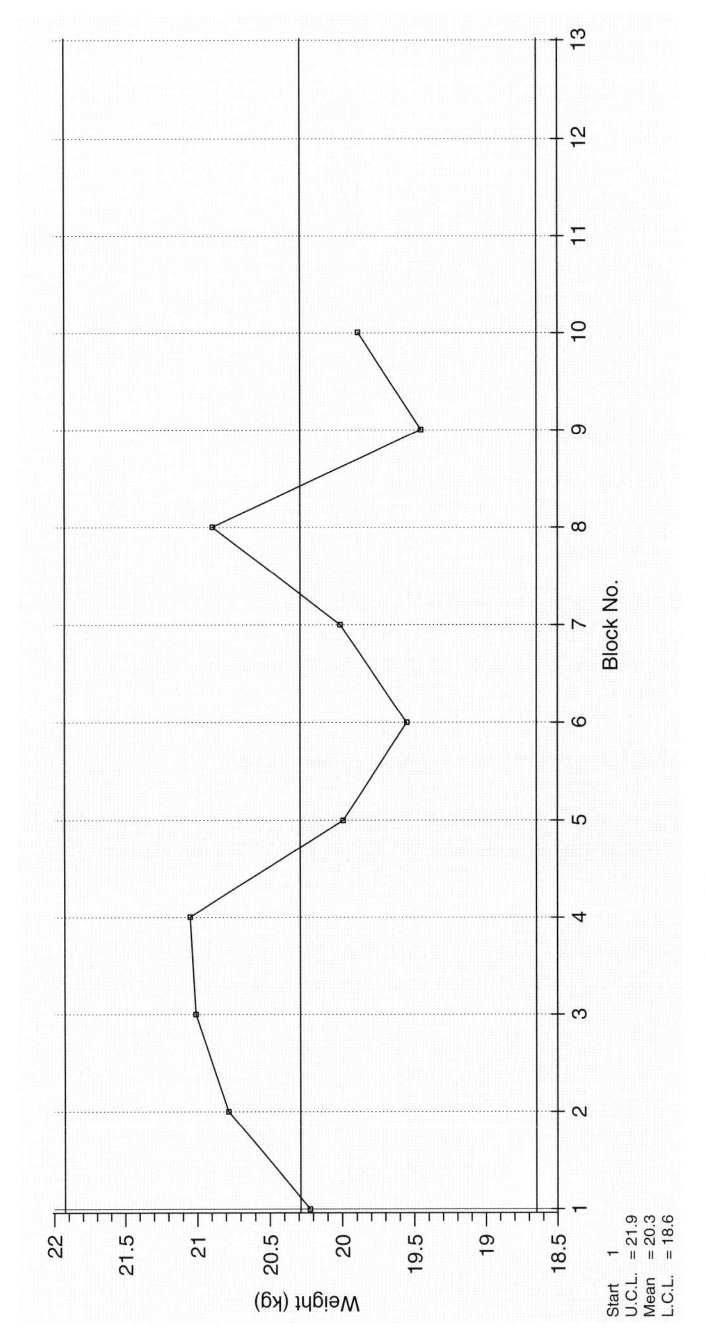

Figure 21.7 Natural process run chart for the block weight data.

upper and lower process limits would be at $\pm 3 \times 1.5811 = \pm 4.743$ of the mean. This will yield incorrect results and moreover the results would be the same for the same numbers in any order.

Now consider the case when the same numbers are used but in a different order as may be represented by a different process.

14, 10, 13, 11, 15, 11, 14, 12, 13, 12

For this second set of numbers (which are the first set rearranged) the mean is still 12.5 and the mathematical standard deviation is still 1.5811. Yet the two processes are clearly different (Figure 21.8).

To resolve this and to obtain measures that are representative of the process the natural process limits are based on the mean of the moving range data. And it is the moving range data that characterises the two different processes.

The mathematical derivation of the standard deviation considers the data as a whole. Whereas for a process the data are time dependent and hence the results will be different. By using the formulas from the previous page one obtains:

$$\text{Process 1 - mean} = 12.5, \overline{mR} = 1.111, \text{UNPL} = 15.456, \text{ and LNPL} = 9.544$$

$$\text{Process 2 - mean} = 12.5, \overline{mR} = 2.667, \text{UNPL} = 19.593, \text{ and LNPL} = 5.407$$

It can also be seen from the chart, that while the means are the same for both processes, the common cause variation in process 2 is much greater. This implies that process 2 will produce a wider range of more variable products that process 1.

21.6.4
Use of a Natural Process Run Chart in Process Improvement

In this section we show a practical example of the use of a NPRC in product and process improvement (Figure 21.9). The chart is in three parts: the existing process on the left hand section of the chart; the change process where the product was being altered for an end purpose; and on the right the new product and process situation. The interesting part about this chart (from a liquid conversion process) is that it shows both a process that is barely predictable and the change to produce a predictable process.

This process is one of converting a liquid to a solid. The plot is not a direct measurement, but is a difference plot of the amount of liquid used over or below the theoretical volume needed for that particular batch of liquid. The *Y* axis represents the amount of liquid needed for the process to produce 1 t of solid product as a difference from theoretical. Positive numbers are a beneficial result, that is, the process made more solid material with less liquid than theoretically required.[7]

7) This is possible due to preselected conversion factors in the formula for theoretical conversion. Provided these are not changed during the study, realistic and usable data can be obtained.

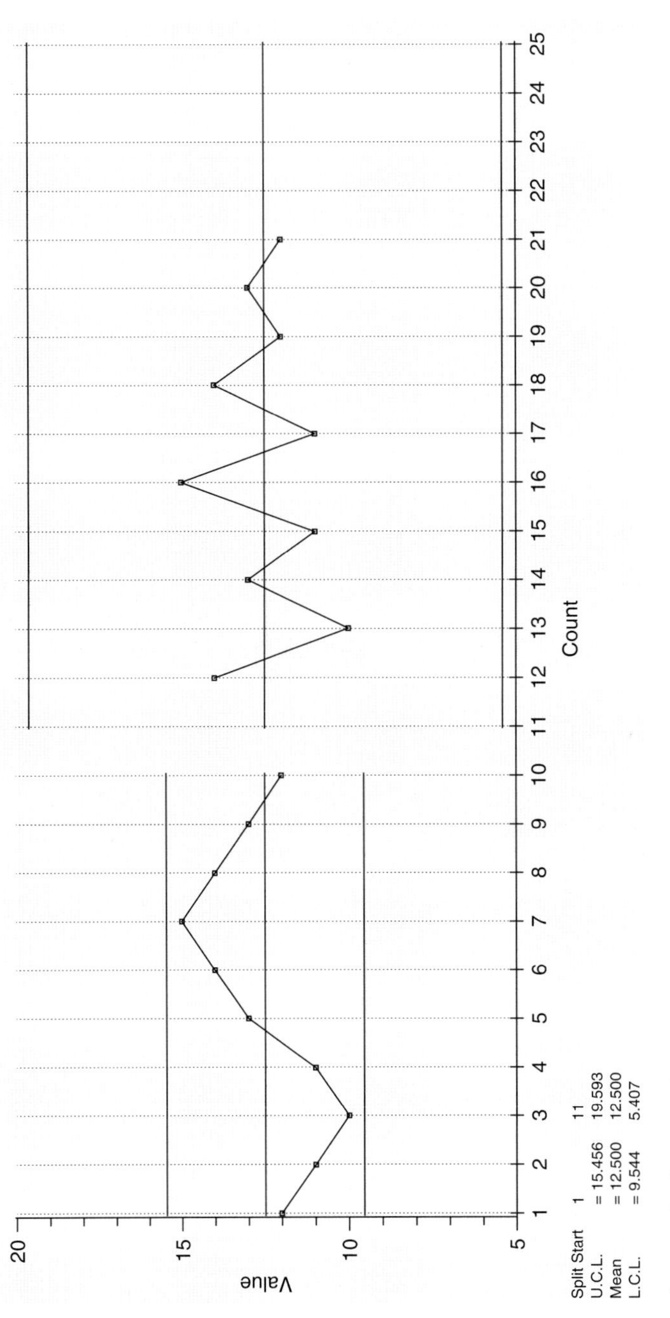

Figure 21.8 Same numbers – two processes.

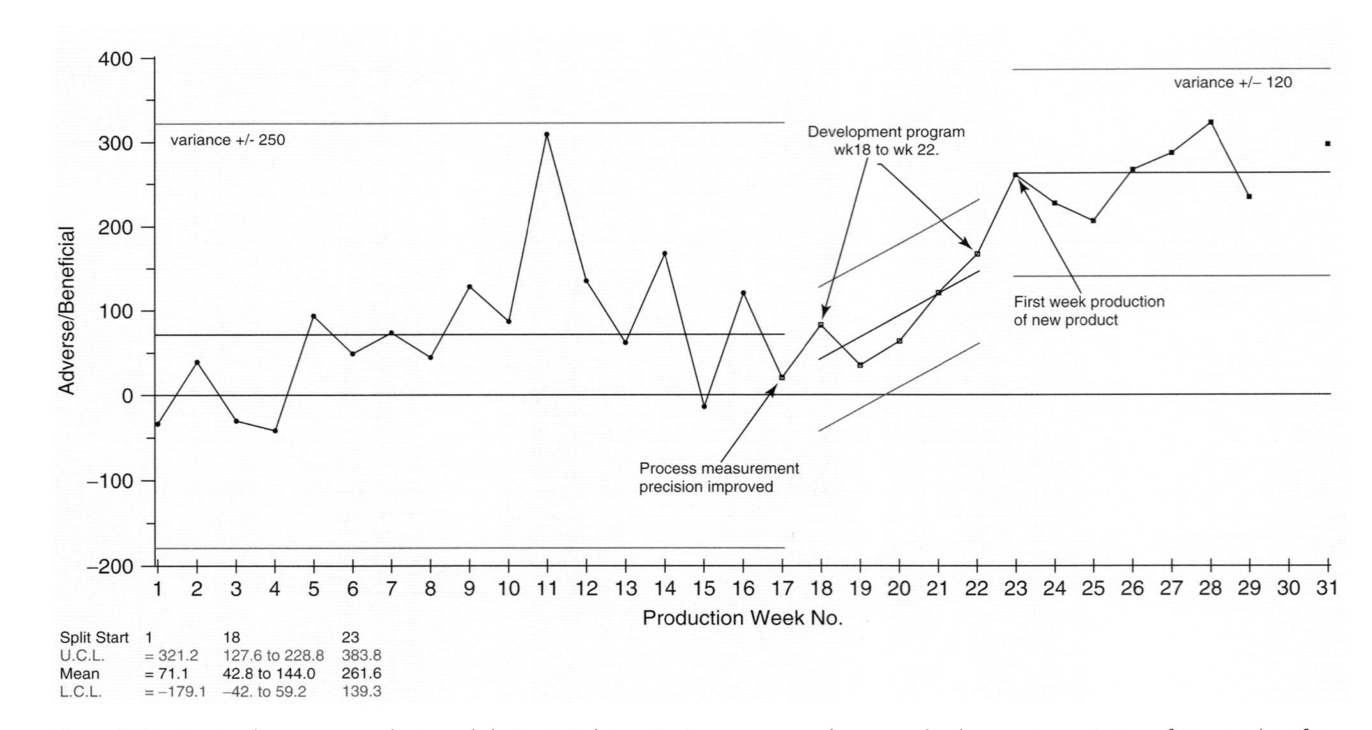

Figure 21.9 A natural process run chart used during initial investigations, process change, and subsequent monitoring of improved performance.

The story starts with a process that was exhibiting both unpredictability (overall shape of the data set) and an event which, although within the UNPL, was not part of the common process but the result of a specific event, itself originally outside the control of the manufacturing plant.

This single chart does not represent all the investigatory work undertaken, nor does it on its own show all the data that contributed to the process improvements, but it serves as the starting point for those investigations and the works to bring about the significant increase in both the financial and process reliability for the business.

Taking elements in turn, we see from the first part of the chart that it has an overall shape that starts low, goes high and then falls low again. The starting point for plant investigation was to ask "what happened at point 11 to create such a difference in yield?" It transpired that during this week the plant was running at around half capacity and some of the parallel batch stages were not being used. More specifically and on investigation it was found that a mechanical device controlling the filling of some batch units in one bank of reactors was not shutting off in time to control flow. More detailed investigations revealed that other mechanical components were sub-optimal in a set of batch units. This level of detail is of course not directly obvious from the chart. In fact it cannot be seen or deduced. But the chart does one vital piece of work; it leads the investigator to ask a better question about why. The result and benefits were then down to good process investigation and engineering.

This is a vital part of the use of this technique. These charts are the pictorial representation of the voice of the process (Wheeler); listening to it will lead one to ask better questions for better understanding.

At week 17 the business set about reformulating the product. With the tool in place and other process abnormalities addressed, the business was in a position to see what effect reformulation would have on the overall process variation and financial viability. During the process it soon became clear that the actual yield of the process was improving. During the five weeks of trials the plant could clearly see the overall direction the product changes were taking the process. Clearly, the process could have become worse. Yields could have come down. Even if these were associated with a substantial improvement in, say, eating quality and texture, at least the business would have a quantifiable basis on which to determine the additional cost of manufacture for the changed product. And this would be a much better basis on which to take business decisions, at least an informed one.

Finally, with a few more process refinements to product handling and plant operation the process arrived at the third section of the plot. This short section of run, some nine weeks, clearly reveals a significant change over the position of the process in weeks 1–7. The mean of the process has moved from 71.1 $l\ t^{-1}$ beneficial to 260 $l\ t^{-1}$ beneficial. This improvement in yield alone is substantial. But it is accompanied by a significant reduction in process variability from ± 250 to ± 120 $l\ t^{-1}$, a dramatic reduction in variability within the process. The whole journey involved many people making measurements and carefully inspecting process equipment. However, the journey was worthwhile, not only for the direct improvements seen here, but also for the integration of the tools into the everyday operation of the plant which will help future improvements and provide early indication of any shift in process performance.[8]

21.7 Brief Introduction to Lean and Waste

Lean is a way of life. It is not simply about waste or improvement tools but about the organisation being customer centric. That means putting value for the customer at the heart of every action, plan, and operation whether customer facing or not. It is also about respect for people and open, blame-free cultures.

21.7.1 Lean

Much has been written about lean: what adopting a lean approach can do for a business, how to build lean ideals into a process, how to change to efficiently reduce waste and more. Indeed, this last point specifically, on reducing waste, has often become the focus of many initiatives to the exclusion of other equally important aspects. A strategy that focuses on waste reduction alone will achieve only some waste reduction but will not achieve other necessary aspects of whole business improvement.

This is probably well illustrated by considering what happened when manufacturing plants in the Western world thought that they had copied the core elements (processes and data handling) of manufacturing plants in Japan, only to subsequently discover that they were not realising the expected improvements.

What went awry was that the Western observers who went to Japan[9] probably only saw what was on the surface – the processes and data handling; in other words, the symptoms of the process, but not the core drivers of the process. What was missing from their observations was the cultural aspect of the way the employees felt about what they did and for whom they were doing it. Latterly it has been speculated that, this cultural approach was probably outside the experience of the Western observers at that time and hence not observable at all.

8) The reader is reminded that throughout the whole process the data plotted were all based on a translation of the raw data through a theoretical model (with variable elements for type of solid) for liquid to solid conversion. Importantly, the model used was not altered during the whole process. And provided the model holds true for the range of input liquid concentrations, then daily changes in raw material feedstock will be normalised and hence the resultant data will represent process changes. It is also important to stabilise the process before making any specific changes to product quality or formulation. Otherwise changes seen in the data cannot be ascribed to any particular event.

9) As a side note Deming forewarns of the pitfalls of simply copying: "*To copy an example of success without understanding it, with the aid of theory, may lead to disaster*" Deming [13]

And that relates back to lean implementation. Lean is a whole way of life. It is not simply a set of rules, which if followed with sufficient thoroughness in the right order will lead to success. Nor is it simply a waste reduction exercise. For lean to be truly rewarding to the business the implementation has to encompass the whole business, and that means starting at the top of the company. Practitioners embarking on this journey need to understand both what drives performance and the culture of the business if systemic change efforts are to be absorbed.

Lean is a journey and all journeys take time. While the basic principles of lean are straightforward, the application takes persistence. After all, if the organisation took time to be where it is, it is not unreasonable for it to take time to learn a new way of doing things and to move along that journey. While there are no quick fixes or magic silver bullets, the journey itself is rewarding and the business gains can be substantial.

Lean is founded on a systems approach to organisations. Traditionally, many businesses are organised on a departmental basis which are mini-organisations in their own right. Left alone, these functions will naturally evolve to be self-serving (or protecting) often to the determent of the business as a whole. Indeed cost/profit center accounting does much to drive this type of existence. Conversely, lean requires that businesses operate as a system, with sub-systems each regarding the recipients (other internal departments) of their outputs as customers. Customer-centric thinking and activity is not only aimed at the final customer or consumer. It starts in the organisation itself.

In 1950, W. Edwards Deming in his work titled *Out of the Crisis* [14] developed the diagram in Figure 21.10 to illustrate a systems view of production in which customers, both internal and external, influence quality and outcome.

This was later expanded by Myron Tribus in *The Germ Theory of Management* [21] and Peter Scholtes [12] in *The Leader's Handbook.*

For many in the West a journey into lean involves a cultural shift. Many in businesses live daily with a target-setting culture. We are used to believing and

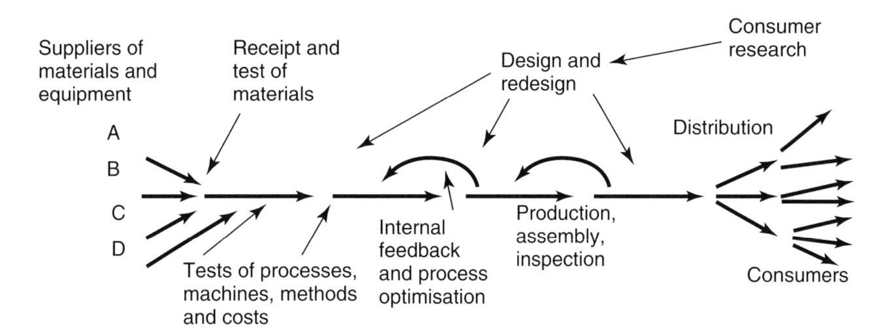

Figure 21.10 Reproduction of Deming's "Production viewed as a system" diagram. From page 4 of *Out of the Crisis*, with customer (both internal and external) feed back.

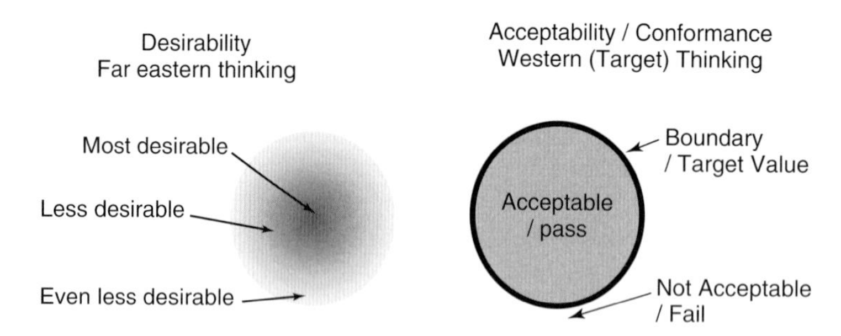

Figure 21.11 Desirability vs. acceptability.

managing in a way that once one meets a specific criterion (often wrongly set by others from outside the system) we have achieved all that was required. This both belies the journey that is lean and encourages focus on another target, while the system for the first target probably remains broken. In lean there is no end point in the traditional Western sense. Professor Kosaku Yoshida summed this up in his thinking on "Desirability vs Acceptability [22]." Figure 21.11 shows a diagrammatic representation of this concept.

While an organisation continues to work with a boundary-constrained model, a lean journey will be difficult. If embarking on a lean transformation journey this change in thinking is an essential early step. The Toyota Production Model describes the pillars of lean:

- elimination of all non-value-adding actions and work;
- respect for people;
- continuous improvement;
- customer satisfaction;
- employee empowerment.

21.7.2
Waste and Culture or the Elimination of Non-Value-Added Activities

A company setting out to become more efficient and customer focused will be able to optimise its business through a structure based on lean thinking. And if waste is the predominant focus then a business will gain benefits quickly from a single pursuit on waste alone. But the pursuit of waste alone leads to a common pitfall and the lean transformation becomes sub-optimal and rapidly runs out of momentum once some easy gains have been achieved. What is more important is that the company focuses on the long-term goals. In doing so, companies adopting a long-term strategic approach gain much larger benefits and cost advantages. Lean itself is not simply about waste but also includes a focus on continuous learning, innovation, and improvement in a customer-centric manner.

Reduction of waste is a key element, but it is the reduction of waste in all forms that is important for true lean organisations. For example, it is not just the waste of overproduction or waste from leaving a resource running (i.e., a water tap dripping away) but from removing unnecessary steps in a process – more the excluding of everything that does not add value for the customer.

The original work on waste elimination in all forms that did not add value for the customer was by Taiichi Ohno [23] of Toyota. He defined seven forms of waste:

- the waste of overproduction – probably the biggest waste;
- the waste of waiting;
- the waste of unnecessary movements;
- the waste of transportation;
- the waste of overprocessing;
- the waste on unnecessary inventory;
- the waste of defects.

To this list Womack and Jones [24] add an eighth waste, that of making products (goods or services) which do not meet customer needs. Today we can also add the wastes of energy, materials, and untapped human potential.

In defining waste Ohno describe waste as having two forms:

- type 1 wastes are activities that create no value;
- type 2 wastes are activities that not only create no value, but actually destroy value.

Looking at list of the wastes above it is easy to find many such wastes in any organisation. Some may be found to have the potential for large savings. However, it is worth remembering that although it is tempting to start focusing on waste elimination rather than embarking on a whole lean strategy, and this might result in some short-term savings, the organisation will lose out on transformation and the lean initiative will eventually fall by the wayside.

Culture transformation in an organisation often means that the existing culture needs to change. This is never easy and demands that the organisation develops a top level commitment to necessary change in style and management practices. Managers will need to change the way they think. Indeed the organisation needs to plan for long-term leadership and stability, as these things will not be achieved overnight. Those leading the change will need to recognise the way the existing organisational structure and culture as a system causes most of the problems the business encounters. Existing cultures can and do de-rail initiatives, so it may be that the culture of the organisation has to be changed first.

In developing a lean strategy there are other known practices that de-rail initiatives with ease. These are:

- blame cultures and where command and control is the order of the day;
- inconsistent goals and objectives handed down from senior management;
- business decisions that are based on guesses, assumptions, and not real unmodified data; and lastly
- those organisations which optimise departmental or silo cultures that are treated as cost/profit centers.

One very good antidote is a business-wide respect for people. This is at the heart of lean.

Lean methodologies were originally developed and successfully exploited in Japan. In their book on lean (*Lean Thinking*) Womack and Jones [24] distilled into five essential strategies the practices and business design criteria of leading lean companies:

Value - Value Streams - Flow - Pull - Perfection.

Understanding the value (or the value you want to achieve) for the customer and the stream by which it is delivered are the starting points for understanding the transformation needed. How often do you simply stand and watch simply what is going on? How often do you take a customer process and map it end-to-end in some detail to see just what happens to the customer interface and the work flow? Undertaking these exercises with actual business activities is often both illuminating as it is rewarding. If the transformation is to be lean, then start with where you are and move forward in measured and adaptable fashion.

In summary, lean is not just waste reduction or a set of tools. It is about value to the customer, which necessitates the reduction of all actions and processes that do not add benefit for the customer. Lean should also be thought of as a journey with the appreciation for a system at its heart. As with all journeys, progress will not always be smooth. Importantly, quality is uppermost and (an anathema to many) monetary gain is not a principal driver. Paradoxically, it is the pursuit of quality and customer-centric deliverables that lead to a reduction in costs and waste, and mean that the business achieves more with considerably less.

To fully do justice to a description of a lean program and its effect on business and understanding of Deming's System of Profound Knowledge (SoPK) is beyond the scope of this chapter. Readers wishing to explore further can usefully start with texts in the list of references but specifically:

- John Bicheno, *The New Lean Toolbox* [25]
- Peter Hines and David Taylor, *Going Lean* [26]
- David Clift, *Lean World* [27].

21.8 Tools for Continuous Improvement

Fortunately the main tools for getting started on the analysis of a process are graphical, well tried and few in number. The seven process improvement tools are:

- process flow diagram (flow charts);
- cause and effect diagram;
- Pareto diagram;
- check sheet;
- histogram;
- scatter plot;
- natural process run chart.

These are divided into two main groups: mapping the process (flow charts and cause and effect (Ishikawa) diagrams) and analysing the numerical data (Pareto plots, check sheets, histograms, scatter plots, and NPRCs).

21.8.1
Mapping the Process

All processes, whether business based or manufacturing based, comprise a series, generally sequential, of interrelated actions and/or tasks. These tasks, as a whole, transform the input (request or raw material) into the output (deliverable or product) and form the basis of all the value added to the business. However, they are also the source of all the added costs, waste, and efficiency losses that reduce the business profitability.

The purpose of mapping the process is to gain an insight into the individual stages of a process. And, within each stage the detailed steps of how the product is processed. This includes all the general process steps of preparation, mixing, heating, cooling cutting, packing, and so on. Once the process elements are identified, they can be analysed individually to asses their impact (value generating or cost addition) on the process and suitable action taken.

One further use of flow charts is during process design. Once a process has been mapped out, product measurements, machinery services, and similar can be added to the diagrams to build up the factory design.

Process mapping needs to be undertaken by the staff running the process. It is not an exercise that can be adequately carried out in the absence of those who have direct knowledge of the line and the interventions that take place.

Typical convention is that the process starts at the top of the page and works downward. There are a set of standardised symbols, which when used convey the same message to any reader of the chart (Figure 21.12).

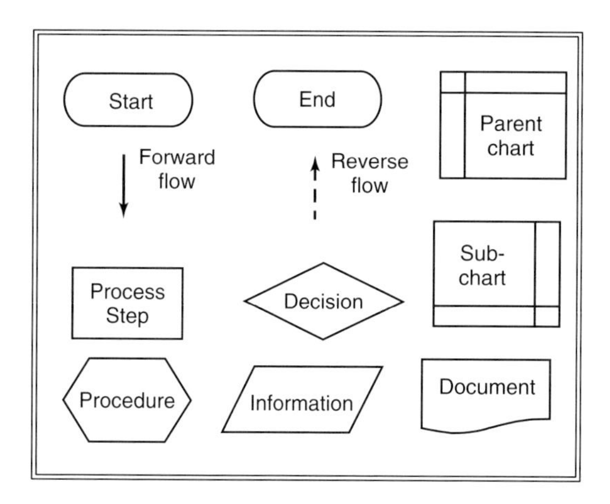

Figure 21.12 Symbols typically used in flow charting and deployment flow charts.

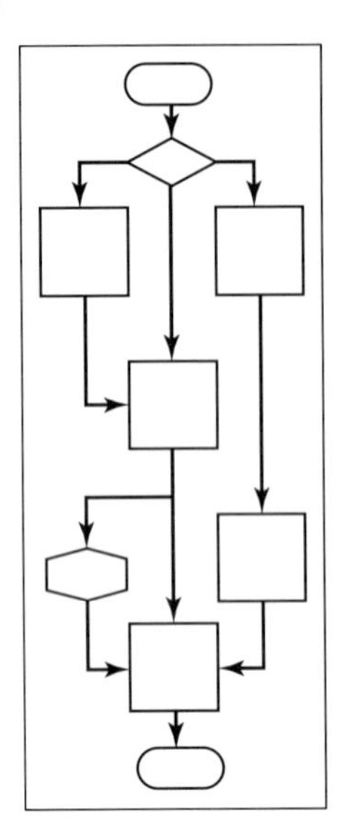

Figure 21.13 Typical block form for a process flow diagram.

21.8.1.1 Process Flow Diagrams

A process flow diagram (Figure 21.13):

- expresses detailed knowledge of the process;
- identifies process flow and interaction among the process steps;
- identifies potential control points;
- when expressed as a deployment flow chart shows time and time delays (non-contributory tasks) in a system.

21.8.1.2 Deployment Process Flow Charts

An extension of the basic process or system flow chart is the deployment flow chart (Figures 21.14 and 21.15). This is a powerful technique for capturing the essential and influential details of a process whether business or physical. As it maps activities across parallel functions, it is a powerful way to represent knowledge about a process that makes it easier to work on improving the process.

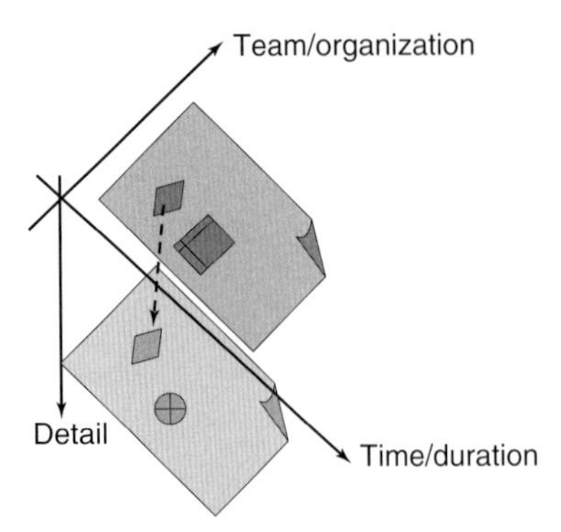

Figure 21.14 Principle of deployment flow charts.

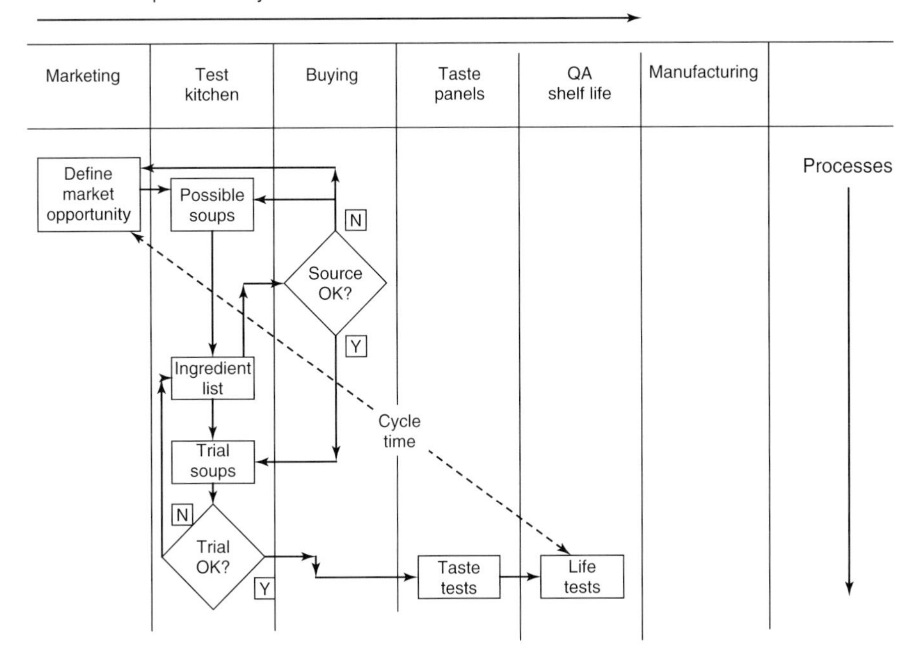

Figure 21.15 Typical (simple) deployment flow chart for a product development program.

21.8.2
Cause and Effect Diagrams

Cause and effect diagrams were originally developed by Kaoru Isikawa as a quality management tool. The cause and effect diagram is a way of visualising all the relevant causes and underlying reasons that contribute to a particular effect. The structure is to place the effect at the head of an arrow (Figure 21.16) and then to put the major group causes as contributory to the effect.

In practice, the causes can be anything, but it often found that in manufacturing examples major group causes are materials, methods, machines; and in service or administrative processes the major group causes are equipment, policies, procedures. The major group causes are a convenient way to collect thoughts under

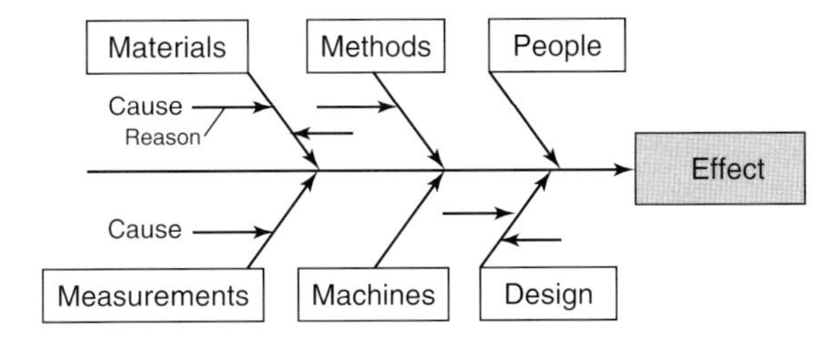

Figure 21.16 A typical cause and effect diagram.

individual elements, but the diagrams should not be constrained by this and any cause, however small it may seem at the time, should be added to the diagram. Reasons for a particular cause need to be added in order to identify the actual or potential problem area.

As with process flow maps, these diagrams are most comprehensive and effective when all the people involved in a process contribute. They do not work as a desk exercise away from the process being studied.

In summary, cause and effects diagrams:

- display all contributing factors and their relationships;
- identify problem areas where data can be collected and analysed;
- reveal minor but influential factors;
- need to be constructed by all workers in the system.

The procedure for construction:

1) Choose the effect to be studied and write it at the end of the arrow.
2) List all the factors and influences under consideration.
3) Arrange and stratify these factors and choose principle factors for the main branches.
4) Draw sub-branches for various sub-factors.

Helpful guidelines:

- Get everyone involved to contribute.
- Do not criticise any suggestion.
- Do not restrict to one's own work area.
- Concentrate on how to eliminate the problem and do not get sidetracked on justifications.

21.8.3 Pareto Diagram

The Pareto principle was formulated by Dr Joseph Juran based on work by the nineteenth-century economist Wilfredo Pareto, who reported that (at that time) 80% of the wealth in Italy was held by just 20% of the population. Indeed it was later found that the proportion of 80 to 20 was applicable to most management systems. Specifically, the Pareto principle states that a small number of causes are responsible for a large proportion of the effect, usually an 80 : 20 ratio.

The usefulness of this technique is that it helps to visualise and identify the important factors which most contribute to the output. It helps to focus efforts on the few elements that will bring the most cost-effective solution to reducing the effect.

A typical Pareto diagram is a bar chart of the percentage frequency of, say, cause of a defect (Figure 21.17). The leftmost bar represents the most frequent cause and subsequent bars to the right represent causes in decreasing frequency order. The chart is completed with a cumulative frequency line.

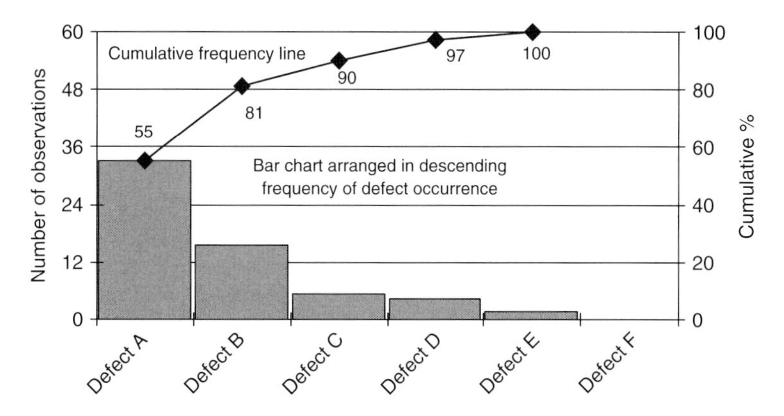

Figure 21.17 A typical Pareto chart.

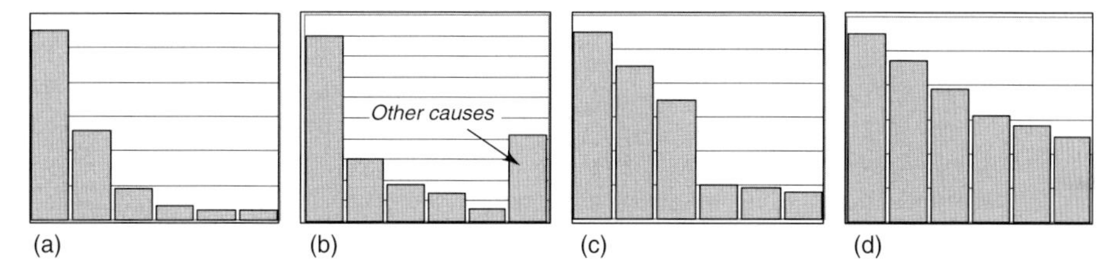

Figure 21.18 (a) Classic Pareto, one or two predominant causes. (b) As the classic chart, but where the catch-all category of "other causes" is also dominant, suggesting that there are numerous smaller events which need separating out. (c) A common result showing multiple causes where several classes need further investigation. Results of this nature do not support the Pareto principle. (d) A common first result showing multiple dominant causes. This can suggest that the defect categories are drawn too broadly or that the process has multiple issues.

When constructing bar charts of, say, frequency of defect type it is important to ensure that there are just one or two causes for the technique to work effectively. Figure 21.18 shows examples of bar charts where the data do not fulfill the criteria.

In summary, the Pareto effect is that 80% of problems are due to the 20% of the factors. A Pareto chart:

- identifies most significant problem to be tackled first;
- shows the factor percentage contribution;
- can show when multiple factors are at work or when special cause variation is predominant.

21.8.4 Check Sheets and Histograms

Check sheets are a useful tool for grouping count data on defects, location, or frequency in order to determine the most common cause of event. Check sheets are simple to construct, robust, and very effective in organising data. There are three basic forms of check sheet:

- count or tally chart;
- location sheet;
- histogram or process distribution sheet.

21.8.4.1 Check Sheet

An example of a check sheet is given in Figure 21.19. Here a product has six recognised forms of defect. Whenever a product is examined and found to have a specific defect, the count for that type of defect is increased by one. A typical result will look like the example. It shows that defect D is the most frequently occurring followed by defect C. The least commonly occurring is defect A. For defects of equal magnitude in cost or product rejection the check sheet would indicate that elimination of the cause of defects C and D would be the next course of action.

This type of check sheet gives no indication of where on the product the defect occurred. Such knowledge may lead to secondary process elements (such as transit damage), being the most frequent cause of defect. A location plot can be useful in these situations.

21.8.4.2 Location plot

Sometimes called a *concentration diagram,* this type of check sheet includes a drawing of the object and the number and types of defect are marked on the sheet. Figure 21.20 is an example of such a sheet for a ready meal.

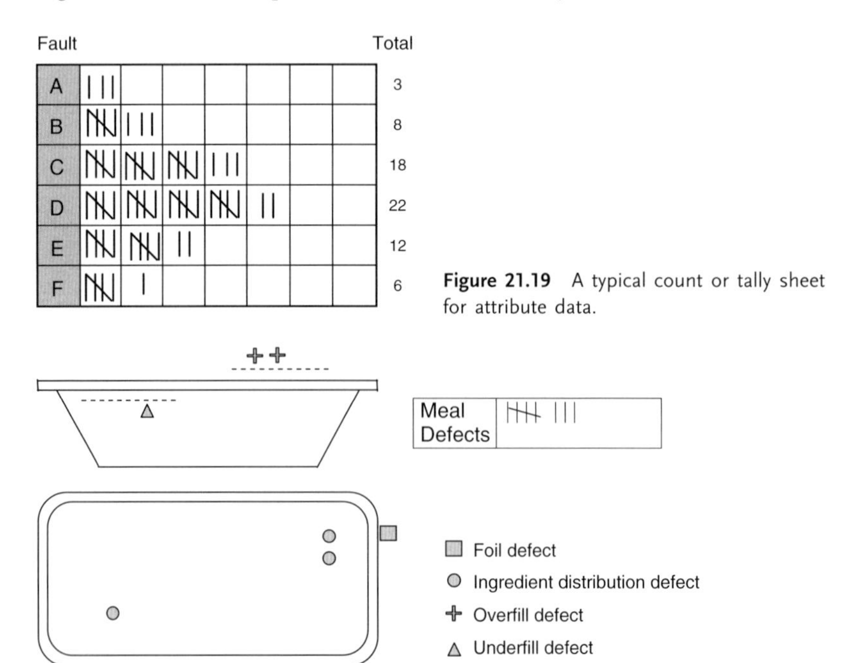

Fault	Total
A	3
B	8
C	18
D	22
E	12
F	6

Figure 21.19 A typical count or tally sheet for attribute data.

Figure 21.20 A typical location plot for a ready meal.

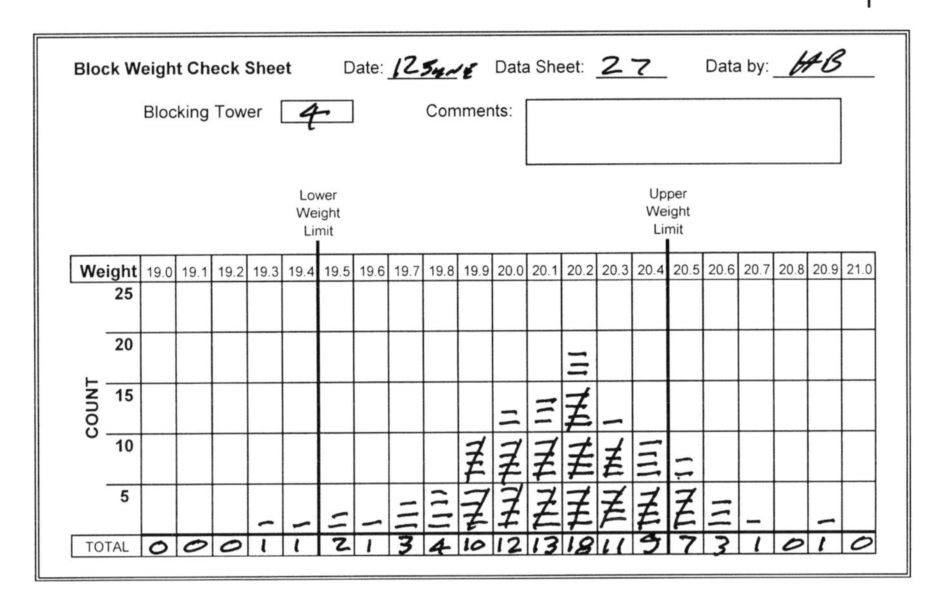

Figure 21.21 Process distribution check sheet for a block forming process.

21.8.4.3 Process Distribution Check Sheet

This type of check sheet is used to build a histogram of the process as it is running (Figure 21.21). It allows the user to see a how the process is performing in terms of distribution of, say, weights as the process is running. It gives a real time picture of the process performance without the need for fuller data analysis or specialist software. We will see later that specialist software for NPRCs can be used as the process is running. Many software packages also plot the histogram or frequency chart (Figure 21.22).

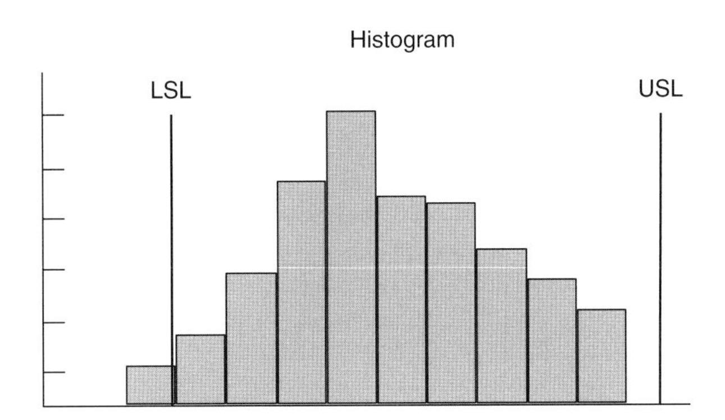

Figure 21.22 Typical histogram for a process. LSL, lower specification limit; USL, upper specification limit.

In summary, check sheets:

- simplify data collection and analysis;
- highlight problem areas by frequency of location, type, or causes;
- are the basis for building a histogram;
- show the extent of the distribution of the data;
- show the central frequency (average) and variability;
- the inclusion of specification limits that can be used to display the capability of the process (and there are mathematical techniques for determining capability).

21.8.5 Scatter Plots

Scatter diagrams are used to show how two variables are related to one another (Figure 21.23). Specifically they show the cause and effect relationship and how strong that relationship is through the correlation function (R^2) of least or normalised squares.

In summary, scatter plots:

- identify the relationship between two variables;
- allow a positive, negative, or no relationship to be readily deduced;
- show degree of association;
- can be expressed mathematically – R_{xy} (least or orthogonal squares).

21.8.6 Natural Process Behavior Charts

The last of the seven tools discussed here is probably the most useful, robust, and flexible of all (Figure 21.24). The core objective is applying and maintaining good quality control; which is not a natural state. Process improvement and quality control only work effectively when the business as a whole focuses on the required outcomes and the tools needed. The tools alone are not a substitute for good

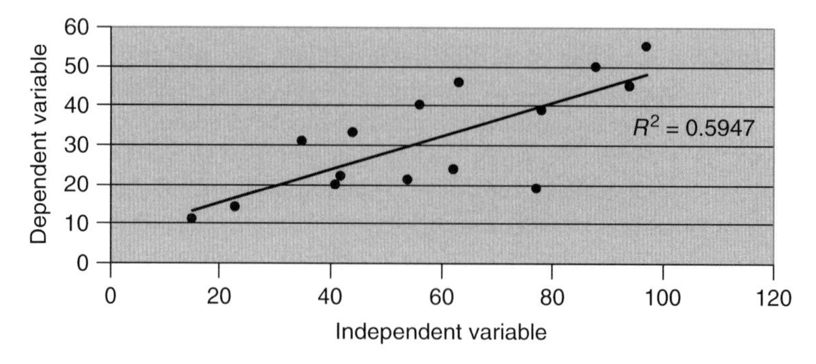

Figure 21.23 Scatter diagram.

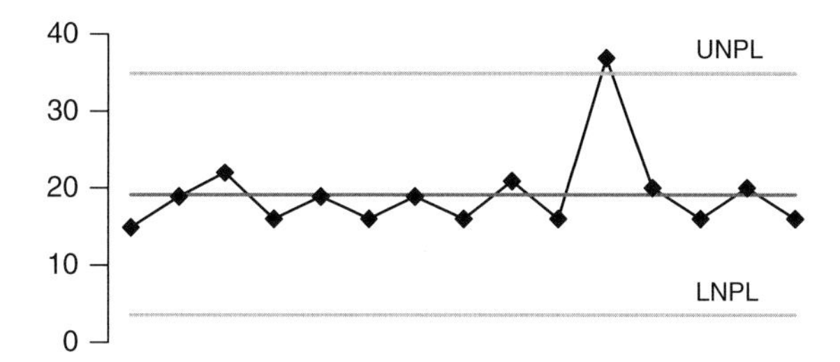

Figure 21.24 A typical natural process run chart showing a special cause event. UNPL, upper natural process limit; LNPL, lower natural process limit.

management where the whole business understands and contributes to the pursuit of quality. Indeed, simply using the tools alone will not bring about good quality control and reduced costs.

The concept of quality being a way of life is described in more detail in the section on lean.

For quality control to work effectively it requires the clear differentiation of products and processes. For products, the measurement is one of conformance to specification (i.e. Quality). For processes, the measure is one of stability or predictability. Wheeler [10] defines the four possibilities that can exist for the combination of these two classifications:

- conforming and predictable – the ideal state;
- nonconforming and predictable – the threshold state;
- conforming but unpredictable – the brink of chaos;
- nonconforming and unpredictable – the state of chaos.

Successful interpretation of how a process is performing requires the use of a natural process run chart and one constructed according to Shewhart's rules. The most useful of the charts is the XmR chart, or measured variable and moving range chart.

There are many papers and documents published on NPRCs, their use, data preconditioning, statistical conformity of data, suitability of the process to this type of analysis, use of specific charts for count data, and much more [6, 8–11]. In practice, data preconditioning is not needed and indeed may do more to obscure than enlighten. First, the charts are built on an empirical model which produces limits based on the variation in the data. It is the data one is testing. Second, the chart with limits at $\pm 3\sigma$ are a reasonable compromise to include all common cause events and reveal special cause events. Indeed, this is one reason for analysing the data. The chart is robust.

Lastly, count (or attribute) data; yes there are specific charts for count data as shown in Table 21.3. These charts are built on an assumption that the data conform to a specific statistical distribution. Table 21.3 gives classifications for data, both measurement and attribute data. However if the attribute data do not conform to the appropriate statistical distribution on which the chart construction is based, then

Table 21.3 Natural process run charts for types of data.

Data type	Measurements		Attributes			
	Individual	Sub-group	Nonconforming units defectives (with upper limit) Binomial		Number of nonconformity defects (no upper limit) Poisson	
Examples	Pressures,	Pressures,	Proportion of defective items	Number of broken biscuits	Number of faults per week or per mile.	Number of broken biscuits in a variable count pack
	weights,	weights,	(variable sample size)	(fixed sample size)	Number of broken biscuits in a fixed count pack	
	times,	lengths,				
	percentage changes in value	times to do something				
Chart type	Individual values	Sample means and sample ranges	Proportions	Number of defectives	Number of events per fixed interval	Number of events per variable interval
	moving ranges					
	(XmR chart)	($\overline{X}$ and $\overline{R}$)	(p-chart)	(np-chart)	(c-chart)	(u-chart)

the calculations will yield inappropriate limits. But all is not lost; rather than using a specific chart based on statistical conformity of data, one can still use the XmR chart. If the attribute data conform to the appropriate statistical distribution for the type of chart to be used, then the chart will yield the correct limits. If the attribute data set do not conform to the appropriate statistical distribution for the type of attribute chart to be used, then while the attribute chart will produce inappropriate limits, the XmR chart will produce a chart with appropriate limits which is of practical use for anayising the process. If in doubt about the statistical distribution of the attribute data, simply use the XmR chart.

In summary, process behaviour charts:

- can be used to help to reduce variability;
- monitor performance over time;
- allow graphical representation to prevent misinterpretation;
- display trends and out-of-control conditions immediately.

References

1. Obeng, E. (1994) *All Change! The Project Leaders Secret Handbook*, Pearson Education Ltd, London.
2. Clark, A.C. (2005) *The Gist of Process Mapping: How to Record, Analyse and Improve Work Processes*, Word4Word, Evesham.
3. Tribus, M. (1989 and 2003) *Deployment Flowcharting, Improving the Interaction of people and Processes*, The Deming Fourm.
4. Goldratt, E.M. (1994) *The Goal*, Gower Publishing Ltd, Aldershot.
5. Goldratt, E.M. (1993) *It's Not Luck*, Gower Publishing, Aldershot.
6. Wheeler, D.J. (1993) *Understanding Variation: The Key to Managing Chaos*, SPC Press, Knoxville, TN.
7. Wheeler, D.J. (2003) *Making Sense of Data: SPC for the Service Sector*, SPC Press, Knoxville, TN.
8. Wheeler, D.J. (1998) *Avoiding Man-Made Chaos*, SPC Press, Knoxville, TN.
9. Wheeler, D.J. (2009) *Twenty Things You Need to Know*, SPC Press, Knoxville, TN.
10. Wheeler, D.J. (1997) The Four Possibilities for Any Process. Quality Digest. *http://www.qualitydigest.com/dec97/html/spctool.html* (accessed June 2011).
11. Wheeler, D.J. (1996) Four Foundations of Shewhart's Charts. Quality Digest. *http://www.qualitydigest.com/oct96/spctool.html* (accessed June 2011).
12. Scholtes, P.R. (1998) *The Leaders Handbook*, McGraw-Hill, New York.
13. Deming, W.E. (1993) *The New Economics – for Industry, Government, Education*, MIT-CAES, Cambridge, MA.
14. Deming, W.E. (1986) *Out of the Crisis*, MIT Press, Cambridge, MA.
15. Neave, H. (1990) *The Deming Dimension*, SPC Press, Knoxville, TN.
16. Gaster, D. (2010) *Visualising Transformation*, HotHive Books, Blackminster, Worcs.
17. Shewhart, W. (1939 and 1986) *Statistical Method from the Viewpoint of Quality Control*, Graduate School of the Department of Agriculture, Washington, DC.
18. Shewhart, W.A. (1980) *Economic Control of Quality of Manufactured Product*, ASQC Quality Press, Milwaukee, WI.
19. Balestracci, D. (2009) *Data Sanity: A Quantum Leap to Unprecedented Results*, Medical Group Management Association, Englewood, CO.
20. SSMT (Society of Motor Manufactures and Traders Ltd) (1986) *Handbook on Guidelines to Statistical Process Control*, SSMT, London.

21. Tribus, M. (1998) The germ theory of management, in *Avoiding Man-Made Chaos Wheeler*, SPC Press Inc., Knoxville, TN, pp. 1–16.

22. Yoshida, K., (1985) Sources of Japanese productivity: competition and cooperation. *Rev. Bus.*, 7 (3), 1820.

23. Taichi, O., (1988) *The Toyota Production System: Beyond Large Scale Production*, Productivity Press, Portland.

24. Womack, J.P. and Jones, D.T. (2003) *Lean Thinking: Banish Waste and Create Wealth in Your Corporation*, Simon and Schuster, New York.

25. Bicheno, J. (2004) *The New Lean Toolbox, Production and Inventory Control*, Systems and Industrial Engineering Books (PICSIE Books), Buckingham.

26. Hines, P. and Taylor, D. (2000) *Going Lean*, Lean Enterprise Research Centre, Cardiff Business School, Cardiff.

27. Clift, D.J. (2007) *Lean World*, Lean World Ltd, Oldbury.

22
Microscopy Techniques and Image Analysis for the Quantitative Evaluation of Food Microstructure

Maria de Jesús Perea-Flores, Angélica Gabriela Mendoza-Madrigal, José Jorge Chanona-Pérez, Liliana Alamilla-Beltrán, and Gustavo Fidel Gutierrez-López

22.1
Introduction

Food materials have different levels of structural organization, such as the atomic, nanometric, molecular, microscopic, and macroscopic levels, and their structure depends primarily on the interactions between the atomic and molecular components. The most important structure-holding factors are van der Waals forces, hydrogen and polar bonds, electrostatic forces, and quantum interactions [1]. At the molecular level, steric effects determine the structural organization of materials and at molecular and microscopic levels the chemical and electrostatic interactions have an important role in the architecture of proteins, polysaccharides, and lipids. Furthermore, the size, shape, surface, and morphology of the microcomponents of a food material are fundamental for the bulk structure of biomaterials, while the interfaces and the geometry of the products have important influences on transport phenomena at the macro- and microscales.

The structures of foods originate from nature and are influenced by processing and storage. These structures are formed by more than one material or phase which may form clear and discernible structural arrangements, giving rise to the geometry and architecture of the foodstuff. In other words, *structure* is defined as the spatial arrangement of various elements of a food and their interactions. Food structure is also defined as the organization of a number of similar or dissimilar elements, their binding into a unit, and the interrelationship between the individual elements and their groupings [2]. Structure can be related to physical properties and to the quality and acceptability of food materials [3, 4].

Handling of foods induces modification of structure which can be controlled to some degree by means of the processing conditions. Novel strategies for production are based on a proper understanding of the organization and architecture of materials. It is necessary to design new food materials with specific functionality [5]. Recently, new approaches for the design of novel food structures with improved functionality have been suggested. Bottom-up and top-down techniques are traditionally used in nanoscience and nanotechnology for building nanostructures, and

Food Processing Handbook, Second Edition. Edited by James G. Brennan and Alistair S. Grandison.

these strategies are starting to be used in food science to create new food materials with better structural characteristics. The characterization of macro-, micro-, and nanostructure is essential for the control of the structure and functionality of food materials [6].

The development of microcomputers and advanced electronic systems has allowed the use of a number of instruments and tools to evaluate the structure of foods on the macro-, micro-, and nanometric scale. At the macrostructural level, several instruments can be used to obtain digital images and other structural properties from food materials, such as charge coupled device (CCD) cameras, multi-spectral and hyper-spectral imaging systems, ultrasound, magnetic resonance imaging (MRI), computed tomography (CT), and electrical tomography (ET) [7]. External attributes such as size, shape, color, surface texture, and external defects can be evaluated using CCD cameras. However, internal structures are difficult to detect with relatively simple and traditional imaging techniques, which cannot provide enough information for detecting internal features, such as water-core, internal breakdown, and hollow heart. It is necessary to apply special image acquisition techniques for food quality evaluation [7]. Ultrasound, MRI, CT, and ET technologies may be used to inspect the internal attributes of food products. Digital images can be used to construct three-dimensional representations of the object by stacking several slices of scanned object.

There are many microstructural characterization systems, such as light and fluorescence microscopes. Confocal laser scanning microscopes (CLSMs) are used to study the internal microstructure of biological materials and to obtain topography and 3D images. Conventional scanning electron microscopy (SEM) and environmental scanning electron microscopy (ESEM) are used to obtain the structure and morphology at micrometric levels and the advantage of ESEM is that it enables the observation of nonconductive and wet samples without pretreatment. These microscopes frequently use a tungsten filament to produce the electron beam and have a resolution of around 1 μm, depending on the sample. On the other hand, transmission electron microscopes (TEMs) with tungsten filaments permit micrometric and nanometric resolution [3, 8–11].

In scanning and transmission modes, a good resolution on the submicrometric and nanometric scales can be obtained with high resolution scanning and TEMs that use field emission filaments. Atomic force microscopes (AFMs) are helpful in studying structures at the micrometric and nanometric scales, including the topography and other properties of food materials such as edible films, cellular organelles, food surfaces, polymeric structures, and proteins [12, 13].

Recently, the analytical and manipulation capabilities of instruments have increased dramatically and it is possible to couple to the microscopes several instruments and devices, such as spectroscopic systems, heating and chilling elements, automatic devices for molecular manipulation, nano-dissection rasping, and milling of the samples, multiple electron and X-ray detectors, among other systems that permit better characterization and design of biomaterials on the micro- and nanoscale. Along with the development of capturing systems and microscopy equipment, image processing methodologies and software have progressed significantly.

It is now possible to obtain detailed quantitative information of the structure of foods using such digital images.

This chapter discusses the concept of structure–property–function relationships [3, 14–17] in the macro-, micro-, and nanoscale and briefly discusses some applications of microscopic techniques and image analysis (IA) in food science, technology, and engineering.

22.2
Microstructure, Nanostructure, and Levels of Structure

Food scientists define levels of structures to identify the scale (or scales) of critical elements of a product and to select techniques for their characterization (Figure 22.1). The relevant scale is the dimensional level at which the effects of certain phenomena are evaluated. The first evaluation of any structure is carried out at the macroscale and then, at microstructural levels [3, 18]. Nowadays, structures can be resolved down to the atomic level (nanostructure), but a major challenge is

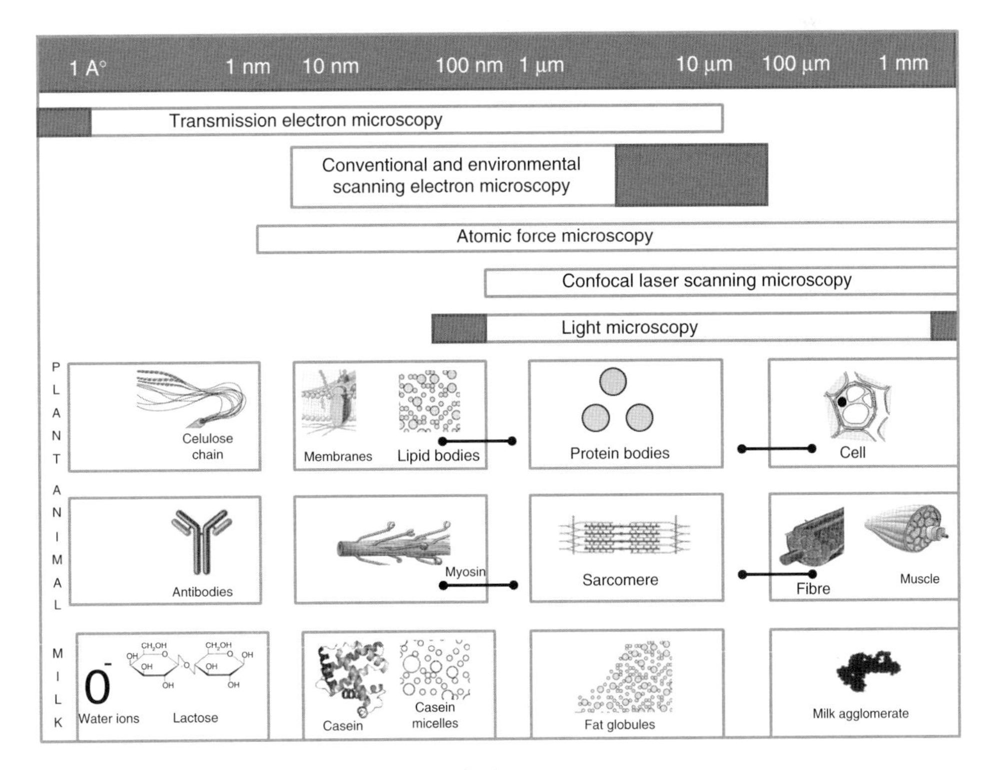

Figure 22.1 Structural levels of biological and food materials and resolution of microscopes used in food science and technology.

to identify the relevant scales responsible for the desired functionality, which may vary from molecules at interfaces to macroscopic particles.

22.2.1 Structure–Property–Function Relationships

Food technology involves a controlled attempt to preserve, transform, create, or destroy a structure that has been produced by nature or by the use of different technologies. The impact of processing on structure has been largely a result of the adaptation of unit operations and the design of processing equipment to transform and preserve foods in ways that were not possible in the past.

Further improvements in the quality of existing foods and the creation of new products to satisfy expanding consumer demands during this century will be based, largely, on manipulations at the microscopic and nanoscopic levels. This is because most of the elements that critically participate in transport properties, physical and rheological behavior, and textural and sensory properties of foods are found within the 100 μm range. The study of food structure at macro-, micro-, and nanoscale levels provides information that helps the understanding of the properties and functionality of entire food systems (structure–property–function relationships). At these three structural levels the food elements are formed by specific and nonspecific interactions that take place successively or simultaneously to yield the nonequilibrium structures of real food systems [2, 3].

Structure–property relationships imply a causal connection between structure and the way the product behaves. So far, most structure–property–function relationships in foods have been assessed on an empirical basis; yet, they provide evidence that structure does play a key role in transport phenomena, physical properties, texture, chemical stability, and bioavailability of nutrients [19]. Nutritional data for foods are determined by laboratory analysis, and the availability of those nutrients for absorption in the gut depends on mastication and release of nutrients during digestion from the food matrices formed by processing. For instance, air drying of carrot severely restricts the availability of carotene, which becomes entrapped within the cell walls of dried cells and is only very slowly released by gastric juices. Enzyme action on starch in the gut depends on the porosity and permeability of the starch matrix formed during processing. The effect of food structures on nutrient availability deserves a good deal of attention because consumers are increasingly concerned about the effects of foods on health and wellbeing.

As mentioned earlier, the relationships between structure and transport properties have also received marginal attention, the problem being circumvented by food engineers through the use of "apparent" transport coefficients that are process- and product-specific. Efforts are being directed at quantifying the microstructure and the properties of microstructural elements intervening in the architecture of foods, while modeling the relationship between structure and transport phenomena will remain largely empirical and qualitative [10, 20–22].

An understanding of the role of structure in dehydrated products is required in order to understand transport mechanisms and to design functional properties. Dehydration of biological material is a controlled effort to preserve the structure or create a new one that serves a functional purpose. Thus, dehydration should be viewed not only as a unit operation but also as a method of producing unique food structures even from the same raw material [10, 23].

Consumer demands have led to improvements in the quality of powders with various properties [24], including rehydration, wettability, solubility, and dispersibility, which are strongly related to food structure [19, 25]. The dependence of morphology and moisture content on drying temperature during spray drying has been studied, especially on those aspects related to breakage and inflation at intermediate and high air drying temperatures, and with collapse of structures during low-temperature drying. According to Alamilla *et al.* [26], when droplets were spray dried at low temperatures, shrinkage was greater than at high spray drying temperatures. This behavior could be associated with more drastic shrinkage and deformation mechanisms under these processing conditions, causing the water diffusion in the material to be slower and allowing the deformation processes to arise in those particles subjected to low processing temperatures.

Product properties and structure depend strongly on the type of drying, and on the original structural elements of the raw material. The quality of dried fruits is strongly dependent on the severe structural changes that arise during processing including discoloration effects and loss of valuable nutrients [23]. Prothon *et al.* [27] evaluated the effect of combining osmotic dehydration with traditional air drying and found that osmotic pretreatment had a positive effect on the final overall quality of the product, including beneficial effects on the firmness of the rehydrated samples. Microstructural studies showed a thickening of the cell wall after the osmotic treatment.

In postharvest technology, one of the major objectives is to preserve the structure of plant materials, since alterations in their structure may lead to detrimental changes in texture, flavor, nutritional properties, and wholesomeness. This is also true for the handling of meat. Similarly, the quality of structured semi-solid and liquid foods, such as gels and emulsions, needs to be maintained during distribution and storage [28]. Bloom on chocolate, for example, is a surface condition resulting in a white, powdery appearance that consumers interpret as a sign of poor quality. Bloom results from incorrect processing or storage [29]. Many innovative methods have been developed and used to investigate the morphology and composition of bloomed chocolate. Briones and Aguilera [30] evaluated the changes in surface color during storage of milk chocolate tablets and related them to the migration of fat to the surface of the sample where major color changes in the surface occurred after 33 days of storage. The presence of surface bloom was observed in two ways. First (around day 33) a few round white specks appeared on the original brown background, probably due to rapid migration of liquid fat through flaws or pores in the surface. Later (around day 36), a whitish homogeneous background gradually set in, slowly replacing the original brown background of the chocolate tablets and superimposed on the white specks.

Macroscopic modifications occurring in frozen gelatinized starch systems, such as syneresis and textural changes, can be related to starch retrogradation and ice crystal formation [28, 31]. James and Smith [29] used ESEM to image the morphology of the surface of fresh and bloomed chocolate and to observe the changes in the morphology of this product in real time, during heating and cooling. The nodular morphology of sugar bloom was distinguished from the blade-like crystals of fat bloom. Surface details for the fat blades indicated an extrusion mechanism during their formation on the surface of chocolate.

Aguilera [19] mentioned that control of biochemical activity within a cell is often achieved by physical separation of reactants in microstructural locations. Disruption of cell microstructure influences the generation of off-flavors and induces browning. A well-known example is *Allium* species where flavor precursors are converted to flavor compounds when cells are disrupted by chewing or other means of mechanical injury. Changes in structure mediated by freezing and drying can activate enzymes in unheated plant tissues, leading to off-flavor development and textural changes. The physical structure of a food can have an effect on the local chemical environment perceived by the microbial cells and during their proliferation.

22.3 Microscopy Techniques

Microscopy and other imaging techniques can be used to observe food structures and, in combination with microanalytical methods and image processing, may provide useful data at the required scale, and hence be used to derive physical models on which basic equations and computational methods may be applied to describe the observed structure. Advanced microprobe techniques combining microscopy and spectroscopy allow localization and mapping of the composition and state of individual phases in a particular area within the field of view (FOV). Raman microprobes can detect spatial resolution of about 1 μm and also, proteins, lipids, and carbohydrates in the presence of water as well as different phases of the same component (e.g., polymorphs, amorphous, crystalline states, etc.). Infrared microprobes have been used to study plant cell walls and crystallization of amylopectin [4].

Examining food microstructure is always a difficult task because of the complexity of materials. One useful approach for differentiating structural features from artifacts is correlative microscopy. New methods available, such as SEM, ESEM, TEM, CLSM, MRI, and AFM, constitute a battery of powerful techniques with good flexibility, minimal intrusion, and simplicity of sample preparation within the whole dimensional scale from the molecular to the macrostructural level [8, 29, 32, 33]. Some of these techniques offer the possibility of simultaneously performing chemical, thermal, or mechanical analyses in simulated and actual processing conditions [10, 19].

The microstructural information generated can be used to understand transport mechanisms during drying and to assess the functionality of finished products. Efforts should be made to derive appropriate scaling laws for results obtained at the microstructural and nanostructural levels. A variety of microscopic techniques are available for studying different structures and components of food systems [21]. The following section describes the different microscopes used in food science and technology-related fields.

22.3.1 Light Microscopy

Light microscopy (LM) has been available for a long time and is used when the structure and microscopic organization is an important issue (food materials). The technique has a resolution about 10^3 times greater than the human eye [34]. Optical or LM involves passing visible light (transmitted through or reflected) from the sample through a single lens or multiple lenses, allowing a magnified view of the sample [35]. In conventional LM using blue light (wavelength of 470 nm) and an oil immersion objective lens with a numerical aperture of 1.40, the resolution would be about 200 nm. LM is able to produce a magnification of about 1000×. Resulting images can be detected directly by the eye or can be captured digitally.

Sample preparation for LM is less complex than for other microscopic techniques. Food materials can sometimes be examined whole, but more usually studies of cellular structure require the cutting of sections. Microtomes enable sections of uniform thickness to be cut from samples. In order to maintain structural integrity, tissues are often embedded in materials such as paraffin wax, plastics, or resins; alternatively, sectioning can be performed on frozen material. In applying LM, it is common to apply a specific stain or dye to improve visual contrast or differentiate tissues. The specimen is placed on a glass microscope slide, usually a suitable mounting medium is applied, and a cover glass is also used [21].

22.3.2 Confocal Laser Scanning Microscopy

CLSM is a relatively new technique for the structural analysis of biological and food materials. In contrast to conventional LM, the light source is replaced by a laser beam. A scanning unit and a pinhole in the back focal plane, which improves the limited depth of focus, are also incorporated in this apparatus [36] Advanced computer imaging technologies, fluorescent probe developments, and computer-designed optics which may be coupled to this microscope enhance the capacity of this instrument for high resolution volumetric imaging [8].

The most important feature of CLSM to research is its ability to produce optical sections of a three-dimensional (3D) specimen. An optical section contains information from one of the focal planes only. Therefore, by moving the focal plane of the instrument within defined distances of micrometer-order through the depth of the specimen, a stack of optical sections can be captured. This property of the

CLSM is fundamental for solving 3D problems where information from regions which are distant from the plane of focus can blur the image of such objects. For imaging in the CLSM, either the epifluorescence or epireflection mode is generally used. As a valuable by-product, the computer-controller CLSM produces digital images which are amenable to IA and processing and can also be used to compute surface or 3D reconstructions of the specimen [5, 36]. CLSM not only provides excellent resolution within the plane of the section (>0.25 mm in *x*- and *y*-axis), but also yields good resolution between section planes (>0.25 mm in the *z*-axis) [8, 20, 34, 36, 37].

The application of CLSM to food materials has been particularly fruitful in the area of lipid components because optical sectioning overcomes the tendency of fats to smear and migrate. Lipids are also well suited to fluorescent staining. The possibility of combining CLSM with rheological measurements, light scattering, and other physical analytical techniques in the same experiments with specially designed stages offers the opportunity to obtain detailed structural information of complex food systems [36].

22.3.3 Electron Microscopy

In electron microscopy, a beam of electrons rather than light is used to form a magnified image of specimens since electrons provide as much as a 1000-fold increase in resolving power. As an electron beam falls on a sample, interactions generate a variety of signals that can be captured to obtain images. The two basic types of electron microscopes are the TEM, in which transmitted electrons are captured, and the SEM, in which secondary electrons are captured. Nowadays, electron microscopy is a useful tool for investigating the microstructure of cereal grains, dough, beef, and others. This technique is particularly in demand for 3D investigations in food science [4, 37].

22.3.3.1 Conventional and Environmental Scanning Electron Microscopy

Microscopic techniques provide useful and informative methods to observe and understand the micrographic details of a food system. The high vacuum in the sample chamber that is needed for most electron microscopic techniques has limited direct observation of dynamic processes of moisture movement. In addition, dimensional change somehow overcomes these barriers and permits the observation of food microstructures when exposed to different relative humidities [38].

22.3.3.2 Transmission Electron Microscopy

The TEM operates on the same basic principles as the light microscope but uses electrons instead of light. The limitation of what can be seen in the LM is the wavelength of light. In TEM, electrons are used as the "light source" and their low wavelength makes it possible to obtain a resolution a 1000 times better than is possible with a light microscope. For instance, it is possible to see objects of about

10^{-10} μm and to study small details in cells or different materials down to near atomic levels. The possibility of high magnification has made the TEM a valuable tool in medical, biological, and materials research.

TEM is a technique in which an electron gun (tungsten filament) is heated and emits a narrow beam of electrons traveling at high speed. The voltage applied to achieve this acceleration is within the 40–300 kV range, the magnification goes from 50 to 1 000 000×, and the resolution goes from 2.4–1.5 to >0.5 Å. In focusing, the electrons will be reflected by a magnetic field and a magnetic lens can be employed analogous to a converging glass lens used in LM. TEM seems to be the most applicable technique for observing certain food components, such as proteins, and how they interact with other components [33, 34, 39].

22.3.4 Atomic Force Microscopy

AFM provides a method for obtaining nanoscale structural information. The images are obtained by measuring changes in the magnitude of the interaction between the probe and the surface of the sample as it is scanned. This instrument measures the current of electrons that tunnels from atoms at the probe tip to the surface with atomic resolution as it scans the surface. The resolution of AFM images includes lateral (X, Y) resolution. Vertical (Z) resolution is determined by the resolution of the vertical scanner movement, that is <0.1 nm. There are three primary imaging modes in AFM operation: contact mode, noncontact mode, and tapping mode. In food science applications, the "tapping" mode is used to reduce artifacts. In tapping mode, the cantilever on which the tip is mounted oscillates while separating from the surface. This has allowed AFM to be applied to soft, hydrated materials such as those found in some food specimens, since it overcomes problems associated with friction, adhesion, and electrostatic forces. Applications of AFM have been reported in food science and technology research, including qualitative macromolecule and polymer imaging, complicated or quantitative structure analysis, molecular interaction, food molecular manipulation, surface topography, and nanofood characterization [5, 13, 40].

Table 22.1 shows a brief description of different microscopes used in food science.

Other important techniques used to study the internal structure of foods are MRI and soft X-ray microscopy. These techniques are discussed next.

22.3.5 Magnetic Resonance Imaging

There are many applications of MRI for characterization of dehydrated products. MRI is particularly useful for measuring changes of state of the water in food because MRI parameters are related to water content, the mobility within the structure and its interactions with macromolecules. Since the signal intensity is directly proportional to moisture content, MRI data can be directly converted to moisture

Table 22.1 Comparative description of microscopes more used in food science.

	Light microscopy	TEM	ESEM	SEM	AFM
General use	Surface structures and sections	Internal thin sections	Surface structures	Surface structure	Topography and phases surfaces
Resolution	200–500 nm	24–15 to >5 nm	7 nm at 30 kV	3–4 nm at 30 kV	Atomic
Magnification (×)	10–1500	50 to >1 000 000	50–200 000	50–300 000 and high resolution 1 000 000	–
Illumination	Visible light	High speed electrons	High speed electrons	High speed electrons	–
Preparation	Easy	Difficult	Easy	Easy	Easy
Thickness	Thin	Very thin	Reflectance	Reflectance	Small and thin
Environment	Versatile	Vacuum	Low vacuum	Vacuum	–

Adapted from Aguilera and Stanley [34].

content, and moisture profiles inside a product may be effectively calculated during drying since magnetic resonance can measure directly translational motion which can be converted to *in situ* effective diffusion coefficients [41].

MRI has been used to assess moisture transfer during drying through structures in corn kernels. MRI and its higher resolution version, MRI microscopy, are likely to become major tools in the study of dehydration due to their noninvasive nature, ability to generate 3D structural information, chemical probing capabilities, and the possibility of studying the drying process inside equipment [42].

22.3.6 Soft X-Ray Microscopy

Soft X-rays have an energy level of about 100–1000 eV and a wavelength in the 1–10 nm range, offering the potential for very high resolution imaging. X-ray microscopy or microtomography works in the same way as X-rays used in medicine but with much finer resolution. It has many potential advantages for studying foods during dehydration: the specimen can be fully or partially hydrated, good contrast is possible without staining, resolution can be as low as 30 nm (better than in LM), and objects can be imaged at atmospheric pressure. Internal structures are reconstructed as a set of flat cross-sections that are used to analyze the two and three-dimensional morphological parameters of the object. The process is nondestructive and requires no special preparation of the specimen [43, 44].

22.4 Image Analysis

IA techniques have been applied for the study of food quality and provide valuable qualitative and quantitative insights into the microstructure of the observed object. Dimensional changes during processing of food materials and increased demand for objectivity, consistency, and efficiency have led to the introduction of computer-based image processing techniques. Recently, computer vision systems employing IA techniques have developed rapidly and can quantitatively characterize complex size, shape, color, and textural properties of foods. Computer vision systems can also be used to acquire and analyze images during a process or to aid in process control. They have been used in the food industry for the detection of defects and identification, grading, and sorting of fruits and vegetables, meat and fish, bakery products, and prepared commodities among others. The automatic pattern recognition process, according to Pedreschi *et al.* [45], Du and Sun [7], and Pérez *et al.* [17] consists of five steps:

- image acquisition;
- image preprocessing;
- image segmentation;

- feature extraction; and
- classification.

22.4.1 Image Acquisition

A digital image of the object is captured and stored in the computer. When acquiring images, high quality can help to reduce the time and complexity of the subsequent processing. It is also important to consider the effect of illumination and the orientation of the specimen relative to the illumination source because the gray level of the pixels is determined not only by the physical features of the surface but also by these two parameters. In recent years, there have been attempts to develop nondestructive, noninvasive sensors for assessing the composition and quality of food products. Various sensors such as the CCD camera, ultrasound, MRI, CT, ET, TEM, SEM, ESEM, and CLSM, among others, are widely used to obtain images of food products.

22.4.2 Image Preprocessing

This step improves image data by suppressing distortions or enhancing some image features that are important for further processing and creates a more suitable image than the original for a specific application. Using digital filtering, the noise of the image, which may degrade the quality of an image and subsequent image processing, can be removed and the contrast can be enhanced. In addition, in this step the color image is converted to a grayscale image, called the *intensity image*.

22.4.3 Image Segmentation

Techniques for image segmentation developed for the study of food quality can be divided into four different approaches: threshold-based, region-based, gradient-based, and classification-based segmentation. The intensity of image is used to identify disjointed regions with the purpose of separating the part of the image under study from the background. This segmented image (S) is a binary image consisting only of black and white pixels, where "0" (black) and "1" (white) mean background and object, respectively. The region of interest within the image corresponds to the area where the material being studied is located.

22.4.4 Feature Extraction

Feature extraction is mainly concentrated around the measurement of geometric properties (perimeter, shape factors, Fourier descriptors, invariant moments, etc.)

and on the intensity and color characteristics of regions (mean value, gradient, second derivative, image texture, features, etc.). The geometric features are computed from the segmented image (S), the intensity features are extracted from the intensity of the image, and the color features from the RGB characteristic of the images. It is important to know which features provide relevant information for the classification. For this reason, a selection of features must be performed by trial and error. The texture of an image is characterized by the spatial distribution of gray levels in a neighborhood. *Image texture* is defined as the repeating patterns of local variations in image intensity that are too fine to be distinguished as separate objects at the resolution used, that is, the local variation of brightness from 1 pixel to the next (or within a small region). Image texture can be used to describe image properties such as smoothness, coarseness, and regularity. Three approaches may be taken to determine features of the image texture: statistical, structural, and spectral. The image texture of some food surfaces has been described quantitatively by fractal methods. Besides fractal properties, other important characteristics can be obtained from an image such as statistical, geometrical, and signal-processing properties, among others.

22.4.5 Classification

The extracted features of each region of the image are analyzed and assigned to one of the defined classes of the IA process which represent all possible types of regions expected in the image. Simple classifiers may be implemented by comparing measured features with threshold values. Nonetheless, it is also possible to use more sophisticated classification techniques such as statistical, fuzzy logic, and neural network that have as a common objective to simulate a human decision-maker's behavior, and have the advantage of consistency and, to a variable extent, explicitness. Figure 22.2 shows a diagram of the basic steps used in automatic pattern recognition and IA which also apply to the characterization of food structures from microscopy images.

22.5 Applications of Microscopy and Image Analysis Techniques

In this section different applications of microscopy and image analysis techniques used in food science are described.

One application of microscopy techniques is the study of the microstructure of immobilized spheres containing bacteria when they are treated under gastrointestinal conditions [46, 47]. This is important for the understanding of the extent of damage to the capsules which may be associated with a decrease in viability of bacteria.

Mendoza [48] evaluated the survival of *Bifidobacterium bifidum* under gastrointestinal *in vitro* simulation using microscopy and IA. The fractal dimensions and entropy of the image depicted in Figure 22.3 indicated that the surfaces of the

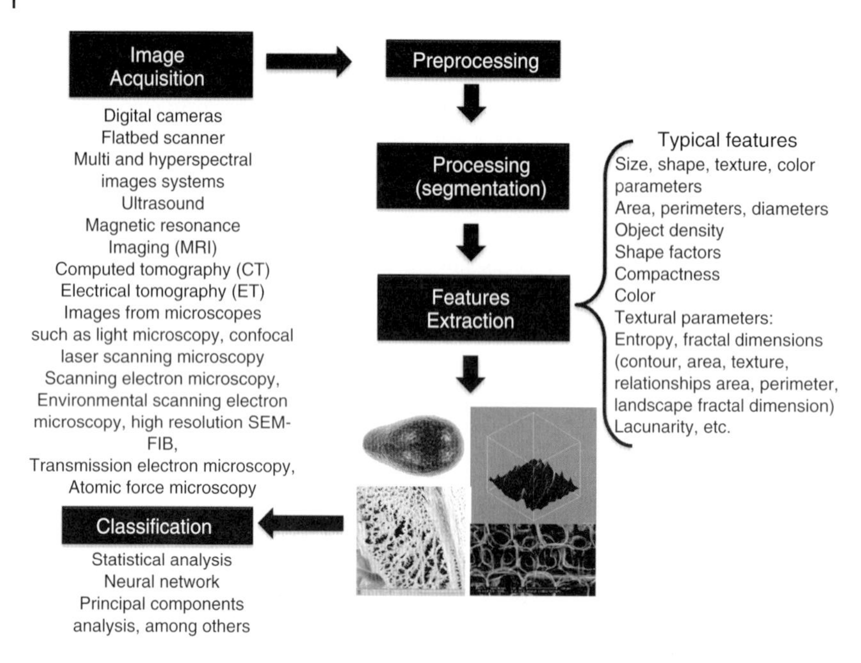

Figure 22.2 Basic steps used in automatic pattern recognition process and image analysis of microstructure images.

spheres were more heterogeneous and rough than their interiors, which were smoother and more homogeneous. ESEM and IA were helpful in evaluating changes that occurred in the immobilization process which is used as a protection system during the exposure to gastric and intestinal *in vitro* simulated conditions. The techniques were also useful tools for evaluating the morphology, as well as for monitoring the morphological and microstructural changes on the immobilization support.

Another example related to the application of microscopy techniques is the drying of coffee. One of the most important processing stages involves the removal of water from the coffee parchment beans into the surrounding environment until the beans reach equilibrium moisture content. However, ultrastructural changes during drying are not well understand, therefore Borém *et al.* [49] evaluated and compared the alterations in the structure of coffee seed endosperm subjected to different temperatures (40, 50, and 60 °C) and drying conditions (airflow of 0.33 $m^3\ m^{-2}$ s). The dried seeds were randomly selected and prepared for the histochemical tests with Sudan IV and SEM and TEM according to traditional techniques. The histochemical results (light, SEM, and TEM micrographs) showed that for the coffee parchment beans dried at 40 °C there was no change in the cellular integrity of the plasma membrane and vesicles. In contrast, in the endosperm of parchment coffee beans dried at 60 °C, fused oil bodies that gave rise to large droplets in the intercellular space were observed, indicating rupture of vesicles and plasma membrane. SEM and TEM micrographs showed that for the parchment

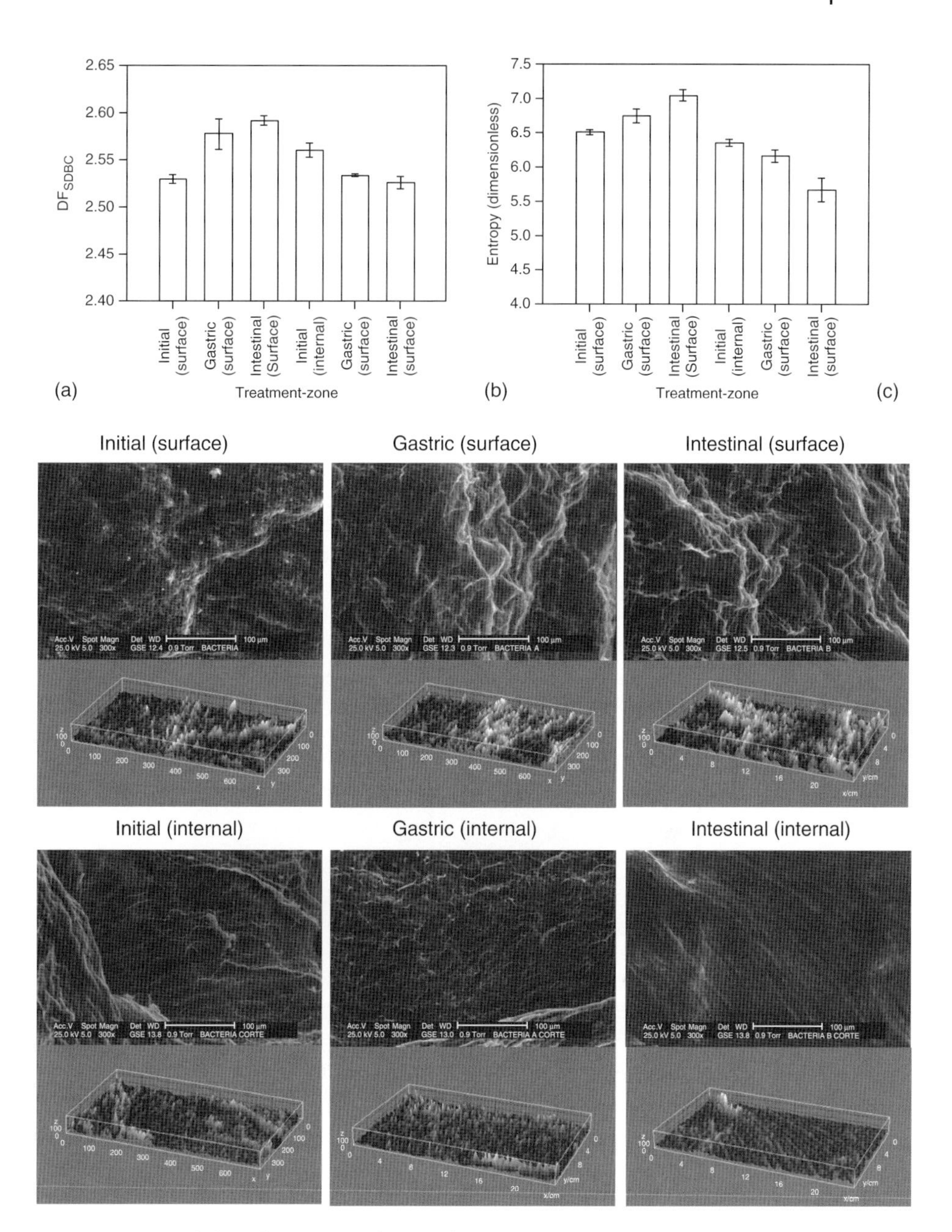

Figure 22.3 Morphological parameters for immobilized spheres under gastrointestinal conditions: (a) fractal dimension, (b) entropy, and (c) environmental scanning electron microscope images with their corresponding gray level intensity plots.

Arabica coffee beans dried at 40 °C, the internal cellular content remained intact and full of cellular material, and the space between the plasma membrane and the cell wall was empty. However, in seeds dried at 60 °C, rupture of the cell was observed, represented by occluded intercellular spaces, indicating leakage of part of the protoplasm. These observations are essential to the preservation of the quality of coffee, as the rupture of the membranes can expose the oils to oxidation and rankness, which in turn leads to the formation of undesirable compounds that alter the aroma and flavor.

When biomaterials are subjected to drying, the moisture that is part of the structure of the solid, pore, or fiber, is removed, affecting the structure and properties of the biomaterial. Micro- and macrostructural studies could be useful to quantify the changes during drying and could help with the understanding of the mass and heat transfer mechanisms and their effect on the functionality of the final product (color, texture, etc.). The evaluation of structural changes can be accomplished through the use of innovative tools such as microscopy techniques and IA. The drying of oilseeds is one of the most important operations in oil extraction, because the total oil yield depends on the extraction time and temperature, particle size of the oil-bearing material, and, most importantly, the moisture content [50]. The higuerilla oilseed, also known as *Ricinus communis*, contains around 40–50% oil, and is widely used in the chemical, pharmaceutical, cosmetic, and energy industries. The drying conditions have clearly been shown to affect the extraction process from this oilseed. For example, it has been demonstrated (Figure 22.4) that at 80 °C

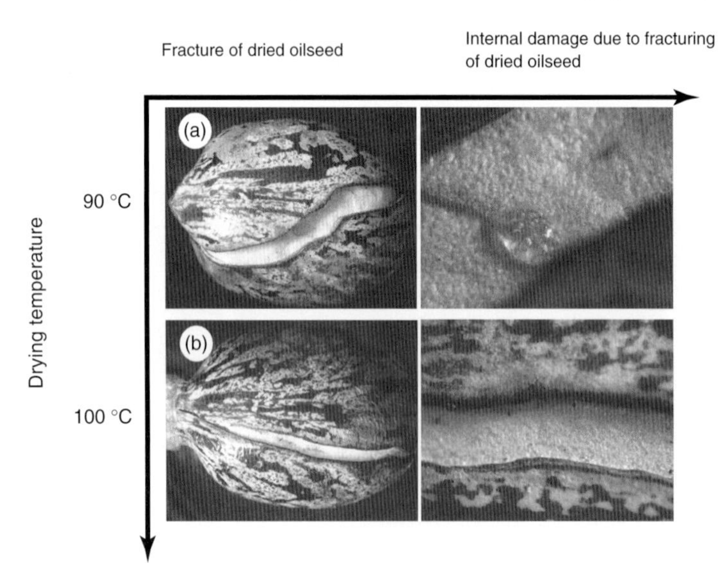

Figure 22.4 Light stereomicrographs of shell fracture of higuerilla oilseed (*Ricinus communis*) (left) and effect on parenchyma (right) due to fluidized bed drying at 90 °C (a) and 100 °C (b); 8.7 m s^{-1} airflow. Observed at 10× and 50×, respectively.

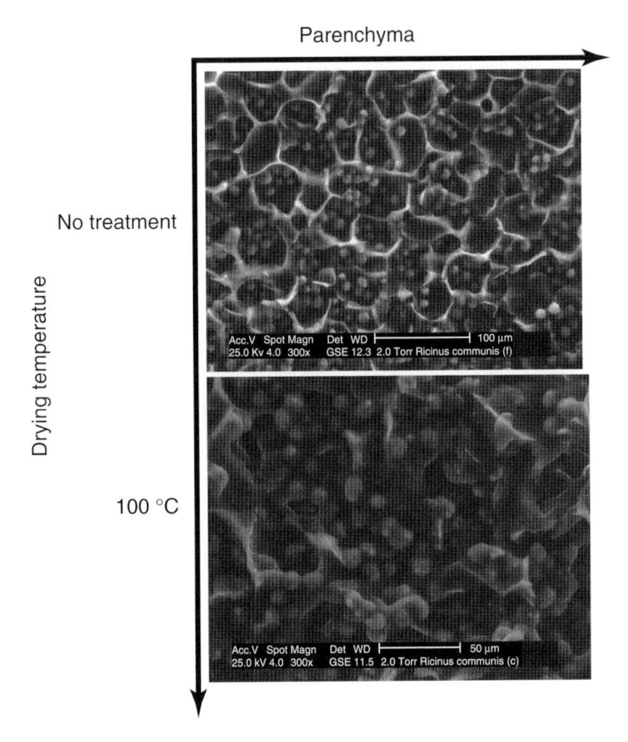

Figure 22.5 Environmental scanning electron micrographs of parenchyma oilseed of higuerilla dried at 100 °C (bottom) and untreated (top).

the proportion of damaged oilseed is low and the fracture is higher and transverse, following a nonregular pattern, in comparison with the seeds dried at 90 °C in which the fracture is often on the lateral side of the seed. In addition, the number of damaged seeds is increased at 90 °C.

Finally, in the seeds dried at 100 °C, the number of damaged seeds increased again, with fractures on the lateral side. In the fractured seeds dried at 90 °C lipid bodies were present in the parenchyma. Changes in the parenchyma are evident (Figure 22.5) in seeds dried at 100 °C, with cell shrinkage and coalescence of protein bodies. These structural changes were observed by means of microscopy techniques and IA could explain the effect of the drying conditions on the structure, which has an influence on the oil extraction and yield.

The microstructure of Marcona almonds using different degrees of roasting has been studied by means of quantitative IA, and the relationship between their microstructure and mechanical fracture studied by macro-IA [51]. IA methods have been successfully used to quantify the features of both macro- and microphotographs related to the fracture of roasted almonds. The particle size distribution of the fractured samples, calculated by IA, would appear to be a very useful tool for characterization of the fracture features of crispy/crunchy foods such as other nuts,

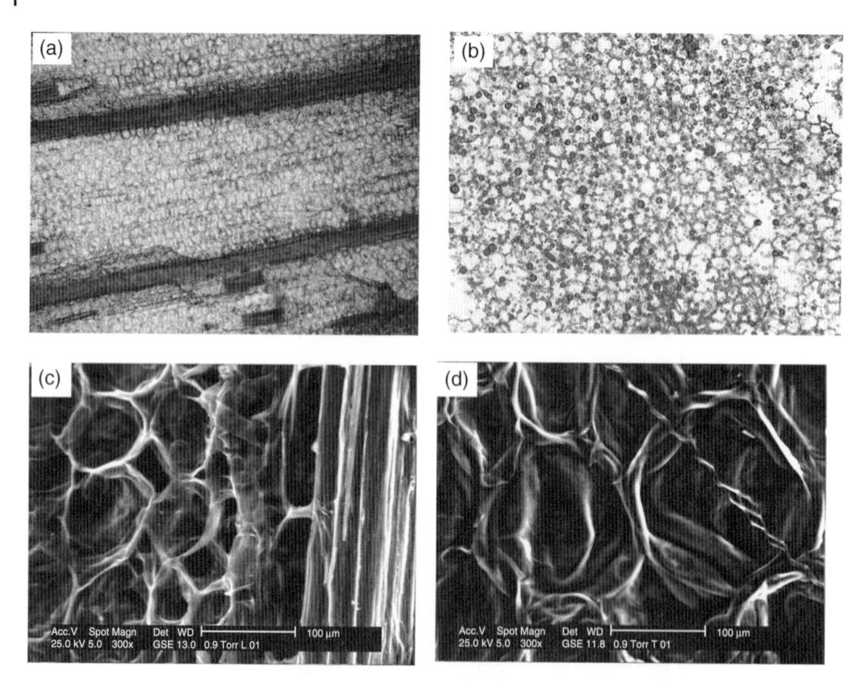

Figure 22.6 Micrographs of different plant tissues used in the food industry. (a) Light stereomicroscopy of *Agave* tissue showing the different structures and arrays in the material. Magnification 40×. (b) Light micrograph of parenchyma tissue of higuerilla showing the oil bodies. Magnification 150×. (c, d) Environmental scanning electron micrographs of fresh Agave tissue: (c) longitudinal cut, (d) transversal cut. Magnification 300×.

snacks, biscuits, and so on. Also LM and SEM micrographs can reveal the thermal damage to samples during the roasting process; this damage can be quantified efficiently [49].

Figure 22.6 shows different plant tissues observed by means of microscopy techniques (ESEM and LM). These micrographs allow the microstructural characterization of different components (lipids, proteins, carbohydrates, water, and fiber, among others).

Drying efficiency and the final characteristics of a dried product are strongly dependent on the composition and structure of the material before it is subjected to modification or processing [2]. Researchers have proved that the structural changes to food during convective dehydration modifies the heat and mass transport properties, reducing the drying rate [52]. The reduction in drying rate is a result of many phenomena that occur during vegetable dehydration, including shrinkage and reduction in the diameter of capillaries in porous solids [53]. Water flow from the interior of the material causes redistribution of compounds that are deposited on the surface, helping with the hardening and the formation of crust, hence reducing water migration from the inside to the outside, resulting in shrinkage of the surface that is exposed to the air, and modifying the mass transfer from the interface to the air [54]. The shrinkage of vegetables is the result of cell water loss and contraction of the cell walls. These phenomena can cause cell and structural collapse, causing irreversible deformation [54–56].

Studies on the dehydration of *Agave atrovirens* evaluated the influence of type of cut characteristics on the structure of the material by means of microscopy techniques (stereomicroscopy and SEM) and IA; such data provide qualitative and quantitative descriptions of the drying effect on the vegetal tissue microstructure that will determine the functionality of its components for use in the food industry. During the drying, it was shown that the microstructural arrangement of the agave tissue and the orientation that the rigid structures adopted (fibers) at a specific cut angle, influenced the shrinkage, deformation, and collapse of its structure. This in turn influences the resistance to the structural or deformation changes occurring during the drying process (Figure 22.7). These phenomena are likely to influence the diffusion of water within the material and its final functionality. Taking advantage of the extracted parameters by means of IA (fractal dimension and entropy), it is possible to quantify these phenomena and understand more fully the effect of convective dehydration on the microstructure, properties, and functionality.

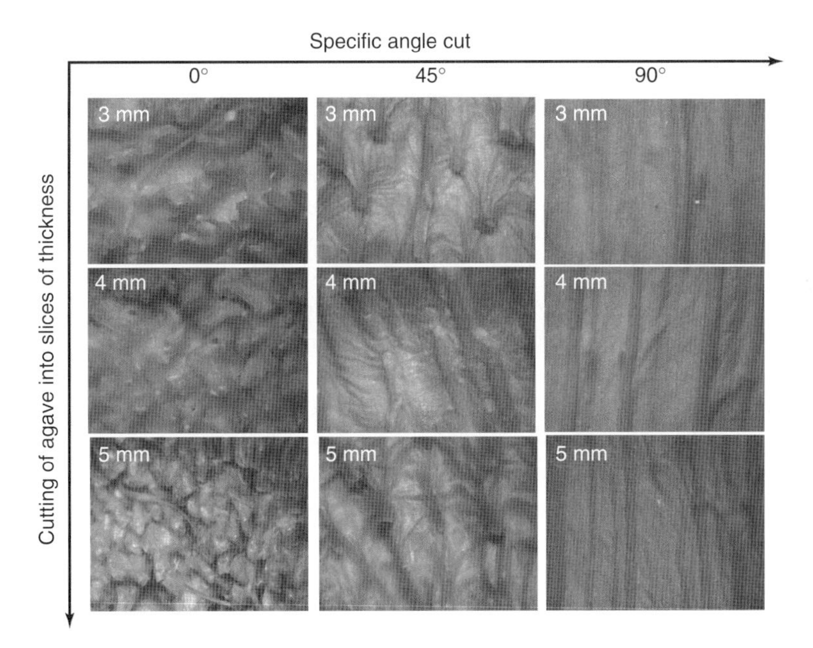

Figure 22.7 Gallery of light stereomicroscopy micrographs of microstructure of *Agave* tissue dried convectively at 60 °C, 2 m s^{-1}. The resistance to the fiber deformation due to the influence on the arrangement and orientation that the rigid structures adopt in a specific angle cut and thickness of slices can be seen.

IA has been useful for describing the enzymatic degradation of particulate food material. Devaux *et al.* [57] correlated the physical changes to the chemical transformation of beet pulp particles by enzymatic degradation. Chemical and physical changes in particles during degradation were interpreted, taking into account the current model of molecular arrangements of primary cell walls. This enabled better interpretation of enzymatic degradation for its application in the food industry for the extraction of functional compounds.

Physical and rehydration properties (also called *functional* or *instant properties*) of food powders have a high dependence on their structure at macro-, micro-, and nanometric levels. The authors have demonstrated the morphology and microstructure of different samples of milk powder (Figure 22.8) and their relationship with rehydration. Light stereomicroscopy (LS) and ESEM were applied to obtain images of milk powder and IA was performed to evaluate their morphology and microstructure. The range of particle sizes (agglomerates) was 100–450 μm. Wettability (8.95–8019 s) decreased as the particle size increased; this behavior is shown in Figure 22.8. According to the literature [58], the lowest values of Fractal dimension (FD) are associated with lower tortuosity (more spherical agglomerates), less roughness on the surface, and lower rehydration capabilities. There are clear relationships between the microstructure and morphology of milk powder agglomerates and their rehydration capability. IA was an adequate tool to evaluate the quality and the agglomeration level of milk powders.

Villalobos *et al.* [15] evaluated the optical (gloss and transparency) and microstructural properties of nine edible films prepared with hydroxypropyl methylcellulose (HPMC) and surfactant mixtures (Span 60 and sucrose ester P-1570) with different hydrophilic–lipophilic balance (HLB). Film microstructure was analyzed using LM, AFM, and SEM. Fractal analysis was applied to the microscopy images to quantify the complexity of the film microstructure. The study demonstrated the strong influence of the film structure (internal and surface arrangement of the different phases) on film optical properties, which define the film quality attributes.

The current authors have also shown that microscopy techniques and IA can help to characterize the structure of edible films at different levels and to establish relationships with their physical properties. The aim of this study was the characterization of the structure of alginate film by means of microscopy (LM, ESEM, and AFM) and IA (Figure 22.9). Alginate films were prepared by varying the concentration of NaCl (0.01, 0.05, and 0.09 M) and alginate (0.1, 0.5, and 0.9%). Textural features from images were extracted using the gray level co-occurrence

Figure 22.8 Study of the physical and rehydration properties of commercial milk powders and relationships with their morphology and microstructure. (a) Environmental scanning electron microscopy of commercial powdered milk showing particles with an agglomerated structure. Magnification 400×. (b) Agglomerated images of powdered milk obtained by means of light stereomicroscope (Nikon SMZ1500) at 60× showing the fractal dimension of contour (FD). (c) Graphs showing the influence of particle size on rehydration capacity and morphology.

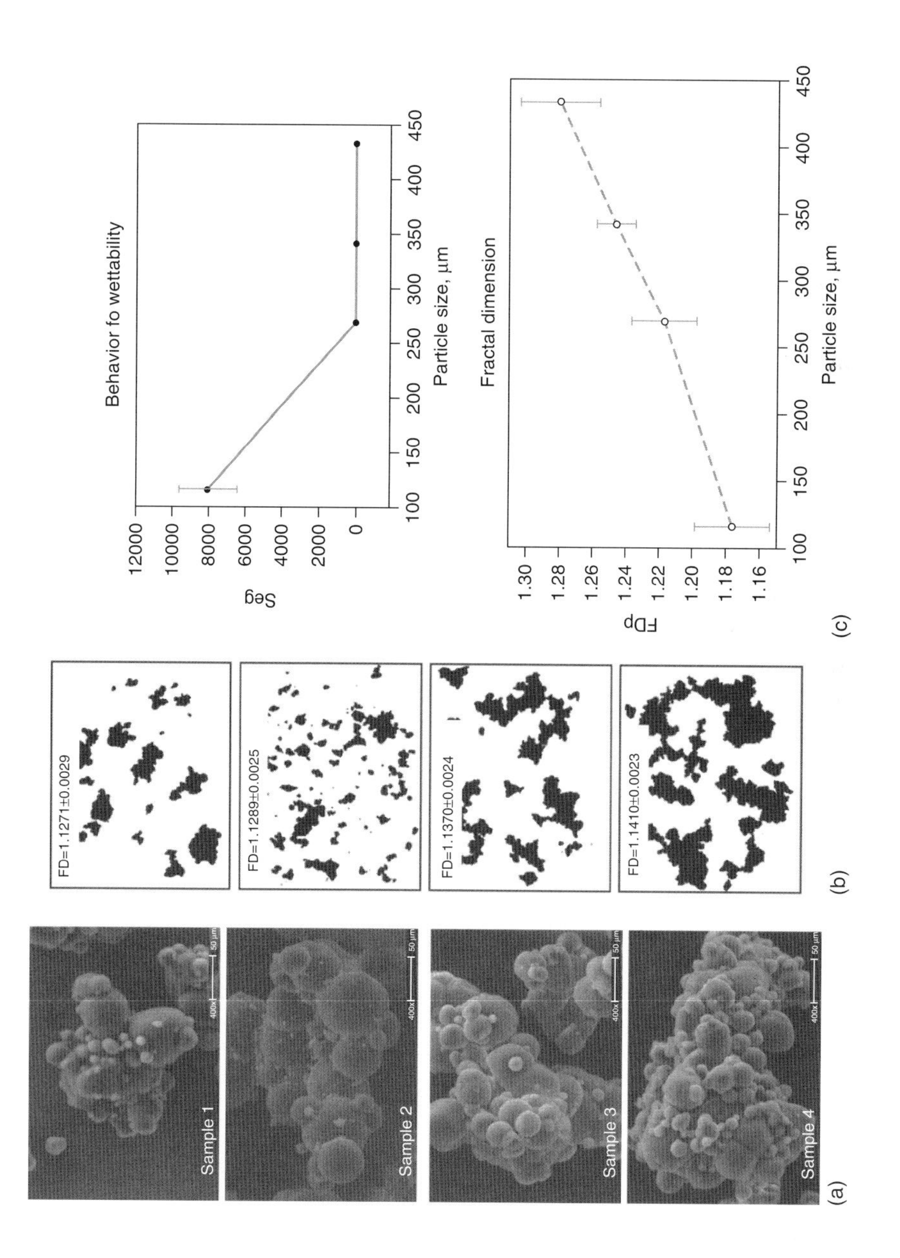
(a)
Sample 1
Sample 2
Sample 3
Sample 4
(b)
FD=1.1271±0.0029
FD=1.1289±0.0025
FD=1.1370±0.0024
FD=1.1410±0.0023
(c)
Behavior fo wettability
Seg
Particle size, µm
Fractal dimension
FDp
Particle size, µm

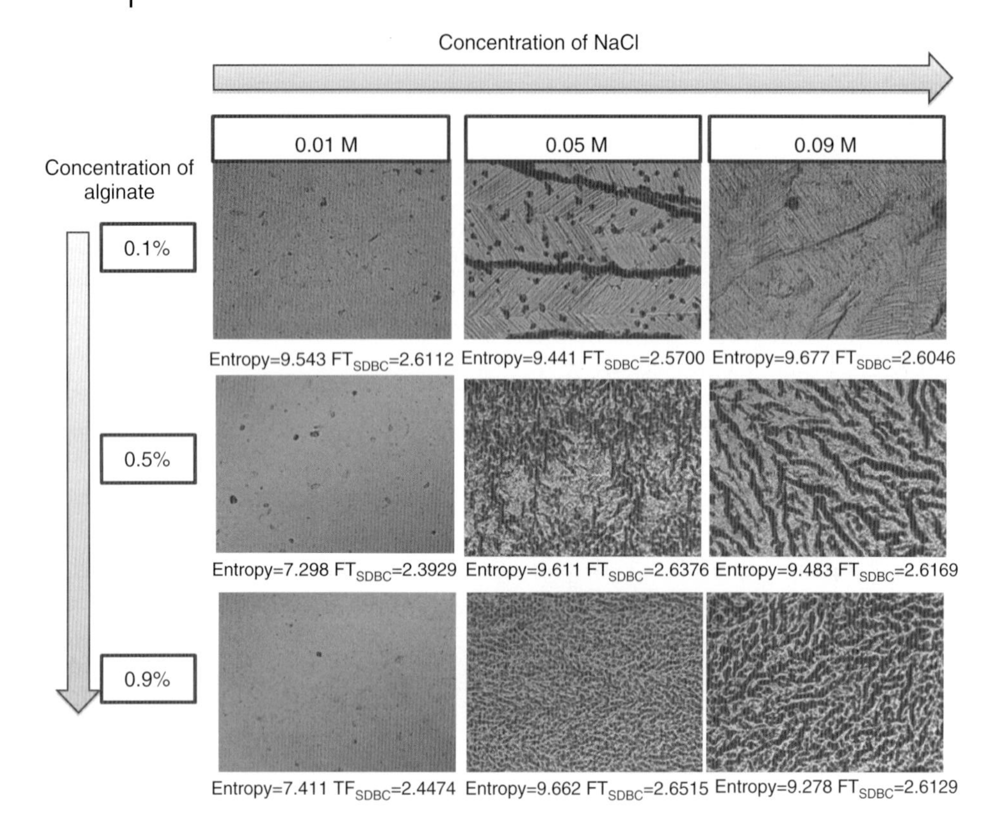

Figure 22.9 Light micrographs of alginate films at different NaCl and alginate concentration with respective entropy and fractal texture (FT) parameters. Magnification 10×.

matrix (GLCM) and fractal texture (FT) using the scattering differential box counting algorithmic [59]. It was shown that when the alginate and NaCl concentration increased, the films had a different structure and these changes influenced the FT.

Figure 22.10 shows images of the topography of the films with NaCl concentrations of 0.1 and 0.3 M and 0.5 and 0.9% of alginate. Figure 22.10 also shows that the structures of the films are formed by agglomerates. It can be seen that when using a concentration of 0.9% of alginate, the agglomerates that form crystallization networks are usually <1 μm in length.

Rodrigues and Fernandes [60] studied the effect of the osmotic dehydration on fruit shrinkage and on the sugar/salt penetration in the fruit, as well as its impact on the overall dehydration rates. IA was a useful technique for the understanding of many aspects related to the osmotic dehydration process.

Fracture of dry cereal foods is important since it can be related to texture and physical performance during transport and storage. The moisture content is a variable key factor affecting these properties and the type of fracture. Castro and

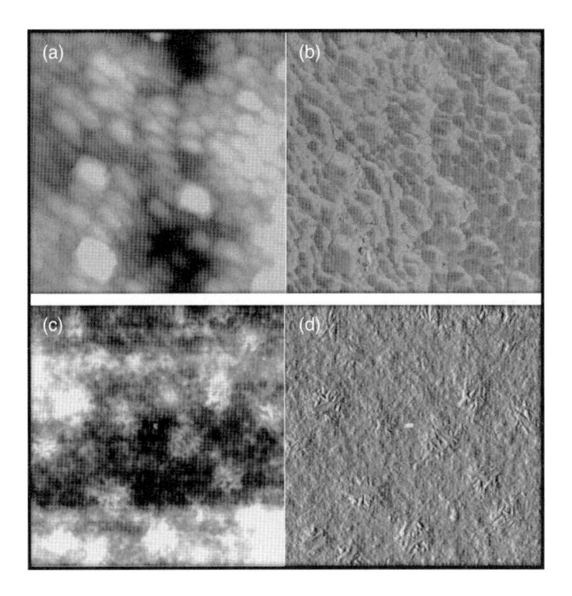

Figure 22.10 Topographic images of the alginate films with 0.05 M NaCl and 0.9% alginate (a and b) and a 0.01 M NaCl and 0.9% alginate (c and d). These images were taken using atomic force microscopy.

Aguilera [61] assessed the effect of air drying and the resulting microstructure on the physical properties of a starch model system as a function of water activity.

22.6 Concluding Remarks

The microstructure of food and food-related materials is a very important feature in production, transport, and design of food ingredients. IA has been developed and proved to be a powerful tool for the description of such features. The development of CVS could be an aid to improving existing processes or could help in the development of engineering-based foodstuffs.

References

1. Peukert, W. (2005) General concepts in nanoparticles technology and their possible implication on cultural science and philosophy. *Powder Technol.*, **158**, 133–140.
2. Aguilera, J.M. (2000) Structure–property relationships in foods, in *Trends in Food Engineering*, Food Preservation Series, (eds J.E. Lozano, C. Añón, E. Parada-Arias, and G.V. Barbosa-Cánovas), Technomic Publishing Co., Lancaster, PA, pp. 1–14.
3. Ko, S. and Gunasekaran, S. (2007) Error correction of confocal microscopy images for in situ microstructure evaluation. *J. Food Eng.*, **79**, 935–944.
4. Witek, M., Weglarz, W.P., De Jong, L., van Dalen, G., Blonk, J.C.G.,

Heussen, P., Van Velzen, E., Van As, H., and van Duynhoven, J. (2010) The structural and hydration properties of heat-treated rice studied at multiple length scales. *Food Chem.*, **120**, 1031–1040.

5. Aguilera, J.M., Stanley, D.W., and Baker, K.W. (2000) New dimensions in microstructure of food products. *Trends Food Sci. Technol.*, **11**, 3–9.
6. Quintanilla-Carvajal, M.X., Camacho-Díaz, B.H., Meraz-Torres, L.S., Chanona-Pérez, J.J., Alamilla-Beltrán, L., Jimenéz-Aparicio, A., and Gutiérrez-López, G.F. (2009) Nanoencapsulation: a new trend in food engineering processing. *Food Eng. Rev.*, **2**, 39–50.
7. Du, C.-J. and Sun, D.-W. (2004) Recent developments in the applications of image processing techniques for food quality evaluation. *Trends Food Sci. Technol.*, **15**, 230–249.
8. Amos, W.B. and White, J.G. (2003) How the confocal laser scannig microscope entered biological research. *Biol. Cell.*, **95**, 335–342.
9. Stokes, D.J. (2003) Recent advances in electron imaging, image interpretation and applications: environmental scanning electron microscopy. *Philos. Trans. R. Soc. London, Ser. A*, **361**, 2771–2787.
10. Chen, X.D., Chiu, Y.L., Lin, S.X., and James, B. (2006) In situ ESEM examination of microstructural changes of an apple tissue sample undergoing low-pressure air-drying followed by wetting. *Drying Technol.*, **24**, 965–972.
11. James, B. (2009) Advances in "wet" electron microscopy techniques and their application to the study of food structure. *Trends Food Sci. Technol.*, **20**, 114–124.
12. Pishkenari, H.N., Jalili, N., and Meghdari, A. (2006) Acquisition of high-precision images for non-contact atomic force microscopy. *Mechatronics*, **16**, 655–664.
13. Shimoni, E. (2008) Using AFM to explore food nanostructure. *Curr. Opin. Colloid Interface Sci.*, **13**, 368–374.
14. Chanona, P.J.J., Alamilla, B.L., Farrera, R.R.R., Quevedo, R., Aguilera, J.M., and Gutiérrez, L.G.F. (2003) Description of convective air-drying of a food model by means of the fractal theory. *Food Sci. Technol. Int.*, **9** (3), 207–213.
15. Villalobos, R., Chanona, J., Hernández, P., Gutiérrez, G., and Chiralt, A. (2005) Gloss and transparency of hydroxypropyl methylcellulose films containing surfactants as affected by their microstructure. *Food Hydrocoll.*, **19**, 53–61.
16. Pescador-Piedra, J.C., Garrido-Castro, A., Chanona-Pérez, J., Farrera-Rebollo, R., Gutiérrez-López, G., and Calderon-Dominguez, G. (2009) Effect of addition of mixtures of glucose oxidase, peroxidase and xylanase on rheological and breadmaking properties of wheat flour. *Int. J. Food Prop.*, **12**, 748–765.
17. Antonio, P.-N., Jorge, C.-P., Reynold, F.-R., Gustavo, G.-L., and Georgina, C.-D. (2010) Image analysis of structural changes in dough during baking. *LWT - Food Sci. Technol.*, **43**, 535–543.
18. Aguilera, J.M. (2000) Microstructure and food product engineering. *Food Technol.*, **54** (11), 56–65.
19. Aguilera, J.M. (2005) Why food microstructure? *J. Food Eng.*, **67**, 3–11.
20. Blonk, J.C.G. and van Aalst, H. (1993) Confocal scanning light microscopy in food research. *Food Res. Int.*, **26**, 297–311.
21. Autio, K. and Salmenkallio-Marttila, M. (2001) Light microscopic investigations of cereal grains dough's and breads. *Lebensm. Wiss. Technol.*, **34**, 18–22.
22. Aguilera, J.M. (2006) Structure–property relationships in low moisture products, in *Water Properties of Food, Pharmaceutical, and Biological Materials*, Food Preservation Technology Series (eds M.P. Buera, J. Welti-Chanes, P.J. Lilford, and H.R. Corti), Taylor and Francis, Boca Raton, FL, pp. 115–132.
23. Aguilera, J.M., Chiralt, A., and Fito, P. (2003) Food dehydration and product structure. *Trends Food Sci. Technol.*, **14**, 432–437.
24. Teunou, E., Fitzpatrick, J.J., and Synnott, E.C. (1999) Characterization of food powder flowability. *J. Food Eng.*, **39** (1), 31–37–15.

25. Schoonman, A., Mayor, G., Dillmann, M.L., Bisperink, C., and Ubbink, J. (2001) The microstructure of foamed maltodextrin/sodium caseinate powders: a comparative study by microscopy and physical techniques. *Food Res. Int.*, **34** (1), 913–929.
26. Alamilla, B.L., Chanona, P.J.J., Jiménez, A.A.R., and Gutiérrez, L.G.F. (2005) Description of morphological changes of particles along spray drying. *J. Food Eng.*, **67**, 179–184.
27. Prothon, F., Ahrné, L.M., Funebo, T., Kidman, S., Langton, M., and Sjóholm, I. (2001) Effects of combined osmotic and microwave dehydration of apple on texture, microstructure and rehydration characteristics. *Lebensm. Wiss. Technol.*, **34**, 95–101.
28. Hindmarsh, J.P., Russell, A.B., and Chen, X.D. (2007) Fundamentals of the spray freezing of foods-microstructure of frozen droplets. *J. Food Eng.*, **78**, 136–150.
29. James, B.J. and Smith, B.G. (2009) Surface structure and composition of fresh and bloomed chocolate analysed using X-ray photoelectron spectroscopy, cryo-scanning electron microscopy and environmental scanning electron microscopy. *LWT – Food Sci. Technol.*, **42**, 929–937.
30. Briones, V. and Aguilera, J.M. (2005) Image analysis of changes in surface color of chocolate. *Food Res. Int.*, **38**, 87–94.
31. Zaritzky, N.E. (2000) Physical and microstructural properties of frozen gelatinized starch suspensions, in *Food Preservation Technology Series: Trends in Food Engineering* (eds J.E. Lozano, C. Añón, E. Parada-Arias, and G.V. Barbosa-Cánovas), Technomic Publishing Company, Inc., Pennsylvania, pp. 15–12.
32. Knowles, A. and FitzGerald, S. (2005) Biopharma imaging and analysis: advancing towards a more detailed picture of chemistry. *Eur. BioPharm. Rev.*, Spring (5).
33. Wirth, R. (2009) Focused ion beam (FIB) combined with SEM and TEM: advanced analytical tools for studies of chemical composition, microstructure and crystal structure in geomaterials on a nanometre scale. *Chem. Geol.*, **261** (3–4), 217–229.
34. Aguilera, J.M. and Stanley, D.W. (1999) *Microstructural Principles of Food Processing and Engineering*, Chapter 1, 2nd edn, Aspen Publishers, New York, pp. 1–70.
35. Abramowitz, M. and Davidson, M.W. (2007) Introduction to Microscopy, Molecular Expressions. *http://micro.magnet.fsu.edu/primer/anatomy/introduction.html* (accessed June 2011).
36. Dürrenberger, M.B., Handschin, S., Conde-Petit, B., and Escher, F. (2001) Visualization of food structure by confocal laser scanning microscopy (CLSM). *Lebensm. Wiss. Technol.*, **34**, 11–17.
37. Peighambardoust, S.H., Dadpour, M.R., and Dokouhaki, M. (2010) Application of epifluorescence light microscopy (EFLM) to study the microstructure of wheat dough: a comparison with confocal scanning laser microscopy (CSLM) technique. *J. Cereal Sci.*, **51**, 21–27.
38. Ma, Q. and Rudolph, V. (2007) Dimensional change behavior of caribbean pine using an environmental scanning electron microscope. *Drying Technol.*, **24**, 1397–1403.
39. Van Driel, L.F., Knoops, K., Koster, A.J., and Valentijn, J.A. (2008) Fluorescent labeling of resin-embedded sections for correlative electron microscopy using tomography-based contrast enhancement. *J. Struct. Biol.*, **161**, 372–383.
40. Yang, H., Wang, Y., Lai, S., An, H., Li, Y., and Chen, F. (2007) Application of atomic force microscopya as a nanotechnology tool in food science. *J. Food Sci.*, **72** (4), R65–R75.
41. McCarthy, M.J. and McCarthy, K.L. (1996) Applications of magnetic resonance imaging to food research. *Magn. Reson. Imaging*, **14** (7–8), 799–802.
42. Frías, J.M., Foucat, L., Bimbenet, J.J., and Bonazzi, C. (2002) Modeling of moisture profiles in paddy rice during drying mapped with magnetic resonance imaging. *Chem. Eng. J.*, **86** (1–2), 173–178.
43. Lott, J.N.A., Greenwood, J.S., and Vollmer, C.M. (1982) Mineral reserves

in castor beans: the dry seed. *Plant Physiol.*, **69**, 829–833.

44. Habibi, Y., Heus, L., Mahrouz, M., and Vignon, M.R. (2008) Morphological and structural study of seed pericarp of Opuntia ficus-indica prickly pear fruits. *Carbohydr. Polym.*, **72**, 102–112.

45. Pedreschi, F., Mery, D., and Aguilera, J.M. (2004) Classification of potato chips using pattern recognition. *J. Food Sci.*, **69**, E264–E270.

46. Annan, N.T., Borza, A.D., and Truelstrup, H.L. (2008) Encapsulation in alginate-coated gelatin microspheres improves survival of the probiotic *Bifidobacterium adolescentis* 15703T during exposure to simulated gastro intestinal conditions. *Food Res. Int.*, **41**, 184–193.

47. Allan-Wojitas, P., Truelstrup, H.L., and Paulson, A.T. (2008) Microstructural studies of probiotic bacteria loaded alginate microcapsules using standard electron microscopy techniques and anhydrous fixation. *LWT – Food Sci. Technol.*, **41**, 101–108.

48. Mendoza, M.A.G. (2009) Determinación de la viabilidad de *Bifidobacterium bifidum* inmovilizado, tratado bajo condiciones gastrointestinales humanas simuladas *in vitro*. Tesis de Maestría en Ciencias en Bioprocesos, realizada en la Unidad Profesional Interdisciplinaria de Biotecnología del Instituto Politécnico Nacional. México, Distrito Federal.

49. Borém, F.M., Marques, E.R., and Alves, E. (2008) Ultrastructural analysis of drying damage in parchment *Arabica* coffee endosperm cells. *Biosyst. Eng.*, **99**, 62–66.

50. Sirisomboon, P. and Kitchaiya, P. (2009) Physical properties of Jatropha curcas L. kernel after heat treatments. *Biosyst. Eng.*, **102**, 244–250.

51. Varela, P., Aguilera, J.M., and Fiszman, S. (2008) Quantification of fracture properties and microstructural features of roasted *Marcona* almonds by image analysis. *Lebensm. Wiss. Technol.*, **41**, 10–17.

52. Wang, N. and Brennan, J.G. (1995) Changes in structure, density and porosity of potato during dehydration. *J. Food Eng.*, **24**, 61–76.

53. Aguilera, J.M. and Stanley, D.W. (1999c) *Microstructural Principles of Food Processing and Engineering*, Chapter 9, 2nd edn, Aspen Publishers, New York, pp. 373–397.

54. Prothon, F., Ahrné, L., and Sjóholm, I. (2003) Mechanisms and prevention of plant tissue collapse during dehydration: a critical review. *Crit. Rev. Food Sci. Nutr.*, **43** (4), 447–479.

55. Gekas, V. (1992) Transport phenomena in solid foods, in *Transport Phenomena of Foods and Biological Materials* (ed. V. Gekas), CRC Press, Boca Raton, FL, pp. 133–166.

56. Mayor, L. and Sereno, A.M. (2004) Modelling shrinkage during convective drying of food materials: a review. *J. Food Eng.*, **61**, 373–386.

57. Devaux, M.F., Taralova, I., Levy-Vehel, J., Bonnin, E., Thibaulth, J.F., and Guillon, F. (2006) Contribution of image analysis to the description of enzymatic degradation kinetics for particulate food material. *J. Food Eng.*, **77**, 1096–1107.

58. Perea-Flores, M.J., Chanona-Pérez, J.J., Terrés-Rojas, E., Calderón-Domínguez, G., Garibay-Febles, V., Alamilla-Beltrán, L., and Gutiérrez-López, G.F. (2010) Microstructure characterization of milk powders and their relationship with rehydration properties, in *Spray Drying Technology* (eds M.W. Woo, W.R. Wan Daud, and A.S. Mujumdar), Transport Processes Research (TPR) Group National University of Singapore, Singapore, pp. 197–218.

59. Haralick, R., Shanmugam, K., and Dinstein, I. (1973) Textural features for image classification. *IEEE Trans. Syst. Man Cybern.*, **3** (6), 610–621.

60. Rodrigues, S. and Fernandes, F.A.N. (2006) Image analysis of osmotically dehydrated fruits: melons dehydration in a ternary system. *Eur. Food Res. Technol.*, **225**, 685–691.

61. Castro, L. and Aguilera, J.M. (2007) Fracture properties and microstructure of low-moisture starch probes. *Drying Technol.*, **25**, 147–152.

23
Nanotechnology in the Food Sector

Christopher J. Kirby

23.1
Introduction

Nanotechnology is a broad and multidisciplinary area of science and technology that is seeing explosive growth in many different sectors. It is attracting considerable interest because of the many potential benefits that it offers, and it is predicted that it will be the most important driver for economic growth in the twenty-first century [1]. At the same time, concerns are increasingly being expressed about possible detrimental effects.

A very general definition for nanotechnology is that it is concerned with the identification, characterization, manipulation, and application of materials, structures, and phenomena that take place in the nanometer size range, where a nanometer is 1 billionth (10^{-9}) of a meter. Individual molecules themselves fall into this size range, and it is known that the properties of materials in a nanoparticulate form are generally very different from those exhibited by bulk quantities of the same substances. This can result in novel properties and functionalities which can open up entirely new possibilities for exploiting the use of such materials in many different fields. Thus, rather than being a specific technology in its own right, it is actually a diverse range of different technologies.

To give a feel for the dimensions involved at the nanoscale, we can start by considering the size range of particles suspended in a liquid. For particles of one thousandth of a meter in diameter, that is, at the millimeter size level, the individual particles are readily discernable, providing a coarse suspension in which, in the absence of other forces, the particles would rapidly sediment. A further, 1000-fold decrease in diameter would provide particles at the micron (micrometer) range, where individual particles can no longer be resolved without the aid of a microscope. For example in homogenized milk, the typical opaque, whiter than white appearance is largely due to fat globules with a median diameter of approximately 1 μm, which is large enough to scatter the entire visible spectrum of light. It is just about possible to resolve the individual particles using an optical microscope. A solid phase example in the micron range is a human hair, which has an average diameter in the region of 90 μm.

Food Processing Handbook, Second Edition. Edited by James G. Brennan and Alistair S. Grandison.

With a further 1000-fold reduction in size, we then arrive at the nanometer level. Since the wavelength spectrum of visible light is from 380 to 740 nm, particles below this size range have very limited ability to scatter light and will appear clear or translucent in suspension. In the milk example, besides the fat globules, another population of particles is present, the casein micelles, and with a median diameter of around 100 nm these are natural nanoparticles. Thus in skimmed milk where the fat droplets have largely been removed, though still white, the visual appearance is influenced by these micelles which by virtue of their smaller size, can only scatter the smaller wavelengths (blue region) of the visible spectrum, giving a slightly bluish tinge to the milk. The term "*nanoscale*" has often to date been used to describe particles of between 1 and 100 nm though it is now being suggested that the upper level of this size range should be increased.

23.2 The Driving Force for Nanotechnology Development

As it exists today, the nanotechnology concept evolved from separate developments in two completely unrelated fields: electronics and biomedicine (see the following sections). The remarkable advances in each of those areas were taken up and developed further by other sectors, leading to a diverse range of novel technologies. This in turn led to a cross-fertilization of ideas that are helping to unite the otherwise disparate disciplines of electronics and biology to create new hybrid technologies, such an nanobiosensors, "lab on a chip" devices, and biocomputers. In this way, nanotechnology promised to revolutionize virtually every aspect of human life.

23.2.1 Electronics Drivers

In the electronics field, the driving force has been the truly astonishing rate of progress in the semi-conductor industry. The ever-increasing rate of miniaturization of semi-conductor technology since the late 1950s has led to a doubling every 2 years in the number of transistors and other components that can be fabricated at reasonable cost onto a silicon chip. Thus for 60 years, the miniaturization rate has literally been increasing exponentially. This phenomenon was first described in 1965 by Gordon E. Moore, a cofounder of Intel, and has since become known as *Moore's law*. At that time it was based purely on past observations, but since then it has been adopted by industry as a developmental goal, and so has become a self-fulfilling prophesy. Because of factors such as continuing improvements in photolithographic techniques, subsequently to be overtaken by technologies with still greater resolving power such as extreme ultraviolet light lithography, it is widely accepted that Moore's law will hold true for some years to come. Continuing developments in miniaturization are being accompanied by parallel improvements in component density, processing speed, memory capacity, and cost per unit, all of which are encouraging further innovation and uptake by different disciplines.

23.2.2 Biomedical Drivers

In sharp contrast to the inert, silicon-based chemistries being employed by the electronics industry, the major applications for nanotechnology in the biomedical field have necessarily been based on organic, biodegradable constituents. At the beginning of the twentieth century, Paul Ehrlich conceived of the concept of "magic bullets," whereby toxic agents might be delivered selectively to harmful microorganisms by associating them with an agent having an affinity for that organism, thus enabling the microorganism to be destroyed without harming the patient. This eventually gave rise to the "drug delivery" concept which has since been employing micro- and nanostructures for that purpose over several decades. The most common delivery approach involves packaging the drug inside a microscopic carrier vehicle and then manipulating the properties of that carrier in order to optimize drug behavior. Using this extremely powerful approach, not only can the eventual target of a drug within the body be accurately specified, whether it be a tissue, an organ, or an individual cell type, but the timing, the duration and the extent of exposure of the target to the drug can also be controlled, thus maximizing the chances of a successful outcome.

For these reasons, a tremendous amount of time and energy has been spent on developing and characterizing a wide range of different carrier types suitable for different biomedical applications. As will be discussed later in this chapter, many of the principles of the drug delivery concept as well as the carriers that have been developed, can also be applied within complex food systems, for example, to improve food safety, quality, and nutritional content as well as to increase the effectiveness of different food processes. Obviously for food use, the carrier would also need to be edible.

23.3 Manufacture of Nanosystems: General Principles

A wide range of procedures have been developed for making nanoparticles and nanostructures, but these can generally be sub-divided into two categories which are often described as the "top-down" and "bottom-up" approaches. The "top-down," attrition-based procedures commence with macroscale material and then break this down to smaller units. This may involve processes with high input of mechanical energy such as high pressure homogenization or ball milling, or the use of ultrasonic radiation. Such methods tend to be high throughput but far less amenable to fine control. Alternatively highly sophisticated procedures such as photolithography and electron beam lithography may be employed, as described above, enabling very precise control of the finished product. As discussed, these latter technologies have seen an exponential growth in the rate of miniaturization and breadth of application over the past few decades.

In the "bottom-up" approach, atomic or molecular building blocks are used which are encouraged to self-assemble to form nanostructures. For example, individual atoms or molecules can be made to crystallize under controlled conditions which prevent them from forming larger crystals. Molecules with surface active properties can be allowed to self-assemble into a range of different polymorphic forms. Self-assembly with regard to lipidic components is discussed in greater detail in Section 23.5.3. A graphic example using DNA has been described recently [2]. Based on the molecular recognition properties of DNA and the fact that it has a predictable and programmable secondary structure, a section of DNA has been designed which was then able to self-assemble into a box of nanoscale dimensions ($42 \times 36 \times 36\ nm^3$) and which included an openable lid, which can be opened using a DNA "key." Suggested applications include controlled release of an encapsulated cargo. This DNA origami approach had previously been used to manufacture a wide range of two-dimensional structures but has only recently been successful in 3D. Similar approaches can be used to precisely guide other molecules and nanoparticulates into predetermined spatial arrangements in order to manufacture functional nanoscale devices.

23.4 Nanotechnology and Food

Modern food processing technology makes use of a diverse range of individual processes, for example, heat treatment, fermentation, filtration, curing, and drying. Moreover, a vast range of functional ingredients are used which fulfil a variety of different roles. These include antioxidants, preservatives, enzymes, stabilizers, flavors, micronutrients, and many others. The great majority are harmless but there is still a market-led trend to reduce the levels of some of these and, where possible, to replace chemically derived ingredients with ones that are perceived to be of natural origin. Other drivers include the need to improve food safety, quality, and traceability, to increase nutritional value and of course to reduce cost and increase sustainability of food production. Many of these needs can be met by developing new technologies or embracing those that have been developed in other, non-food sectors.

Food nanotechnology is a rapidly developing field which promises to revolutionize a great many aspects of food science, ranging from improvements in safety and quality of foodstuffs, to enhanced rheological and organoleptic properties and increased bioavailability of essential nutrients. It has been proposed that by 2010, the overall market for food-related nanotechnologies will be worth US$20.4 billion [3]. Global food corporations such as Kraft, Nestle, Unilever, Heinz, and Hershey Foods are investing heavily in the area [4] and a great many other companies are involved in nanotechnology research, with the market leader being the United States, followed by Japan and China. Factors such as climate change, population growth, and the need for cost efficiencies are expected to make significant contributions to the rate of development of these technologies [3].

Nanotechnology in its various manifestations is being applied in a multitude of different ways within the food sector. The different areas of application can be divided into three main categories:

- Incorporation of nanostructures comprising food compatible materials, or of food components themselves, where the intention is that those nanostructures will be consumed as a part of the food product. These "*in vivo*" approaches generally make use of "delivery systems." This category can also include formulations for administration of micronutrients and nutraceuticals.
- Use of nanoparticulates in areas of food packaging, or for incorporation into other food contact surfaces.
- Development of nanoscale devices for use as biosensors, processing tools, nanotracking devices, nanocatalysts, and so on. In this respect, these applications can be described as "*in vitro*" since many will not be part of the marketed food product.

23.5 Delivery Systems for Functional Food Ingredients

Among the first deliberate uses for nanotechnology in food processing was the development of delivery systems for functional food ingredients. More than 20 years ago we introduced the idea that the principles underlying the drug delivery concept, for example, stabilization, targeting, and controlled release, might also be used to enhance the performance of functional ingredients in the food sector [5–7]. At that time, the word nanotechnology was not widely used in this context, and "microencapsulation" was generally accepted as an overriding term to describe incorporation into microscopic carriers. In fact, the carriers already being used at that time tended to be relatively large. A typical application was for encapsulation of flavor oils within particles of a food grade hydrocolloid such as a polysaccharide or gum, using a process such as spray drying. This could improve handling and flow properties of the oil by converting it from a liquid to a solid form [8–10].

Other potential benefits included protection of vulnerable materials from heat, light or oxidation, and masking of unpleasant tastes and odors, for example, in the case of fish oils. In most cases the capsule was water soluble and its role ended when the ingredient was added to the food or when water was added to a dry food mix containing the microencapsulated ingredients. Although first coined in 1974 by Norio Taniguchi, a Japanese university professor [11], the word nanotechnology did not become widely used until the mid-1990s.

The basis of our putative delivery approach was that active ingredients would be encapsulated in a stable form inside microscopic carrier systems. However the delivery systems that we envisaged were of more defined structure than had previously been used by the food sector. Obviously they would need to be able to accommodate and retain the substance of interest, would preferably be of micro- or nanoscale dimensions, and with properties such as permeability, stability, and release characteristics able to be carefully controlled. The structures would need to be stable, food compatible and to be able to survive in an aqueous environment without loss of integrity. The general features of an ideal delivery system are described below. By these means it can be possible to control when, where and how a particular ingredient is released and able to exert its intended action [6, 7].

23.5.1 General Features of an Ideal Delivery System

23.5.1.1 Stabilization

- Segregation of ingredient from potentially harmful conditions such as the presence of inhibitors, metal ions, free radicals, pH extremes, enzymes, moisture, and so on.
- Co-encapsulation with stabilizing agents, for example, pH buffers, antioxidants, chelators, thermostabilizers, cryoprotectants, cofactors, and so on.
- Prevention of dilution (various ingredients are more stable in a concentrated form).

23.5.1.2 Controlled release

- **Time/stage-specific release::** Release controlled by general stability properties of the carrier within a particular food environment or by intrinsic factors within food, for example, enzymes, moisture content, pH, osmotic pressure.
- **Signaled release::** Selection of a carrier that will release the contents in response to a signal or processing variable, for example, microwave irradiation or temperature change.

23.5.1.3 Targeting (site-specific release)

- Selection of nanocarriers with appropriate surface properties can encourage accumulation at a predetermined microscopic location within a complex, multiphase food. When the encapsulated functional ingredient is eventually released, it will then be able to exert its intended action more selectively and effectively rather than being dispersed nonspecifically throughout the food matrix.

A number of review articles relating to nanoparticulate delivery of functional food ingredients have appeared in the literature [7, 12, 13].

Examples of applications based on the principles described above are discussed in Section 23.5.5.

23.5.2 Classification of Delivery Vehicles

Many different types of nano- and microscale structures have been harnessed to act as delivery vehicles for active substances. As discussed previously, the majority of these had originally been investigated for delivery of drugs and other biomedically relevant agents such as enzymes, hormones and genetic materials. Their acceptability in those situations suggested that they might also be usable for delivery of functional ingredients in food applications. A large proportion of the delivery systems are lipid-based, though many non-lipid vehicles have also been used. The actual choice of system will be dependent on a number of factors, including the physicochemical properties of the ingredient, the nature of the food product and the pattern of delivery required.

23.5.3 Lipid-Based Systems

The properties of lipid molecules enable them to organize themselves into a diverse range of microstructures. This is made possible because many are amphiphilic, which means that one part of the molecule is polar and hydrophilic (water-loving), whereas another part is hydrophobic, and will try to avoid contact with water. The hydrophobic region typically consists of one or more extended hydrocarbon chains, resulting in an elongated molecule. In the presence of water, amphiphiles will assemble into distinct structures in order to prevent energetically unfavorable interactions from occurring. In general, each molecule will seek to align itself relative to the adjacent ones such that the polar head groups are in proximity to each other and/or the hydrocarbon chains are also in close proximity to themselves. Depending on the effective geometry of the molecule, a large number of structures can form, some of which are shown in Figure 23.1, all having very different properties, and most of which are found naturally within the body. This capability for self-assembly is one of the most important phenomena to be found in living systems. Specific examples of self-assembled microstructures that are being explored or are actually in use, are discussed below.

23.5.3.1 Micelles and Reverse Micelles

In micellar structures, the amphiphiles behave as though they are cone-shaped because one end of the molecule has a larger cross-sectional area than the other, either for steric reasons or because of electrostatic attractive or repulsive forces between adjacent polar head groups. Micelle-forming surfactants tend to have a single hydrocarbon chain which has a narrower cross-section than the polar head group, and when the molecules try to align in water, the conical shape imposes curvature, resulting in formation of small spherical micelles as shown in Figure 23.1. Lipid-soluble substances can be dissolved in the hydrocarbon interior and in this way; micelles can be used as delivery systems for such substances.

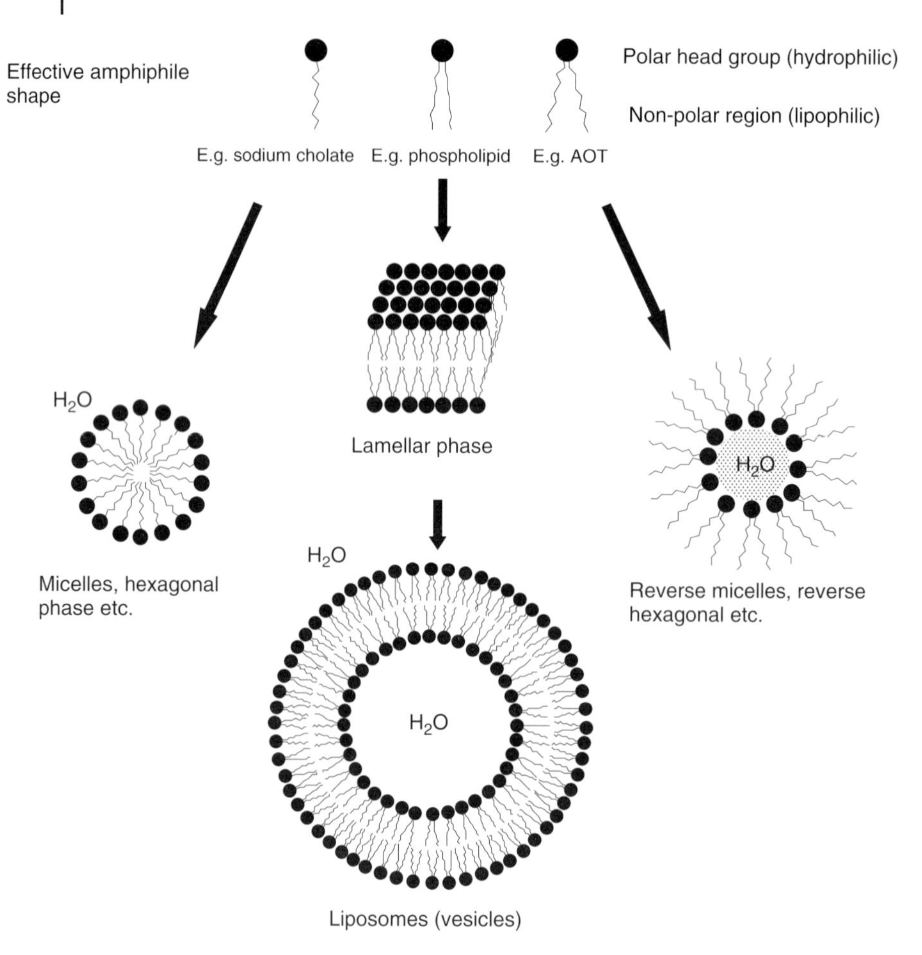

Figure 23.1 Dependence of phase behavior on effective molecular geometry of individual amphiphile molecules.

The converse situation is where the hydrocarbon region has the larger cross-sectional area and these molecules would be unable to associate in water but would often be able to form reverse micelles in oils and solvents as shown in Figure 23.1. In this case, a small amount of water containing dissolved solutes can be accommodated in the center of the structure, and so reverse micelles can be used as delivery systems for water-soluble substances in food oils.

23.5.3.2 **Liposomes**

Liposomes are artificially made, microscopic membrane vesicles comprising one or more concentric bilayers of phospholipid molecules (lamellae), each bilayer resembling the plasma membrane that surrounds plant and animal cells. This arrangement arises because phospholipids have two hydrocarbon chains attached to the polar group and the cross-sectional area is relatively uniform along the length of the molecule, making them effectively cylindrical. When they align in water, the short range order is therefore planar rather than curved. In order to achieve stability in water, two of these planar regions are aligned back to back to form a phospholipid bilayer, and sheets of bilayer then round off to form the spherical liposomes (Figures 23.1 and 23.4).

Liposomes make highly versatile delivery systems and can be designed for a wide variety of different applications. By varying factors such as size and lipid composition, parameters such as permeability, stability, affinity, and surface activity can be controlled, enabling liposomes to be made with widely differing physical properties. For this reason, they have been studied in greater depth than any other carrier and have been used extensively as drug delivery systems to contain and carry a great many types of drugs and related materials (including vaccines, hormones, enzymes, vitamins, and genetic material) around the body [14]. In this respect they have been administered orally, parenterally, topically, and via the respiratory tract. They are ideal as delivery systems for food ingredients for a number of reasons [7, 15]. The phospholipids from which they are normally composed are found in large amounts in many types of food, so they are completely food acceptable with no toxicity issues. Many types of active ingredient can be encapsulated with high efficiency and they are particularly well suited to entrapping and retaining hydrophilic molecules in solution. Few other encapsulation systems can maintain good barrier properties in this situation. At the same time, many types of nonhydrophilic (i.e., lipophilic) ingredients can be included, by incorporating these into the lipid bilayer region of the liposome. And as mentioned above, there are many ways of altering their properties to accommodate the needs of different food processes.

Depending on the methods used to prepare them, liposomes range in size from around 25 nm for the smallest single-layered vesicle, to a normal maximum of several micrometers for the multilamellar type. Despite their small size, they are much less fragile than many of the alternative types of vehicle because of their membranous structure. Liposomes are also easy to make and can be stored in a stable freeze-dried form. A preparation method was developed by the author in order to enable very high encapsulation efficiencies using simple conditions suitable for scale-up [16]. It also employs mild conditions which make it especially appropriate for encapsulating sensitive molecules which might otherwise be denatured, such as enzymes and genetic material. Some of the potential applications for liposomes in the food and agricultural sectors are described in a recent review [17].

23.5.3.3 Emulsions, Nanoemulsions, and Derived Systems

Emulsions are metastable dispersions of two immiscible liquids in which one of the liquids, known as the *disperse phase*, is dispersed as droplets within the second liquid which is known as the *continuous phase*. A surfactant or emulsifier is generally employed to prevent phase separation of the two liquids. Emulsions have a long history as food systems in their own right, common examples being milk and mayonnaise in which the oil phase is dispersed within the aqueous phase to form an oil-in-water emulsion, while margarine and butter are examples of water-in-oil

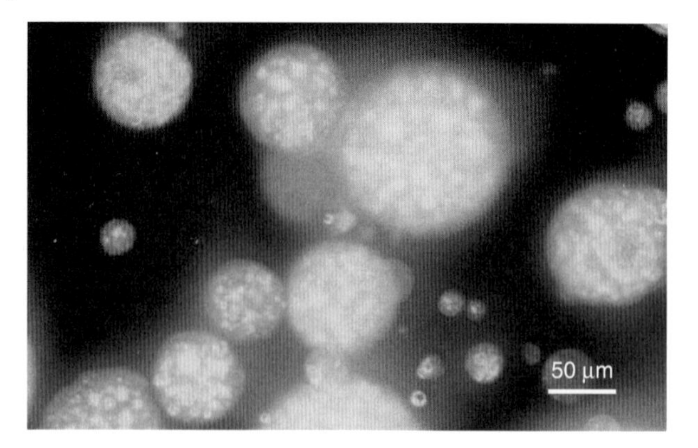

Figure 23.2 Water-in-oil-in-water (w/o/w) emulsions of glycerol in cocoa butter by fluorescence light microscopy. A fluorescent dye was included in the glycerol phase to aid visualization. The large white spheres are cocoa butter microspheres.

emulsions. They have also been used in medicine for many years for intravenous nutrition in order to provide calories where oral feeding is problematic. In this case the droplet size has to be reduced to the nanoscale by high pressure homogenization procedures so that larger droplets do not block the microcapillaries of the vascular system. More recently nanoemulsions are being used to enable solubilization and delivery of poorly soluble drugs into the body. In the same way they can be used in food situations for delivery of lipophilic ingredients such as flavor components. It has been reported that Unilever are developing a low-fat ice cream where the large increase in surface area caused by nano-sizing the fat droplets enables the fat content to be reduced from 17 to 1% [18].

Another way of utilizing emulsions and nanoemulsions is to prepare these using high melting point fats or waxes in the molten state rather than using liquid oils, and then allow the resulting dispersions to solidify by reducing the temperature. In this way, the emulsion droplets solidify to form fat microspheres [19] while the nanoemulsions form solid lipid nanoparticles (SLN) [20]. The latter are also used in drug delivery but both approaches can be used as delivery systems for food applications.

A further refinement of the emulsion approach is to prepare multiple emulsions [21], which can be either water-in-oil-in-water (w/o/w) or oil-in-water-in-oil (o/w/o) systems. We used a combination of a w/o/w multiple emulsion and a high melting fat in order to develop microcapsules of cocoa butter containing dispersed micro-droplets of glycerol as shown in Figure 23.2 [19]. In this case, the objective was to incorporate these into chocolate so that when this came into contact with the warm surface of the fingers, some of the encapsulated glycerol would be released to form a thin film between the chocolate and the skin, thus preventing stickiness. The multiple emulsion approach is also being used commercially to develop low-fat products by replacing some of the fat in the interior of the droplet with water [22].

23.5.3.4 Microemulsions

Microemulsions are very different systems to emulsions and nanoemulsions, and can be defined as thermodynamically stable, isotropic, optically clear dispersions of oil and water, stabilized by an interfacial layer of surfactant, and sometimes a co-surfactant. The actual microstructure is largely determined by the relative proportions of the different components, with an oil-in-water configuration being formed when water is in excess, a water-in-oil configuration when the oil is in excess, and a bi-continuous configuration when the quantities are relatively similar. In contrast to emulsions and nanoemulsions where input of mechanical energy is required, microemulsions form spontaneously simply by mixing the different components in the appropriate ratio. A minimal amount of energy input in the form of simple shaking is generally used simply to speed up the process rather than relying on slow mixing by diffusion. Microemulsions are true nanostructures in that the nanodroplets of disperse phase range from 5 up to a maximum of 200 nm in diameter, substantially smaller than the wavelength of visible light and thus transparent or translucent. Because of their surface properties they have proved highly effective at solubilizing and increasing the bioavailability of poorly soluble drugs. The well-known Neoral formulation of cyclosporin, for control of transplant rejection, is microemulsion-based.

23.5.3.5 Novel Oil-Phase Nanoparticles

One drawback of microemulsions and reverse micelles as a means of incorporating functional ingredients into oils is the fact that the ingredient must be dissolved in the aqueous phase, which significantly limits the amount that can be incorporated, and hence its functional efficacy. The presence of water can also increase instability of some labile substances. We had previously developed a novel technology with which water-soluble substances could be solubilized in a dry form into oils and many organic solvents [23, 24]. In this case the amount of the substance that can be solubilized is not limited by its solubility in water and so much higher concentrations can be achieved. The technology, known also as *Macrosol*™, was originally developed for drug delivery of therapeutic peptides such as insulin and calcitonin, and produced dramatic results in preclinical trials [25]. It was subsequently shown to be applicable to virtually any type of water-soluble substance, and many insoluble ones, with these substances being self-assembled to form stabilized, dehydrated nanoparticles. Table 23.1 lists some of the classes of substance that have been formulated using this approach, and demonstrates the tremendous versatility of the technology.

The procedure for preparing the nanoparticles has been described elsewhere and is simple enough to be readily scaled up, uses mild conditions, and the materials involved are normally food compatible and generally regarded as safe (GRAS)-approved. Figure 23.3 shows an electron micrograph of a typical example comprising a nanoparticulate peptide, with a diagrammatic representation of a single particle shown as inset.

Table 23.1 Examples of classes of hydrophilic materials which have been self-assembled into a nanoparticulate form within oil phases.

Class	Examples
Simple salts	Sodium chloride, calcium chloride
Small organic molecules	Glucose, aqueous dyes, vitamins, micronutrients
Drugs	Antibiotics, antiglaucoma drugs, anti-asthmatics
Complex proteins	Albumin, various enzymes, hormones
Polynucleotides	RNA, genomic DNA, plasmid DNA
Polysaccharides	Dextrans, hydroxypropyl cellulose, hyaluronic acid
Others	Colloidal gold, viruses, vaccines

Figure 23.3 Freeze-fracture electron micrograph to show nanoparticles of a peptide in an oil phase. Inset shows a diagrammatic representation of a single nanoparticle comprising a central core surrounded by molecules of phospholipid. In this case the core is composed of small peptide molecules, but virtually any type of hydrophilic substance can be used. Nanoparticles smaller than 100 nm are difficult to resolve.

23.5.3.6 **Other Lipid-Based Delivery Systems**

Because of the multiple configurations that amphiphilic molecules can adopt, a rich variety of other lipid-based systems have been developed with drug delivery in mind, but which could also be amenable to use in food applications. Nanocochleates [26], like liposomes, are based on phospholipid bilayers, but contain a relatively large proportion of negatively charged phospholipids. In the presence of multivalent cations (positively charged ions) such as calcium ions, sheets of bilayer roll up tightly, with the calcium ions neutralizing the negative charges on each opposing layer and allowing these to lie alongside each other to form cylindrical nanostructures. In one adaption, the BioGeode (Biodelivery Sciences Intl. Inc.), a nano-droplet of oil containing dissolved micronutrients is encapsulated in the center of a crystalline nanocochleate which protects the contents from damaging conditions during food manufacture and storage without affecting the taste or odor of the food product.

Cubic phase lipid is a three-dimensional array of hydrated lipid, for example a monoglyceride, which is biocontinuous with respect to both the water and the lipid phases. Hexagonal phase lipid comprises a two-dimensional array of long rod-shaped or cylindrical micelles. Using appropriate procedures, both of these phases can made into nanosized dispersed forms in excess water to form cubosomes and hexosomes, respectively [27]. As with liposomes, both lipid-soluble and water-soluble substances can be incorporated into the respective phases and used as delivery systems for both medical and food applications. In reality, hydrophilic ingredients are probably best incorporated into liposomes whereas the particularly high surface area of lipid in cubosomes and hexasomes may be more appropriate for carrying small lipophilic and amphiphilic substances.

23.5.4 Non-Lipid-Based Delivery Systems

Of the non-lipid carriers that have been explored for delivery of drugs, a large proportion are based on polymeric systems, using both naturally derived and synthetic polymers. Natural biopolymers include protein and polysaccharide systems. Of the proteins, gelatine microcapsules have a long history of use by the food sector, for encapsulating oils for nutritional (e.g., omega 3 oils) and flavor purposes. These have generally been multi-micron in size but more recently gelatine nanoparticles have been developed with a diameter of 45 ± 5 nm which could in principle be used for food applications [28]. Casein nanoparticles containing functional food ingredients have recently been described [29]. Examples of polysaccharide-based nanoparticles include those based on cross-linked starch [30] and on cellulose [31], including synthetic derivatives such as ethyl cellulose. Another example that is finding increasing use in areas such as nanoparticle manufacture is *chitosan*. This is a deacetylated form of chitin, a linear polysaccharide found widely in the shells of crustaceans as well as insects and fungi, and thus extremely abundant in nature.

In addition to the naturally occurring types, a wide range of synthetic polymers have been used for making nanoparticulate drug delivery systems, and particular attention has been focused on those prepared from biodegradable polylactide and polylactide-co-glycolide [32]. These are especially appropriate for food use since their breakdown products – lactic and glycolic acids – are normal components of food. Depending on the preparation methods used, both water-soluble and oil-soluble substances can be entrapped.

23.5.5 Examples of Potential and Realized Applications for Delivery of Functional Ingredients in the Food Sector

The multiplicity of the types of colloidal carrier systems that are available, and of their properties, has led to their being investigated for a diverse range of food applications. The examples described below have been selected in order to illustrate the variety of the different ways in which the delivery principles might be applied.

23.5.5.1 Solubilization and Protection of Nutritional Supplements and Nutraceuticals

Nano- and microencapsulation of potentially labile materials, including many vitamins, can protect these against a wide range of conditions and substances which might otherwise lead to their destruction during food processing. For example, ascorbic acid (vitamin C) is added extensively to many types of foods for two quite different purposes, as a supplement to reinforce dietary uptake of vitamin C and as an antioxidant to protect the sensory and nutritive quality of the food itself. In the latter respect it is used extensively for stabilization of color and flavor in a wide range of products, including fruit and vegetable juices, canned fruit, beer, wine, and nonalcoholic beverages [33]. Because of its high reactivity, it can degrade rapidly during food processing by a variety of different mechanisms including metal ion-catalyzed oxidative processes, enzymic destruction, adverse pH, and interaction with other food components. As such, it provides an ideal model of a labile food additive with which to test the protective capabilities for nano/microencapsulation.

In order to investigate the potential benefits of this approach, ascorbic acid was encapsulated into lecithin liposomes and its survival measured by enzymic assay under a range of different storage conditions [34]. It was found that the vitamin could be encapsulated with high efficiency and at high concentration which itself can bestow improved stability, since degradation is known to take place more rapidly in a dilute form. Figure 23.4 shows an example of a multilamellar liposome containing concentrated ascorbic acid compared with an empty one. Even in the absence of added antagonists, survival of ascorbic acid in solution inside liposomes was substantially better than in free solution. However, when challenged with copper ions, lysine or the enzyme ascorbic acid oxidase, protection of the encapsulated vitamin was much more dramatic compared with the free ascorbic acid solution. For example, in the presence of ascorbic acid oxidase or lysine, more than 60% of the encapsulated vitamin was still present after 50 days while in free solution, the vitamin was completely destroyed within a day of exposure to the enzyme and only 14% remained after 8 days exposure to lysine. In the presence of copper ions, 40% of the encapsulated vitamin was recovered after 50 days while the free material was destroyed within 1 day.

These results demonstrate substantial improvements in the shelf life of an encapsulated vitamin, both in simple aqueous solution and particularly in the presence of common food components which would otherwise lead to its rapid

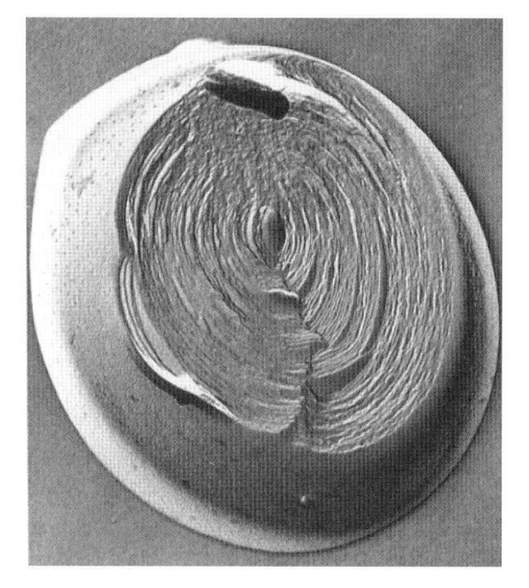

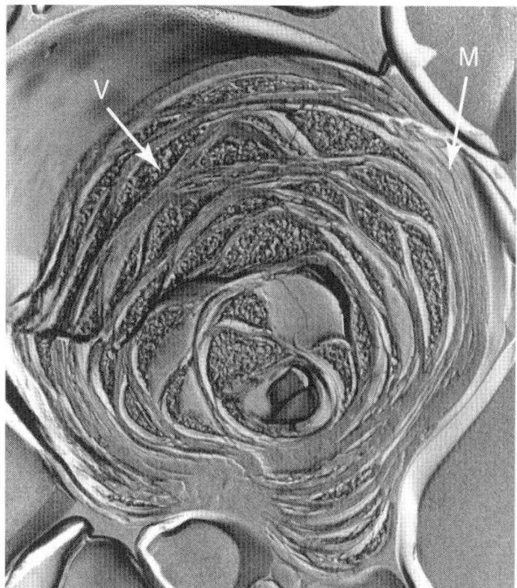

Figure 23.4 Comparison of empty and vitamin C-loaded multilamellar liposomes. The vitamin C-loaded liposome is shown on the right. "M" denotes regions of phospholipid multilayer and "V" denotes vitamin C.

destruction. Vitamin C is also known to have a protective effect on some other food components and co-encapsulation of these with the vitamin could provide further protection. Other workers have demonstrated protection of vitamin A against temperature, pH, and light-induced degradation by encapsulation in liposomes [35]. Alpha amylase was similarly protected against the damaging effects of pepsin, cold temperature storage, and extreme pH conditions [36]. Glucose oxidase was protected against proteolytic and thermal degradation [37]. Also liposome encapsulation provided maximum shielding of immunoglobulin Y from acid and pepsin attack, leading to the conclusion that this approach represented a powerful potential method for oral delivery of immunoglobulins [38]. Thus it seems likely that the same approach could stabilize many other types of vulnerable food additive against adverse effects arising during food processing.

A number of companies are employing nanotechnology commercially to solubilize and improve the bioavailability of nutritional supplements and other ingredients. Several are using micellar approaches to solubilize water-insoluble vitamins and other nutrients, such as flavonoids, carotenoids, herbal extracts, and essential oils, and thus aid their absorption, both for dietary purposes as well as for cosmetic and pharmaceutical applications. Aquanova's Nutri-nano product contains alpha lipoic acid and coenzyme Q10 aimed at weight reduction and satiety. The micelles also improved the stability of these substances. Amongst their range, Solgar also markets lipoic acid and coenzyme Q10 in micellar form. RBC Life Sciences manufacture a range of nanoceutical nutritional and skincare supplements, as well as NanoClusters™, which claims to combine with nutritional supplements to reduce their surface tension and promote absorption. Nutralease have developed a product known as nano-sized self-assembled structured liquids (NSSLs) technology which incorporates ingredients such as lycopene, lutein, coenzyme Q10, and phytosterols. The product is more akin to a microemulsion rather than a micellar phase, but unlike the former it can be diluted to any extent in an appropriate continuous phase without losing its structure. Freedom Plus Corporation market nano calcium/magnesium as a dietary supplement to increase bioavailability of these elements. BASF manufacture nanoparticulate lycopene and other health ingredients embedded in a solid matrix and sold in a powder form.

23.5.5.2 **Enzyme Delivery for Flavor and Texture Development**

Our first investigations into the use of delivery systems in foods entailed using enzymes encapsulated within liposomes to speed up the rate of ripening of Cheddar cheese. This brought together the delivery principles of protection, targeting, and controlled release of a functional ingredient within a complex, multiphase food. Enzymes were already being used to try to reduce the one year-plus ripening time of Cheddar [39], but were proving to be unviable because of premature enzyme action, poor distribution within the curd structure and heavy losses of both the enzyme and the product of its action. A preliminary attempt to reduce losses by adding enzymes which had been microencapsulated inside multilamellar liposomes could only achieve limited results because only very low levels of encapsulation were possible [40]. However, a newly developed liposome manufacturing procedure enabled much higher encapsulation efficiencies to be achieved [16], which reduced the ripening time of Cheddar cheese by half [41]. Figure 23.5 is a micrograph of a section of Cheddar cheese in which liposomes added to the milk at the start of

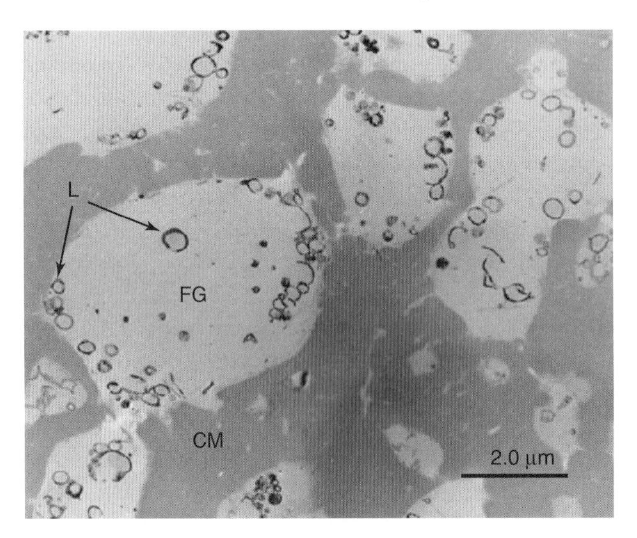

Figure 23.5 Electron micrograph showing distribution of liposomes in Cheddar cheese. "L" denotes liposomes, "FG" is fat globules and "CM" is the casein matrix of the cheese. Liposomes contain horseradish peroxidase to aid visualization. The dark-staining liposomes are trapped between the casein matrix (protein) and the fat globules.

the process become enmeshed in the curd after rennet addition, and eventually break open as a result of lipase action to release the ripening enzymes dispersed uniformly throughout the forming cheese.

The success of this work prompted the realization that similar approaches might provide improved functionalities for many other types of active ingredient, including other enzymic systems, antimicrobial agents, vitamins, and antioxidants [7, 42]. In this way, strategies might be developed by which synthetic food additives might potentially be replaced by naturally derived alternatives. Although liposomes were used initially, this was a stimulus for us to go on to explore the possibilities for using different types of nanostructured vehicles including microemulsions, polymeric nanospheres, and other novel nanoparticle technologies.

23.5.5.3 Delivery of Natural Antimicrobial Agents

The most common route to loss of food quality is microbial contamination. The most serious consequence is transmission of pathogens to the consumer, for example, *Salmonella*, *Campylobacter*, and *Listeria monocytogenes*, but from a commercial viewpoint, deterioration of the quality of the food itself is obviously a major concern. Heat or irradiation can provide adequate protection to some classes of foodstuff, but for others it is impractical or ineffective. For example, complex multiphase foods such as cheeses are not amenable to such measures. However, the number of food-acceptable antimicrobial agents is very limited. There are a number of natural antimicrobial agents which can provide very valuable protection, such as the polypeptide nisin, which is derived from strains of *Lactococcus lactis*, and the enzyme lysozyme, which is isolated from egg white and other sources. On the other hand, the antimicrobial efficacy of both of these can be substantially reduced as a result of adverse interactions with other food components, or due to food processing effects. For example, the effectiveness of lysozyme can be reduced 1000-fold due to unwanted interactions with milk components [43], while nisin can destroy other essential food microorganisms.

As a novel way of improving the effectiveness of such natural ingredients, we conceived the idea of encapsulating them in delivery systems such as liposomes [7, 42]. By this means, the stability and efficacy of the antimicrobials might be greatly increased and unwanted interactions with other food components and organisms could be prevented. We had also demonstrated a targeting effect in that, during our earlier studies on enzymic cheese ripening, we had noticed that liposomes and bacteria become localized in a common microcompartment within the ripening curd, viz. the microscopic spaces between the casein matrix and fat globules. This gave rise to the idea that we might be able to deliver agents such as lysozyme and nisin to the direct vicinity of food spoilage organisms, and indeed we demonstrated this targeting effect using liposomes containing horseradish peroxidase, by which means it was possible to visualize the liposomes microscopically. Figure 23.6 shows a liposome in the direct vicinity of a microbial spore. The success of this approach has since been demonstrated by other groups who have used the same antimicrobial agents encapsulated within liposomes to prevent the growth of food

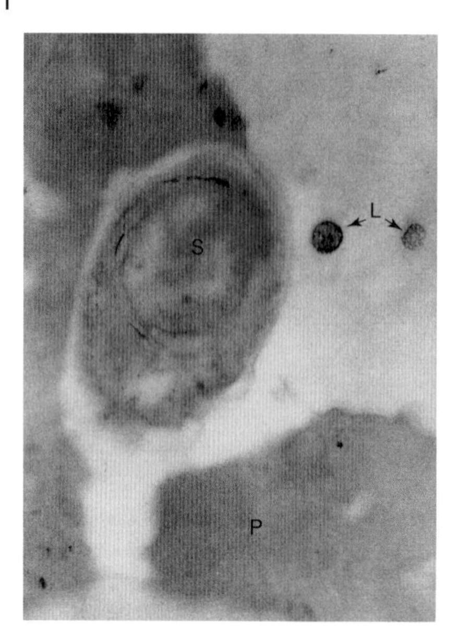

Figure 23.6 Electron micrograph to show liposomes in the vicinity of a contaminating bacterial spore in Gouda cheese. Liposomes, denoted "L" contain horseradish peroxidase to aid visualization. Dark staining areas, "P" are casein matrix. "S" is a contaminating bacterial spore.

spoilage microorganisms, concluding that it provides a powerful tool for this purpose [44–46].

23.5.5.4 **Antioxidant Protection of Polyunsaturated Oils**

Oxidative degradation has already been mentioned with regard to the protective effect on ascorbic acid in aqueous conditions [34]. However, oxidation is more of a problem in lipid-based foods, especially when these contain polyunsaturated fats such as fish oils. These are in demand by virtue of their high content of ω-3 fatty acids, which are known to protect against cardiovascular diseases, inflammatory conditions, and so on. The presence of antioxidants is essential and though natural oils do generally contain endogenous antioxidants, these tend to be used up during storage and processing, so that it is often necessary to add back exogenous ones. The downside is that the most potent antioxidants are chemical-derived and for many years there has been a consumer-led trend to replace these with more natural ones. Natural lipid-soluble antioxidants such as α-tocopherol (vitamin E) are less effective but their effectiveness can be enhanced.

An important property of certain antioxidants is their ability to interact synergistically with each other and interactions such as these are vital for protecting living systems against free radical-induced damage. A well-known example of such synergy is that which exists between vitamins C and E, where the vitamin C behaves as the primary antioxidant to repair organic free radicals, and is then itself repaired by vitamin E [47]. In living systems, vitamin E is present in the intracellular membranes and so is able to interact with synergists in the cell cytosol. In food systems, however, it is dissolved in the bulk oil and so would be inaccessible to vitamin C dissolved in an aqueous phase. The food industry currently approaches this by using lipid-soluble derivatives of vitamin C such as ascorbyl palmitate, but this itself has a poor dissolution rate, requiring heat to dissolve it which may increase the vulnerability of unsaturated fats to oxidation.

We have explored a number of different approaches to solve this problem. Since we had already shown that ascorbic acid can be entrapped in high concentration inside liposomes [34] and it is a simple matter to incorporate α-tocopherol into the membrane bilayer, we anticipated that this might provide an effective, synergistic antioxidant system. This was tested using larger liposomes together with free radical-inducing oxidative chemicals, but we were unable to demonstrate antioxidant activity. That work was discontinued though in retrospect the lack of effect may have been due to interaction between the ascorbate and the free radical initiators. Instead, other approaches were explored as discussed below.

Microemulsion-based antioxidant activity Microemulsions can allow solubilization of both water-soluble and lipid-soluble substances and thus enable the intimate contact between the two which might form the basis for an effective antioxidant system using ascorbic acid and α-tocopherol. This approach had already been demonstrated using microemulsions stabilized by monoglycerides [48] and phospholipids [49]. In the latter case, the induction time for oxidation of fish oil was increased to 40 days under oxidizing conditions but in both cases, relatively large amounts of emulsifier were required to achieve a satisfactory level of aqueous phase solubilization. With this as background, we set out to try to develop an emulsifier which would enable higher levels of solubilization.

Previously a mixture of polyglycerol esters of unsaturated fatty acids had been shown to allow relatively high levels of water incorporation into triglyceride oils in the form of microemulsions [50] and we chose to try to refine this approach. An enzymic method was developed using immobilized lipase and used to synthesize linoleic acid esters of specific types of polyglycerol. One particular fraction was isolated which, when screened against a range of other food acceptable emulsifiers, produced superior levels of water solubilization [51]. This was subsequently used to incorporate ascorbic acid into unsaturated oils at higher levels than had previously been achieved, and was shown to extend their oxidative stability. Using this system, the induction period of oxidation was increased from 2 to 70 days using trilinolein as the unsaturated oil phase, and from 20 to approximately 100 days using sunflower oil.

Vitamin C nanoparticles As discussed in Section 23.5.2.6, one of the drawbacks of microemulsions and reverse micelles for incorporating functional ingredients into oils is the fact that the ingredient must first be dissolved in water, which limits the amount that can be incorporated and can also increase instability of some labile substances. This was overcome by the development of a novel procedure for self-assembling such substances into dehydrated nanoparticles, an example of which is shown in Figure 23.3. Using this approach, ascorbic acid could be solubilized in polyunsaturated oils such as fish oil, in a dry form, and at much higher levels than are possible using microemulsions or reverse micelles, resulting in a crystal clear product. Moreover, the ascorbic acid was able to interact

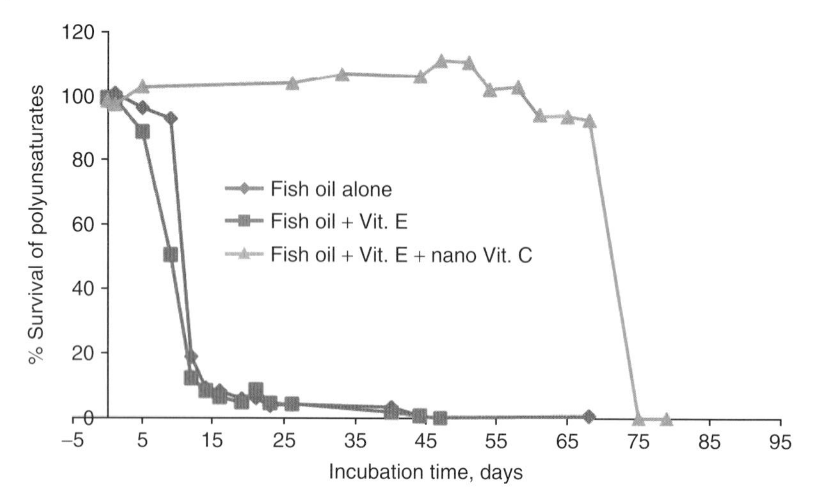

Figure 23.7 Protection of fish oil from oxidative degradation using nanoparticles of vitamin C.

synergistically with vitamin E dissolved in the oil to produce a powerful antioxidant action [52, 53], an effect which cannot occur when using dry ascorbic acid in its normal bulk form. As discussed previously, this increased reactivity is in keeping with the known properties of nanoparticulates.

Antioxidant efficacy was initially demonstrated by monitoring the rate of disappearance of unsaturated fatty acid components of the fish oil triglycerides by gas chromatography, indicating an increase in the induction period of oxidation from 5 days to 68 days under artificially accelerated oxidative conditions (Figure 23.7). The protective effect was subsequently confirmed on a commercial scale using additional procedures for measuring oxidation. The components of the system are all natural ingredients that are abundantly present in a great many food types, viz. ascorbic acid, phospholipids and fish oil, with or without added α-tocopherol.

Oxidative protection using nano-encapsulated ascorbic acid has also been employed more recently by Aquanova AG to form commercial products. In that case the vitamin is incorporated into surfactant micelles based on polysorbates (Tweens). The same approach has been used to solubilize a wide range of other active substances, both in water and fat phases, and has been especially used for increasing the oral bioavailability of nutrients and for cosmetic applications.

23.6
Application of Nanotechnology in Food Packaging and Other Contact Surfaces

The dramatic way in which the materials properties of substances can be altered by fabricating at the nanoscale is being exploited to powerful effect by the food packaging industry [54]. The key aspect is the dispersion of nanoparticulate substances within a polymeric matrix, typically comprising a thermoplastic, to form a so-called *nanocomposite*. The reinforcement of polymers by incorporating different types of fillers, whether these are in a particulate or dissolved form, is a common approach to modifying the properties of plastics. However, by reducing filler particles to a nanoparticulate size, such modifications can open up a wealth of new opportunities. Nanoparticulates are now being included in food and beverage packaging for a whole range of different reasons, and such applications currently represent the largest area of uptake of nanotechnology within the sector. In fact it is claimed that the market in 2008 was US$4.13 billion, and is expected to grow to US$7.30 billion by 2114 [55]. The different ways in which nanotechnology is being used in food and beverage packaging applications can be subdivided into three categories (adapted from [56]):

- improvement of the materials properties of the packaging, especially its mechanical and permeability characteristics;
- incorporation of "active" nanoparticulates of direct benefit to the food; and
- incorporation of nanosensors to monitor and report the condition of the food content.

These areas are discussed in greater detail below.

23.6.1 Improvement of Materials Properties of Packaging

Inclusion of different types of nanoparticulates can alter the mechanical and permeability properties of the packaging material in a variety of ways that can be exploited to great advantage [54]. Many of these applications utilize naturally occurring nanoclays such as montmorillonite, which is derived from volcanic deposits. Chemically, this is a hydrated, mixed silicate of sodium, calcium, magnesium, and aluminum, while physically it exists as plate-like crystals with average diameters of around 1 μm and only 1 nm thick. When incorporated at typical levels of 5%, the spatial characteristics and high surface area of these materials confer unique properties to the host polymer. For example, packaging can be made lighter, yet have greater tensile strength, toughness and hardness than the unmodified polymer. As a result it may be possible to reduce the overall mass of packaging material needed, which also has implications for materials and transportation costs. Other potential benefits include improved flexural strength, thermal stability, flame retardance, adhesion resistance, and optical clarity when compared with conventionally filled polymers.

In order to maintain food quality, it is often desirable to retain an inert environment, so of great relevance to the food sector is the fact that nanocomposites have significantly enhanced barrier properties against gases, moisture, and volatiles such as flavor components. Rather than diffusing directly through the matrix, gaseous molecules must instead follow a tortuous path in order to avoid the intervening nanoparticles. For example, shelf life can be greatly extended by restricting entry of oxygen, which might otherwise contribute to oxidative degradation of vulnerable food constituents. In the case of carbonated beverages, loss of carbon dioxide is substantially reduced. Thus companies such as Nanocor and Honeywell Polymers have produced plastic beer bottles which are lighter than glass, less expensive than metal cans and which can considerably extend the shelf life of bottled beer. Moisture-related damage can also be prevented. These principles are now being used to protect a wide range of food products such as processed meats, cheeses and other dairy products, cereals, confectionary products, fruit juices, carbonated drinks, beer, and "boil-in-the-bag" food products [54]. In addition to nanoclays, other types of nanoparticulates can be incorporated into nanocomposites in order to improve their materials properties. For example, nanoparticulate titanium dioxide and zinc dioxide can prevent UV-mediated degradation of plastics without the opacity that would be present with larger particles whereas nano-silicon dioxide can improve resistance to abrasion.

A major impetus for research into nano-packaging is the drive toward developing sustainable, biodegradable packaging materials. Biodegradable matrices such as cellulose, starch, polylactic acid, and chitosan are being reinforced with biodegradable nanoparticulate inclusions such as cellulose nano-whiskers, to produce greatly enhanced materials properties with greater biodegradability than the matrix polymer itself [57]. Nanoclays have also been incorporated into such non-plastic matrices to improve their properties.

23.6.2 Incorporation of "Active" Nanoparticulates

As discussed previously, reduction of particle size to the nanoscale range can bring about substantial increases in chemical reactivity, even of relatively inert substances, due to factors such as greatly increased surface area, high aspect ratios, and quantum confinement effects. This can result in new functionalities such as antimicrobial behavior and oxygen scavenging activity, both of which can be exploited by the food packaging industry. In this way, the nanoparticulates can exert their beneficial effects on the food in their own right, rather than simply improving the materials properties of the packaging. Many such applications have been developed based on nanoparticulate metals and metal oxides, including silver, gold, titanium and zinc dioxides, iron (ferrous and ferric) oxides, and aluminum oxide (alumina) and a number of these applications are now being commercialized.

Titanium dioxide has a long history of use as a white pigment and opacifier in a wide range of consumer goods including foods, and has been allocated an E number, E 171. It possesses photocatalytic activity and this effect is magnified when it is used in a nanoparticulate form. On exposure to ultraviolet light it can generate free radicals which are able to decompose organic molecules and thus kill bacteria and other microorganisms. This effect is also the basis of self-cleaning coatings since once organic contaminants have been broken down, it is much easier to remove inorganic substances. It has also been suggested that titanium dioxide nanoparticles embedded in food packaging could slow the ripening of fruit and vegetables by breaking down ethylene gas [58].

Silver has long been known to have antimicrobial properties and its use in a nanoparticulate form increases this activity and enables a given amount of the metal to go vastly further. Examples of nanosilver-based antimicrobial packaging products include FresherLonger™ food storage containers and storage bags, marketed by Sharper Image®, USA and Nano Silver food containers and other products from A-DO Korea. Samsung have also developed a range of domestic appliances, including refrigerators, containing nanosilver-impregnated surfaces for antimicrobial control.

Functional properties other than antimicrobial effects include oxygen scavenging activity for oxidative protection. In this respect, Kodak is developing oxygen scavenging condensation copolymers which can be incorporated into plastic bottles. When "active" protection is combined with the "passive" barrier properties achieved using nanoclays, entry of oxygen can be reduced to extremely low levels.

23.6.3 "Smart" Packaging

Nanoscale-sized biosensors, which can detect and measure specific food pathogens, toxins, allergens, and chemicals, are the subject of numerous studies that are expected to revolutionize the food packaging field. The sensors themselves are discussed in greater detail in Section 23.7. These so-called "smart" or "intelligent" devices can be embedded in the packaging or else attached to the packaging interior to enhance the quality, safety, and traceability of foodstuffs. They could, for example, sense and indicate when food is beginning to spoil, and so provide a true indication of when it is approaching the end of its shelf life, thus eliminating a great deal of current food wastage. Some can even help to solve the contamination problem. An example is the BioSwitch encapsulation and delivery system developed by TNO Life Sciences. The basis of the approach is that antimicrobial agents are incorporated in a charged, cross-linked biopolymer matrix which is vulnerable to breakdown by bacterial enzymes. Particles of the matrix are embedded in the food packaging material. If and when bacterial contamination occurs, the gel matrix will be broken down, the antimicrobial released and the bacteria destroyed. A recent configuration utilizes the antimicrobial enzyme lysozyme embedded in cross-linked potato starch which can then be digested by amylase secreted by contaminating bacteria [59].

Of increasing importance in the food packaging industry is radiofrequency identification (RFID) technology. Electronic tags that can carry a wealth of data relating to a particular food product are attached to the packaging and can then be read wirelessly by a reader, with that information being passed to a computer for processing. This provides major benefits for food monitoring and traceability, resulting in increased speed and efficiency of stock control and improved safety and quality. According to the US Food and Drug Administration (FDA), a very large number of product recalls during previous years could have been avoided using technologies such as RFID [60].

Continued advances in miniaturization mean that tags can be made the size of powder particles and can be embedded within the packaging. For example, Hitachi's Mu chip is just 0.4 mm^2 and 50 μm thick and there are plans to reduce this size further. The tags could eventually communicate directly with nanosensors to provide a detailed analysis of food quality to the reader.

23.7 Other Areas of Application

In addition to delivery of functional ingredients and packaging, a variety of miscellaneous applications for nanotechnology in food and related areas are under development. Nanosensors have already been mentioned in relation to food packaging, but many applications are being developed for general use in food safety and quality control. For example, in microbiological monitoring, culture-based approaches are complex, very time consuming and need to be carried out in a laboratory environment, whereas biosensor technologies can provide rapid throughput, portability, sensitivity, and specificity, yet still be inexpensive.

Sensors typically comprise a recognition component, which in the case of a biosensor might be an antibody, an enzyme or a complementary DNA sequence, each of which can interact with its corresponding target in the food, a transducer component which is able to convert the result of that interaction into a measurable signal (e.g., an optical signal or an electric current), and a detector to process and display the results. Because of their special properties such as increased surface area, chemical reactivity and electrical conductance, nanoparticulates are proving to be uniquely valuable in terms of their ability to interface between the biological molecule and the transduction event. Typically the biomolecule will be adsorbed onto the surface of the nanoparticulate, which might be a gold nanoparticle, a carbon nanotube or a quantum dot.

Numerous examples of biosensors under development for food application can be found in the scientific literature. For example, a gold/silicon nanorod array coated with anti-salmonella antibodies and fluorescent organic dye molecules was used to detect *Salmonella* in food with high sensitivity [61]. Staphylococcal enterotoxin B (SEB) was targeted by immobilizing anti-SEB antibodies onto gold nanoparticles [62] or carbon nanotubes [63]. In each case the nanoparticles were immobilized on a polycarbonate surface and SEB in various food samples (mushrooms, tomatoes, and baby food) was detected by ELISA assay using an enhanced chemiluminescence immunosensor. In the former case, sensitivity was increased 10-fold, and in the latter, 6-fold, compared with traditional ELISA techniques.

Anti-*Escherichia coli* antibodies immobilized on gold nanorods could be used to detect *E. coli* 0157:H7 with high sensitivity and selectivity based on the two photon Rayleigh scattering (TPRS) properties of the nanorods, the intensity of which increased 40 times in the presence of the bacteria [64]. Liposomes have been used for immunoassay-based simultaneous detection of *E. coli*, *Salmonella*, and *Listeria monocytogenes* with high sensitivity [65] and to detect peanut allergen proteins [66].

Specificity can also be provided using nucleotides such as RNA and DNA rather than antibodies. DNA microarrays can be used to simultaneously analyze different DNA sequences via polymerase chain reaction (PCR), thus enabling identification of multiple pathogens and genes in a single assay [67]. A comprehensive review of food pathogen detection using bioconjugated nanomaterials can be found in Ref. [68].

An alternative analytical approach to biosensors is to use electronic tongue technology which is also the subject of numerous research initiatives. In this case, arrays of nonspecific detectors are used to measure electrochemical changes within food, beverage, or water samples. Sophisticated data processing techniques are then used to analyze the sensor outputs and to try to recognize patterns that can be correlated with human sensory assessment of food flavors, thus enabling these to be quantified. A recent review of electrochemical approaches can be found in Ref. [69].

Another application for nanotechnology in food, beverage, and water processing is nanofiltration. This is already a well established area using conventional polymer-based membranes, with applications such as water softening and whey processing. More recently, filters have been developed based on co-aligned carbon nanotubes [70] which promise to revolutionize the water purification and desalination industries. Their pore diameters, at 2 nm, are an order of magnitude smaller than conventional membranes yet, counter-intuitively, their permeability is several orders of magnitude higher. These are likely to significantly reduce the cost of desalination compared to conventional approaches, and may also have other applications such as for separation of sugars.

Agricultural products such as fertilizers, pesticides, and herbicides are sometimes used in a nanoparticulate form for various reasons. Some biologically active materials are made more potent by designing them to have increased lipid solubility in order to aid penetration into cells, and as such they may have very low water solubility. Conversion to a nanoparticulate form can increase water solubility as a result of increased surface area and reactivity. Formulation as microemulsions can also achieve this; for example, Syngenta's MAXX® growth regulator product is marketed in this form.

Other products are incorporated into controlled release microcapsules which can break open in contact with leaves (Karate with Zeon Technology®) or in the alkaline environment of an insect's stomach (Gutbuster), or as nanoemulsions for improved administration.

23.8
Potential Health and Safety Concerns Involved with Ingestion of Nanoparticulates

Despite its rapid growth, the food nanotechnology field is still in its infancy and as such there are major gaps in our knowledge concerning the consequences of long-term dietary exposure to nanoparticulates, whether these are intentionally added to the food or enter inadvertently, for example, by migration from food contact surfaces such as packaging. These gaps arise partly through the lack of knowledge about the body's responses to ingested nanoparticles, much of the current information being derived from inhalation studies. There is also limited understanding concerning the dynamics of nanoparticulates within food contact materials when these are in contact with different foods. This lack of knowledge was highlighted in the 2004 Royal Society/Royal Academy of Engineering report [71] discussed in Section 23.9, and reiterated in several subsequent reports. It is known that particle size has an important effect on the fate of particulates within the gut. For example, nanoparticulates can penetrate the mucous layers more effectively [72], and are also taken up more readily by the enterocytes than are larger particles. They might then accumulate in those or in other intestinal cells, or in certain circumstances, pass across the cellular barrier to ultimately reach the bloodstream. Other factors such as particle shape can also influence cellular interactions.

The majority of nanoparticulates that are added purposely into the food will be there as delivery systems, usually adapted from the drug delivery field where they will already have been selected for their lack of toxicity. Many will be composed of normal food components and as such are unlikely to have any toxic effect, especially at the concentrations that are likely to be present. On the other hand, nutritional supplements containing nanoparticulate silver that have been marketed in the European Union for several years may be withdrawn from sale because of the difficulty in proving that they are safe.

Those nanoparticles incorporated into food contact materials are often inorganic materials that have been activated by the process of conversion to the nanoscale. Titanium dioxide has a long history of use as a white pigment and opacifier in a wide range of consumer goods including foods, and has been allocated an E number, E 171. In its normal form it has generally been considered to be chemically inert. However, it is known to be photoreactive and rats have been shown to develop respiratory tract cancers after inhalation or intratracheal instillation of powdered and ultrafine grades [73]. *In vitro* studies in mice showed that nanoparticulate titanium dioxide caused DNA damage, probably as a result of free radical generation, and indicated the possibility of it being carcinogenic in humans. This is of increasing concern because of its use in nanocomposites for UV protection.

Silver has long been known to have antimicrobial properties and its use in a nanoparticulate form increases this effect as well as enabling a given amount of the expensive metal to go vastly further. Various companies have incorporated it into a variety of different food utensils in order to exploit this property. For example, Samsung use it for coating of appliances such as refrigerators. Other nanoparticulate substances are also being explored for incorporation into food contact surfaces. In some cases the shape of the particle might also be a safety factor. There are fears that carbon nanotubes might behave in a similar way to asbestos fibers because of their high aspect ratios.

The question in all of these cases is to what extent the particles might be able to migrate from the polymer matrix into the food. This is exacerbated by the paucity of sensitive methodologies for detecting and quantifying any released nanoparticles. An early study using nanoclays embedded in a biodegradable matrix suggested a limited tendency of the nanoclay to migrate into a vegetable food model [74]. A more recent study using conventional plastic nanocomposites [75] showed that, except where the particles were very small (approximately 1 nm radius) and the polymer matrix had a low dynamic viscosity (e.g., polyolefines), no significant migration of nanoparticles occurred. Nanoparticulate silver is of this size range and could migrate from such matrices as long as there is no interaction with the polymer. The study concluded that further investigations need to be carried out using alternative polymers. One point worth noting is that after a very extensive consultation exercise, a recent enquiry carried out by the House of Lords [76] stated that it had "received no evidence, however, of instances where *ingested* nanomaterials have harmed human health."

Another point in partial mitigation is that the epithelial cells of the gastrointestinal tract are known to be in a constant state of renewal such that there is a complete turnover of gut endothelia every few days [77]. This would limit any problems that might occur due to accumulation of nanoparticles within the gut endothelium. On the other hand, nanoparticles that have been absorbed rapidly and transported to other parts of the body might still be a problem if they accumulate in sufficient amount. It is likely that if migration is very low, it could be extremely difficult to detect those particles that that do get through. It is possible that absorption might be increased in conditions such as inflammatory bowel disease, where inflammation might compromise the normal barrier properties of the gut [56].

23.9 Regulatory Aspects

Issues relating to regulation of the use of nanoparticulates in food have been summarized effectively by others [56] and will be discussed only briefly here. In the United Kingdom and other European member states, regulation in the food area is largely decided at the European Commission level, and is outlined in a range of legislations covering various aspects of food law. These regulations are intended to ensure that all food products and components that come to market have been evaluated with respect to their potential risk to human health. The main ones relevant to nanoparticulates are those relating to food additives (Directive EC/1333/2008), novel foods and food ingredients (Regulation EC/258/97), and food contact materials (Regulation EC/1935/2004). In addition, general safety is covered by the General Principles of Food Law Regulation (EC/178/2002). The directive for novel foods and food ingredients applies to foods or ingredients (other than food additives) not consumed within the European Union before May 15, 1997, and is currently being revised. Whereas nanoparticulates are not specifically mentioned in the existing version, they will be included in the new directive.

Additives are defined as substances that are added to foods for a technological purpose (e.g., preservatives) rather than being a food in their own right. A basic tenet of the regulations is that additives must be explicitly authorized before they can be used in food. Applications for authorization must be made to the European Food Safety Authority (EFSA) which is also responsible for safety evaluation and risk assessment of the additive. Thus each additive is considered on a case-by-case basis before being permitted for use. Where doubt still exists, the "precautionary principle" may be applied, which is based on the notion of "better safe than sorry."

Within the United Kingdom, the Food Standards Agency (FSA) is responsible for maintaining food safety and protecting consumer interests. In this respect, it is advised by the Advisory Committee on Novel Foods and Processes (ACNFP), an independent body of scientific experts which carries out rigorous safety assessments on additives and processes which are submitted for evaluation. At the same time, issues such as technological needs, ethical considerations, and consumer concerns are taken into account. There is a general consensus that the existing regulatory approach is adequate to identify any risks associated with newly developed products of nanotechnology, and that the development of specific laws to regulate these is unnecessary.

The FDA in the United States have taken a similar view except that nanoparticulates developed from a GRAS ingredient may be able to bypass the regulatory process. On the other hand, the FDA claims that where any uncertainties exist, expert opinions would be used to clarify the issue. In the EU, nanoparticulate forms of a known ingredient would in any case be viewed as novel with changed properties, and so would be assessed anyway.

A number of reviews and reports have contributed to shaping the current regulatory position in relation to nanotechnology and foods. The Royal Society and the Royal Academy of Engineering carried out a study which culminated in a report in 2004 [71] entitled "Nanoscience and Nanotechnologies: Opportunities and Uncertainties" which highlighted the uncertainties arising from the lack of evidence concerning the risks posed by manufactured nanoparticles and recommended that the adequacy of existing regulations be reviewed and any regulatory gaps be addressed by the appropriate regulatory agencies. In 2005, the government issued a response, stating that the regulations would be reviewed, recognizing that free engineered nanoparticles should be treated as new chemicals from a regulatory perspective and should undergo thorough safety assessment before being placed on the market. A series of regulatory reviews covering general aspects of nanoparticulates followed, including a 2006 report for the Department of the Environment, Food and Rural Affairs (Defra) [78] which gave some focus on foods and food contact materials. A report was also presented by the FSA in 2008 [79] which focused entirely on food-related areas. All were in general agreement that the lack of knowledge concerning the potential effects of exposure to nanoparticles on human health and the environment was a major limitation to drafting effective regulations. Additional research is thus required, including the development of methods for quantifying exposure limits in order to aid risk assessment. At the same time it was considered that the existing regulations could be adapted to encompass the impact of nanoparticles without the need to draft new regulations. Improved definitions and nomenclature relating to nanoparticles would help in this respect.

A report by the House of Lords Select Committee on Science and Technology in 2010 [76] on nanotechnology in the food sector also gave some consideration to regulatory issues. This concurred with several of the conclusions of the previous reports while making a number of additional recommendations. These included the suggestion that more emphasis be given to functionality rather than size, and that the upper size limit of a nanoparticle might be 1000 nm rather than the popularly accepted 100 nm limit. Also that for regulatory purposes, the definition used for nanomaterials "should exclude those created from natural food substances, except for nanomaterials that have been deliberately chosen or engineered to take advantage of their nanoscale properties." This report is discussed further in the following section.

23.10
Recent Initiatives

An ever-increasing awareness of the impact that nanotechnology will have on everyday life, not least in the food sector, has led to the rolling out of numerous national and international initiatives intended to promote greater understanding, appreciation and further development of the field. In the United States, the National Nanotechnology Initiative (NNI) was established in 2001 with the specific aim of coordinating Federal nanotechnology research and development. In Europe, the framework programmes are responsible for supporting general research and technological development over a broad range, but over the past decade, have seen an increasing focus on nanotechnology. The Seventh Framework Programme runs from 2007 to 2013 and has an overall budget of around €50.5 billion with almost €3.5 billion allocated for research on nanosciences, nanotechnology, knowledge-based materials and new production technologies (NMP).

A large number of initiatives have also been conducted at the national level. The 2006 RS/RAE report recognized that "public attitudes play a crucial role in the realization of the potential of technological advances" and recommended that "the Government initiates adequately funded public dialogue around the development of nanotechnologies." This was echoed by the various regulatory reviews that followed, and has led to a range of different nanotechnology dialogue initiatives such as the Nanotechnologies Engagement Group (NEG) organized by the Government Office of Science and Innovation, the Nano Stakeholder Forum chaired by Defra and the NanoJury, which was set up by Greenpeace together with the Guardian newspaper and several UK universities. An interactive web site entitled *www.Nano&me.com* also provides a useful information resource for the public.

An enquiry by the House of Lords Select Committee on Science and Technology to look at the potential offered by nanotechnology to the food sector and to consider issues such as public awareness, communication, health and safety, and regulatory aspects, culminated in a report which was finally submitted on January 8, 2010 [76]. Its findings and recommendations were based on an extensive number of written submissions, oral testimonies, and meetings with a wide range of experts and stakeholders. The report acknowledges the many potential benefits that nanotechnology can provide, while being fully appreciative of consumer concerns that might arise. It starts out with recommendations for support in various areas of research, such as in the development and commercialization of new scientific advances in food nanotechnology. This could be encouraged, for example, by increased funding and by improved knowledge transfer between industry and academia. Research initiatives should also be encouraged in aspects relating to health and safety, for example, to provide better understanding of how nanoparticulates from food or from food contact materials, interact with and are absorbed by the gut, thus enabling more effective risk assessment. This would also include investigations into how the natural gut flora is affected by the antimicrobial properties of some nanoparticulates. Support for this could be aided by the development by the FSA, in collaboration with the food industry, of a confidential database listing information about ongoing research on nanomaterials in the food sector.

Another area which received substantial coverage in the report concerned the perceived lack of openness and transparency by the food industries with regard to work on nanoparticulates. There is a widely held view that the industry is unwilling to talk about ongoing work in this area, presumably because of fear of consumer negativity. Public acceptance of food nanotechnology is paramount if the potential benefits are to be realized, and lessons must be learned from previous instances such as the GM foods and food irradiation debacles, if suspicion and distrust is to be avoided. Strenuous efforts must be made to maintain consumer confidence by keeping them informed about ongoing research and development, and allowing them to be involved in the decision making process. Recommendations made regarding regulatory issues have already been discussed in the preceding section.

References

1. European Committee for Standardization (2009) Nanotechnologies. *http://www.cen.eu/cen/Sectors/Sectors/Nanotechnologies/index.asp* (accessed June 2011).
2. Anderson, E.S., Dong, M., Nielsen, M.M., Jahn, K., Subramani, R., Mamdouh, W., Golas, M.M., Sander, B., Stark, H., Oliveira, C.L.P., Pedersen, J.S., Birkedal, V., Besenbacher, F., Gothelf, K.V., and Kjems, J. (2009) Self-assembly of a nanoscale DNA box with a controllable lid. *Nature*, **459**, 73–76.
3. Helmut Kaiser Consultancy Report (2008) Nanotechnology in Food and Food Processing Industry Worldwide 2008-2010-2015. *http://www.hkc22.com/nanofood.html* (accessed March 9, 2010).
4. Scrinis, G. and Lyons, K. (2007) The emerging nano-corporate paradigm: nanotechnology and the transformation of nature, food and agri-food systems. *Int. J. Sociol. Food Agric.*, **15** (2), 28.
5. Kirby, C.J. and Law, B.A. (1986) Recent developments in cheese flavour technology: application of enzyme microencapsulation. Proceedings of the Biotechnology in the Food Industry Conference, December, 1986, London Online Publications, Pinner, Middlesex.
6. Kirby, C.J. and Law, B.A. (1987) Developments in the microencapsulation of enzymes in food technology, in *Chemical Aspects of Food Enzymes* (ed. A.T. Andrews), Royal Society of Chemistry, London, pp. 106–119.
7. Kirby, C.J. (1993) Controlled delivery of food ingredients: opportunities for

liposomes in the food industry, in *Liposome Technology*, 2nd edn, vol. II (ed. G. Gregoriadis), CRC Press, London, pp. 215–232.

8. Balassa, L.L. and Fanger, G.O. (1971) Microencapsulation in the food industry. *Crit. Rev. Food Technol.*, **2**, 245–252.

9. Risch, S.J. and Reineccius, G.A. (eds) (1988) *Flavour Encapsulation*, ACS Symposium Series, Vol. 370, American Chemical Society, Washington, DC.

10. Dziezak, J.D. (1988) Microencapsulation and encapsulated ingredients. *Food Technol.*, **42**, 136–148.

11. Taniguchi, N. (1974) On the basic concept of 'nano-technology'. Proceedings of the International Conference on Production Engineering Tokyo, Part II, Japan Society Precision Engineering, pp. 18–23.

12. Weiss, J., Takhistov, P., and McClemants, D.J. (2006) Functional materials in food nanotechnology. *J. Food Sci.*, **71** (9), R107–R116.

13. Chen, H., Weiss, J., and Shahidi, F. (2006) Nanotechnology in nutraceuticals and functional foods. *Food Technol.*, **60** (3), 30–36.

14. Kirby, C.J. and Gregoriadis, G. (1999) Liposomes, in *Encyclopaedia of Controlled Drug Delivery* (ed. E. Mathiowitz), John Wiley & Sons, Inc., New York, pp. 461–492.

15. Kirby, C.J. (1991) Microencapsulation and controlled delivery of food ingredients. *Food Sci. Technol. Today*, **5** (2), 74–78.

16. Kirby, C.J. and Gregoriadis, G. (1984) Dehydration–rehydration vesicles: a simple method for high yield drug entrapment in liposomes. *Biotechnology*, **2**, 979–984.

17. Taylor, T.M., Davidson, P.M., Bruce, B.D., and Weiss, J. (2005) Liposomal nanocapsules in food science and agriculture. *Crit. Rev. Food Sci. Nutr.*, **45**, 587–605.

18. Renton, A. (2006) Welcome to the World of Nanofoods, Guardian Unlimited UK December 13, 2006, *http://observer.guardian.co.uk/foodmonthly/futureoffood/story/0,,1971266,00.html* (accessed March 5, 2010).

19. Kirby, C.J. (1992) Development of Cocoa Stearine Microcapsules Containing Encapsulated Glycerol, AFRC final report.

20. Muller, R.H., Mader, K., and Gohla, S. (2000) Solid lipid nanoparticles (SLN) for controlled drug delivery-a review of the state of the art. *Eur. J. Pharm. Biopharm.*, **50**, 161–177.

21. Garti, N. and Benichou, A. (2004) Recent developments in double emulsions for food applications, in *Food Emulsions*, 4th edn (eds S. Friberg, K. Larsson, and J. Sjoblem), Marcel Dekker, New York, pp. 352–412.

22. Groves, K. (2009) Nanotechnology and the food industry. *Food Sci. Technol.*, **23** (2), 22–24.

23. Kirby, C.J., New, R.R.C., Gay, C., Tynen, W., and Satchdev, A. (1995) "Macrosol": a new oil-based carrier vehicle. *Proc. Int. Symp. Control Rel. Bioact. Mater.*, **22**, 588–589.

24. Kirby, C.J. and New, R. (1994) Hydrophobic preparations of hydrophilic species and process for their preparation. Patent No. WO 95/13795, filed Nov. 14, Published May 26, 1995.

25. Kirby, C.J. (2000) Oil-based formulations for oral delivery of therapeutic peptides. *J. Liposome Res.*, **10** (4), 391–407.

26. Jin, T., Zarif, L., and Mannino, R. (1999) Nanocochleate formulations, process of preparation and method of delivery of pharmaceutical agents. US Patent 6153217, filed 1999.

27. Larsson, K. (1999) Colloidal dispersions of ordered lipid–water phases. *J. Disp. Sci. Technol.*, **20** (1), 27–34.

28. Moharty, B., Aswal, V.K., Kohlbrecher, J., and Bohidar, H.B. (2005) Syntheses of gelatin nanoparticles via simple coacervation. *J. Surf. Sci. Technol.*, **21** (3–4), 149–160.

29. Aimi, M., Nemori, R., and Ogiwara, K. (2007) Casein nanoparticles. US Patent 20090280148, filed 2007.

30. Chakraborty, S., Sahoo, B., Teraoka, I., and Gross, R.A. (2005) Solution properties of starch nanoparticles in water and DMSO as studied by dynamic light scattering. *Carbohydr. Polym.*, **60** (4), 475–481.

31. Zhang, J., Elder, T.J., Yunqiao Pu, Y., and Ragauskas, A.J. (2007) Facile synthesis of spherical cellulose nanoparticles. *Carbohydr. Polym.*, **60** (3), 607–611.
32. Jalil, R. and Nixon, J. (1992) Microencapsulation with biodegradable materials, in *Microencapsulation of Drugs* (ed. T.L. Whately), Harwood Academic Publishers, London, pp. 177–188.
33. Klaui, H. (1974) Technical uses of vitamin C, in *Vitamin C, Recent Aspects of its Physiological and Technological Importance* (eds G.G. Birch and K.J. Parker), Applied Science Publishers Ltd, London, pp. 17–29.
34. Kirby, C.J., Whittle, C.J., Rigby, N., Coxon, D.T., and Law, B.A. (1991) Stabilisation of ascorbic acid by encapsulation in liposomes. *Int. J. Food Sci. Technol.*, **26**, 437–449.
35. Lee, S.C., Yuk, H.G., Lee, D.H., Lee, K.E., Hwang, Y.I., and Ludescher, R.D. (2002) Stabilization of retinol through incorporation into liposomes. *J. Biochem. Mol. Biol.*, **35** (4), 358–363.
36. Hsieh, Y.F., Chen, T.L., Wang, Y.T., Chang, J.H., and Chang, H.M. (2002) Properties of liposomes prepared with various lipids. *J. Food Sci.*, **67**, 2808–2813.
37. Rodriguez-Nogales, J.M. (2004) Kinetic behavior and stability of glucose oxidase entrapped in liposomes. *J. Chem. Technol. Biotech.*, **79**, 72–78.
38. Chang, H.M., Lee, Y.C., Chen, C.C., and Tu, Y.Y. (2001) Microencapsulation protects immunoglobulin in yolk (IgY) specific against *Helicobacter pylori*. *J. Food Sci.*, **67**, 15–20.
39. Law, B.A. and Wigmore, A.S. (1982) Accelerated cheese ripening with food grade proteinases. *J Dairy Sci.*, **49**, 137–146.
40. Law, B.A. and King, J. (1985) Use of liposomes for proteinase addition to Cheddar cheese. *J. Dairy Res.*, **52**, 183–188.
41. Kirby, C.J., Brooker, B.E., and Law, B.A. (1987) Accelerated ripening of cheese using liposome-encapsulated enzymes. *Int. J. Food Sci. Technol.*, **22**, 355–375.
42. Kirby, C.J. (1990) Delivery systems for enzymes. *Chem. Br.*, **26** (9), 847–850.
43. Thapon, J.L. and Brule, G. (1986) Effets du pH et de la forme ionize sur raffinit lysozymes-caseines. *Lait*, **66**, 19–30.
44. Benech, R.O., Kheadr, E.E., Lacroix, C., and Fliss, I. (2002) Antibacterial activities of nisin Z encapsulated in situ by mixed culture during Cheddar cheese ripening. *Appl. Environ. Microbiol.*, **68**, 5607–5619.
45. Benech, R.O., Kheadr, E.E., Laridi, R., Lacroix, C., and Fliss, I. (2002) Inhibition of *Listeria innocua* in Cheddar cheese by addition of nisin Z in liposomes or by in situ production in mixed culture. *Appl. Environ. Microbiol.*, **68**, 3683–3690.
46. Were, L.M., Bruce, B.D., Davidson, P.M., and Weiss, J. (2003) Size, stability and entrapment efficiency of phospholipid nanocapsules containing polypeptide antimicrobials. *J. Agric. Food Chem.*, **51**, 8073–8079.
47. Tappel, A.L. (1968) Will antioxidant nutrients slow ageing processes? *Geriatrics*, **23**, 97–105.
48. Moberger, L., Larsson, K., Buchheim, W., and Timmen, H. (1987) A study of fat oxidation in a microemulsion system. *J. Disp. Sci. Technol.*, **8**, 207–215.
49. Han, D., Yi, O.-S., and Shin, H.-K. (1991) Solubilisation of vitamin C in fish oil and synergistic effect with vitamin E in retarding oxidation. *J. Am. Oil Chem. Soc.*, **68**, 740–743.
50. El-Nokaly, M., Hiler, G., and McGardy, J. (1991) Solubilization of water and water-soluble compounds in triglycerides, in *Microemulsions and Emulsions in Foods* (eds M. El-Nokaly and D. Cornell), American Chemical Society, Washington, DC, pp. 26–43.
51. Kirby, C.J. and Needs, E.C. (1995) Microemulsions containing functional substances. UK Patent Application 9502745.4, filed Feb. 24, 1995.
52. Kirby, C.J., Mullen, A.N., New, R.R.C., and Mallinson, C.B. (1996) The use of Macrosol technology for highly effective antioxidant protection in unsaturated oils. *Proc. Int. Symp. Control Rel. Bioact. Mater.*, **23**, 130–131.
53. Kirby, C.J. (1995) Antioxidant compositions. Patent No. WO/1996/017899, filed Dec. 8, 1995, published June 13, 1996.

54. Akbari, Z., Ghomashchi, T., and Aroujalian, A. (2006) Potential of nanotechnology for food packaging industry. Proceedings of the Institute of Nanotechnology, Amsterdam on "Nano Micro Technologies in the Food Health Food Industries", October 25–26.
55. Report by Innovative Research and Products Inc. (2010) Nano-Enabled Packaging for the Food and Beverage Industry – A Global Technology, Industry and Market Analysis, *http://www.innoresearch.net/Press_Release.aspx?id=18* (accessed March 29, 2010).
56. Chaudhry, Q., Scotter, M., Blackburn, J., Ross, B., Boxall, A., Castle, L., Aitken, R., and Watkins, R. (2008) Applications and implications of nanotechnologies for the food sector. *Food Addit. Contam.*, **25** (3), 241–258.
57. de Mesquita, J.P., Donnici, D.L., and Pereira, F.V. (2010) Biobased nanocomposites from layer-by-layer assembly of cellulose nanowhiskers with chitosan. *Biomacromolecules*, **11** (2), 473–480.
58. Maneerat, C. and Hayata, Y. (2008) Gas-phase photocatalytic oxidation of ethylene with TiO2-coated packaging film for horticultural products. *Trans. ASABE*, **51**, 163–168.
59. Li, Y., Vries, R., Slaghek, T., Timmermans, J., Cohen Stuart, M.A., and Norde, W. (2009) Preparation and characterization of oxidized starch polymer microgels for encapsulation and controlled release of functional ingredients. *Biomacromolecules*, **10** (7), 1931–1938.
60. Brody, A.L., Bugusu, B., Jung, H., Han, J.H., Sand, C.K., and McHugh, T.H. (2008) Innovative food packaging solutions. *J Food Sci.*, **73** (8), R107–R116.
61. Fu, J., Park, B., Siragusa, G., Jones, L., Tripp, R., Zhao, Y., and Cho, Y.-J. (2008) Au/Si hetero-nanorod-based biosensor for salmonella detection. *J. Nanosci. Nanotechnol.*, **19**, 1–7.
62. Yang, M.H., Kostov, Y., Bruck, H.A., and Rasooly, A. (2009) Gold nanoparticle-based enhanced chemiluminescence immunosensor for detection of staphylococcal enterotoxin B in food. *Int. J. Food Microbiol.*, **133**, 265–271.
63. Yang, M.H., Kostov, Y., and Rasooly, A. (2008) Carbon nanotubes based optical immunodetection of Staphylococcal Enterotoxin B (SEB) in food. *Int. J. Food Microbiol.*, **127**, 78–83.
64. Singh, A.K., Senapati, D., Wang, S.G., Griffin, J., Neely, A., Candice, P., Naylor, K.M., Varisli, B., Kalluri, J.R., and Ray, P.C. (2009) Gold nanorod-based selective identification of *Escherichia coli* bacteria using two-photon Rayleigh scattering spectroscopy. *ACS Nano*, **3**, 1906–1912.
65. Chen, C.-S. and Durst, R.A. (2006) Simultaneous detection of *Escherichia coli* O157:H7, *Salmonella* spp. and *Listeria monocytogenes* with an array-based immunosorbent assay using universal protein G liposomal nanovesicles. *Talanta*, **69** (1), 232–238.
66. Wen, H.W., Borejsza-Wysocki, W., DeCory, T.R., Baeumner, A.J., and Durst, R.A. (2005) A novel extraction method for peanut allergenic proteins in chocolate and their detection by a liposome-based lateral flow assay. *Eur. Food Res. Technol.*, **221**, 564–569.
67. Rasooly, A.K. and Herold, K.E. (2008) Food microbial pathogen detection and analysis using DNA microarray technologies. *Foodborne Pathog. Dis.*, **5**, 531–550.
68. Yang, H., Li, H.P., and Jiang, X.P. (2008) Detection of foodborne pathogens using bioconjugated nanomaterials. *Microfluid. Nanofluid.*, **5**, 571–583.
69. Vaddiraju, S., Tomazos, I., Burgess, D.J., Jain, F.C., and Papadimitrakopoulos, F. (2010) Emerging synergy between nanotechnology and implantable biosensors: a review. *Biosens. Bioelectron.*, **25** (7), 1553–1565.
70. Holt, J.K., Park, H.G., Wang, Y., Stadermann, M., Artyukhin, A.B., Grigoropoulos, C.P., Noy, A., and Bakajin, O. (2006) Fast mass transport through sub-2-nanometer carbon nanotubes. *Science*, **312** (5776), 1034–1037.
71. The Royal Society and The Royal Academy of Engineering (2004) Nanoscience and Nanotechnologies: Opportunities and Uncertainties, *http://www.nanotec.org.uk/finalReport.htm* (accessed June 2011).

72. Tang, B.C., Dawson, M., Lai, S.K., Wang, Y.-Y., Suk, J.S., Yang, M., Zeitlin, P., Boyle, M.P., Fu, J., and Hanes, J. (2009) Biodegradable polymer nanoparticles that rapidly penetrate the human mucus barrier. *Proc. Natl Acad. Sci. U.S.A.*, **106**, 19268–19273.

73. Trouiller, B., Reliene, R., Westbrook, A., Solaimani, P., and Schiestl, R. (2009) Titanium dioxide nanoparticles induce DNA damage and genetic instability in vivo in mice. *Cancer Res.*, **69** (22), 8784–8789.

74. Avella, M., De Vlieger, J.J., Errico, M.E., Fischer, S., Vacca, P., and Volpe, M.G. (2005) Biodegradable starch/clay nanocomposite films for food packaging applications. *Food Chem.*, **93**, 467–474.

75. Simon, P., Chaudhry, Q., and Bakos, D. (2008) Migration of engineered nanoparticles from polymer packaging to food – a physicochemical view. *Food Nutr. Res.*, **47**, 105–113.

76. House of Lords Science and Technology Committee – First Report, Nanotechnologies and Food, *http://www.publications.parliament.uk/pa/ld200910/ldselect/ldsctech/22/2202.htm* (accessed January 10, 2010).

77. Creamer, B., Shorter, R.G., and Bamforth, J. (1961) The turnover and shedding of epithelial cells. *Gut*, **2**, 110–116.

78. Choudhry, Q., Blackburn, J., Floyd, P., George, C., Nwaogu, T., Boxall, A., and Aitken, R.A. (2006) A Scoping Study to Identify Gaps in Environmental Regulation for the Products and Applications of Nanotechnologies, A report for Defra.

79. Report of FSA Regulatory Review (2008) A Review of Potential Implications of Nanotechnologies for Regulations and Risk Assessment in Relation to Food, *http://www.food.gov.uk/multimedia/pdfs/nanoregreviewreport.pdf* (accessed January 12, 2010)

24
Fermentation and the Use of Enzymes

Dimitris Charalampopoulos

24.1
Introduction

Fermented foods have been produced for many centuries as the means for increasing the storage stability of processed foods and modifying the organoleptic and textural properties of raw materials. Fermented foods are very popular even today, although the main reason for their popularity is their specific organoleptic properties, rather than their preservation abilities. Notable examples of fermented foods include dairy products, such as yoghurts, cheeses, and sour milks; alcoholic drinks, such as wine and beer; fermented vegetables, such as sauerkraut and pickles; and fermented meats, such as sausages.

A more recent development to the use of microorganisms in foods is the use of enzymes isolated either from microorganisms or plant and animal sources. Enzymes are used extensively nowadays in the food processing industry, for the manufacturing of cheese, beer, wine, and bread. Depending on the application, the enzymes can be either in a relatively pure form or in a crude form. Important factors influencing the activity of the enzymes include the environmental conditions, such as temperature, substrate concentration, and pH. These need to be taken into account when selecting an enzyme for a food use.

This chapter covers food fermentation, including fermentation kinetics, bioreactor systems, and types of fermented foods, and enzymes, including enzyme nomenclature and kinetics, as well as their applications in foods.

24.2
Fermentation Theory

Fermentations can be carried out as batch or continuous processes. This section will consider the microbial growth kinetics in these processes and present the equations that can be used to describe them.

Food Processing Handbook, Second Edition. Edited by James G. Brennan and Alistair S. Grandison.

24.2.1 Growth Kinetics

24.2.1.1 Batch Growth

In batch culture, the cells are grown in a closed system that contains the appropriate nutrients. A typical growth curve is depicted in Figure 24.1. It usually consists of five phases: lag, exponential, deceleration, stationary, and death phase. During the lag phase the cells adapt to the new environment, because during their transfer from the inoculation medium to the fermentation medium several parameters can be different. Typical differences include a higher pH, an increased availability of nutrients, and a decrease in the levels of growth inhibitors [1]. Following the lag phase the cells enter the exponential (or logarithmic) phase, where they grow at a constant, maximum rate. The exponential phase is described by the equation:

$$\mathrm{d}X/\mathrm{d}t = \mu X \tag{24.1}$$

where X is the cell concentration, t is the time, and μ is the specific growth rate. Integration of Equation (24.1) from cell concentration X_0 at time $t = 0$ to X_t at time t results in:

$$\ln X = \ln X_0 + \mu t \tag{24.2}$$

Graphically, as shown in Figure 24.1, $\ln X$ versus time produces a straight line, the slope of which is the specific growth rate. Equation (24.2) can be rearranged mathematically to:

$$X_t = X_0 e^{\mu t} \tag{24.3}$$

Cell growth rates are often expressed in terms of doubling time (t_d), which can be derived from Equation (24.3) by substituting $2X_0$ for X_t, and t_d for t. This gives:

$$2 = e^{\mu t} \tag{24.4}$$

After taking the natural logarithm of both sides, Equation (24.4) results in:

$$t_d = \ln 2/\mu \tag{24.5}$$

As nutrients (e.g., carbon, nitrogen, phosphorus) in the growth medium become depleted and/or inhibitory products accumulate (e.g., organic acids, ethanol), the growth slows down and the cells enter the deceleration phase, as shown in Figure 24.1. This is followed by the stationary phase during which the cell concentration remains constant. However, during this phase the cells are still metabolically active and may produce various secondary metabolites, such as antibiotics and enzymes, which are of great commercial interest. Following the stationary phase, the cells enter the death phase, during which cell viability decreases. The death rate of a microorganism varies depending on the environmental conditions (e.g., temperature), the species, and their age and size [2].

During the growth and deceleration phases of batch culture, the specific growth rate depends on the concentration of the nutrients in the medium. It is a common feature of fermentations that a single substrate, such as the carbon or nitrogen

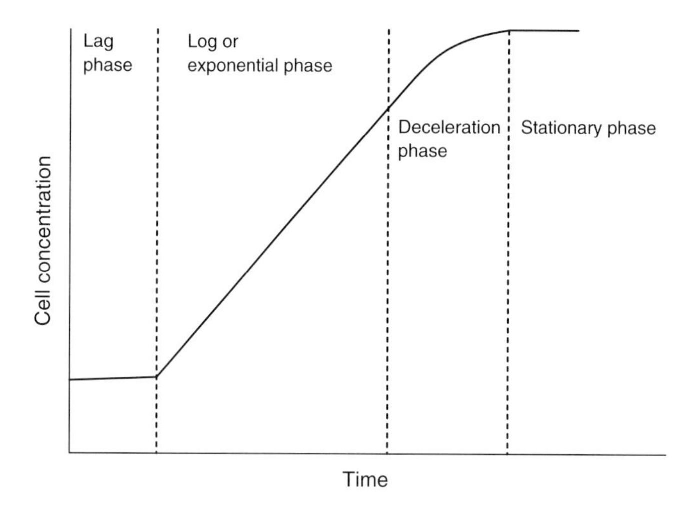

Figure 24.1 A typical microbial growth curve.

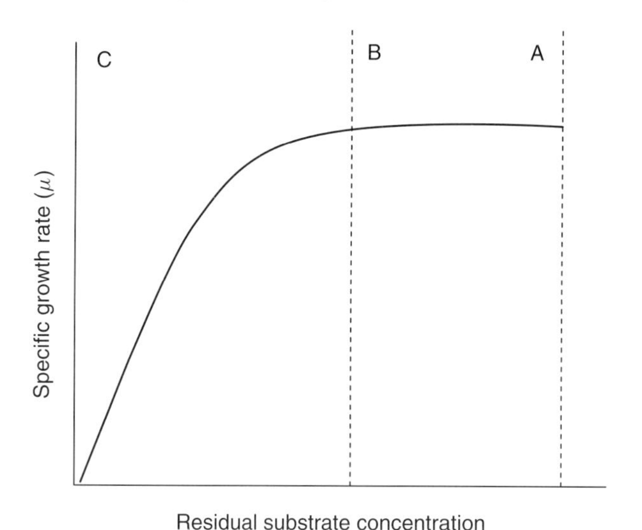

Figure 24.2 Effect of sugar concentration on the microbial growth rate.

source, influences to a large extent the growth rate; this substrate is known as the *growth limiting substrate*. The relationship between the specific growth rate and the concentration of the growth limiting substrate is described by Equation (24.6), the Monod equation [3], and is depicted schematically in Figure 24.2.

$$\mu = \mu_{\max} S/(K_S + S) \tag{24.6}$$

where μ is the specific growth rate, $\mu_{\max}$ is the maximum specific growth rate, S is the concentration of the growth limiting substrate, and K_S is the substrate constant.

In Figure 24.2, the zone A to B represents the exponential phase of the cell growth, during which the substrate concentration is in excess, and the growth rate is at its maximum (μ_{max}). The zone B to C represents the deceleration phase of the growth curve, during which the cells grow at a lower rate than μ_{max}. When the microorganism has a high affinity for the limiting substrate, (thus a low K_S), the growth rate will start decreasing when the substrate concentration has reached a very low value, and thus the deceleration phase will be most likely short [4]. The values for K_S are usually very small, that is, in the order of milligrams per liter of carbohydrates and micrograms per liter for other nutrients, such as amino acids [5].

24.2.1.2 **Continuous Culture**

In continuous culture, fresh medium containing the growth limiting substrate is continuously added in the bioreactor, and spent medium (containing cells and products) is continuously removed. If the medium is added and removed at the same rate, then a steady state is eventually achieved. The rate at which the growth limiting substrate is added is given by:

$$D = {}^{F}\!/_{V} \tag{24.7}$$

where D is the dilution rate, F is the flow rate, and V is the volume of the bioreactor. Under steady state conditions the specific growth rate is controlled by the dilution rate, therefore:

$$D = \mu \tag{24.8}$$

Combining Equations 24.6 test and 24.8 gives:

$$D = \mu_{max}{}^{\overline{S}}\!/(K_S + \overline{S}) \tag{24.9}$$

where $\overline{S}$ is the concentration of the growth limiting substrate at steady state. Rearrangement of this equation gives:

$$\overline{S} = {}^{K_S D}\!/(\mu_{max} - D) \tag{24.10}$$

The concentration of the cells in the bioreactor at steady state is described by:

$$X = Y\left(S_0 - \overline{S}\right) \tag{24.11}$$

where S_0 is the concentration of the growth limiting substrate in the feeding medium.

Equations 24.8–24.11 can be used to describe the effect of dilution rate on the steady state cell concentrations and on the residual substrate concentration. These relationships are schematically depicted in Figure 24.3. It can be observed that at low dilution rates the substrate is almost completely consumed by the cells. As D increases, S increases slowly and then more rapidly as D approaches μ_{max}. Meanwhile X decreases very slowly as the dilution rate increases, but drops rapidly as D approaches μ_{max}.

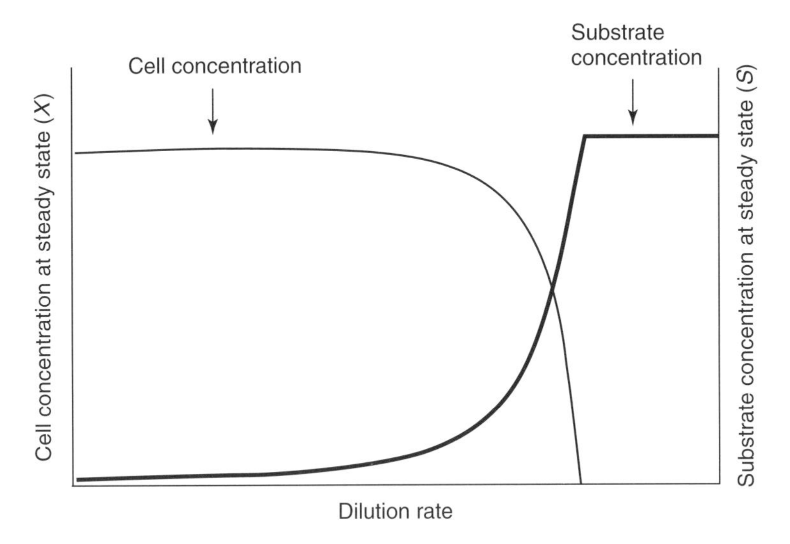

Figure 24.3 Effects of dilution rate on the cell concentration and substrate concentration in a continuous culture. (⁻⁻) Substrate concentration at steady state (*S*). (⁻⁻) Cell concentration at steady state (*X*).

24.2.2 Fermentation Systems

The fermentations are carried out in bioreactors (or fermenters). The sections below provide details on the main design configurations used. The sizes vary depending on the application, and range from laboratory size bioreactors (1–10 l) used for research and development purposes, up to 300 m^3 industrial bioreactors, used for commercial production.

24.2.2.1 Stirred Tank Reactor

The stirred tank reactor (STR) is the most commonly used bioreactor. Figure 24.4 illustrates a basic design. The main features include a stirrer, baffles, and a sparger (for aerobic processes). Mixing and bubble dispersion is provided by mechanical stirring using a variety of impeller sizes and shapes (e.g., disk turbines, open turbines, propellers), each providing different flow patterns. The baffles are vertically placed steel sheets, and are used to break the laminar liquid flow and reduce vortexing [5]. Usually, only 70–80% of the volume of the reactor is filled, which provides enough headspace for the disengagement of droplets from the exhaust gas and for any foam that might be generated during the fermentation. The geometry of the reactors depends to large extent on the application, but in most cases the height-to-width ratio ranges from 2 : 1 to 6 : 1. When high aeration is required then a high height-to-width ratio is preferred [5].

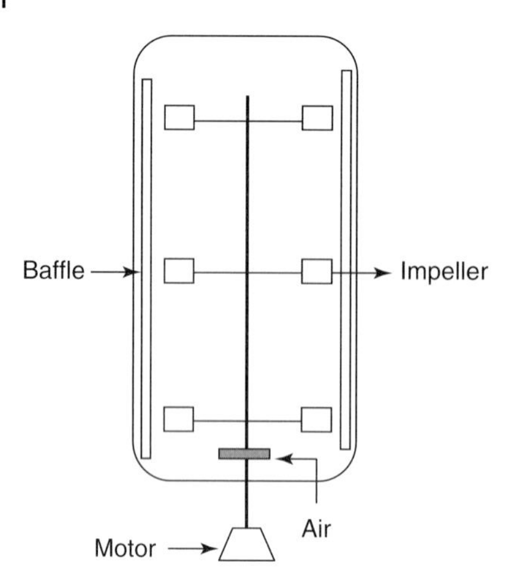

Figure 24.4 A typical stirred tank reactor (STR).

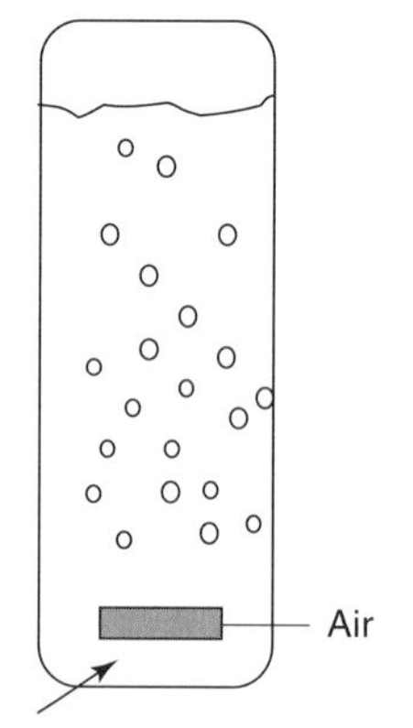

Figure 24.5 A bubble column bioreactor.

24.2.2.2 **Bubble Column**

The bubble column is a tall cylindrical vessel in which the air is dispersed by the sparger at the bottom (Figure 24.5). Gas sparging is the driving force for mixing in bubble columns. The height-to-diameter ratio ranges from 3 : 1 to 6 : 1, depending on the application [5]. The oxygen transfer capacity is lower than in STR, but it is considerably cheaper to operate the bubble column, as it does not use mechanical stirring. Bubble columns are used industrially for the production of baker's yeast, beer, and vinegar [5].

24.2.2.3 **Airlift Reactor**

In the airlift reactor the vessel is divided into two sections: the riser and the downcomer. The gas is sparged into the riser. The bulk density of the gas–liquid dispersion in the riser is higher than in the downcomer, therefore the liquid moves up in the riser. At the top, the gas disengages, leaving bubble-free liquid moving in the downcomer. Figure 24.6 illustrates the most common configurations. In the internal-loop vessels, the riser and the downcomer are separated by a draft tube

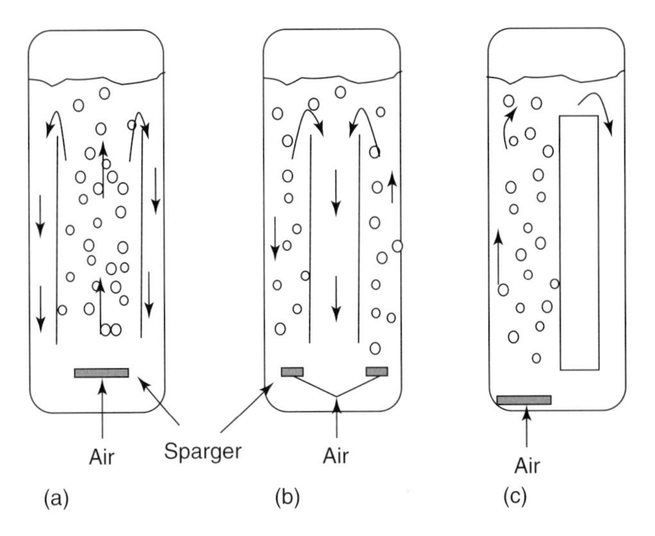

Figure 24.6 (a–c) Configurations of airlift bioreactors.

or a baffle (Figure 24.4a,b). In external loop vessels the riser and the downcomer are two separate pipes that are interconnected by horizontal sections at the top and the bottom (Figure 24.4c). Due to the higher distance between the riser and the downcomer in the case of the external loop reactors, gas disengagement is more effective than in internal loop reactors, and thus the mixing of the liquid is usually better. Airlift reactors in general provide better mixing than bubble columns [5], and are more energy efficient than STRs. They are mainly used to cultivate shear-sensitive cells, such as mammalian and plant cells, for the production of biopharmaceuticals, such as recombinant therapeutic proteins [6].

24.2.2.4 **Packed Bed Reactor**

The packed bed reactor consists of a tube packed with solid particles, as shown in Figure 24.7. The particles can be from compressible polymeric or more solid material [6]. The biocatalyst (e.g., cells, enzymes) is supported on or within the matrix of solids. The medium can be introduced in the reactor either from the bottom or the top, with the latter being the most commonly used method. Due to poor mixing, the environment of a packed bed reactor is not homogeneous, and for this reason pH control is very difficult. Packed bed reactors are used commercially as immobilized enzyme reactors, for example, for the production of aspartic acid from fumaric acid, and the conversion of penicillin to 6-aminopenicillanic acid [5].

24.2.2.5 **Fluidized Bed Reactor**

The fluidized bed reactor (Figure 24.8) is geometrically similar to the bubble column. The feed is introduced from the bottom of the reactor at high flowrates, and as a result the bed expands due to the upward motion of the particles. Because the particles in a fluidized bed reactor are continuously moving, there is much

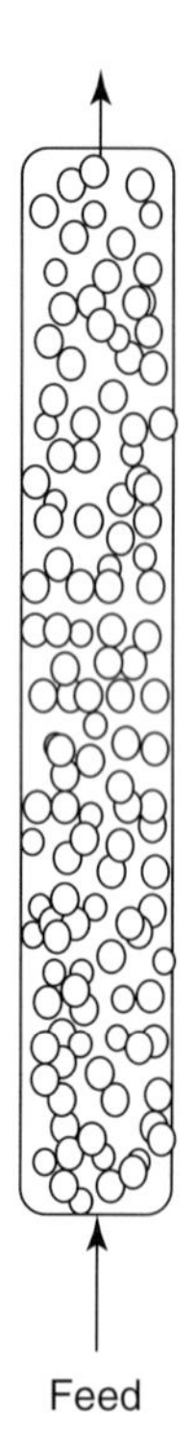

Figure 24.7 Packed bed bioreactor.

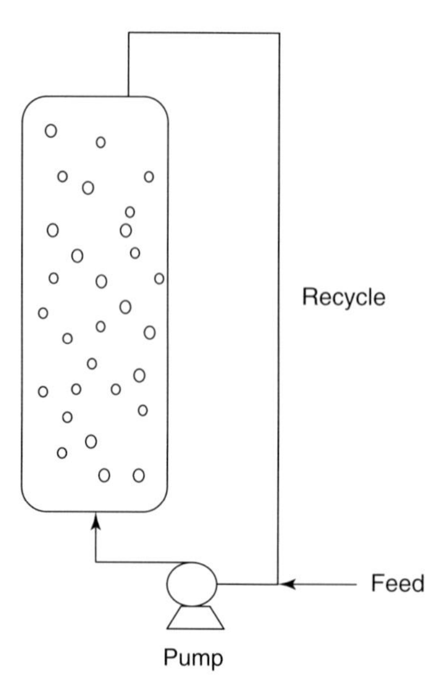

Figure 24.8 Fluidized bed bioreactor.

Table 24.1 Fermented foods and types of biochemical reactions.

Food product	Fermentation
Cheese	Enzymatic hydrolysis, lactic acid fermentation, mold fermentation
Yoghurt	Lactic acid fermentation
Fermented milks	Lactic acid fermentation
Sauerkraut	Lactic acid fermentation
Olives	Lactic acid fermentation
Bread	Ethanol fermentation
Beer	Enzymatic hydrolysis, ethanol fermentation
Wine	Ethanol fermentation, malolactic fermentation
Cider	Ethanol fermentation, malolactic fermentation
Soy products	Enzymatic hydrolysis by molds, lactic acid fermentation, ethanol fermentation
Vinegar	Ethanol, acetic acid fermentation
Sausage	Lactic acid fermentation

better mixing and oxygen transfer compared to packed bed reactors. Fluidized bed reactors are used in beer and vinegar production [5].

24.3 Fermented Foods

Fermented foods have been produced for many centuries. Table 24.1 presents a list of the most common fermented foods and the types of fermentation taking place. In most cases this is either lactic acid fermentation or ethanol fermentation. The remainder of this section focuses on specific fermented foods and provides more information about them.

24.3.1 Beer

Beer is a product of the alcoholic fermentation of malted barley. Figure 24.9 presents a schematic outline of the brewing process, whereas detailed information on brewing science and technology is provided elsewhere [7]. The first step in the brewing process is malting, which is the production of wort, a soluble malt extract. The process starts by soaking the barley grains in water at 10–15.6 °C; they are left there to germinate for 3–4 days. During germination, hydrolytic enzymes, such as α- and β-amylases, hemicellulases, and proteases, are activated and start to break down the complex polymers into simple molecules. The germination process is arrested by kilning, which uses heating to reduce the moisture content from about 45 to 4%. During kilning nonenzymatic browning reactions take place, which

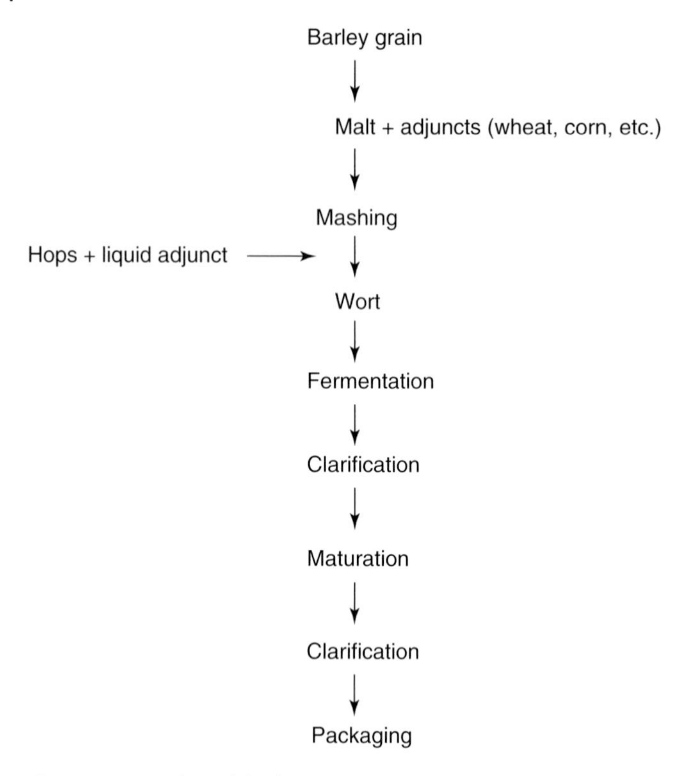

Figure 24.9 Outline of the brewing process.

contribute to the flavor and color of the final beer. Following kilning, the malt is ground and mixed with hot water and kept at 65 °C, a process called *mashing*. During mashing, starch, proteins, and enzymes are extracted from the grains in the hot aqueous solution. In addition, starch is degraded through the action of α- and β-amylases to fermentable sugars (e.g., maltose, maltotriose, glucose, fructose) and nonfermentable dextrins, whereas malt proteins are broken down to peptides and amino acids, which are very important for the growth of the yeast during the subsequent fermentation process.

The product of mashing is called *wort*. This is boiled in order to sterilize it and inactivate all the enzymes. After clarification the wort is fermented by a yeast strain, which can be either a top fermenting yeast, *Saccharomyces cerevisiae*, or a bottom fermenting, *Saccharomyces uvarum* (*Saccharomyces carlsbergenesis*). Top fermentation is typical for ale and stout beers, and is usually carried out at 15–22 °C for 3–5 days. Bottom fermentation is typical for lager and pilsner beers, and is usually carried out at 5–10 °C for 8–10 days [2]. Following fermentation the beer is transferred into wooden or stainless steel tanks for maturation, where it is kept at around 0 °C for 2–6 weeks. Then the beer is clarified, usually by filtration, and packaged in bottles, cans, or kegs.

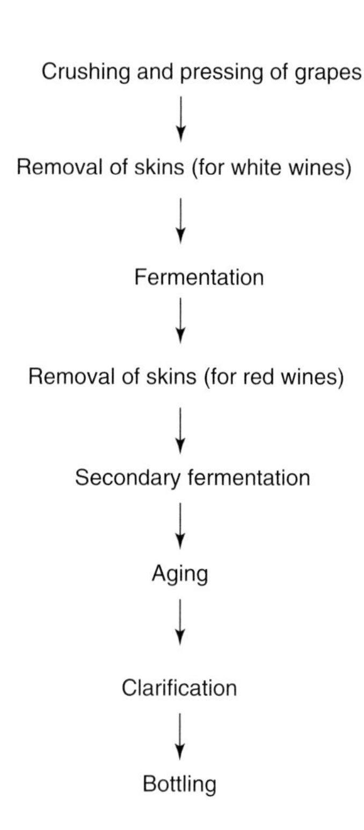

Figure 24.10 Outline of the wine making process.

24.3.2 Wine

The production of wine is schematically presented in Figure 24.10. Detailed information on wine science and technology is provided elsewhere [8]. Briefly, wine making starts with the collection and crushing of the grapes and the separation of the liquid (must). For the production of white wines, only the liquid of the white grapes is used, whereas for red wines the grape skins are allowed to remain in contact with the liquid in order to transfer their color and flavor. The next step is the fermentation step, which can be performed either using defined starter cultures, for example, strains of *Saccharomyces cerevisiae*, or mixtures of undefined cultures, usually containing a combination of lactic acid bacteria and yeast. For red wines the fermentation takes 3–5 days at 25–30 °C, whereas for white wines it takes 7–14 days at around 20 °C [2]. Depending on the tolerance of the yeast strains, the final ethanol concentration ranges between 10 and 18% ethanol [9]. During fermentation, the simple sugars present in the juice, mainly glucose, fructose, arabinose, and rhamnose, are transformed into ethanol, with the concomitant production of a variety of volatile compounds, such as higher alcohols, aldehydes, and esters, all of which contribute to the wine flavor.

Ethanolic fermentation is followed by a secondary fermentation, called *malolactic fermentation*, during which malic acid is converted to lactic acid. This results in an increase in pH and an improvement in flavor stability. The next processing step is aging, which is carried out in barrels and during which the bouquet of the wine is developed. Subsequently, the wine is clarified from various precipitated materials either using fine agents (e.g., bentonite, egg white, gelatin) or filtration. Finally, the wine is transferred into bottles, where it is left to age for one to several years [2].

24.3.3 Cheese

More than 400 cheese varieties are produced throughout the world. This large diversity of cheeses is attributed to the variety of milk sources, starter cultures, and manufacturing processes. Although the protocols for production differ, Figure 24.11 illustrates a general processing scheme for unripened and ripened cheeses. Detailed information on the science and technology of cheese can be found elsewhere [10]. Common processes for most varieties include the acidification of milk (due to lactic acid fermentation), coagulation of milk proteins, and curd dehydration and salting. Typical starters include mesophilic starters (temperatures ~30 °C), such as *Lactococcus lactis* (e.g., for Cheddar cheese) or thermophilic starters (temperatures of 38–45 °C), such as *Lactobacillus helveticus, Lactobacillus delbruecki*

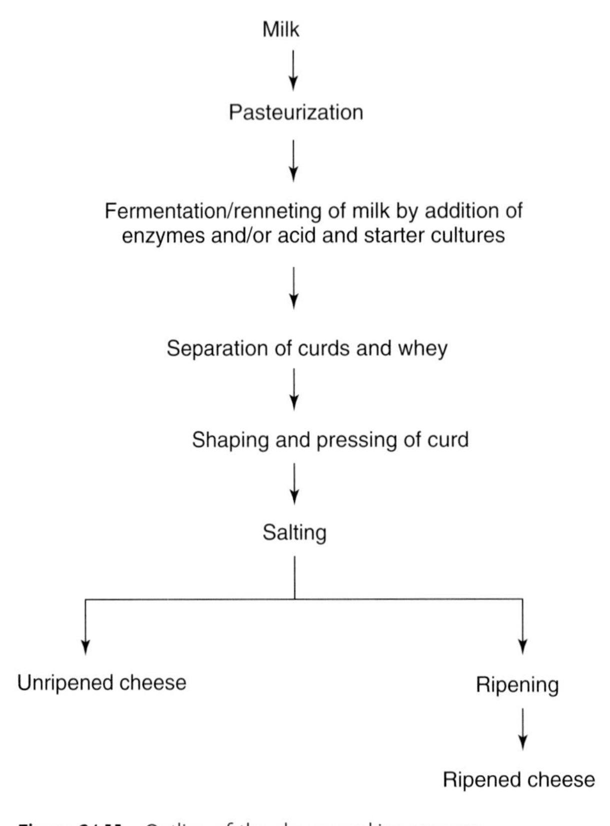

Figure 24.11 Outline of the cheese making process.

subsp. *bulgaricus*, or *Streptococcus thermophilus* (e.g., for Emmental and Parmesan). Coagulation of casein is initiated through the addition of rennet (or chymosin), a protease, and the high levels of lactic acid, and results in the formation of the curd. The coagulated curd forms a gel that entraps fat, water molecules, enzymes, and starter bacteria [11]. Subsequent steps include cutting and pressing of the curd, resulting in the release of whey, and salting, which aims to reduce the moisture content of the curd, and inhibit the growth of spoilage microorganisms. Finally, ripened cheeses undergo a ripening period, which can last from a few weeks up to 2 years, and which contributes to the flavor and texture of the cheeses.

24.3.4 Yoghurt

Yoghurt is one of the oldest fermented milk products. The manufacturing protocols and raw materials vary widely from country to country. In general, two types of yoghurt are made, the stirred type and the set type. In the former, the fermentation is carried out in a reactor, and then the yoghurt is transferred to containers, whereas in the latter, the fermentation is carried out within individual containers. Figure 24.12 depicts a generic process for the production of set-style yoghurt. Detailed information on the science and technology of yoghurt can be found elsewhere [12].

Briefly, the process starts with the standardization of milk to the desired fat and milk solids contents (usually 12–14%). The latter is achieved by adding non-fat milk powder, and aims to reduce syneresis. The milk is then homogenized in order to

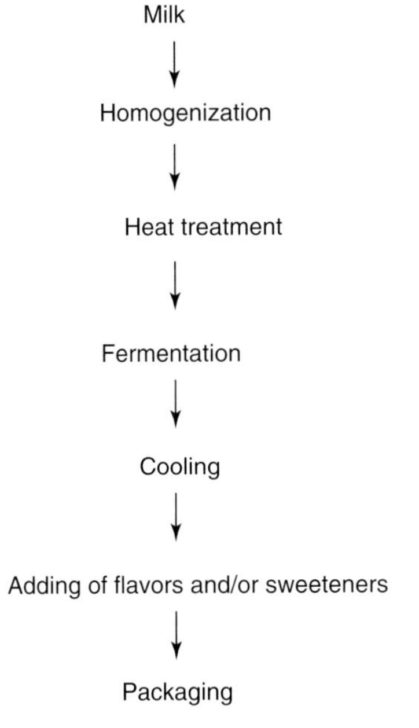

Figure 24.12 Outline of the set-style yoghurt making process.

decrease the size of the fat globules, reduce syneresis, and increase the firmness of the yoghurt. Pasteurization of the milk then follows, usually at 85 °C for 30 min, in order to eliminate spoilage microorganisms, inactivate the milk enzymes, denature the whey proteins, and reduce the levels of oxygen. Subsequently, the milk is cooled to 43–45 °C and inoculated with the starter culture, which consists of *Lactobacillus delbrueckii* spp. *bulgaricus* and *Streptococcus thermophilus*. The inoculated milk is incubated at 40–45 °C for 2–5 h, during which the acids are produced through the metabolism of the starter cultures, which result in a drop in pH and consequently in the aggregation of the casein micelles, and the formation of a gel.

The two bacteria used for yoghurt making have a synergistic effect on each other's growth [12–14]. On one hand, *Lactobacillus delbrueckii* spp. *bulgaricus* has protease activity and thus is able to break down the milk proteins to peptides and amino acids. These in turn can be taken up by *Streptococcus thermophilus*, which has limited proteolytic activity, and enhance its growth. On the other hand, *Streptococcus thermophilus* produces formic acid and carbon dioxide, which stimulate the growth of *Lactobacillus delbrueckii* spp. *bulgaricus*. As a result, the growth profiles are as follows. *Streptococcus thermophilus* grows faster initially due to the availability of nitrogenous compounds and then slows down due to the increased acidity. *Lactobacillus delbrueckii* spp. *bulgaricus* dominates in the final stages of the fermentation, as it is stimulated by formic acid.

24.3.5 Fermented Milks

More than 400 different fermented milks are produced throughout the world [9, 15]. These include cultured buttermilks and sour creams, which are produced using mesophilic starter cultures, including *Lactococcus lactis* ssp. *lactis*, *Lactococcus lactis* biovar. *diacetylactis* and *Leuconostoc mesenteroides* ssp. *cremonis*. These bacteria are able to produce diacetyl and contribute to the buttery flavor of the fermented products. The fermentations are carried out between 20 and 30 °C, and are stopped by cooling [16].

Another fermented milk product is *kefir*, which is very popular in Eastern Europe. It is prepared by inoculating cow, goat, or sheep's milk with kefir grains; these are complex matrices of polysaccharides and proteins that contain lactic bacteria, yeast, and acetic acid bacteria [17]. Kefir is sour, contains up to 2% ethanol, and is considered to have beneficial health effects [18, 19].

Other types of fermented drinks with a considerable market share are fermented milks with probiotic bacteria, such as *Lactobacillus acidophilus*. This is called *acidophilus milk*. According to their definition probiotics are "Live microorganisms which when administered in adequate amounts confer a health benefit on the host" [20]. Probiotic lactobacilli have been associated with the prevention and treatment of gastrointestinal disorders, such as rotavirus diarrhea, antibiotic-associated diarrhea, and traveler's diarrhea, and have also been suggested as potential therapeutic agents against irritable bowel syndrome and inflammatory bowel disease [21–23].

24.3.6 Fermented Meats

Fermentation followed by drying of meats has been used since ancient times as a method to extend the shelf life of meat products. Nowadays, a wide variety of fermented sausages are produced, depending on the raw materials used, the starter cultures and the processing conditions. Fermented sausages are classified on the basis of the moisture content and range from dry (moisture content 25–45%) to semi-dry (moisture content 40–50%). The ingredients of dry and semi-dry sausages usually include [24]:

- lean meat, 55–70%;
- fat, 25–40%;
- curing salts, 3%;
- fermentable carbohydrate, 0.4–2%;
- spices and flavorings, 0.5%;
- starter, acidulant, ascorbic acid, and so on, 0.5%.

Typically, the manufacturing process includes grinding and mixing the meat with fat at cold temperatures. Then, spices, flavorings, curing salts, carbohydrates, microbial starter cultures, and sodium ascorbate are added, the mixture is homogenized under vacuum and then stuffed into casings, which can be either natural (e.g., collagen based) or synthetic [2, 11]. The sausages are then placed in air-conditioned and temperature-controlled rooms. The temperature and time of fermentation vary depending on the type of sausage being made. High fermentation temperatures (35–40 °C) are typical for sausages from the United States, intermediate temperatures (25–30 °C) for those from northern European countries, and milder temperatures (18–24 °C) for those from Mediterranean countries [11]. Fermentation times range between 20 and 60 h. During the fermentation, acids are produced as a result of microbial metabolism and the pH drops to below 5.2. This promotes the coagulation of the meat proteins and the development of the desired texture and flavor [24]. Fermentation is followed by a ripening/drying stage, during which the water activity decreases and the flavor is developed. This stage can take from 7 to 90 days depending on the product's desirable characteristics [11].

The fermentation characteristics depend to a large extent on the type of microorganisms that are present. Nowadays, in order to standardize the process, ensure the overall quality of the product, and improve its safety, most manufacturers use starter cultures. These are single or mixed cultures of well-characterized strains with defined attributes that are beneficial for sausage manufacturing. The main components of these starters are lactic acid bacteria, including *Lactobacillus, Lactococcus*, and *Pediococcus* strains, and nitrate-reducing bacteria, mainly *Micrococcus* species. Certain yeasts are also used, such as *Debaryomyces hansenii* and *Candida famata*, as well as molds, such as *Penicillium* species [24]. During fermentation the added carbohydrate is converted into lactic acid, and to a lesser extent to acetic acid and acetoin [11].

24.3.7 Fermented Vegetables

Vegetables, such as cabbage, cucumbers, and olives are used as substrates for the production of fermented products, such as sauerkraut, pickles, and fermented olives, respectively. Detailed information on the production of such products can be found elsewhere [24]. Typically, the process includes submerging the vegetables in brine (2–6%) under anaerobic conditions, and at temperatures between 18 and 30 °C, depending on the product [2]. This environment inhibits the growth of spoilage bacteria and promotes the growth of lactic acid bacteria. Starter cultures are available, but the above method of controlling the environmental conditions is preferred by manufacturers. During the fermentation process, the microbial population, which is naturally present in the raw material changes, and as a result a succession of species is observed. In the case of sauerkraut, in the beginning of the fermentation lactic acid bacteria comprise less than 1% of the total microbial population, but at the end of it they account 90% of the total population [24]. The fermentation is initiated by *Leuconostoc mesenteroides*, a heterofermentative lactic acid bacterium producing lactic acid, carbon dioxide, acetic acid, and ethanol. As the fermentation progresses *Leuconostoc mesenteroides* is replaced by heterofermentative lactobacilli, and eventually by *L. plantarum*, an homofermentative *Lactobacillus* producing mainly lactic acid, as this particular species can survive high lactic acid concentrations. Overall, the fermentation takes between 4 and 8 weeks, and the final product contains 1.7–2.3% lactic acid, with a pH between 3.4 and 3.6.

24.3.8 Vinegar

Vinegar is produced from ethanol-containing raw materials, such as wine, cider, spirit, or malt liquor, and involves the oxidation of ethanol to acetic acid. The most common strains used to catalyze the oxidation process belong to *Acetobacter* and *Gluconobacter* species. The process takes place in specially designed bioreactors under high aerobic conditions. At the end of the fermentation the acetic acid concentration ranges between 40 and 150 g l^{-1} [25].

24.4 Enzyme Technology

24.4.1 Enzyme Nomenclature

Enzymes are proteins that are produced from living systems, and are able to catalyze either synthesis or degradation biochemical reactions. Historically, they have been used extensively by the food processing industry, for example, for the production of cheeses, beers, wines, breads, and many more food products and food ingredients.

There has been significant growth in recent years in the use of enzymes as the means for utilizing more effectively natural raw materials and improving their handling, as well as reducing the processing costs of food products and improving their shelf life and organoleptic properties. This stems from the experience of the food manufacturers in using enzymes in their processes, but more importantly from the significant improvements in the technology of their production and the advances in recombinant DNA technology and protein engineering, and has led to the generation of novel foods and food ingredients with improved properties.

Enzymes are classified based on the type of reaction that they catalyze. The following classes exist [26]:

- oxidoreductases (catalyze oxidation/reduction reactions);
- transferases (catalyze the transfer of a group from one molecule to another);
- hydrolases (catalyze the hydrolytic cleavage of C–C, C–N, C–O, and O–P bonds);
- lyases (catalyze elimination reactions resulting in the cleavage of C–C, C–O, C–N, and other bonds);
- isomerases (catalyze structural rearrangements);
- ligases (catalyze bond formation coupled with hydrolysis of a high energy phosphate bond).

The systematic name of enzymes is given by the International Union of Biochemistry and Molecular Biology (IUBMB) and consists of a numerical classification hierarchy of the form "EC i,j,k,l," in which "i" represents the class of the reaction, "j" the subclass, "k" the sub-subclass, and "l" the serial number of the enzyme within the sub-subclass [27]. Frequently however, instead of their official name, other names are used to describe enzymes; these are derived from the truncated names of the substrate with the suffix "ase" added. Examples include lipases, which break down lipids into fatty acids and glycerol, proteinases (or proteases), which break proteins into peptides and amino acids, carbohydrases, which break carbohydrates, such as pectin, starch, cellulose, hemi-cellulose, into mono- and oligosaccharides.

24.4.2 Enzyme Kinetics

Detailed discussions on the mechanisms of enzyme action, including mathematical expressions of enzyme kinetics can be found elsewhere [26, 28, 29]. In brief, the kinetics of most enzyme catalyzed reactions are described by the Michaelis–Menten equation. The reaction is modeled as a two-step reaction:

$$\mathrm{E} + \mathrm{S} \rightleftharpoons \mathrm{ES} \rightleftharpoons \mathrm{E} + \mathrm{S} \quad (24.12)$$

where E is the free enzyme, S is the free substrate, ES is the enzyme–substrate complex, and P is the product. The Michaelis–Menten equation is:

$$v = v_{\max}[S] / (K_m + [S]) \quad (24.13)$$

where v is the volumetric rate of reaction, $[I]$ is the concentration of the substrate, v_{max} is the maximum rate of reaction, and k_m is the Michaelis constant. A key feature of Michaelis–Menten kinetics is that the catalyst becomes saturated at high substrate concentrations, and therefore the rate of reaction (v) reaches a limit (v_{max}). This is diagrammatically described in Figure 24.13a.

The factors that influence the reaction, besides the substrate concentration, are pH and temperature. As temperature increases, the rate of reaction increases (Figure 24.13b). However, after a certain temperature, which is characteristic for each enzyme, the activity drops. This is due to the fact that the folded structure of the enzyme, which is critical for its activity, is disrupted. The enzyme is then said to be denatured [27]. In the case of pH, the relationship between enzyme activity and pH is depicted in Figure 24.13c. The curve is similar to the temperature curve, as there is an optimum pH, either side of which the enzyme activity is lower. The lower activities probably reflect changes in three-dimensional fold of the enzyme that distorts the active site, or a reduction in bond dipoles in active site functional groups [27].

24.4.3 Production of Enzymes

Traditionally, enzymes have been obtained by extracting them from plant and animal tissues. Examples include the protease papain, which is extracted in crude form from plant material, and is used for meat tenderization and haze reduction in beer, and the protease chymosin, which is extracted from calf stomach and is used in cheese processing to promote the coagulation of milk. However, the extraction processes suffer from a number of drawbacks [30]. The raw materials often are not readily available and the enzyme concentration in the raw material is very low. In the latter case, large amounts of the raw materials need to be processed in order to extract the required amounts of enzyme, which makes the process uneconomical. In addition to the above, the extraction methodologies depend greatly on the type of raw materials used, therefore, a new process needs to be designed and developed when a new enzyme needs to be produced. This increases the costs and reduces the scalability of the process. For the above reasons, the production of enzymes by extraction from animal and plant sources is nowadays limited.

The main alternative for producing enzymes is through microbial fermentation. The advantages of this method are that the enzymes are produced in a more cost effective way, scale-up is easier, the process is more reproducible and robust, and can be used for different enzymes with certain modifications [2, 30]. Important criteria for selecting a particular microbial strain to produce an enzyme include that the strain must give high yields and productivities, it must secrete the enzyme into the medium, as it is significantly more cost effective than to extract it from the cells, and should have a GRAS (generally regarded as safe) status. Information on the screening approaches, including molecular approaches, for the discovery of novel enzymes can be found elsewhere [31].

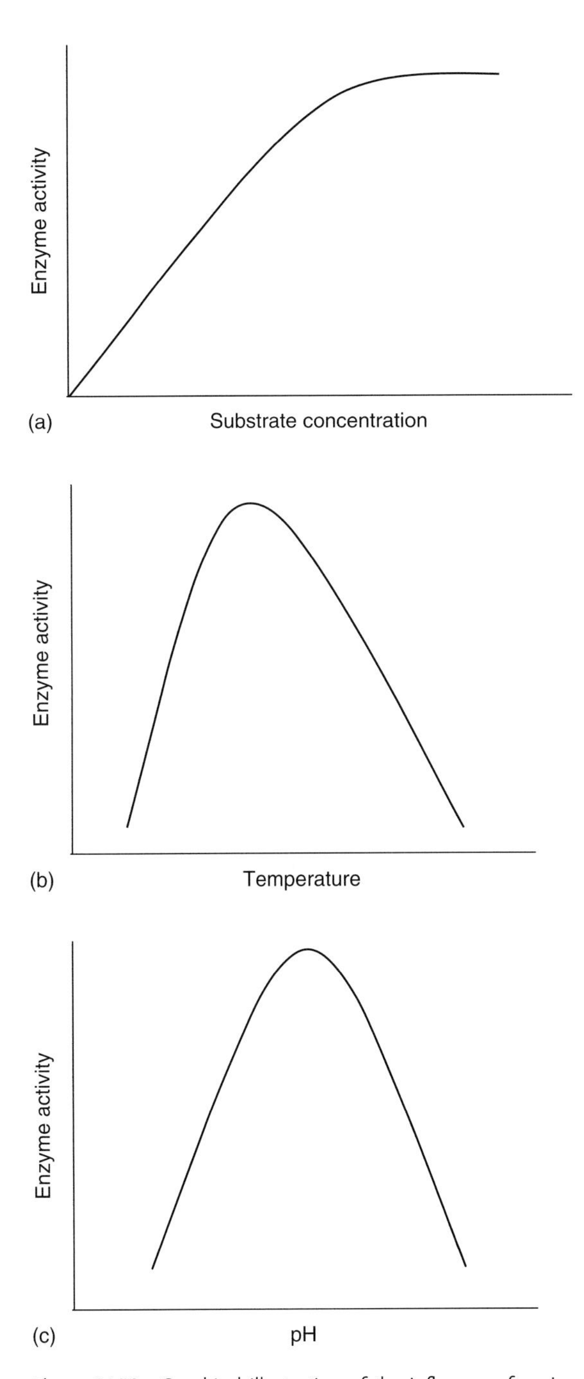

Figure 24.13 Graphical illustration of the influence of environmental factors on enzyme activity. (a) Substrate concentration, (b) temperature, and (c) pH.

The industrial production of enzymes from microorganisms started after the end of World War II. The main microorganisms that were initially used included *Aspergillus* species, such *A. oryzae* and *A. niger*, and *Bacillus licheniformis*. These microorganisms produce significant amounts of enzymes. Before the advent of recombinant DNA technology, strain improvements were carried out using classical mutagenesis methods, in an effort to increase yields and productivities. The advances in molecular microbiology and DNA technologies in the late 1980s have provided manufacturers with the possibility of transferring genes from microorganisms that are not suitable for industrial production to industrial host microorganisms, and are thus able to produce high levels of enzymes. The main industrial host microorganisms include among the fungi *Aspergillus* species, such as *A. oryzae*, *A. awamori*, and *A. foetidus*. *Trichoderma reesei*, although used extensively for the production of industrial enzymes, is not generally used for food applications due to the high protein background [31]. Among yeasts, *Kluyveromyces lactis* has been mainly used for food enzymes production and to a lesser extent *Pichia pastoris*, *Hansenula polymorpha*, and *S. cerevisiae* [31, 32]. Among bacteria, the main ones used are *Bacillus subtilis* and *Bacillus licheniformis*. *E. coli* is not used extensively for food application [31].

Detailed information on the genetics and molecular biology of the above mentioned microbial expression systems can be found elsewhere [33]. In addition to optimizing the microbial host as the means for increasing yields and productivities, the performance of enzymes, for example, their stability and functionality within a food system, can be improved by protein engineering [34].

Food enzymes are produced using the same processes that are used for other enzymes, such as industrial enzymes. A generic production process is diagrammatically described in Figure 24.14. It consists of several steps, including fermentation, enzyme recovery, concentration, drying, and formulation. The fermentation process can be carried out either as submerged fermentation or as surface fermentation. In the latter case the cells grow on the surface of solid media, for example, wheat bran, either in trays or drum reactors. However, the majority of industrial fermentations nowadays are carried out as submerged fermentations. The main reason for that is that surface fermentation is less controllable, more difficult to automate, more labor intensive, and the downstream processing is more complicated [30]. In submerged fermentation the enzyme is produced during growth of the microorganism in an industrial scale fermenter.

For commercial production, the size of fermenters used ranges between 20 000 and 200 000 l [35]. The process starts by inoculating a vial of the production strain into a flask, which after growth is transferred into a seed fermenter. The cells from the seed fermentation are harvested and are used to inoculate the production fermenter. Industrial growth media include cheap carbon and nitrogen sources, such as whey, molasses, and soybean and corn steep liquors. The fermentations can be carried out in batch, fed-batch, or continuous mode. The fermentation conditions, such as oxygen, pH, and temperature are controlled so that maximum enzyme production is achieved.

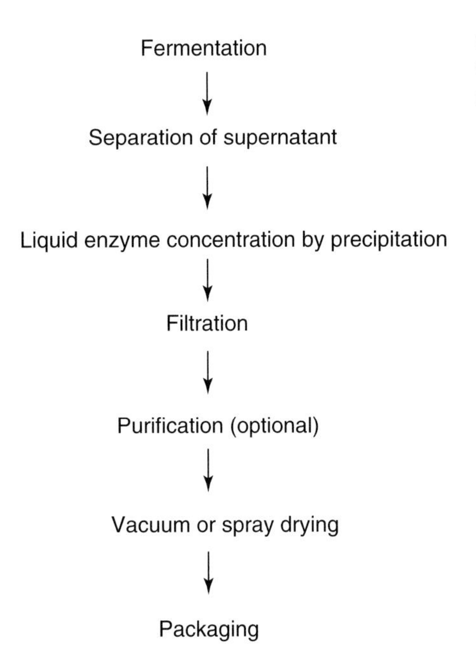

Figure 24.14 Outline of the process for the production of food enzymes from microbial sources.

When pure enzymes rather than crude formulations are required, chromatographic steps can be included as part of the downstream processing, although these are usually vey expensive [30]. If the enzyme is produced intracellularly, the first step in the process involves breaking up the cells in order to release its content into the extracellular medium. This can be carried out either by mechanical or physical treatment. Detailed information on the technologies used for the separation and purification of enzymes can be found elsewhere [36].

24.4.4 Application of Enzymes in Food Processing

A wide variety of enzymes are used in food manufacture (Table 24.2). The section below focuses on the most important ones and provides information regarding their applications.

24.4.4.1 Amylases

α-Amylase is an endo-amylase that randomly cleaves the α-1,4 linkages in pre-gelatinized starch, which consists of amylose and amylopectin. This results in a decrease in the starch viscosity, and the production of maltose, glucose, and low molecular weight dextrins [37]. Thermostable α-amylases are produced from *Bacillus licheniformis* and *Bacillus stearothermophillus* (temperature range 95–105 °C) and *Bacilllus subtilis* (80–85 °C), whereas mesophilic α-amylases are produced from fungi, such as *Aspergillus niger* and *Aspergillus oryzae*. The latter two produce high amounts of maltose [37].

Table 24.2 Applications of important food enzymes.

Enzymes	Source	Applications
α-Amylase	*Bacillus* spp.[a], *Aspergillus* spp.	Starch liquefaction, alcohol production
β-Amylase	Malt	Maltose production, alcohol production
Cellulase	*Aspergillus* spp., *Trichoderma* spp.	Fruit and vegetable processing
Chymosin	Calves, *Mucor michei*, *Aspergillus* spp.[a], *Kluyveromyces* spp.[a]	Cheese manufacture
β-Galactosidase	*Aspergillus* spp., *Kluyveromyces* spp.	Dairy products
Glucoamylase	*Aspergillus* spp., *Rhizopus* spp.	Starch saccharification, brewing
Glucose isomearse	*Bacillus* spp.	High fructose corn syrup production
Hemicellulase	*Aspergillus* spp.[a], *Bacillus* spp.[a], *Trichoderma* spp.[a]	Bread making
Lipase/esterase	*Aspergillus* spp.[a], *Candida* spp., *Penicillium* spp.	Cheese manufacture, milk fat modification
Pectinase	*Aspergillus* spp., *Penicillium* spp.	Extraction and clarification of fruit juices
Pentosanase	*Humicola* spp., *Trichoderma* spp.	Bread making
Pullulanase	*Bacillus* spp.[a], *Klebsiella* spp.[a]	Starch saccharification, brewing
Protease	*Aspergillus* spp.[a], *Penicillium* spp., *Rhizopus* spp., *Bacillus* spp[a]	Cheese manufacture, bread making, meat processing

[a] These enzymes are produced using recombinant strains.

β-Amylase is an exo-enzyme that is able to hydrolyze α-1,4 linkages and produce maltose from the nonreducing end of amylose. In the case of amylopectin the cleavage stops two or three units from the nonreducing ends [37]. β-Amylases are mainly produced commercially from soybean and barley malt, although various *Bacillus* species can also produce them [38].

Glucoamylase (also called *amyloglucosidase*) hydrolyzes α-1,4 and to a lesser extent α-1,6 linkages in liquefied starch. Its action is to split off glucose from the nonreducing ends of the substrate molecules. Maltose and maltotriose are hydrolyzed more slowly than higher saccharides, whereas α-1,4 linkages are hydrolyzed much faster than α-1,6 linkages [37]. For this reason glucoamylase is often used in conjunction with pullulanase (pullulan-6-glucanohydrolase) and isoamylase (glycogen-6-glucanohydrolase), which are able to hydrolyze the α-1,6 linkages efficiently. Glucoamylases are mainly formed by fungi, including *Aspergillus* or *Rhizopus* species.

The amylases, along with the 1,6-glycoside-splitting enzymes, have several applications in the food industry, including starch processing, ethanol production, and baking.

24.4.4.2 Lactase

Lactase (or β-galactosidase) catalyzes the conversion of lactose to galactose and glucose. It can be used to increase the digestibility of dairy products, as many individuals are lactose-intolerant, increase their sweetness, and prevent lactose crystallization in concentrated dairy products. Most applications of lactase are in milk and milk powders. Lactases are also used for the production of galacto-oligosaccharides, which are considered to be prebiotics [39]. A *prebiotic* is defined as "a selectively fermented ingredient that allows specific changes, both in the composition and/or activity in the gastrointestinal microbiota that confers benefits upon host well-being and health" [40]. Lactases are produced commercially from yeasts, mainly *Kluyveromyces lactis* and *Kluyveromyces fragilis*, and fungi, for example, *Aspergillus niger* and *Aspergillus oryzae*. In the case of *Kluyveromyces* spp. the optimum pH range is between 6 and 7, whereas in the case of *Aspergillus* species it is between 4 and 5 [3].

24.4.4.3 Lipase

Lipase hydrolyzes fats (triglycerides) into di- or mono-glycerides and fatty acids. Lipases are produced from fungi (*Aspergillus, Mucor, Penicillium, Geotrichum, Rhizopus*), yeasts (*Torulopsis, Candida*), and bacteria (*Pseudomonas, Achromobacter, Staphylococcus*) [3]. Of these, *Aspergillus* and *Rhizopus* spp. are used commercially [38]. Bacterial lipases have a pH optimum between the neutral and alkaline range, whereas fungal lipase has optimum in the neutral to slightly acidic range [38]. Lipase is mainly used for improving or enhancing the development of flavor in cheeses during the ripening process, and for producing free fatty acids, such as butyric, caproic, caprylic, capric, and longer chain acids from butter fat [3]. These fatty acids are used in cheese flavored dips, processed cheeses, confectionery, and snack foods to give a creamy/buttery flavor [3, 41].

24.4.4.4 Pectinase

Pectins are complex polysaccharide polymers consisting of a linear chain of α-1,4-linked galacturonic acid, which has more than 50% of its carboxyl groups partly esterified with methanol. Pectins are hydrolyzed by two main types of enzymes: polygalacturonases and pectin esterases. The former are endo- and exo-galacturonase, which hydrolyze the α-1,4 linkages randomly or from the reducing end of the polysaccharide, respectively. Pectin esterase hydrolyzes the methoxyl group, producing methanol and low methoxyl pectin. The main applications of pectinases include increasing the yield in juice extraction and clarifying fruit juices and grape must. Commercial pectinases are primarily derived from fungi, such as *Aspergillus* and *Rhizopus* species.

24.4.4.5 Glucose Isomerase

Glucose isomerase (D-glucose ketoisomerase) causes the isomerization of glucose to fructose. The product of the reaction is a mixture of glucose and fructose, with their ratio being dependent on the enzyme concentration and temperature. Commercially, glucose isomerase is produced by *Bacillus* and *Streptomyces* spp. [1].

Its main application is for the production of high-fructose corn syrups (HFCSs), which are used as sugar substitutes in various food products.

24.4.4.6 **Proteases**

Proteases are commonly classified on the basis of their pH optima into alkaline, neutral, and acid proteases. Acid proteases are mainly produced by fungi (*Aspergillus* spp.), whereas neutral and alkaline proteases are produced by both fungi and bacteria (*Bacillus* spp.). Proteases are used in foods in order to modify the molecular structure of proteins and thus improve their functional properties, such as solubility, emulsification, gelling, and foaming properties, as well as nutritional properties [42]. A well-known and established food process that employs proteases is the conversion of milk into cheese. This is carried out using chymosin, a protease that coagulates milk proteins, to form a casein curd. Traditionally, rennet has been isolated from calves, although nowadays it is produced using recombinant DNA technologies. Other proteases that are used for food applications include proteases in baby foods aiming to reduce the risk of allergenicity in formula milks, the production of meat hydrolysates that are used as flavor enhancers, the tenderization of meat, and the modification of wheat gluten [42]. Novel applications of proteases include the enzymatic cross-linking of proteins to improve the textural properties of foods [43], the production of bioactive peptides from milk and whey proteins [44], and the enhancement of the fat-mimicking properties of proteins by enzymatic modification [45].

References

1. Crueger, W. and Crueger, A. (1984) *Biotechnology. A Textbook of Industrial Microbiology*, Science Tech Inc., Madison.
2. Lee, B.H. (1996) *Fundamentals of Food Biotechnology*, Wiley-VCH Verlag GmbH, Weinheim.
3. Monod, J. (1942) *Recherches sur les Croissances des Cultures Bacteriennes*, Hermann and Cie, Paris.
4. Stanbury, P.F., Whitaker, A., and Hall, S.J. (1995) *Principles of Fermentation Technology*, Butterworth-Heinemann, Oxford.
5. Doran, P.M. (1995) *Bioprocess Engineering Principles*, Academic Press, London.
6. Chisti, Y. (2006) in *Basic Biotechnology* (eds C. Ratledge and B. Kristiansen), Cambridge University Press, Cambridge, pp. 181–200.
7. Priest, F.G. and Stewart, G.G. (2006) *Handbook of Brewing*, CRC Press, Boca Raton, FL.
8. Jackson, R.S. (2008) *Wine Science. Principles and Applications*, Elsevier, London.
9. Prescot, L.M. (2005) *Microbiology*, McGraw-Hill, New York.
10. Fox, P.F., McSweeney, P., Cogan, T.M., and Guinee, T.P. (2000) *Fundamentals of Cheese Science*, Aspen Publishers, Gaithersburg.
11. Hui, Y.H. (2006) *Food Biochemistry and Food Processing*, Blackwell Publishing, Oxford.
12. Tamime, A.Y. and Robinson, R.K. (2007) *Tamime and Robinson's Yoghurt: Science and Technology*, CRC Press, Boca Raton, FL.
13. Rajagopal, S.N. and Sandine, W.E. (1990) Associative growth and proteolysis of *Streptococcus thermophilus* and *Lactobacillus bulgaricus* in skim milk. *J. Dairy Sci.*, **73**, 894–899.
14. Driessen, F.M., Kingma, F., and Stadhouders, J. (1982) Evidence that *Lactobacillus bulgaricus* in yogurt is stimulated by carbon dioxide produced by

Streptococcus thermophilus. *Neth. Milk Dairy J.*, **36**, 135–144.
15. Oberman, H. and Libudzisz, Z. (1998) Fermented milks, in *Microbiology of Fermented Foods* (ed. B.J. Wood), Thomson Science, London, pp. 308–350.
16. Tamime, A.Y. (2005) *Fermented Milks*, Blackwell, Oxford.
17. Farnworth, E.R. and Mainville, I. (2003) Kefir: a fermented milk product, in *Handbook of Fermented Functional Foods* (ed. E.R. Farnworth), CRC Press, pp. 77–112.
18. Thoreux, K. and Schmucker, D.L. (2001) Kefir milk enhances intestinal immunity in young but not old rats. *Nutr. Aging*, **131**, 807–812.
19. Vinderola, C.G., Duarte, J., Thangavel, D., Perdigon, G., Farnworth, E., and Matar, C. (2005) Immunomodulating capacity of kefir. *J. Dairy Res.*, **72**, 195–202.
20. FAO/WHO (2001) Joint Report of Expert Consultation: HEALTH and Nutritional Properties of Probiotics in Food Including Powder Milk with Live Lactic Acid Bacteria. *http://www.who.int/foodsafety/publications/fs_management/en/probiotics.pdf* (accessed June 2011).
21. Lomax, A.R. and Calder, P.C. (2009) Prebiotics, immune function, infection and inflammation: a review of the evidence. *Br. J. Nutr.*, **101**, 633–658.
22. Parkes, G.C., Sanderson, J.D., and Whelan, K. (2009) The mechanisms and efficacy of probiotics in the prevention of *Clostridium difficile*-associated diarrhea. *Lancet Infect. Dis.*, **9**, 237–244.
23. Ruemmele, F.M., Bier, D., Marteau, P., Rechkemmer, G., Bourdet-Sicard, R., Walker, W.A., and Goulet, O. (2009) Clinical evidence for immunomodulatory effects of probiotic bacteria. *J. Pediatr. Gastroenterol. Nutr.*, **48**, 126–141.
24. Adams, M.R. and Moss, M.O. (2008) *Food Microbiology*, The Royal Society of Chemistry, Cambridge.
25. Ebner, H., Follmann, H., and Sellmer, S. (1995) in *Biotechnology* (eds G. Reed and T.W. Nagodawithana), Wiley-VCH Verlag GmbH, Weinheim, pp. 579–591.
26. Smith, G.M. (1995) The nature of enzymes, in *Biotechnology* (eds G. Reed and T.W. Nagodawithana), Wiley-VCH Verlag GmbH, Weinheim, pp. 1–72.
27. Law, B.A. (2002) The nature of enzymes and their action in foods, in *Enzymes in Food Technology* (eds R.J. Whitehurst and B.A. Law), CRC Press, Sheffield, pp. 1–18.
28. Price, N.C. and Stevens, L. (1999) *Fundamentals of Enzymology: Cell and Molecular Biology of Catalytic Proteins*, Oxford University Press, Oxford.
29. Buchholz, K., Kasche, V., and Bornscheuer, U.T. (2005) *Biocatalysts and Enzyme Technology*, Wiley-VCH Verlag GmbH, Weinheim.
30. Hjort, C. (2007) Industrial enzyme production for food applications, in *Novel Enzyme Technology for Food Applications* (ed. R. Rastall), Woodhead Publishing, Cambridge, pp. 43–59.
31. Schaffer, T. (2007) Industrial enzyme production for food applications, in *Novel Enzyme Technology for Food Applications* (ed. R. Rastall), Woodhead Publishing, Cambridge, pp. 3–15.
32. Berka, R.M. and Cherry, J.R. (2006) Enzyme biotechnology, in *Basic Biotechnology* (eds C. Ratledge and B. Kristiansen), Cambridge University Press, Cambridge, pp. 477–498.
33. Gellissen, G. (2005) *Production of Recombinant Proteins: Novel Microbial and Eukaryotic Expression Systems*, Wiley-VCH Verlag GmbH, Weinheim.
34. Machielsen, R., Dijkhuizen, S., and van der Oost, J. (2007) Industrial enzyme production for food applications, in *Novel Enzyme Technology for Food Applications* (ed. R. Rastall), Woodhead Publishing, Cambridge, pp. 16–42.
35. Piggott, R. (2002) The nature of enzymes and their action in foods, in *Enzymes in Food Technology* (eds R.J. Whitehurst and B.A. Law), CRC Press, Sheffield, pp. 229–244.
36. Foster, K.A., Frackman, S., and Jolly, J.F. (1995) The nature of enzymes, in *Biotechnology* (eds G. Reed and T.W. Nagodawithana), Wiley-VCH Verlag GmbH, New York, pp. 73–120.
37. Olsen, H.J. (2002) Enzymes in starch modification, in *Enzymes in Food*

Technology (eds R.J. Whitehurst and B.A. Law), CRC Press, Sheffield, pp. 200–228.

38. Enfors, S.O. and Haggstrom, L. (2000) *Bioprocess Technology Fundamentals and Applications*, Royal Institute of Technology, Stockholm.
39. Tzortzis, G. and Vulevic, J. (2009) Galacto-oligosaccharide prebiotics, in *Prebiotics and Probiotics Science and Technology* (eds D. Charalampopoulos and R. Rastall), Springer, New York, pp. 207–244.
40. Gibson, G.R., Probert, H.M., Van Loo, J., Rastall, R.A., and Roberfroid, M. (2004) Dietary modulation of the human colonic microbiota: updating the concept of prebiotics. *Nutr. Res. Rev.*, **17**, 259–275.
41. Fellows, P.G. (2000) *Food Processing Technology*, Woodhead Publishing Limited, Cambridge.
42. Nielsen, P.M. and Olsen, H.S. (2002) Enzymic modification of food protein, in *Enzymes in Food Technology* (eds R.J. Whitehurst and B.A. Law), Sheffield Academic Press, Sheffield, pp. 109–143.
43. Buchert, J., Selinheimo, E., Kruus, K., Mattinen, M.L., Lantto, R., and Autio, K. (2007) Using crosslinking enzymes to improve textural and other prperties of food, in *Novel Enzyme Technology for Food Applications* (ed. R. Rastall), Woodhead Publishing, Cambridge, pp. 101–139.
44. Ortiz-Chao, P.A. and Jauregi, P. (2007) Enzymatic production of bioactive peptides from milk and whey proteins, in *Novel Enzyme Technology for Food Applications* (ed. R. Rastall), Woodhead Publishing, Cambridge, pp. 160–182.
45. Leman, J. (2007) Enzymatically modified whey protein and other protein-based fat replacers, in *Novel Enzyme Technology for Food Applications* (ed. R. Rastall), Woodhead Publishing, Cambridge, pp. 140–159.

Index

Food Processing Handbook, Second Edition. Edited by James G. Brennan and Alistair S. Grandison.

C

e

g

i

j

k

l

m

n

o

p